£50.60
each.

MARINE MYCOLOGY
The Higher Fungi

MARINE MYCOLOGY
The Higher Fungi

Jan Kohlmeyer
Erika Kohlmeyer

Institute of Marine Sciences
University of North Carolina at Chapel Hill
Morehead City, North Carolina

ACADEMIC PRESS   New York   San Francisco   London   1979

A Subsidiary of Harcourt Brace Jovanovich, Publishers

ACADEMIC PRESS, INC.
111 Fifth Avenue, New York, New York 10003

United Kingdom Edition published by
ACADEMIC PRESS, INC. (LONDON) LTD.
24/28 Oval Road, London NW1 7DX

Library of Congress Cataloging in Publication Data

Kohlmeyer, Jan.
    Marine mycology.

    Includes bibliographical references and indexes.
    1. Marine fungi. I. Kohlmeyer, Erika, joint author.
II. Title.
QK618.K59   589'.2'09162   79–14703
ISBN 0–12–418350–6

PRINTED IN THE UNITED STATES OF AMERICA

79 80 81 82   9 8 7 6 5 4 3 2 1

To Willy Höhnk
and
to the memory of Frederick K. Sparrow, Jr.
in recognition of their pioneering work
in marine mycology

Contents

Preface

Growing general interest in marine and estuarine habitats in recent years has also led to an increase in studies on marine fungi which were often overlooked as participants in ecological processes. Because of this rapid growth of knowledge in marine mycology, the first comprehensive treatise on marine fungi by Johnson and Sparrow (1961) is partly out-of-date. G. C. Hughes (1975), in a very informative review, has summarized papers published since 1961. However, extensive descriptions and complete keys are not available for all marine fungi. The "Icones Fungorum Maris" and "Synoptic Plates of Higher Marine Fungi" (Kohlmeyer and Kohlmeyer, 1964–1969, 1971b) include diagnoses for only 90 species and keys for the identification of all taxa described up to January 1971, respectively. An updated treatise was clearly needed.

This book deals with the higher marine fungi, i.e., Ascomycotina, Basidiomycotina, and Deuteromycotina. This book combines features of a monograph with those of a text. It includes sections on ecological groups of fungi and other topics, such as phylogeny, ontogeny, physiology, and vertical and geographical distribution, providing information on known facts and open questions. The taxonomic–descriptive part contains complete descriptions of each genus and species, together with substrates, range, etymology of generic and specific names, and literature. There are keys for all species within a given genus, and a general illustrated key leads to the individual species. The taxonomic section is based on our examinations of almost all of the filamentous marine fungi, and unpublished data on new hosts and geographical distributions are included for many species. The filamentous higher marine fungi are represented by 149 Ascomycetes, 4 Basidiomycetes, and 56 Deuteromycetes. The majority, namely 191(91%) of the filamentous fungi, are obligately marine species, whereas the remainder are facultatively marine. One new species and seven new combinations are proposed (see Table I). Excluded

---

TABLE I
New Species and Combinations Proposed

Ascomycotina
  1. *Didymella gloiopeltidis* (Miyabe et Tokida) Kohlm. et Kohlm.
     (≡*Guignardia gloiopeltidis*)
  2. *Haligena amicta* (Kohlmeyer) Kohlm. et Kohlm.
     (≡*Sphaerulina amicta*)
  3. *Halosphaeria galerita* (Tubaki) Schmidt ex Kohlm. et Kohlm.
     (≡*Remispora galerita*)
  4. *Leiophloea pelvetiae* (Sutherland) Kohlm. et Kohlm.
     (≡*Dothidella pelvetiae*)
  5. *Massarina cystophorae* (Cribb et Cribb) Kohlm. et Kohlm.
     (≡*Otthiella cystophorae*)
  6. *Pontogeneia codiicola* (Dawson) Kohlm. et Kohlm.
     (≡*Sphaerulina codiicola*)
  7. *Turgidosculum complicatulum* (Nylander) Kohlm. et Kohlm.
     (≡*Leptogiopsis complicatula*)
Deuteromycotina
  1. *Camarosporium palliatum* Kohlm. et Kohlm. sp. nov.

---

are 49 *nomina confusa, dubia,* or *nuda*. The yeasts are treated in a separate chapter and comprise 177 species or varieties. Authorities on fungal names and hosts are given in the Organism Index. We have considered all literature except for unpublished reports or theses.

Jan Kohlmeyer
Erika Kohlmeyer

Acknowledgments

The investigations have been supported in part by grants GB 27-265, EMS 74-18539, and DEB 74-18539 of the U.S. National Science Foundation and by the Brown–Hazen Fund of the Research Corporation. We are most grateful for this support which made possible the revisions presented in this treatise. We acknowledge gratefully loan of type and other collections by curators of herbaria* ABD, AHFH, B, BM ex K, BPI, C, H, HBG, L, LD, LE, LPS, MA, MPU, PAD, PC, UC, and the University of Sheffield. The following colleagues and friends kindly collected for us or made their own collections of marine fungi from many parts of the world available: J. Acosta, A. A. Aleem, R. A. Barkley, R. D. Brooks, A. B. Cribb, O. Eriksson, G. Feldmann, M. H. Hommersand, R. G. Hooper, D. E. and C. M. Hoss, P. W. Kirk, Jr., L. Malacalza, R. T. Moore, A. Munk, M. J. Parsons, C. Pujals, M. Salmon, I. Schmidt, F. J. Schwartz, R. B. Setzer, G. R. South, A. J. Southward, S. Udagawa, I. M. Wilson, E. M. Wollaston, and H. Yabu. We express our gratitude to these people who gave invaluable information on taxonomic and other matters: J. A. von Arx, R. K. Benjamin, G. H. Boerema, I. M. Brodo, B. Bruce, G. S. Daniels, T. Eckardt, W. E. Fahy, M. L. Farr, J. Gerloff, W. Haupt, M. H. Hommersand, B. Kaussmann, R. P. Korf, I. M. Lamb, N. D. Latham, E. Müller, G. F. Papenfuss, J. Poelt, B. Sahlmann, D. E. Shaw, J. Walker, I. M. Wilson, and J. S. ZoBell. Thanks are due the staffs of the Laboratoire de Biologie Marine, Collège de France at Concarneau (France) and of the Biologische Anstalt Helgoland (Federal Republic of Germany); to I. Gamundi de Amos and staff of the Instituto de Botanica Spegazzini of the Universidad Nacional de la Plata (Argentina); and Instituto Argentino de Oceanografia, Bahia Blanca (Argentina) for kind assistance during our stay at their institutions. Grateful acknowledgment is made to the follow-

---

* Abbreviations follow Holmgren and Keuken (1974).

ing colleagues for permission to reproduce illustrations: G. E. Cole (Fig. 34), J. W. Fell (Fig. 129), R. V. Gessner (Fig. 22), and L. Vitellaro Zuccarello (Fig. 31). Stearn's (1966) "Botanical Latin" was invaluable in clarifying the etymologies of many generic and specific names. T. L. Herbert and B. J. Lamore ably assisted us in the laboratory, B. B. Bright and J. W. Lewis provided us with quick library service and administrative assistance, respectively, and J. B. Garner carefully typed the manuscript. We are greatly indebted especially to L. S. Olive, R. V. Gessner, and J. W. Fell for reading parts of the manuscript and making helpful suggestions. Many thanks go also to T. L. Herbert for collecting in Trinidad, for proof-reading and indexing; to B. J. Lamore for help in the preparation of indexes; to P. E. Boyd for typing; and to V. A. Zullo for giving advice on authorities of host species.

1. Introduction

One single publication has probably influenced the development of marine mycology more than any other paper, namely, "Marine Fungi: Their Taxonomy and Biology" by Barghoorn and Linder (1944). These authors demonstrated that there was an indigenous marine mycota, showing growth and reproduction on submerged wood after defined periods of time. The existence of true marine saprobic fungi was often questioned, for instance, by Bauch (1936), who wrote: "Saprobic Ascomycetes which play an important role in forest and soil in the deterioration of organic material, especially of wood, appear to be completely absent in seawater" (translation from the German). Investigations by Barghoorn and Linder and later authors have proven without a doubt that fungi, including Ascomycetes, do contribute to the decomposition of organic substrates in the oceans, although the extent of biodeterioration caused by fungi in the sea is still not fully understood.

The first facultative marine fungus, *Phaeosphaeria typharum*, was described by Desmazières (1849) as *Sphaeria scirpicola* var. *typharum* from *Typha* in freshwater. Durieu and Montagne (1869) discovered the first obligate marine fungi on the rhizomes of the sea grass, *Posidonia oceanica*, and marveled at the most remarkable life-style of *Sphaeria posidoniae* (=*Halotthia posidoniae*), which spends all parts of its cycle at the bottom of the sea. Figure 1 summarizes the numbers of species of higher marine fungi described per decade since 1840. If numbers of new descriptions can be used to indicate the activity dealing with a particular group of organisms at a certain time, this figure shows that up to the period of 1930–1939 there was only sporadic interest in marine fungi. Descriptions of marine species during these first hundred years were supplied mostly by authors working on a wide variety of fungi with no particular interest in the marine habitat. Exceptions may be the Crouan brothers (Kohlmeyer, 1974a), who described five marine fungi in 1867 in

1

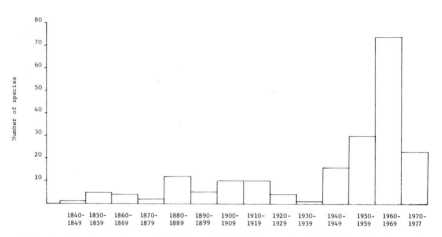

Fig. 1. Numbers of accepted filamentous higher marine fungi described per decade since 1840. Research in marine mycology was stimulated by a paper of Barghoorn and Linder in 1944. The numbers of newly described species show a decreasing trend since 1970.

their *Florule du Finistère* and prepared eleven others for publication (remaining unpublished), and particularly Sutherland (1915a–c, 1916a,b), who published a series of papers dealing exclusively with marine fungi. Almost three-quarters of all recognized higher marine fungi have been described in the last four decades (Fig. 1). This sudden increase in new descriptions after 1940 can be attributed mostly to the publication of Barghoorn and Linder (1944), which stimulated worldwide research in marine mycology. A proliferation of literature on the higher marine fungi occurred, as I. M. Wilson (Great Britain) in 1951, W. Höhnk (Germany) in 1952, S. P. Meyers (United States) in 1953, A. B. Cribb and J. W. Herbert (Australia) in 1954, T. W. Johnson (United States) in 1956, G. Feldmann (France) in 1957, J. Kohlmeyer (Germany) in 1958, and G. Doguet (France) in 1962 and E. B. G. Jones (Great Britain) also in 1962 published their first papers in this field. The first monograph on marine mycology by Johnson and Sparrow appeared in 1961.

I. DEFINITION OF MARINE FUNGI

Past definitions of marine fungi have often been based on their ability to grow at certain seawater concentrations (Johnson and Sparrow, 1961;

Tubaki, 1969). It has been shown that marine fungi cannot be defined on a strictly physiological basis (see discussions in Kohlmeyer, 1974a; G. C. Hughes, 1975), and we use a broad, ecological definition, namely: *Obligate marine fungi* are those that grow and sporulate exclusively in a marine or estuarine habitat; *facultative marine fungi* are those from freshwater or terrestrial milieus able to grow (and possibly also to sporulate) in the marine environment (Kohlmeyer, 1974a).

Marine or estuarine habitats are the oceans and ocean-associated smaller bodies of water that contain salt or brackish water, including river mouths, sounds, lagoons, tidal creeks, salinas, étangs, and the like. Fungi from inland salt lakes are not well known, but the mycota of these habitats appears to be identical with that of the oceans (Anastasiou, 1961, 1963b; Davidson, 1974b).

Some authors feel that attempts to define marine fungi are premature (Schaumann, 1975b), but, in our opinion, such efforts are necessary to understand the role of fungi in a particular environment such as the oceans. Fungi are often obtained by culture methods, for example, plating techniques or incubation, from marine substrates. Most investigators using such methods have failed to demonstrate conclusively that the species isolated in the laboratory can actually grow (and, possibly, reproduce) in the marine habitat. The mere isolation of a species from marine samples does not prove that this fungus is active in the marine environment. Such species are possibly dormant in the form of spores or hyphal fractions until the surrounding conditions become favorable for germination and growth. Research by Tyndall and Kirk (1973) and preliminary experiments carried out by Dr. R. V. Gessner and ourselves indicate that a mycostatic factor in fresh seawater inhibits germination of spores of terrestrial fungi but does not affect certain marine fungi. Mycostasis is a well-known phenomenon in terrestrial soils and on leaf surfaces (Lockwood, 1977). The unidentified mycostatic principle in raw seawater is destroyed by heat sterilization. Conidia of terrestrial fungi, for example, *Penicillium* sp., germinate in sterilized or aged seawater or in distilled water, but they do not germinate in the natural habitat. Therefore, fungi isolated from seawater by means of culture methods cannot be considered to be marine species unless their growth in the marine environment is demonstrated. A valid criterion for the definition of a marine fungus might be its ability to germinate and to form a mycelium under natural marine conditions. These conditions may vary, depending on the species. Some fungi might require permanent submergence, whereas others, such as species occurring in beach sands, may need a greater supply of oxygen for germination.

II. NUMBERS OF MARINE FUNGI

About 50,000 fungal species are known from terrestrial habitats (Ainsworth, 1968), but, in contrast, less than 500 species have been described from oceans and estuaries, which cover a much larger area, namely, three-quarters of the world. The higher filamentous marine fungi include 209 species, the marine-occurring yeasts comprise 177 species, and the lower marine fungi comprise probably less than 100 species. The decrease of new descriptions of marine fungi during the period 1970–1977 (Fig. 1) indicates that the most common species have been named and that considerable additions of new species in the future are unlikely. Undescribed fungi can be expected in some less explored habitats, for example, in temperate salt marshes, in mangrove forests, particularly of the Pacific and Indian oceans, and in the deep sea.

The oceans, compared to the land masses, provide stable environments with small changes in temperatures and salinities. Organic substrates, such as algae, marsh plants, and plant litter, are concentrated mostly along the shores, where they provide nutrients for fungi. The open ocean is a fungal desert where only yeasts or lower fungi may be found attached to planktonic organisms or pelagic animals. The existence of such organisms in a free-living state has not been demonstrated convincingly, in our opinion. In contrast to terrestrial habitats, the incubatorlike stable environment of the oceans and the comparable small number of hosts and substrates in the marine environment have probably not exerted enough selective pressure during the course of evolution to induce the formation of a high number of different types of fungi. Ainsworth (1968) estimates the number of named and undescribed terrestrial fungi to be 250,000 or more. We expect the number of marine fungi to be about 1% of the terrestrial species, namely, the same ratio as is found between the presently known marine and terrestrial fungi.

III. SIZES OF MARINE FUNGI

Most of the fungi found in marine habitats are microscopic. The largest ascocarps occur in *Amylocarpus encephaloides,* which do not exceed 3 mm, and the Basidiomycetes *Digitatispora marina* and *Nia vibrissa* have fruiting bodies 4 mm in length and 3 mm in diameter, respectively. Obviously, the marine environment does not permit the development of large, fleshy fruiting bodies, because abrasion by waves and grains of sand impedes formation of such structures. Macromycetes growing in the leaf litter of forests need an extended nutrient-supplying mycelium and an undisturbed habitat. Similarly, soft fruiting bodies of large marine species (*A. encephaloides, Eiona tunicata, D. marina, N. vibrissa,* and *Halocyphina villosa*) develop mostly in sheltered habitats, namely, on firmly anchored wood at or above the high-tide line or protected in cracks of the wood or under bark. The deep sea appears to be another environment where large fruiting bodies could develop because water currents are weak, and, indeed, the Ascomycete *Oceanitis scuticella*, occurring at a depth of about 4000 m, has fleshy ascocarps up to 2 mm in height.

IV. THE MODE OF LIFE AND DISTRIBUTION OF MARINE FUNGI

The higher marine fungi occur as parasites on plants and animals, as symbionts in lichenoid associations with algae and as saprobes on dead organic material of plant or animal origin. Examples of the occurrence of marine fungi in different substrates and habitats are given in the following chapters. These organisms are found in all marine environments from the high-water line down to the deep sea. Their distribution is mainly limited by dissolved oxygen and temperature of the water, less so by salinities. Besides ubiquitous species, there are fungi restricted to temperate waters and others restricted to tropical or subtropical waters.

V. UNSOLVED MAJOR PROBLEMS

Among the topics to be solved in marine mycology, we consider the following three to be of foremost importance.

A. Quantification

So far we have no means to measure the biomass, numbers or activity of marine fungi in a particular habitat. Microscopic observation of decaying marsh plants or marine wood indicates that fungi play a major role in the decomposition of these plants and wood, but fungal hyphae and substrate cannot be separated.

B. Definition of Marine Fungi

To obtain a better understanding of the role of fungi in the marine environment, we need to separate the indigenous marine species (obligate and facultative) from the "contaminants." The latter are terrestrial or freshwater species [sometimes called "transients" (Park, 1972b)] that are dormant in marine habitats. A way of separating indigenous from nonindigenous species appears to be a test of their germination ability in the natural environment, as discussed above.

C. Fossil Records

In the absence of fossil higher marine fungi, discussions on the phylogeny of this group are more or less speculative. A search for parasitic fungi on fossil marine algae, for instance, on calcified Rhodophyta, might supply data on the origin of marine fungi.

2. Methods

Techniques used to collect, preserve, and culture higher marine fungi do not differ greatly from those applied in research on fungi from other habitats, and recent handbooks (Norris *et al.*, 1971; Hawksworth, 1974; Stevens, 1974) should be consulted for general information. We explain only those methods that are particularly relevant to research on the filamentous higher marine fungi. Detailed descriptions of methods used in marine mycology have been published by E. B. G. Jones (1971b) and Höhnk (1972), whereas G. C. Hughes (1975) reviewed the literature appearing between 1961 and 1974.

I. COLLECTING TECHNIQUES

As discussed in Chapter 1, collecting techniques in marine mycology are crucial to determine if a fungus can be considered marine, that is, an active inhabitant of marine habitats. Therefore, the collecting methods based on direct observations are always preferred over the incubation (or "indirect") techniques.

A. "Direct" Examination Methods

Material collected according to the methods described in the following paragraphs is examined under the dissecting microscope for the presence of ascocarps, basidiocarps, pycnidia, or conidia. Fruiting bodies are transferred with a needle to a microscope slide, torn apart in a drop of water to expose the spores, and carefully squeezed under a cover glass. For the identification of an unknown species, the observation of ascospores, basidiospores, or conidia usually suffices. In some cases it is necessary to search for asci and sterile elements of the ascocarps, such as paraphyses and pseudoparaphyses. Additional information leading to the identification of higher marine fungi is given in the introduction to the key (Chapter 23).

1. Wood and Other Cellulosic Matter

Fungi found growing on cellulosic materials in marine habitats belong to the Ascomycetes, Basidiomycetes, and Deuteromycetes. Substrates include breakwaters, pilings, and other shoreline fortifications, driftwood, ropes, and mangrove roots, all of which are submerged in, or wetted by, seawater for some part of the day. Specimens are collected by cutting or sawing off pieces of the substrate and placing them in sterile plastic bags.

In the laboratory, the material is either examined at once under a dissecting microscope or stored in a refrigerator at about 5°C if the investigation can be made during the following days. If a longer storage period is necessary, the specimens should be frozen at about −18°C.

a. Driftwood. Some pieces of driftwood, ropes, or other matter washed ashore may not yet contain any marine fungi because they have been in seawater for only a short period. Terrestrial fungi growing on such material may then be mistakenly considered as marine. Therefore, care should be taken to collect only specimens that have been in the marine habitat for several weeks, as indicated by the attack of marine fouling or boring organisms, such as barnacles, algae, or shipworms. Wood wedged between jetties or natural rocks is often inhabited by marine fungi. G. C. Hughes (1968) distinguishes between "driftwood" *sensu stricto,* namely, wood that is actually drifting or capable of drifting, and "intertidal wood," that is, wood that has become wedged between rocks or partially buried in sand for long periods of time. Intertidal wood also includes submerged parts of permanently fixed wooden structures such as pilings and bulkheads.

b. Test Panels and Poles. Marine mycologists often use wooden panels to trap fungi. In this method, first described by Meyers (1953) and later modified by T. W. Johnson *et al.* (1959) and by other researchers, panels are placed on a string or chain and submerged for several weeks until recovered and examined. Wood pieces, about $15 \times 13 \times 2$ cm, are arranged in a sandwich fashion on a nylon or polyethylene cord 6 to 10 mm in diameter, separated from each other by brass washers. Smaller wood specimens (e.g., $3 \times 2 \times 0.5$ cm) separated by pieces of plastic tubes have also been used successfully (Kohlmeyer, 1963c). The optimal length of exposure depends on location and the species of wood. It is to be expected that in tropical or subtropical areas untreated wood is deteriorated much faster by boring animals than it is in colder regions. After the wood is removed from the water, it must be examined as soon as possible to disclose fungal structures present at the time of recovery. If desired, the wood can be incubated for a certain time (Meyers and Reynolds, 1958; see p. 15 on "indirect" methods) after the initial examination has been performed. Some investigators (e.g., Johnson and Sparrow, 1961) recommend scraping the test panel with a putty knife to remove the fouling organisms. This procedure usually damages or even removes fungal fruiting structures present in or on the softened outer layers of the wood. If growth of bryozoa, algae, or tunicates impedes handling of the panel under the dissecting microscope, it is preferable to cut these organisms carefully off with a sharp knife without damaging the wood surface. Complete cleaning of the wood by scraping should be done only if the specimen must be incubated in order to obtain a secondary growth of fungi (p. 15).

Our experience in two decades of collecting marine fungi has shown that submerged wood panels yield a smaller number of species than do wooden substrates found in the natural habitat, such as pilings, driftwood, lobster pots, eroded trees, or roots of mangroves. The cause for the greater variety of fungal species on the latter substrates is probably related to the different age, origin of growth, and diversity of tree species that produced these materials. Panels used in submergence experiments are restricted to a few species of wood that usually come from one source of supply, may even be derived from one tree, and are decorticated. Bark that is often found on driftwood branches is host to a variety of higher marine fungi (Table XVI), some of which may not be found if only dressed wood panels are used in ecological surveys. G. C. Hughes (1975) also concluded that studies of intertidal wood give a better estimate of species diversity and distribution of lignicolous fungi in a certain area than do trapping experiments with wood panels. On the other hand, an advantage of the panel method over the collecting of driftwood and other substrates is that the species of wood and the time of submergence of the panels are known, whereas these data are usually missing for material washed up on the beach or for shoreline fortifications.

Untreated wooden poles of at least 5-cm diameter are valuable substrates for determining the vertical distribution of marine fungi. Schaumann (1968) described the zonation of fungi along a partly corticated oak pole from Bremerhaven, Germany. Experiments made by us in North Carolina with fir and pine poles (5 × 5 cm in diameter, 1.5 m long) placed in salt marshes showed the suitability of such poles for experimental studies. Optimal length of exposure varies with the locality and depends largely on the degree of attack by wood-boring animals. The poles must be recovered shortly before destruction by borers causes fracture of the wood at the intertidal zone. Zonation of fungal species below, at, and above the high-tide line can be determined.

c. Roots and Submerged Branches of Mangroves and Other Trees. Partly submerged trees along eroding shorelines and roots of mangroves are promising substrates for wood- and bark-inhabiting marine fungi (Cribb and Cribb, 1955, 1956; Kohlmeyer and Kohlmeyer, 1965). Fungi grown on permanently or intermittently submersed roots and branches are doubtlessly aquatic. Salinity determinations in the vicinity of the trees must establish whether the habitat is marine, brackish, or freshwater. Mangrove trees (*Avicennia, Rhizophora* spp., and others) are able to grow in full salinity seawater as well as in freshwater, even developing best in the latter. Overlapping marine, freshwater, and terrestrial mycotas in the mangrove habitat offer fascinating opportunities for ecological studies.

Submersed roots of mangroves or other trees are cut immediately above the sediment with a saw or knife. Parts of roots embedded in the soil are generally useless for mycological investigations because the surrounding substrate is low in oxygen and full of anaerobic bacteria, as indicated by the black color and smell of H_2S. Only those parts of roots and branches that are submerged for some part of the day in water are suitable for collecting aquatic fungi. Exposed parts above the water line are inhabited by terrestrial fungi. Thus, the distribution of terrestrial and aquatic fungi overlaps on mangrove roots and branches at the high-water mark (Kohlmeyer, 1969d). Injured, partly decorticated roots should be collected because fruiting bodies and conidia of higher marine fungi develop more readily on such wounded areas. Roots are placed in sterile plastic bags and further treated as explained above (Section I, A, 1, a and b).

Large root systems of mangrove trees can also be collected for the study of vertical distribution of fungi along the roots. Pieces of horizontal *Avicennia* roots up to 1 m long, with their attached vertical pneumatophores, can be cut, folded, and placed in a plastic bag. In the laboratory the roots are unfolded and examined under a dissecting microscope. Young damaged *Rhizophora* seedlings with prop roots, up to about 1.5 m in length, can also be collected *in toto* for further examination.

2. Algae

Representatives of all fungal groups except Basidiomycetes have been found on marine algae. Living algae are generally collected at low tide from the shore by dredging or by skin diving. Algae washed ashore and lying on the beach for a certain time are often contaminated by terrestrial molds, and critical examination has to prove which of the fungi are indigenous marine species. In general, collecting algicolous marine fungi is not easy because it requires time-consuming examinations of great numbers of algae in order to detect a few fungal parasites, symbionts, or saprobes. Fungal infestations occur particularly on Rhodophyta and Phaeophyta, rarely on Chlorophyta. Discolorations of the plants indicate attack by fungi, but similar spots may be caused by parasitic algae or bacteria.

   a. Ascomycetes. Among the 39 Ascomycetes occurring on marine algae, 31 have been collected on living plants (see Table V, p. 56). The majority of algicolous fungi can be found on algae washed up on the beach. *Haloguignardia irritans* is a distinct parasitic, gall-forming species that can be easily collected along California beaches on *Cystoseira* and *Halidrys*. Other gall-causing Ascomycetes occur on a small number of plants of *Cystophora* and *Sargassum*. Black, stalked ascocarps of *Thalassoascus tregoubovii* (see Fig. 100) develop on the stalk near the holdfasts of *Aglaozonia* and *Cystoseira* in the Mediterranean and on the Canary Islands. This fungus is also uncommon and grows predominantly in habitats with stagnant or contaminated water. *Phycomelaina laminariae,* forming black patches on stalks of *Laminaria* spp., is easily gathered with drift or attached algae, especially in New England, but in northern European countries as well. At low tide, this fungus can be seen *in situ* on *Laminaria* stalks and holdfasts. An endemic Ascomycete in brown algae, *Mycosphaerella ascophylli*, forms ascocarps in receptacles of *Ascophyllum nodosum* and *Pelvetia canaliculata*. A common Ascomycete in *Chondrus crispus,* namely, *Didymosphaeria danica* (see Fig. 88d), is collected either *in situ* or with plants washed ashore. This fungus is readily detected by the black spots formed in the tips of the thalli, in the region of the cystocarps. Perthophytic *Didymella fucicola* develops in damaged lower side branches of *Fucus* spp. with ascocarps embedded in the central midrib. Many specimens of *Corallina, Halimeda,* and *Jania* must be examined under the dissecting microscope before one of the small epiphytic ascocarps of *Mycophycophila corollinarum* will be found.

   Collected specimens are placed in plastic bags on ice in a cooling box during transportation. Yeasts, bacteria, and terrestrial molds develop rapidly on the surface of algae and tend to grow over the marine fungi, unless the material is kept cool. Therefore, the algae should be examined immediately or stored in a freezer at about −18°C.

   b. Deuteromycetes. Direct observation of algicolous Deuteromycetes is more or less restricted to Coelomycetes, whereas conidia of Hyphomycetes appear mainly after incubation of the substrate and, in most cases, cannot be considered to be true marine fungi. Ascocarps of several Ascomycetes are associated in nature with pycnidia or spermogonia (Table

XXI), for instance, in *Didymella fucicola* (on *Fucus*), *Didymosphaeria danica* (on *Chondrus*), *Mycosphaerella ascophylli* (on *Pelvetia* and *Ascophyllum*), *Phycomelaina laminariae* (on *Laminaria*), and *Thalassoascus tregoubovii* (on *Cystoseira*).

3. Marine Phanerogams

Collecting techniques for fungi occurring in marine phanerogams are similar to those described above for algicolous and lignicolous species.

a. Permanently Submerged Plants. Host plants of marine fungi include sea grasses, such as *Posidonia oceanica, Ruppia maritima, Thalassia testudinum,* and *Zostera marina.* Rhizomes and leaves of these plants are often washed ashore and are collected along the beach or gathered by dredging or skin diving. The large ascocarps of *Halotthia posidòniae* on *P. oceanica* are easy to detect on bare parts of the rhizomes near old leaf bases. The foliicolous Ascomycete on *Thalassia, Lindra thalassiae,* occurs on leaves washed up on the beach. Brown and whitish spots on the tips of the leaves often indicate attack by the fungus. Fruiting bodies can be found *in situ,* or they develop after incubation of damaged leaves.

b. Partly or Intermittently Submerged Plants. Higher marine fungi are easily collected in salt marshes on *Juncus, Salicornia, Spartina,* and other hosts. Old rhizomes, culms, or leaves of these plants lying on the sediment are deteriorated by a number of fungi, predominantly Ascomycetes (Table XI). Dead standing upper parts that have never been submerged may be inhabited by terrestrial fungi. Ascocarps and pycnidia of marine fungi are often embedded in the substrate without being visible from the outside. The cuticle or outer cell layers of stems or leaves are cut off in a tangential section with a razor blade under the dissecting microscope, or the specimens are split lengthwise, in order to make visible the fruiting bodies, which are often arranged in rows parallel to the long axis of the culm. Black spots on stems and leaf sheaths of *Spartina* spp. indicate the location of hidden ascocarps. Salt-marsh plants offer a wide field for new discoveries of marine fungi; however, descriptions of new species should be made only after a careful literature search because some of the marsh fungi may have been described before from other hosts of freshwater or terrestrial habitats.

c. Litter-Bag Experiments. Most research on marine phanerogams has been carried out with plants collected *in situ,* except for some litter-bag experiments. In such tests, sterilized and weighed plant material is submerged in estuarine or marine habitats in noncorroding plastic containers, usually made of nylon mesh. Anastasiou and Churchland (1969) submerged leaves of terrestrial plants (*Arbutus menziesii* and *Prunus laurocerasus*), whereas Fell and Master (1973) used mangrove leaves

(*Rhizophora mangle*). In both of these investigations leaves were recovered at regular intervals and fungi were isolated after incubation. Gessner and Goos (1973a) employed the litter-bag method to investigate the mycota of *Spartina alterniflora* in a Rhode Island tidal marsh. These authors examined the grass microscopically after recovery, determined the loss in dry weight, and prepared dilution plates with the decomposing plants. The pour-plate technique proved to be selective for terrestrial molds, a fact that points to the importance of direct microscopic examination after collection. Finally, Newell (1976) utilized untreated and injured seedlings of *Rhizophora mangle* submerged in nylon mesh bags. The extensive sampling program of Newell included direct observation at the time of collection and after incubation, as well as plating, baiting, and use of the wash water.

Most litter-bag studies did not consider the influence of tides, since the plant material was submerged at all times. Certain fungi of the salt marshes and mangroves may require an alternation between immersion and drying out, and future investigations should take this possibility into account.

4. Sea Foam Containing Fungal Spores

The environment between sand grains in marine beaches, the endopsammon, harbors a rich microfauna and microflora (see Chapter 8). Higher marine fungi occur in the interstitium, and ascospores and conidia of marine fungi are released from fruiting bodies and conidiophores, respectively, and carried away with the waves. Ascospore appendages or tetraradiate shapes of spores and conidia favor their entrapment in foam between air bubbles. Foam is easily collected along the shore with a spoon-type skimmer. This slightly convex spoon, about 11 cm in diameter, with about 100 holes, each 2 mm in diameter, is used to scoop up foam, which is poured into a widemouthed bottle. Care should be taken not to collect any sand or water with the foam. Bottled samples are refrigerated at about 5°C for several hours to allow foam to "dissolve" and spores to settle. Storage at low temperatures is necessary to keep spores from germinating. Finally, a droplet of the sediment is taken out with a pipette, placed on a slide, and covered with a large cover glass (25 × 25 mm). Fungal spores are sought out by scanning the slide under a 40× objective, possibly in dark field or phase contrast, where ascospores, basidiospores, or conidia stand out by their refraction between nonrefractive amorphous particles (see Fig. 12). If samples cannot be examined the same day, formalin (45%) is added to the dissolved foam to yield a 5% solution.

On certain days, when foam formation on the shore is heavy,

receding water leaves thick patches of foam along the upper edge of the high-tide line. Eventually, this foam collapses and dries, forming greyish-green spots on the sand and covering plant and animal debris. The dry foam can be carefully scraped off the sand or debris and shaken in water, and the sediment examined for fungi. Fungal spores found in foam give a good indication of the mycota present in the sand of a particular beach (Kohlmeyer, 1966a).

5. Marine Animals

The Ascomycete *Pharcidia balani* has a worldwide distribution on shells of living balanids, as well as limpets and other mollusks growing in the intertidal zone. Ascocarps and pycnidia of this fungus appear as black dots on the surface of the shells.

Other substrates of animal origin attacked by marine fungi are the cellulose mantle of tunicates and calcareous linings of empty shipworm tubes in rotten wood. Ascocarps of *Halosphaeria quadricornuta* occur in both materials. Imperfect fungi may also be found in the tube linings, forming conidia inside the substrate and breaking through the surface. Perithecia and hyphal aggregations appear as dark spots in the white calcareous tubes. These $CaCO_3$-inhabiting fungi are more frequent in tropical and subtropical waters than in temperate zones.

B. Indirect Collecting Methods (Incubation)

1. Sediment and Water Samples

Sediment or soil samples are collected on beaches, or in mangrove or salt-marsh communities, placed in sterile plastic bags (e.g., Nasco Whirl-Pak), and refrigerated at about 4°C until used. Soil from offshore locations is gathered with a bottom grab or coring tube lowered from shipboard. Several types of sampling devices have been used successfully by mycologists. Höhnk (1972) describes the use of the van Veen bottom grab, which was applied in the Indian Ocean to depths of 3450 m. Another device used by Höhnk is the 2.5-m-long coring tube devised by Pratje. Parts of the sediment can be taken for examination from any depth of the core. This type of tube works satisfactorily in every location except when it hits bare rock.

Water specimens from below the surface can be collected with the Cobet–Weyland sampler (Höhnk, 1972) or other devices, for example, Nansen bottles. Sediment or water specimens are used in dilution plates to obtain fungal cultures after incubation.

2. Natural Substrates

At this point, one must be cautioned again that the incubation method does not exclude isolation of fungi of nonmarine origin. Nevertheless, some widely used techniques are as follows: Pieces of wood, living or dead algae, or marine phanerogams collected in marine habitats are incubated in moist chambers, for example, petri dishes or larger containers lined with wet filter paper. Mycelium or fruiting bodies developing on the surface of the substrates can be used for the isolation of fungi. Incubation of algae or other plants in sterilized seawater from the collecting site is another method employed.

Marine animals also serve as a source of fungi. The contents of the intestines of fishes, sea cucumbers, sea urchins, and starfishes or the fluid of sponge colonies are used for dilution plate isolations. Yeasts can be obtained from surface smears of fishes, algae, or other organisms.

3. Baits

Depending on the fungal species, different substrates should be used for trapping. Wood panels as a source of higher fungi have been mentioned in Section I, A. Some authors (e.g., Johnson and Sparrow, 1961) advise submerging the panels for certain periods of time, followed by removal of the fouling organisms after recovery and incubation of the wood in moist chambers. Another method uses cores cut with a borer from pilings or other wooden substrates and subsequent incubation of the specimens. Again, it is emphasized that nonmarine fungi may develop in addition to marine species after incubation in the laboratory.

II. PRESERVATION

In view of the large number of doubtful and rejected marine species (Chapter 27), we recommend strongly not only that type material be deposited in recognized herbaria, but also that other specimens which are the bases for check lists or other research should be deposited there. Such thoroughly labeled material will permit later investigators to confirm or correct earlier identifications. As Dennis (1968) so aptly stated: "Lists of records that cannot be verified are mere waste paper."

Collected specimens can be preserved for future examination in the following ways: alive in a pure culture, frozen, dried, in liquids, or on a microscope slide. There is no rule for the most suitable preservation technique of marine fungi. Every species may have its own requirements, but freezing, freeze-drying, and microscope mounts are probably the best preservative methods.

A. Freezing

Algae- and wood-inhabiting marine fungi are well preserved if stored at temperatures around −18°C (Kohlmeyer, 1968e, 1969c). The morphology of Ascomycetes and Deuteromycetes remains unaltered after repeated freezing and thawing. Conidia of *Robillarda rhizophorae* survived this treatment and germinated at room temperature in distilled water. To prevent drying of the specimens in the freezer, the substrate is stored in seawater in a plastic bag. Fungi kept in frozen seawater will probably remain morphologically unchanged indefinitely.

B. Drying

Drying techniques depend on the substrate on which the fungus has developed. Bulky, hard material such as wood, bark, rhizomes, or calcareous algae can be air-dried (not in the sun!). Soft algae containing fungi are handled like other algal collections, that is, dried in a plant press between pieces of cloth (Dawson, 1956). Leaf-inhabiting fungi are dried with the substrate between drying papers in the plant press. Pure cultures are made into dried reference specimens according to the method described by Pollack (1967). The partially dried agar culture is removed from the petri dish, 15 ml of hot 2.5% glycerol agar is poured into the lid of the plate, and the culture is floated on the hot agar until the lower layer is solidified. The culture is left uncovered and is then peeled off the plate after it is completely dry. The pliable agar can be mounted in a cardboard folder for filing.

Freeze-drying of marine fungal specimens is recommended if a freeze-dryer is available. This technique guarantees the best possible preservation of the material and prevents, for instance, collapsing of fruiting bodies, which often occurs during slow air-drying.

C. Liquid Storage

Preservation of marine fungi in the most widely used preservatives, namely, ethyl alcohol (about 70%) or 5% formaldehyde in seawater, is not always satisfying. Delicate structures, such as asci, interthecial tissues, or spore appendages, may shrink and become useless for future work. Identification of species preserved in alcohol or formalin is often difficult or even impossible if spore appendages have lost their natural shape. The

use of alcohol or formaldehyde as fixative is suitable if the material is to be sectioned after dehydration and embedding in paraffin wax. However, sectioning the living material in a cryostat is preferable.

Liquids containing fungal spores, such as seawater resulting from "liquified" foam, should be examined as soon as possible after they have been collected. These specimens can be preserved by adding formalin to yield a 5% solution.

D. Permanent Microscope Slides

Squash mounts or cryostat sections of living specimens are examined in water and then made into permanent slides according to the double–cover-glass method (Kohlmeyer and Kohlmeyer, 1972c). For the initial examination of a fungus, a square cover glass (25 × 25 mm, No. 1) is placed on a slide (76 × 26 mm) and tiny droplets of distilled water are placed with a dissecting needle at two sides of the cover glass to make it adhere to the slide (Fig. 2a). A larger droplet of distilled water is dropped in the center of the cover glass, and the specimen is placed in the water and covered with a small cover glass (18 × 18 mm, No. 1; Figs. 2b and 2c). Now the living material is ready for microscopic investigation, including under oil immersion if required. After the necessary observations are made and drawings or photographs are prepared or measurements are taken, a drop-

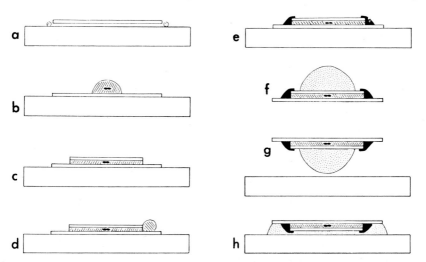

Fig. 2. Double–cover-glass method for permanent microscopic mounts. Preparation in vertical section; drawings not to scale. See the text for explanation. From Kohlmeyer and Kohlmeyer, *Mycologia* **64,** 667 (1972), Figs. 1–8. Reprinted with permission.

let of concentrated glycerin is added to the water from the side (Fig. 2d). The slide is stored horizontally in a dust-proof container for several days to permit the water to evaporate. After cleaning the edges, the mount is sealed twice with a ring of clear fingernail polish (Fig. 2e). The large cover glass is detached from the slide and a drop of mounting medium (e.g., synthetic Canada balsam "Caedax," E. Merck) is placed on the small cover glass (Fig. 2f). The preparation is turned over with the drop of mounting medium hanging down and then placed on the slide (Fig. 2g). The drop of balsam flattens out and surrounds the edges of the small cover glass, thereby permanently covering the dried ring of nail polish (Fig. 2h). A slowly hardening mounting medium is more suited for permanent preparations than fast-drying media, which tend to become brittle and crack over the years.

A reference collection of permanent slides is probably more valuable than other kinds of preserved specimens, because such slides keep indefinitely and are immediately available for comparison.

III. SECTIONING

The preparation of sections is essential for the investigation of fruiting bodies of the higher marine fungi. Squash mounts are often sufficient for the identification of known species, but sections must be made for thorough examinations of critical or new fungi. Hard substrates, such as wood or stems of marsh plants, can be hand-sectioned with a razor blade. Small specimens or fruiting bodies embedded in algae are sectioned on a microtome, preferably in a cryostat, where fixation of the fungi is not required. We have had excellent results in sectioning frozen algicolous fungi enclosed in the host tissue and have been able to obtain sections as thin as 2–5 μm in some species by using an International Cryostat (Model CTI) of the International Equipment Company. Delicate internal structures of ascocarps and pycnidia remain unaltered during the freezing, sectioning, and thawing process. The sections are transferred into a drop of distilled water on a slide, covered with a cover glass, and examined under the microscope. Stains can be added from one side of the preparation by drawing water out from the other side with blotting paper.

IV. MICROSCOPIC EXAMINATION

Many higher marine fungi are characterized by morphological adaptations to the aquatic habitat, in particular, the ascospores (Chapter 3).

Gelatinous or mucilaginous sheaths or appendages of the spores are often invisible in the bright-field microscope; therefore, phase-contrast optics, differential interference contrast, or stains must be used to detect their presence. The following stains are useful for this purpose: acid fuchsin–lactophenol, aqueous nigrosin, Delafield's hematoxylin, picric acid–anilin blue, or violamin. India ink added to an ascospore or conidial suspension will make those sheaths visible that do not take up stains, for example, in *Haligena amicta* and *Leptosphaeria pelagica*. If the chemical composition of ascospores and their appendages is to be examined, we recommend consulting Kirk's (1966) extensive treatise on the cytochemistry of marine Ascomycetes, where tests and procedures are described in detail.

Studies with the transmission electron microscope have been carried out on only a few higher marine fungi (Lutley and Wilson, 1972a,b). The scanning electron microscope (SEM) has been employed by Brooks *et al.* (1972) and Gessner *et al.* (1972) to investigate surfaces of wood and *Spartina,* respectively, and Moss and Jones (1977) have examined the ascosporogenesis of *Halosphaeria mediosetigera.* Basidiocarps of *Halocyphina villosa* have been illustrated in SEM pictures by Kohlmeyer and Kohlmeyer (1977). The application of transmission as well as scanning electron microscopy in marine mycology will probably increase in the near future, because both instruments have proven to be valuable in detecting previously unknown features in marine fungi.

V. ISOLATION AND CULTURE

Substrates collected as suggested in the preceding sections usually contain several species of marine fungi. In addition, the surface of the material recovered from a marine habitat may be contaminated by bacteria and dormant propagules of terrestrial fungi. Therefore, care must be taken not to isolate such contaminants in place of the usually slower growing marine fungi. Wood, leaves, or other substrates are blotted with sterile filter paper and externally dried at room temperature. Superficial fruiting bodies or conidia are located under a dissecting microscope. Immersed ascocarps or pycnidia are found by making thin sections with a sterile razor blade parallel to the surface. Some ascospores or conidia are picked out with a sterile glass or metal needle and transferred onto a slide for identification, others from the same fruiting body are used to prepare a spore suspension in 2 ml of sterile distilled water for dilution plates, or the spores are spread directly on the surface of agar. The superficial drying of the natural substrate compresses the fruiting bodies and often presses the spores out, causing them to accumulate at the orifice, where they can be taken up with a needle. Conidia of Hyphomycetes developing on the surface are transferred onto a slide, directly onto a solidified agar medium, or into water to be poured into dilution plates. After inoculation, agar plates are examined regularly under a dissecting microscope, if possible in dark field, to observe germination. Germinating single spores are transferred into test-tube cultures for permanent maintenance.

Media used for isolation should contain a bacterial inhibitor, such as 0.03% streptomycin sulfate (Kirk, 1969), potassium tellurite (20–50 mg/ liter), or tetracycline (500 mg/liter) (G. C. Hughes, 1975). Streptomycin sulfate (100 mg/liter) plus penicillin-G (100 mg/liter) is used routinely with good success by R. V. Gessner (personal communication). Glucose–yeast extract agar (Johnson and Sparrow, 1961) is generally employed for isolating marine fungi.* Natural seawater must be aged, that is, stored for at least 1 week before use, or artificial seawater can be used, for instance, made up with Rila Marine Salt Mixture (Rila Products, Teaneck, New Jersey). For general maintenance of cultures, we have also used Emerson's YpSs agar,† available from Difco Laboratories (Detroit, Michigan). The addition of a cellulose source [e.g., birch applicator sticks—*Betula papyrifera* (Kirk, 1966, 1969)] to the agar enhances the fruiting of cellulose-decomposing fungi. We have also had good results in growing cellulolytic species on 10-cm-long strips of filter paper or cellophane or on birch sticks in test tubes half filled with seawater plus yeast extract (0.1%). Such cultures can be maintained over long periods without drying out.

A liquid medium particularly recommended for experimental use is that employed by Sguros *et al.* (1962), later modified by Meyers and Simms (1965) and Meyers (1966).** These authors developed a procedure to supply uniform inocula and to measure gravimetrically the growth of marine fungi. This standardized method yields data with a high degree of reproducibility, and it is easily modified. As G. C. Hughes (1975) pointed out, the reproducibility is highly significant statistically ($P < 0.001$) with only three replicates.

A closed system used in nutritional studies by most investigators has the disadvantage that nutrients become exhausted, pH and osmotic changes occur, and toxic metabolites accumulate. Therefore, a continuous culture system, such as that developed by Churchland and McClaren (1976), appears to be advantageous over the closed (or batch culture)

---

* Glucose, 1 g; Bacto-yeast extract, 0.1 g; agar, 18 g; aged seawater, 1 liter.

† Bacto-yeast extract, 4 g; soluble starch, 15 g; dipotassium sulfate, 1 g; magnesium sulfate, 0.5 g; Bacto-agar, 20 g; seawater, 1 liter.

** Glucose, 5 g; $MgSO_4 \cdot 7H_2O$, 2.4 g; NH_4NO_3, 2.4 g; yeast extract (Difco), 1 g; Tris buffer, 1.21 g (pH 7.5); natural seawater, 1 liter.

system. The growth of three marine Hyphomycetes tested by these authors was significantly greater in continuous culture than in batch culture. Churchland and McClaren (1976) conclude that "the continuous system can possibly be regarded as a closer approximation to the marine environment than the batch culture system." Another continuous culture chamber was devised by Padgett and Lundeen (1977) in which the salinity can be varied in a cyclic pattern, simulating the conditions in estuaries. The continuous flow systems require a large apparatus with 6-liter reservoirs, which limit the use of this excellent method when extensive experiments are required.

Temperatures for isolating and maintaining pure cultures of marine fungi depend on the source of the species. The majority of ubiquitous fungi can be grown at temperatures between 15° and 25°C [e.g., *Nia vibrissa* (Doguet, 1968)]. Some tropical species, for example, *Halosphaeria quadricornuta,* require temperatures around 28°C for germination (e.g., Kohlmeyer, 1968c), whereas some fungi from temperate waters, such as the Ascomycete *Herpotrichiella ciliomaris,* germinate only at about 16°C (Kohlmeyer, 1960), and growth of the Basidiomycete *Digitatispora marina* ceases above 23°–25°C (Doguet, 1964).

Small laboratories are usually unable to maintain pure cultures of microorganisms in perpetuity, and it is suggested that cultures of marine fungi be made generally available by depositing subcultures of them in large collections, for instance, in the American Type Culture Collection (Rockville, Maryland) or in the Centraalbureau voor Schimmelcultures (Baarn, The Netherlands).

3. Release, Dispersal, and Settlement of Ascospores, Basidiospores, and Conidia

This chapter deals with the different types of morphological adaptations of higher marine fungi to the aquatic habitat and discusses the importance of these features for the release, dispersal, and settlement of propagation units.

I. SPORE RELEASE

A. Ascomycetes

In this group, spores are released actively or passively. Mechanisms that eject spores forcibly away from the fruiting body have probably evolved in terrestrial habitats where spore dispersal is most efficient if a spore is catapulted into the air stream. In water, however, forceful ejection of spores is of no apparent value, because spores are easily washed away from the orifice of a fruiting body as soon as they reach this opening. Therefore, we consider the "jack-in-the-box mechanism" of bitunicate asci a character of terrestrially adapted Ascomycetes, whereas dissolving, thin-walled asci are more likely to be found among originally aquatic species (Barghoorn and Linder, 1944; Kohlmeyer, 1974a, 1975a; see also Chapter 22).

The forcible discharge of ascospores is found in the unitunicate as well as the bitunicate ascus. In both cases the increasing osmotic pressure at maturity causes an elongation of the ascus. The unitunicate ascus stretches into the ostiole, ruptures at the apex, and releases the ascospores, one after the other, into the water. This mechanism occurs in *Chaetosphaeria chaetosa* and other members of Sphaeriales. In the bitunicate ascus of the Loculoascomycetes, the outer wall, or ectoascus, breaks at the apex; the inner wall, or endoascus, elongates into the ostiole, ruptures apically, and forces the ascospores through the opening (Fig. 3). Ascospores may be released into the air or under water, for example, in *Herpotrichiella ciliomaris*, where ascocarps often develop on permanently submerged bark (Kohlmeyer, 1960). An aerial spore release is found in *Mycosphaerella ascophylli*, when the host, *Ascophyllum*

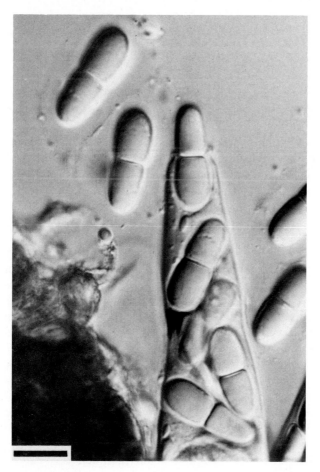

Fig. 3. *Mycosphaerella salicorniae,* tip of ascus, releasing ascospores; spores surrounded by a gelatinous sheath (Herb. J.K. 3772; Nomarski interference contrast; bar = 10 μm).

nodosum, is exposed at low tide and dries up superficially, thereby squeezing the ascocarps. Ascospores of *M. ascophylli* can be shot upward into the air to a distance of about 5 mm (F. C. Webber, 1967), where wind currents may carry them farther away.

In some marine Ascomycetes the ascus becomes detached from the ascogenous tissue before the ascospores are discharged. In *Gnomonia marina* the asci move singly up the neck of the ascocarp and eject the ascospores through the ostiole (Cribb and Cribb, 1956). The shriveled ascus follows the spores into the water and regains its original shape. A similar ascospore release occurs in terrestrial fungi [e.g., *Ceratostomella ampullasca* (Ingold, 1971)]. The asci of *Halosarpheia fibrosa* also separate from the ascogenous tissue before releasing the ascospores, but the spore discharge has not been observed (Kohlmeyer and Kohlmeyer, 1977). In *Lignincola laevis,* asci are the actual dispersal units, because they break loose at the base, float freely in the ascocarp venter, and are eventually released through the ostiole into the water (Kohlmeyer and Kohlmeyer, 1971a). The free-floating ascus swells in the central part (Fig. 4), finally ruptures, and releases the ascospores.

The most common case of ascospore release in marine Ascomycetes, especially in the Halosphaeriaceae, is by dissolution of the ascus, shortly before or at ascospore maturation. At this stage the ascocarp is still closed at the apex. Ascospores accumulate in the venter of the ascocarp and are liberated after the formation of an ostiolum, probably by increased pressure in the ascocarp through gelatinization of the pseudoparenchyma or swelling of catenophyses. After leaving the fruiting body, either the spores are carried away by water currents or, when ascocarps are exposed at low tide, they collect over the ostiole. Fazzani and Jones (1977) examined the spore release of *Halosphaeria appendiculata,* in which a rapid initial discharge of spores occurred during the first hour (up to 50 per minute), with only a few following subsequently. About 2000 spores were released from one ascocarp with an average rate of 11 spores per minute.

A turgescent swelling cushion in the ostiole of *Turgidosculum ulvae* (Fig. 5) prevents the penetration of water into the venter of mature ascocarps, where asci develop in succession. The host, *Blidingia minima* var. *vexata,* dries out during low tides and the shrinking ostiolar pulvillus permits release of the ascospores, which first accumulate at the ostiolum and are dispersed at high tide (Kohlmeyer and Kohlmeyer, 1972b). A similar dispersal mechanism and protection for immature asci appear to occur in *Turgidosculum complicatulum,* in which the venter of the ascocarp is filled with a gelatinous matrix, surrounding the asci. The discharged asci of other fungi, for example, *Didymella fucicola* and *Myco-*

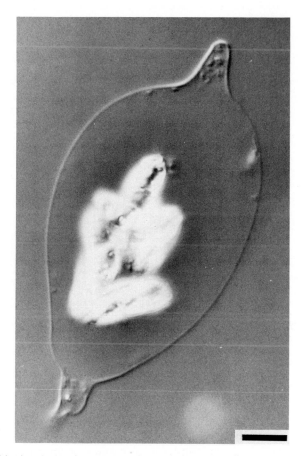

Fig. 4. *Lignincola laevis,* ascus with central part swelling in water after being released intact from the ascocarp (Herb. J.K. 2694; Nomarski interference contrast; bar = 10 μm).

sphaerella ascophylli, seem to have a similar function, by forming a dense "palisade" in the ostiole of the ascocarp, preventing the entrance of water.

A special case of adaptation appears to occur in the marine yeast *Metschnikowia bicuspidata* var. *australis*. The needle-shaped ascospores are forcibly discharged and, thus, may be inserted into vital elements of the host. This yeast is pathogenic to brine shrimp (*Artemia salina*), and the spore release may constitute an active mechanical predation (Lachance *et al.*, 1976).

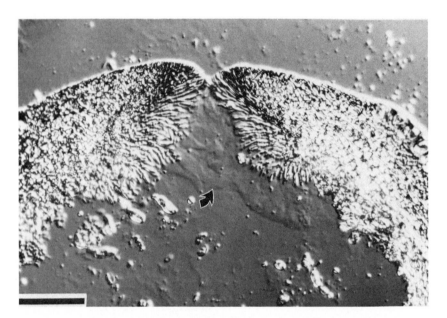

Fig. 5. *Turgidosculum ulvae*, 4-μm longitudinal section through ascocarp apex; ostiolar canal filled with a turgescent pulvillus (arrow) of gelatinous material that dries upon exposure at low tides and permits dispersal of ascospores (Herb. Setzer 3750; Nomarski interference contrast; bar = 20 μm).

B. Basidiomycetes

Release of basidiospores of marine Basidiomycetes has rarely been described, although the method of discharge can be inferred from the morphology of the basidiocarps. Fruiting bodies of *Digitatispora marina* are flat or crustlike, with the hymenium covering the outer side. Since the basidia are exposed to the water, the spores are probably detached from the basidium and washed away as soon as they become mature. Cupulate or funnel-shaped basidiocarps of *Halocyphina villosa* bear mature basidiospores in a cluster on the top. Spore dispersal of *H. villosa* is most likely passive by water currents. Also *Nia vibrissa* has no active release mechanism of basidiospores. Its basidiocarps are globose, puffball-like structures without openings. At maturity the peridium ruptures irregularly and the appendaged basidiospores are passively released. This breaking of the wall is probably caused by bacterial action and not by abrasion or other mechanical modes (Fazzani and Jones, 1977).

By morphological adaptations to the aquatic habitat, basidiocarps of *N. vibrissa* may have a function in their own right for dispersal of the fungus. They develop superficially, become easily detached from the substrate, and float on the surface of the water because air is trapped among the numerous superficial hairs (Kohlmeyer, 1963d). Fazzani and Jones (1977) determined that basidiocarps of *N. vibrissa* placed in seawater float for about 7 days. In this respect, *N. vibrissa* resembles *Limnoperdon incarnatum,* the Gasteromycete that was described from wood in freshwater (Escobar *et al.*, 1976). Basidiocarps of *L. incarnatum* are superficially similar to those of *N. vibrissa* and float on water as well. These two species are not closely related and have apparently developed in analogous habitats by convergent evolution.

C. Deuteromycetes

Conidial release in most marine Hyphomycetes appears to be passive. No release mechanism is known, but conidia probably fall off the conidiogenous cells as soon as they become mature. Developmental stages of *Varicosporina ramulosa* show that mature conidia sever easily from the point of attachment (Fig. 6). A similar conidial release occurs in other

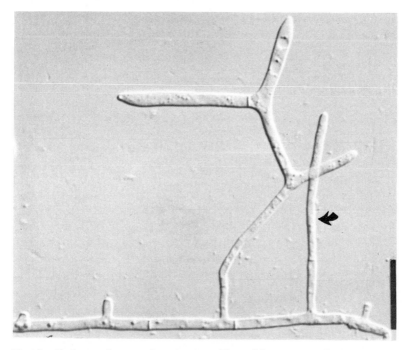

Fig. 6. *Varicosporina ramulosa,* hypha with conidiogenous cells; conidia severed from the left two and the right one; immature conidium with the first-formed primary axis (arrow) (Herb. J.K. 3907; Nomarski interference contrast; bar = 20 μm).

marine Hyphomycetes, for example, *Cirrenalia* and *Zalerion* spp. Conidia of *Z. maritimum* may separate from the conidiogenous cell, or the latter may become detached from the vegetative hypha and remain connected to the conidium (Cole, 1976). In *Asteromyces cruciatus* the question of conidial secession is not as clear, depending on the definition of "conidium" in this species. Based on observations in nature where the propagule consists of a cluster of smaller cells ("conidia") attached with denticles to a larger bulbous central cell ("conidiogenous cell"), we tend to consider this whole aggregate as one conidium, comparable to conidia of *Orbimyces spectabilis* or *Clavariopsis bulbosa*. This view is in contrast to the opinion of most other mycologists (e.g., Cole, 1976), who regard each cell as one conidium. Propagules of *A. cruciatus* are commonly trapped in foam along the shores of temperate zones, but rarely are the single cells separated from the central "conidiogenous" cell. When a secession occurs, the denticle wall fractures at any point between the base of the "conidium" and the surface of the conidiogenous cell (Cole, 1976). The secession of the conidial aggregate as it usually happens in nature is by simple separation of the conidiogenous cell from the point of attachment to the hypha.

Conidial release of most marine Sphaeropsidales has not been observed. In some species, for example, *Cytospora rhizophorae* on roots of mangroves, conidia ooze out of the pycnidial orifice at low tide and form yellow to reddish beads or curled cirrhi. These conglomerates "dissolve" with rising water and conidia are dispersed.

II. SPORE MORPHOLOGY, DISPERSAL, AND SETTLEMENT

Morphological characters that enhance dispersal of spores occur in marine Ascomycetes, Basidiomycetes, and Deuteromycetes. In particular, those formations that enlarge the spore surface and thereby minimize the settling rate assist in keeping the spores suspended in the water and increase their chance of touching a substrate. Barghoorn and Linder (1944) were first to observe the dual function of ascospore appendages of marine fungi as floating and attachment devices.

A. Ascomycetes

The greatest variability of appendages is found in the Ascomycetes, in particular, in the family Halosphaeriaceae. Such ascospore appendages are thorn-, spine-, tube-, cap-, veil-, or fiberlike, and their consistence is tough, gelatinous, or mucilaginous, and, consequently, they are dry or sticky. Soft appendages or sheaths attach the spores to submerged substrates (Fig. 7) until the germ tube penetrates the substratum. Tubelike appendages of *Corollospora tubulata* and *Halosphaeria tubulifera* and apical chambers in ascospores of *Lulworthia* spp. contain mucilage that adheres to submerged objects. In *Halosarpheia fibrosa, Haligena viscidula,* and others, the apical ascospore caps evolve into long filamentous threads that aid in spore dispersal by byssoid drifting (Kohlmeyer and Kohlmeyer, 1977) and anchor the spores to substrates.

Appendages that appear almost identical (e.g., those of *Halosphaeria appendiculata* and *Ceriosporopsis calyptrata*) may develop by quite different processes. In particular, T. W. Johnson (1963a–d) and T. W. Johnson and Cavaliere (1963) examined ascospore ontogeny and development of appendages in several species of marine Ascomycetes. Other publications on appendage formation are by Kohlmeyer (1966b) and Lutley and Wilson (1972a,b). Kirk (1966) did pioneer work in exploring the cytochemistry of ascospores in eight marine species. Ascospore appendages are usually chitinized membranous or nonchitinized mucilaginous processes of exosporic nature (Kirk, 1966, 1976). They develop by

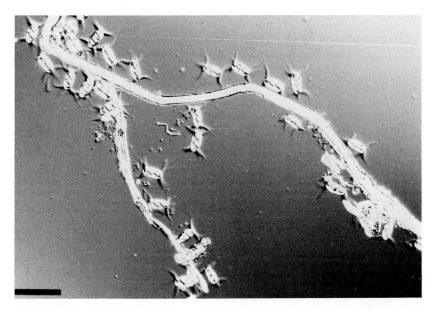

Fig. 7. *Halosphaeria salina,* ascospores attached by sticky appendages to a cellulose fiber (Herb. J.K. 2725; Nomarski interference contrast; bar = 50 μm).

outgrowth or by fragmentation of the outer, exosporic layer, which separates in different ways from the epispore (Fig. 8). An additional sheath, derived from the epispore, may enclose the ascospore and the appendages, for example, in *Ceriosporopsis halima* (Kirk, 1976). Striplike ascospore appendages are found in most species of *Corollospora,* developing from a thin layer of chitinous exospore (Kirk, 1966; Kohlmeyer, 1966b; Kohlmeyer *et al.*, 1967; Lutley and Wilson, 1972a; E. B. G. Jones, 1976). Additional apical spines are formed as extensions of the spore body at an early stage of ascospore development in *Corollospora intermedia, C. lacera,* and *C. maritima* (T. W. Johnson, 1963b). Ascospore appendages of *Torpedospora radiata, Halosphaeria quadriremis,* and *H. stellata* contain parallel fibrils (Kohlmeyer, 1960; Lutley and Wilson, 1972a), which possibly transport polysaccharides (Kirk, 1976) and add strength to

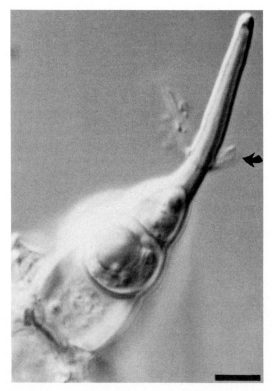

Fig. 8. *Corollospora lacera,* tip of ascospore with an apical thorn from which an exosporic layer is peeling off (arrow) (Herb. Brooks; Nomarski interference contrast; bar = 5 μm).

the appendages, like iron rods in reinforced concrete. Ascospore sheaths may also be reinforced by parallel fibrils, for example, in *Carbosphaerella leptosphaerioides* and *C. pleosporoides* (Schmidt, 1969a,b).

Ascospore appendages, such as those discussed above, are not restricted to marine species. Similar appendages occur among terrestrial, coprophilous Ascomycetes and have probably evolved as adaptations to a different environment while serving a similar purpose. Sticky ascospore appendages of coprophilous Ascomycetes ensure that the propagules adhere to plants, where they are ingested by herbivores and pass through the digestive tract.

B. Basidiomycetes

Two of the four filamentous marine Basidiomycetes are morphologically adapted to the marine habitat. The Hymenomycete *Digitatispora marina* has tetraradiate basidiospores (Fig. 9) that are comparable to similar conidia among freshwater Hyphomycetes, which are known for their radiate shapes (Ingold, 1975a, 1976). The one-celled basidiospores of the Gasteromycete *Nia vibrissa* are provided with four to six filamentous

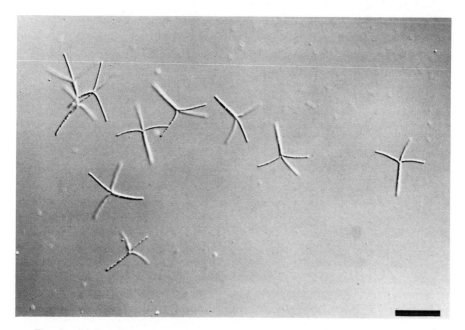

Fig. 9. *Digitatispora marina,* tetraradiate basidiospores (Herb. J.K. 905; Nomarski interference contrast; bar = 50 μm).

appendages, also bearing a resemblance to conidia of aquatic Hyphomycetes, in particular to *Actinospora megalospora* Ingold (Ingold, 1975a).

C. Deuteromycetes

Appendaged or tetraradiate conidia are rare among the marine Fungi
Imperfecti, and only eight such species have been found so far. *Robillarda
rhizophorae* is a member of Coelomycetes, with ellipsoidal conidia usually bearing three radiating appendages at the apex. *Camarosporium
metableticum* belongs to the same group of fungi and is characterized by
muriform conidia with a gelatinous cap at each end. One bristlelike
appendage is attached to each conidial apex of *Dinemasporium marinum*
(Coelomycetes).

In the Hyphomycetes, there is *Varicosporina ramulosa* with tetraradiate conidia. Conidia of both, *Orbimyces spectabilis* and *Clavariopsis bulbosa,* consist of a subglobose basal cell with several septate, radiating arms. The dispersal unit of *Asteromyces cruciatus* is composed of
several ellipsoidal cells (conidia?) connected by thin denticles to the
central "conidiogenous cell." Finally, *Clavatospora stellatacula* produces
staurosporous conidia with four short projections, a typical caltrop shape
(Savile, 1973; Bandoni, 1975).

D. General Considerations

Ingold (1975a, 1976) discussed the theories of biological significance of
appendages and tetraradiate shapes in aquatic fungi. He concluded that a
reduced rate of sinking caused by a large surface to volume ratio is of no
particular advantage in streaming water. However, Ingold accepts the
second theory that tetraradiate shapes increase the efficiency of impaction
of aquatic fungal spores. The attachment of conidia to a firm object in
running water has been examined among freshwater Hyphomycetes by
Webster (1959). Most efficiently trapped were tetraradiate conidia, followed by filamentous conidia, whereas ellipsoidal conidia were the least
successful in becoming attached to the substrate.

The sinking rate of spores of marine fungi has not been measured thus
far and, therefore, we do not know if an appendaged ascospore of, for
instance, *Ceriosporopsis calyptrata* sinks more slowly than a nonappendaged ascospore of *Nais inornata*. A reduced sedimentation rate in appendaged spores can be expected, since sinking ascospores of *Halosphaeria mediosetigera* rotate along the transverse axis because of the
lateral appendages (Kohlmeyer and Kohlmeyer, 1966). Ascospores of
both *H. mediosetigera* and *Ceriosporopsis circumvestita* became attached

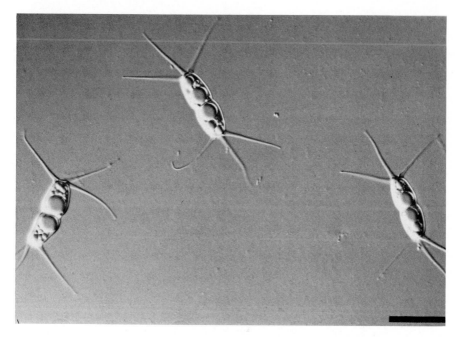

Fig. 10. *Corollospora trifurcata,* ascospores with three elongated appendages at each apex (Herb. Brooks; Nomarski interference contrast; bar = 25 μm).

upon contact with a substrate, and it was concluded that the primary functions of ascospore appendages are to respond to unorganized eddy diffusion currents and to anchor the spore (Kohlmeyer and Kohlmeyer, 1966). Ascospore appendages of *Torpedospora radiata* function as parachutes, keeping the spores in a vertical position while they slowly sink to the bottom. This process can be examined when ascospores are placed in a cuvette with water and observed with a horizontal microscope (J. Kohlmeyer, unpublished). Appendages and tetraradiate shapes are particularly advantageous in arenicolous fungi (Fig. 10), the propagules of which are trapped between air bubbles and deposited with foam on substrates along the beach (Kohlmeyer, 1966a). *Corollospora* species and the Hyphomycetes *Asteromyces cruciatus* and *Varicosporina ramulosa* are representative of this habitat. The accumulation of their spores in foam ensures that large numbers of propagules are kept on the shore in an environment suitable for growth of sand-inhabiting fungi (see Chapter 8).

In the absence of experimental data for marine fungi, we are unable to support or to reject Ingold's (1975a, 1976) conclusion that the reduced sinking rate of tetraradiate spores as compared with globose spores of the same volume is of no advantage in streaming water. The manifold spines and other structures that increase the surface to volume ratio in marine phyto- and zooplankton are known to be advantageous in keeping passively drifting organisms afloat, and we assume that they serve a similar purpose in propagules of marine fungi. Experimental work is needed to clarify the role of spore appendages and tetraradiate shapes of marine fungi and to determine whether they are merely attachment devices or also decrease the sinking rate, as well as how they function in the trapping of spores on the surface of air bubbles.

4. Geographical Distribution

Data on the biogeography of marine fungi are still scarce. Distributional records of species mostly coincide with the location of marine laboratories or with travel routes of marine mycologists or their friends who collect specimens for them. Based on the limited number of studies on marine fungi, we are not able to establish phytogeographic (or, rather, mycogeographic) regions, as they are prepared for algae (van den Hoek, 1975). Most probably, the distribution of fungi cannot be compared with that of benthic algae because the fungal occurrence depends not only on physical conditions of the environment, but also on the availability of certain substrates or hosts. In addition, many marine fungi may survive extremely unstable temperature regimes that constitute a barrier for the distribution of benthic algae. As van den Hoek (1975) points out, extreme temperature fluctuations during glaciations have caused an impoverishment of the benthic marine flora and fauna, for instance, in the Boreal Atlantic area. Latitudinal barriers, such as sediment coasts of the Carolinas, prevented southward or northward dispersals of algal species after the glaciations.

Marine fungi, on the other hand, do not seem to form a variety of discrete floristic units, as they are found among algae. The distribution of host-specific fungi, of course, depends entirely on their respective hosts. Mangrove-inhabiting fungi, for example, *Keissleriella blepharospora* on submerged roots and seedlings of *Rhizophora* spp., usually occur wherever their hosts are able to live (see Chapter 12). The same is generally true for fungi on marsh plants (e.g., Gessner and Kohlmeyer, 1976) or for obligate algae-inhabiting species (see also Chapters 9 and 10), although in some cases the algal host may have a wider distribution than its parasite. For instance, the red alga, *Ballia callitricha,* occurs in Antarctic waters, but also around New Zealand and Southern Australia. The parasitic Ascomycete *Spathulospora antarctica* is found on *B. callitricha* only below the 10°C isotherm (surface water temperature in August), whereas four other *Spathulospora* species that parasitize the same host occur at or above this isotherm. Apparently, the temperature controls the distribution of *Spathulospora* species (Kohlmeyer and Kohlmeyer, 1975).

Saprobic marine fungi have a much wider geographical distribution than most parasites, and van den Hoek's (1975) statement about algal species applies also to marine fungi, namely: "The most important environmental factor, past and present, which is responsible for the distribution over very large areas is temperature." This was recognized by G. C. Hughes (1974) when he divided the oceans into five biogeographic temperature-determined regions, namely, arctic, temperate, subtropical, tropical, and antarctic, and supplied a map outlining the proposed areas.* This scheme permits a precise classification and terminology for the distribution of marine fungi and should form a good basis for future biogeographical studies. In this treatise, we have not assigned individual species to any of Hughes' zones because distributional data for many species are too scanty to make meaningful generalizations possible. However, the known areas of occurrence (countries with states) are listed for each species (see descriptive part, Chapter 26), and general patterns of distribution can be recognized. It seems to be clear, for instance, that *Spathulospora antarctica* has an antarctic distribution, *Halosphaeria quadricornuta* is tropical and subtropical, *Asteromyces cruciatus* is temperate, and *Corollospora maritima* is cosmopolitan. Physiological studies, such as that by Doguet (1964) on *Digitatispora marina,* are important in explaining the distributional patterns of single species. Doguet (1964), based on the behavior of *D. marina* in culture, predicts that this psychrophilic fungus would not occur in areas where the water temperature reaches 30°C, even for a short time, because the species has an optimum temperature of 15°C and all development ceases at about 25°C. In fact, *D. marina* has not been recorded from tropical or subtropical regions, and Brooks (1975) collected the species in Rhode Island only during winter and spring, when water temperatures there are 15°C or below.

G. C. Hughes (1974) plotted the distributions of 20 marine wood-inhabiting fungi on zonal maps based on his proposed classification. Such distributional maps are useful in demonstrating the occurrence of "out of

---

* Delimitation of the zones: arctic–antarctic, the 10°C isotheres of sea surface temperatures for the warmest calendar month; temperate, the 10°C isotheres for the warmest calendar month (toward the pole) and the 17°C isocrymes for the coldest calendar month (toward the equator); subtropical, between 17° and 20°C isocrymes for the coldest calendar month; tropical, the 20°C isocrymes for the coldest calendar month, north and south of the equator (G. C. Hughes, 1974).

place" records in the literature [e.g., *Halosphaeria quadriremis* in the tropics (G. C. Hughes, 1974, Fig. 3)], which could be based on wrong identifications or on possible physiological races with different temperature requirements (G. C. Hughes, 1974). Of special interest are areas of overlapping of different biogeographic regions, for instance, in North Carolina (Cape Hatteras), where temperate and tropical–subtropical species meet.

Collections of higher marine fungi have been made mostly along coastlines of continents, whereas only few data are available on the occurrence of fungi in the open ocean, in the deep sea, or on the shores of islands. The marine mycota of Icelandic islands was studied by Cavaliere (1968) and T. W. Johnson (1968), who collected 25 and 16 higher fungi, respectively. As can be expected, these species are common representatives of temperate waters, *sensu* G. C. Hughes (1974).

Although the Canary Islands are included in the northern subtropical zone by G. C. Hughes (1974), the higher marine mycota consists of temperate-water species (Kohlmeyer, 1967). Studies on Tenerife yielded 17 Ascomycetes and 4 Deuteromycetes, all members of temperate waters or cosmopolitan species, whereas tropical or subtropical fungi were not found at all (Kohlmeyer, 1967). These findings indicate that the zones proposed by G. C. Hughes (1974) may require modifications in certain areas.

Another island study was by Kohlmeyer (1969c) on the Hawaiian Islands, where 19 Ascomycetes and 8 Deuteromycetes were collected. These are tropical, subtropical, and cosmopolitan species. *Helicascus kanaloanus* and *Robillarda rhizophorae* are known only from Hawaii, but it would be premature to consider them endemic species, because data from the tropics are scanty. Two species, *Keissleriella blepharospora* and *Cytospora rhizophorae*, were introduced in Oahu in 1902 together with their host plant, *Rhizophora mangle,* from Florida. All other marine fungi collected along the Hawaiian Islands may have invaded the archipelago with driftwood or wooden boats or as air- or waterborne spores.

Marine fungi of the island of Helgoland were studied by Schaumann (1969, 1975a,b), who found 33 Ascomycetes and 38 Deuteromycetes in 241 samples. The mycota of Helgoland does not differ from the continental marine mycota of northern Europe and is composed of temperate and cosmopolitan species.

The latest island research was conducted in Bermuda, where we collected 15 marine Ascomycetes, 1 Basidiomycete, and 6 Deuteromycetes (Kohlmeyer and Kohlmeyer, 1977). The fungi occurred on mangroves (*Avicennia, Conocarpus,* and *Rhizophora*) and other shoreline trees (*Tamarix),* as well as on marsh plants (*Salicornia*) and sea grasses

(*Thalassia*). All species found around Bermuda are tropical, subtropical, or cosmopolitan and correspond to the marine mycota of Florida and other areas of the Caribbean.

In conclusion, we would like to support G. C. Hughes' (1975) statement that most of the objectives of biogeography as applied to marine fungi are not fulfilled, and we are probably a long way from their fulfillment. A major problem, for instance, is our present inability to quantify the numbers of fungi. For the time being, we have to be content to register the occurrence of fungal species and to fill the blank spots on the maps where no collections of marine fungi have been made before. Information on the geographical distribution of marine fungi is missing particularly in the Southern Hemisphere, in the Arctic, Antarctic, and Indian oceans, and in the eastern Mediterranean and the Black Sea.

5. Vertical Zonation

The vertical distribution of marine fungi in intertidal and subtidal habitats has rarely been studied. The most suitable substrates for such observations are stationary wood (e.g., pilings) or standing spermatophytes of the intertidal zone, for instance, mangrove trees or salt-marsh plants. Reproductive organs (ascocarps, basidiocarps, pycnidia, and conidia) that develop on the surface can be observed under a dissecting microscope. The purpose of such investigations has been to determine whether the fungi show a pattern of zonation that depends on the depth of the water and on changes between submergence and emergence.

Schaumann (1968, 1969) recorded the occurrence of wood-inhabiting fungi on stationary wooden substrates in the Weser Estuary and Helgoland in relation to their growth below the low-tide line, between the tidal marks, and above the mean high-tide mark. Although these observations seem to indicate zonation patterns for certain species within each study, they must be analyzed critically, and more observations are necessary before any generalizations can be made. In the Weser Estuary (Schaumann, 1968, Fig. 2) *Halosphaeria hamata* occurred only in the littoral zone between the tide marks ("Tidenhubzone"), whereas the same species was found only above the mean high-tide mark around the island of Helgoland (Schaumann, 1969, Fig. 4). *Cirrenalia macrocephala, Dictyosporium pelagicum,* and *Monodictys pelagica* extended from below the mean low-tide mark to above the mean high-tide mark in the Weser Estuary (Schaumann, 1968), but the same species occurred only around the mean high-tide line in Helgoland (Schaumann, 1969). Similar disagreements in the two studies can be noted in *Halosphaeria maritima* and *Lignincola laevis*. Schaumann (1969) separated the fungi found in Helgoland into seven distributional types according to their vertical location. He stated that the number of species increased from the low-water mark toward the mean high-tide mark, but was not certain if this diversity in the upper zone was caused by larger numbers of samples taken in this area or by conditions of the particular habitat or if the higher diversity represented a general principle of distribution. These uncertainties demonstrate the need for further investigations to determine if true vertical distributional patterns occur and how they are influenced by biological and environmental factors, such as fouling and boring organisms, water temperature and current, salinity, species of wood, and composition of sediments.

No vertical zonation pattern of marine fungi was found by Kohlmeyer (1969d) on roots of mangroves. Prop roots of *Rhizophora* spp. and pneumatophores of *Avicennia* spp. often bear ascocarps and pycnidia. These tropical and subtropical fungi do not show any preference for certain zones along the roots. However, they grow only up to the high-tide mark, where they overlap with terrestrial fungi. An exception is a parasite of *Rhizophora* spp., *Cytospora rhizophorae,* which can be found in dead parts of the plant, as well below as above the water line. Mangrove roots surrounded by soil or mud are free of fungi, probably because of anaerobic conditions in the sediment. Ascocarps of *Keissleriella blepharospora,* for instance, occur only on immersed parts of *Rhizophora* roots and seedlings, from the soil surface to the high-water mark.

Senescent plants in tidal salt marshes are invaded by marine and terrestrial fungi, which have different vertical distribution patterns. Salt-marsh grasses of the genus *Spartina* are hosts to over 100 fungi (Gessner and Kohlmeyer, 1976). Marine species are found mainly in the lower portion of the plant, which is submerged regularly by the tides, whereas terrestrial fungi inhabit the upper part of the host, which is not immersed during regular high tides (Gessner, 1977). Obligate marine fungi developing in the bases of *Spartina alterniflora* are, for instance, *Halosphaeria* and *Lulworthia* spp., *Leptosphaeria albopunctata, L. obiones,* and *L. pelagica* (see Fig. 22 and Table XI). A similar separation between marine and terrestrial fungi can be observed in *Juncus roemerianus,* a salt-marsh plant common along the east coast of the United States (J. Kohlmeyer, unpublished). Marine species found in leaves close to the rhizomes of *J. roemerianus* include *Leptosphaeria marina, L. neomaritima,* and *Lulworthia* sp. To our present knowledge, marine fungi on *Spartina* and *Juncus* spp. show no distinct vertical distributional patterns along the standing plants. *Lulworthia* sp. may be an exception because its ascocarps are found mostly in heavily deteriorated, wet culms or leaf bases near the rhizomes. Further studies are necessary to determine if other marine fungi on salt-marsh plants exhibit zonation patterns.

The vertical distribution of marine fungi in offshore waters is discussed in Chapter 6.

6. Deep-Sea Fungi

The oceans can be divided into several vertical zones (Fig. 11) (Bruun, 1957; Wolff, 1970). The upper epipelagic, or photic, zone includes the littoral and sublittoral and extends down to a depth of about 200 m. The bathyal zone has an upper limit at about 200 m and is followed, between 2000 and 4000 m, by the abyssal zone. The latter extends down to about 6000 m. The deepest zone, beginning at 6000 m, is the hadal, which includes the submarine trenches with the greatest depth known thus far in the Pacific Ocean, near Guam, namely, 10,899 m.

Although some authors restrict the definition of the deep sea to waters below 3800 m (e.g., ZoBell, 1968), we draw an arbitrary line at 500 m and include in the deep-sea mycota all fungi growing and reproducing below this depth at hydrostatic pressures of 50 atm and more. This appears to be acceptable because the fungi found at depths below 500 m are different from the epipelagic mycota. An adaptation to high pressures and low temperatures can be expected in deep-sea fungi.

About half of the earth's surface is abyssal, that is, covered by water with depths of 3000 to 6000 m (Bruun, 1957). The mycota of this environment is almost unknown because research has concentrated on littoral and sublittoral zones. Only five species of filamentous higher marine fungi have been described from the deep sea (Table II and Fig. 11), a mere 2.4% of all higher fungi known in the marine environment. The poor knowledge of deep-sea fungi results from the high expense of submergence and retrieval of test samples in deep water, or from trawling in great depths. Most attempts at obtaining deep-sea fungi have been based on incubation of deep-sea sediments or water samples in the laboratory under ambient pressure. With this technique, lower and higher fungi have been isolated from 2500 m (Gaertner, 1968), 4450 m (Roth et al., 1964), and 4610 m (Höhnk, 1961), but it is uncertain if these organisms are indigenous deep-sea fungi or merely "contaminants" from terrestrial or

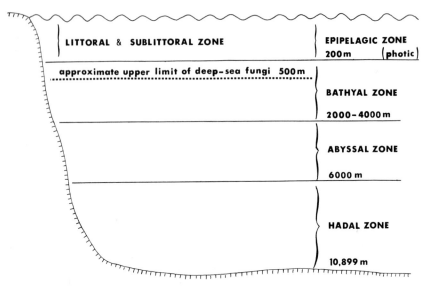

Fig. 11.   Vertical zones of the oceans.

freshwater habitats. It is known that barophobic and obligately thermophilic bacteria occur in a dormant state at depths between 7000 and
10,750 m; however, they "are believed to be passive alien microorganisms which have settled to the sea floor" (ZoBell, 1968). Similarly, fungi
isolated from deep-sea materials under normal laboratory conditions may
derive from dormant propagules of terrestrial or freshwater origin. Deep-
sea pressures, combined with low temperatures and low nutrient concentrations, may even have preservative effects on certain microorganisms
(ZoBell, 1968). Tests for the tolerance of high pressures and low temperatures can indicate whether the isolated fungal species are indigenous
deep-sea forms or aliens from other habitats. No such experiments with
fungi have been conducted thus far. The other method to obtain indigenous deep-sea fungi is to search for them directly on substrates that had
been submerged in the deep sea at known depths (Kohlmeyer, 1968b,
1969a, 1977). If substrates are collected by trawling (Kohlmeyer, 1970,
1977), they should contain other marine organisms as indicators of a
prolonged submergence in the deep sea.

Four of the five named deep-sea fungi (Table II) as well as two
unidentified species (Kohlmeyer, 1977) grow on wood. These species
cause a type of wood decomposition known as "soft-rot" (see Chapter
15). Fungal mycelia have been found in every piece of wood collected in
the deep sea, unless a low oxygen content of the surrounding water had
impeded development of fungi (Kohlmeyer, 1969a, 1977). Wood is com-

TABLE II

Indigenous Deep-Sea Fungi

Species	Depth (m)	Location	Substrate
Ascomycotina			
Abyssomyces hydrozoicus	631–641	Atlantic (near South Orkney Islands)	Chitin (hydrozoa)
Bathyascus vermisporus	1615 + 1720	Pacific (near California)	Wood
Oceanitis scuticella	3975	Atlantic (Gulf of Angola)	Wood
Deuteromycotina			
Allescheriella bathygena	1720	Atlantic (near Bahama Islands)	Wood
Periconia abyssa	3975 + 5315	Atlantic (Gulf of Angola and Iberian Sea)	Wood

mon in the deep sea, although irregularly distributed (Turner, 1973), and wood infested by marine borers usually contains fungal mycelia. Judging from experiences with shallow-water fungi, it can be concluded that deep-sea fungi play a similar role as decomposers of cellulose and possibly also aid in the larval settlement of bivalve borers (Kampf *et al.*, 1959) by "predigesting" the wood surface.

The only other substrate known to be attacked by a deep-sea fungus is chitin of hydrozoa (Kohlmeyer, 1970, 1972a). Hyphae of *Abyssomyces hydrozoicus* grew in hydrorhiza and hydrocaulus of a hydrozoan colony and developed ascocarps on the surface. The mycelium covered the exoskeleton of the hydrozoa, particularly under and near the ascocarps. The hyphae formed channels on the surface and penetrated the substrate. This and other species found in hydrozoan tubes (Kohlmeyer, 1972a) indicate that fungi may be important decomposers of chitin in the deep sea as well. It is probable that chitin accumulates in large amounts on the bottom of the ocean and is broken down by chitinoclastic bacteria and fungi.

Indigenous higher fungi from the deep sea have not yet been cultured, and only preserved specimens have been available for examination. Therefore, we have no information on possible physiological adaptations of fungi to environmental conditions of the deep sea, namely, high pressures, low temperatures, and constant darkness. It can be assumed that deep-sea fungi are barophilic and psychrophilic or that they show tolerance toward high pressures and low temperatures. Transformation rates of organic material by bacteria in the deep sea are slow (Jannasch and Wirsen, 1973; Jannasch *et al.*, 1976), and the fungi also develop slowly (Kohlmeyer, 1969a). Sterile hyphae were found in wood panels submerged for 13 months at 2073 m at 2.5°C, but ascocarps and more mycelia appeared only after a submergence of 2 to 3 years on panels at the same site. Slow developmental rates are probably caused by low temperatures, because Meyers and Reynolds (1960a) have shown that also shallow-water Ascomycetes required between 126 and 191 days for fruiting when wood was exposed at sites with water temperatures of 8°C or less. Many samples of wood from the deep sea contain hyphae without propagules and it was speculated (Kohlmeyer, 1977) that the time for development of fruiting bodies in this environment is long and that rapidly growing boring animals (Knudsen, 1961; Turner, 1973) break the wood down into small fragments before the fungi can reproduce. Possibly, most fruiting of marine fungi in the deep sea occurs on small pieces of wood that may escape collection during trawling.

Higher marine fungi appear to require high rates of dissolved oxygen for growth. Wooden panels submerged off the California coast at 722 m in a "minimum oxygen zone," containing 0.30 ml/liter of dissolved oxygen, were free of fungal mycelia on recovery after 13 months (Kohlmeyer, 1969a). These panels were attacked by cellulolytic bacteria and the wood borer *Xylophaga washingtona*, which were both, apparently, not affected by the low oxygen content of the environment. Wood submerged at a site close to the aforementioned one, but with a higher content of dissolved oxygen (1.26 ml/liter), was attacked by fungal mycelia, although the temperature was lower and the atmospheric pressure higher than at the low oxygen site (Kohlmeyer, 1969a).

Two of the deep-sea species are morphologically adapted to life in an aquatic habitat, namely, *Abyssomyces hydrozoicus* and *Oceanitis scuticella*. Ascospores of both species are provided with mucilaginous appendages, caplike in the first species, whiplike in the second. The advantages of such appendages to marine fungi are discussed in detail in Chapter 3. Although wood and other cellulose-containing substrates are common in the deep sea, they occur in constantly shifting "islands" (Turner, 1973). The chance for isolated pieces of newly submerged wood to become infested by marine fungi is small unless fungal propagules occur in large numbers in water currents close to the deep-sea sediment.

The geographical distribution of higher fungi in the deep sea cannot be determined as yet. Only cooperation between workers in different fields of deep-sea research can provide the necessary data for our studies. A beginning has been made by several zoologists and sorting centers, who kindly supplied deep-sea specimens for mycological research.

7. Fungi Isolated from Marine and Estuarine Waters, Sediments, and Soils

In 1930 Elliott reported on a large number of fungi isolated from the soil of a tidal salt marsh. Most of the species were ubiquitous terrestrial soil fungi. Sparrow (1937) isolated numerous fungi from marine sediments by using a plating method and, again, the species all belonged to genera commonly found in terrestrial soils. Sparrow (1937) concluded that no characteristically marine species were recovered by the methods used, but there was "strong evidence for believing that certain soil and dust-borne fungi can exist in the surface muds although it has not been shown in what form they occur" and "it is doubtful in the present state of our knowledge whether these organisms play an active part in the disintegration of organic materials present in the mud."

Considerable efforts have been made since 1937 to isolate fungi from marine waters, intertidal soil, and benthic sediments, and these efforts are reflected in the large number of publications that have appeared on this topic. Most investigators used standard isolation techniques, such as soil plating or dilution-plate methods, or baiting with a variety of substrates. These experiments resulted in the isolation of hundreds of terrestrial fungi, with a few exceptional marine or facultative marine species, such as *Asteromyces cruciatus, Dendryphiella arenaria, D. salina,* and, possibly, *Drechslera halodes.* Sparrow's (1937) doubts are still as valid as ever, namely: Are the fungi obtained with traditional isolation methods active in the particular sediment, soil, or even in the water column, or are they only present in the form of dormant propagules that come to life in the laboratory? Such doubts were shared or discussed by only a few authors, for instance, Swart (1958), Kohlmeyer (1963e, 1974a), Pugh (1968), and Subramanian and Raghukumar (1974). The isolation methods used to obtain fungi from sediments and waters are selective and give heavily biased results (Pugh, 1968). The possibility exists that some of the isolated fungi are active in the particular habitat, but we do not know which species. Two considerations appear to support our view that the majority of species isolated so far from sediments and waters originate from dormant propagules and not from actively growing fungi and that isolation methods excluded almost all true marine inhabitants.

First, there are more than 100 species of higher fungi known from *Spartina* spp. (Gessner and Kohlmeyer, 1976), most of them non-host-specific decomposers, but only a few of these species have been recovered by culture methods from *Spartina* salt-marsh soil. The same is true for submersed parts of mangroves (see Chapter 12) or other halophytes of intertidal areas (e.g., *Salicornia* spp.) supporting large numbers of saprobic fungi that do not appear in the lists of fungi isolated from sediments of the corresponding habitats. However, these saprobes should also occur on plant fragments in the soil under the vascular halophytes, or their propagules should be present in the soil.

Second, preliminary experiments in our laboratory indicate that terrestrial molds do not germinate under natural marine conditions, but they do *in vitro*. If propagules of terrestrial fungi derived from land via air or water cannot germinate in a marine habitat, obviously they cannot become active in this environment but may remain dormant for unknown periods of time in the sediment. A mycostatic effect of seawater on terrestrial fungi was also postulated by Tyndall and Kirk (1973).

Park (1972a) wrote, about freshwater fungi: "Clearly the ecological role of fungal or other species detected in water should not be assumed, but must be determined by investigation." Furthermore: "Many fungi are commonly accepted as having a role in aquatic habitats solely on the basis of frequent isolation from or detection in such habitats and their real role or significance has not been established." The same is also true for marine fungi. Pawar and Thirumalachar (1966) compared the growth of pure cultures of marine and terrestrial isolates of the same species of soil fungi and concluded that most of the marine isolates grew better on seawater agar than on a distilled-water medium, whereas the terrestrial isolates of the same species showed the reverse reaction. It has to be demonstrated if such culture experiments can give conclusive evidence for the existence of marine-adapted forms or races of terrestrial fungi. Pawar and Thirumalachar (1966) maintain "that the only differentiation between marine and terrestrial fungi is that the former have been adapted to grow and tolerate saline conditions," and they speculate that the present marine species might have migrated from terrestrial habitats and became adapted to saline conditions. The question of fungal activity in marine habitats is more fully discussed in connection with the definition of obligate and facultative marine fungi (see Chapter 1).

In view of the uncertainty of the marine activity of most of the fungi isolated from marine waters and sediments by plating or dilution plates, we do not list these species but refer the reader to the following original publications, or to Johnson and Sparrow (1961, Appendix A), who have compiled lists of these organisms. Few authors have isolated higher fungi from seawater (Höhnk, 1959; Meyers *et al.*, 1967b; Roth *et al.*, 1964; Schaumann, 1974a). The following investigators used marine sediment or soil for the isolation of fungi: Abdel-Fattah *et al.* (1977), Apinis and Chesters (1964), Borut and Johnson (1962), Cowley (1973), Dabrowa *et al.* (1964), Elliott (1930), Höhnk (1956, 1958, 1959, 1962b, 1967), Moustafa (1975), Moustafa and Al-Musallam (1975), Moustafa *et al.* (1976), Nicot (1958a,c), Pitts and Cowley (1974), Pugh (1960, 1962b, 1966, 1968, 1974), Pugh *et al.* (1963), Saitô (1952), Schaumann (1974a, 1975c), Siepmann (1959a,b), Sparrow (1937), Steele (1967), and Subramanian and Raghukumar (1974). Further literature on fungi isolated from marine soils and sand may be found in the chapters on mangroves (Chapter 12) and sand-inhabiting fungi (Chapter 8).

8. Fungi in Sandy Beaches and Sea Foam

Sandy beaches harbor interesting organisms that are adapted to living in the water-filled spaces between grains of sand. These spaces are small, mostly microscopic, and, accordingly, the organisms are minute and provided with adaptations to life in such a habitat. An extraordinary microfauna, named "interstitial fauna" or "thalassopsammon" (Zinn, 1968) lives in intertidal and subtidal areas of sandy beaches. Most invertebrate classes are represented in the thalassopsammon, such as flagellates, ciliates, echinoderms, tunicates, bryozoa, annelids, polychaetes, oligochaetes, and gastrotrichs, but mainly nematodes and flatworms (Zinn, 1967). Interest in the interstitial fauna developed about 1904 in Europe, and a large group of zoologists is now working on all aspects of marine sand-inhabiting animals (Jansson, 1968).

Scanty information is available on the interstitial flora, even though bacteria, diatoms, and filamentous algae, especially blue–greens, commonly cover grains of sand in intertidal and subtidal parts of sandy beaches (Meadows and Anderson, 1966, 1968; R. G. Johnson, 1974; Wicks, 1974; Weise, 1976; Rheinheimer, 1977). The mycota of such habitats is also not well known and investigations of sand-inhabiting or arenicolous fungi have been restricted mainly to taxonomic and morphological studies, whereas ecological observations are rare and have started only recently (Kohlmeyer, 1966a; Wagner-Merner, 1972; Koch, 1974). The number of indigenous sand-inhabiting fungi is small (Tables III and IV), although these lists will be augmented by some still undescribed arenicolous species from temperate waters in the future. Arenicolous fungi can be simply defined as fungi living among or on grains of sand,

TABLE III

Sand-Inhabiting Fungi with Propagules Found in Sea Foam

Ascomycotina	Basidiomycotina	Deuteromycotina
Corollospora comata	Nia vibrissa	Alternaria sp.
C. intermedia		Asteromyces cruciatus
C. lacera		Dendryphiella arenaria
C. maritima		Varicosporina ramulosa
C. pulchella		
C. trifurcata		
Lindra marinera		

without implying that they obtain any nutrients from the sand itself.*

Marine fungi in beach sands have the same function as litter-inhabiting fungi in forests and other terrestrial habitats. They degrade organic material, which serves as their source of nutrition. Depending on the geographic location, such substrates consist of washed-up algae, leaves, and rhizomes of sea grasses (*Posidonia, Thalassia,* and *Zostera* spp.), parts of salt-marsh plants (*Juncus* and *Spartina* spp.), or driftwood. Most of the fungi found on these buried substrates are able to break down cellulose of higher plants (Koch, 1974), alginates or laminarin of Phaeophyta (Chesters and Bull, 1963a–c), or agar of Rhodophyta. At maturity, arenicolous fungi release their propagules into the water and these are trapped by air bubbles in the foam (Fig. 12). Foam contains only nongerminated ascospores, basidiospores, and conidia. Apparently, the constant movement in water prevents germination, but germ tubes are formed within a few hours after the propagules come to rest in the collecting bottle or in a microscope preparation. Similar observations have been made for conidia of freshwater Hyphomycetes (Ingold, 1976; Iqbal and Webster, 1973). Propagules of marine fungi contained in foam are deposited by the waves on washed-up substrates, such as sea grasses, algae, or animal remains (Kohlmeyer, 1966a). There, the spores germinate and penetrate the medium, which may be covered by sand or washed away by the rising tide. Ascospores deposited at the high-tide line are able to survive in a

---

* Fungi isolated from coastal sands by Dabrowa *et al.* (1964), Steele (1967), Kishimoto and Baker (1969), and Bergen and Wagner-Merner (1977) have been omitted from this discussion because the organisms obtained by these authors do not meet our criteria for obligate or facultative marine species. In addition, all the species isolated by soil plate techniques by Moreau and Moreau (1941), Brown (1958b), and Nicot (1958a,c) from tide-washed beaches are not considered, except for *Asteromyces cruciatus* and *Dendryphiella arenaria* (see also Chapter 7).

TABLE IV
Ascomycetes with Ascocarps Attached to Grains of Sand or Other Hard Substrates

Corollospora	Carbosphaerella
C. lacera	C. leptosphaerioides
C. intermedia	C. pleosporoides
C. maritima	
C. pulchella	
C. trifurcata	
C. tubulata	

dried state if subsequent high tides do not reach them for several days. We collected dried foam from a Florida beach and kept it dry for 10 days. Ascospores of *Corollospora lacera* contained in this foam germinated overnight after immersion in distilled water (J. Kohlmeyer, unpublished). This remarkable resistance of ascospores to prolonged periods of drying is an adaptation of arenicolous marine fungi to extreme conditions occurring in the upper layers of beach sands.

Arenicolous fungi are probably important links in the food web because they break down cellulose, which cannot be digested by interstitial ani-

Fig. 12. *Corollospora lacera*, ascospore from foam, surrounded by detritus (Herb. J.K. 3908; Nomarski interference contrast; bar = 10 μm).

mals. Schuster (1966), for instance, indicated that certain mites (Oribatei) living in washed-up plant material do not feed on this deposit but instead on the fungi (in particular their spores) occurring in this habitat.

Fruiting bodies of arenicolous Ascomycetes (*Corollospora* and *Carbosphaerella* spp.) are generally attached to grains of sand or shell fragments (Table IV and Fig. 13). Höhnk (1954a) was first to observe that there is a distinct affinity of the ascocarps to siliciferous and calcareous substances. Other authors confirmed Höhnk's observation and illustrated ascocarps attached to grains of sand (Fize, 1960; Kohlmeyer and Kohlmeyer, 1964–1969; Tubaki, 1968; Koch, 1974; Schmidt, 1974). Even in pure culture the fruiting bodies are often attached to the hard surface of culture vessels. It must be pointed out that the fungi are merely mechanically fixed to the inorganic substrate without breaking it down. This was demonstrated when ascocarps of *Corollospora maritima* developed on cover glasses that were added to agar cultures. The glass was not corroded by the subicula (J. Kohlmeyer, unpublished).

Ascocarps of *Corollospora* and *Carbosphaerella* species are morphologically adapted to the interstitial habitat. The walls are usually thick, hard, and carbonaceous. Necks are short or absent, and the ostiole is

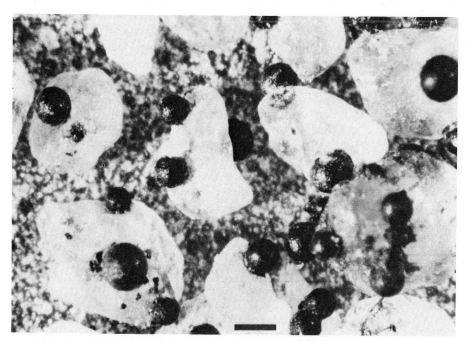

Fig. 13. *Corollospora maritima,* ascocarps attached to grains of sand (Herb. J.K. 3909; bar = 300 μm).

inconspicuous and usually situated near the base, close to the place of attachment to the substrate (Fig. 13). This way the weakest part of the ascocarp, the ostiole, is protected against abrasion by being located near the subiculum. Fruiting bodies of *Carbosphaerella* and *Corollospora* species are the hardest of all marine fungi and appear to be well adapted to a habitat of constantly shifting grains of sand.

The propagules of most fungi found in beach sands are also adapted to this particular environment. Appendaged ascospores and basidiospores, and tetraradiate conidia, are regularly found in sea foam along sandy beaches (Fig. 12 and Table III) (Kohlmeyer, 1966a), mostly together with algae and protozoa (Schlichting, 1971). The shapes of the fungal propagules apparently aid in attaching them to air bubbles, thereby concentrating them in foam along the shore. Since Ingold (1942) published the first paper on fungi from submerged decaying leaves, almost 100 freshwater Hyphomycetes have become known (Ingold, 1975b). Conidia of most of them accumulate in foam, which acts as a spore trap, and "the examination of a foam sample can very quickly give a picture of the species present in a particular stream above the point of collection" (Ingold, 1975b). The same is true for arenicolous marine fungi, because a foam sample from an oceanic shore can serve as an indicator for the species occurring in this beach (Kohlmeyer, 1966a; Schaumann, 1969; Wagner-Merner, 1972). Most of the arenicolous fungi have characteristic spores or conidia that can be readily identified (Table III). However, foam samples may also contain spores of fungi from other habitats (Kohlmeyer, 1966a; Wagner-Merner, 1972). Most of such extraneous propagules are morphologically indistinctive and can be identified to genus at the most. In some cases it is even uncertain if a propagule belongs to the Ascomycetes or to the Deuteromycetes. Therefore, identifications of species based on spores alone, except for the fungi listed in Table III, should be discouraged.

The efficiency of air bubbles in trapping microorganisms has been demonstrated for bacteria by Carlucci and Williams (1965) and by Blanchard and Syzdek (1970, 1974) and for freshwater Hyphomycetes by Iqbal and Webster (1973). No such experiments have been reported for marine fungi, but marine foam samples from many geographic locations make it obvious that propagules of these organisms are efficiently trapped by air bubbles. It would take several liters of seawater collected along the shore to find the same number of fungal spores that are present in a few milliliters of dissolved foam from the same area. E. B. G. Jones (1973), for instance, found five to eight spores per liter of filtered seawater. So far, there is no method to quantify the fungi occurring in beach sand. The number of propagules in a given sample of foam is no indication of the frequency of the species in this habitat. The structure of sea foam varies considerably, because its formation depends on wind velocity and direction, the tides, precipitation, and possibly other factors. Consequently, the spore-trapping efficiency of foam from the same location changes from day to day, or even from hour to hour. It is possible, however, to compare the abundance of different species in one and the same sample of foam. It was observed, for instance, that ascospores of *Corollospora maritima* occur about two to ten times more often than those of *C. trifurcata* (Kohlmeyer, 1966a, 1969c). This observation agrees with collections of ascocarps on wood, where the first species is also much more frequently found than the second (Koch, 1974).

Information on the vertical distribution of arenicolous fungi is scarce. Most substrates with ascocarps have been collected on the surface of the sediment, but some collections from greater depths have also been made. An ascocarp of *Corollospora* sp. (as *Arenariomyces* sp.) was found in sand at 15 cm below the sediment–water interface in Massachusetts (R. G. Johnson, 1974). Koch (1974) buried wood panels at a depth between 25 and 40 cm in a Danish beach and recovered them after 2 years, at which time he found the wood attacked by seven species of fungi, among them *Carbosphaerella* and *Corollospora*. These observations indicate that arenicolous fungi can grow at considerable depths, probably anywhere between the sediment surface and the anoxic black layers of sand (Fenchel and Riedl, 1970) or the groundwater level. Mycelial strands may be abundant and may bind grains of sand together (Koch, 1974). Again, it must be emphasized that the fungi do not attack the sand, but their hyphae are always connected to some organic substrate. Representatives of the shore mycota may also occur in deeper water if the conditions are suitable, as shown by E. B. G. Jones and LeCampion-Alsumard (1970a,b), who report *Corollospora maritima* from depths down to 280 m on polyurethane panels anchored offshore.

Arenicolous fungi can be divided into three geographical groups: (1) species of temperate waters, (2) tropical and subtropical species, and (3) cosmopolitan species. A representative of the first group is *Asteromyces cruciatus*, which is known only from temperate waters of the Atlantic Ocean (Europe; New Jersey to Massachusetts) and the Pacific Ocean (California). Spores of several other, still undescribed species are found regularly in foam samples from the same temperate waters. *Varicosporina ramulosa* belongs to the second group and appears to occur only in tropical and subtropical regions of the Atlantic Ocean (Mexico to North Carolina) and the Pacific Ocean (Japan, Hawaii). The third group includes such cosmopolitans as *Corollospora maritima* and *C. trifurcata*. Ascospores of these last two species can be expected in foam collections from beaches all around the world. The geographical distribution of arenicolous fungi seems to be mainly controlled by temperature, as also indicated for other marine fungi (G. C. Hughes, 1974).

9. Algae-Inhabiting Fungi

Almost one-third of all known higher marine fungi are associated with algae. Most algicolous marine fungi are Ascomycetes. Filamentous Basidiomycetes have not been recorded from marine algae, and only eleven Deuteromycetes are known from such hosts. The algicolous fungi are symbionts, parasites, or saprobes, but only the latter two are treated in this chapter, as symbiotic marine fungi are dealt with under submarine lichens and lichenlike associations (Chapter 10). The first descriptions of algae-inhabiting fungi appeared at the end of the nineteenth century, but it was Sutherland (1915a–c, 1916a,b) who started a thorough survey of these organisms and described 24 species. In the following decades most marine mycologists concentrated on wood-inhabiting fungi, and algicolous fungi were considered by only a few investigators, for instance, Jean and Geneviève Feldmann in France, A. B. Cribb and Joan W. Cribb in Australia, Irene M. Wilson in Great Britain, and by us in the United States. Obviously, the preference for lignicolous fungi as research objects over algicolous species can be explained by the ubiquitous nature of most fungi living on wood and by the easier accessibility of such substrates. Wooden panels can be easily submerged all around the world to serve as traps for cellulolytic fungi. On the other hand, algae-inhabiting fungi are relatively rare and limited in their geographic distribution to the range of their hosts. We know, therefore, much more about the morphology, activities, and distribution of wood-inhabiting fungi than about algicolous species. The sporadic occurrence of algicolous fungi may be explained by antibiotic substances produced by healthy algae (Sieburth, 1968). Recent reviews of the literature on algicolous fungi have been published by Andrews (1976), E. B. G. Jones (1976), and Kohlmeyer (1974b). As pointed out by Andrews (1976), studies of these fungi have been undertaken mostly from a taxonomic approach, whereas few data on the host–parasite interaction, pathogenicity, predisposition, and epidemiology have become available. A major cause for the neglect of these research areas is the relative scarcity of most algicolous fungi in nature and the difficulties in growing the host plants in laboratory cultures for infection experiments. Thus, most investigations of algae-inhabiting fungi have been based on herbarium specimens. There is no need, however, to deplore the past descriptive and systematic-oriented research in this field, because any work on plant pathology should be based on sound taxonomics (see, e.g., Walker, 1975).

Algae-inhabiting marine fungi are listed in Tables V (parasites) and VIII (saprobes). We refrain from separating pathogens, i.e., disease-causing species, from other parasites, because disease symptoms in an alga cannot be easily recognized in many cases. Andrews (1976) defines disease as "a continuing disturbance to the plant's normal structure or function such that it is altered in growth rate, appearance, or economic importance." For example, *Spathulospora* species generally do not alter the outer appearance of their hosts, *Ballia* spp., but chloroplasts of some host cells are abnormally contracted (Kohlmeyer and Kohlmeyer, 1975). If one were to follow Andrews' (1976) definition of a disease, probably all parasites listed in Table V could be considered to be pathogens in a broad sense. We accept the definition of a parasite by Ainsworth *et al.* (1971) as "an organism living on or in, and obtaining its food from, its host, another living organism." A saprobe, on the other hand, is an organism "using dead organic material as food, and commonly causing its decay" (Ainsworth *et al.*, 1971).

I. PARASITES

A. Taxonomic Groups

At the present time, we recognize 32 higher fungi parasitic on marine algae (Table V). All but one belong to the Ascomycetes. The one exception, *Sphaceloma cecidii,* is a Deuteromycete in the order Melanconiales. The Ascomycetes are members of the Pyrenomycetes and Loculoascomycetes (Table VI). All the pyrenomycetous genera are composed of obligate marine species, whereas three of the loculoascomycetous genera, namely, *Didymella, Didymosphaeria,* and *Massarina,* also contain terrestrial species.

B. "Imperfect States"

Of the 31 parasitic Ascomycetes, ten have pycnidialike fructifications, usually produced near the ascocarps (Table VII). These "pycnidia" contain

TABLE V
Parasitic Fungi on Marine Algae

Fungus	Hosts	Host classes	Range
Ascomycotina			
Chadefaudia balliae	*Ballia callitricha*	Rhodophyta	Pacific Ocean
Chadefaudia gymnogongri	*Curdiea, Gigartina, Gymnogongrus, Laurencia, Microcladia, Ptilonia* spp.	Rhodophyta	Antarctic, Atlantic, Indian, and Pacific oceans; Mediterranean Sea
Chadefaudia marina	*Rhodymenia palmata*	Rhodophyta	Atlantic Ocean
Chadefaudia polyporolithi	*Polyporolithon* spp.	Rhodophyta	Pacific Ocean
Didymella fucicola[a]	*Fucus spiralis, F. vesiculosus, Pelvetia canaliculata*	Phaeophyta	Atlantic Ocean
Didymella gloiopeltidis	*Gloiopeltis furcata*	Rhodophyta	Pacific Ocean
Didymella magnei	*Rhodymenia palmata*	Rhodophyta	Atlantic Ocean
Didymosphaeria danica	*Chondrus crispus*	Rhodophyta	Atlantic Ocean
Haloguignardia decidua	*Sargassum daemelii, Sargassum* sp.	Phaeophyta	Pacific Ocean
Haloguignardia irritans	*Cystoseira osmundacea, Halidrys dioica*	Phaeophyta	Pacific Ocean
Haloguignardia oceanica	*Sargassum fluitans, S. natans*	Phaeophyta	Atlantic Ocean
Haloguignardia tumefaciens	Several *Sargassum* spp.	Phaeophyta	Atlantic and Pacific oceans
Lindra thalassiae[a]	*Sargassum* sp. (also in turtle grass, *Thalassia testudinum*)	Phaeophyta	Atlantic and Pacific oceans
Lulworthia fucicola[a]	*Fucus vesiculosus*	Phaeophyta	Atlantic and Pacific oceans; Baltic Sea
Lulworthia kniepii	*Lithophyllum, Porolithon, Pseudolithophyllum* spp.	Rhodophyta	Atlantic and Pacific oceans; Mediterranean Sea

TABLE V *(Continued)*

Fungus	Hosts	Host classes	Range
Massarina cystophorae	*Cystophora retroflexa, C. subfarcinata*	Phaeophyta	Pacific Ocean
Orcadia ascophylli[a]	*Ascophyllum, Fucus, Pelvetia* spp.	Phaeophyta	Atlantic Ocean
Phycomelaina laminariae	*Laminaria* spp. (also *Alaria* sp.?)	Phaeophyta	Atlantic Ocean
Pontogeneia calospora	*Castagnea chordariaeformis*	Phaeophyta	Atlantic Ocean
Pontogeneia codiicola	*Codium fragile, C. simulans*	Chlorophyta	Pacific Ocean
Pontogeneia cubensis	*Halopteris scoparia*	Phaeophyta	Atlantic Ocean
Pontogeneia enormis	*Halopteris scoparia*	Phaeophyta	Atlantic Ocean
Pontogeneia padinae	*Padina durvillaei*	Phaeophyta	Pacific Ocean
Pontogeneia valoniopsidis	*Valoniopsis pachynema*	Chlorophyta	Pacific Ocean
Spathulospora adelpha	*Ballia callitricha*	Rhodophyta	Pacific Ocean
Spathulospora antarctica	*Ballia callitricha*	Rhodophyta	Atlantic, Indian, and Pacific oceans
Spathulospora calva	*Ballia callitricha*	Rhodophyta	Pacific Ocean
Spathulospora lanata	*Ballia hirsuta, B. scoparia*	Rhodophyta	Pacific Ocean
Spathulospora phycophila	*Ballia callitricha, B. scoparia*	Rhodophyta	Pacific Ocean
Thalassoascus tregoubovii	*Aglaozonia, Cystoseira, Zanardinia* spp.	Phaeophyta	Atlantic Ocean; Mediterranean Sea
Trailia ascophylli	*Ascophyllum nodosum, Fucus* sp.	Phaeophyta	Atlantic Ocean
Deuteromycotina			
Sphaceloma cecidii	*Cystoseira, Halidrys, Sargassum* spp.	Phaeophyta	Atlantic and Pacific oceans

[a]Species probably perthophytes, attacking damaged tissues of the host.

TABLE VI

Classification of Genera of Parasitic Marine Ascomycetes

Class	Order	Family	Genus
Pyrenomycetes	Spathulosporales	Spathulosporaceae	*Spathulospora*
			Trailia?
			Chadefaudia
	Sphaeriales	Halosphaeriaceae	*Lindra*
			Lulworthia
			Haloguignardia?
		Polystigmataceae	*Phycomelaina*
		Sphaeriaceae	*Pontogeneia*
Loculoascomycetes	Dothideales	Mycosphaerellaceae	*Didymella*
			Didymosphaeria
		Pleosporaceae	*Massarina*
			Thalassoascus
Ascomycotina incertae sedis			*Orcadia*

small one-celled, hyaline spores and, most probably, these constitute spermogonia and spermatia, respectively. However, no information exists on the function of the spores, whether they serve as dispersal units (conidia) or as male sex cells (spermatia) in the fertilization process.

C. Host Specificity

Fungi and their hosts listed in Table V indicate the limitations of certain fungal species to certain hosts and disclose the indiscriminate nature of other fungi. It is notable that each parasite is restricted to one class of algae, namely, Chlorophyta, Phaeophyta, or Rhodophyta. However, some fungi attack a variety of genera within one class, as do, for instance, the Ascomycetes *Chadefaudia gymnogongri, Didymella fucicola, Haloguignardia irritans, Lulworthia kniepii, Orcadia ascophylli,*

TABLE VII

Algae-Inhabiting Parasitic Ascomycetes with Pycnidialike Fructifications (Spermogonia?)

Didymella fucicola	*Haloguignardia oceanica*
Didymella gloiopeltidis	*Haloguignardia tumefaciens*
Didymosphaeria danica	*Massarina cystophorae*
Haloguignardia decidua	*Phycomelaina laminariae*
Haloguignardia irritans	*Thalassoascus tregoubovii*

Thalassoascus tregoubovii, and *Trailia ascophylli* and the Deuteromycete *Sphaceloma cecidii.* Other fungal species occur on several host species within one genus, and some are host-specific *sensu stricto* in parasitizing only one algal species. *Lindra thalassiae* is the most indiscriminate species of all, as it occurs in leaves of a spermatophyte (*Thalassia*) as well as in air vesicles of *Sargassum* sp. It may be a perthophyte, attacking only host tissues that are weakened or damaged by environmental conditions or other causes.

D. Life Cycles

We do not know the complete life cycle of a single parasitic algicolous fungus. The ability to culture the host plant would be a prerequisite for such investigations, and cooperative efforts between phycologists and mycologists could clarify the life cycles of obligate parasites on marine algae. The only case in which many developmental stages of a parasite have been found is *Spathulospora lanata* (Kohlmeyer and Kohlmeyer, 1975). This fungus occurs on *Ballia hirsuta* and *B. scoparia* in New Zealand waters. Its one-celled ascospores are provided with two sub-gelatinous caps, which probably cause attachment of the spores to the host. The ascospore forms a short germ tube ending in a lobed appressorium (Fig. 14a). Further proliferation results in a crustlike multicelled thallus on the algal surface. Peglike cells perforate the wall and penetrate the host cell (Fig. 14d). An intracellular stroma composed of "assimilating" or "foraging" cells is formed along the peripheral wall of the alga. Although the fungal thallus eventually covers about five or six cells of the filamentous host, only one algal cell is penetrated by pegs and contains a fungal stroma (Fig. 14e). The chloroplast in the parasitized cell is severely damaged, and the adjoining uninvaded cells have abnormally contracted chloroplasts. It can be surmised that nutrients pass from other cells of the filament into the infected cell, and plugs in the algal pit connections obviously do not obstruct an exchange of nutrients between the cells.

When the thallus of *S. lanata* has reached a diameter of about 50 μm, it forms hairlike, branched protrusions, bearing numerous bottle-shaped antheridia at their tips (Fig. 14b). Rod-shaped spermatia with a mucilaginous cap are produced in succession in the necks of the antheridia.

The next stage in the development is a thickening of the fungal thallus by centrifugal development of the cells. Eventually the thallus is composed of seven to ten layers of radiating cells (Fig. 14e). During this growth several simple, septate filaments are formed at one end of the thallus. Up to seven of these receptive hyphae, or trichogynes, can be

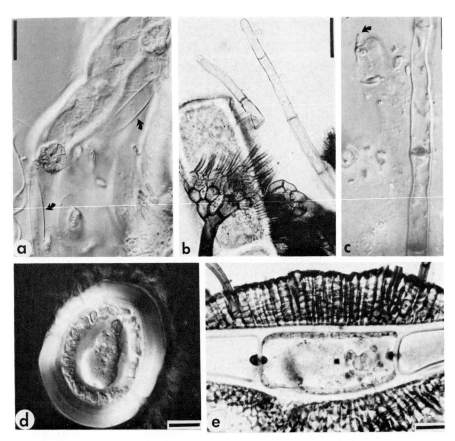

Fig. 14. *Spathulospora lanata* from *Ballia hirsuta*. (a) Germinating ascospores (arrows) forming lobed appressoria (Herb. J.K. 3360; bar = 20 μm). (b) Two trichogynes and tip of an antheridia-bearing hair (Herb. J.K. 3349; bar = 25 μm). (c) Septate trichogyne and tip of a trichogyne forming a contact papilla upon attachment of a spermatium (arrow) (Herb. J.K. 3360; bar = 10 μm). (d) Cross section (16 μm) through thick-walled algal cell surrounded by the dark fungal stroma; penetration pegs perforate the wall and give rise to an intracellular stroma of assimilating cells along the algal wall (Herb. J.K. 3052; bar = 10 μm). (e) Longitudinal section (10 μm) through algal filament surrounded by fungal stroma; only the central cell is penetrated by pegs and contains assimilating cells (acetocarmine; Herb. J.K. 3360; bar = 25 μm). (b and e in bright field, the others in Nomarski interference contrast.)

found on one thallus (Fig. 14b), and at the same time one or two semi-globose ascocarp initials bulge out at the base of the trichogynes.

Spermatization, namely, the attachment of spermatia to the trichogynes, has also been observed in *S. lanata*. After contact with a spermatium, the trichogyne forms a small, thin-walled "contact papilla," which probably aids in the transfer of the male nucleus (Fig. 14c). This reaction of the receptive hypha indicates that the spermatia function as gametes and are not merely conidia. However, migration of the male nucleus and karyogamy have not been observed so far. Spermatia were found attached to the apical or penultimate cells of trichogynes, near the tip or even 50 μm below it.

The complete ascocarp ontogeny of *S. lanata* is not known because some early stages of development have not become available. Immature ascocarps are thick-walled, with antheridia-bearing hairs deeply immersed in the wall. The center is filled with a thin-walled pseudoparenchyma, and large ascogenous cells originate at the base and sides of the ventral cavity. Thin-walled asci grow upward, compressing the pseudoparenchyma and releasing ascospores into the venter before an ostiolar canal is formed. A short papilla originates at the apex of the ascocarp and its canal is initially filled by a small-celled pseudoparenchyma. After disintegration of these cells, ascospores are released into the water.

E. Symptoms of Fungal Attack

Parasitic fungi on marine algae may cause obvious changes in the host or they may not affect the outer appearance of the plant at all. In the following, some fungal diseases and their causal organisms are discussed according to the symptoms.

1. Host Externally Unaltered

The nutritional relationships between host and parasite are poorly understood in most cases of higher marine fungi. However, some of the species appear to be weak parasites that form sparse mycelia supporting an immersed or superficial ascocarp, without discoloring the host. The only visible signs of parasitism are the dark fruiting bodies, which are often difficult to observe in or on the algal thallus. All species of *Pontogeneia* in Chlorophyta (Fig. 15a) and Phaeophyta (Fig. 15b) (Kohlmeyer, 1975b) are examples of such weak parasites, as are *Chadefaudia balliae* in *Ballia callitricha* and *C. polyporolithi* in species of

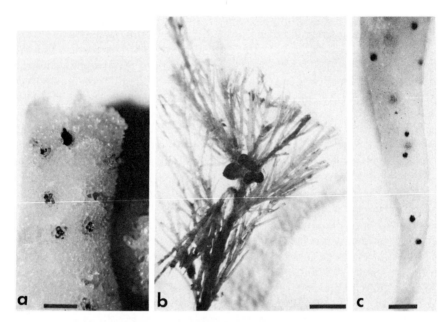

Fig. 15. Ascocarps of parasitic fungi in marine algae. (a) *Pontogeneia codiicola* in *Codium mucronatum* (holotype Dawson 4766b in AHFH; bar = 750 μm). (b) *P. cubensis* on *Halopteris scoparia* (type in PC; bar = 1 mm). (c) *Didymella gloiopeltidis* in *Gloiopeltis furcata*, causing "black-dots disease" (type in AHFH; bar = 600 μm).

Polyporolithon. Other instances are some species of *Spathulospora*, namely, *S. antarctica* and *S. lanata* on *Ballia callitricha,* which do not alter the appearance of the host, whereas other species in the genus cause a wild growth of algal filaments around the fungal thalli. Hyaline ascocarps of *Orcadia ascophylli* are most difficult to detect in the brown host tissues of *Ascophyllum, Fucus,* and *Pelvetia* spp.

2. Discolorations

This group includes fungi causing light or dark areas in the host, but no abnormal growth, such as galls or other proliferations. Discolorations may consist of faded zones in which the algal cells are damaged, as observed in calcareous red algae infected by *Lulworthia kniepii* (Bauch, 1936). Mycelium grows in the middle lamellae of *Lithophyllum, Porolithon,* and *Pseudolithophyllum* spp. and destroys the protoplasts of adjoining cells, whereas the calcified cell walls of the algae are not attacked. The embedded ascocarps and surrounding hyphae appear as dark spots in the light-colored areas.

Another fungus associated with pale spots is *Didymella gloiopeltidis* in *Gloiopeltis furcata* in Japan. The disease was named "black-dots disease" because of the black ascocarps immersed in the infected areas of the host (Fig. 15c) (Miyabe and Tokida, 1948).

Chadefaudia marina causes yellow-greenish spots with dark ascocarps in the red thalli of its host, *Rhodymenia palmata,* in France (G. Feldmann, 1957).

A blackening of the host thalli may be caused by a general darkening of the algal tissues during senescence or by the formation of blackish fungal material, such as stromata, on or in the host. The first case is found in air vesicles of *Sargassum* sp. that become infected by *Lindra thalassiae.* These vesicles turn dark brown, become soft and wrinkled, and resemble raisins, for which reason this disease was named "raisin disease" (Kohlmeyer, 1971a). The fungus is restricted to the air vesicles and does not penetrate the stalks of the air bladders or the stipes or blades of the host. Hyphae and ascocarps develop in the walls of the air vesicles, softening the tissue until the bladders break open and fall off. It was speculated (Kohlmeyer, 1971a) that epiphytic animals, such as tube-forming annelids and polychaetes, or others, damage the air vesicles and make them susceptible to fungal attack (Fig. 16). The disease occurs in the Sargasso Sea, but only a small number of plants appear to be affected.

Most other cases of discoloration consist of blackenings resulting from growth of black fungal stromata or hyphae. The most widespread and conspicuous disease is "stipe blotch of kelps" [a term coined by Andrews (1976)] of *Laminaria* species. The tarlike spots on the stalks are caused by *Phycomelaina laminariae* (Sutherland, 1915b; Kohlmeyer, 1968e). Ascocarps and spermogonia (?) are embedded in a black pseudostroma that forms circular or oblong spots on algal stipes, rarely on holdfasts, but never on the blades. At first these necrotic spots are isolated, small, and light brown. Later they merge and turn black, sometimes covering large parts of the stalks. The mycelium is mostly restricted to the cortex and grows between the mucilaginous walls of the host but also penetrates the cells with haustoria. Groups of algal cells are enclosed by hyphae and by the walls of the fruiting bodies; thus, a pseudostroma, composed of fungal and algal material, is formed. Eventually, the parasitized tissues break down and expose the inner part of the stipe, making it susceptible to attack by perthophytes and saprobes. Usually the gross morphology of the parasitized algae is unaltered, but one *Laminaria* was found in which adventitious hapteres had developed on the stalk around a stroma, about 35 cm above the holdfast (Kohlmeyer, 1968e). The cause for the formation of these hapteres is unknown, but, obviously, the fungus had induced

Fig. 16. *Lindra thalassiae* causing softening and collapsing of air vesicles (arrow) of *Sargassum* sp.; fouling organisms (e.g., *Spirorbis* spp.) may make the vesicles susceptible to fungal attack (Herb. J.K. 2729; bar = 1 mm).

their growth. A report of *P. laminariae* from *Alaria esculenta* is doubtful (Kohlmeyer, 1968e) and we have been unable to find any diseased *Alaria* in areas where the fungus was abundant on species of *Laminaria*.

Ascocarps of *Phycomelaina laminariae* are often parasitized by an Oomycete that fills the locules with lobed thalli and extends discharge tubes through the ostiole (Fig. 17a). This peculiar hyperparasite is similar to representatives of *Petersenia* (Kohlmeyer, 1968e) or *Atkinsiella* (e.g., Sparrow, 1973) and prevents formation of asci in *P. laminariae*. Propagules have not been observed.

Didymosphaeria danica is another common Ascomycete in marine algae (Wilson and Knoyle, 1961). The fungus is tissue specific, that is, restricted to cystocarps of *Chondrus crispus*, causing a blackening of the host tissue. Ascocarps and spermogonia are embedded in the stroma, which prevents reproduction of the alga. Often holes can be observed in tips of *C. crispus* where cystocarps had been located. Either infected algal fruiting bodies fall out after senescence of the fungus or marine animals feed preferentially on the parasitized tissues.

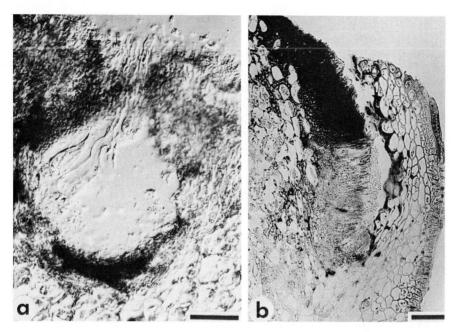

Fig. 17. Hyperparasitic fungi. (a) Section (16 μm) through *Laminaria digitata;* Oomycete with lobed thalli in empty ascocarp of *Phycomelaina laminariae;* discharge tubes extending through the ostiole (Herb. J.K. 1949). (b) Section (6 μm) through gall caused by *Haloguignardia irritans* in *Cystoseira osmundacea;* acervulus of *Sphaceloma cecidii* ruptures the outer layers of the gall (Herb. J.K. 2646). (a in Nomarski interference contrast, b in bright field; both bars = 50 μm.)

The only named parasitic Deuteromycete, *Sphaceloma cecidii,* is found exclusively in galls caused by species of *Haloguignardia* in *Cystoseira, Halidrys,* and *Sargassum* spp. (Kohlmeyer, 1972c). The fruiting bodies of *S. cecidii* become black at maturity and give the galls a blackish-brown color. The Deuteromycete damages the gall tissues by rupturing the outer cell layers (Fig. 17b) and closing the ostioles of ascocarps and spermogonia of the primary parasites, namely, *Haloguignardia* spp. The intimate mingling of algal and fungal cells does not allow any clues whether *S. cecidii* obtains its nutrients from the algal tissue alone or also from the primary parasite. Because of its restriction to diseased tissues, we consider *S. cecidii* to be a hyperparasite.

A final example of a parasite causing distinct black marks on its hosts is *Thalassoascus tregoubovii,* which occurs on the stipes of *Aglaozonia, Cystoseira,* and *Zanardinia* spp. in Europe (Ollivier, 1926; Kohlmeyer,

1963b). Black stromatic crusts are restricted to the base of the stalks and give rise to elongated, stalked ascocarps and sessile pycnidia.

3. Malformations

The number of fungi causing malformations in marine algae is relatively small and the ones we know are restricted to the genera *Spathulospora*, *Haloguignardia*, and *Massarina*. Some species of *Spathulospora* do not alter the outer appearance of the host, whereas others induce wild growths of hair in the algae. Such proliferating hairs have been found regularly around ascocarps of *S. calva* on *Ballia callitricha* (Kohlmeyer, 1973a) and occasionally under fruiting bodies of *S. adelpha* or around ascocarps of *S. phycophila* (Kohlmeyer and Kohlmeyer, 1975). The adventitious algal hairs are even produced in young tips of *Ballia,* where such hairs normally do not occur. This wild growth of hairs can be compared to witches'-broom, namely, galls induced by parasitic insects or fungi in vascular terrestrial plants.

Several genera of brown algae are infected by *Haloguignardia* species, which cause the formation of galls on the stipes or, more rarely, on vesicles and blades of the hosts. Estee (1913) was first to describe such galls on *Cystoseira* and *Halidrys* from California. Ferdinandsen and Winge (1920) found galls on *Sargassum* spp., and other authors described additional malformations on members of the same host genus (Cribb and Cribb, 1956, 1960b; Cribb and Herbert, 1954; Kohlmeyer, 1971a, 1972c). Galls induced by species of *Haloguignardia* are usually subglobose or ellipsoidal outgrowths of the algal thallus and contain one or more immersed ascocarps or spermogonia, or both, near the surface. The gall tissue consists mainly of algal cells with interspersed fungal hyphae. The mode of infection and the physiological changes in the host leading to gall formation are unknown. Diseased *Cystoseira* and *Halidrys* are often washed up in California, but gall-bearing *Sargassum* plants occur much less frequently. *Haloguignardia* species appear to be weak parasites that do not seriously harm the hosts or kill them.

The last example of a gall-inducing Ascomycete is *Massarina cystophorae* in *Cystophora retroflexa* in Tasmania (Cribb and Herbert, 1954). The galls are rounded or irregular, about 1.5 cm in diameter, and have a warty surface caused by the projecting black ascocarps and spermogonia. In this case also, pathological aspects are unknown.

To our knowledge, infectious fungal diseases on algae have never occurred as epidemics. This situation may change if algae become predisposed to infection by stress caused, for instance, by thermal or chemical pollution (Andrews, 1976).

II. SAPROBES

Fungi observed on dead algae on the shore or isolated from such substrates are listed in Table VIII. Additional facultative marine fungi, in particular, Deuteromycetes and yeasts, may occur and will be added to the list in the future. Algae washed up on the beach and incubated in the laboratory in a damp chamber almost regularly yield obligate or facultative marine fungi. Arenicolous species such as *Corollospora* often develop if grains of sand or shell fragments are attached to the algae. However, terrestrial and facultative marine fungi also may develop if the

TABLE VIII
Saprobic Fungi on Marine Algae

Fungus	Hosts	Host classes
Ascomycotina		
Corollospora intermedia	*Fucus vesiculosus*	Phaeophyta
Corollospora maritima	*Ceramium, Fucus, Macrocystis, Sargassum* spp.	Rhodophyta and Phaeophyta
Corollospora pulchella	*Fucus vesiculosus*	Phaeophyta
Crinigera maritima	*Fucus vesiculosus*	Phaeophyta
Lulworthia spp.	*Fucus vesiculosus, Laminaria hyperborea, L. saccharina, Saccorhiza polyschides*	Phaeophyta
Nectriella laminariae	*Laminaria* sp.	Phaeophyta
Orbilia marina	*Ascophyllum, Fucus, Halidrys* spp.	Phaeophyta
Pleospora pelvetiae	*Ceramium* sp., *Chondrus, Furcellaria, Laminaria, Pelvetia* spp.	Rhodophyta and Phaeophyta
Deuteromycotina		
Asteromyces cruciatus	*Cystoseira osmundacea, Egregia menziesii*	Phaeophyta
Cladosporium algarum	*Laminaria digitata*	Phaeophyta
Dendryphiella arenaria	*Sargassum* sp.	Phaeophyta
Dendryphiella salina	*Chondrus, Furcellaria, Laminaria, Sargassum* spp.	Rhodophyta and Phaeophyta
Epicoccum sp.	*Laminaria* sp.	Phaeophyta
Phoma laminariae	*Laminaria* sp.	Phaeophyta
Phoma spp.	*Fucus vesiculosus, Macrocystis integrifolia, Porolithon onkodes*	Rhodophyta and Phaeophyta
Stagonospora haliclysta	*Pelvetia canaliculata*	Phaeophyta
Stemphylium sp.	*Sargassum muticum*	Phaeophyta
Varicosporina ramulosa	*Hypnea charoides, Sargassum* sp.	Rhodophyta and Phaeophyta

algae are cast up at the high-water mark (Figs. 18a–18e). There are no reports available on the role of filamentous fungi in the breakdown of detached and rotting algae. In some geographic areas with a rich vegetation of macroalgae, large banks of decaying seaweeds can be found along the shore. Bacteria, yeasts (Seshadri and Sieburth, 1975), and filamentous higher fungi (Figs. 18a–18e) are involved in the decomposition, but we do not know how much each group of organisms contributes to this process. It is likely, however, that the fast-growing bacteria and yeasts are more important here than the filamentous fungi, in particular, members of

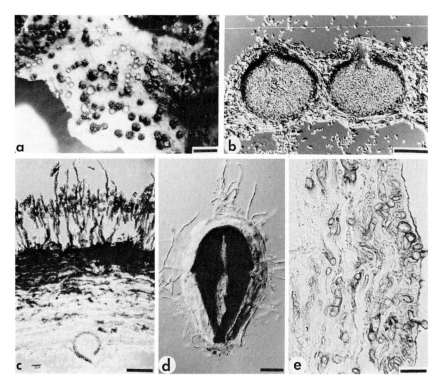

Fig. 18. Saprobic fungi in deteriorating marine algae. (a) Pycnidia of *Phoma laminariae* on *Laminaria* sp. (holotype K; bar = 300 μm). (b) 16-μm section through pycnidia of *P. laminariae* (holotype K; bar = 50 μm). (c) 16-μm section through heavily infested *Laminaria* sp. with conidiophores of *Cladosporium* sp. on the top and a pycnidium of *P. laminariae* on the bottom (from Ayrshire, K; bar = 100 μm). (d) Old egg of *Pelvetia canaliculata* with hyphae (Herb. J.K. 2965; bar = 50 μm). (e) 6-μm section through thallus of *Laminaria digitata* with hyphae of *Cladosporium algarum* (type NY; bar = 25 μm). (a = close-up photograph, b and d in Nomarski interference contrast, c in phase contrast, e in bright field.)

Ascomycetes. Phaeophyta are the main substrate for saprobic fungi (Table VIII and Fig. 18), probably because the tough brown algae are more resistant to decay than Chlorophyta and Rhodophyta and permit slow-growing filamentous fungi to form mycelia and propagules, whereas the more delicate green and red algae are decomposed by bacteria and yeasts before the filamentous fungi have time to reproduce.

III. GEOGRAPHICAL DISTRIBUTION

In general, parasitic marine fungi are found throughout the range of their particular host plants. Species of *Spathulospora,* parasites on *Ballia* spp. in the Southern Hemisphere, appear to be an exception to this rule (Kohlmeyer, 1973a; Kohlmeyer and Kohlmeyer, 1975). The distribution of these species is temperature dependent, as *S. antarctica* occurs only in antarctic and subantarctic regions below the 10°C isotherm (surface water temperature in August), whereas the other four species grow above that isotherm. A temperature barrier appears to limit the growth of *Spathulospora* species to certain areas, in spite of the possibility that water currents could transport ascospores across this barrier to healthy host plants.

Algal exudates may be another limiting factor for the occurrence of fungi. The number of fungal species as well as the total number of infected algae (*Sargassum* spp.) are small in the Sargasso Sea (Kohlmeyer, 1971a). It can be speculated that the cause for the paucity of fungi in this area is related to the production of antibiotic tannin by *Sargassum* spp. Conover and Sieburth (1964) reported that epiphytic algae, bacteria, and animals on *Sargassum* were rarer in the distribution center than outside the gyre of the Sargasso Sea and that the antibacterial activity of algae was strongest in the center of the gyre. These findings may explain the rare occurrence of parasitic fungi in the center of the Sargasso Sea and a relatively higher abundance at its periphery, namely, on the coast of North Carolina (Kohlmeyer, 1972c).

10. Submarine Lichens and Lichenlike Associations

Lichens may be defined in a variety of ways. Henssen and Jahns (1974), for instance, include in the lichens all fungi that are obligately bound in their nutrition to certain algae, forming with them a morphological–physiological unit. The habit of a lichen may or may not differ from the habit of both free-living partners. In addition, the formation of lichen products or bipartite vegetative reproductive units does not necessarily characterize a true lichen, according to these authors. The lichen association is not necessarily mutually beneficial to both partners (Hawksworth, 1976). Consequently, there is a wide array of symbiotic partnerships. At one end, there are loose associations of primitive lichens and at the other end we have algal–fungal associations that border on parasitism:

primitive lichens ↔ "true" lichens ↔ mycophycobioses ↔ parasites

We have listed these groups in Tables V, IX, and X. Marine lichenoid symbioses are poorly understood and all associations in Table X are in need of thorough examination. In particular, the physical contact and transport of nutrients between algal and fungal partners must be clarified. None of the organisms has been cultured and the algae must be checked for their ability to occur free-living in nature.

"True" marine lichens included in Table IX usually form obligate morphological–physiological units in which the fungal partner determines the habit of the association. Generally, the thallus of true lichens differs from the habit of both free-living partners. Littoral lichens belong to the genera *Arthopyrenia, Lichina, Stigmidium,* and *Verrucaria.* For the treatment of true marine lichens (Table IX) we refer to the lichenological and ecological literature (Zschacke, 1925; Santesson, 1939; Lamb, 1948; Klement and Doppelbaur, 1952; J. R. Lewis, 1964; Dodge, 1973; Fletcher,

---

TABLE IX
Marine Lichens of the Littoral Zone Associated with *Littorina*, Barnacles, or Algae[a]

Arthopyrenia halodytes	*V. erichsenii*
A. sublitoralis	*V. maura*
Arthopyrenia sp.	*V. microspora auct. non* Nyl.
Lichina confinis	*V. microspora* Nyl.
L. pygmaea	*V. mucosa*
Lichina sp.	*V. prominula*
Stigmidium marinum	*V. psychrophila*
Verrucaria amphibia	*V. sandstedei*
V. degelii	*V. serpuloides*
V. ditmarsica	*V. striatula*

[a] From Fletcher (1975a), Lamb (1948), and Santesson (1939).

1975a,b, 1976). The following discussion includes only borderline cases of submarine lichenoid associations, namely, the primitive lichens and mycophycobioses, which are usually not treated in books on lichens (but see Henssen and Jahns, 1974).

I. PRIMITIVE MARINE LICHENS

Primitive marine lichens (Kohlmeyer, 1967) or lichenoids are loose associations in which the algal partners are able to occur in a free-living state. No new structures originate from such associations. The algae are usually microscopic and the fungi belong to such genera as *Chadefaudia, Leiophloea,* and *Pharcidia* (Table X). In this context we understand the term "lichen" in a wide sense, meaning a symbiosis between algae and fungi.

A. *Chadefaudia corallinarum*

The association between this Ascomycete and epiphytic, crustose red algae of the genera *Dermatolithon* and *Epilithon* was discussed by Kohlmeyer (1973b). The fact that the fungus may live with different phycobionts indicates that the symbiosis is rather loose. In addition, the algal partners may occur without the fungus. The association is only found under submerged conditions and the fungus grows on macroalgae (Chlorophyta, Phaeophyta, and Rhodophyta) and leaves of sea grasses. Hyphae of *C. corallinarum* grow between the cells of the phycobiont without apparent damage to the partner. The supporting macroalgae or

TABLE X
Submarine Lichenoid Associations

Fungus	Algal partner or substrate	Type of association
Chadefaudia corallinarum	*Dermatolithon pustulatum, Dermatolithon* sp., *Epilithon membranaceum* on diverse macroalgae and sea-grass leaves	Primitive lichen
Leiophloea pelvetiae	Unidentified blue–green algae on *Pelvetia canaliculata*	Primitive lichen
Pharcidia balani (lichen: *Arthopyrenia sublitoralis*)	Various species of microscopic algae in calcareous shells of mollusks and cirripeds	Primitive lichen
Pharcidia laminariicola	*Ectocarpus fasciculatus* on *Laminaria digitata*	Primitive lichen
Pharcidia rhachiana	Unidentified blue–green and probably brown algae on *Laminaria digitata*	Primitive lichen
Blodgettia bornetii	*Cladophora caespitosa, C. fuliginosa*	Mycophycobiosis
Mycosphaerella ascophylli	*Ascophyllum nodosum, Pelvetia canaliculata*	Mycophycobiosis
Turgidosculum complicatulum	*Prasiola borealis, P. tessellata*	Mycophycobiosis (or parasitism)
Turgidosculum ulvae	*Blidingia minima* var. *vexata*	Mycophycobiosis (or parasitism)

leaves are never penetrated by the fungus. The ascocarps are often partially covered by the calcareous thallus of the algal partner. This lichen has been collected in the Atlantic, Indian, and Pacific oceans and in the Black and Mediterranean seas.

B. *Leiophloea pelvetiae*

This fungus has been recently investigated (Kohlmeyer, 1973b, *sub Dothidella pelvetiae*) and is being transferred (see p. 376) to the genus *Leiophloea*. Initially the lichenized character of *L. pelvetiae* had not been recognized and the fungus was considered to be a parasite of *Pelvetia canaliculata*. Sections through this alga show that ascocarps and pycnidia of *L. pelvetiae* are superficially seated on the cortical cells of *Pelvetia* without penetrating it (Fig. 19a). Both types of fruiting bodies are surrounded by colonies of at least two species of one-celled blue–green algae

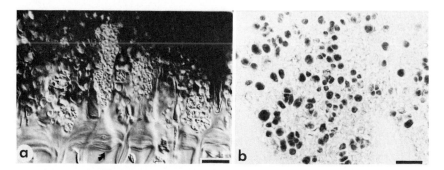

Fig. 19. *Leiophloea pelvetiae,* a lichenized Ascomycete. (a) 4-μm longitudinal section through the supporting host, *Pelvetia canaliculata* (arrow), with the dark base of the ascocarp above and colonies of blue–green algae in between (bar = 10 μm). (b) Section under and parallel to the base of the ascocarp through the layer of symbiotic blue–green algae (bar = 15 μm). (a in Nomarski interference contrast, b in bright field; both Herb. J.K. 2951.)

(Fig. 19b). Single algal cells are usually embedded in the peridium. Thus, there is a close contact between phycobiont and fungus, but haustoria have not been observed. So far it has not been possible to isolate or identify the blue–green algae. Because the supporting host, *P. canaliculata,* is exposed at low tides, *L. pelvetiae* and associated algae dry out frequently. The fungus occurs in France and Great Britain and on the Spanish Atlantic coast. We have been unable to find it on *Pelvetia* in Norwegian waters.

C. *Pharcidia balani*

This species is a facultative symbiont of various species of microscopic algae in the shells or tests of live or dead intertidal mollusks and cirripedes, respectively, but it also occurs in calcareous rocks. Lichenologists usually include this association under the lichen name *Arthopyrenia sublitoralis,* as does Santesson (1939), who states, "I have only after a lengthy search succeeded in finding gonidia; sometimes the result has been quite negative." The algal partners in this association belong to the Cyanophyceae. Ascocarps and pycnidia (or spermogonia) are immersed in the calcareous substrate, but the mycelium does not penetrate deeply into the shell, and the animals are usually not damaged by the fungus. However, Santesson (1939) remarks that the lichen may occasionally be a dangerous parasite. It grows in intertidal areas, infrequently in the *Ascophyllum* and *Pelvetia* zones, but has its optimum of distribution in the *Fucus vesiculosus* and *F. spiralis* horizons. *Pharcidia balani* is a cosmopolitan species found in the tropics, subtropics, temperate zones, and even arctic seas. It is one of the most common fungal species of rocky intertidal areas and occurs on shells of more than 60 animal species (Santesson, 1939).

D. *Pharcidia laminariicola*

This submarine fungus lives in lichenoid associations with epiphytic *Ectocarpus fasciculatus* on the stalks of *Laminaria digitata* (Kohlmeyer, 1973b). Ascocarps and pycnidia (or spermogonia) are superficially seated on the cortex of *Laminaria*. Hyphae of the fungus and filaments of *Ectocarpus* are intimately entwined and there is no indication that the mycelium penetrates the supporting host, *Laminaria*. This symbiosis between a filamentous marine brown alga and a fungus is unique among lichenoid associations. The only other cases of marine Phaeophyta living together with an Ascomycete are *Ascophyllum* and *Pelvetia,* which will be treated in Section II (mycophycobioses). *Pharcidia laminariicola* has been collected only once on the Norwegian coast. Because of its inconspicuous habit it may be easily overlooked.

E. *Pharcidia rhachiana*

Ascocarps and pycnidia (or spermogonia) are found associated with epiphytic blue–green and possibly brown algae on the surface of hapteres of *Laminaria digitata* (Kohlmeyer, 1973b). Superficially, *P. rhachiana* resembles *P. laminariicola,* but ascospores and dimensions of fruiting bodies and asci are distinctly different. The association of *P. rhachiana* with microscopic algae consists of a thin lichenoid crust around the hapteres of the supporting host, *Laminaria* (Fig. 20a). There is no penetration of hyphae into the host. So far, *P. rhachiana* is known only from a single collection from the Norwegian coast.

An Ascomycete on the holdfast of *Padina durvillaei,* namely, *Pontogeneia padinae,* may possibly be associated with epiphytic, microscopic algae (Kohlmeyer, 1975b). However, the nutritional relationships of this fungus are still unresolved.

II. MYCOPHYCOBIOSES

Mycophycobioses are obligate symbiotic associations between a systemic marine fungus and a marine macroalga in which the habit of the alga dominates (Kohlmeyer and Kohlmeyer, 1972a). Fungal partners in

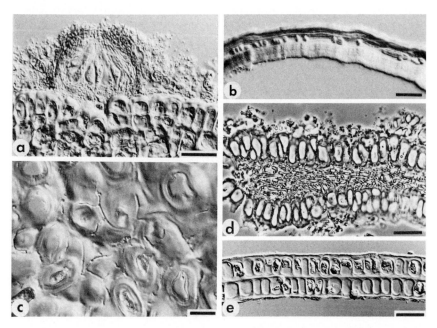

Fig. 20. Lichenized marine fungi. (a) Longitudinal section (6 μm) through as-cocarp of *Pharcidia rhachiana* on *Laminaria digitata;* the fungus is superficially seated on the macroalga and is surrounded by one-celled microscopic algae (Herb. J.K. 3119; bar = 25 μm). (b) Cross section (8 μm) through wall of *Cladophora* sp., hyphae of *Blodgettia bornetii* between the outer and inner walls (from PC; bar = 15 μm). (c) Section (6 μm) through medulla of *Ascophyllum nodosum* with intercellular hyphae of the symbiont, *Mycosphaerella ascophylli* (Herb. J.K. 2481; bar = 15 μm). (d) Cross section through thallus of *Blidingia minima* var. *vexata* with hyphae of *Turgidosculum ulvae* pushing the two algal cell layers apart (Herb. Setzer 3750; bar = 30 μm). (e) Cross section through thallus of *Ulva* sp. as a comparison of a fungus-free chlorophyte with an infested one in d (bar = 25 μm). (d in phase contrast, the others in Nomarski interference contrast.)

mycophycobioses belong to *Blodgettia* and *Mycosphaerella.* Some cases of mycophycobioses are questionable and may possibly be classified as parasitism, for instance, the association of *Turgidosculum* spp. with *Blidingia* or *Prasiola* (Table X). If the host plants are able to live independently, the fungus could be considered a parasite. However, the identity of fungus-free *Blidingia* and *Prasiola* with similar fungus-infested plants appears not to be established with certainty. Distinct cases of parasitism are discussed in Chapter 9 on algicolous fungi. According to D. H. Lewis (1973, Table 3), mycophycobioses may possibly belong to his group 5, namely, "Ecologically Obligately Symbiotic Biotrophs (Obligate

Biotrophs),'' a group also embracing fungi responsible for ''symptomless parasitism.''

A. *Blodgettia bornetii*

This Hyphomycete forms a dense network of anastomosing hyphae with hyaline chlamydospores inside the wall of the chlorophytes *Cladophora caespitosa* and *C. fuliginosa* (J. Feldmann, 1938). The mycelium grows throughout the plant (Fig. 20b) but never penetrates the algal cells. Hyphae are only absent in the very tips and in the cross walls of the host. The algal partner clearly determines the habit of this association and does not seem to be damaged by the presumably symbiotic fungus. *Cladophora* species with the endophyte appear to be restricted to subtropical and tropical areas of the Atlantic and Pacific oceans.

B. *Mycosphaerella ascophylli*

Ascophyllum nodosum and *Pelvetia canaliculata* regularly contain this systemic Ascomycete (F. C. Webber, 1967; Kohlmeyer and Kohlmeyer, 1972a). The constant occurrence of the endosymbiotic fungus in both species of Phaeophyta is generally not mentioned in the phycological literature. A network of fungal hyphae grows intercellularly in the cortex and medulla of the host (Fig. 20c). The mycelium follows the growing tip of the alga, usually six to ten cells behind the apical cell. Haustoria have never been observed. Ascocarps and spermogonia develop in the receptacles of *A. nodosum* and *P. canaliculata* and are totally immersed in the outer three cell layers of the host. Trichogynes protrude from the ostiole of immature ascocarps, and occasionally spermatia attached to a trichogyne have been observed (F. C. Webber, 1967). Although fruiting bodies are always restricted to the receptacles of attached plants of *A. nodosum* and *P. canaliculata,* they have also been found in the vegetative tips of the free-floating form, *A. nodosum* f. *mackaii.*

Some authors characterize *M. ascophylli* as a parasite (e.g., Henssen and Jahns, 1974), whereas others regard the fungus as a symbiont (J. Feldmann, 1938; Mattick, 1953). We consider the symbiosis between *M. ascophylli* and its hosts, *A. nodosum* and *P. canaliculata,* to be a typical example of a mycophycobiosis in which the thallus of the alga determines the habit of the association. It may be compared to an ectotrophic mycorrhiza. The algae and the fungus never occur separately in nature. This indicates that the fungus depends on its hosts. *Ascophyllum,* in turn, appears to require the presence of its mycobiont, *Mycosphaerella,* because F. C. Webber (1967) found that uninfected sporelings of *A.*

nodosum do not survive in nature beyond a certain age. The infection probably occurs at a very early stage. Thorough studies with axenic cultures of the fungus and the algal hosts should be made to clarify this biologically interesting phenomenon. The location of *Ascophyllum* and *Pelvetia* on rocks at the high-tide line indicates their ability to resist desiccation at low tides, and it is probable that the fungal partner provides a protection against drying of the thallus. The fungus also appears to profit from its association with *Ascophyllum* and *Pelvetia* by being protected against desiccation inside the algal tissues. Ascocarps of *M. ascophylli* are perfectly preserved after years of dry storage in the herbarium. The fungus–alga association seems to be as advantageous to both organisms in the upper intertidal zone as it is in terrestrial habitats, where most lichens are so ably adapted to dry conditions. Because of these considerations, a parasitic relationship between *M. ascophylli* and its phaeophyte hosts is unlikely.

Penot and Penot (1977) determined by means of autoradiography that long-distance transports in *Ascophyllum nodosum* are of a diffusive type, involving the ionic transfer by an apoplastic pathway. This translocation differs from that of other marine algae where it is symplastic, using the cytoplasm of cells and plasmodesmata as means of transportation. The possibility that ions move through the mycelium of the systemic fungus has not been mentioned by Penot and Penot (1977), but may be a hypothesis worth testing.

C. *Turgidosculum complicatulum*

The chlorophytes *Prasiola borealis* and *P. tessellata* are associated with the Ascomycete *T. complicatulum*. Hyphae grow throughout the algal thallus and separate it into groups of cells. These groups consist of tetrads or rows of algal cells. Large dark ascocarps and smaller spermogonia are embedded in this thallus, giving it a warty appearance. The question is unresolved whether *T. complicatulum* is a symbiont or a parasite. The host, *Prasiola,* also occurs in a fungus-free form on the same rocks with the infested plants (Reed, 1902). Possibly, the symbiosis with a fungus renders the alga more resistant to exposures at low tides, similar to other cases of marine macroalgae, namely, *Ascophyllum* and *Pelvetia,* which, however, live in obligate symbioses. Therefore, we tend to consider the *Prasiola–Turgidosculum* association a mycophycobiosis rather than a case of parasitism. It has to be proven that the free-living form of *Prasiola* is indeed identical with the lichenized alga.

Ascocarps of *Turgidosculum complicatulum* are perfectly preserved in dried herbarium material. The centrum is filled with a gelatinous matrix surrounding the asci and apparently protecting them against desiccation. Also, the ostiolar canal is occupied by a similar turgescent matrix that probably has a function in the dispersal of ascospores. The *Prasiola–Turgidosculum* association is commonly found in arctic and antarctic waters but also occurs in British Columbia.

D. *Turgidosculum ulvae*

The green alga *Blidingia minima* var. *vexata* (= *Ulva vexata*) grows on rocks in high intertidal to splash zones and is often exposed to heavy surf (Norris, 1971). In contrast to *B. m.* var. *minima,* the variety *vexata* is characterized by its usual association with the Ascomycete *Turgidosculum ulvae* (Reed, 1902; Kohlmeyer and Kohlmeyer, 1972b). The question may be raised whether a separation of the host, based solely on the infection by a symbiont (or parasite?), is justified or *B. minima* var. *minima* should be regarded as the host. This doubt is supported by the fact that the range of the variety *minima* covers that of var. *vexata,* and both varieties occur in the same habitats of the upper intertidal zone (Norris, 1971).

The mycelium of *T. ulvae* forms dense mats between the two layers of the host (Figs. 20d and 20e), and ascomata and spermogonia also develop between these layers and push them apart. As in *T. complicatulum,* the ascocarps contain a gelatinous matrix that fills the ostiolar canal. This pulvillus closes the ostiole of wet fruiting bodies and shrinks in exposed algae at low tide. By the shrinking process, mature ascospores are pressed out of the ostiole, accumulate on top of the opening, and are washed away by the rising water. The renewed imbibition makes the pulvillus swell and close the ostiole, thereby protecting immature asci against exposure to water.

We regard the association between *T. ulvae* and *Blidingia* as mycophycobiosis because the alga determines the habit. However, this facultative symbiosis is close to parasitism, especially if the host, *Blidingia minima* var. *vexata,* should prove to be identical with the non-lichenized *B. m.* var. *minima.* Possibly, infection experiments of uninfested *B. minima* with *Turgidosculum* could clarify the question of host identity. No damage to the alga is noticeable under the light microscope.

In conclusion, we would like to point out again that mycophycobioses are not lichens in a strict sense, but that they consist of macroalgae that, in all likelihood, profit rather than suffer from the association with a fungus. Terrestrial lichens are extremely well protected against desiccation and frost (Kappen and Lange, 1972). Similarly, mycophycobioses, which all occur at the upper intertidal zone, are exposed to, and protected against, desiccation and frost.

11. Fungi in Halophytes of Tidal Salt Marshes

Maritime salt marshes are coastal land areas which are more or less covered by vegetation and which become periodically inundated by the tides (Chapman, 1960, 1974; Reimold and Queen, 1974). The vegetation in these marshes is composed of halophytes, that is, plants capable of tolerating high amounts of salts, usually 0.5% or more NaCl (Chapman, 1974; Waisel, 1972). Salt marshes are among the most productive ecosystems on earth and their detritus supports a complex food web in adjacent estuaries that harbor many commercially important animals. The primary production in a Louisiana *Spartina alterniflora* marsh, for instance, ranges between 750 and 2600 g/m²/year (Kirby and Gosselink, 1976). Odum and De La Cruz (1967) determined the rate of decomposition of salt-marsh plants and found that losses after 300 days were 35% in *Juncus,* 53% in *Distichlis,* 58% in *Spartina,* and 94% in *Salicornia.* The nutritive value of the detritus increases, as small suspended particles contain 24% protein, compared to 6% in dead *Spartina*; thus, the detritus becomes a better food source for animals than the standing plants. Odum and De La Cruz (1967) attribute this enrichment to the action of bacteria, but recent mycological investigations indicate that fungi may be as important decomposers in salt marshes as are the bacteria (e.g., Gessner, 1977; Gessner and Goos, 1973a,b; Meyers *et al.*, 1970b).

This chapter deals with saprobic fungi on plants of temperate and subtropical salt marshes, whereas the mangrove-inhabiting fungi of the tropics and subtropics are treated in Chapter 12. Fungi isolated by culture techniques from soils and sediments of salt marshes are discussed in Chapter 7. The plants of tidal salt marshes become completely exposed at most low tides, and normally they are only partly covered by water, even at regular high tides. This is the reason we find different fungal colonizers on different parts of the plants. The always-emersed portions of stems and leaves of tall plants bear terrestrial fungi, whereas submersed organs of the plants, namely, roots, rhizomes, and lower parts of stems and leaves, are attacked by marine fungi (see Table XI). Our discussion will be restricted mainly to those fungi found on submerged parts of salt-marsh plants. There is, however, an overlapping of marine and terrestrial fungi at the water–air interface. A clear-cut line between these two groups cannot always be drawn and fungi of this intermediate zone may be termed facultative marine fungi (Kohlmeyer and Kohlmeyer, 1964) or amphibious species (Pugh, 1974). In questionable cases we prefer to include a species in this treatment even if it cannot be considered to be a marine fungus *sensu stricto*. Marine fungi are sometimes collected on Gramineae growing at the base of dunes. Such plants are always exposed to salt spray and are occasionally reached by high tides. Fungi found on standing dune plants, or more often on broken-off parts along the high-water mark, are listed in Table XII.

Fungi and their hosts in the salt-marsh vegetation are often exposed to extreme environmental conditions, such as high temperatures at midday, drying out at low tides, changes between high salt concentrations and freshwater conditions during and after rains, and freezing during the winter months in northern areas. The mycota of salt marshes can be expected to be physiologically adapted to such conditions and probably differs from fungi of permanently submerged habitats.

I. HOST SPECIFICITY

The list of halophytes with their fungal partners (Table XI) shows that many fungi are host specific or at least restricted to one family of host plants. The upper parts of halophytes often support fungi specific for certain host families, for example, *Ascochyta salicorniae, Camarosporium roumeguerii, Mycosphaerella salicorniae,* and *Tubercularia pulverulenta* in Chenopodiaceae. These species can probably be classified as facultative marine fungi. In contrast, basal parts of the halophytes that are always or mostly submerged usually harbor ubiquitous obligate marine fungi, such as *Aniptodera chesapeakensis, Haligena elaterophora, H. spartinae, Halosphaeria hamata, Lignincola laevis,* and *Lulworthia* spp.

II. TAXONOMY

In all, the compilation in Table XI includes 33 Ascomycetes, 1 Basidiomycete, and 23 Deuteromycetes. We are still at the beginning of our research on fungi from salt marshes and the list will be expanded

---

TABLE XI

Higher Fungi on Halophytes of Tidal Salt Marshes and Salt Meadows (Exclusive of Mangroves)[a]

Armeria pungens (Plumbaginaceae)
  Ascomycotina: *Mycosphaerella staticicola*
Arundo donax (Gramineae)
  Ascomycotina: *Ceriosporopsis circumvestita*
               C. halima
               Halosphaeria appendiculata
               Leptosphaeria oraemaris
Atriplex halimus (Chenopodiaceae)
  Deuteromycotina: *Camarosporium roumeguerii*
Atriplex sp.
  Deuteromycotina: *Camarosporium roumeguerii*
Crithmum maritimum (Umbelliferae)
  Deuteromycotina: *Phoma* sp.
Distichlis spicata (Gramineae)
  Deuteromycotina: *Drechslera halodes*
                Septoria thalassica
D. "thalassica"
  Deuteromycotina: *Septoria thalassica*
Glaux maritima (Primulaceae)
  Ascomycotina: *Leptosphaeria oraemaris*
Halimione portulacoides (Chenopodiaceae)
  Ascomycotina: *Didymosphaeria maritima*
               Leptosphaeria obiones
               Pleospora gaudefroyi
  Deuteromycotina: *Alternaria* sp.
                Ascochytula obiones
                Camarosporium roumeguerii
                Coniothyrium obiones
                Dendryphiella salina
                Phoma sp.
Juncus maritimus (Juncaceae)
  Ascomycotina: *Leptosphaeria albopunctata*
               L. neomaritima
J. roemerianus
  Ascomycotina: *Aniptodera chesapeakensis*
               Leptosphaeria marina
               L. neomaritima
               L. paucispora?
               L. typhicola
               Lulworthia sp.
  Deuteromycotina: *Alternaria* sp.   *(Continued)*

---

[a] This preliminary list is by no means complete because relatively few halophytes have been surveyed thoroughly for fungi thus far. Most of the fungi grow on submerged parts of the plants, but some characteristic species from emersed parts have also been included. Host names are used as they appear in the original literature, and therefore a host species may occasionally be listed under more than one synonym.

TABLE XI (Continued)

Juncus sp.
   Deuteromycotina: *Monodictys pelagica*
Limonium sp. (Plumbaginaceae)
   Ascomycotina: *Mycosphaerella staticicola*
Phragmites communis (Gramineae)
   Ascomycotina: *Leptosphaeria albopunctata*
   Deuteromycotina: *Cirrenalia fusca*
Puccinellia maritima (Gramineae)
   Deuteromycotina: *Dendryphiella salina*
Salicornia ambigua (Chenopodiaceae)
   Ascomycotina: *Mycosphaerella salicorniae*
                Pleospora gaudefroyi
   Deuteromycotina: *Camarosporium palliatum*
                Tubercularia pulverulenta
S. europaea
   Deuteromycotina: *Ascochyta salicorniae*
S. cf. *fruticosa*
   Ascomycotina: *Mycosphaerella salicorniae*
                Pleospora gaudefroyi
S. herbacea
   Ascomycotina: *Mycosphaerella salicorniae*
                Pleospora gaudefroyi
   Deuteromycotina: *Ascochyta salicorniae*
                Camarosporium roumeguerii
                C. palliatum
                Phoma sp.
                Tubercularia pulverulenta
S. h. var. *procumbens*
   Ascomycotina: *Mycosphaerella salicorniae*
S. patula
   Deuteromycotina: *Ascochyta salicorniae*
S. perennis
   Ascomycotina: *Mycosphaerella salicorniae*
S. peruviana
   Ascomycotina: *Mycosphaerella salicorniae*
   Deuteromycotina: *Camarosporium roumeguerii*
                Tubercularia pulverulenta
S. stricta
   Deuteromycotina: *Dendryphiella salina*
                Phoma sp.
S. subterminalis
   Ascomycotina: *Mycosphaerella salicorniae*
S. virginica
   Ascomycotina: *Mycosphaerella salicorniae*
                Pleospora sp. I
   Deuteromycotina: *Camarosporium palliatum*
Salicornia sp.
   Ascomycotina: *Pleospora gaudefroyi*

TABLE XI (Continued)

Deuteromycotina: *Camarosporium palliatum*
       C. roumeguerii
Salsola kali (Chenopodiaceae)
 Deuteromycotina: *Alternaria* sp.
        Stemphylium sp.
Spartina alterniflora (Gramineae)
 Ascomycotina: *Aniptodera chesapeakensis*
      Buergenerula spartinae
      Corollospora maritima
      Haligena elaterophora
      H. spartinae
      Halosphaeria hamata
      Leptosphaeria albopunctata
      L. halima
      L. marina
      L. neomaritima
      L. obiones
      L. oraemaris
      L. paucispora
      Lulworthia spp.
      Mycosphaerella sp. I
      Nais inornata
      Phaeosphaeria typharum
      Pleospora pelagica
 Basidiomycotina: *Nia vibrissa*
 Deuteromycotina: *Alternaria* spp.
        Cirrenalia macrocephala
        Dendryphiella arenaria
        D. salina
        Dictyosporium pelagicum
        Drechslera halodes
        Epicoccum spp.
        Monodictys pelagica
        Phoma spp.
        Stagonospora sp.
S. cynosuroides
 Ascomycotina: *Leptosphaeria obiones*
      Lulworthia sp.
S. cf. *densiflora*
 Ascomycotina: *Mycosphaerella* sp. I
S. patens
 Deuteromycotina: *Stagonospora* sp.
S. cf. *pectinata*
 Ascomycotina: *Mycosphaerella* sp. II
S. townsendii
 Ascomycotina: *Ceriosporopsis halima*
      Gnomonia salina
      Haligena spartinae

(Continued)

TABLE XI (*Continued*)

         Halosphaeria hamata
         Leptosphaeria albopunctata
         L. marina
         L. neomaritima
         L. obiones
         L. oraemaris
         L. paucispora
         Lignincola laevis
         Lulworthia spp.
         Phaeosphaeria typharum
         Pleospora spartinae
  Deuteromycotina: *Alternaria* spp.
         Dendryphiella salina
         Phoma spp.
Spartina sp.
  Ascomycotina: *Buergenerula spartinae*
         Gnomonia salina
         Leptosphaeria albopunctata
         L. marina
         L. neomaritima
         L. obiones
         L. paucispora
         L. typhicola
         Mycosphaerella sp. II
         Phaeosphaeria typharum
  Deuteromycotina: *Asteromyces cruciatus*
         Stagonospora sp.
         Stemphylium sp.
Spergularia salina (Caryophyllaceae)
  Deuteromycotina: *Cladosporium algarum*
Suaeda australis (Chenopodiaceae)
  Ascomycotina: *Mycosphaerella suaedae-australis*
S. fruticosa
  Ascomycotina: *Mycosphaerella salicorniae*
  Deuteromycotina: *Camarosporium roumeguerii*
S. maritima
  Ascomycotina: *Pleospora gaudefroyi*
  Deuteromycotina: *Cladosporium algarum*
         Dendryphiella salina
         Phoma suaedae
Triglochin maritima (Juncaginaceae)
  Ascomycotina: *Pleospora triglochinicola*
  Deuteromycotina: *Stemphylium triglochinicola*
Typha sp. (Gramineae)
  Ascomycotina: *Haligena spartinae*
         Halosphaeria hamata
  Deuteromycotina: *Monodictys pelagica*

TABLE XII
Higher Fungi on Gramineae in Dunes Exposed to Salt Spray

Agropyron junceiforme
  Ascomycotina: *Haligena spartinae*
                Leptosphaeria pelagica
A. pungens
  Ascomycotina: *Haligena spartinae*
                Halosphaeria hamata
                Leptosphaeria obiones
                L. paucispora
                Lignincola laevis
                Lulworthia spp.
Agropyron sp.
  Deuteromycotina: *Asteromyces cruciatus*
Ammophila arenaria
  Ascomycotina: *Phaeosphaeria ammophilae*
                Sphaerulina oraemaris
  Deuteromycotina: *Asteromyces cruciatus*
                   Camarosporium metableticum
                   Cirrenalia macrocephala
A. baltica
  Deuteromycotina: *Camarosporium metableticum*
Uniola paniculata
  Deuteromycotina: *Camarosporium metableticum*

considerably in the future, because many fungal species from these habitats await identification and description. From *Spartina* spp. alone over 100 fungi have been described (Gessner and Kohlmeyer, 1976) and *Juncus roemerianus* harbors at least 30 higher fungi (J. Kohlmeyer, unpublished). The mycota of halophytes is a conglomerate of species of many unrelated genera. However, there seems to be a separation between fungal taxa attacking the basal parts of marsh plants and fungi of the emersed parts. Ascohymeniales, in particular, representatives of Halosphaeriaceae, predominate in the submerged stems and rhizomes, whereas the exposed parts of the hosts harbor mostly Ascoloculares and pycnidial Deuteromycetes. Examples of Ascohymeniales in the plants' bases are, for instance, the genera *Ceriosporopsis, Haligena, Halosphaeria, Lignincola, Lulworthia,* and *Nais*. Genera of Ascoloculares, often found in aerial organs of halophytes, are *Leptosphaeria, Mycosphaerella,* and *Pleospora*. The phylogenetic implications for this separation of Ascoloculares above the high-tide line, and of Ascohymeniales below this line, are discussed in Chapter 22.

III. ACTIVITIES OF FUNGI IN SALT MARSHES

Live halophytes may be attacked by parasitic fungi, and saprobes start
decomposing senescent plants at the end of the growing season.

A. Parasites

Emersed parts of live halophytes are sometimes invaded by parasites,
for example, *Claviceps purpurea* (Fig. 21a), *Puccinia seymouriana, P.
sparganioides, Uromyces acuminatus,* and *U. argutus* on *Spartina* spp.
(Gessner and Kohlmeyer, 1976). Among marine fungi there are few
parasitic species known thus far from salt-marsh plants. One of them, the
Ascomycete *Buergenerula spartinae* in *Spartina* spp., is one of the first
fungi to appear on the green standing plant, and it produces runner hyphae
and lobed hyphopodia (Fig. 21b) on basal parts of culms of *Spartina* spp.
below the leaf sheath (Gessner *et al.*, 1972; Goos and Gessner, 1975). It is
probably a weak parasite that forms ascocarps only in senescent tissues of

Fig. 21. Fungi in salt-marsh halophytes. (a) Sclerotia of *Claviceps purpurea* on
Spartina brasiliensis (LPS No. 18649; bar = 5 mm). (b) Hyphopodia and hyphae of
Buergenerula spartinae in coleoptile of *S. alterniflora* (Herb. Gessner; bar = 50 μm). (c)
Stem of senescent *S. alterniflora* with erumpent ascocarps of *B. spartinae* (Herb. J. K.
3498; bar = 1 mm).

the host (Fig. 21c). Leaf browning of *S. alterniflora* is possibly caused by *Leptosphaeria typharum* and a *Phoma* sp. (Gessner and Goos, 1973b).

A serious dieback of *Spartina townsendii* in England had physiological causes, and fungi isolated from moribund plants in dieback areas were exclusively saprobic species (Goodman, 1959). Sivanesan and Manners (1970) also concluded in observations of *S. townsendii* in healthy and dieback sites that fungi are probably insignificant even as secondary parasites in relation to the dieback. These authors used four different sampling techniques and isolated only terrestrial fungi.

B. Saprobes

Many halophytes have fleshy leaves and stems that are not invaded by saprobic fungi unless they have lost their turgor and have dried up. An exception may be *Halimione portulacoides,* the living leaves of which support vegetative hyphae of *Ascochytula obiones* (Dickinson, 1965). Eventually all parts of dead marsh plants are attacked by fungi. The first ascocarps and pycnidia on *Salicornia* spp., for instance, appear on the lower, desiccated side branches and on the dry stem bases of otherwise healthy plants. Fruiting bodies of *Mycosphaerella salicorniae, Ascochyta salicorniae,* and *Camarosporium roumeguerii* develop subepidermally in dried parts of *Salicornia* spp. and are scarcely visible as tiny black dots. Other fungi cause a distinct blackening of large areas of host tissues, for example, several species of *Leptosphaeria* that form black stromata in the epidermis of dead stems and leaves of *Spartina* spp.

Few observations have been made on the ecology of fungi in marsh plants. Wagner (1965) thoroughly investigated *Leptosphaeria obiones,* a saprobe in several halophytes. She described the growth of the germ tube through the cuticle and pits of epidermal cells of *Spartina alterniflora.* This fungus did not seem to dissolve cell walls, but rather entered neighboring cells through the pits. Ascocarps of *L. obiones* occurred twice as frequently at the nodes as at the internodes and developed only on basal parts of culms submerged at high water (Wagner, 1969). Furthermore, *L. obiones* was more frequent and had greater densities of ascocarps near the low-tide line. Wave action decreased the height at which the fungus occurred on the stems. Detritus and algae around the base of *Spartina* stems prevented development of *L. obiones,* probably because of lack of oxygen in this zone. Wagner's (1969) studies showed that *L. obiones* in North Carolina required about 6 to 9 months to develop, after the plants became infected between July and November.

Other ecological surveys in Rhode Island (Gessner, 1977; Gessner and Goos, 1973a,b) also concerned fungi active in the decomposition of

Spartina alterniflora. Direct microscopic observations of living and dead standing plants (Gessner, 1977; Gessner and Goos, 1973b) and a litter-bag method (Gessner and Goos, 1973a) were used to characterize and enumerate the fungal populations on this host. The direct observations, combined with damp-chamber incubations, yielded 14 marine and 18 terrestrial fungi. The litter-bag study, in which ground parts of *S. alterniflora* were added to pour-plates, resulted in the isolation of 14 terrestrial fungi, a sterile mycelium, and three marine species (*Dendryphiella arenaria, D. salina,* and *Dictyosporium pelagicum*). A control consisted of the direct microscopic examination of *S. alterniflora* from litter bags and yielded nine marine fungi, all different from those obtained in the culture experiments. Gessner and Goos (1973a) emphasize the selectiveness of the dilution-plate technique, which favors fast-growing terrestrial species, whereas the upper, noninundated parts are mainly inhabited by terrestrial discussed in Chapters 1 and 2, the role of terrestrial fungi in the marine habitat is unknown and such species isolated with culture techniques in the laboratory may derive from dormant propagules of fungi that are inactive in this environment.

Gessner (1977) examined the seasonal occurrence and distribution of fungi on the aerial parts of *Spartina alterniflora* in a Rhode Island estuary between June 1972 and July 1974. Marine Ascomycetes and Deuteromycetes occur on the lower, twice daily submerged parts of the plants, whereas the upper, noninundated parts are mainly inhabited by terrestrial species (Fig. 22). *Buergenerula spartinae* and *Phaeosphaeria typharum* invade the host early in the season, and some Deuteromycetes (*Alternaria alternata, Epicoccum nigrum*) are most frequent during the time of seed production, since they grow primarily on inflorescences and seeds. In general, the greatest number of fungal species per plant occur from September till January, that is, during seed formation, senescence, and death of the host. Dead culms of *S. alterniflora* left over from the previous year are mostly inhabited by *Leptosphaeria obiones, L. pelagica, Lulworthia* sp., *Pleospora pelagica,* and *P. vagans.* Some fungi occur throughout the seasonal growth–death cycle of the host, for example, *Buergenerula spartinae, Halosphaeria hamata, Phaeosphaeria typharum,* and *Stagonospora* sp. Gessner (1977) emphasized that the successional scheme observed is mainly a seasonal pattern of appearance of reproductive structures, not necessarily a replacement of one fungus by another.

Apinis and Chesters (1964) studied the Ascomycetes colonizing various halophytes of salt marshes and sand dunes in England. These authors found 17 species of marine Ascomycetes in the intertidal zone, mainly on standing plants or on litter of *Spartina townsendii* and *Agropyron* spp.

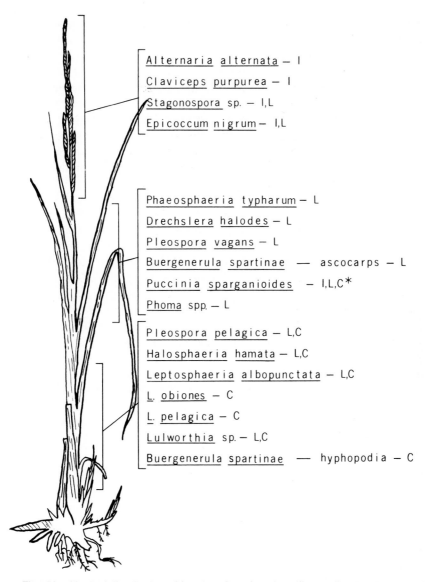

Fig. 22. Vertical distribution of fungi on *Spartina alterniflora* indicating the area of the plant where each fungus was most commonly observed. I, inflorescence and seeds; L, leaves; C, culm; asterisk, throughout plant. From Gessner, *Mycologia* **69**, 488 (1977), Fig. 5. Reprinted with permission.

Debris of *S. townsendii* transported by high tides from the intertidal to higher grassland lost the marine mycota, which was replaced by terrestrial species within a short time.

A variety of fungi, mostly terrestrial species, can be isolated with culture techniques from the root surface (rhizoplane) of marsh plants (Pugh, 1960, 1974). The role of these fungi on the roots is not clear, and they may be merely dormant terrestrial species, or they may be facultative marine fungi that are involved in the decomposition of senescent or dead roots. *Dendryphiella salina* is an example of such a facultative marine species, which is frequently isolated from the root surface, rhizosphere, and seeds of *Salicornia stricta* (Pugh, 1960) but occurs commonly also on dead plant material of the drift line (Pugh, 1962a).

Another woody perennial from developing salt marshes, namely, *Halimione portulacoides,* was studied in England by Dickinson and Pugh (1965a–c) and Dickinson and Morgan-Jones (1966). These authors used plating techniques to observe the mycota on root surfaces and dispersal propagules and found *Dendryphiella salina* and *Ascochytula obiones* as dominant initial colonists, besides a number of terrestrial fungi. Root surfaces of mature plants were mostly invaded by sterile hyphae, whereas sporulating fungi, among them *D. salina,* were rarely isolated. The mycota of green and moribund leaves of *H. portulacoides* included again *D. salina* and *A. obiones* (Dickinson, 1965), which both can be regarded as facultative marine species because they grow in submersed as well as emersed parts of marsh plants. Two other Deuteromycetes, *Camarosporium obiones* and *Coniothyrium obiones,* occur in the cortex of moribund stems of *H. portulacoides* and are obviously involved in the latter stages of decomposition of this host. These two fungi grow even on plants that are seldom if ever inundated; therefore, host species and organs determine their distribution rather than the frequency of inundation by seawater (Dickinson and Morgan-Jones, 1966).

C. Symbionts

Studies of possible mycorrhizal associations in salt-marsh plants in England demonstrated a vesicular endophyte in *Puccinellia maritima* but the absence of mycorrhizas in *Salicornia stricta, Spartina townsendii,* and *Suaeda maritima* (Pugh, 1962b). Mason (1928) found mycorrhizas in species of *Agrostis, Armeria, Aster, Cochlearia, Glaux, Glyceria,* and *Plantago,* but did not see them in *Juncus maritimus, J. gerardi, Salicornia europaea, Spergularia marginata,* and *Triglochin maritima.*

IV. GEOGRAPHICAL DISTRIBUTION

Most of the studies on fungi on salt-marsh plants have been carried out in England and in the United States and, therefore, our knowledge on the geographical distribution of these fungi is limited. A compilation of literature on 101 fungi occurring in *Spartina* spp. (Gessner and Kohlmeyer, 1976) indicates that certain common species such as *Buergenerula spartinae, Claviceps purpurea, Leptosphaeria obiones,* and *Phaeosphaeria typharum* are ubiquitous and occur wherever the host grows. Others have a worldwide distribution on other substrates but have been collected only infrequently on *Spartina,* for example, *Corollospora maritima* and *Cirrenalia macrocephala.* Salinity and latitude appear to have less influence on the occurrence of certain common fungi than does the condition of the host plants, because the same species have been found to occur on *Spartina alterniflora* from Canada to Argentina under a variety of salinity conditions (Gessner and Kohlmeyer, 1976). The same observations were made with fungi on *Salicornia* spp. in Europe, North America and South America, and Bermuda (Kohlmeyer and Kohlmeyer, 1977).

In concluding this chapter on fungi in halophytes of tidal salt marshes we would like to emphasize the importance of thorough studies in this area. The role of "terrestrial" species frequently isolated in the laboratory from marsh plants has to be assessed. So far, only direct microscopic observations of fungi on different organs of the hosts can indicate with certainty which species are growing and reproducing on these substrates. Regular isolations with culture methods do not prove that the fungi obtained are active in the habitat, as discussed more fully in Chapter 2.

12. Fungi on Mangroves and Other Tropical Shoreline Trees

Mangrove vegetation (Walsh, 1974; Chapman, 1976) or "mangal" is the tropical counterpart of tidal salt marshes of temperate regions. In the subtropics, both types of communities may mingle. The mangal is composed of a wide variety of shoreline trees and bushes belonging to numerous, often unrelated families. Chapman (1976, Tables 1 and 3) lists over 30 families with more than 100 mangrove species which can be divided into Old World and New World mangroves. The greatest diversity of species is found in the Old World mangrove vegetation, that is, in the Indian Ocean and the West Pacific. In comparison, the New World mangal, restricted to the African west coast, the shores of the Americas, and the West Indies, is poor and includes only some ten species. Many mangrove species have interesting adaptations to the aquatic habitat, such as prop roots, negatively geotropic aerating roots (pneumatophores), and viviparous seedlings (Chapman, 1976). Mangrove plants usually develop in areas that are protected from wave action, such as estuaries, bays, and lagoons and on the leeward side of islands and spits. The land-building activities of mangroves have been overemphasized, according to Chapman (1976), but there is no doubt that they play an important role in the production of organic detritus and, thus, support a large animal community in the mangal (Heald and Odum, 1969; Fell and Master, 1973; Fell et al., 1975).

The higher marine mycota or manglicolous fungi on submerged parts of mangroves include 42 species (Table XIII) and are the fourth largest ecological group after the wood-, salt-marsh-, and algae-inhabiting species. These mangrove fungi are almost exclusively saprobes and belong to the Ascomycetes (23 spp.), Deuteromycetes (17 spp.), and Basidiomycetes (2 spp.). The most frequently encountered fungi on man-

TABLE XIII

Obligate Marine Fungi on Submerged Parts of Mangroves and Other Tropical Shoreline Trees and Bushes

Avicennia africana
  Ascomycotina: *Didymosphaeria enalia*
                    Leptosphaeria avicenniae[a]
                    Lulworthia spp.
                    Mycosphaerella pneumatophorae[a]
  Deuteromycotina: *Phoma* and *Macrophoma* spp.
                    Rhabdospora avicenniae[a]

A. germinans
  Ascomycotina: *Didymosphaeria enalia*
                    Gnomonia longirostris
                    Halosarpheia fibrosa
                    Hydronectria tethys
                    Kymadiscus haliotrephus
                    Leptosphaeria australiensis
                    Leptosphaeria avicenniae[a]
                    Lulworthia spp.
                    Mycosphaerella pneumatophorae[a]
                    Paraliomyces lentiferus
                    Torpedospora radiata
  Deuteromycotina: *Phoma* spp.
                    Rhabdospora avicenniae[a]
                    Trichocladium achrasporum

A. marina var. *resinifera*
  Ascomycotina: *Gnomonia longirostris*
                    Gnomonia marina
                    Halosphaeria quadricornuta
                    Leptosphaeria australiensis
                    Lulworthia spp.
                    Ophiobolus australiensis
  Deuteromycotina: *Phialophorophoma litoralis*

Casuarina sp.
  Ascomycotina: *Corollospora maritima*
                    Leptosphaeria australiensis
                    Lulworthia grandispora

Conocarpus erecta
  Ascomycotina: *Halosphaeria quadricornuta*
                    Hydronectria tethys
                    Leptosphaeria australiensis
                    Lulworthia spp.
  Basidiomycotina: *Halocyphina villosa*

Hibiscus tiliaceus
  Ascomycotina: *Halosphaeria quadricornuta*
                    Hydronectria tethys
                    Kymadiscus haliotrephus
                    Leptosphaeria australiensis

[a] Host-specific fungus.   *(Continued)*

TABLE XIII *(Continued)*

         Lignincola laevis
         Lulworthia spp.
  Deuteromycotina: *Phoma* spp.
Pachira aquatica
  Ascomycotina: *Lignincola laevis*
         Lulworthia spp.
  Deuteromycotina: *Macrophoma* spp.
Pluchea x *fosbergii*
  Ascomycotina: *Halosphaeria cucullata*
         Lignincola laevis
  Deuteromycotina: *Humicola alopallonella*
         Periconia prolifica
Rhizophora mangle
  Ascomycotina: *Didymosphaeria enalia*
         D. rhizophorae[a]
         Halosarpheia fibrosa
         Halosphaeria quadricornuta
         Helicascus kanaloanus
         Hydronectria tethys
         Keissleriella blepharospora[a]
         Kymadiscus haliotrephus
         Leptosphaeria australiensis
         Leptosphaeria neomaritima
         Lignincola laevis
         Lulworthia spp.
         Manglicola guatemalensis
         Paraliomyces lentiferus
         Torpedospora radiata
  Deuteromycotina: *Cirrenalia macrocephala*
         Cirrenalia pseudomacrocephala
         Cirrenalia pygmea
         Cytospora rhizophorae[a]
         Dendryphiella salina
         Dictyosporium pelagicum
         Humicola alopallonella
         Papulaspora halima
         Periconia prolifica
         Phoma and *Macrophoma* spp.
         Robillarda rhizophorae[a]
         Trichocladium achrasporum
         Varicosporina ramulosa
         Zalerion varium
  Basidiomycotina: *Halocyphina villosa*
         Nia vibrissa
R. racemosa
  Ascomycotina: *Didymosphaeria enalia*
         Haligena viscidula

TABLE XIII (Continued)

	Leptosphaeria australiensis
	Lulworthia spp.
	Trematosphaeria mangrovis[a]
Deuteromycotina:	*Cirrenalia pygmea*
	Cirrenalia tropicalis
	Cytospora rhizophorae[a]
	Phoma and *Macrophoma* spp.
	Trichocladium achrasporum
Tamarix gallica	
Ascomycotina:	*Halosarpheia fibrosa*
	Hydronectria tethys
	Leptosphaeria australiensis
Basidiomycotina:	*Halocyphina villosa*

groves are *Lulworthia* spp., which occurred in about 20% of our collections, *Leptosphaeria australiensis* (15%), and *Phoma* sp. (10%; Kohlmeyer, 1969d). Cribb and Cribb (1955, 1956) in Australia were the first mycologists to collect marine fungi on mangroves, and other reports on higher and lower species from different tropical areas followed soon after (Kohlmeyer, 1966c, 1968a,c, 1969c,d, 1976; Kohlmeyer and Kohlmeyer, 1964–1969, 1965, 1971a, 1977; Maxwell, 1968; Lee and Baker, 1972b, 1973; Fell and Master, 1973, 1975; Fell *et al.*, 1975; Newell, 1976).

Mangrove trees are fascinating study objects for the mycologist because the bases of their trunks and aerating roots are permanently or intermittently submerged, whereas the upper parts of roots and trunks are never reached by the salt water, although they sometimes are by salt spray. Thus, terrestrial fungi and lichens occupy the upper part of the trees and marine species the lower part. At the edge of the intertidal area along trunks and roots there is an overlap between marine and terrestrial fungi (Kohlmeyer, 1969d). Literature on terrestrial mangrove fungi has been compiled by Kohlmeyer (1969d). Our knowledge of marine manglicolous fungi is limited in spite of the wide distribution and importance of mangrove trees in the tropics. Of over 100 mangrove species listed by Chapman (1976), only eight have been examined for the occurrence of marine fungi. These hosts belong to the families Avicenniaceae (*Avicennia africana, A. germinans,* and *A. marina* var. *resinifera*), Combretaceae (*Conocarpus erecta*), Compositae (*Pluchea* x *fosbergii*), Malvaceae (*Hibiscus tiliaceus*), and Rhizophoraceae (*Rhizophora mangle* and *R. racemosa*). In addition, we found marine fungi in the tropics on two shoreline trees that are not included in Chapman's list, because they do not belong to the mangal, namely, on *Pachira aquatica* (Bombacaceae) and *Tamarix gallica* (Tamaricaceae).

I. PARASITIC FUNGI IN MANGROVES

Among the species listed in Table XIII, only the Deuteromycete, *Cytospora rhizophorae,* appears to be parasitic. This opinion is based on observations in the field and is not supported by experimental evidence. We have classified this widely distributed fungus as a halotolerant and facultative marine species, because it was growing in submerged parts of *Rhizophora* spp. next to obligate marine organisms, as well as in emersed branches above the waterline (Kohlmeyer and Kohlmeyer, 1971a). The fungus appears to cause a dying-back of young *Rhizophora* plants and seedlings and a destruction of aerial roots and prop roots. Pycnidium initials develop under the leathery bark, which becomes wrinkled in diseased roots and separates from the soft underlying fibrous tissue. At maturity the multichambered brown pycnidium is surrounded by a white mycelium and produces a central ostiole that pierces the bark or emerges through a lenticel, and numerous tiny conidia are released. Experiments with pure cultures of *C. rhizophorae* and uncontaminated seedlings of *Rhizophora* spp. should be undertaken to substantiate the pathogenicity of the fungus.

II. HOST SPECIFICITY

The majority of manglicolous marine fungi are omnivorous and occur mostly on dead cellulosic substrates all around the tropics. There are just a few host-specific fungi that are limited to one host genus or species. *Didymosphaeria rhizophorae, Keissleriella blepharospora,* and *Robillarda rhizophorae* are restricted to *Rhizophora mangle. Trematosphaeria mangrovis* was found only in *R. racemosa,* and *Cytospora rhizophorae* occurs in both *R. mangle* and *R. racemosa.* Finally, *Leptosphaeria avicenniae, Mycosphaerella pneumatophorae,* and *Rhabdospora avicenniae* live in *Avicennia africana* and *A. germinans.* Additional species, for example, *Helicascus kanaloanus, Manglicola guatemalensis,* and *Ophiobolus australiensis,* might also be host specific, but they are each known from only a single collection. Mangroves in freshwater habitats have a different mycota, and these fungi may also include host-specific species, such as *Psathyrella rhizophorae* on *R. mangle* in Hawaii (Kohlmeyer, 1969d).

III. FUNGI ON SUBMERGED ROOTS, TRUNKS, AND BRANCHES

Most of the literature of manglicolous fungi concerns organisms that live in the bark or wood of mangroves. Bark of many mangrove trees contains high amounts of tannin (Walsh, 1974; Chapman, 1976), which could protect them against biodeterioration. Extracts of mangroves have been patented for antifouling compounds (Sieburth and Conover, 1965), but such compounds apparently do not prevent attack by marine fungi and wood-boring animals on mangrove trees. Barkless wood of most mangrove species does not have any natural resistance to boring mollusks (Southwell *et al.*, 1970). Our observations suggest that undamaged, bark-covered tissues are not invaded by marine fungi. Exceptions may be those fungi that are found exclusively on the surface of the bark. Three fungi, namely, the Ascomycetes *Keissleriella blepharospora* and *Mycosphaerella pneumatophorae* and the Deuteromycete *Rhabdospora avicenniae*, are restricted to the bark of living mangrove parts. These species can be considered saprobes as they grow among the superficial dead cork cells of the plant without penetrating the live tissue. Because of their restriction to bark it can be assumed that *K. blepharospora, M. pneumatophorae,* and *R. avicenniae* have an affinity for certain contents of the bark, for example, suberin.

Wood of submerged mangrove branches, trunks, and roots is attacked by most of the species listed in Table XIII. These fungi usually invade the wood after the protective bark has been damaged or torn off. Such damage can result from attack by wood-boring animals [e.g., *Sphaeroma terebrans* (Rehm and Humm, 1973)] or from storms or human activities. Hyphae grow mainly in the less lignified secondary layers of the cell walls and cause a type of decomposition known as "soft rot" (see Chapter 15 on lignicolous fungi). Ascocarps generally develop immersed in the substrate, whereas conidia of Deuteromycetes and basidiocarps of *Halocyphina villosa* and *Nia vibrissa* originate on the surface, often in cracks of the wood or partly sheltered by bark. Prop roots of *Rhizophora mangle* can regenerate easily and form adventitious roots after the original tips are damaged and attacked by marine fungi (Kohlmeyer, 1969d).

Most literature concerned with higher fungi on roots and other submerged wood of mangroves is taxonomic and contains predominantly descriptions of new species, new host records, or data on geographic distribution (e.g., Cribb and Cribb, 1955, 1956; Kohlmeyer, 1966c, 1968a,c, 1969c, 1976; Kohlmeyer and Kohlmeyer, 1964–1969, 1965, 1971a, 1977). While these studies are based on direct observations of fungal species as they developed in nature, Lee and Baker (1973) used plating techniques to isolate fungi from the surface of roots of *Rhizophora mangle*, from macerated root tissue, and from rhizosphere soil in Hawaii. It is interesting to note that none of the 15 fungi isolated from roots belongs to the obligate marine manglicolous fungi listed in Table XIII. The *Phoma* sp. of Lee and Baker (1973) may or may not be identical with the *Phoma* reported as growing on roots in the natural habitat. As discussed in Chapters 1 and 2, the possibility exists that fungi isolated by culture methods may belong to the indigenous marine mycota; however, this assumption needs confirmation. In the case of the 15 fungi isolated from red mangrove roots, it has not been proven that all, or even some, are facultative marine species, that is, that they are active members of the mangrove mycota. Lee and Baker did not find any mycorrhizal fungi in cross and longitudinal sections of *R. mangle* roots.

IV. FUNGI ON MANGROVE SEEDLINGS

Rhizophora mangle produces viviparous seedlings; that is, the fruit germinates on the tree and forms an elongated, torpedolike seedling. When they are mature the seedlings fall from the tree and some settle under the mother tree, but most of them are washed away by the tides and may float for several weeks, or even months, over long distances (Gunn and Dennis, 1971, 1976; Chapman, 1976). Drifting seedlings may serve as vehicles for transoceanic dispersal of marine fungi. *Keissleriella blepharospora* and *Lulworthia* sp., for instance, are transported by the Gulf Stream over hundreds of miles on seedlings of *R. mangle*, which are often washed up on the outer banks of North Carolina (Gunn and Dennis, 1971, 1976; Kohlmeyer, 1968c; Kohlmeyer and Kohlmeyer, 1971a). The only extensive study of the microbial colonization of mangrove seedlings is that by Newell (1973, 1976), who investigated the succession of fungi on submerged seedlings of *R. mangle*. He used the method of direct observation of fungi fruiting at the time of collection, as well as the observation of species developing on the seedlings after damp-chamber incubation. In addition to direct observations, Newell also applied culture techniques to detect species not sporulating on incubated seedlings, and he found altogether 84 species of fruiting filamentous fungi. These species belong to the following groups: Ascomycetes (13 spp.), Basidiomycetes (1 sp.), Hyphomycetes (42 spp.), Sphaeropsidales (19 spp.), Melanconiales (3 spp.), Oomycetes (3 spp.), Zygomycetes (1 sp.), and ectoplasmic-net fungi (2 spp.). Only 25 species exhibited a frequency of occurrence of 5% or more and were considered "prevalent fungi" (Newell, 1976, Table 2.2). Five of the prevalent species can be considered to be obligate marine fungi, namely, *Keissleriella blepharospora, Lulworthia grandispora, L. medusa* var. *biscaynia, Cytospora rhizophorae,* and *Zalerion varium.* The remaining 20 species, of which 11 have been identified to genus only, may be either facultative marine or terrestrial fungi. As explained in Chapters 1 and 2, it cannot be decided with certainty if species isolated in the laboratory but not directly observed in the natural habitat are active members of the marine mycota. Future experiments, such as reintroducing laboratory isolates into the marine habitat and observing their behavior under natural conditions, have to clarify this point.

Newell (1976) described four overlapping stages of fungal succession on mangrove seedlings. The first stage was characterized by 16 common plant-surface (phylloplane) Hyphomycetes developing superficially on about 80% of the examined seedlings attached to the tree. Most frequently isolated were species of the genera *Alternaria, Aureobasidium, Cladosporium, Pestalotia,* and *Zygosporium.* No fungal structures were found under the epidermis. The next stage of fungal development occurred during the first 2 months during which the seedlings were submerged in the water at the experimental stations. This second mycota included all the species found on the attached seedlings, as well as some new ones—altogether 27 species, of which 16 were prevalent. During the second through fifth month of submergence, the original preabscission fungi declined, and eventually fungal structures occurred for the first time within the subepidermal seedling tissue. Injured seedlings softened, turned brown, and showed a rapid increase in nitrogen content, whereas uninjured seedlings remained firm and green. Bacteria, protozoa, nematodes, and other meiofauna penetrated the damaged seedlings through the wound. *Pestalotia* sp. was the most common member of the second community.

After 5 or 6 months the second successional stage graded into the third one, and, besides terrestrial and facultative marine species, the first obligate marine fungi appeared (*Keissleriella blepharospora, Lulworthia grandispora, L. medusa* var. *biscaynia, Cytospora rhizophorae,* and *Zalerion varium*). The latter five species were initially found on dying or dead parts of viable seedlings, but they soon invaded all areas, as the seedlings senesced and died. More and more cortex cells contained hyphae or fungal pseudoparenchyma and the nitrogen content of injured seedlings had tripled, to about 1.1%. In a postulated fourth and last stage of succession the frequency of prevalent species of the preceding stage increased markedly and some new members, namely, *Papulaspora, Penicillium,* and *Trichoderma* spp., appeared. In comparing the mycota of other mangrove parts, Newell (1976) concluded that seedlings, leaves, and wood of *Rhizophora mangle* "all appear to harbor quantitatively and qualitatively unique mycoflorae."

The major fungal groups showed the following succession in Newell's investigations: Hyphomycetes and Melanconiales were dominant on seedlings attached to the tree. Both of these groups decreased after submergence of seedlings. During the third stage of succession, Ascomycetes and Sphaeropsidales predominated. Finally, Hyphomycetes and Ascomycetes became dominant on the senescent and dead seedlings. Oomycetes and Zygomycetes occurred infrequently in the course of this study. Basidiomycetes did not appear in Newell's investigations and may have been excluded by the method used, namely, total submergence of the seedlings in litter bags. The Basidiomycetes *Halocyphina villosa* and *Nia vibrissa* occur on *Rhizophora mangle,* but their fruiting bodies are only found in the intertidal zone and may need intermittent exposure and submergence for development. In conclusion, we would like to direct attention to the small number of obligate marine fungi, about 15, found by Newell (1976, Table 2.1) among a total of 84 species isolated, or only five obligate marine species among the 25 prevalent species (Newell, 1976, Table 2.2). Although a number of facultative marine species can be expected, Newell's excellent study does not prove with conclusive evidence that *all* 84 species isolated from red mangrove seedlings are active in the marine habitat, namely, growing or reproducing, or both.

V. FUNGI ON MANGROVE LEAVES

Leaves of the red mangrove, *Rhizophora mangle,* contribute a considerable amount of biomass to the tropical and subtropical environment. Heald and Odum (1969) and Odum and Heald (1972) determined the detritus production as being over 3 metric tons (dry weight) per acre per year from mangrove leaf-fall alone in a South Florida estuary. According to these authors, mangrove twigs, bark, and leaf scales are less important contributors than leaves to the food web. In view of the importance of mangrove leaves as a source of detritus it is of great interest to investigate the role of microorganisms in the breakdown of the leaves. The only research in this area is by Fell and Master (1973, 1975) and Fell et al. (1975), who examined the activities of higher and lower fungi in the degradation of *R. mangle* leaves. One paper by Fell and Master (1975) deals with *Phytophthora* and *Pythium* spp. exclusively, whereas the other two include both higher and lower fungi on mangrove leaves. Attached but senescent leaves harbor a number of parasitic and saprobic terrestrial fungi (listed in Kohlmeyer, 1969d), some of which persist after the leaves fall into the water (e.g., *Pestalotia* sp.). During the first week of submergence the lower fungi are prevalent, and a few other primary invaders appear, mostly Hyphomycetes. Within the second and third week the first obligate marine fungi, *Lulworthia* sp. and *Zalerion varium,* were observed, while at the end of this period most of the lower fungi have disappeared. These studies, which resulted in isolations of 53 genera of fungi, were based on incubation of leaf discs on agar plates in the laboratory (Fell and Master, 1973). Among 43 Deuteromycetes and 3 Ascomycetes were the following obligate marine fungi: *Cirrenalia macrocephala, Dictyosporium pelagicum, Varicosporina ramulosa, Zalerion varium,* and *Lulworthia* spp. As Fell and Master (1973) point out, the majority of the fungi isolated from mangrove leaves are ubiquitous saprobes that are often associated with the decomposition of terrestrial plant material. In our opinion, the same reservation has to be made in this case, as in the studies of Lee and Baker (1973) on mangrove-root fungi and of Newell (1976) on mangrove-seedling fungi, namely, that the activity of most of the isolated species in nature has not been demonstrated (see also Chapters 1 and 2) and that some of the isolated fungi derived from dormant propagules of terrestrial species. There is no doubt that obligate and facultative marine fungi play an important role in the breakdown of leaves in the natural habitat, as abundant mycelia can be observed in decomposing mangrove leaves. After 6 weeks of submergence the leaves become fragile, and they break up into small particles after 10 weeks if they are exposed to wave action, while they may remain intact for up to 1 year in sheltered areas (Fell and Master, 1973).

VI. FUNGI IN SOIL OF THE MANGAL

Swart (1958, 1963) did the first comprehensive studies on fungi of soil under an East African mangrove vegetation. Additional investigations on Indian mangal soils were conducted by Pawar and Thirumalachar (1966), Padhye *et al.* (1967), Rai *et al.* (1969), and Rai and Chowdhery (1975, 1976). Ulken (1970, 1972, 1975) isolated Phycomycetes from mangrove sediments in Brazil and Hawaii, and Lee and Baker (1972a,b, 1973) investigated soil microfungi from a Hawaiian mangrove swamp. Other papers dealt with descriptions of single species isolated from mangrove soils, for example, that by Stolk (1955) on an *Emericellopsis* and *Westerdykella ornata* from East Africa and the paper by Swart (1970) on a *Penicillium* from Australia. Further publications from India were by Rai and Tewari (1963) on *Preussia* isolates, by Pawar *et al.* (1963) on a *Monosporium,* by Pawar *et al.* (1965) on a *Cladosporium,* and by Pawar *et al.* (1967) on *Phoma* spp.

The difficulty in evaluating investigations of mangal soil fungi was expressed by Swart (1958), who stated: "Hardly any method makes it possible to differentiate between fungi that are actively growing in the soil and others that are present in a dormant state. Only direct observation is entirely reliable in distinguishing between these two groups." The investigators of mangrove soil mycota used standard isolation techniques, and the usually extensive lists of organisms obtained did not include a single obligate marine fungus. These lists consist of genera and species that are common in terrestrial habitats and contain, for instance, large numbers of *Aspergillus* and *Penicillium* spp. Lee and Baker (1973) pointed out that the mycota they found by direct observation on mangrove roots differed almost completely from that obtained by culture methods. It is unknown which of the species isolated from mangal mud are facultative marine species, that is, whether they play a role in this habitat or not, and we refer the reader to Chapters 1 and 2 for a general discussion.

VII. VERTICAL AND HORIZONTAL ZONATION OF MANGLICOLOUS FUNGI

Mangrove prop roots or dead branches hanging into the water display diverse microhabitats of marine and terrestrial organisms. The intertidal and subtidal parts are often covered by fouling organisms, such as red and green algae, barnacles, and worms with calcareous tubes. Damaged roots are invaded by shipworms (e.g., *Lyrodus pedicellatus*) and isopods (*Limnoria* and *Sphaeroma* spp.). The exposed parts of prop roots bear lichens and often house insects, such as ants, or support nests of termites. Marine fungi occur in the submerged parts of the roots, but they are absent in substrates embedded in the mud. No distinct pattern of vertical distribution of Ascomycetes was observed (Kohlmeyer, 1969d). The Basidiomycetes *Halocyphina villosa* and *Nia vibrissa* appear to occur in the intertidal parts of roots and branches, and basidiocarps of the former are usually found near barnacles or even inside empty barnacle tests (Kohlmeyer and Kohlmeyer, 1977). Terrestrial fungi develop on roots and branches above the high-tide line (Lee and Baker, 1973), and an overlapping between marine and terrestrial species may occur at the water–air interface (Kohlmeyer, 1969d). Pycnidia of the facultative marine Deuteromycete, *Cytospora rhizophorae,* are found on submerged as well as emerged parts of *Rhizophora* plants. Small black ascocarps of *Keissleriella blepharospora* occur on the bark of prop roots, stems, and seedlings of *Rhizophora* spp. These fruiting bodies are restricted to the submerged parts of the trees above the soil surface, and on seedlings they do not develop in the green part of the hypocotyl, which is covered by the cuticle (Kohlmeyer, 1969c,d).

Mangroves such as *Rhizophora* and *Avicennia* grow in seawater as well as in freshwater sites. Little is known about the horizontal distribution of manglicolous fungi from saltwater habitats toward brackish and freshwater locations. The only observations made are by Kohlmeyer (1969c,d) in Hawaii, where obligate marine fungi occur on *Rhizophora mangle* at salinities ñear 35%/$_{oo}$, and they are replaced by a different mycota in the freshwater part of the swamp. *Keissleriella blepharospora,* a host-specific Ascomycete on *Rhizophora,* is found only in marine localities, not in the freshwater sites. On the other hand, a small mushroom, *Psathyrella rhizophorae,* grew on dead *Rhizophora* seedlings in the upper part of the swamp, but not in the saline habitats.

VIII. GEOGRAPHICAL DISTRIBUTION OF MANGLICOLOUS FUNGI

Our knowledge of the distribution of obligate marine fungi on mangroves is incomplete because collections have been made only in a few areas, and the distribution records reflect efforts of a few investigators who collected in Africa (Liberia), in America [United States (Florida), Mexico, Guatemala, Colombia, Venezuela, and Brazil], on Atlantic islands (Bermuda, Bahamas, and Trinidad), in Australia, and in Hawaii. Large parts of the tropics are unexplored; for instance, no manglicolous fungi are known from the Indian Ocean, and only a few fungal species have been collected in the eastern mangrove in general.

Omnivorous species, such as *Didymosphaeria enalia, Halosphaeria quadricornuta, Kymadiscus haliotrephus, Leptosphaeria australiensis, Lulworthia* spp., and others, occur throughout the tropics on mangroves and other substrates. Only some host-specific species appear to be restricted in their distribution; for instance, *Trematosphaeria mangrovis* is found only in West Africa, whereas *Didymosphaeria rhizophorae* occurs only in the American tropics and subtropics, although their hosts have a wider geographical distribution. *Keissleriella blepharospora* is also restricted to America and West Atlantic islands, except for its presence in Hawaii. This disjunct occurrence can be explained by the fact that the host, *Rhizophora mangle,* was introduced in Hawaii in 1902 by seedlings from Florida (Walsh, 1967), and *K. blepharospora* most probably accompanied these seedlings to the new habitat (Kohlmeyer, 1969c).

In concluding this chapter on mangrove-inhabiting fungi, we would like to mention a few topics that need investigation. The diverse Old World mangrove vegetation with almost 100 species has barely been considered by mycologists. The Indian Ocean in particular should be explored for manglicolous fungi. Those fungi isolated from mangrove soils need to be explored for their role in this habitat and the question must be answered: Are they active decomposers in these soils or are they merely dormant stages? Furthermore, the pathogenicity of certain species should be investigated, for instance, that of *Cytospora rhizophorae,* which is often found in dead parts of *Rhizophora.* Also, the physiology of most manglicolous fungi is unknown, and questions, such as the resistance of these organisms to mangrove tannin, need explanation.

13. Leaf-Inhabiting Fungi

Cast leaves of higher plants reach the marine environment from a variety of sources. Indigenous submerged plants, such as sea grasses of the genera *Cymodocea, Posidonia, Thalassia,* and *Zostera,* shed old leaves, which may eventually wash up on the beach (Den Hartog, 1970). At times they accumulate in great numbers, forming large banks on the seashore or floating in rafts through the ocean. Old leaves may also drift to the bottom of the ocean, even far away from their site of origin (Menzies *et al.*, 1967).

Mangrove trees of the tropics and subtropics shed considerable amounts of leaves each year and contribute to the detritus production of estuaries (Heald and Odum, 1969). Heald and Odum state that the production of mangrove-leaf material exceeds 2.9 dry tons per acre per year in the Florida estuary they investigated. Also, Anastasiou and Churchland (1969) assume "that in most parts of the world the amount of leaf material that enters the ocean far exceeds the amount of wood."

Although leaves represent an important source of food in the marine environment, little is known about their occurrence and breakdown by marine microorganisms. Leaf-inhabiting or foliicolous marine fungi are comparable to freshwater Hyphomycetes or to litter fungi in terrestrial habitats. Most of them are saprobes living on dead loose leaves. Only a few Ascomycetes, namely, *Lindra thalassiae, L. marinera,* and *Lulworthia* sp., are known to develop in attached but moribund leaves of sea grasses. The majority of the marine foliicolous fungi belong to the Deuteromycotina and lower fungi. Possibly, the frail leaves break up before most Ascomycetes are able to reproduce and are, therefore, mainly attacked by the faster growing and sporulating imperfect and lower fungi.

Anastasiou and Churchland (1969) submerged autoclaved leaves of *Prunus laurocerasus* and *Arbutus menziesii* in marine waters in British

Columbia. They found 14 species of fungi, among them two Ascomycetes and nine Deuteromycetes, of which *Zalerion maritimum* was most abundant and occurred on 59% of the leaves. Later experiments in the same area with leaves of *Alnus rubra* and *Arbutus menziesii* yielded 12 fungal species, including one Ascomycete and seven Deuteromycetes (Churchland and McClaren, 1973). Again, *Z. maritimum* was a pioneer and an abundant species. Another record of foliicolous marine fungi from temperate waters was published by Schmidt (1974), who found *Corollospora trifurcata, Halosphaeria mediosetigera, Alternaria tenuis,* and *Zalerion maritimum* on leaves ("Laub") in the Baltic Sea.

Fungi associated with submerged leaves of the red mangrove, *Rhizophora mangle,* have been studied by Fell and Master (1973, 1975) and Fell *et al.* (1975). Initial invaders are lower fungi, and the only significant secondary invaders among the higher obligate marine fungi are *Lulworthia* spp. and *Z. varium.* The majority of the fungi isolated were ubiquitous saprobes known also from nonmarine habitats. Fell *et al.* (1975) conclude that there is a fungal succession on the submerged mangrove leaves. These authors suggest that simple carbon compounds are used first, whereas the more complex compounds of the leaves, such as cellulose and lignin, are attacked after 2 or 3 weeks. Studies of mangrove-leaf breakdown are discussed in more detail in Chapter 12.

Only a small number of marine fungi are known from sea grasses. The rhizome inhabitants are treated in Chapter 14 where fungi occurring on *Posidonia oceanica* and *Ruppia maritima* are included. Submerged and washed-up leaves of *Zostera marina* are attacked by the Ascomycetes *Lulworthia* sp. [Mounce and Diehl, 1934 (as *Ophiobolus*)] and *Corollospora maritima* and the Deuteromycetes *Alternaria* sp., *Phoma* sp., and *Varicosporina ramulosa* (Kohlmeyer, 1963d, 1966a). When leaves of the tropical turtle grass, *Thalassia testudinum,* are cast on the beach they are decomposed by *Corollospora lacera, C. maritima, Lindra marinera, L. thalassiae, Lulworthia* sp., *Dendryphiella arenaria,* and *Varicosporina ramulosa* (Orpurt *et al.*, 1964; Meyers and Kohlmeyer, 1965; Meyers *et al.*, 1965; Meyers, 1969; Kohlmeyer and Kohlmeyer, 1977). Meyers *et al.* (1965) and Meyers (1968b) isolated some of these species from incubated leaf segments of live *T. testudinum* in Florida. They frequently isolated other Deuteromycetes also, but the isolation techniques used do not permit a judgment if representatives of ubiquitous genera, such as *Aspergillus, Penicillium,* and others, are actual invaders of submerged *Thalassia* leaves. The dominant species in leaves of *T. testudinum* is *Lindra thalassiae,* which occurs in the host throughout the year (Orpurt *et al.*, 1964; Meyers, 1968b). This Ascomycete and *L. marinera* have been used in numerous investigations (e.g., Meyers and Simms, 1965, 1967; Meyers

Fig. 23. *Lindra thalassiae* in leaf of the sea grass *Thalassia testudinum*; upper part of leaf is discolored by the action of the fungus; ascocarps (arrow) in the zone of growth and reproduction (Herb. J.K. 1708; bar = 3 mm).

et al., 1965; Meyers, 1969) because they grow and fruit readily in pure culture. In nature these fungi usually penetrate *Thalassia* leaves from the tip (Fig. 23). Ascocarps form in the mesophyll and push the upper and lower epidermis apart (Orpurt *et al.*, 1964; Meyers, 1969). The neck perforates the epidermis and releases ascospores into the surrounding water. *Lindra thalassiae* and *L. marinera* are probably perthophytes attacking senescent leaves of *T. testudinum,* but their predominant role is that of saprobes on shed leaves. *Lindra thalassiae* is also found in air vesicles of *Sargassum* sp. (see Fig. 16), causing "raisin disease" (Kohlmeyer, 1971a). This Ascomycete does not occur on submerged wood, but it is able to degrade cellulose noticeably in culture (Meyers and Simms, 1967). Filamentous ascospores of *L. marinera* may be found in large numbers in foam along beaches in the neighborhood of *Thalassia* beds (Kohlmeyer and Kohlmeyer, 1971a).

Sea grasses and algae washed up above the high-tide line by spring tides are covered by sand and are eventually decomposed by bacteria, yeasts, and terrestrial molds of genera such as *Cephalosporium, Hormodendrum,* and *Penicillium* (Suckow and Schwartz, 1960). Obligate marine fungi are probably absent in such habitats.

None of the foliicolous fungi discussed above is host specific. All are decomposers of senescent or dead leaves. Information on their role in the breakdown of leaf material in the marine habitat is missing. We know that higher fungi occur on submerged leaves but we have no quantitative data of their activities and how they compare with bacteria in decomposing leaves. The vertical distribution of foliicolous fungi is also unknown, but it is likely that such organisms occur also in the deep sea because *Thalassia* leaves have been found at 3160 m (Menzies *et al.*, 1967), and other saprobes, namely, lignicolous fungi, are common even at greater depths (see Chapter 6 on deep-sea fungi). Many of the sea grasses (Den Hartog, 1970) have not been examined for fungi and it can be expected that the list of foliicolous fungi will be considerably expanded in the future.

14. Rhizome-Inhabiting Fungi

Fungi are not often found on sea grasses, and they are mostly reported from leaves rather than from rhizomes, the rootlike stems of these plants. Rhizomes of sea grasses (Den Hartog, 1970) contain tannins, which may protect them from microbial attacks. A few fungi, however, seem to be resistant to such inhibitors. Some of the leaf-inhabiting saprobes occur on rhizomes of the same hosts as well, for instance, *Lulworthia* sp. on rhizomes of *Zostera marina* [Mounce and Diehl, 1934 (as *Ophiobolus*); Kohlmeyer, 1963d] and *Thalassia testudinum* (Kohlmeyer and Kohlmeyer, 1977). Washed-up rhizomes of *Z. marina* may be attacked by *Cladosporium algarum* (J. Kohlmeyer, unpublished). Only two higher fungi are known from rhizomes of *Ruppia maritima,* namely, the saprobe *Lulworthia* sp. (Kohlmeyer, 1962a) and the parasite *Melanotaenium ruppiae* (G. Feldmann, 1959). The smut, *M. ruppiae,* develops in air spaces of rhizomes and leaf bases, without causing noticeable deformations. It is known only from the type location, a salt lagoon in southern France.

Posidonia oceanica, a Mediterranean sea grass, harbors two host-specific Ascomycetes, *Halotthia posidoniae* and *Pontoporeia biturbinata* (Kohlmeyer, 1963a). Both species grow on the defoliated ends of rhizomes, apparently without harming the host. Conical or semiglobose ascocarps and spermogonia of *H. posidoniae* develop in the substrate and break through the surface at maturity. Ascocarps of *P. biturbinata* are globose and superficial and reach diameters of more than 1.3 mm. They are seated in the substrate with an inconspicuous hypostroma. *Halotthia posidoniae* develops on live attached plants and can be found frequently on washed-up *Posidonia* rhizomes on the shore. *Pontoporeia biturbinata,* on the other hand, is rare and has been collected only twice, in Algeria and in Spain (Kohlmeyer, 1963a). Thus far, it has not been possible to culture *H. posidoniae* or *P. biturbinata,* possibly because the fungi require live, albeit senescent parts of, rhizomes for development.

The only fungus reported from manatee grass, *Cymodocea manatorum,* is *Lulworthia* sp., found in a floating rhizome in the Sargasso Sea (Kohlmeyer and Kohlmeyer, 1971a).

Scanty information on rhizome-inhabiting fungi demonstrates that this area is particularly worthy of mycological investigations. Other sea grasses (see Den Hartog, 1970) should be examined for fungi and their role in the breakdown of rhizomes investigated. Of special interest is the possible resistance of these fungi to tannins or other growth-inhibiting substances present in rhizomes of sea grasses.

15. Fungi on Wood and Other Cellulosic Substrates

The young history of marine mycology began in 1944 with a paper by Barghoorn and Linder on wood-inhabiting (or lignicolous) marine fungi. This publication inspired mycologists all over the world to search for marine fungi, and it is one of the reasons for the predilection of research on wood-inhabiting species over the past 35 years. Literature on lignicolous marine fungi surpasses all other topics, such as algae-, mangrove-, or marsh plant-inhabiting fungi (Johnson and Sparrow, 1961; Kohlmeyer and Kohlmeyer, 1971b; G. C. Hughes, 1975). The easy availability of wood as a bait or as a naturally occurring substrate of marine fungi is another reason for the preference that this material has enjoyed. Every untreated piece of wood that is submerged for a certain period in marine or estuarine waters is attacked by higher marine fungi (Figs. 24a and 24b). This attack is faster and more severe in tropical waters than in temperate, near-pole, or deep-sea habitats. Only a low level of dissolved oxygen in the water may prevent development of lignicolous fungi (Kohlmeyer, 1969a).

Table XIV includes 107 species of higher marine fungi that have been collected on submerged wood and other cellulosic substances in marine habitats. Wood-inhabiting fungi from mangrove wood are listed separately in Table XIII (p. 93). The Ascomycetes are the largest group, with 76 spp., followed by Deuteromycetes (29 spp.) and Basidiomycetes (2 spp.). None of the fungi is host specific. In fact, some species that have been considered lignicolous in the past also occur on marsh plants (Gessner and Kohlmeyer, 1976) or submerged seeds of *Spartina* (R. V. Gessner, unpublished). Therefore, the terms "wood-inhabiting" and "lignicolous" may not be applicable in many cases and a more general expression, such as "cellulolytic," might be more appropriate. The large number of species found on wood of *Fagus sylvatica* and *Pinus sylvestris*

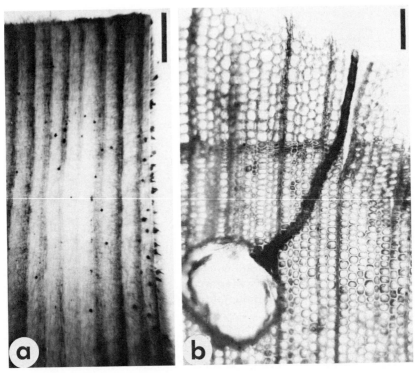

Fig. 24. Attack of pine wood by marine Ascomycetes. (a) Ascocarps and mycelium of *Lulworthia* sp. on the wood surface; the late wood is preferentially darkened and covered by hyphae (bar = 5 mm). (b) Cross section through wood and a submerged ascocarp of *Lulworthia* sp. with long neck reaching the surface (bar = 100 μm).

(Table XIV) can be explained by the fact that panels of these trees are used preferentially in tests by British scientists (see, e.g., E. B. G. Jones, 1968b).

I. SOURCES OF WOOD AND OTHER CELLULOSIC SUBSTRATES IN THE MARINE ENVIRONMENT

A. Untreated Wood of Natural Origin

Forests bordering on the ocean are rich sources of dead wood, which falls into the sea, or rivers may carry trees and branches into the ocean (G. C. Hughes, 1975). This wood can become buried in sandy beaches or wedged between rocks of the intertidal zone. Some of the drifting trees

TABLE XIV

Wood-Inhabiting Marine Fungi[a]

Abies alba
  Deuteromycotina: *Papulaspora halima*
Abies firma
  Ascomycotina: *Ceriosporopsis halima*
                Corollospora maritima
                Heleococcum japonense
  Deuteromycotina: *Cirrenalia macrocephala*
                   Papulaspora halima
Abies sp.
  Ascomycotina: *Carbosphaerella leptosphaerioides*
                Ceriosporopsis cambrensis
                C. halima
                Corollospora comata
                C. lacera
                Halonectria milfordensis
                Halosphaeria maritima
                H. pilleata
                H. stellata
Acer sp.
  Ascomycotina: *Amylocarpus encephaloides*
                Lulworthia grandispora
                Lulworthia sp.
Aesculus hippocastanum
  Ascomycotina: *Amylocarpus encephaloides*
Alnus glutinosa
  Ascomycotina: *Ceriosporopsis cambrensis*
Alnus sp.
  Ascomycotina: *Amylocarpus encephaloides*
                Ceriosporopsis halima
                Halosphaeria stellata

Araucaria sp.
  Ascomycotina: *Ceriosporopsis halima*
                Lulworthia sp.
  Deuteromycotina: *Zalerion maritimum*
Arbutus menziesii
  Ascomycotina: *Amylocarpus encephaloides*
                Corollospora trifurcata
                Halosphaeria appendiculata
  Deuteromycotina: *Papulaspora halima*
Betula papyrifera
  Deuteromycotina: *Papulaspora halima*
Betula pubescens
  Ascomycotina: *Amylocarpus encephaloides*
                Ceriosporopsis tubulifera
                Haligena spartinae
                Halosphaeria appendiculata
                H. hamata
                Leptosphaeria contecta
                L. obiones
                Lulworthia sp.
                Trematosphaeria britzelmayriana
  Basidiomycotina: *Digitatispora marina*
  Deuteromycotina: *Dendryphiella salina*
                   Humicola alopallonella
                   Monodictys pelagica
                   Zalerion maritimum
Betula sp.
  Ascomycotina: *Amylocarpus encephaloides*
                Ceriosporopsis halima
                Halosphaeria cucullata

[a] Mangrove-inhabiting fungi are included in Table XIII.

(Continued)

TABLE XIV (Continued)

Betula sp. continued
  Deuteromycotina: *Alternaria* spp.
                   Humicola alopallonella
Bombax malabaricum
  Ascomycotina: *Halosphaeria quadricornuta*
Cocos nucifera (endosperm)
  Ascomycotina: *Halosphaeria quadricornuta*
                H. salina
                Torpedospora radiata
  Deuteromycotina: *Humicola alopallonella*
                   Periconia prolifica
Cryptomeria japonica
  Ascomycotina: *Halosphaeria appendiculata*
                H. quadriremis
  Deuteromycotina: *Cirrenalia macrocephala*
                   Zalerion maritimum
                   Z. varium
Dicorynia paraensis
  Ascomycotina: *Ceriosporopsis halima*
                Corollospora cristata
                C. maritima
                C. trifurcata
                Halosphaeria appendiculata
                H. hamata
                H. mediosetigera
                Lulworthia sp.
                Nais inornata
  Deuteromycotina: *Humicola alopallonella*
                   Monodictys pelagica
                   Zalerion maritimum
Dipterocarpus sp.
  Ascomycotina: *Halosphaeria quadricornuta*

Fagus crenata
  Ascomycotina: *Ceriosporopsis halima*
                Corollospora maritima
  Basidiomycotina: *Nia vibrissa*
Fagus sylvatica
  Ascomycotina: *Ceriosporopsis calyptrata*
                C. cambrensis
                C. circumvestita
                C. halima
                C. tubulifera
                Corollospora comata
                C. maritima
                C. trifurcata
                Gnomonia longirostris
                Haligena elaterophora
                H. unicaudata
                Halonectria milfordensis
                Halosphaeria appendiculata
                H. hamata
                H. maritima
                H. mediosetigera
                H. quadricornuta
                H. torquata
                Leptosphaeria obiones
                L. oraemaris
                L. pelagica
                Lignincola laevis
                Lulworthia fucicola
                Lulworthia spp.
                Nais inornata
                Nautosphaeria cristaminuta
                Savoryella lignicola

Fagus sylvatica continued

    Sphaerulina oraemaris
Basidiomycotina: *Digitatispora marina*
    Nia vibrissa
Deuteromycotina: *Alternaria* spp.
    Asteromyces cruciatus
    Cirrenalia macrocephala
    Dendryphiella salina
    Dictyosporium pelagicum
    Humicola alopallonella
    Monodictys pelagica
    Papulaspora halima
    Phoma spp.
    Sporidesmium salinum
    Stemphylium sp.
    Trichocladium achrasporum
    Zalerion maritimum

Fagus sp.
Ascomycotina: *Amylocarpus encephaloides*
    Ceriosporopsis halima
    Corollospora maritima
    Lulworthia sp.
    Sphaerulina oraemaris

Fraxinus sp.
Ascomycotina: *Amylocarpus encephaloides*
    Halosphaeria stellata

Laphira procera
Ascomycotina: *Ceriosporopsis halima*
    Corollospora cristata
    C. maritima
    C. trifurcata
    Halosphaeria appendiculata

Laphira procera continued

    H. hamata
    H. mediosetigera
    Lulworthia sp.
    Nais inornata
Deuteromycotina: *Monodictys pelagica*
    Zalerion maritimum

Larix decidua
Deuteromycotina: *Papulaspora halima*
Larix sp.
Ascomycotina: *Ceriosporopsis calyptrata*
    Corollospora comata

Liquidambar styraciflua
Ascomycotina: *Halosphaeria mediosetigera*
    Lignincola laevis

Liriodendron tulipifera
Ascomycotina: *Leptosphaeria halima*
    Lulworthia sp.

Deuteromycotina: *Dictyosporium pelagicum*
    Zalerion maritimum

Liriodendron sp.
Ascomycotina: *Lulworthia* sp
Mangifera indica
Ascomycotina: *Corollospora pulchella*
    Didymosphaeria enalia
    Halosphaeria cucullata
    Leptosphaeria australiensis

Deuteromycotina: *Cirrenalia pygmea*
Ochroma lagopus
Deuteromycotina: *Papulaspora halima*
Ocotea rodiaei
Ascomycotina: *Ceriosporopsis halima*
    Corollospora cristata

(Continued)

TABLE XIV (Continued)

Ocotea rodiaei continued

        Halosphaeria appendiculata
        H. mediosetigera
        H. stellata
        Nais inornata
        Savoryella lignicola
Deuteromycotina: *Cirrenalia macrocephala*
        Dictyosporium pelagicum
        Humicola alopallonella
        Monodictys pelagica
        Zalerion maritimum
Olea europaea
Deuteromycotina: *Papulaspora halima*
Phoenix sp.
Ascomycotina: *Halosphaeria appendiculata*
        Lulworthia sp.
Phyllostachys pubescens
Ascomycotina: *Halosphaeria appendiculata*
        Leptosphaeria oraemaris
Picea sp.
Ascomycotina: *Carbosphaerella leptosphaerioides*
        Ceriosporopsis calyptrata
        Corollospora comata
        C. lacera
        C. maritima
        C. trifurcata
        C. tubulata
        Haligena amicta
        H. elaterophora
        Halosphaeria maritima
        H. pilleata

Picea sp. continued

        H. stellata
        Lindra inflata
        Lulworthia spp.
Basidiomycotina: *Nia vibrissa*
Deuteromycotina: *Cirrenalia macrocephala*
Pinus densiflora
Ascomycotina: *Halosphaeria appendiculata*
Pinus echinata
Ascomycotina: *Corollospora pulchella*
Pinus monticola
Ascomycotina: *Leptosphaeria oraemaris*
        Lulworthia sp.
Deuteromycotina: *Trichocladium achrasporum*
        Monodictys pelagica
Pinus palustris
Deuteromycotina: *Cremasteria cymatilis*
        Zalerion maritimum
Pinus pinaster
Deuteromycotina: *Papulaspora halima*
Pinus ponderosa
Deuteromycotina: *Clavatospora stellatacula*
Pinus sylvestris
Ascomycotina: *Ceriosporopsis cambrensis*
        C. circumvestita
        C. halima
        C. tubulifera
        Corollospora comata
        C. maritima
        Haligena elaterophora
        H. viscidula
        Halonectria milfordensis

Pinus sylvestris continued
    Halosphaeria appendiculata
    H. hamata
    H. maritima
    H. mediosetigera
    H. pilleata
    H. quadricornuta
    H. quadriremis
    H. stellata
    Leptosphaeria oraemaris
    Lindra inflata
    Lulworthia fucicola
    Lulworthia spp.
    Microthelia linderi
    Nais inornata
    Savoryella lignicola
    Torpedospora radiata
    Trematosphaeria britzelmayriana
Basidiomycotina: *Digitatispora marina*
Deuteromycotina: *Alternaria spp.*
    Cirrenalia macrocephala
    Cremasteria cymatilis
    Dendryphiella salina
    Dictyosporium pelagicum
    Humicola alopallonella
    Monodictys pelagica
    Orbimyces spectabilis
    Phoma spp.
    Sporidesmium salinum
    Stemphylium sp.
    Trichocladium achrasporum
    Zalerion maritimum

Pinus taeda
Ascomycotina: *Halosphaeria cucullata*

Pinus sp.
Ascomycotina: *Carbosphaerella leptosphaerioides*
    Ceriosporopsis halima
    Corollospora comata
    C. lacera
    C. maritima
    C. trifurcata
    C. tubulata
    Haligena elaterophora
    H. viscidula
    Halosphaeria maritima
    H. mediosetigera
    H. pilleata
    H. quadriremis
    H. salina
    H. stellata
    Lulworthia grandispora
    Lulworthia spp.
    Massariella maritima
    Nais inornata
    Torpedospora radiata
Basidiomycotina: *Nia vibrissa*
Deuteromycotina: *Dinemasporium marinum*
    Stemphylium sp.
    Zalerion maritimum

Platanus occidentalis
Ascomycotina: *Halosphaeria appendiculata*
    H. mediosetigera

Populus alba
Deuteromycotina: *Papulaspora halima*
Populus sp.
Ascomycotina: *Amylocarpus encephaloides*
    Corollospora lacera
    C. maritima

(Continued)

TABLE XIV (Continued)

Populus sp. continued

        Haligena amicta
        Halosphaeria appendiculata
        H. mediosetigera
        Leptosphaeria oraemaris
   Deuteromycotina: *Cirrenalia macrocephala*

Prosopis sp.
   Deuteromycotina: *Zalerion varium*

Prunus persica
   Ascomycotina: *Ceriosporopsis halima*

Pseudotsuga douglasii
   Ascomycotina: *Lulworthia sp.*
   Deuteromycotina: *Cirrenalia macrocephala*
                  Zalerion maritimum

Pseudotsuga menziesii
   Ascomycotina: *Corollospora maritima*
                  Lulworthia sp.
   Deuteromycotina: *Monodictys pelagica*
                  Zalerion maritimum

Pseudotsuga taxifolia
   Ascomycotina: *Ceriosporopsis halima*
   Deuteromycotina: *Cirrenalia macrocephala*

Pseudotsuga sp.
   Ascomycotina: *Ceriosporopsis halima*
                  Halosphaeria quadricornuta
                  H. stellata

Quercus virginiana
   Ascomycotina: *Didymosphaeria enalia*
                  Kymadiscus haliotrephus

Quercus sp.
   Ascomycotina: *Amylocarpus encephaloides*
                  Carbosphaerella leptosphaerioides

Quercus sp. continued

        Ceriosporopsis calyptrata
        C. halima
        C. tubulifera
        Corollospora maritima
        C. trifurcata
        Crinigera maritima
        Halonectria milfordensis
        Halosphaeria appendiculata
        H. stellata
        Leptosphaeria obiones
        L. oraemaris
        Lignincola laevis
        Lulworthia fucicola
        Lulworthia sp.
        Pleospora gaudefroyi
   Basidiomycotina: *Digitatispora marina*
   Deuteromycotina: *Dictyosporium pelagicum*
                  Monodictys pelagica
                  Papulaspora halima
                  Phialophorophoma litoralis
                  Zalerion maritimum

Salix sp.
   Ascomycotina: *Amylocarpus encephaloides*
                  Ceriosporopsis halima
                  C. tubulifera
                  Halosphaeria stellata
   Basidiomycotina: *Nia vibrissa*

Tamarix aphylla
   Ascomycotina: *Corollospora maritima*
                  C. pulchella
                  Lulworthia spp.

Tamarix aphylla continued
  Deuteromycotina: *Papulaspora halima*
                   Zalerion varium

Tamarix gallica
  Ascomycotina: *Halosarpheia fibrosa*
                Hydronectria tethys
                Leptosphaeria australiensis
  Basidiomycotina: *Halocyphina villosa*

Tamarix sp.
  Ascomycotina: *Corollospora trifurcata*
                Halosphaeria mediosetigera

Tilia americana
  Deuteromycotina: *Humicola alopallonella*
                   Trichocladium achrasporum

Tilia sp.
  Ascomycotina: *Halosphaeria appendiculata*
                Lulworthia spp.
                Massariella maritima
                Torpedospora radiata

Tsuga heterophylla
  Ascomycotina: *Halosphaeria torquata*

Ulmus sp.
  Ascomycotina: *Ceriosporopsis halima*
Unidentified wood (including "balsa")
  Ascomycotina: *Amylocarpus encephaloides*
                Aniptodera chesapeakensis
                Banhegyia setispora
                Bathyascus vermisporus
                Biconiosporella corniculata
                Carbosphaerella leptosphaerioides
                C. pleosporoides
                Ceriosporopsis calyptrata
                C. cambrensis

C. circumvestita
C. halima
C. tubulifera
Chaetosphaeria chaetosa
Corollospora comata
C. cristata
C. lacera
C. maritima
C. pulchella
C. trifurcata
C. tubulata
Didymosphaeria enalia
D. rhizophorae
Eiona tunicata
Gnomonia longirostris
G. salina
Haligena amicta
H. elaterophora
H. spartinae
H. viscidula
Halonectria milfordensis
Halosphaeria appendiculata
H. cucullata
H. galerita
H. hamata
H. maritima
H. mediosetigera
H. pilleata
H. quadricornuta
H. quadriremis
H. salina
H. stellata

(Continued)

TABLE XIV (Continued)

Unidentified wood *continued*

H. torquata
H. trullifera
Heleococcum japonense
Hydronectria tethys
Kymadiscus haliotrephus
Leptosphaeria albopunctata
L. australiensis
L. contecta
L. halima
L. marina
L. obiones
L. oraemaris
L. paucispora
L. pelagica
Lignincola laevis
Lindra inflata
Lulworthia fucicola
Lulworthia spp.
Microthelia linderi
Nais inornata
Nautosphaeria cristaminuta
Oceanitis scuticella
Paraliomyces lentiferus
Pleospora gaudefroyi
Savoryella lignicola
Sphaerulina albispiculata
S. oraemaris
Torpedospora ambispinosa
T. radiata
Zopfiella latipes

Unidentified wood *continued*

Basidiomycotina: Digitatispora marina
Nia vibrissa
Deuteromycotina: Allescheriella bathygena
Alternaria spp.
Botryophialophora marina
Cirrenalia macrocephala
C. pseudomacrocephala
C. pygmea
C. tropicalis
Clavariopsis bulbosa
Cremasteria cymatilis
Dendryphiella salina
Dictyosporium pelagicum
Diplodia oraemaris
Humicola alopallonella
Macrophoma sp.
Monodictys pelagica
Orbimyces spectabilis
Papulaspora halima
Periconia abyssa
P. prolifica
Phialophorophoma litoralis
Phoma sp.
Stemphylium sp.
Trichocladium achraspermum
Zalerion maritimum
Z. varium

will eventually become waterlogged and sink into the deep sea, where "islands" of wood can be found frequently (Turner, 1973). Initially, most mycologists followed Barghoorn and Linder's (1944) technique of collecting fungi on test panels (e.g., Meyers, 1953, 1954; T. W. Johnson, 1956a,b; Gold, 1959). This trapping method, however, is limiting and a wider variety of species can be obtained when naturally occurring intertidal wood is collected, as demonstrated, for example, by Höhnk (1954a, 1955) and Kohlmeyer (1959, 1960). G. C. Hughes (1968) proposed to distinguish between "intertidal wood" and "driftwood." The former is permanently fixed material in the intertidal zone, such as wooden structures along the shore and pieces of wood partially buried in the sand or wedged between rocks. "Driftwood," on the other hand, is wood that is floating or found loose on the shore. We have adopted these definitions herein.

Other naturally occurring cellulosic substrates of the intertidal zone, namely, stones of fruits and endocarps and mesocarps of coconuts, are also included in this chapter.

B. Man-Made Wooden Structures

Harbor and shore fortifications, such as pilings, seawalls, dikes, breakers, and jetties, are excellent substrates for marine lignicolous fungi (e.g., Höhnk, 1954a). The upper, always-emersed parts may harbor terrestrial fungi, and an overlapping between terrestrial and marine species may occur at the sea–air interface, similar to the zonation of fungi on mangrove roots (see Chapter 12). The mycota on such man-made, chemically treated structures is comparable to that of untreated wood, although certain bark-inhabiting fungi (see Chapter 16) are missing on lumber.

The destruction of timber in the ocean is of considerable economic concern (Trussel and Jones, 1970), as the replacement cost of wooden structures is about $50 million per year in North America alone, and, as an example from Great Britain, £15,000 in Blyth Harbour (E. B. G. Jones and Irvine, 1971). The main deterioration of wood in the sea is caused by wood borers, but fungi and bacteria may be important in the life cycles of these animals (see Chapter 19).

C. Wooden Panels Used as "Traps"

Since Barghoorn and Linder's (1944) pioneering publication, the immersion of wood panels in marine habitats has become very popular. Meyers (1953) and Meyers and Reynolds (1958) first described a method of exposing series of wooden panels (attached sandwichlike) on an anchored chain, and others followed, employing similar arrangements (e.g., T. W. Johnson *et al.*, 1959; E. B. G. Jones, 1971b). As mentioned above, this method is selective but could be expanded by testing wood species that have not been used before, for instance, those not included in Table XIV. Wooden-panel experiments are advantageous because the exact duration of exposure, the location, and the species of tree are known, whereas these data are unknown when naturally occurring marine wood ("driftwood") is collected.

D. Manila Cordage

Barghoorn (1944) submerged untreated manila hemp rope for 6 months in Woods Hole Harbor and other Massachusetts sites and observed a softening and blackening of the material, accompanied by a decline in tensile strength. Microscopic observations showed innumerable hyphae and ascocarps of a *Lulworthia*. Meyers (1968b), Meyers *et al.* (1960), and Meyers and Reynolds (1960b) employed the loss of tensile strength of manila rope to test the cellulolytic activities of higher marine fungi in the laboratory. Meyers and Reynolds (1963) observed that the first fungal infestations of manila twine submerged in Biscayne Bay (Florida) occurred within 5 days following exposure. These authors isolated a number of marine Ascomycetes from the submersed cordage, for example, species of *Corollospora, Lulworthia,* and *Torpedospora,* but also several terrestrial Deuteromycetes, which were probably derived from extraneous surface material on the ropes.

In view of the rapid attack of fungi on untreated rope in the marine environment, it can be assumed that marine fungi were active decomposers of fishing nets and ropes and caused considerable losses before the introduction of artificial fibers.

E. Treated Cellulose (Cotton Filters, Filter Paper, Solka Floc, and Cellophane)

Chemically treated cellulose is more easily broken down by fungi than native cellulose. Therefore, altered cellulose is rarely used in field experiments but is often used in the laboratory to test the cellulolytic abilities of marine fungi (see Chapter 21).

A cotton filter (kier-boiled, bleached, dry-textile processed) used in the seawater system of the Institute of Marine Science (University of Miami) was thoroughly decomposed by *Lulworthia* sp. after about 3 months of exposure (Meyers and Hopper, 1967; Meyers, 1968a). Only a small portion of the original matrix remained and the Ascomycete reproduced vigorously. Noticeable in the degraded cotton filter was a rich population of nematodes belonging to four different genera.

Filter paper, solka floc cellulose, and cellophane are often employed in determining the utilization of cellulose by fungal pure cultures (e.g., E. B. G. Jones and Irvine, 1972; Kohlmeyer, 1958a,b, 1960; Kohlmeyer and Kohlmeyer, 1966). The only record of a marine fungus growing and reproducing on paper in a natural habitat is by Kohlmeyer (1971b), who found ascocarps of *Lulworthia* sp. in a piece of newspaper that was lodged between a cement weight and the wood of a submerged lobster pot.

Regenerated cellulose (e.g., Phriphan, cellophane, and dialysis tubing) is an excellent substrate for culturing cellulolytic fungi in pure culture (Kohlmeyer, 1956, 1958a,b; T. Nilsson, 1974a). Different growth patterns caused by different fungal species can be noticed on and in the foil (Kohlmeyer and Kohlmeyer, 1966). Vargo *et al.* (1975) submerged dialysis membranes for up to 17 days in Narragansett Bay (Rhode Island) and observed the settlement by diatoms and cellulolytic bacteria. No fungi were mentioned or illustrated by Vargo *et al.* (1975), but it can be assumed that the exposure was too short for the settlement of fungi. They would have appeared eventually on the membranes, just as every submerged piece of wood is eventually colonized by higher fungi. Settlement of fungi on newly submersed surfaces takes probably longer than that of bacteria because fungal propagules are less numerous in the water column than bacterial cells. It is possible that, in nature, cellulose membranes disintegrate by the action of bacteria before higher fungi have a chance to colonize the substrate.

Fungi occurring on leaves in marine habitats are discussed in Chapter 13.

II. DEGRADATION OF WOOD

A. Micromorphological Symptoms of Wood Decay

Many microscopic Ascomycetes and Deuteromycetes cause a decomposition of wood called "soft rot" (Savory, 1954a,b). This term* designates characteristic patterns of penetration and growth of hyphae within the secondary cell wall of wood, and considerable literature has accumulated on this subject (e.g., Levy, 1965; Liese, 1970; Wilcox, 1970; T.

---

* The term "soft rot" was also applied earlier in phytopathological literature to the decomposition of fruits, roots, stems, and other plant parts that become soft by the action of fungi and bacteria (Ainsworth *et al.*, 1971).

Nilsson, 1973, 1974a–c). Soft rot in terrestrial wood was known since the middle of the nineteenth century, and this type of wood decay was later described and illustrated for fungi of marine habitats (Barghoorn, 1944; Kohlmeyer, 1958b). To understand the symptoms of wood decay, the structure of the cell walls must be considered (Fig. 25) (Liese and Cote, 1960). Surrounding the cell is the middle lamella, composed mainly of lignin and pectopolyuronides. This intercellular substance is bordered by the primary wall. Next is the secondary wall, which is composed of the S_1 (or outer) layer, the S_2 (or central) layer, and, adjoining the lumen, the thin S_3 layer (tertiary wall). The S_2 layer is the thickest of all and is composed of cellulose microfibrils arranged in the form of a steep spiral almost parallel to the longitudinal axis of the cell.

The micromorphological symptoms of wood decay caused by soft-rot fungi depend on the fine structure of the cell wall (Courtois, 1963). Fungal hyphae grow first in the lumina of the cells. They then pass through the tertiary wall and grow into the less lignified middle layer (S_2) of the secondary cell wall. The hyphae follow the orientation of the microfibrils, thereby forming characteristic spirally arranged, elongated chains of cavities with conical ends within the wall (Fig. 26a). In cross sections these cavities appear as round or oval holes, which are especially distinctive in thick-walled late wood (Fig. 26b). Leightley and Eaton (1977) found a mucilaginous sheath around the hyphae of *Halosphaeria mediosetigera,* which supposedly is the carrier of enzymes. They also discovered that the hyphae are actively moving within the wall cavity.

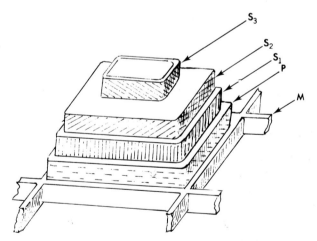

Fig. 25. Diagram of the different layers composing the cell wall of a wood tracheid or fiber. M, middle lamella; P, primary wall; S_1, outer layer of secondary wall; S_2, central layer of secondary wall; S_3, inner layer of secondary wall or tertiary wall.

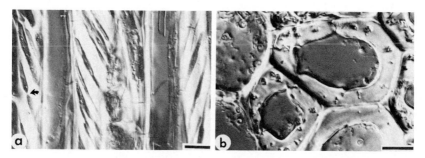

Fig. 26. Soft-rot decay of wood caused by hyphae of marine fungi. (a) Longitudinal section through wood showing chains of cavities with conical ends (arrow) within the wall (Herb. J.K. 1646; bar = 10 μm). (b) Wood from the deep sea (3710 m), cross section (6 μm) through cells with cavities in the central (S_2) layer of the secondary walls (Herb. J.K. 2470; bar = 10 μm). (Both in Nomarski interference contrast.)

These authors suggest that the hyphal sheath together with the hyphal movement within the cavity "serve to facilitate the movement of extracellular enzymes from the hypha to the wood cell wall substrate."

Although the general pattern of degradation is similar in different species of wood, Courtois (1963) found 14 types of cavities, the shapes of which depend on the wood anatomy, fungal species, age of the hyphae, temperature, and water content of the wood. After decomposition of the secondary wall, the tertiary wall (S_3), which adjoins the lumen, and the lignified middle lamellae are attacked.

A second type of wood decay may be caused by soft-rot fungi in hardwoods (e.g., *Betula* and *Fagus* spp.). Simultaneously with the formation of cavities in the secondary walls, erosions are produced by hyphae appressed to the S_3 layer along the cell lumen (Courtois, 1963; Corbett, 1965).

Only one lignicolous marine Ascomycete is known that does not cause the typical cavities in the S_2 layer of the secondary wall. Hyphae of *Amylocarpus encephaloides* erode the cell wall from the cell lumen, by dissolving first the tertiary wall, then the secondary wall, and finally leaving only a skeleton of middle lamellae (Lindau, 1899; Kohlmeyer, 1960). This pattern of wood decay appears to exclude *A. encephaloides* from the soft-rot fungi and makes the species comparable to some terrestrial Basidiomycetes that cause brown rot.

Soft-rot decay is mostly restricted to the wood surface, and hyphae of marine Ascomycetes and Deuteromycetes normally penetrate only a few millimeters deep into the wood, probably being restricted by a high oxygen requirement. Some lightweight tropical woods may be colonized throughout by fungal hyphae even far away from the surface if large vessels permit penetration of air into the interior of the wood. Such deep-penetrating soft rot is found in Indian fishing crafts and causes a rapid decay of these catamarans (Becker and Kohlmeyer, 1958a,b).

Pure cultures of marine fungi have been tested for their ability to degrade wood and to cause soft rot (Barghoorn, 1944; Kohlmeyer, 1963e; T. Nilsson, 1973, 1974a). E. B. G. Jones (1971a) has summarized information on marine fungi that have been shown to cause soft rot under laboratory conditions, and 13 Ascomycetes and five Deuteromycetes are listed in his paper. Their number could be increased considerably by isolating and testing additional lignicolous species.

The Basidiomycete *Nia vibrissa* causes a typical pattern of white-rot decay when grown in pure culture on balsa (*Ochroma lagopus*), beech, and Scots pine wood (Leightley and Eaton, 1977). In contrast to soft-rot decomposition, this fungus forms erosion troughs and gradually thins the cell walls. The cellulolytic enzymes diffuse readily from the hyphae into the substrate. The hyphae of *N. vibrissa* are surrounded by a delicate sheath, and irregular, loose strands lead to the outside of the lysis zone (Leightley and Eaton, 1977).

B. Wood Colonization in the Marine Environment

1. Identification of Colonizers

Microscopic examination of the wood surface immediately after removal of the substrate from the water shows the presence of reproductive structures that have developed *in situ*. The wood may then be incubated in a moist chamber in the laboratory to give those fungal species present in the wood in a vegetative state a chance to fruit. However, these secondary fungi must be regarded critically because some may have derived from dormant terrestrial propagules that have been superficially attached to the wood surface.

2. Frequencies of Fungal Species

Numerous studies have dealt with the colonization of wood in marine habitats, in particular, with the influence of salinities, temperatures, and wood species. Most of these experiments have been performed in geographically restricted areas such as single estuaries [Gold, 1959; G. C. Hughes, unpublished thesis (see Johnson and Sparrow, 1961); Schaumann, 1968, 1975b; Shearer, 1972b] or an island (Schaumann, 1969, 1975a,b). E. B. G. Jones *et al.* (1972) conducted a worldwide test with wooden panels of *Pinus sylvestris* submerged at 23 sites. *Cirrenalia macrocephala* was the most frequent species and occurred on 45% of all test panels. This species was followed by *Lulworthia* spp. (44%), *Zalerion maritimum* (32%), *Dendryphiella salina* (16%), *Ceriosporopsis halima* (15%), and *Humicola alopallonella* (13%). *Lulworthia* sp. and *Z. maritimum* were also the most common species in a survey in British waters (Byrne and Jones, 1974). Schneider (1976) submerged test panels of five wood species in the western Baltic Sea and found the following species to be most common: *Ceriosporopsis halima, Halosphaeria appendiculata, H. mediosetigera, Lulworthia* sp., *Nais inornata, Monodictys pelagica,* and *Zalerion maritimum*. Experiments with four different species of wood in the Mediterranean resulted in the colonization of the panels by *Cirrenalia macrocephala, Z. maritimum, Ceriosporopsis halima,* and *Lulworthia* sp. (Kohlmeyer, 1963c). These are the same species prevalent in the survey made by E. B. G. Jones *et al.* (1972). Studies by Schaumann (1975b) on fungi of fixed wooden structures (e.g., groins) of the German North Sea coast generally agree with the results of E. B. G. Jones *et al.* (1972), because most of the fungi mentioned above are among the frequent species in Schaumann's list also. However, the frequencies are much lower than in the worldwide study. *Cirrenalia macrocephala* was present on 15.7% of 344 samples, *Z. maritimum* on 14.2%, *Lulworthia* spp. on 6%, *C. halima* on 2.8%, and *H. alopallonella* on 8.2% (Schaumann, 1975b). *Dendryphiella salina* did not occur at all in Schaumann's wood collections, but, instead, he found the species in foam and sediments. *Monodictys pelagica* (20.4%) and *Lignincola laevis* (18.6%) were the two prevalent fungi on wood of the German coast, along with unidentified Ascomycetes (23%) and Deuteromycetes (20.4%). These two examples from the literature (E. B. G. Jones *et al.*, 1972; Schaumann, 1975b) show that frequencies of fungal species cannot be generalized. Results of studies made in different geographical areas are difficult to compare because of variations in temperatures, salinities, and other factors. The wood species, in particular, may have an influence on the fungal colonizers. E. B. G. Jones *et al.* (1972) used pinewood panels, whereas Schaumann (1975b) collected samples from a variety of unidentified stationary wood, most likely derived from several wood species. A major problem in recording numbers of fungi is the possibility that some fungi may be present in vegetative form in the substrate but are not recognized because they do not fruit.

3. Successions of Fungal Species

There is no general agreement among mycologists on successional patterns of marine fungi attacking wood. Several authors reported successions of fungal species on wooden panels, for example, Meyers (1954), Meyers and Reynolds (1960a), T. W. Johnson and Sparrow (1961), E. B. G. Jones (1963c, 1968a,b), and R. D. Brooks [unpublished thesis (see G. C. Hughes, 1975)]. The observed patterns apparently depend on the wood species, length of exposure, salinity, temperature, and geographical location of the test site. T. W. Johnson (1967) believes that a putative succession of fungi "is an expression of differential fruiting time, rather than replacement of one species by another." So-called successional patterns may also depend on the relative frequency of fungal species in certain locations. The likelihood of a wood specimen to become colonized by a common fungus such as *Ceriosporopsis halima, Lulworthia* spp., *Humicola alopallonella,* or *Zalerion maritimum* is much greater than by a fastidious species, which does not produce as many propagules. The first settler on newly submerged wood is most probably the first one to fruit and to be identified.

4. Preference for Wood Species, Natural Resistance of Wood, and Wood Preservatives

Some fungi appear to favor certain species of wood; for instance, beechwood (*Fagus sylvatica*) is preferentially colonized by *Ceriosporopsis calyptrata, C. tubulifera, Corollospora maritima, Halosphaeria appendiculata, H. hamata, Nautosphaeria cristaminuta,* and *Digitatispora marina* (E. B. G. Jones, 1968b; Byrne and Jones, 1974; Schneider, 1976). Pinewood (*Pinus sylvestris*) is preferred by *Halosphaeria circumvestita, Cirrenalia macrocephala,* and *Humicola alopallonella* according to E. B. G. Jones (1968b), whereas Schneider (1976) found *Corollospora comata, Halosphaeria maritima,* and *H. stellata* exclusively on this wood species. T. W. Johnson *et al.* (1959) counted the number of ascocarps on panels of 35 wood species after 2 months of submergence in seawater and found the least number of fruiting bodies on hardwoods, with the exception of maple. Ascocarps were absent on Philippine narra (*Pterocarpus indicus*) and primavera (*Cybistax donnell-smithii*), and their occurrence was negligible on bald cypress (*Taxodium distichum*), black cherry (*Prunus serotina*), and mahogany (*Swietenia mahagoni*). T. W. Johnson *et al.* (1959) point out that ascocarp numbers are probably not proportional or related to mycelial abundance within the wood. In our opinion, the suppression of fungal reproduction also indicates that a wood species may have some antimicrobial properties. Microscopic examination of such wood species after submersion should be made to determine the presence or absence of fungal hyphae.

Little is known about the natural resistance of wood species against the attack by marine fungi. Ritchie (1968) submerged wood of eight tropical trees in the Panama Canal Zone. He examined wood sections and found greenheart (*Ocotea rodiaei*) and red mangrove (*Rhizophora mangle*) practically fungus-free after about 6 weeks of exposure. The other wood species showed variable degrees of infestation by fungal hyphae. The only obligate marine fungus isolated in Ritchie's (1968) test was a *Lulworthia* sp., in addition to some 20 Deuteromycetes of apparently terrestrial origin. It is not certain whether these species originated from mycelium present in the wood or from dormant terrestrial propagules that grew out from the incubated wood splinters in the laboratory.

Among 113 tropical trees tested in the Canal Zone, Cocobolo (*Dalbergia retusa*) was the most resistant against marine borers and completely resistant to attack by marine fungi (Bultman and Southwell, 1972; Bultman *et al.*, 1973). In a cooling tower using brackish water for cooling, *D. retusa* heartwood showed some resistance to fungi and was infrequently colonized by marine soft-rot fungi (S. E. J. Furtado and Jones, 1976). Cocobolo contains in its heartwood an orange pigment, obtusaquinone, which shows a mycostatic effect on a *Chaetomium* and a *Pestalotia* isolated from marine wooden panels (Bultman *et al.*, 1973).

Another product of *D. retusa,* obtusastyrene, is even more effective in suppressing fungal growth and reproduction (Bultman and Ritchie, 1976; S. E. J. Furtado *et al.*, 1977). It appears that obtusastyrene derivatives act as fungicides against obligate marine fungi (Bultman and Ritchie, 1976); however, under high-hazard conditions of cooling towers, obtusastyrene does not offer complete protection against attack by marine fungi (S. E. J. Furtado and Jones, 1976).

Although marine fungicides may be of great economic importance because of possible relationships between wood borers and marine fungi (see Chapter 19), it is surprising to see how little information is available on the reaction of marine fungi to chemically treated wood. Schaumann's (1969, 1975a) investigations on fungi from pilings of Helgoland show that these structures are regularly attacked by marine fungi. Wood used for man-made structures in the marine environment is normally treated with preservatives, mostly creosote, but after a certain time leaching from the surface (Irvine *et al.*, 1972) permits settlement of fouling organisms, borers, and microorganisms, including fungi. E. B. G. Jones and Irvine (1971), E. B. G. Jones (1972), and Irvine and Jones (1975) tested the tolerance of some marine fungi on agar plates to a copper–chrome-arsenate preservative and found them to be more tolerant to this compound than are many terrestrial species. Field tests with copper–chrome-arsenate (2.5–7.5%)-treated pine panels, however, showed the absence of fungi after 40 weeks, or even up to 24 months, of exposure (E. B. G. Jones and Irvine, 1971; Irvine *et al.*, 1972; Eaton and Dickinson, 1976). Further experiments with other species of marine lignicolous fungi and other wood preservatives should help to explore ways to protect marine timber against attack by fungi.

5. Influence of Pollution on Fungal Wood Colonization

Information on the effect of pollution on the development of marine fungi is scarce. Preliminary experiments with wooden panels submerged in two different locations in the Mediterranean at Banyuls-sur-Mer (France) showed that the wood in polluted water was covered with a thicker mycelium and more ascocarps than panels tested in clean water (Kohlmeyer, 1963c). Nutrients present in polluted environments obviously favor the growth of fungi and accelerate wood destruction, but quantitative data are lacking.

6. Marine Cellulolytic Bacteria

Wood in the marine environment is colonized more or less simultaneously by bacteria and fungi, although Brooks *et al.* (1972) report that initial colonization is dominated by fungi, whereas Cundell and Mitchell (1977) and Leightley and Eaton (1977) state that treated or untreated wood is initially attacked by bacteria. Under certain conditions, for instance, in waters with a low content of dissolved oxygen, higher fungi may be absent, but cellulolytic bacteria may still be able to attack the wood (Kohlmeyer, 1969a). Preservative-treated timber is initially colonized by bacteria, whereas fungi follow later (Irvine and Jones, 1975). The symptoms of wood decay caused by marine bacteria differ from those produced by soft-rot fungi. Bacteria start degradation of the cell walls at the cell lumina and along the rays. Irregular depressions are formed in the tertiary wall, cavities extend into the S_2 layer of the secondary wall, and finally all layers of the wall are transformed into amorphous masses (Fig. 27). Leightley and Eaton (1977) describe colonies of marine Actinomycetes causing networks of shallow erosion channels in the S_3 layer of the cell wall. Similar colonies also occur on wood and cellophane submerged in the deep sea at depths between 253 and 5000 m (J. Kohlmeyer, unpublished). Marine bacteria were also reported to cause a second type of wood deterioration, namely, a swelling of the S_2 layer and folding of the S_3 layer (Kohlmeyer, 1969a). In general, there is very little known about wood-decomposing marine bacteria and their relationship to soft-rot fungi, and, apparently, the field of cellulolytic marine bacteria is wide open for research in the future.

C. Tests for Wood Degradation

Wood decay caused by marine fungi can be examined by measuring weight losses of test panels, in addition to making visual observations of micromorphological symptoms. The microscopic examinations of soft-rot

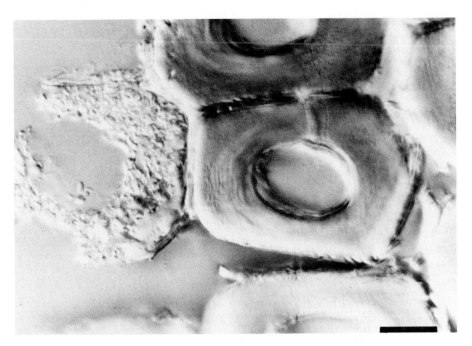

Fig. 27. Cross section (6 μm) through fir wood submerged in the deep sea at 2000 m; all layers of the left cell are decomposed by bacteria (Herb. J.K. 2480; Nomarski interference contrast; bar = 10 μm).

patterns cannot be accurately quantified, and subjective grades of decomposition have to be assigned [e.g., "slight," "mild," and "heavy" soft rot (E. B. G. Jones, 1971a)]. Weight losses of wood after fungal attack in nature have been measured by E. B. G. Jones and Irvine (1971), who submerged wooden panels in British coastal waters. After 40 weeks of exposure, beech showed a weight loss of 27.6%, while Scots pine had lost 19.8%. Byrne and Eaton (1972) also exposed panels for 42 weeks in the sea and obtained weight losses of about 20% in both beech and Scots pine. The same authors tested the wood decomposition in tanks with various concentrations of running seawater and found the greatest decay of beechwood (14.7% weight loss) in the lowest salinity, namely, in 10% seawater concentration. In comparison, Scots pine deteriorated faster in higher salinities and suffered a weight loss of 13% in 75% seawater concentration. Byrne and Eaton (1972) concluded that, under low salinity conditions, Scots-pine wood may be more resistant to attack by soft rot than beechwood.

Only few quantitative data are available on wood decomposition by individual species of marine fungi. E. B. G. Jones (1971a), Eaton and Jones (1971a), and Leightley and Eaton (1977) tested six marine Ascomycetes, one Basidiomycete, and five Deuteromycetes for their abilities to decompose beech- and pine wood (Table XV). Weight losses in Scots pine were negligible or at least smaller than in beechwood. Beech, on the other hand, suffered considerable losses, especially through attack by

TABLE XV

Weight Losses of Wooden Panels Caused by Pure Cultures of Marine Fungi

	Reference	Average weight loss (%)		Time (weeks)
		Fagus sylvatica	*Pinus sylvestris*	
Ascomycotina				
Corollospora maritima	E. B. G. Jones (1971a)	25.7	Not tested	18
Halosphaeria mediosetigera	Leightley and Eaton (1977)	2.1	2.4	13
H. quadriremis	Leightley and Eaton (1977)	0.0	0.0	13
Lulworthia sp.	E. B. G. Jones (1971a)	9.8	Not tested	18
Nais inornata (R131)	Eaton and Jones (1971a)	9.2	0.0	15
Savoryella lignicola (R176A)	Eaton and Jones (1971a)	14.3	4.0	15
S. lignicola (R175)	Eaton and Jones (1971a)	0.0	0.5	15
S. lignicola	Leightley and Eaton (1977)	9.0	5.3	13
Basidiomycotina				
Nia vibrissa	Leightley and Eaton (1977)	0.0	0.0	13
Deuteromycotina				
Alternaria "*maritima*"	Leightley and Eaton (1977)	2.8	2.9	13
Cirrenalia macrocephala	Leightley and Eaton (1977)	7.6	4.1	13
Humicola alopallonella (R72)	Eaton and Jones (1971a)	29.8	1.1	15
H. alopallonella	Leightley and Eaton (1977)	11.3	0.0	13
Trichocladium achrasporum	Leightley and Eaton (1977)	4.0	2.0	13
Zalerion maritimum	Leightley and Eaton (1977)	1.7	1.4	13

Corollospora maritima and *Humicola alopallonella.* However, different isolates of the same species yielded divergent results. The Basidiomycete *Nia vibrissa* did not cause any weight losses within 13 weeks. These tests with pure cultures corresponded only partly to results of field experiments (E. B. G. Jones and Irvine, 1971; Byrne and Eaton, 1972) that produced weight losses of 27.6% in beechwood and about 20% in pinewood. The greater decomposition of pine under natural conditions than *in vitro* is in part due to longer exposure times but may also be caused by mixed mycota, possibly accompanied by cellulolytic bacteria. This assumption is supported by the fact that E. B. G. Jones (1972) obtained weight losses of 32.8% in Scots pine after 48 weeks by using mixed cultures in laboratory experiments. The development of laboratory test methods that simulate natural conditions (E. B. G. Jones, 1972) will be a prerequisite for testing the effectiveness of fungicides against marine soft-rot fungi.

Henningsson (1976a) tested 14 species of Ascomycetes and Deuteromycetes, mostly isolated from wood submerged in the Baltic Sea, for their ability to decompose wood. All species produced soft-rot decay and the majority caused measurable losses of dry weight in hardwoods. Weight losses of dry matter were considerably greater in hardwoods (*Betula pubescens* and *Quercus robur*) than in softwoods (*Picea abies* and *Pinus sylvestris*) and reached, for instance, 12.2% in oak attacked by *Zalerion maritimum* after 3 months, whereas heartwood of spruce and pine did not suffer any losses during this time. The highest rate of loss in dry mass was caused by *Monodictys pelagica,* namely, 25.9% in birchwood after 6 months of growth. Tests for changes in the chemical composition of birchwood after attack by soft-rot fungi showed that the fungi preferably utilize cellulose, hemicelluloses, and other carbohydrates of lower degrees of polymerization. Lignin is also degraded but to a lesser degree. Therefore, a great part of the lignin remains in the wood after attack by higher marine fungi.

16. Bark-Inhabiting Fungi

Bark, the tough external covering of woody perennial stems or roots, is found in the marine environment on live roots of mangrove trees, on dead submerged wood of terrestrial origin, or separated from the wood in intertidal or subtidal habitats. Bark is a complex tissue composed mainly of dead cork cells that are made impervious to water by high amounts of suberin. Cork is quite resistant to decomposition, and information on decay of cork is scarce (DeBaun and Nord, 1951; Dapper, 1967, 1969). The majority of the 35 fungi found on bark in the marine environment are plurivorous species, which usually occur on submerged wood as well (Table XVI). The only known exceptions are *Herpotrichiella ciliomaris, Keissleriella blepharospora, Mycosphaerella pneumatophorae,* and *Rhabdospora avicenniae,* which all appear to be restricted to bark. The last three species are host specific on mangrove roots in the tropics and subtropics, whereas *H. ciliomaris* is limited to temperate waters of the Pacific and Atlantic oceans. Fruiting bodies of all four obligate bark inhabitants are superficial or partly immersed in the substrate. Hyphae extend between the cork cells, and it is likely that these species obtain nutrients by decomposing the bark tissue. However, we are still ignorant about the processes involved and about the cause for the restriction of the four species to bark. Only *K. blepharospora* has been isolated thus far (Kohlmeyer, 1969c), but poor growth in culture has prevented further experiments on the nutritional requirements of this species. Ascocarps of *K. blepharospora* develop between cork cells of roots or submerged seedlings of *Rhizophora* spp. The upper green part of seedlings, which is covered by a cuticle, is never attacked by the fungus (Kohlmeyer, 1969d).

Herpotrichiella ciliomaris fruits in clusters in small depressions of the bark surface. These ascocarps are the prettiest of all marine fungi, displaying whitish to bluish colors and brown hairs around the ostiole. The

---

TABLE XVI
Bark-Inhabiting Marine Fungi

Ascomycotina
  Ceriosporopsis halima
  Corollospora comata
  C. maritima
  C. trifurcata
  Crinigera maritima
  Didymosphaeria rhizophorae
  Haligena elaterophora
  H. viscidula
  Halosphaeria appendiculata
  H. mediosetigera
  H. quadricornuta
  H. stellata
  Herpotrichiella ciliomaris[a]

  Hydronectria tethys
  Keissleriella blepharospora[a]
  Kymadiscus haliotrephus
  Leptosphaeria australiensis
  L. obiones
  Lignincola laevis
  Lulworthia sp.
  Microthelia linderi
  Mycosphaerella
    pneumatophorae[a]
  Nais inornata
  Torpedospora radiata
  Trematosphaeria mangrovis

Deuteromycotina
  Camarosporium metableticum[b]
  Cirrenalia macrocephala
  C. pygmea
  Dictyosporium pelagicum
  Humicola alopallonella

  Monodictys pelagica
  Phialophorophoma litoralis
  Rhabdospora avicenniae[a]
  Trichocladium achrasporum
  Zalerion maritimum

---

[a] Species known from bark only.
[b] Doubtful record (Schmidt, 1974).

fruiting bodies are easily overlooked because of their small size (77–226 μm in diameter), especially when the bark is covered by fouling organisms.

    In view of the scarce information available on bark-inhabiting (or corticolous) fungi, it would be particularly important to clarify nutritional requirements of these organisms.

17. Fungi on Man-Made Materials

Fouling and biodegradation of man-made structures in the ocean have aroused considerable interest. Tests with a large number of materials have yielded growth of fungi in submerged wood, but not in plastics and other man-made specimens (e.g., Muraoka, 1966a,b). The only reports on fungal deterioration of synthetic materials are by Le Campion-Alsumard (1970) and by E. B. G. Jones and Le Campion-Alsumard (1970a,b), who found three Ascomycetes and one Deuteromycete in polyurethane submerged for up to 4 years in the Mediterranean. Fungal mycelia had penetrated up to 1 mm into the 3-mm-thick layer of polyurethane. The substrate was swollen and torn by the action of the fungi, but the investigators were unable to determine if the corrosion of polyurethane was purely mechanical or chemical, or both. Fungi growing on the polyurethane were *Corollospora maritima, Haligena unicaudata, Lulworthia purpurea,* and *Zalerion maritimum.*

The only other fungus known to derive from a man-made substrate is *Varicosporina ramulosa,* which was isolated from a synthetic sponge, the composition of which was not indicated (Meyers and Kohlmeyer, 1965).

18. Fungi in Animal Substrates

Fungal diseases of marine animals have been discussed by Alderman (1976). Higher marine fungi infesting animals are rare, and most of them are saprobes restricted to exoskeletons, shells, or protective tubes. Such substrates of animal origin consist of cellulose, chitin, keratin, and calcium carbonate with an organic matrix.

Tunicin is an animal cellulose that resembles plant cellulose and occurs in the test of tunicates. Ritchie (1954) isolated *Halosphaeria quadricornuta* [as "Form No. 2" of Meyers (1953)] from a tunicate. *Ceriosporopsis circumvestita* and *Pleospora gaudefroyi* produced a good growth in pure culture on pieces of tunicin from *Phallusia mammillata* in addition to decomposing balsa wood and cellulose foil (Kohlmeyer, 1962c; Kohlmeyer and Kohlmeyer, 1966). These observations indicate that fungi may play a role in deteriorating tunicate cellulose in nature, and additional species will probably be detected on such substrates.

Chitin, a nitrogen-containing polysaccharide of considerable mechanical strength, is found in the exoskeletons of many marine animals. Higher fungi have rarely been recorded from chitinous substrates in the ocean, and most of the chitin produced in marine habitats is probably decomposed by chitinoclastic bacteria (see, e.g., Harding, 1973). A rare record was made by Alderman (1973), who found septate hyphae of an unidentified fungus in the exoskeleton of a crawfish, *Palinurus elephas* [as *P. vulgaris* in the chapter by Alderman (1976)], from the western coasts of the British Isles. The mycelium had damaged the shell extensively, even penetrating it to expose the muscle tissue beneath. Bacterial necrosis followed the fungal infection.

Alderman (1976) reports on a fungal infection of the crab *Carcinus maenas* from the Dorset coast. The fungus, tentatively identified as *Periconia prolifica*, caused "burn spots" on the surface of the exoskeleton and black nodules in the hepatopancreas. Gills and other tissues contained hyphae, conidia, and aleuriospores. It was not proven whether the fungus was a parasite or secondary invader of the crabs. Detailed morphological information on the fungus is not available and the identification of *P. prolifica* needs confirmation. This species occurs in other animal substrates, namely, calcareous shipworm tubes (Kohlmeyer, 1969b,c).

The third record of a chitinoclastic higher fungus is the Ascomycete *Abyssomyces hydrozoicus* from the chitinous hydrorhiza and hydrocaulus of hydrozoa that were attached to a stony coral collected near the South Orkney Islands (Kohlmeyer, 1970, 1972a). The mycelium of *A. hydrozoicus* covers the hydrozoan exoskeleton (Fig. 28a), forming channels on the surface or penetrating the substrate and laminating it (Fig. 28b). Ascocarps are formed on the outside of the tubelike exoskeleton, and their diameter is comparatively large in relation to the thin hydrozoan colonies.

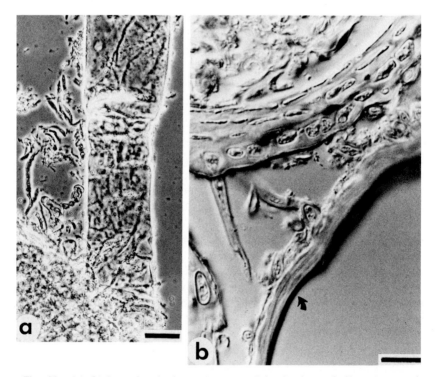

Fig. 28. (a) Stolon of a hydrozoan, covered by hyphae of *Abyssomyces hydrozoicus* (bar = 30 μm). (b) Section through base of ascocarp of *A. hydrozoicus*, hyphae attached to the wall of a hydrozoan exoskeleton (arrow) (bar = 10 μm). (Herb. J.K. 2754; a in phase contrast, b in Nomarski interference contrast.)

Fungal hyphae have also been found in other collections of hydrozoa in the Atlantic Ocean, on and off the North Carolina coast (Kohlmeyer, 1972a). These hyphae displayed four different types of corrosion of the substrate, differing in the diameter of hyphae and in their mode of branching. Identifications could not be made because reproductive structures were absent. It is also not known whether the fungi occurring in hydrozoa attack the living hosts or infest the exoskeletons after the animals have died.

The last known record of a chitin-inhabiting higher fungus in the marine habitat is *Laboulbenia marina* (Picard, 1908), an Ascomycete living on the beetle *Aepus robini* of the *Laminaria* zone along the French Atlantic coast. Ascocarps of the fungus develop on hairs or the bases of the anterior wings. There is no visible corrosion of the chitin and the nutrition of the fungus is unknown, as are the exact nutritional requirements of all other Laboulbeniales (Benjamin, 1971). Examination of other marine insects (e.g., Cheng, 1976) will probably disclose additional marine Laboulbeniales.

Keratin, a tough fibrous protein containing sulfur, is produced particularly in the epidermis of vertebrates. Keratinophilic fungi from salt marshes, beaches, and dunes have been investigated by Pugh and Mathison (1962) and Pugh and Hughes (1975), who isolated several terrestrial species. Although they found no obligate or facultative marine fungi, and keratinophilic species occurred only above the high-water mark, these authors mention that *Arthroderma curreyi* and *Ctenomyces serratus* are capable of growing and sporulating on autoclaved dogfish egg cases and decalcified cuttlefish "bone." Knowledge about the decomposition of keratin in marine habitats is negligible. The only report on the deterioration of keratinlike substances by marine fungi is by Kohlmeyer (1972a), who found Ascomycetes in tubes of the polychaete *Chaetopterus variopedatus*. This sand-inhabiting worm lives in mud flats of many parts of the world. The leathery protective tubes of the animal are often washed ashore. Tubes collected along California beaches contain hyphae and immersed ascocarps of *Lulworthia* sp. (Figs. 29a and 29b), which cause the substrate to laminate into many thin layers. The mycelium spreads throughout the tube, always in close contact with the substrate, which indicates that enzymes act only in immediate contact between hyphae and matrix. Eventually the ascocarps are exposed. The chemical composition of the tubes of *C. variopedatus* was unknown until recently, when N. D. Latham (personal communication) analyzed the tubes and found the solid fraction to be mainly inorganic, containing calcium, iron, manganese, magnesium, phosphorus, silicon, and sulfur. The organic fraction, which has not been analyzed, may be the source of nutrients for higher marine fungi.

Fig. 29. *Lulworthia* sp. in tubes of the polychaete, *Chaetopterus variopedatus*. (a) Section (8 μm) through the base of an ascocarp with hyphae penetrating the substrate in all directions (bar = 35 μm). (b) 6-μm section through the tube, branching hyphae in the layered wall of the substrate (bar = 25 μm). (Herb. J.K. 2644; both in phase contrast.)

Baiting experiments with sand-inhabiting fungi yielded *Corollospora trifurcata,* which developed on submerged snake skin (J. Kohlmeyer, unpublished).

These observations demonstrate that obligate marine fungi are able to attack keratinlike substances, but further research is needed to clarify how important they are in the breakdown of such materials in the marine environment.

The penetration of calcium carbonate substrates by algae, fungi, and invertebrates has attracted much interest in recent years (Carriker *et al.*, 1969; Golubic *et al.*, 1975). Early records of shell-boring microorganisms have been dubious as to the identity of the borers, and both algae and fungi could have been involved (e.g., Kölliker, 1860a,b). Most of the fungi reported from shells belong to the lower fungi, or else they have not been identified with certainty and are excluded from the following discussion (Zebrowski, 1936; Porter and Zebrowski, 1937; T. W. Johnson and Anderson, 1962; Höhnk, 1967, 1968, 1969; Cavaliere and Alberte, 1970; Cavaliere and Markhart, 1972; Golubic *et al.*, 1975; Alderman, 1976).

Only a few higher marine fungi have been found as invaders of calcareous substances such as shells of mollusks, tests of barnacles, or linings of burrows. Höhnk (1967, 1969), illustrates septate hyphae and fruiting bodies of unidentified higher fungi from shell fragments. The beaked "pycnidia" in his photographs could be ascocarps, because the fruiting bodies were apparently empty when collected (Höhnk, 1967). Calcareous linings of empty shipworm tubes, especially in tropical and subtropical waters, are frequently attacked by Ascomycetes and Deuteromycetes (Kohlmeyer, 1969b). Fungal hyphae from the surrounding wood penetrate the tubes after death of the animals, and the hard substrate becomes soft, brittle, and brown. Eventually, conidia or ascocarps develop within the calcareous matrix and break through the surface. Sometimes, fruiting bodies develop in the wood under the tubes, forming long necks that pierce the calcareous lining. The Ascomycetes *Halosphaeria quadricornuta* and *H. salina* and the Deuteromycetes *Cirrenalia pygmea, Humicola alopallonella,* and *Periconia prolifica* are known from such substrates (Kohlmeyer, 1969b). Hyphae and conidia of *C. pygmea* and *H. alopallonella* have been found only on the surface of the calcareous tubes, whereas the other fungi also developed inside this material.

An Ascomycete, *Pharcidia balani,* often associated with various species of microscopic algae, is regularly found all over the world in shells of living intertidal mollusks and tests of cirripedes (Fig. 30). In the lichenological literature the association is treated as a lichen, *Arthopyrenia sublitoralis* (e.g., Santesson, 1939). Black ascocarps and pycnidia (or spermogonia?) of *P. balani* are half embedded in the calcareous substrate (Kohlmeyer and Kohlmeyer, 1964–1969). When old fruiting bodies disintegrate, they leave a crater in the surface of the shell, giving it a spongy and pitted appearance. Hyphae grow throughout the outer layers of shells and tests and cause a softening of the calcareous substrate. As

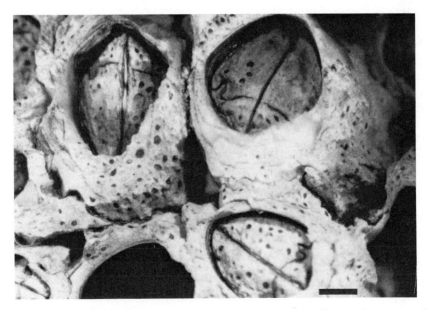

Fig. 30. Fruiting bodies of *Pharcidia balani* in the balanid, *Chthamalus montagui,* giving the shells a pitted appearance (Herb. J.K. 3837; bar = 1 mm).

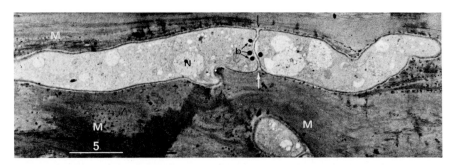

Fig. 31. Hypha of a higher marine fungus within the byssus thread of *Mytilus galloprovincialis*. M, byssus matrix; N, nucleus; b, Woronin bodies; unmarked arrows indicate cross septum (bar = 5 μm). From Vitellaro-Zuccarello, *Marine Biol.* **22**, 226 (1973), Fig. 2. Reprinted with permission.

the fungus occurs only embedded in the outside of the shells, it appears not to damage the animals.

The mechanism and chemistry of fungal infestation of calcareous animal shells and tubes are unknown. Probably, the fungi utilize conchyolin, the organic matrix of the shells, whereas the $CaCO_3$ crystals or the amorphous calcium carbonate are not attacked. Investigations are needed to clarify the biodeterioration of these calcareous substrates. Such research is impeded by inadequate knowledge about these materials, their structure and chemical composition. Important techniques have been developed recently for the preservation of fungal and algal structures *in situ* by impregnating the substrate with resin or plastic, followed by removal of the carbonate matrix (Golubic *et al.*, 1970; Perkins and Halsey, 1971). These methods could be employed by using identifiable marine fungal species growing in calcareous substrata.

Byssus threads of *Mytilus galloprovincialis* from the Mediterranean regularly contain fungal hyphae (Vitellaro-Zuccarello, 1973). The byssus matrix consists of proteic substances, including some collagen, and is rather resistant to chemical and physical decomposition. Therefore, a penetration by fungal hyphae (Fig. 31) is of great interest. The ultrastructure of the hyphae suggests an ascomycetous affinity of the fungus (Vitellaro-Zuccarello, 1973). Fruiting bodies have not been observed; therefore, a relationship with Deuteromycetes also appears to be possible.

The preceding examples show the occurrence of fungi in a variety of substrates of animal origin. The importance of higher fungi in the breakdown of such materials in the oceans is not clear because the number of observations is limited and no quantitative evaluations have been made. Thorough searches for microorganisms in animal substrates should be undertaken to clarify their deteriorative activities and to make identifications possible.

19. Fungal–Animal Relationships

Although many invertebrates and fungi occur in the same marine microhabitats, little is known about interactions between them. Fungi grow, for instance, on organic matter buried in sandy beaches (see Chapter 8), and it is likely that members of the meiofauna, which graze on bacteria and microscopic algae from grains of sand (Meadows and Anderson, 1968), also feed on hyphae or ascocarps found in the interstices (R. G. Johnson, 1974). No data on the ingestion of sand-inhabiting marine fungi by the interstitial fauna are available. The only experiments using marine fungi as a food source were conducted with wood-boring animals, salt-marsh amphipods, and nematodes.

I. MARINE WOOD BORERS

A. Boring Mollusks

Bivalve mollusks of the families Teredinidae (shipworms) and Pholadidae (Turner, 1971) are important decomposers of wood in marine habitats. Besides their role as recyclers of natural submerged wood in the ocean, these "termites of the sea" are considered pests because of their enormous destruction of man-made wooden structures. Reports on damages of ships and shoreline fortifications go back to the seventeenth and eighteenth centuries, and the economic importance of marine wood borers has stimulated a considerable number of publications (Clapp and Kenk, 1963). In spite of the extensive literature on wood borers, basic questions about their biology are still unanswered, or contradictory theories are discussed. It was not clear, for instance, if shipworms produce a cellulase, or if they are unable to digest wood without the help of microorganisms. Rosenberg and Breiter (1969) and Cutter and Rosenberg (1972) found symbiotic cellulolytic bacteria (*Cellulomonas* sp.) in the cecum of *Teredo navalis*. In addition, they demonstrated an endogenous cecal cellulase, but postulated an apparent requirement for symbiotic bacterial cellulase in the digestive process of this borer. The same authors (Rosenberg and Cutter, 1973) stated later that the bacterial cellulase is the dominating factor in cellulose decomposition and utilization by *T. navalis*. B. Sahlmann (personal communication), however, examined the ultrastructure of the digestive tract of *T. navalis* and was unable to detect any bacteria. According to Sahlmann, wood particles are digested with an endogenous cellulase by secretion–resorption cells. Dean (1976) did not find a significant degradation of native crystalline cellulose by cell-free digestive tract extracts of *Bankia gouldi,* a fact that makes the involvement of microorganisms in the breakdown of wood by this borer doubtful. This author speculated on the possibility of an oxidative degradation of native crystalline cellulose, possibly coupled with lignin degradation. Whatever the final outcome in this controversy, it seems to be clear that fungi are not involved in the digestive process of adult shipworms because fungal hyphae are not found in their digestive tracts and do not occur in the wood near the boring shells. However, fungi may play a role in the settlement of larvae.

Larvae of wood-boring mollusks are ejected through the exhalant syphon from the maternal animal and dispersed by currents (Lane, 1961). They swim aimlessly with the help of an umbrella-shaped velum composed of long cilia and rest on any substrate they come in contact with. The foot serves to explore the surface in search of a suitable place for penetration. At this stage, the shells of the larvae are not calcified and are unable to penetrate fresh wood. If the wood surface is softened by the action of cellulolytic fungi or bacteria, or both, the larva starts scraping the substrate, forming within minutes a mound of debris around it. Metamorphosis begins about 12 hr after penetration of the wood. The shells become calcified, teeth develop on the shell, siphons appear, and the long foot changes into a short, stout organ of attachment

Experiments by Kampf *et al.* (1959) with *Lyrodus pedicellatus* (=*Teredo pedicellata*) have shown that the larvae are not chemotactically attracted by certain substances, as some authors had suggested earlier. Larvae were not attracted by wood or fungal extracts, sodium hydroxyde, malic and hydrochloric acids, or by mycelia or wood with or without fungi. However, they settled on fungus-infested wood when they came in contact with it. Other series of experiments of choice with *L. pedicellatus* gave conclusive evidence that the larvae accumulated preferably on wood that was "predigested" by fungi when the larvae reached it by chance and then settled on it. They rarely remained on fresh wood without microorganisms, on wood softened by acids or alkalies, or on wood covered with agar. Apparently, the softness of the substrate is not the only factor inducing the larvae to settle. Wood attacked by marine Ascomycetes and Deuteromycetes and by terrestrial Basidiomycetes was distinctly favored over wood without fungi. The process of choice consists of a "testing" of the surface by the crawling larva, which leaves the substrate if it is unsuitable for settlement. Cellulolytic bacteria also appear to improve the wood surface for penetration by larvae. Wooden panels submerged in a zone of low oxygen content were attacked by bacteria and deep-sea borers (*Xylophaga*), whereas fungi were absent (Kohlmeyer, 1969a).

The experiments by Kampf *et al.* (1959) suggest that the wood surface is "conditioned" by the growth of microorganisms, which partly decompose the cellulose, soften the outer layers, and permit penetration of the soft-shelled larvae. It is unknown if larvae at this stage use the softened wood as food or if they just use it for shelter until metamorphosis is completed. The softer spring wood is usually preferred over the denser late wood for settlement by larvae of *L. pedicellatus*.

B. Boring Crustacea

The most important marine wood destroyers in this group belong to the Isopoda in the families Limnoriidae and Sphaeromatidae (Kühne, 1971). Especially the Limnoriidae or gribbles are known to cause considerable economic losses in marine wooden structures. The gribbles form tunnels closely under the wood surface, produce respiratory pits, and, thus, come in contact with cellulolytic fungi and bacteria growing in the same outer wood layers. For many years there has been a controversy over the question whether *Limnoria* bores in wood solely for shelter or uses the wood in its diet, and is possibly assisted by microorganisms in the breakdown of the cellulose. Yonge (1927) postulated the absence of cellulase production in *Limnoria,* but later authors (Ray and Julian, 1952; Ray, 1959a,b) demonstrated an endogenous cellulase in the gribble and thus its ability to digest wood without the assistance of symbiotic microorganisms. Finally, Ray (1959c) and Sleeter *et al.* (1978), the latter by using electron microscopic techniques, did not find any evidence of bacterial symbionts in the digestive tract of *Limnoria*. It appears to be clear, therefore, that *Limnoria* species digest wood with their own enzymes. They may, however, require additional nutrients that are not available in this substrate but could be derived from cellulolytic microorganisms living in the wood.

Results of preliminary experiments with *Limnoria tripunctata* (Becker *et al.*, 1957; Becker, 1959) indicated that this species was unable to live on sterile wood, but depended on cellulolytic marine fungi to digest wood. Repeated series of extensive experiments (Kohlmeyer *et al.*, 1959) showed that the gribble could indeed reach its maximum lifetime on a diet of sterile wood but did not reproduce under these conditions. We would like to point out that the earlier publications by Becker and co-workers (Becker *et al.*, 1957; Becker, 1959) are superseded by the paper of Kohlmeyer *et al.* (1959), but data from the former publications are, unfortunately, still the only ones discussed (e.g., Johnson and Sparrow, 1961; Eltringham, 1971). The following will summarize the results of our final experiments to clarify any misunderstandings.

Mature *L. tripunctata* were kept on a variety of substrates, with and without fungi, and transferred once or twice weekly onto new media. The animals were observed until they died naturally, which occurred in some individuals after more than 1.5 years. Substrates used were softwoods (pine and spruce), hardwoods (alder, beech, and oak), filter paper, cellophane, sunflower pith, date endosperm, starch, coconut fat, mycelia of terrestrial and marine fungi, and wood attacked by representatives of both groups of fungi. The survival rate of *L. tripunctata* on these nutrients was tested and compared with starving animals that survived without food at 20°C for 7.5 weeks (Fig. 32). The animals reached their natural age on sterile hardwoods, and the lifetime was not extended when they were given fungus-infested wood of the same kind. Mycelium of marine and terrestrial fungi without wood was eaten, but was not sufficient as a source of food over a longer period. Sterile softwood extended the survival of the animals two or up to five times compared with the starved controls. A considerable extension of the lifetime occurred on softwood attacked by marine fungi, but the longest survival was found on softwood decomposed by the terrestrial white-rot fungus *Pleurotus ostreatus* (Fig. 32). The results demonstrate without doubt that *Limnoria* can live on fungus-free wood; however, the lifetime is extended if fungi are present in the wood. The most important finding is that *L. tripunctata* does not reproduce on any of the substrates unless marine fungi are included in the diet. Most probably, the fungi supply proteins, vitamins, and oils, which may be required for reproduction. Dietrich and Höhnk (1958) found oil and vitamins D and E in a marine isolate of *Ceratostomella* sp., but obligate marine fungi have not been analyzed thus far.

The failure of Stevenson (1961) to demonstrate the presence of chitinase in *Limnoria* indicates that the animal is unable to digest chitin in fungal cell walls. This supposition is supported by the results of our experiments in which *L. tripunctata* was kept on an exclusive diet of

Start of experiment	Significance of mean value compared with starved control (%)	Average lifetime compared with starved control = 1
XI 1957		7.5 (11) Starved control
XI 1957	> 99.9	15.1 (22) Zalerion marit.-mycelium, dead
XI 1957	> 99.9	12.5 (24) Zalerion marit.-mycelium, living
II 1958	95...99	12.6 (30) Corollospora marit.-mycelium, living
XI 1957	99...99.9	10.9 (20) Coniophora-mycelium
XI 1957	95...99	11.3 (23) Lenzites-mycelium
XI 1957	99...99.9	16.0 (40) pine without fungi, sterile
XI 1957	> 99.9	29.8 (62) pine + Dictyosporium pelagicum, living
XI 1957	-	7.6 (14) pine + Lentinus, sterile
XI 1957	95...99	11.7 (24) pine + Coniophora, sterile
XI 1957	95...99	12.3 (25) pine + Poria, sterile
XI 1957	> 99.9	14.1 (23) pine + Lenzites, sterile
XI 1957	99...99.9	21.4 (35) pine + Paxillus, sterile
XI 1957	> 99.9	43.7 (>96) pine + Pleurotus, sterile
XI 1957	99...99.9	18.0 (45) pine + Lent., dead, Dictyspor. pel., living
XI 1957	99...99.9	19.2 (42) pine + Con., dead, Dictyospor. pel., living
XI 1957	> 99.9	25.5 (>96) pine + Pax., dead, Dictyospor. pel., living
XI 1957	> 99.9	25.7 (>96) pine + Por., dead, Dictyospor. pel., living
XI 1957	> 99.9	29.9 (>96) pine + Lenz., dead, Dictyospor. pel., living
XI 1957	95...99	32.6 (81) pine + Fleur., dead, Dictyospor. pel., living
XI 1957	> 99.9	31.0 (80) beech without fungi, sterile
XI 1957	> 99.9	30.0 (63) beech + Dictyospor. pelagicum, living
XI 1957	> 99.9	25.4 (>96) beech + Polystictus, sterile
XI 1957	> 99.9	29.3 (89) beech + Pol., dead, Dictyospor. pelagicum, living

Average lifetime compared to starved control = 1
Longest survival in weeks
Average lifetime in weeks

Fig. 32. Survival of the gribble, *Limnoria tripunctata,* in feeding experiments with different substrates at 20°C. See the text for explanation. After Kohlmeyer *et al.* (1959).

fungal mycelia and spores (Kohlmeyer, 1958a; Kohlmeyer *et al.*, 1959). The fecal pellets consisted of compressed, undigested hyphae and co-nidia and ascospores, which were both able to germinate after passing through the digestive tract. The fungal cell walls appear not to be affected by the digestive process, but cell contents of hyphae damaged during feeding can probably be used by the gribble. In addition, the wood cellulose is only incompletely digested after passing through the gut of *Limnoria,* as can be observed in fecal pellets under polarized light; the pellets still contain refractive cellulose particles (Kohlmeyer *et al.*, 1959). Seifert (1964) determined that *L. tripunctata* utilized only 30% of the consumed wood and decomposed 40 to 50% of the cellulose.

The controversy over possible relationships between marine fungi and *Limnoria* is not solved, but some of the open questions (Eltringham, 1971) appear to be answered. Ray (1959c) and Ray and Stuntz (1959), who refuted such a relationship, stated that a fundamental premise for such an association would be their general joint occurrence in the normal environment. These authors wrote that they did "not find that marine fungi are universal inhabitants of submerged or floating wood" and "that there are occasional hyphae cannot be denied, but we find the evidence grossly insufficient to claim a universal or even a common occurrence of marine fungi on or in submerged wood." Ray (1959c) submerged test panels of fresh Douglas fir and western yellow pine in the Puget Sound (Washington) and in the Mediterranean (Naples) and could not verify the invasion by marine fungi after 8 and 5 months, respectively, although *Limnoria* was found. There is strong reason to believe that marine fungi were present in the wood without being recognized. All wood specimens submerged in marine habitats are eventually attacked by cellulolytic fungi, except in rare areas of extremely low oxygen content (Kohlmeyer, 1969a). This fact is supported, for instance, by research on wood-inhabiting fungi carried out in the Puget Sound (Kohlmeyer, 1960, 1961a) and in the Mediterranean (Kohlmeyer, 1963c; Corte, 1975). Possibly, Ray and Stuntz (1959) searched for fungal reproductive structures and superficial hyphae and overlooked the decay pattern caused by soft-rot fungi, namely, tunnels within the secondary cell walls. In our experience, wood containing populations of *Limnoria* always harbors fungi, even though free mycelium is often not present because of the browsing activities of the gribble. However, soft-rot decay can be observed regularly inside wood cell walls next to *Limnoria* galleries.

The other hypotheses, that *Limnoria* is unable to survive in sterilized wood and does attack wood in nature only after it has been "preconditioned" (Becker *et al.*, 1957; Becker, 1959), are no longer tenable, as explained above. However, fungi seem to supplement the cellulose diet of *Limnoria* with some nutrients that permit reproduction. Perhaps these nutrients can also be contributed by bacteria, a possibility suggested by Ray (1959c). The experiments by Kohlmeyer *et al.* (1959) were aimed at elucidating the role of fungi in the nutrition of *Limnoria,* and other microorganisms were not considered at that time. The regular transfers of the gribbles onto fresh substrates during the experiments eliminated the slower growing fungi but not bacteria and protozoa, which thrive after 3 days on the fecal pellets of *Limnoria* (Kohlmeyer *et al.*, 1959). Eltringham

(1971) speculated that such experiments were possibly concentrating on the wrong group and that not fungi but bacteria may be the source of essential nutrients for borers. We agree that the role of bacteria in the nutrition of marine borers is unknown and should be clarified. However, there is no doubt that fungi have a positive influence on the development of the borers. This conclusion was also drawn by Schafer (1966), who found a prolonged survival time of *Limnoria* spp. on fungi-infested wood (>20 weeks), as compared to the maximal survival time of animals kept on sterile pinewood (15 weeks).

Some reports indicate that *Limnoria* has a positive effect on the growth of marine fungi. Corte (1975) speculated that ascocarps of *Lulworthia* sp. develop only after the wood is perforated by gribbles. Conidia of *Cirrenalia macrocephala* are produced preferably in abandoned tunnels of *Limnoria* (Kohlmeyer, 1958a, 1962a). An endozoic dispersal of conidia by *Limnoria* and other wood-inhabiting invertebrates (Fig. 33) is possible, as they are able to germinate after passing through the digestive tract (Kohlmeyer *et al.*, 1959).

The clarification of nutritional relationships between wood borers and marine fungi is of basic biological interest, as well as of great economic importance. Research in this field has ceased for more than a decade, and

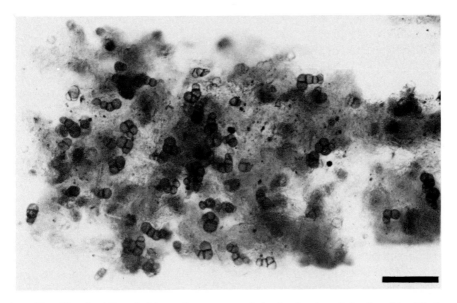

Fig. 33. Conidia of *Cirrenalia macrocephala* in fecal pellet of unidentified surface-browsing invertebrates from wood; specimen collected after animals had left the substrate (Herb. J.K. 3697; bright field; bar = 50 μm).

discussions of the problems involved have not added any new data (e.g., T. W. Johnson and Sparrow, 1961; Eltringham, 1971). Nutritional experiments, such as those carried out in Berlin–Dahlem (Kampf *et al.*, 1959; Kohlmeyer *et al.*, 1959), are very time-consuming, but it is desirable that similar experiments should be conducted using other species of wood borers and fungi as well as bacteria.

II. SALT-MARSH AMPHIPODS

Investigations on freshwater-detritus feeders have shown that amphipods assimilate higher percentages of dry weight, protein, and energy of ingested fungal mycelium than when feeding on elm or maple leaves. The animals prefer leaves colonized by microorganisms, especially leaf areas with high hyphal concentrations, over freshly fallen or sterile leaves (Bärlocher and Kendrick, 1973, 1975a,b). A similar situation can be expected in tidal salt marshes, where large amounts of plant detritus accumulate and serve as nutrients for detritivores. The nutritional value of marsh-plant detritus is increased by fungi (see Chapter 11) and these microorganisms are grazed upon by invertebrates, for example, the amphipod *Gammarus mucronatus* (R. V. Gessner, unpublished). In preliminary experiments, Gessner was able to keep these amphipods for 2–3 months on an exclusive diet of 17 different species of marine fungi. The animals ingested hyphae and fungal spores, and spores were found in the fecal pellets. The fungal diet appears to provide all essential nutrients because the amphipods reproduce in these cultures.

III. NEMATODES

Marine benthic ecosystems contain large numbers of nematodes, which are probably the most abundant animals in sublittoral sediments. The nutritional requirements for many species are unknown, but Meyers *et al.* (1963a, 1964a) have demonstrated that a stylet-bearing nematode, *Aphelenchoides* sp., actively feeds on fungal hyphae by draining the cytoplasm from the hyphae. This animal was able to grow and reproduce in culture on a sole diet of mycelia deriving from a wide variety of fungi. Among the fungi tested, marine Ascomycetes and Deuteromycetes (e.g., *Halosphaeria mediosetigera, Lulworthia* sp., *Dendryphiella arenaria, Trichocladium achrasporum,* and *Zalerion maritimum*) were extremely effective as a nutrient source, whereas isolates of terrestrial fungi (e.g., genera *Aspergillus, Cladosporium, Fusarium,* and *Pestalotia*) resulted in a very low utilization factor (Meyers *et al.*, 1964a). In experiments with other nematodes, Meyers and Hopper (1966, 1967) did not observe any feeding on the fungi, but the animals were attracted by fungal mats growing on cellulose. As fungal mycelium alone is ineffective, the authors postulated that the attraction is probably a response to substances emitted from the fungal–cellulose complex or from other associated microorganisms such as bacteria, diatoms, ciliates, or protozoa. At any rate, nematodes accumulate and reproduce in substrates covered by fungal hyphae, for instance, cotton–cellulose filters (Meyers and Hopper, 1967) or filter-paper disks submersed in natural habitats (Meyers and Hopper, 1966). The nematode populations in these materials consisted almost entirely of gravid females of mostly one species, *Metoncholaimus scissus*.

Accumulations of nematodes can be observed occasionally in decaying fruiting bodies of marine fungi. Nematode colonies developed, for example, in old ascocarps of *Lulworthia* sp. and probably contributed to the decomposition of the fungal material (Kohlmeyer, 1958a).

IV. MITES

Mites are known as nuisances, causing contaminations in fungal and bacterial cultures in the laboratory. Their role in marine habitats is not fully recognized and only a few cases are reported where mites appear to feed on fungal mycelium, for instance, on detritus along beaches (Schuster, 1966) and on decaying *Spartina* stems and leaves (Gessner *et al.*, 1972). Dead mites may be decomposed by fungi, for example, *Periconia prolifica*, which grew and sporulated on and in these animals attached to old shipworm tunnels (Kohlmeyer and Kohlmeyer, 1977).

V. MOLLUSCA

Observations on fungus-infested algae in nature indicate that marine mollusks graze preferentially on spots containing fungal hyphae, stromata, and fruiting bodies (J. Kohlmeyer, unpublished). Examples of such food sources are *Didymosphaeria danica* attacking *Chondrus crispus* and *Phycomelaina laminariae* producing black spots on *Laminaria* spp. R. Hooper (Newfoundland; personal communication) sent us mechanically damaged specimens of the calcareous red alga, *Clathromorphum compactum,* also infected by *Lulworthia kniepii* and wrote that chitons (*Ischnochiton* and *Tonicella*), gastropods (*Littorina*), and other animals often graze selectively on dead regions of the host, causing rapid and severe erosions of infected areas. In this case, the algae also appear to be grazed upon after attack by the fungus.

VI. FUNGI USED AS FEEDS IN MARICULTURE

Preliminary experiments by Newell and Fell (1975) have shown that marine fungi can convert agricultural by-products (wheat bran, sugar cane bagasse, and straw) into inexpensive microbial–detrital complexes, which can be used to rear shrimp in aquaculture.

The few examples of animal–fungal relationships show how little is known about marine fungi as possible food sources for benthic animals. Extensive searches for fungal particles in digestive tracts of the interstitial fauna or of salt-marsh invertebrates would probably reveal further instances of nutritional connections.

20. Ontogeny

The whole course of development from spore to mature fruiting body is known for only a few species of marine fungi. Special attention has been paid to certain aspects of life histories, for instance, to the development of ascospores and their appendages (e.g., T. W. Johnson, 1963a–d; T. W. Johnson and Cavaliere, 1963; Kirk, 1966; Moss and Jones, 1977) and to the morphology of mature ascocarps (Cavaliere, 1966a–c; Cavaliere and Johnson, 1966b). The ontogeny of ascocarps, ascospores, basidiocarps, and basidiospores of some marine fungi is described in the following sections.

I. ASCOCARP ONTOGENY

Ontogenetic studies of ascocarps of marine fungi require the availability of all stages of development, from ascocarp initials to mature, ascospore-containing fruiting bodies. Wood or other dead organic matter in natural habitats seldom harbors both young and mature stages of one fungus. In addition, several species often occur together on one substrate, and it cannot be determined with certainty to which fungus a certain ascocarp initial belongs. Therefore, material is usually obtained by isolating saprobic species in unifungal culture or by collecting in nature host-specific plant parasites or symbionts, which mostly occur by themselves, often in characteristic spots or stromata.

A. Pyrenomycetes

G. Feldmann (1957) was first to describe the ascocarp ontogeny of a marine fungus, *Chadefaudia marina*. This parasite on the red alga *Rhodymenia palmata* forms a young ascocarp consisting of a wall surrounding nutritional, pseudoparenchymatous cells and a central, ascogonial tissue. In a later stage the ascogenous cells are arranged at the periphery of the ascocarp center, and the asci develop in a hymenium along the lower two-thirds of the inner wall (G. Feldmann, 1957; Kohlmeyer, 1973b). The central pseudoparenchyma disappears as soon as the asci originate, and the thin ascus walls dissolve early and release their spores into the cavity of the fruiting body. A similar type of ascocarp development occurs in the wood inhabitants *Halosphaeria mediosetigera* and *Ceriosporopsis circumvestita* (Kohlmeyer and Kohlmeyer, 1966) and probably in all other members of Halosphaeriaceae. Reproductive hyphae of *H. mediosetigera* and *C. circumvestita* coil up to form simple protoperithecial initials, which develop into protoperithecia. These pseudoparenchymatous balls grow into young ascocarps with a differentiated peridium enclosing a sterile pseudoparenchyma. Hyphae making up the wall form a *textura intricata,* which later develops into a *textura angularis.* Ascogonial cells originate in the center of the pseudoparenchyma and can be distinguished from the surrounding sterile cells only by a denser cytoplasm staining deeply in hematoxylin. These ascogonial cells give rise to indistinguishable ascogenous cells, which in turn form the asci. The sterile pseudoparenchymatous cells of the center are compressed by the growing asci and dissolve. Eventually, the neck develops from a meristematic zone in the apical area of the ascocarp between the pseudoparenchyma and the peridium. Croziers were not observed in *H. mediosetigera* and *C. circumvestita.*

The ascocarp ontogeny of *Lulworthia medusa,* a saprobe in *Spartina townsendii,* was examined by Lloyd and Wilson (1962). These observations were made on ascocarps developing in both culture and host tissue from nature. Essentially, the ontogeny is the same as in the foregoing species, and *Ceriosporopsis halima,* a ubiquitous saprobe on cellulosic substrates in the sea, also has the same type of development (Wilson, 1965). All species of *Ceriosporopsis, Halosphaeria,* and *Lulworthia* examined thus far show a *Diaporthe*-type (Luttrell, 1951) ascocarp ontogeny in which the central pseudoparenchyma derives from the inner wall layer.

B. Loculoascomycetes

Wagner (1965) investigated the life history of *Leptosphaeria obiones,* a common saprobe in culms of *Spartina* spp. and more rarely in wood. Ascocarp initials, subglobose masses of pseudoparenchymatous cells, develop between the epidermis and the air chamber of the host. The next stage consists of young fruiting bodies with pseudoparaphyses surrounded by a peridium in the lower half and meristematic cells at the top. The meristem gives rise to the ostiolar pseudoparenchyma. Ascogenous hyphae appear at the time of ostiolar inception, and ascus formation initiates with the crozier. Young bitunicate asci, which mature successively, grow up between the pseudoparaphyses. Wagner (1965) supposed that pseudoparaphyses originated from ascogenous hyphae and concluded that *L. obiones* follows Luttrell's (1951) *Pleospora* type of development. Spermogonia occur before the ascocarps, both in nature and in culture, and Wagner (1965) observed that ascocarps occurred only in cultures infested by mites. The author assumed that mites were the vectors of spermatia, bringing them in contact with receptive hyphae.

Didymosphaeria danica, a parasite on *Chondrus crispus*, was examined by Wilson and Knoyle (1961). These authors were able to observe only a few developmental stages of spermogonia and ascocarps in the host tissue. The youngest recognizable stage of ascocarps consists of a compact mass of interwoven hyphae between cortex and fruiting area of the alga. This small stroma enlarges to form pseudoparaphyses, ascogenous hyphae, a peridium of more or less loosely bounded hyphae, and a clypeus. A neck is formed in the upper part of this globose ascocarp and the mature fruiting body assumes a flasklike shape. Bitunicate asci grow up between the pseudoparaphyses, which form wavy strands around the asci or hang down from the top and sides of the locule.

The last ontogenetic studies of ascocarps are those by F. C. Webber (1967) on *Mycosphaerella ascophylli,* the endophyte of *Ascophyllum nodosum,* and include observations on the life history and biology of the fungus. At first a weft of hyphae can be observed in the cortex of the host receptacle. The weft enlarges to form a solid plectenchymatous stroma with three to five ascogonia and trichogynes. The ascogonia are probably spermatized by filiform spermatia, which originate in simultaneously developing spermogonia. The central stromatic tissue breaks down and makes room for the ascogonial coils, ascogenous hyphae with croziers, and finally the young bitunicate asci. Pseudoparaphyses and periphyses were not observed by F. C. Webber (1967), but they occurred in specimens examined by Kohlmeyer (1968e) (see Figs. 86b–86d).

II. ASCOSPORE ONTOGENY

Ascospores are the major structures for identification of marine Ascomycetes, as evidenced by the keys included in this treatise. However, identification should not be confused with classification. Early describers of new marine fungi used mainly ascospore characters to classify different taxa, often neglecting, for instance, ascocarp morphology. In the meantime, it is well recognized by most workers in this field that *all* characters of a taxon should be used to make meaningful classifications possible. Therefore, when ascospores are concerned, it is necessary to consider not only the morphology of mature ascospores and their appendages, but also the mode of origin of these spores, because analogous structures may arise from basically different material. Examples are *Halosphaeria appendiculata* and *Ceriosporopsis calyptrata,* which both have ascospores with similar-appearing apical and lateral appendages (see Figs. 48 and 58a); however, these appendages originate by different processes.

Linder (1944) was the first to indicate the origin of appendages of a marine fungus when he described the rupturing and shedding of the "perispore" from ascospores of *Peritrichospora (Corollospora) lacera,* resulting in the formation of lateral and apical cilialike processes. Some of the later studies concentrated mainly on the mode of appendage formation in marine Ascomycetes (Kohlmeyer, 1966b; Lutley and Wilson, 1972a,b), but extensive investigations by T. W. Johnson (1963a–d) and T. W. Johnson and Cavaliere (1963) considered the complete ascosporogenesis of 18 marine Ascomycetes. The usefulness of stains, for example, Delafield's hematoxylin and others, to demonstrate appendage origin and morphology was emphasized by Kohlmeyer and Kohlmeyer (1964), and Wilson (1965) applied cytochemical methods to demonstrate development of spore ornamentations in *Ceriosporopsis halima.* However, Kirk's (1966, 1976) important research on the cytochemistry of eight marine Ascomycetes was the first thorough investigation revealing the chemical composition of the different parts of spores, in particular, the appendages. Methods to show appendage origin and morphology became even more refined with the use of the transmission electron microscope (Lutley and Wilson, 1972a,b; E. B. G. Jones, 1976) and finally with the scanning electron microscope (Moss and Jones, 1977).

The earliest recognizable signs of ascospore formation consist of aggregates of refractive bodies, guttules, or granulated matter in the cytoplasm of the ascus. These first stages of ascosporogenesis, termed "ascospore rudiments" by T. W. Johnson (1963a), do not show any distinct walls. Kirk (1966, 1976) demonstrated that glycogen content within the spores decreased during development, whereas neutral oil accumulated. Apparently, glycogen is transformed into oil, which coalesces into guttules. These droplets are surrounded by Schiff-positive, alcohol-soluble substances.

The next stage is characterized by the delimitation of developing ascospores from the ascus cytoplasm by two unit membranes, which include the wall proper (Lutley and Wilson, 1972a,b). The outer membrane becomes the spore membrane and the inner one the plasmalemma. Such phases with completed walls are "ascospore initials" *sensu* T. W. Johnson (1963a). The first wall layer, or epispore, developing between the unit membranes, is composed of chitin, acid mucopolysaccharides, protein, and lipid (Kirk, 1966). The mesospore develops between epispore and plasmalemma (Lutley and Wilson, 1972b), and the final outer wall layer, or exospore, is secreted toward the outside by the epispore. The chemical composition of the mesospore is not clear (Kirk, 1966; Lutley and Wilson, 1972b), but protein, neutral Periodic acid Schiff (PAS)-negative carbohydrates, and sometimes uronic acids are contained in the exospore. Ascosporogenesis is concluded with nuclear divisions, in some species followed by centripetal (Kirk, 1976) or centrifugal (Wagner, 1965) formation of a septum by the mesospore, and finally with appendage formation. Mature ascospores of *Ceriosporopsis halima* and *Corollospora maritima* contain glycogen, large oil globules, and volutin bodies (Lutley and Wilson, 1972a,b). It should be pointed out that the foregoing descriptions of ascospore ontogeny and cytochemistry are based on only a few species. Other types of development may well exist among marine Ascomycetes.

Ascospore appendages of marine fungi can be formed by two main processes: by fragmentation of the exospore or epispore or as outgrowths of the epispore (e.g., T. W. Johnson, 1963a–d; T. W. Johnson and Cavaliere, 1963; Lutley and Wilson, 1972a,b; Kirk, 1966; Kohlmeyer, 1961a; Wilson, 1965). A supposed third way of appendage formation, namely, the retention of ascus cytoplasm on ascospores of *Corollospora maritima* (T. W. Johnson, 1963b), was not confirmed (Kirk, 1966; Kohlmeyer, 1966b). Examples of two well-examined and common species shall suffice to explain the different types of appendage origin. Mature ascospores of *Corollospora maritima* are two-celled and are provided at each end with a spinelike appendage (see Fig. 54a). The tips of these spines bear thin processes, at first caplike and then fiberlike, and similar processes are arranged around the equator of each spore. The apical spines are outgrowths of the wall at both ends of the ascospore initial (Kirk, 1966). A cytoplasmic membrane at the base of the spine separates it from the cytoplasm of the spore, and mesospore deposition fills the cavity with chitinous material. The outer wall layer, or exospore, covers the whole ascospore, even the apical spines, and at maturity this exospore ruptures at the base of the spine and peels back toward the equator and toward the tips of the spines (see Fig. 8), thereby forming the characteristic lateral and apical irregular strips, which appear as fibers under the light microscope (Kirk, 1966; Kohlmeyer, 1966b; Lutley and Wilson, 1972a; E. B. G. Jones, 1976).

The apical ascospore processes of *Ceriosporopsis halima* are mucilaginous because chitin is absent. They are modified exosporic remnants that usually grow straight out from apical caps of the epispore (Kirk, 1966, 1976; Lutley and Wilson, 1972b). It should be pointed out that the apical appendages are not secretions exuded from the ascospore cell. Late in development there is a sheath produced that surrounds the spore except for the tips, where it forms pouches containing the appendages. Kirk (1966) and Lutley and Wilson (1972b) have also shown that the spore sheath derives from the epispore by deposition of polysaccharides surrounding a network of proteinaceous tubules. While appendages and sheath are chemically similar, they are not identical. Cytochemical evidence and electron microscopy have revealed numerous fibrillar structures in the apical processes (Kirk, 1966; Lutley and Wilson, 1972b). These tubules probably transport raw materials from the cytoplasm of the spore, transforming the materials into chitin and mucopolysaccharides (Kirk, 1976).

III. BASIDIOCARP ONTOGENY

Most work on the morphology, basidiocarp ontogeny, and origin of basidiospores of filamentous marine Basidiomycetes has been carried out by Doguet (1962a,b, 1963, 1967, 1968, 1969). Detailed information on the development of fruiting bodies exists only for the Gasteromycete *Nia vibrissa*. Doguet (1969) described the following pattern of growth, starting with an embryonic stage and followed by a filamentous stage, a protocyst stage, a basidial and spore-forming stage, and a final stage.

The development begins with the appearance of a minute white tuft of delicate hyphae, borne on a small mycelial cushion. This embryonic basidiocarp of less than 100 μm in height grows and becomes round, but remains filamentous. The superficial hairs are already present. Next, the protocysts appear in the center, that is, inflated bulbous tips of ramified hyphae, 15–25 μm in diameter, containing a large vacuole. Fruiting bodies of the protocyst stage have a small pedicel, a peridium of the *textura angularis* type covered by external hairs followed toward the centrum by loose tangential hyphae forming a *textura intricata,* and in the centrum an irregular network of hyphae and protocysts. When the basidiocarp reaches about 1 mm in diameter, basidia originate from new ramifications of the central hyphae and are characterized by a dense cytoplasm turning dark in a variety of stains. Basidia never form a hymenium, but are always irregularly dispersed throughout the center. While basidia produce the first basidiospores, the protocysts loose their turgor, become empty, and collapse. Eventually they disappear and make room for the spores. In the final stage, the center of the basidiocarp consists only of basidiospores, with hyphae and basidia rarely present. Except for the spores, all parts of the gleba vanish, probably by dissolution of the cell walls.

IV. BASIDIOSPORE ONTOGENY

Doguet (1962b, 1969) described in detail the development of basidia and spores in *Nia vibrissa* and *Digitatispora marina*. Young basidia of *N. vibrissa* differ from the other elements of the gleba by their dense, strongly staining cytoplasm and binucleate condition. Mature basidia are clavate, 1.5 μm in diameter and up to 50 μm long. After karyogamy, the diploid nucleus migrates toward the tip of the basidium, which inflates, followed by meiosis that results in the formation of four haploid nuclei. A vacuole develops at the base of the bulbous part of the basidium, pushing the cytoplasm upward and later into the spores. At the same time, four to eight tubelike sterigmata are produced on the apex of the basidium, and their tips inflate to produce the spores. Each spore then forms the five characteristic slender appendages. The four nuclei divide simultaneously and some of the resulting eight nuclei move through the narrow sterigmata into the spores. During this passage they become elongate, but regain their globose shapes inside the spore. Most basidiospores receive one nucleus; they rarely contain two or even three nuclei. The remaining nuclei probably perish within the spent basidium. Sporogenesis is concluded with the formation of a septum between basidiospore and sterigma. When the spores are released from the basidium, part of the sterigma remains attached to the base of the spore.

The second Basidiomycete, examined thoroughly by Doguet (1962a,b, 1963), is *Digitatispora marina*. The resupinate basidiocarps of this species form grayish or white cushions on the wood surface. They are covered with the hymenium, consisting of cylindrical basidia and tetraradiate basidiospores. Apical cells of binucleate hyphae on the surface of the fruiting body develop into nonseptate basidia. They swell slightly and, after the second meiosis, apically form four cylindrical extensions, which are not sterigmata but constitute the basal cells of the future basidiospores. The four haploid nuclei are pushed toward the apex of the basidium by a large vacuole and mitosis occurs, resulting in eight nuclei. Each spore receives one nucleus and the leftover ones degenerate. After the main basidiospore cells have reached a length of about 30–35 μm, three papillae appear at their apices. These papillae elongate synchronously until they almost equal the basal cell. The nucleus remains in the lower cell of the basidiospore, and the side arms do not become closed off by septa, according to Doguet (1963). No trace of the spore insertion remains on the rounded tip of the basidium after the spores are released. In culture, and sometimes in nature, basidiocarps of *D. marina* produce abnormal basidiospores with more than three side arms or with bifurcate projections (Doguet, 1963; Kohlmeyer, 1963d).

V. CONIDIAL ONTOGENY

In recent years conidium development has become more and more important in the classification of imperfect fungi (Kendrick, 1971a). The conidiogeneses of marine Deuteromycetes are listed in the descriptive part of this treatise (pp. 464–545) as far as they are known. There is only one paper dealing with detailed ontogenetic investigations of marine Fungi Imperfecti, namely, that by Cole (1976) on *Asteromyces cruciatus* and *Zalerion maritimum*. This author used time-lapse photography with the light microscope, as well as scanning electron microscopy to observe conidial development.

Vegetative hyphae of *A. cruciatus* form conidiogenous cells as lateral branches (Fig. 34, 1). A terminal primary conidium develops, and, about 3.5 hr after maturation of this conidium, an area of the conidiogenous cell below the base of the conidium inflates and becomes the new fertile region (Fig. 34, 4). A transverse septum delimits the upper fertile area from the rest of the conidiogenous cell (Fig. 34, 5). The following conidium develops 1 hr later as a lateral cylindrical outgrowth from the fertile region, and its tip grows into a conidium during the next 4 hr (Fig. 34, 5–8). The cylindrical bases between conidia and the conidiogenous cell become denticles. The next conidium is initiated 2.75 hr later on the opposite side of the preceding conidium (Fig. 34, 9), and this conidium is fully developed after 6.25 hr. More secondary conidia originate in the same manner, and, eventually, one or more lateral whorls of mostly four conidia are produced. Cole (1976) suggests that conidia secede by a fracture through the denticle wall. However, conidia found in nature, for instance, in foam along the shore, usually occur in aggregates attached to the conidiogenous cell. Therefore, we offer a different interpretation of "conidia" in *A. cruciatus,* namely, that these aggregates are actually multicelled conidia, comparable to those of *Clavariopsis bulbosa* and *Orbimyces spectabilis*. Cole (1976) speculated on the possible presence of two transverse septa in each denticle without actually observing them, since thin sections would have been required to confirm their occurrence. If such septa should be found, we would favor the notion even more that the

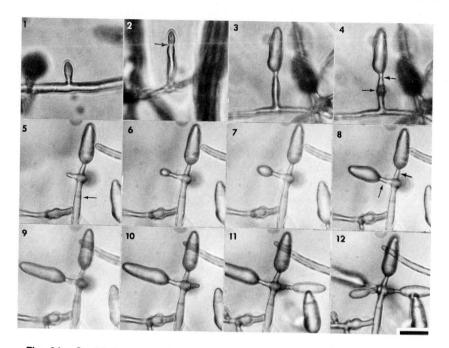

Fig. 34. Conidial ontogeny of *Asteromyces cruciatus*, taken with 35-mm time-lapse photomicrography. Arrowhead in 2 indicates constriction below conidium initial; arrowheads in 4 and 8 point to denticle of primary conidium; arrow in 4 indicates conidiogenous cell; arrow in 5 locates transverse septum delimiting upper fertile area of conidiogenous cell; arrow in 8 points to denticle of secondary conidium. Further explanations in the text (bar = 10 μm). From Cole, *Marine Biol.* **38,** 149 (1976), Fig. 1. Reprinted with permission.

"conidia" of *A. cruciatus* be considered multicellular dispersal units with a large central, bulbous cell to which smaller lateral and apical cells are attached. However, the present generally accepted interpretation of conidium ontogeny in *A. cruciatus* assumes a blastic, basipetal, and whorled pattern (Cole, 1976), as described above.

Cole (1976) also investigated conidiogenesis in *Zalerion maritimum*. The vegetative hyphae produce short erect conidiogenous cells that swell at the apex. After 1.75 hr, the tip of the conidium initial starts to elongate and continues to expand for 1.75 hr. At the same time, septation starts and a change in the direction of elongation from left to right takes place over the following 6 hr. Cole (1976) assumes that the shift in direction of growth corresponds to the position of apical clusters of secretory vesicles, which add wall components to the growing tip. Cells of almost equal length are produced by septa that are formed at the same distance behind the growing apex of the young conidium. Finally, the cells become slightly inflated. Secession occurs by separation of the conidium from the conidiogenous cell, but sometimes the latter may break off the hypha and remain connected to the conidium. Cole (1976) determined that *Z. maritimum* produces solitary, terminal, blastic, helicoid conidia at the tips of determinate conidiogenous cells.

In concluding this chapter on the ontogeny of marine fungi we would like to point out again the scanty knowledge in this area. In particular, ascocarp ontogeny, cytology, and conidiogenesis have been examined in only a few species. One should also be cautioned against overemphasizing a single character in the generic placement of species, for instance, the origin of ascospore appendages; instead, all characters, such as ascocarp and ascus morphology, as well as chemical composition of appendages, should be considered.

21. Physiological Processes and Metabolites

Numerous papers dealing with physiological processes of marine fungi have been published since Barghoorn (1944) conducted the first experiments on the behavior of eleven marine wood-inhabiting Ascomycetes and Deuteromycetes in pure culture. Nevertheless, we are still unable to explain which physiological characters enable a fungal species to germinate, grow, and reproduce in a marine habitat. Results of physiological studies are incomplete and often confusing, "making a comparative review not only difficult but possibly meaningless" (G. C. Hughes, 1975). In view of these difficulties and considering the fact that processes occurring in nature often cannot be explained from laboratory studies, we discuss in the following only a selection of pertinent papers on the physiology and biochemistry of higher marine fungi.

I. PRODUCTION OF ENZYMES

A. Cellulase

In the past, experimental work with marine fungi has been restricted mainly to wood-inhabiting species, and a large number of papers have dealt with the production of enzymes involved in the breakdown of native and other celluloses. Barghoorn (1944) was first to use marine Ascomycetes and Deuteromycetes to demonstrate their ability to grow on wood flour and regenerated cotton cellulose by measuring the rate of radial growth on agar medium. Barghoorn emphasized that "the ability of an organism to attack cellulose is not always clearly demonstrable by its development on synthetic media containing regenerated cellulose." This reservation must be kept in mind when the results of other experiments with marine fungi grown on a variety of chemically treated cellulose products are assessed.

Extensive experiments on the cellulolytic activity of marine fungi have been carried out, in particular, by S. P. Meyers and co-workers at the University of Miami. Meyers and Reynolds (1959b,c, 1960b, 1963), grew marine Ascomycetes and Deuteromycetes on cellulosic substrates and obtained cell-free filtrates from these liquid cultures. The liquid was added to various cellulose materials (e.g., treated cotton, ashless cellulose powder, ground balsa wood, and sodium carboxymethyl cellulose), and the reducing sugars were determined by the Nelson–Somogyi method. Other tests measured the reduction in tensile strength of cordage or the weight loss of cellulose substrates after attack by marine fungi (e.g., E. B. G. Jones and Irvine, 1972; Meyers, 1968a; Meyers *et al.*, 1960; Meyers and Scott, 1968). Schaumann (1974c) compared a viscosimetric method using sodium carboxymethyl cellulose with the clearing of cellulose-containing agar plates to demonstrate the production of cellulase by 20 marine fungi. Schaumann (1974c) was able to determine considerable differences in the cellulolytic activities between the species tested, but concluded that only the C_x component (β-1,4-glucanases) of the total cellulase complex could be estimated with these methods and that no correlation between the activity of the fungi *in vitro* and their frequency on wood in nature could be found. The clearing of cellulose-containing agar by 14 marine fungi was also used by Henningsson (1976a) as a measure of cellulase and xylanase production. T. Nilsson (1974b) employed several methods to assay the enzymatic activities of 36 wood-inhabiting fungi, among them one marine species, namely, *Humicola alopallonella*. Twelve of these fungi were unable to degrade pure cellulose substrates in culture, but produced characteristic soft-rot patterns, namely, cavities in the secondary cell walls of wood. T. Nilsson (1974b) demonstrated the advantages and disadvantages of different methods for determining cellulase activity and showed that growth on a cellulose–agar medium is an uncertain criterion for such activity. It can be concluded that the numerous experiments carried out on cellulase production by marine fungi have not supplied a quantitative method that permits correlation between behavior of a species in culture and its activity in nature. The most reliable qualitative test of wood decomposition is still the microscopic observation of soft-rot decay produced by a species in pure culture (see Chapter 15).

B. Hemicellulases and Pectinase

The major components of wood, besides cellulose, are xylan, glucomannan, and pectin. T. Nilsson (1974b) demonstrated a xylanase in

Humicola alopallonella, whereas mannase was absent. Barghoorn (1944) found that *d*-xylose, produced by the hydrolysis of xylan, was used by the eight species of marine fungi tested. Pectin was used as a carbon source by the same species (Barghoorn, 1944). Leightley and Eaton (1977) determined the ability to degrade wood cell wall components of several marine fungi belonging to the genera *Cirrenalia, Culcitalna, Halosphaeria, Humicola, Nia,* and *Zalerion.* They compared them with freshwater and terrestrial fungi and found production of cellulase, xylanase, and mannanase in all species tested.

C. Amylase

The decomposition of starch by marine fungi was demonstrated by Barghoorn (1944) for representatives of the genera *Ceriosporopsis, Corollospora, Lulworthia, Microthelia, Phialophorophoma,* and *Zalerion* and by T. Nilsson (1974b) for *Humicola alopallonella.*

D. Laminarinase

Laminarin, a reserve polysaccharide in many Phaeophyta, is barely attacked by filamentous marine fungi, according to Chesters and Bull (1963a). An exception in this study is *Dendryphiella salina,* which occurs saprobically on *Laminaria* spp. and shows considerable laminarinase activity in culture. Tubaki (1969) found that laminarin was an adequate carbon source for *D. arenaria, Lindra thalassiae,* and *Varicosporina ramulosa,* causing luxuriant growth on a yeast extract–seawater medium with 1% laminarin.

E. Proteases

Pisano *et al.* (1964) screened 14 marine fungi for their gelatinase activities and found such activity in the culture filtrates of 13 isolates. *Halosphaeria mediosetigera* produced the highest levels of gelatinase.

F. Other Enzymes

Sguros and co-workers examined the enzyme systems of several marine fungi, in particular, those of *Halosphaeria mediosetigera* and its imperfect state, *Trichocladium achrasporum.* Strong dehydrogenase activities connected with the citric acid cycle were found in culture extracts of these species (Sguros *et al.,* 1970; Rodrigues *et al.,* 1970; Vembu and Sguros, 1972). Chitinases are probably present in certain marine fungi, as indicated, for instance, by the growth of *Abyssomyces hydrozoicus* on the chitinous tubes of hydrozoa (Kohlmeyer, 1970), but chitinase production in culture has not been demonstrated so far. The use of other organic nutrients by marine fungi is discussed below.

II. METABOLITES

A. Amino Acids

Schafer and Lane (1957) demonstrated 12 amino acids in *Lulworthia* sp., and Peters *et al.* (1975) found the following compounds common to 10 species of marine fungi: alanine, aspartic acid, cysteine, cystine, glutamic acid, glycine, histidine, hydroxyproline, isoleucine, leucine, lysine, methionine, ornithine, phenylalanine, proline, serine, tryptophan, threonine, tyrosine, and valine. The species examined by Peters *et al.* (1975) belong to the ascomycetous genera *Corollospora, Haligena, Halosphaeria, Leptosphaeria, Lignincola,* and *Nais* and to the Deuteromycetes *Culcitalna* and *Zalerion*. These authors concluded that higher marine fungi could be an important source of amino acids for detritus-feeding animals.

B. Amines

Choline sulfate (ester) was found to be the principal amine produced by 10 marine Ascomycetes and the Deuteromycete *Zalerion maritimum* (Catalfomo *et al.*, 1972–1973; Kirk and Catalfomo, 1970; Kirk *et al.*, 1974).

C. Lipids

Triglyceride fatty acids, in particular, oleic, palmitic, and linoleic acids, were isolated from *Corollospora maritima* and *Zalerion maritimum* by Block *et al.* (1973) and Kirk *et al.* (1974). A number of these and other fatty acids were also determined in *Buergenerula spartinae* (*sub Sphaerulina pedicellata*) and *Dendryphiella salina* (Schultz and Quinn, 1973). Ergosterol was the most common sterol found in marine Ascomycetes and Deuteromycetes screened by Kirk and Catalfomo (1970) and Kirk *et al.* (1974). This compound occurred in species of *Ceriosporopsis, Corollospora, Halosphaeria, Lignincola, Nais,* and *Zalerion*. It is interesting to note that ergosterol was produced by one isolate but was absent in another isolate of the same species, for example, in *H. appendiculata* and *Z. maritimum* (Kirk *et al.*, 1974).

D. Sugar Alcohols

Mannitol occurred in 17 isolates of 11 species of marine fungi tested by Kirk *et al.* (1974).

E. Diverse Wall and Cell Constituents

Szaniszlo and Mitchell (1971) compared the hyphal wall compositions of marine and terrestrial species of the genus *Leptosphaeria* and found qualitatively identical compositions in both groups. The walls consisted of glucose, mannose, galactose, glucosamine, amino acids, and traces of galactosamine. Capsular or mucoid sheaths around hyphae of *Leptosphaeria albopunctata* and a *Lulworthia* species were found *in vitro* by Szaniszlo *et al.* (1968) and Davidson (1973), respectively. The hyphal envelopes of *L. albopunctata* consist of polysaccharides, composed mainly of glucose and small amounts of mannose. The sheaths, which are remetabolized in old cultures, have not been found in nature.

Kirk (1966, 1976) studied the cytochemistry of ascospores of marine fungi. Ascospore constituents found in these extensive investigations are described in Chapter 20.

III. EFFECT OF NUTRIENTS AND ENVIRONMENTAL PARAMETERS ON GROWTH AND REPRODUCTION

A. Nutrients and Growth

The use of some organic nutrients, such as cellulose, gelatin, hemicelluloses, laminarin, and starch, by marine fungi was mentioned earlier in Section I. The most extensive nutritional studies, using a standardized, gravimetric method, are by Sguros *et al.* (1973). These authors tested 38 inorganic and organic compounds as nitrogen sources and 79 organic compounds as carbon sources in the nutrition of *Culcitalna achraspora, Halosphaeria mediosetigera,* and *Humicola alopallonella.* All species grew significantly on inorganic nitrogen sources (alanine, arginine, asparagine, aspartate, glutamate, glutamine, hypoxanthine, leucine, urea, valine, and xanthine), with glucose as a carbon source. The fungi were more selective in their use of carbon compounds, showing good growth only in the presence of cellobiose, fructose, glucose, mannose, and xylose, whereas other sugars or sugar derivatives were unacceptable.

Sguros *et al.* (1973) concluded that the three species were probably insignificantly proteolytic, lipolytic, nucleolytic, or ligninolytic.

Previous investigations, similar to those of Barghoorn (1944), on carbohydrate and nitrogen source utilization by marine fungi were conducted by Gustafsson and Fries (1956) and T. W. Johnson *et al.* (1959). The extremely variable growth of the isolates on standardized media led T. W. Johnson *et al.* (1959) to the conclusion that the results could not be interpreted and were not indicative of the growth potential of a certain species in nature. Other studies on carbohydrate nutrition and inorganic nutrients of Ascomycetes and Deuteromycetes were carried out by Sguros and Simms (1963a,b, 1964), E. B. G. Jones and Jennings (1964, 1965), Meyers (1966, 1969), Meyers and Hoyo (1966), Meyers and Scott (1967), and Tubaki (1966, 1969).

Wood in the marine habitat is usually inhabited by more than one fungus, but observations on interrelationships between different species *in situ* are absent. Therefore, the following discovery made by Tubaki (1966) is of particular interest: *Corollospora maritima* assimilates sodium nitrite ($NaNO_2$), but *Ceriosporopsis halima* is unable to utilize this nitrogen source. When *C. halima* is inoculated on a nitrite medium next to *Corollospora maritima*, the former is able to grow by utilizing the ammonium produced after nitrite reduction by *C. maritima*.

Most of the other investigations on marine fungi dealt with woodinhabiting species, whereas Gessner (1976) examined gravimetrically the growth and nutrition of *Buergenerula spartinae*, a host-specific Ascomycete from salt-marsh cordgrass. Ten carbohydrates were utilized; only galactose and mannitol were not metabolized. Ammonium salts were the best nitrogen sources, followed by amino acids and $NaNO_3$. The fungus required biotin, thiamine, and pyridoxine and showed reduced growth when iron and zinc were absent. In nature, the vitamins and trace metals are probably supplied by the host.

Former nutritional studies on marine fungi have been conducted in solid agar cultures or in closed liquid systems. Churchland and McClaren (1976) developed a continuous culture system that supplies fresh nutrients, controls pH and osmotic changes, prevents accumulation of toxic metabolites, and, therefore, is closer to conditions in nature. These authors employed this system to determine the effect of inorganic nitrogen sources and L-glutamic acid on the growth of *Zalerion maritimum*. Comparisons between growth in a closed system with that in continuous culture showed a preference of the fungus for the organic nitrogen source in the first, but the absence of such preference in the second. These disagreeing results obtained in the two systems indicate that the continuous culture technique should be used in addition to the closed system, in order to come to meaningful conclusions.

B. Reproduction

Doguet (1964, 1968) examined the conditions under which two marine Basidiomycetes grew and fruited in culture. *Digitatispora marina* forms basidiospores at temperatures below 20°C, whereas *Nia vibrissa* sporulates between 15° and 25°C. These observations appear to explain the geographical distribution of the two species, namely, temperate waters for *D. marina* and temperate as well as tropical waters for *N. vibrissa*. Nutrients required for reproduction in these Basidiomycetes have not been determined. *Nia vibrissa* remains sterile if cultures are exposed to the light for 12 hr per day, but it fruits on the shaded side of pieces of wood. In nature, *N. vibrissa* is often found in protected niches, for example, inside hollow roots of mangroves.

Reproduction by Ascomycetes and Deuteromycetes was studied mainly by S. P. Meyers and co-workers, who found, for instance, that cellulose and cellulose-breakdown products support the formation of ascocarps in *Ceriosporopsis halima, Corollospora maritima, Halosphaeria mediosetigera, Lindra marinera, L. thalassiae, Lulworthia* spp., and *Torpedospora radiata* (Meyers, 1966, 1969; Meyers and Reynolds, 1959a; Meyers and Scott, 1967; Meyers and Simms, 1967). Species that did not fruit in culture, for example, *Halosphaeria quadricornuta, H. salina,* and *Lignincola laevis* on balsa wood in yeast-extract broth, produced ascocarps when the wood was transferred to aquaria with aerated seawater (Meyers and Reynolds, 1959a). Meyers and Simms (1965) established that *Lindra thalassiae* needs amino acids and vitamins to reproduce. Thiamine and biotin in conjunction with leaf sections of *Thalassia testudinum* support reproduction of *L. thalassiae,* but either of the two compounds alone or vitamin combinations without them were ineffective. In addition, Tubaki (1969) pointed to the significance of thiamine and demonstrated that 13 of 21 thiamine-deficient marine Ascomycetes and Deuteromycetes were able to synthesize thiamine by combining the thiazol moiety with the pyrimidine moiety.

The influence of salinity on reproduction was examined by Byrne and Jones (1975b), who showed that five marine Ascomycetes (*Ceriosporopsis halima, Corollospora maritima, Halosphaeria hamata, Lulworthia* sp., and *Microthelia linderi*) produced ascocarps on balsa wood in yeast-extract broth at seawater concentrations from 0 to 100%. *Halosphaeria appendiculata* formed fruiting bodies over the whole range of salinities, except in distilled-water media.

Sporulation in Deuteromycetes has not been examined as much as fruiting in Ascomycetes. Byrne and Jones (1975b) determined that *Dendryphiella salina* formed conidia equally well in distilled water and in 100% seawater, while *Asteromyces cruciatus* showed reduced sporulation in distilled water and in low salinities. Conidium production of *Varicosporina ramulosa* is affected by salinity, pH, and nutrients (Meyers and Hoyo, 1966). In distilled water, at low salt concentrations, or under other adverse conditions this species forms chlamydospores and atypical conidia and mycelia that become encysted. Encystment disappears at 20% seawater, and normal conidium production occurs above 50% seawater. *Orbimyces spectabilis* reacts similarly to the various salinities.

Churchland and McClaren (1976) noted that *Zalerion maritimum* sporulated heavily in continuous culture in a nitrogen-free medium, while the same fungus remained sterile in a closed system.

C. Environmental Parameters

This section deals with the effect of environmental parameters, such as temperature, salinity, and pH, on the vegetative growth of marine fungi. The influence of such factors on reproduction has been dealt with in the preceding section. It is hardly possible to discuss the effect of single parameters separately, since Ritchie (1957) demonstrated the so-called "*Phoma* pattern" of growth, namely, an increase in salinity tolerance with rising temperature. *Phoma* sp., isolated from submerged wood, had optimal growth at 16°C combined with 19%ₒₒ salinity, but the growth optimum at 37°C required a salinity of 47%ₒₒ. No growth occurred if low temperatures were combined with high salinities or high temperatures with low salinities. A *Pestalotia* sp. showed a similar pattern, but a *Lulworthia* exhibited no temperature–salinity relationship. Other species for which a *Phoma* pattern was established are the Basidiomycetes *Digitatispora marina* and *Nia vibrissa* (Doguet, 1964, 1968) and the Deuteromycetes *Robillarda rhizophorae* (Lee and Baker, 1972b) and *Asteromyces cruciatus* and *Dendryphiella salina* (Schaumann, 1974b). Ritchie and Jacobsohn (1963) determined that the *Phoma* pattern in *Zalerion maritimum* was based on an osmotic rather than an ionic effect of the seawater concentrations. The same growth occurred at corresponding osmotic pressures, whether the osmotically active substances (salts, sugars, and alcohols) were ionic or not, metabolizable or not, penetrable or not. Gessner (1976) found the opposite reaction in *Buergenerula spartinae,* namely, an ionic rather than an osmotic effect with increasing salinity.

The primary purpose of laboratory experiments should be to correlate the behavior of fungi *in vitro* with their activity in nature. This goal has been reached in some cases, for instance, when it was demonstrated that fungal strains have maximum growth at salinities approximating those of their natural habitat and that isolates from tropical waters show higher temperature optima than isolates from temperate habitats. *Robillarda rhizophorae* and *Dendryphiella salina* grow best at salinities found at their mangrove site (Lee and Baker, 1972b). Schaumann (1974b) demonstrated that a *Lulworthia* from the West African coast exhibited optimal growth at 25° and 30°C and at salinities around 20%‰, while *Lulworthia* isolates from the North Sea (Helgoland and Bremerhaven) grew best at 20° and 25°C, but poorly at 30°C. Furthermore, the Bremerhaven strain, isolated from brackish water, had a corresponding salinity optimum of 10%‰. Schaumann (1974b) pointed out that such obvious correlations between behavior *in vitro* and *in situ* are not the rule and that other factors besides temperature and salinity should be tested to explain the natural distribution of marine fungi.

Comparisons between the responses of marine and nonmarine fungi to salinity showed in the few species examined that marine fungi usually grew better in seawater than in freshwater media, but nonmarine fungi had reduced vegetative growth, spore germination, and reproduction under saline conditions (Davidson, 1974a; E. B. G. Jones *et al.*, 1971; E. B. G. Jones and Jennings, 1964).

Barghoorn (1944) was first to examine the influence of hydrogen ion concentrations on the growth of marine fungi and determined that five out of six isolates developed best in media with an initial pH above 7.4. Growth was reduced in acidic media, indicating a physiological adaptation of marine fungi to pH levels of seawater, which normally vary between 8.1 and 8.3. E. B. G. Jones and Irvine (1972) observed two pH optima for *Asteromyces cruciatus, Corollospora cristata, Dendryphiella salina,* and *Lulworthia* sp., one in the acidic pH range (6.0–6.6) and the other in the neutral to light alkaline range (7.0–8.0). A similar double peak was also observed by Barghoorn (1944) for a *Lulworthia* sp. (sub *Halophiobolus opacus*).

The effect of pH on growth and sporulation of *Varicosporina ramulosa* was studied by Meyers and Hoyo (1966). Conidia were produced only in media with initial pH values of 6.9 and 7.5, whereas pH 4.0 caused the formation of mycelial cysts and chlamydospores, but conidia were absent. Mycelia grown initially at pH 7.5 and transferred to pH 4.0 showed increased conidial production upon resuspension. In contrast, mycelia from an initial pH 4.0 culture transferred to media with pH values of 6.9 and 7.5 did not initiate conidial formation.

Initial and terminal pH values, but rarely transient pH changes of culture filtrates, are often registered in physiological experiments (Sguros et al., 1973; Sguros and Simms, 1963a). Meyers and Scott (1968) described the growth of *Halosphaeria mediosetigera* in a culture medium with an initial pH of 4.1. The pH rose to 8.0 after 6 days and dropped to pH 4.0 by 21 days. Such changes of pH in a closed system do not reflect natural conditions, and, as discussed above, continuous culture systems such as that developed by Churchland and McClaren (1976) hold out promise for experiments approximating natural situations.

IV. UNSOLVED PHYSIOLOGICAL PROBLEMS

Several important areas of the physiology of higher marine fungi have been grossly neglected, such as the algae-inhabiting fungi in general, the influence of atmospheric pressure and oxygen tension, and spore germination.

A. Algicolous Fungi

Physiological research on the higher marine fungi has been restricted in the past mostly to the saprobic wood-inhabiting species, and nothing is known of the physiology of symbiotic, parasitic, or saprobic marine fungi on algae. The major problem is the prerequisite of being able to grow the host plant in the laboratory in order to cultivate the fungus.

B. Pressure

All marine fungi tested in culture have been isolated from beaches, intertidal or subtidal habitats. These species belong to a mycota that differs morphologically from that of the deep sea (Kohlmeyer, 1977). No information is available on the influence of pressure on the growth and sporulation of marine fungi in general, and native deep-sea species have not been cultured at all. Techniques employed for the isolation and cultivation of bacteria under deep-sea conditions (Jannasch and Wirsen, 1977; Jannasch et al., 1976; Wirsen and Jannasch, 1975) could give answers to the questions whether shallow- and deep-water fungi differ physiologically and whether pressure impedes or favors development of either group.

C. Oxygen Tension

The influence of dissolved oxygen in seawater on marine fungi has not been tested, although observations in the field indicate that the higher marine fungi are unable to grow under low oxygen conditions. A deep-sea area off the California coast known as a "minimum oxygen zone" and having an oxygen content of 0.30 ml/liter did not allow growth of soft-rot fungi, when such fungi occurred on wood at a neighboring site with a content of dissolved oxygen of 1.26 ml/liter (Kohlmeyer, 1969a). Higher fungi are also absent in sediments of some North Carolina estuaries, where submerged wood is softened by cellulolytic bacteria but does not contain any fungi (J. Kohlmeyer, unpublished). Another indication for the high oxygen requirement of higher marine fungi is the fact that wood-inhabiting soft-rot fungi usually penetrate only a few millimeters into the substrate. A deep penetration of Ascomycetes and Deuteromycetes is observed only in light wood containing vessels with wide lumina, which permit adequate aeration of the interior (Becker and Kohlmeyer, 1958a,b).

D. Spore Germination

Experiments employing ascospores, basidiospores, or conidia can be instrumental in testing the viability of fungal species under different environmental conditions. Indeed, the ability to germinate in seawater *in situ* may be a test to separate "true" marine fungi from terrestrial dormant propagules. As explained in Chapter 1, a mycostatic principle in seawater (Borut and Johnson, 1962; Tyndall and Kirk, 1973) inhibits spore germination of certain nonmarine fungi.

Ascospore germination has been described incidentally for some species of marine fungi (e.g., E. B. G. Jones, 1973; Wagner, 1965), but detailed experimental data, such as salinities or temperatures employed, are generally missing. Byrne and Jones (1975a) investigated the effect of salinity and temperature on spore germination of terrestrial, freshwater, and marine fungi. Germination in terrestrial species was inhibited by increasing salinities, while marine species exhibited a broad tolerance to salinity and temperature stresses.

22. The Possible Origin of Higher Marine Fungi

The following discussion deals mainly with the Ascomycetes, because we consider the four filamentous marine Basidiomycetes to be secondary inhabitants of the oceans, as will be explained below. Marine Ascomycetes are a heterogeneous assembly belonging to all classes except the Discomycetes, with the majority in the Pyrenomycetes and Loculoascomycetes (Kohlmeyer, 1974a). Although fossil marine fungi have not been found so far, comparisons of extant marine species indicate that there are two groups: primary and secondary marine fungi. The primary marine fungi are derived from marine ancestors and have never left the marine environment, whereas the secondary marine fungi evolved from terrestrial ancestors and migrated secondarily into the sea (Fig. 35). To understand the difference between these two groups, we must first discuss some phylogenetic principles that determine the antiquity or the evolved state of an organism.

I. PHYLOGENETIC PRINCIPLES

Guiding principles should be the basis of evolutionary schemes, so that discussions about the evolution of organisms are based on rationale rather than on emotion. Among the phylogenetic principles listed by Savile (1968) and quoted below in abbreviated form, the following may be particularly pertinent to the question of the origin of higher marine fungi:

1. Obligate parasitism in the fungi is...a fundamental attribute of primitive groups; and saprophytism arose repeatedly from it (see also Raper, 1968).
2. In strict parasitism the antiquity of the host reflects that of the parasite and vice versa. Host and parasites evolve together . . . .

174

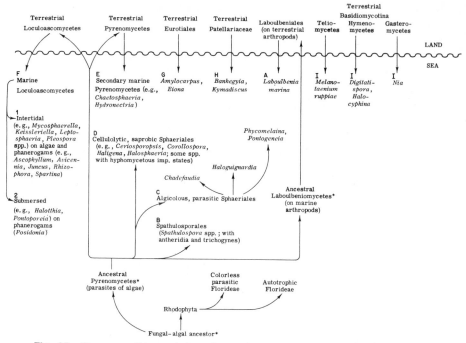

Fig. 35. The possible origin of marine filamentous Ascomycotina and Basidiomycotina (only fungal classes and orders with marine representatives are included). Hypothetical ancestors are marked with an asterisk. Primary marine fungi include groups B, C, and D; all others (A, E–I) are secondary marine fungi. See discussion in the text.

3. Elaborate sexual mechanisms and self-sterility are generally ancient . . . . Simple sexual mechanisms and self-fertility or apomixis occur later . . .
4. In simple organisms biochemistry often proves more reliable than gross morphology in indicating genetic relationships.
5. Preadaptation . . . refers to a structure or mechanism adapted to one function in an ancestral group being, without change, adaptive in some degree for a new function in a derived group.
6. If there are few possible ways of attaining important ends, convergent evolution is probable if not inevitable.

II. CHARACTERS OF ARCHAIC ASCOMYCETES

Based on the phylogenetic principles of the preceding paragraph, we expect an archaic Ascomycete to be a parasite on an archaic host. Recent phylogenetic schemes regard Taphrinales (Mix, 1949) as the oldest, true Ascomycetes (e.g., Savile, 1968; Raper, 1968). These parasites without ascocarps or paraphyses live on ferns and phanerogams (Betulaceae,

Rosaceae, etc.). There is no doubt that marine algae, such as Rhodophyta, are phylogenetically older than plum trees or even terrestrial ferns and, consequently, parasites on these algae can be expected to be more ancient than Taphrinales. The existence of parasitic Ascomycetes on marine algae, or of marine fungi in general, has not found entry in modern textbooks. Therefore, the possibility that Ascomycetes could have evolved from marine ancestors is generally not even considered any more.

In 1966, Denison and Carroll revived and modified the hypothesis of Sachs (1874) that the Ascomycetes derived from Floridean (red algal) ancestors, proposing that the primitive Ascomycetes evolved as saprobes on driftwood in oceans and estuaries. Discovery of the Spathulosporales (Kohlmeyer, 1973a), which we consider to be living fossils, appeared to support our view that this order as well as the Laboulbeniales are closely related to a hypothetical marine ancestor, which in turn was akin to parasitic red algae (Kohlmeyer, 1973a, 1974a, 1975a). Demoulin (1974) suggested that parasitic red algae gave rise to parasitic Ascomycetes and Basidiomycetes, whereas Chadefaud (1975) advanced the idea that Florideae and the higher fungi had parallel, but not identical, evolutions from a common ancestor, a theory termed "para-floridean" origin of fungi. We agree that a direct derivation of Ascomycetes from red algae is unlikely and propose the scheme depicted in Fig. 35. A hypothetical "fungal–algal ancestor" may have given rise to Rhodophyta and to hypothetical ancestral Pyrenomycetes, parasites on red algae. Based on the phylogenetic principles explained above, we regard Spathulosporales as the oldest types known of Ascomycetes, but, of course, not as direct precursors of other extant species. We propose that the hypothetical ancestral Pyrenomycete had the following characters: It was a monoecious parasite on permanently submerged marine red algae; the ascocarps were unilocular and ostiolate and had thick, dark walls; it was provided with functional spermatia and ascogonia with trichogynes; the immature ascocarp was filled with a deliquescing pseudoparenchyma; the asci were unitunicate, thin-walled, and deliquescing at maturity; the ascospores were one-celled, thin-walled, and hyaline; asexual spores were absent. This hypothetical ancestor of extant Ascomycetes is similar to that postulated by Denison and Carroll (1966), except for the parasitic trait and one-celled spores in our hypothesis. Parasitism is more archaic than saprophytism, according to Savile (1968), and red algae are much older, phylogenetically, than wood-producing land plants. Our preference for one-celled ascospores is based on extant algicolous Ascomycetes, almost all of which have nonseptate ascospores.

III. HOMOLOGIES BETWEEN RHODOPHYTA AND ASCOMYCETES

If a scheme proposes that red algae and Ascomycetes have branched off from a common ancestor, a large number of corresponding characters occurring in both groups are needed to support their relationship. Such similarities exist, for instance, in the alternation of generations of certain Rhodophyta and Ascomycetes (Fig. 36), and corresponding structures of

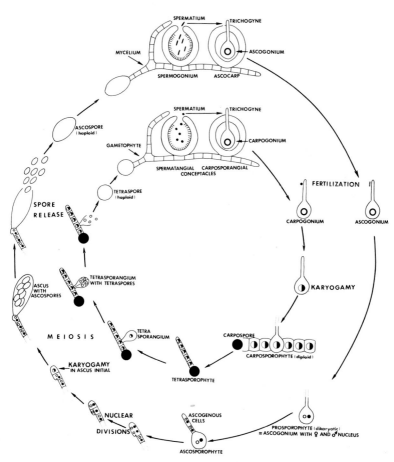

Fig. 36. Diagrammatic representation of alternation of generations and nuclear phase of a red alga (inner circle) and an Ascomycete (outer circle). From Kohlmeyer, *BioScience* **25**, 87 (1975), Fig. 1. Reprinted with permission.

these two groups have been compared recently (Demoulin, 1974; Kohlmeyer, 1975a). The carposporangial conceptacle of Corallinaceae shows a remarkable similarity to the ascocarp of Spathulosporales (see Figs. 41c and 42c), a fact pointed out by Demoulin (1974). In addition, the spermatangial conceptacle of Corallinaceae bears a striking resemblance to spermogonia of marine Ascomycetes (e.g., *Haloguignardia* spp., Figs. 66c and 68c). The conceptacles and spermogonia may be regarded as examples of preadaptations (see the phylogenetic principles listed on pp. 174–175. These fruiting bodies with male reproductive cells, occurring in marine algae and in ancient types of marine Ascomycetes, became pycnidia in the more highly evolved fungi that formed conidia able to germinate and propagate the organism without sexual processes.

Biochemical characters appeared to contradict a common ancestry of Rhodophyta and Ascomycetes, for instance, the composition of cell walls, namely, cellulose in the former and chitin in the latter (Gäumann, 1964). However, recent investigations have demonstrated cellulose also in the cell walls of 35 species of Ascomycetes (Rosinski and Campana, 1964; Jewell, 1974). Furthermore, the ester of choline sulfate occurs in higher marine fungi and in red algae, but is absent in the lower fungi (Catalfomo *et al.*, 1972–1973), a fact that supports relationships between the first two groups, but speaks against the origin of Ascomycetes from the lower fungi, as advocated, for instance, by Gäumann (1964) and Savile (1968).

An important argument against the derivation of higher fungi from the lower ones is the absence of any flagella or cilia composed of microtubules in the (9 + 2) pattern in Ascomycetes, Basidiomycetes, and Rhodophyta, whereas such complex motile organelles occur in the lower fungi, Chlorophyta and Phaeophyta. A putative discovery of flagellar structures in Rhodophyta proved to be erroneous (Young, 1977). For more detailed discussions of relationships between Rhodophyta and Ascomycetes, we refer the reader to Denison and Carroll (1966), Demoulin (1974), and Chadefaud (1975).

IV. POSITION OF EXTANT HIGHER MARINE FUNGI IN THE PHYLOGENETIC SCHEME

The following discussion will try to separate the primary marine higher fungi from those that have evolved from terrestrial ancestors and migrated secondarily into the marine environment. Our phylogenetic ideas for this section, A through I, are illustrated in Fig. 35.

A. Laboulbeniales

Although the Laboulbeniales show archaic traits (Vuillemin, 1912) and are considered to be closely related to Spathulosporales (Kohlmeyer, 1973a), the only marine representative, *Laboulbenia marina* living on a marine beetle among algae of the intertidal zone, is most probably a secondary invader of the sea. Terrestrial arthropods have evolved from marine ancestors (Størmer, 1977) and the extant Laboulbeniales have probably emerged from the sea together with their hosts. No primary marine Laboulbeniales are known, but it is possible that such ancient types may be found on archaic arthropods, such as *Limulus*, the horseshoe crab, or in the fossil record.

B. Spathulosporales

This order has many characters in common with Rhodophyta (Kohlmeyer, 1973a, 1974a, 1975a), as listed in Table XVII. Obligate parasitism on Rhodophyta and the ancient character of the red algae in general indicate that Spathulosporales are on the bottom of the phylogenetic tree, close to the hypothetical ancestral Pyrenomycetes (Fig. 35). Furthermore, the elaborate sexual mechanisms of Spathulosporales with a receptive filament of the ascogonium, which forms a receptive papilla upon contact with the spermatium (Kohlmeyer and Kohlmeyer, 1975), and diversified spermatia with various types of appendages point to the archaic nature of Spathulosporales. This order and ancestral Laboulbeniales probably evolved from a common ancestral Pyrenomycete.

C. Algicolous, Parasitic Sphaeriales

Representatives of this group are mostly obligate parasites on marine algae. Species of *Chadefaudia* occur on Rhodophyta, *Haloguignardia* and *Phycomelaina* on Phaeophyta, and *Pontogeneia* on Chlorophyta and Phaeophyta. We consider these genera to be more highly evolved than Spathulosporales, but still originally marine forms that evolved from marine ancestors. The most ancient genus in this group appears to be *Chadefaudia* because of the restriction to Rhodophyta, as well as because of the central pseudoparenchyma in immature ascocarps and the one-celled ascospores. More highly evolved are probably *Haloguignardia* species because the hosts, namely, Phaeophyta, are phylogenetically younger than Rhodophyta. Morphologically, *Haloguignardia* is an old

Table XVII

Corresponding Structures in Florideae (Rhodophyta and Ascomycetes; see also Fig. 36)[a]

General term	Florideae	Ascomycetes
Central pore in cell walls between adjoining cells	Pit connection	Pore
Haploid gametophyte	Gametophyte	Mycelium and sterile cells of fruiting bodies
Female fruiting body	Carposporangial conceptacle[b]	Ascocarp
Female gametangium	Carpogonium	Ascogonium
Receptive filament of gametangium	Trichogyne	Trichogyne
Male fruiting body	Spermatangial conceptacle[b]	Spermogonium
Male gametangium	Spermatangium	Antheridium or spermatiophore
Male gamete	Spermatium	Spermatium
Prosporophyte	Carposporophyte (diploid; develops from fertilized carpogonium after caryogamy)	Prosporophyte [fertilized ascogonium with male and female nucleus (Chadefaud, 1972)]
Carpospore	Diploid carpospore	Dicaryotic carpospore [rare (Chadefaud, 1972)]
Sporophyte	Tetrasporophyte (diploid)	Ascosporophyte (dicaryotic ascogenous cells or hyphae)
Meiosporangium	Tetrasporangium (meiosis; tetraspores develop by cleavage)	Ascus (caryogamy and meiosis; ascospores develop by free cell formation)
Meiospore	Tetraspore (haploid)	Ascospore (haploid)

[a] From Kohlmeyer (1975a).
[b] In some representatives of Florideae (e.g., Corallinaceae).

type, in consideration of the thin-walled pseudoparenchyma and one-celled ascospores. *Phycomelaina* and *Pontogeneia* are the most evolved genera in this group, since their species are mostly parasites on Phaeophyta, immature ascocarps contain paraphyses, and the ascospores are uni- to multiseptate.

D. Cellulolytic, Saprobic Sphaeriales

This last group of the primary marine Ascomycetes has probably evolved from parasitic ancestors after woody plants became available as substrates in the intertidal zone. Saprophytism is considered a secondary trait that developed after parasitism (see the list of phylogenetic principles, pp. 174–175); therefore, these cellulolytic members of Sphaeriales can be considered younger in an evolutionary sense than the parasites of algae belonging to the same order. However, morphologically, there are close relationships between the saprobic and parasitic marine Sphaeriales. The cellulolytic Sphaeriales (especially in the genera *Ceriosporopsis, Corollospora, Haligena, Halosarpheia,* and *Halosphaeria*) are also characterized by a thin-walled pseudoparenchyma filling immature ascocarps, by early deliquescing asci, and by hyaline ascospores. Hyphomycetous imperfect states with brown conidia or chlamydospores are probably secondary traits that developed as adaptations to intermittent dehydration and for protection from insolation in the intertidal zone.

E. Secondary Marine Pyrenomycetes

We regard certain species of saprobic Pyrenomycetes as secondary invaders of the marine environment, having more relations to terrestrial genera than to marine genera. Examples are the wood-inhabiting *Chaetosphaeria chaetosa* and *Hydronectria tethys,* which both have paraphyses and persistent asci. Most probably such species developed from terrestrial ancestors and migrated into the marine environment via shoreline trees or driftwood. It can be assumed that paraphyses are adaptations to intertidal or terrestrial habitats, where they provided protection for the developing asci and supplied pressure in the ascocarp (Savile, 1968). In addition, asci with persistent walls are advantageous in the terrestrial environment if they can forcibly eject spores away from the ascocarp into the air stream. Fungi with dissolving asci in combination with a dissolving or compressed pseudoparenchyma as found in the primary marine Ascomycetes (especially algicolous Sphaeriales and Spathulosporales) would be at a disadvantage in dry habitats because spore dispersal could not function properly.

F. Marine Loculoascomycetes

We postulated earlier that fungi with thick-walled, bitunicate asci, the Loculoascomycetes, developed when Ascomycetes moved from aquatic to terrestrial habitats (Kohlmeyer, 1974a). The jack-in-a-box mechanism of ascospore discharge most probably evolved when ancestral Pyrenomycetes passed the barrier of the sea–land interface. In addition, pseudoparaphyses originated as protection devices for developing asci in dry habitats, similar to the development of paraphyses in terrestrial Pyrenomycetes. The hypothesis of a terrestrial origin of marine Loculoascomycetes is supported by two facts: (1) the limitation of bitunicate species to intertidal or secondary marine plants; (2) the limitation of dark, thick-walled ascospores to species on such marine plants and their absence in parasites on algae. Dark ascospore pigments are regarded as protective devices against radiation and desiccation of terrestrial fungi (Savile, 1968), whereas thin-walled hyaline ascospores are characteristic of the ancient types of parasitic Ascomycetes on algae. We propose that marine Loculoascomycetes derived from terrestrial ancestors, which kept their main traits when they reentered the marine environment. There are two groups: those on intertidal substrates and others on permanently submersed hosts.

1. Intertidal Loculoascomycetes

To this group belong marine representatives of *Mycosphaerella, Didymosphaeria, Keissleriella, Leptosphaeria,* and *Pleospora,* all genera with numerous species in terrestrial habitats. Host genera are, for instance, *Ascophyllum, Chondrus, Pelvetia, Avicennia, Rhizophora, Juncus, Salicornia,* and *Spartina,* that is, algae, mangroves, and salt-marsh plants that live in intertidal environments exposed to regular periods of emersion. A forceful ascospore ejection through the jack-in-a-box mechanism at low tides is advantageous for these intertidal Ascomycetes, because the propagules are carried by air currents to new, exposed substrates along the shore. It is interesting to note that elaborate dispersal devices are rare in this group, for example, ascospore appendages, which are so common in Halosphaeriaceae. Exceptions are *Keissleriella blepharospora* with ascospore bristles and *Pleospora gaudefroyi* with gelatinous, fingerlike extensions. However, simple gelatinous ascospore sheaths occur in many species and may serve as attachment devices.

2. Submersed Loculoascomycetes

At first sight, the occurrence of two Ascomycetes with bitunicate asci and brown ascospores (*Halotthia posidoniae* and *Pontoporeia bitur-*

binata) in a permanently submersed habitat may seem to contradict our theories of marine ascomycetous ancestry. However, their host, *Posidonia oceanica,* derives from terrestrial ancestors, and it is likely that the two Ascomycetes accompanied the sea grass when it entered the marine habitat.

G. Marine Eurotiales

The two marine representatives of Eurotiales, *Amylocarpus en-cephaloides* and *Eiona tunicata,* show no relationship to the primary marine Ascomycetes, and we propose that they have developed from terrestrial ancestors. This opinion is based on the nonperithecial as-cocarps, on the absence of a central pseudoparenchyma, on the irregular dispersal of the asci throughout the ascocarp, on the saprobic trait, and on the restriction of the species to the high-water line. Both species are morphologically adapted to the aquatic habitat, as ascospores of *A. en-cephaloides* are provided with spines and those of *E. tunicata* with delicate, subgelatinous appendages.

H. Marine Patellariaceae

Patellariaceae are also rare in the marine environment. The two members, *Banhegyia setispora* and *Kymadiscus haliotrephus*, are definitely derived from terrestrial ancestors. As we consider the perithecium of fungal parasites on algae to be an ancestral type of fruiting body, disklike ascocarps of *B. setispora* and *K. haliotrephus* are highly evolved. These two species are the only marine species that have discoid ascocarps. Apparently, the apothecium of "true" Discomycetes with an unprotected hymenium of thin-walled asci was not able to adapt to submersed conditions in the marine environment because representatives of this class (Korf, 1973) are absent in the ocean. *Orbilia marina,* included elsewhere in our treatise (p. 222), is not a marine fungus *sensu stricto* because it grows above the high-tide line.

I. Basidiomycotina

Representatives of this subdivision are rarely found in the marine environment. Only four filamentous marine Basidiomycetes are known, and, in addition, 17 yeastlike species. Several reasons support our view that Basidiomycetes as a group originated on land and not in a marine habitat. The small number of marine species and the monotypic state of the genera speak against an origin of Basidiomycetes in the ocean. Furthermore, the genera are taxonomically unrelated, which also indicates that they have developed independently of each other from terrestrial ancestors. The fact that no parasitic Basidiomycete is known to occur on marine algae also suggests a nonmarine origin of these fungi. Basidiomycetes in general have a very low tolerance for NaCl. Tresner and Hayes (1971) found that only 1.9% of 975 species of terrestrial fungi tested could withstand a solution of 10% NaCl. Terrestrial smuts belong to these exceptional salt-tolerant fungi. The salt resistance of this group may explain the relative frequency of heterobasidiomycetous yeasts in marine locations and the occurrence of the smut, *Melanotaenium ruppiae,* parasite of the sea grass *Ruppia,* in brackish-water habitats.

V. CONVERGENCES IN THE MARINE FUNGI

Certain characters found in the higher marine fungi must definitely be regarded as convergences that have developed independently in unrelated groups. A case in point are tetraradiate shapes of spores and spore appendages. The tetraradiate type of spore occurs commonly in conidia of freshwater Hyphomycetes (Ingold, 1975a) and in basidiospores of *Digitatispora marina* (see Fig. 9). Spore appendages are mostly variations of one theme, namely, structures originating from the outer layer (exospore) of the ascospore. Apical or equatorial fibers occur, for instance, in Loculoascomycetes (*Herpotrichiella ciliomaris* and *Keissleriella blepharospora*), in primary marine Pyrenomycetes (*Corollospora* spp.), in secondary marine Pyrenomycetes (*Chaetosphaeria chaetosa*), in Eurotiales (*Eiona tunicata*), and in Hysteriales (*Banhegyia setispora*). In other species the outer wall layer forms a uniform gelatinous sheath (*Leptosphaeria avicenniae, L. contecta, L. halima,* and *Haligena amicta*) or wide ribbonlike appendages (*Haligena elaterophora*). It is obvious that similar structures in dissimilar fungal groups are convergences (see No. 6 of the phylogenetic principles. p. 175).

VI. CONCLUSIONS

There is no doubt that the higher marine fungi are of polyphyletic origin. Extant marine fungi belong to two groups: primary marine fungi that derived directly from marine ancestors, and secondary marine fungi that originated from terrestrial ancestors (Fig. 35). The most ancient types of Ascomycetes are Spathulosporales and some parasitic Sphaeriales on marine Rhodophyta. Secondary marine fungi are mostly saprobes which belong to Loculoascomycetes, Eurotiales, Hysteriales, Laboulbeniales, and Basidiomycotina. A direct derivation of Ascomycetes from red algae is not likely, but Rhodophyta and Ascomycetes probably arose from a hypothetical common marine ancestor.

Critics of theories relating Ascomycetes to Rhodophyta point out that more biochemical evidence is needed to confirm possible relationships between the two groups. The lysine pathway of red algae, in particular, should be compared with that of Ascomycetes. A search for fossil marine Ascomycetes appears to be promising, as extant marine fungi associate with calcified red algae, and these plants are well preserved in the fossil record. Furthermore, ancient types of Ascomycetes related to terrestrial Laboulbeniales may possibly occur on marine arthropods, in particular, on living fossils such as *Limulus,* the horseshoe crab. Research in these areas could give further clues about the origin of marine fungi and about the higher fungi in general.

23. Identification

A major objective of this treatise is to assist students of marine fungi in the identification of species. The main tool to be used for this purpose is the following illustrated key (Chapter 24), which is based essentially on the morphology of fungal propagules. This key is "artificial" and aims only at the fastest and most reliable way to identify a species, and it is not intended to reveal any relationships among species. The first step in identifying an unknown fungus is to search for the propagules and their mode of formation: inside an ascus of an ascocarp, superficially on the basidium of a basidiocarp, in pycnidia, or on hyphae. Special attention should be paid to appendages or sheaths around ascospores and conidia. These structures are usually hyaline and phase-contrast microscopy or stains are needed to make them visible (see Chapter 2). After a fungal name has been found in the key, the *complete* description of the particular species should be compared thoroughly with the characters of the organism to be identified. If disagreements exist in some morphological features, for instance, in the sizes of spores or asci, or if the unidentified fungus occurs outside the known range of the species in question, it is possible that the name found in the key does not apply to the collection under examination. The use of illustrations only as a base for identification of a species without comparing the detailed diagnosis can lead to wrong determinations. Mis-identifications have, for instance, led to the report that a marine fungus was found in a desert or that a host-specific algal parasite was found growing on dead wood. Such assignments of wrong names are a burden for the mycological literature and should be avoided. If any doubt about the exact identity of the fungus exists, it should be listed under the name of the species it most closely resembles with a preceding "cf." (from *confer* = compare) or under a generic name only. In case a marine fungus cannot be identified by means of the present key, this does not necessarily indicate that it is a new species. It could have been described from a nonmarine habitat or substrate and may have been overlooked in the literature by marine mycologists. On the other hand, our collections show that there are a large number of undescribed species in the marine environment, but insufficient material has precluded us from naming and describing them.

Type or authentic material of marine fungi should be compared if precise identifications are required. The largest collection of higher marine fungi is housed at our Institute [University of North Carolina, Institute of Marine Sciences, official herbarium abbreviation IMS (see Holmgren and Keuken, 1974)]. Pure cultures of many marine fungi are maintained at the American Type Culture Collection (Rockville, Maryland).

24. Key to the Filamentous Higher Marine Fungi*

I. KEY TO SUBDIVISIONS OF EUMYCOTA

1. Spores absent, only sterile hyphae and bulbils formed . . . Mycelia Sterilia, p. 211
1'. Reproduction by sexual spores or asexual reproductive units 2
    2(1') Sexual spores 3
    2'(1') Asexual reproductive units 4
3(2) Spores produced in asci . . . . Ascomycotina, p. 189
3'(2) Spores produced on basidia . . . . Basidiomycotina, p. 204
    4(2') Smut spores, originating inside the host . . . . Basidiomycotina, p. 204
    4'(2') Smut spores absent; conidia or chlamydospores formed
    . . . . Deuteromycotina, p. 205

* Figure letters after the fungal names in the keys refer to the illustrations of spores at the bottom of the same page; the full description of each species is found on the page indicated. The illustrations are not drawn to scale. Spore diagrams are merely aids in the use of this key, and identifications should not be made solely on the basis of the figures.

II. KEY TO ASCOMYCOTINA

1. Ascospores not filiform 2
1'. Ascospores filiform or vermiform, at least 15 times longer than wide (see also *Bathyascus,* lead 5, and *Pontogeneia,* leads 101 and 101') 144
    2(1) Ascospores one-celled at maturity 3
    2'(1) Ascospores several-celled at maturity 25
3(2) Ascospores without appendages, mucilaginous sheaths, or chambered apices . 4
3'(2) Ascospores with distinct appendages, mucilaginous sheaths, or chambered, deciduous, or nondeciduous apices 8
    4(3) In wood 5
    4'(3) In or on algae 6
5(4) Ascospores longer than 40 μm; deep-sea species
. . . . *Bathyascus vermisporus,* Fig. a; p. 249
5'(4) Ascospores shorter than 40 μm; littoral species
. . . . *Halonectria milfordensis,* Fig. b; p. 324
    6(4') Ascocarps discoid with exposed asci (apothecia)
. . . . *Orbilia marina,* Fig. c; p. 222
    6'(4') Ascocarps subglobose, with enclosed asci (perithecia) 7
7(6') Ascospores up to 7 μm in diameter, in *Blidingia*
. . . . *Turgidosculum ulvae,* Fig. d; p. 365
7'(6') Ascospores up to 5 μm in diameter, in *Prasiola*
. . . . *Turgidosculum complicatulum,* Fig. e; p. 361
    8(3') Ascospores over 130 μm long (including apical tubes)
. . . . *Corollospora tubulata,* Fig. f; p. 278
    8'(3') Ascospores shorter, without long apical tubes
. . . . 9
9(8') Hyphae absent; crustose thallus bearing flask-shaped antheridia; ascocarps superficial on *Ballia* spp. (Rhodophyta) 10
9'(8') Hyphae present; antheridia absent; on other substrates 14

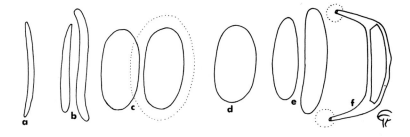

10(9) Ascocarps without hairs; in antarctic and subantarctic waters
. . . . *Spathulospora antarctica*, Fig. a; p. 234
10'(9) Ascocarps with hairs; in Australia or New Zealand 11
11(10') Ascospores longer than 65 μm 12
11'(10') Ascospores usually shorter than 65 μm 13
  12(11) Ascospores less than 14 μm in diameter, spathulate to spoon-shaped at the apices; few antheridia on long hairs
  . . . . *Spathulospora phycophila*, Fig. b; p. 240
  12'(11) Ascospores more than 14 μm in diameter, with a conical appendage at each end; many antheridia on short stalks . . . . . *Spathulospora adelpha*, Fig. c; p. 232
13(11') Ascocarps enclosed by long hairs; ascospore tips spathulate; gelatinous appendages subterminal . . . . *Spathulospora lanata*, Fig. d; p. 238
13'(11') Ascocarps bare, except for a few short hairs around the ostiole; ascospore tips rounded, with terminal appendages . . . . *Spathulospora calva*, Fig. e; p. 236
  14(9') On wood; ascospores with bristles, tufts of fibers, or fingerlike appendages 15
  14'(9') In algae; ascospores with gelatinous apical caps or chambered tips . . 17
15(14) Ascocarps perithecial; ascospore appendages tufts of fibers, 1 at each end and 4 around the equator . . . . *Nautosphaeria cristaminuta*, Fig. f; p. 322
15'(14) Ascocarps cleistothecial; ascospore appendages different 16
  16(15') Ascospores subglobose to ovoidal, with rigid, awl-like appendages all around the surface . . . . *Amylocarpus encephaloides*, Fig. g; p. 223
  16'(15') Ascospores ellipsoidal, with fingerlike, radiating, apical, or subterminal appendages . . . . *Eiona tunicata*, Fig. h; p. 225
17(14') In Phaeophyta; conical ascospore appendages chambered or striate 18
17'(14') In Rhodophyta; gelatinous, rounded appendages without structure 21

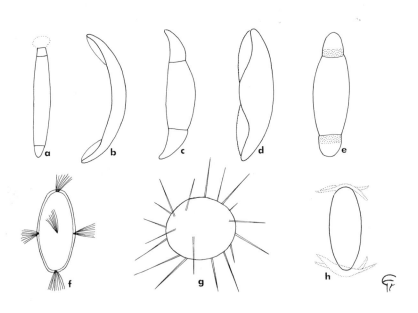

18(17) In *Cystoseira* or *Halidrys* . . . . . . *Haloguignardia irritans*, Fig. a; p. 330
18'(17) In *Sargassum* spp. 19
19(18') Ascospore appendages persistent
. . . . *Haloguignardia tumefaciens*, Fig. b; p. 336
19'(18') Ascospore appendages deciduous 20
    20(19') Galls with fingerlike projections; ascospores, including appendages, longer than
    35 μm; Pacific Ocean . . . . *Haloguignardia decidua*, Fig. c; p. 330
    20'(19') Galls without projections; ascospores, including appendages, commonly
    shorter than 35 μm; Atlantic Ocean . . . *Haloguignardia oceanica*, Fig. d; p. 334
21(17') Ascospores longer than 25 μm (excluding appendages)
. . . . *Chadefaudia balliae*, Fig. e; p. 260
21'(17') Ascospores shorter than 25 μm 22
    22(21') Ascocarps superficial, but often covered by epiphytes 23
    22'(21') Ascocarps immersed in the host tissue 24
23(22) Symbiont with epiphytic algae; ascocarp diameter generally less than 300 μm; upper
peridium less than 40 μm thick; ostiolar projection into ascocarp cavity absent
. . . . *Chadefaudia corallinarum*, Fig. f; p. 261
23'(22) Parasite; ascocarp diameter generally over 300 μm; ostiolar canal forming tubelike
projection into ascocarp cavity . . . . . *Chadefaudia polyporolithi*, Fig. f; p. 265
    24(22') In *Rhodymenia;* ascospores up to 16.5 μm long (excluding appendages)
    . . . . *Chadefaudia marina*, Fig. g; p. 264
    24'(22') In other hosts; ascospores up to 20 μm long
    . . . . *Chadefaudia gymnogongri*, Fig. f; p. 262
25(2') Ascospores never have more than 1 septum at maturity 26
25'(2') Ascospores with 1 and more septa at maturity 92
    26(25) Fruiting body an apothecium; ascospores without appendages
    . . . . *Kymadiscus haliotrephus*, Fig. h; p. 395
    26'(25) Other types of fruiting bodies; if apotheciumlike, ascospores with appen-
    dages 27
27(26') Ascospores septate near the base; large cell dark, small cell light-colored . 28
27'(26') Ascospores septate near the middle or, if asymmetrical, concolorous . . . 30

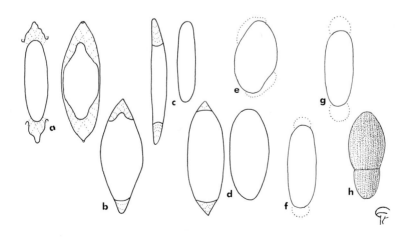

28(27) Ascospores longer than 70 μm
. . . . *Manglicola guatemalensis,* Fig. a; p. 425
28'(27) Ascospores less than 40 μm long   29
29(28') Ascospore with subapical germ pore; basal cell broadly cylindrical, up to 7–8 μm in diameter . . . . *Zopfiella latipes,* Fig. b; p. 343
29'(28') Ascospore with apical germ pore; basal cell elongate cylindrical, up to 4 μm in diameter   . . . . *Zopfiella marina,* Fig. b; p. 345
   30(27') Ascospores hyaline, light grayish, yellow, or yellowish brown at maturity   31
   30'(27') Ascospores dark-colored (brown or blackish) at maturity   86
31(30) Ascospores without appendages or gelatinous or mucilaginous sheaths   . . . 32
31'(30) Ascospores with appendages or sheaths   54
   32(31) Ascospores longer than 33 μm, in *Chondrus*
. . . . *Didymosphaeria danica,* Fig. c; p. 398
   32'(31) Ascospores shorter than 33 μm; if longer, in other substrates 33
33(32') Ascospores thick-walled 34
33'(32') Ascospores thin-walled 35
   34(33) Ascospores more than 18 μm in diameter; in algae (*Cystoseira*)
. . . . *Thalassoascus tregoubovii,* Fig. d; p. 448
   34'(33) Ascospores less than 18 μm in diameter; in wood or higher plants
. . . . *Aniptodera chesapeakensis,* Fig. e; p. 248
35(33') On animals   36
35'(33') In plant material 37
   36(35) In shells of mollusca and balanidae . . . . . *Pharcidia balani,* Fig. f; p. 356
   36'(35) On chitin of insects   . . . . *Laboulbenia marina,* Fig. g; p. 227
37(35') In wood or higher plants   38
37'(35') In or on algae 48
   38(37) Ascospores distinctly ridged longitudinally; ascocarps orange-yellowish, ostiolate   . . . . *Hydronectria tethys,* Fig. h; p. 327
   38'(37) Ascospores without longitudinal ridges; ascocarp color different, ostiolate or nonostiolate 39
39(38') Ascospore diameter 10 μm and more 40
39'(38') Ascospore diameter below 10 μm   41

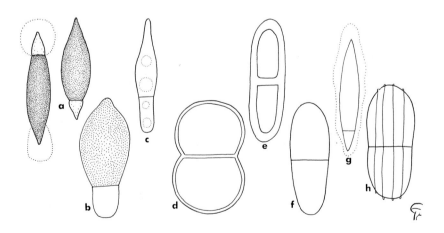

40(39) Asci scattered irregularly through the center of a cleistothecium; ascospores shorter than 22 μm . . . . *Heleococcum japonense,* Fig. a; p. 326

40'(39) Asci in a hymenial, basal layer of a perithecium; ascospores 22 μm and longer . . . . *Nais inornata,* Fig. b; p. 320

41(39') In bark of living pneumatophores of *Avicennia*
. . . . *Mycosphaerella pneumatophorae,* Fig. c; p. 388

41'(39') In other substrates 42

42(41') Asci bitunicate 43

42'(41') Asci unitunicate 47

43(42) In Plumbaginaceae . . . . *Mycosphaerella staticicola,* Fig. d; p. 391

43'(42) In other hosts 44

44(43') In Chenopodiaceae 45

44'(43') In *Spartina* spp. 46

45(44) Ascocarps 100 μm or more in diameter; ascospores 18 μm long or longer . . . . *Mycosphaerella suaedae-australis,* Fig. d; p. 392

45'(44) Ascocarps less than 100 μm in diameter; ascospores up to 18 μm long (in fresh material in a gelatinous sheath) . . . . . *Mycosphaerella salicorniae,* Fig. d; p. 390

46(44') Ascospore length 23 μm or more . . *Mycosphaerella sp. I,* Fig. d; p. 392

46'(44') Ascospore length 20 μm or less . . *Mycosphaerella sp. II,* Fig. d; p. 393

47(42') Asci with apical pore; ascospores 13–19.5 × 4–7.5 μm; catenophyses absent
. . . . *Gnomonia longirostris,* Fig. e; p. 242

47'(42') Asci without apical pore; ascospores mostly 16–24 × 6–8 μm; catenophyses present . . . . *Lignincola laevis,* Fig. e; p. 307

48(37') Symbiont; ascocarps in receptacles of *Ascophyllum* and *Pelvetia*
. . . . *Mycosphaerella ascophylli,* Fig. f; p. 386

48'(37') Parasites or saprobes; ascocarps in stipes of algae, not in receptacles . 49

49(48') Ascospores septate near the base; in *Gloiopeltis*
. . . . *Didymella gloiopeltidis,* Fig. g; p. 382

49'(48') Ascospores septate near the middle; in other algae 50

50(49') Symbionts of epiphytic algae; ascocarps on the surface of macroalgae . 51

50'(49') Parasites or saprobes; ascocarps embedded in the hosts 52

51(50) On *Laminaria;* asci less than 13 μm in diameter
. . . . *Pharcidia rhachiana,* Fig. h; p. 360

51'(50) On *Pelvetia;* ascus diameter 13 μm or more
. . . . *Leiophloea pelvetiae,* Fig. i; p. 376

52(50') In *Rhodymenia;* asci shorter than 45 μm; ascospore diameter less than 5 μm (inconspicuous gelatinous sheath) . . . . *Didymella magnei,* Fig. j; p. 384

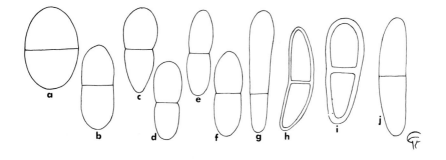

52'(50') In other algae; asci longer than 45 μm; ascospores more than 5 μm in diameter   53

53(52') In *Fucus;* ascocarps depressed, ellipsoidal, wall black above, almost hyaline below; asci thick-walled . . . . *Didymella fucicola,* Fig. a; p. 379

53'(52') In *Laminaria;* ascocarps ampulliform or subglobose, wall pale brown to red; asci thin-walled   . . . . *Nectriella laminariae,* Fig. b; p. 329

54(31') Ascospores without appendages, but enclosed in uniform gelatinous sheath   55

54'(31') Ascospores with appendages; at maturity without uniform sheath . . . 56

55(54) In *Salicornia;* asci thick-walled   . . . *Mycosphaerella salicorniae,* Fig. c; p. 390

55'(54) In *Rhodymenia;* asci thin-walled   . . . . . *Didymella magnei,* Fig. d; p. 384

56(54') Ascospores with one caplike appendage   57

56'(54') Ascospores with more than one appendage 58

57(56) 1 lenticular, lateral cap on the ascospore septum (hematoxylin!); no additional septa formed during germination   . . . . *Paraliomyces lentiferus,* Fig. e; p. 432

57'(56) Ascospore with 1 deciduous apical cap; 1–3 additional septa may be formed during germination . . . . *Halosphaeria cucullata,* Fig. f; p. 290

58(56') Ascospores with terminal radiating deciduous setae (thin bristles) . . . 59

58'(56') Ascospore appendages different 62

59(58) 5 setae on one end of ascospore . . . *Keissleriella blepharospora,* Fig. g; p. 407

59'(58) Setae on each end of the ascospore   60

60(59') Ascospores shorter than 14 μm   . . . . *Crinigera maritima,* Fig. h; p. 452

60'(59') Ascospores longer than 14 μm 61

61(60') Fruit body a globose perithecium with brown setae around the ostiole . . . . *Herpotrichiella ciliomaris,* Fig. i; p. 375

61'(60') Fruit body a disk-shaped apothecium; hymenium exposed . . . . *Banhegyia setispora,* Fig. j; p. 394

62(58') Ascospore appendages only terminal or subterminal (an additional gelatinous cover around the spore may occur) 63

62'(58') Ascospore appendages terminally and laterally attached 80

63(62) Mature ascospores enclosed in a gelatinous sheath that is pierced at each apex by an elongated outward growing, tapering appendage   64

63'(62) Mature ascospores without sheath 65

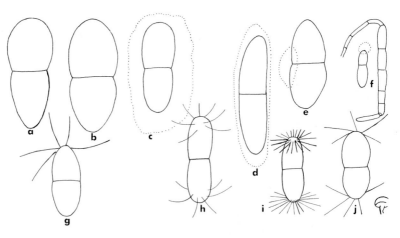

64(63) Ascospores usually longer than 28 μm, wider than 10 μm; centrum of ascocarp containing persistent, wide filaments, similar to pseudoparaphyses or catenophyses . . . . *Ceriosporopsis cambrensis,* Fig. a; p. 254

64'(63) Ascospores usually shorter than 28 μm; diameter 6–12 μm; centrum of ascocarp without persistent filaments . . . . . . *Ceriosporopsis halima,* Fig. a; p. 256

65(63') Ascospore appendages slender, rigid, tapering, 2–9 at each end 66

65'(63') Ascospore appendages different, not more than 1 at each end 70

66(65) 2 subterminal appendages at each ascospore apex, these pairs arranged at right angles to one another (anomalous spores with 3 to 4 appendages on both ends may occur) . . . . *Halosphaeria quadricornuta,* Fig. b; p. 299

66'(65) 3–9 appendages at each ascospore apex 67

67(66') 3 appendages at each apex of the ascospore; ascocarps developing predominantly on the surface of grains of sand and shells . . . *Corollospora trifurcata,* Fig. c; p. 277

67'(66') 3–9 appendages at each apex; in wood, occasionally within teredinid tubes 68

68(67') 3–4 (rarely 5) subterminal ascospore appendages; species of tropical and subtropical waters . . . . *Halosphaeria salina,* Fig. d; p. 302

68'(67') 4 or more terminal ascospore appendages; species of temperate zones 69

69(68') Generally 4 radiating appendages at each ascospore apex . . . . *Halosphaeria quadriremis,* Fig. e; p. 301

69'(68') Generally 6 appendages at each apex . . . *Halosphaeria stellata,* Fig. f; p. 303

70(65') Parasites or symbionts of Phaeophyta 71

70'(65') Saprobes in wood or higher plants 73

71(70) Symbiont of *Ectocarpus,* on the surface of *Laminaria;* ascospores rhomboid, almost pointed . . . . *Pharcidia laminariicola,* Fig. g; p. 359

71'(70) Parasites, embedded in the hosts; ascospores ellipsoidal with blunt tips . . 72

72(71') Ascospores shorter than 30 μm; in *Laminaria* (and *Alaria?*) . . . . *Phycomelaina laminariae,* Fig. h; p. 339

72'(71') Ascospores longer than 30 μm; in *Cystophora* . . . . *Massarina cystophorae,* Fig. i; p. 427

73(70') Appendages initially enclosing the ascospores completely, finally detached around the septum; polymorphous, yoke-shaped, or veil-like 74

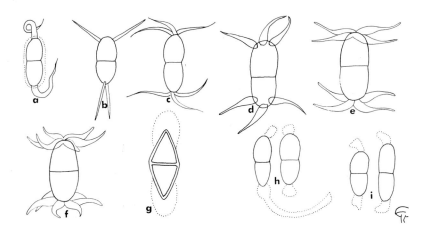

73'(70') Base of appendages at or near the apices of the mature ascospores only . 75

74(73) Ascospores thin-walled, ovoid or ellipsoidal; no distinct striae in the appendages . . . . **Halosphaeria maritima,** Fig. a; p. 294

74'(73) Ascospores thick-walled, rhomboid; distinct striae in the base of appendages (hematoxylin!) . . . . **Halosphaeria pilleata,** Fig. b; p. 298

75(73') Ascospore appendages thin sheaths, at first completely attached subapically; finally appearing hamate, tapering; asci thin-walled, early deliquescing
. . . . **Halosphaeria hamata,** Fig. c; p. 292

75'(73') Ascospore appendages thick apical caps; ascus walls apically thickened, persistent, or thin-and deliquescing 76

76(75') Immature ascospores enclosed in a gelatinous sheath; each apex of mature ascospores is surrounded by a large, subglobose, subgelatinous cap with delicate radiating striae (stain!) . . . . **Halosphaeria galerita,** Fig. d; p. 291

76'(75') Immature ascospores not completely enclosed in a sheath; appendages small and apical, without radiating striae 77

77(76') Asci thin-walled, deliquescent, without apical apparatuses; catenophyses present . . . . **Halosphaeria trullifera,** Fig. e; p. 305

77'(76') Asci thick-walled, persistent, with or without apical plate; with or without catenophyses 78

78(77') Ascospores shorter than 28 µm; diameter less than 15 µm; ascus with apical plate (hematoxylin!) . . . . **Gnomonia marina,** Fig. f; p. 243

78'(77') Ascospores longer than 28 µm; diameter over 15 µm; ascus without apical plate 79

79(78') Ascospores forcibly ejected from ascus; catenophyses absent; ascospore appendages not transformed into a coil of filaments; temperate species
. . . . **Gnomonia salina,** Fig. f; p. 244

79'(78') Ascospores not ejected; catenophyses present; ascospore appendages at maturity transformed into a coil of filaments (hematoxylin!); tropical and subtropical species . . . . **Halosarpheia fibrosa,** Fig. g; p. 285

80(62') Appendages consisting of a spore sheath enclosing mucilage; both surrounding the ascospore completely at maturity 81

80'(62') Appendages not enclosing the ascospore completely at maturity . . . 82

81(80) Ascospore appendages consisting of a distinct tube at each apex and a rupturing ring around the septum . . . . **Ceriosporopsis tubulifera,** Fig. h; p. 258

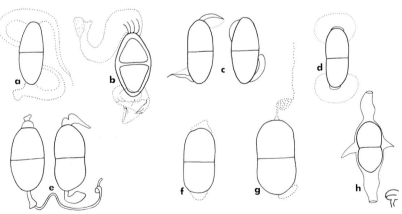

81'(80) Ascospores without distinct tubes or ring; appendages inconspicuous, irregular, extended into lobes around the septum and into 1 process at each end of the ascospore; core of apical process different from outer part (staining dark with hematoxylin)
. . . . *Ceriosporopsis circumvestita,* Fig. a; p. 255
    82(80') A tubular annulus around septum
    . . . . *Halosphaeria torquata,* Fig. b; p. 304
    82'(80') Equatorial appendages different 83
83(82') 2–3 lunate, rigid appendages, attached with the middle part to the ascospore over the septum; in addition a small apical cap, which may become reversed, at each end
. . . . *Halosphaeria mediosetigera,* Fig. c; p. 296
83'(82') Ascospore appendages different 84
    84(83') Numerous flexuous cilialike appendages around the septum
    . . . . *Corollospora maritima,* Fig. d; p. 273
    84'(83') 3–4 lateral appendages, rigid, thick, not cilialike 85
85(84') Ascospore appendages spoon-shaped at the bases, their tips without refractive bodies or small caps; without persistent paraphysoid chains in the ascocarp center
. . . . *Halosphaeria appendiculata,* Fig. e; p. 288
85'(84') Base of ascospore appendage cylindrical, its tip with refractive body and covered with a small cap; persistent paraphysoid chains in the ascocarp center
. . . . *Ceriosporopsis calyptrata,* Fig. f; p. 252

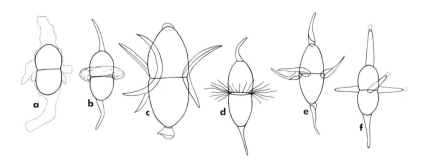

86(30′) Ascospores surrounded by a gelatinous sheath, which may be deliquescing; wall smooth . . . . *Helicascus kanaloanus,* Fig. g; p. 406

86′(30′) Ascospores without gelatinous sheath; wall smooth or sculptured . . . 87

87(86′) Ascospores verrucose to verruculose
. . . . *Didymosphaeria enalia,* Fig. h; p. 399

87′(86′) Ascospores smooth or striate, not verrucose 88

88(87′) Ascospores distinctly striate; on *Rhizophora*
. . . . *Didymosphaeria rhizophorae,* Fig. i; p. 402

88′(87′) Ascospores not striate; on other substrates 89

89(88′) Ascospores shorter than 22 μm; diameter less than 10 μm
. . . . *Microthelia linderi,* Fig. j; p. 430

89′(88′) Ascospores longer than 22 μm; diameter over 10 μm 90

90(89′) Ascospores shorter than 35 μm; diameter less than 16 μm; in *Atriplex*
. . . . *Didymosphaeria maritima,* Fig. k; p. 400

90′(89′) Ascospores longer than 35 μm; diameter over 16 μm; in *Posidonia* . . 91

91(90′) Ascospores shorter than 61 μm; dark band around the septum; without distinct germ pores . . . . *Halotthia posidoniae,* Fig. l; p. 404

91′(90′) Ascospores longer than 65 μm; without band around the septum; hyaline, apical germ pores . . . . *Pontoporeia biturbinata,* Fig. m; p. 446

92(25′) Ascospores with transverse septa only 93

92′(25′) Ascospores muriform (with transverse and longitudinal septa) 138

93(92) Ascospores without appendages or gelatinous sheaths 94

93′(92) Ascospores with appendages or gelatinous sheaths 117

94(93) Ascospores hyaline at maturity 95

94′(93) Ascospores colored (light yellow, yellow, brown, or olivaceous), at least some of the cells 106

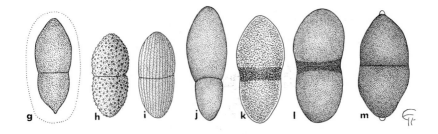

95(94) Ascospores consistently have 3 septa, never attenuate at one end 96
95'(94) Ascospore septation varying (2–12 septa), attenuate or not at one end . . . 100
    96(95) Ascospores longer than 50 μm, diameter 20 μm or more; in *Codium*
    . . . . *Pontogeneia codiicola,* Fig. a; p. 350
    96'(95) Ascospores shorter than 50 μm or diameter less than 15 μm; in other sub-
    strates 97
97(96') Asci unitunicate 98
97'(96') Asci bitunicate 99
    98(97) Asci opening with an operculus; ascospores longer than 33 μm
    . . . . *Orcadia ascophylli,* Fig. b; p. 453
    98'(97) Asci without operculus; ascospores shorter than 33 μm
    . . . . *Sphaerulina oraemaris,* Fig. c; p. 456
99(97') Ascospores 28 μm and longer; species of temperate waters
    . . . . *Leptosphaeria pelagica,* Fig. d; p. 422
99'(97') Ascospores shorter than 28 μm; species of tropical and subtropical waters
    . . . . *Leptosphaeria australiensis,* Fig. d; p. 411
    100(95') Ascospores longer than 60 μm; in algae 101
    100'(95') Ascospores shorter than 60 μm; in wood or *Spartina* 105
101(100) Ascospores longer than 200 μm 102
101'(100) Ascospores shorter than 200 μm 103
    102(101) Ascospore length 280 μm or more; 4–5 septa
    . . . . *Pontogeneia enormis,* Fig. e; p. 352
    102'(101) Ascospore length less than 280 μm; 12 or 13 septa
    . . . . *Pontogeneia cubensis,* Fig. f; p. 352
103(101') Ascospore diameter 20 μm or more; in *Valoniopsis*
    . . . . *Pontogeneia valoniopsidis,* Fig. g; p. 355
103'(101') Ascospore diameter below 20 μm; in other hosts 104
    104(103') Ascospores longer than 90 μm; in *Padina*
    . . . . *Pontogeneia padinae,* Fig. h; p. 353
    104'(103') Ascospores shorter than 90 μm; in *Castagnea*
    . . . . *Pontogeneia calospora,* Fig. i; p. 349

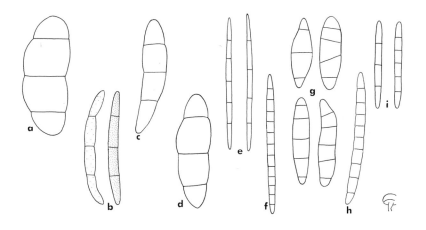

105(100') Ascospores distinctly attenuate at the base, longer than 35 μm
. . . . **Buergenerula spartinae,** Fig. a; p. 229
105'(100') Ascospores not attenuate at one end, shorter than 35 μm
. . . . **Sphaerulina albispiculata,** Fig. b; p. 455
   106(94') Apical cells of ascospores hyaline or lighter colored than the brown middle
   cells 107
   106'(94') Ascospores concolorous throughout 110
107(106) Ascospores unequally biconical, with 7–8 tubercules around the equator
. . . . **Biconiosporella corniculata,** Fig. c; p. 341
107'(106) Ascospores ellipsoidal to fusoid, without tubercules around the middle . . 108
   108(107') Asci unitunicate, 8-spored; pseudoparaphyses absent
   . . . . **Savoryella lignicola,** Fig. d; p. 370
   108'(107') Asci bitunicate, 8- or 4-spored; pseudoparaphyses present 109
109(108') Ascospore length 36 μm or less; asci typically 8-spored, longer than 110 μm
. . . . **Leptosphaeria obiones,** Fig. e; p. 418
109'(108') Ascospore length 36 μm or more; asci typically 2-spored, shorter than 110 μm
. . . . **Leptosphaeria paucispora,** Fig. e; p. 421
   110(106') Ascospores have 1–3 septa 111
   110'(106') Ascospores have 4 and more septa 116
111(110) Ascospores have 1–3 septa 112
111'(110) Ascospores consistently have 3 septa 113
   112(111) Ascospores longer than 33 μm, diameter 8 μm or more; subhyaline to yellowish . . . . **Leptosphaeria marina,** Fig. f; p. 415
   112'(111) Ascospores shorter than 33 μm, diameter (4 to) 5–8 μm; pale brown or
   darker . . . . **Leptosphaeria oraemaris,** Fig. g; p. 420
113(111') Ascocarps thick-walled, carbonaceous; asci longer than 150 μm
. . . . **Trematosphaeria mangrovis,** Fig. h; p. 451
113'(111') Ascocarps thin-walled, coriaceous; asci shorter than 150 μm 114
   114(113') Asci unitunicate, thin-walled, opening with an opercule; ascospores hyaline,
   possibly yellowish at maturity; in algae
   . . . . **Orcadia ascophylli,** Fig. i; p. 453
   114'(113') Asci bitunicate, thick-walled, without opercule; ascospores brown; in other
   substrates 115

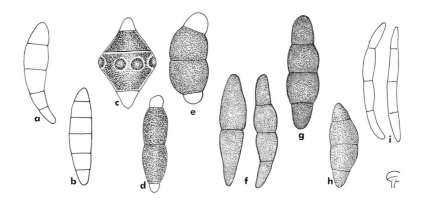

115(114') Ascospores longer than 20 μm, diameter over 7 μm
. . . . *Phaeosphaeria typharum*, Fig. a; p. 435
115'(114') Ascospores shorter than 20 μm, diameter less than 7 μm
. . . . *Leptosphaeria peruviana*, Fig. b; p. 422
  116(110') Ascospores usually have 5 septa (rarely up to 7 septa); third cell mostly the largest . . . . c . . . . . *Leptosphaeria albopunctata*, Fig. c; p. 410
  116'(110') Ascospores usually have 8 septa (rarely 4–7 or 9–10 septa); fourth cell mostly the largest . . . . *Trematosphaeria britzelmayriana*, Fig. d; p. 450
117(93') Mature ascospores completely enclosed in a uniform gelatinous sheath; apical cells may be free in *Carbosphaerella* (use India ink!) 118
117'(93') Mature ascospores without gelatinous sheath, but with other appendages . 125
  118(117) Asci thin-walled, early deliquescent; without pseudoparaphyses . . . 119
  118'(117) Asci thick-walled, not deliquescent; pseudoparaphyses present . . . 120
119(118) Ascospore diameter more than 12 μm; central cells dark
. . . . *Carbosphaerella leptosphaerioides*, Fig. e; p. 250
119'(118) Ascospore diameter less than 12 μm; hyaline throughout
. . . . *Haligena amicta*, Fig. f; p. 280
  120(118') Ascospores 30 μm and longer 121
  120'(118') Ascospores shorter than 30 μm 124
121(120) Ascospores hyaline, 3 (rarely 4) septa
. . . . *Leptosphaeria contecta*, Fig. g; p. 414
121'(120) Ascospores yellow–brown, 3 and more septa 122
  122(121') Ascospores have 3–5 septa; third cell from apex largest
. . . . *Leptosphaeria neomaritima*, Fig. h; p. 416
  122'(121') Ascospores usually have 6 or more septa; fourth or fifth cell largest 123
123(122') Ascospores usually have 6 septa (rarely have 5, 7, or 8 septa); diameter 12 μm or more; smooth; mostly in *Ammophila* . . . *Phaeosphaeria ammophilae*, Fig. i; p. 433
123'(122') Ascospores have 7–11 septa; diameter rarely over 10 μm; verrucose in age; in *Juncus, Spartina, Phragmites,* and *Typha* . . . *Leptosphaeria typhicola*, Fig. j; p. 423
  124(120') Ascospores hyaline; in pneumatophores of *Avicennia*
. . . . *Leptosphaeria avicenniae,* Fig. k; p. 412
  124'(120') Ascospores yellow–brown; in *Spartina* and wood
. . . . *Leptosphaeria halima*, Fig. l; p. 415
125(117') Ascospore appendages only lateral or apical 126
125'(117') Apical as well as lateral appendages on the ascospore 134

126(125) Lateral spore appendages only, cilialike
. . . . *Chaetosphaeria chaetosa,* Fig. a; p. 348
126'(125) Apical spore appendages only, not cilialike 127
127(126') Ascospore appendages on one end only 128
127'(126') Ascospore appendages on both ends 130
    128(127) Ascospores with 3–4 radiating, acuminate appendages
. . . . *Torpedospora radiata,* Fig. b; p. 373
    128'(127) Ascospores with one cap or sheathlike appendage 129
129(128') Ascospores have 1 septum (upon germination up to 4 septa); diameter 6 μm or more; appendage subglobose, terminal . . . . *Halosphaeria cucullata,* Fig. c; p. 290
129'(128') Ascospores have 3 (rarely 4–5) septa; diameter 5 μm or less; appendage an elongate irregular sheath around the apex or along the upper side
. . . . *Haligena unicaudata,* Fig. d; p. 283
    130(127') 4–7 (normally 5) radiating, acuminate, subterminal appendages on each ascospore apex . . . . *Torpedospora ambispinosa,* Fig. e; p. 372
    130'(127') Ascospore appendages single, different 131
131(130') Appendages very long, enclosing the ascospore completely within the ascus, later on expanded, elaterlike . . . . *Haligena elaterophora,* Fig. f; p. 281
131'(130') Appendages short apical caps or processes not enclosing the ascospore . 132
    132(131') Ascospores shorter than 30 μm, with 3 septa; on animal substrates
. . . . *Abyssomyces hydrozoicus,* Fig. g; p. 346
    132'(131') Ascospores longer than 30 μm, 4 or more septa; on plant material . 133
133(132') Ascospores more than 10 μm in diameter, predominantly 5 septa
. . . . *Haligena spartinae,* Fig. h; p. 282
133'(132') Ascospores less than 10 μm in diameter, predominantly 11 septa
. . . . *Haligena viscidula,* Fig. i; p. 284
    134(125') Ascospores bearing a rigid, tapering appendage at each apex 135
    134'(125') Ascospores bearing a tuft of bristles at each apex 136

135(134) Ascospores longer than 37 μm (excluding apical thorns); predominantly 5 septa
. . . . *Corollospora lacera,* Fig. a; p. 271
135'(134) Ascospores shorter than 37 μm (excluding apical thorns); 3 septa
. . . . *Corollospora intermedia,* Fig. b; p. 270
136(134') Ascospores with 7 or more septa, hyaline throughout
. . . . *Corollospora pulchella,* Fig. c; p. 275
136'(134') Ascospores with 6 or less septa, central cells fuscous 137
137(136') Mature ascospores typically have 3 septa; asci less than 100 μm long
. . . . *Corollospora cristata,* Fig. d; p. 268
137'(136') Mature ascospores typically have 5 septa; asci longer than 100 μm
. . . . *Corollospora comata,* Fig. e; p. 268
138(92') Ascospores with striate gelatinous sheath or subconical appendages . 139
138'(92')Ascospores without sheaths or appendages; if a simple sheath present, not
striate 140
139(138) Ascospores with gelatinous sheath without striae; subconical appendages at each
apex; asci thick-walled, persistent . . . . *Pleospora gaudefroyi,* Fig. f; p. 438
139'(138) Ascospores with a gelatinous, striate sheath, without appendages; asci thin-
walled, deliquescing . . . . *Carbosphaerella pleosporoides,* Fig. g; p. 251
140(138') In algae . . . . *Pleospora pelvetiae,* Fig. h; p. 441
140'(138') In higher plants 141
141(140') In *Spartina;* ascospores without sheaths 142
141'(140') In other hosts; ascospores enclosed in gelatinous sheaths 143
142(141) Ascospores typically with 5 transverse septa; up to 38μm long
. . . . *Pleospora spartinae,* Fig. i; p. 442
142'(141) Ascospores with 7–9 transverse septa; up to 52 μm long
. . . . *Pleospora pelagica,* Fig. h; p. 440
143(141') In *Triglochin;* ascospores with 7 transverse septa
. . . . *Pleospora triglochinicola,* Fig. j; p. 442
143'(141') In *Salicornia;* ascospores with 8–10 transverse septa (see remarks about similar
Pleospora spp. on other hosts under this species) . . *Pleospora* sp. I, Fig. k; p. 443

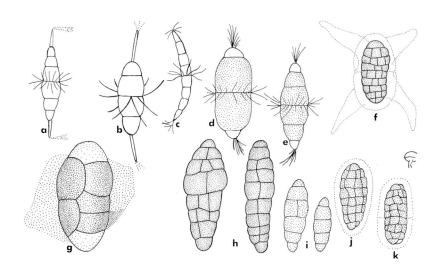

144(1') Ascospores with a whiplike appendage at the upper apex; a deep-sea species . . . . *Oceanitis scuticella,* Fig. a; p. 367
144'(1') Ascospores without such appendages; species from shallow water . . 145
145(144') Ascospores tapering on one or both ends to narrow long apices 146
145'(144') Ascospore diameter approximately equal throughout (note: the very tips on both sides may form either a tapering end chamber or a globose appendage) 148
146(145) Ascospores 3–3.5 μm wide on one end, tapering to the very narrow other end, shorter than 150 μm . . . . *Trailia ascophylli,* Fig. b; p. 323
146'(145) Ascospores tapering on both ends, longer than 150 μm 147
147(146') Ascospores mostly longer than 230 μm . . *Lindra thalassiae,* Fig. c; p. 311
147'(146') Ascospores mostly shorter than 230 μm . . *Lindra marinera,* Fig. c; p. 310
148(145') Ascospores with a globose gelatinous appendage on each apex; 30 to 50 septa . . . . *Lindra inflata,* Fig. d; p. 309
148'(145') Ascospores without such appendages, but other apical processes may occur; usually nonseptate 149
149(148') Ascospores without processes . . . . *Ophiobolus australiensis,* Fig. e; p. 369
149'(148') Ascospore apices tapering to mucilage-containing, cell-like processes (use stains, e.g., hematoxylin) 150
150(149') Ascospores longer than 500 μm; species on mangroves and wood in tropical and subtropical regions . . . . *Lulworthia grandispora,* Fig. f; p. 314
150'(149') Ascospores shorter than 500 μm 151
151(150') Ascospore length usually 110 μm or less; species of temperate regions . . . . *Lulworthia fucicola,* Fig. f; p. 313
151'(150') Ascospores longer than 110 μm; in temperate or tropical regions 152
152(151') Parasitic in calcified Rhodophyta . . . *Lulworthia kniepii,* Fig. f; p. 315
152'(151') Saprobic in other substrates . . . . . . *Lulworthia sp.,* Fig. f; p. 315

III. KEY TO BASIDIOMYCOTINA

1. Parasitic smut in the stem of *Ruppia* . . . . . *Melanotaenium ruppiae,* Fig. g; p. 464
1'. Saprobic on the surface of wood or *Spartina* 2
2(1') Basidiospores consisting of 4 radiating arms; formed on the surface of a flat, cushionlike basidiocarp . . . . *Digitatispora marina,* Fig. h; p. 460

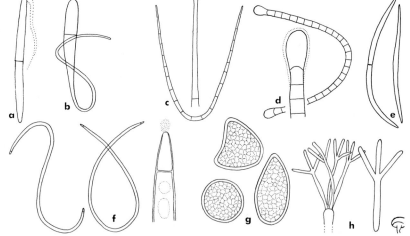

2′(1′) Basidiospores subglobose; in a globular or funnel-shaped basidiocarp . .   3
3(2′) Basidiocarp subglobose, puffball-like, irregularly dehiscing at maturity; basidiospores
with 5 filamentous appendages . . . . *Nia vibrissa,* Fig. a; p. 458
3′(2′) Basidiocarp funnel-shaped, releasing the nonappendaged basidiospores apically as a
ball through a wide aperture . . . . *Halocyphina villosa,* Fig. b; p. 462

IV. KEY TO DEUTEROMYCOTINA

1. Conidia borne on or in fruit bodies (acervuli, pycnidia, and sporodochia) (Coelomycetes) 2
1′. Conidia not produced on or in fruit bodies, but on hyphae (Hyphomycetes) . . 28
 2(1) Conidia on sporodochia, on the surface of the substrate 3
 2′(1) Conidia enclosed in pycnidia or acervuli 4
3(2) Sporodochia brown, on wood in the deep sea
 . . . . *Allescheriella bathygena,* Fig. c; p. 509
3′(2) Sporodochia brightly colored, on marsh plants (*Salicornia* spp.)
 . . . . *Tubercularia pulverulenta,* Fig. d; p. 512
 4(2′) Acervuli with thin upper wall, rupturing irregularly; hyperparasitic in galls formed
 by *Haloguignardia* spp. in Phaeophyta . . . . *Sphaceloma cecidii,* Fig. e; p. 543
 4′(2′) Pycnidia with thick upper wall, ostiolate; not hyperparasitic 5
5(4′) Pycnidia with a long beak, almost equal in diameter with the venter; on wood
 . . . . *Halonectria milfordensis,* Fig. f; p. 324
5′(4′) Pycnidia without beak or with a short papilla; in wood, algae, mangroves, or marsh
plants 6
 6(5′) Conidia always nonseptate 7
 6′(5′) Conidia zero- to multiseptate 16
7(6) Conidia with apical setae; produced in cupulate, setose acervuli
 . . . . *Dinemasporium marinum,* Fig. g; p. 542
7′(6) Conidia without setae; produced in pycnidia without setae 8
 8(7′) Conidia filiform (at least 20 times longer than thick); in *Ascophyllum*
 . . . . *Septoria ascophylli,* Fig. h; p. 537
 8′(7′) Conidia globose, ellipsoidal, or elongate cylindrical (not more than 3 to 4 times
 longer than thick) 9
9(8′) Composite pycnidia, 1.5–1.7 mm in diameter, with a central ostiole; under the bark of
Rhizophora . . . . *Cytospora rhizophorae,* Fig. i; p. 525
9′(8′) Pycnidia different, smaller 10

10(9′) Pycnidia on the bark of pneumatophores and tree trunks of *Avicennia*
. . . . *Rhabdospora avicenniae,* Fig. a; p. 534
   10′(9′) Pycnidia on other substrates 11
11(10′) Conidia brownish . . . . *Coniothyrium obiones,* Fig. b; p. 523
11′(10′) Conidia hyaline 12
   12(11′) Conidia produced endogenously in phialides (oil immersion!)
. . . . *Phialophorophoma litoralis,* Fig. c; p. 528
   12′(11′) Conidia produced exogenously on conidiophores 13
13(12′) On algae 14
13′(12′) On other substrates 15
   14(13) Saprobe on decaying *Laminaria;* pycnidia separate, not in a stroma
. . . . *Phoma laminariae,* Fig. d; p. 531
   14′(13) Parasite on *Chondrus crispus;* pycnidia in a stroma (imperfect or spermogonial
state of *Didymosphaeria danica*) . . . . *Phoma marina,* Fig. e; p. 532
15(13′) On *Suaeda* . . . . *Phoma suaedae,* Fig. f; p. 532
15′(13′) On other substrates . . . . *Phoma* and *Macrophoma* **spp.,** Fig. f; pp. 533, 528
   16(6′) Conidia muriform 17
   16′(6′) Conidia with transverse septa only 19
17(16) Conidia predominantly have 3 septa, usually not longer than 20 μm; without gelatinous sheaths or appendages . . . . . . *Camarosporium roumeguerii,* Fig. g; p. 522
17′(16) Conidia with 5 or more transverse septa; with distinctive gelatinous sheaths or appendages 18
   18(17′) Conidia with a gelatinous caplike appendage at each end; mostly on Gramineae in dunes . . . . *Camarosporium metableticum,* Fig. h; p. 518
   18′(17′) Conidia completely surrounded by a gelatinous sheath; on salt-marsh *Salicornia* spp. . . . . *Camarosporium palliatum,* Fig. i; p. 519
19(16′) Conidia filiform or elongate cylindrical, zero- to multiseptate 20
19′(16′) Conidia ellipsoidal or ovoid, always 1 septum when mature 24
   20(19) Conidia longer than 35 μm, with 5 or more septa; in higher plants (Gramineae) 21
   20′(19) Conidia shorter than 35 μm, 0–3 septa; in algae 23
21(20) Conidia without appendages
. . . . imperfect state of *Pleospora spartinae,* Fig. j; p. 442
21′(20) Conidia with apical gelatinous caps 22

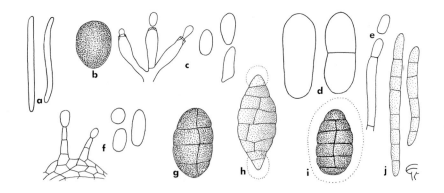

22(21') Conidia mostly longer than 60 μm; caps at both ends; in *Distichlis*
. . . . ***Septoria thalassica,*** Fig. a; p. 537
22'(21') Conidia mostly shorter than 60 μm; caps at one end only; in *Spartina*
. . . . ***Stagonospora* sp.,** Fig. b; p. 540
23(20') In *Ascophyllum;* conidia have 0–2 septa, diameter less than 2 μm
. . . . imperfect state of ***Mycosphaerella ascophylli,*** Fig. c; p. 386
23'(20') In *Pelvetia;* conidia have 3 septa, diameter more than 2 μm
. . . . ***Stagonospora haliclysta,*** Fig. d; p. 539
24(19') Conidia with appendages, in *Rhizophora*
. . . . ***Robillarda rhizophorae,*** Fig. e; p. 536
24'(19') Conidia without appendages, in other substrates 25
25(24') Conidia hyaline, in *Laminaria* . . . . ***Phoma laminariae,*** Fig. f; p. 531
25'(24') Conidia colored; in wood or phanerogams 26
26(25') Conidia shorter than 9 μm; in wood . . ***Diplodia oraemaris,*** Fig. g; p. 526
26'(25') Conidia 9 μm or longer; in marsh plants 27
27(26') In *Salicornia;* conidia up to 19 (to 20) μm long, up to 7 μm in diameter
. . . . ***Ascochyta salicorniae,*** Fig. h; p. 514
27'(26') In *Halimione;* conidia up to 11.5 μm long, up to 5 μm in diameter
. . . . ***Ascochytula obiones,*** Fig. i; p. 517
28(1') Conidia or chlamydospores 1-celled, but single cells may form long chains that
break up eventually 29
28'(1') Conidia or chlamydospores several-celled; or 1-celled conidia remaining attached to, and distributed with, a central sporogenous cell 37
29(28) Hyaline conidia 30
29'(28) Dark conidia or chlamydospores 32
30(29) Conidia with radiating projections (staurospores)
. . . . ***Clavatospora stellatacula,*** Fig. j; p. 470
30'(29) Conidia simple 31
31(30') Conidia globose, 2–3.3 μm in diameter, produced in phialides
. . . . ***Botryophialophora marina,*** Fig. k; p. 468
31'(30') Conidia ellipsoidal, 8–12 μm long, produced on conidiophores
. . . . imperfect state of ***Heleococcum japonense,*** Fig. l; p. 326
32(29') Conidia borne on conidiophores, often forming dense pustules 33
32'(29') Conidia without distinct conidiophores or pustules 34

33(32) Conidia 16 μm in diameter or more; deep-sea species
. . . . *Periconia abyssa*, Fig. a; p. 500
33'(32) Conidia generally 13 μm in diameter or less; littoral species
. . . . *Periconia prolifica*, Fig. b; p. 501
   34(32') Chlamydospores deep olivaceous to black
   . . . . secondary spores of *Clavariopsis bulbosa*, Fig. c; p. 485
   34'(32') Chlamydospores more or less brownish 35
35(34') Chains straight, unbranched . . . . *Cremasteria cymatilis*, Fig. d; p. 485
. . . . or imperfect state of *Halosphaeria mediosetigera*, Fig. d; p. 296
35'(34') Chains more or less curved, branching occurs 36
   36(35') Chains straight or curved, same diameter throughout, single cells 3.5–7.5 μm
   wide . . . . imperfect state of *Ceriosporopsis circumvestita*, Fig. e; p. 255
   36'(35') Chains predominantly curved, sometimes increasing in diameter from base to
   apex, single cells 6–17 μm wide
   . . . . imperfect state of *Ceriosporopsis halima*, Fig. e; p. 256
37(28') Conidia with three or more arms, often with a large bulbous cell 38
37'(28') Conidia without radiating arms or appendages 41
   38(37) Conidia hyaline, without bulbous cell
   . . . . *Varicosporina ramulosa*, Fig. f; p. 471
   38'(37) Conidia dark, with bulbous cell 39
39(38') Conidia 1-celled, often distributed in aggregates of 5–9 cells, attached to the central
sporogenous cell . . . . *Asteromyces cruciatus*, Fig. g; p. 475
39'(38') Conidia composed of 1 basal, bulbous, dark cell with 1 or 2 crowns of radiating,
nondeciduous arms 40
   40(39') Basal cell hyaline to light olive, 6–20 μm in diameter
   . . . . *Clavariopsis bulbosa*, Fig. h; p. 485
   40'(39') Basal cell dark brown, 23–42 μm in diameter
   . . . . *Orbimyces spectabilis*, Fig. i; p. 498
41(37') Conidia 1- or 2-celled, rarely 3-celled
. . . . *Humicola alopallonella*, Fig. j; p. 494
41'(37') Conidia or chlamydospores multicellular, with more than 2 cells 42
   42(41') Conidia with transverse septa only 43
   42'(41') Conidia muriform 57

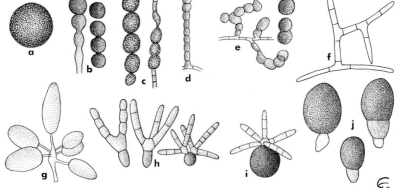

43(42) Symbiotic fungus with chlamydospores, forming chains inside the walls of living Chlorophyta (*Cladophora*) . . . . ***Blodgettia bornetii,*** Fig. a; p. 466

43'(42) Saprobic fungi with true conidia, on decaying algae, sea grasses, marsh plants, or wood 44

  44(43') Conidia borne in chains on the conidiophore 45

  44'(43') Conidia borne singly on the conidiophore 46

45(44) Conidia have 0–2 (sometimes 3) septa
. . . . . ***Cladosporium algarum*** (see also *Cladosporium* spp.), Fig. b; pp. 482, 484

45'(44) Conidia have 1–9 (sometimes 10 or 11) septa, predominantly with 3–5 septa
. . . . ***Dendryphiella salina,*** Fig. c; p. 488

  46(44') Conidia more or less straight 47

  46'(44') Conidia coiled or curved (helicoid) 51

47(46) Apical cell not broader than basal cell 48

47'(46) Apical cell broader than basal cell 50

  48(47) Conidia thick-walled, with 6–12 pseudosepta; apical cells light-colored, separated from adjoining cells by dark septa . . . ***Drechslera halodes,*** Fig. d; p. 492

  48'(47) Conidia thin-walled, with less than 6 true septa, concolorous throughout
. . . . 49

49(48') Conidia 1–3 septate; shorter than 20 μm ***Dendryphiella arenaria,*** Fig. c; p. 487

49'(48') Conidia usually have 3–5 septa; rarely with fewer septa or up to 11 septa; usually longer than 20 μm, up to 75 μm . . . . ***Dendryphiella salina,*** Fig. c; p. 488

  50(47') Conidia constricted at the septa, shorter than 100 μm
. . . . ***Trichocladium achrasporum,*** Fig. e; p. 504

  50'(47') Conidia not constricted at the septa, longer than 100 μm
. . . . ***Sporidesmium salinum,*** Fig. f; p. 502

51(46') Apical cell considerably broader than basal cell 52

51'(46') Apical cell not conspicuously broader than basal cell (see also imperfect state of *Ceriosporopsis halima*) 56

  52(51) Mature conidia black, shiny, fist-shaped, not constricted at the obscured septa
. . . . ***Cirrenalia pygmea,*** Fig. g; p. 480

  52'(51) Mature conidia brown, all septa distinct, constricted at the septa . . . 53

53(52') Conidia have 6–12 septa, slightly constricted at the septa
. . . . ***Cirrenalia tropicalis,*** Fig. h; p. 481

53′(52′) Conidia mostly with less than 6 septa, distinctly constricted at the septa   .   54
54(53′) Conidia reddish-brown; height of apical cell less than 13.5 μm
. . . . ***Cirrenalia macrocephala***, Fig. a; p. 477
54′(53′) Conidia without reddish tint; height of apical cell more than 13 μm . . 55
55(54′) Conidia have 3–5 (sometimes 6) septa, apical cell subglobose to ellipsoidal; a
tropical species . . . . ***Cirrenalia pseudomacrocephala***, Fig. b; p. 479
55′(54′) Conidia have 3 septa, rarely 2 or 4 septa, apical cell often sausage-shaped; a
temperate species   . . . . ***Cirrenalia fusca***, Fig. c; p. 476
  56(51′) Conidial filament turns to produce a terminal, regular spiral; no additional
  complex spores formed . . . . ***Zalerion maritimum***, Fig. d; p. 506
  56′(51′) Conidial filament turns to a lateral, variable spiral; additional complex spores
  composed of up to several hundred cells formed in the substrate
  . . . . ***Zalerion varium***, Fig. e; p. 508
57(42′) Conidia borne acropetally in chains   . . . . . . ***Alternaria*** spp., Fig. f; p. 473
57′(42′) Conidia single on the conidiophore 58
  58(57′) Conidia black at maturity, septa obscured
  . . . . ***Monodictys pelagica***, Fig. g; p. 496
  58′(57′) Conidia fuscous, gray or brown at maturity, septa distinct 59
59(58′) Conidia consisting of 3–8 parallel branches arising from a single cell
. . . . ***Dictyosporium pelagicum***, Fig. h; p. 490

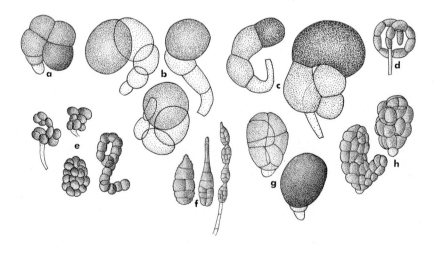

59'(58') Cells of the conidium not in parallel branches 60
  60(59') Conidiophores united into sporodochia
      . . . . *Epicoccum* **spp.,** Fig. i; p. 511
    60'(59') Conidia not on sporodochia 61
61(60') Conidia trigonous, on dead leaves of *Triglochin*
      . . . . *Stemphylium triglochinicola,* Fig. j; p. 503
  61'(60') Conidia not trigonous, on algae or wood
      . . . . *Stemphylium* **spp.,** Fig. k; p. 503

V. MYCELIA STERILIA

Only one representative . . . . *Papulaspora halima,* Fig. l; p. 465

25. Classification

As we have previously emphasized (Kohlmeyer and Kohlmeyer, 1964, 1971b), marine fungi are not a natural, taxonomic group, but consist of a heterogeneous assemblage of species growing in marine habitats. The first attempt at a suprageneric classification of the filamentous higher marine fungi was begun in our "Icones" (Kohlmeyer and Kohlmeyer, 1964–1969) and was later supplemented (Kohlmeyer, 1974a). The classification in the following chapter, containing descriptions of genera and species, and in Tables XVIII and XIX is based on Müller and von Arx (1973) for the Pyrenomycetes, on von Arx and Müller (1975) for the Loculoascomycetes, and on Ainsworth *et al.* (1971) for the Deuteromycetes. It is evident that true marine fungi occur in all classes of higher fungi, except in the class Discomycetes (Tables XVIII and XIX). The one discomycetous exception, *Orbilia marina,* is not marine in the strict sense, since it develops on decaying algae above the high-water mark. Apparently, Discomycetes are more highly evolved than Pyrenomycetes and Loculoascomycetes. Probably the Discomycetes are unable to live in submerged conditions in salt water and, therefore, have not succeeded in colonizing the marine environment.

The number of Plectomycetes is small in marine habitats, and the only two representatives, *Amylocarpus encephaloides* and *Eiona tunicata*, occur at the high-water mark, which indicates that members of Plectomycetes are also secondary invaders of marine localities (see Chapter 22 on phylogeny).

Pyrenomycetes and Loculoascomycetes include the largest number of marine species. Two orders, one in each of these two classes, contain the majority of marine Ascomycetes. The Sphaeriales contain 84 species and the Dothideales 51 species. Together, these two orders embrace 84% of the genera with marine representatives and 90% of all described filamentous marine Ascomycetes (Table XIX).

TABLE XVIII

Classification of Filamentous Higher Marine Fungi[a]

Ascomycotina

Discomycetes
  Helotiales
    Orbiliaceae
      Orbilia (1 supralittoral sp.)
Plectomycetes
  Eurotiales
    Eurotiaceae
      Amylocarpus (1 sp.)
      Eiona (1 sp.)
Pyrenomycetes
  Laboulbeniales
    Laboulbeniaceae
      Laboulbenia (1 marine sp.)
  Phyllachorales
    Physosporellaceae
      Buergenerula (1 marine sp.)
  Spathulosporales
    Spathulosporaceae
      Spathulospora (5 spp.)
  Sphaeriales
    Diaporthaceae
      Gnomonia (3 marine spp.)
    Halosphaeriaceae
      Aniptodera (1 sp.)
      Bathyascus (1 sp.)
      Carbosphaerella (2 spp.)
      Ceriosporopsis (5 spp.)
      Chadefaudia (5 spp.)
      Corollospora (8 spp.)
      Haligena (5 spp.)
      Halosarpheia (1 sp.)
      Halosphaeria (13 spp.)
      Lignincola (1 sp.)
      Lindra (3 spp.)
      Lulworthia (at least 3 spp.)
      Nais (1 sp.)
      Nautosphaeria (1 sp.)
      Trailia? (1 sp.)
    Hypocreaceae
      Halonectria (1 sp.)
      Heleococcum (1 marine sp.)
      Hydronectria (1 marine sp.)
      Nectriella (1 marine sp.)

[a] Genera preceded by an asterisk contain marine species exclusively. A question mark indicates the uncertain position of a genus under the preceding higher taxon.

(Continued)

TABLE XVIII *(Continued)*

Ascomycotina (continued)

Polystigmataceae
  *Haloguignardia? (4 spp.)
  *Phycomelaina (1 sp.)
Sordariaceae
  *Biconiosporella? (1 sp.)
  Zopfiella (2 marine spp.)
Sphaeriaceae
  *Abyssomyces (1 sp.)
  Chaetosphaeria (1 marine sp.)
  *Pontogeneia (6 spp.)
Verrucariaceae
  Pharcidia (3 marine spp.)
  *Turgidosculum? (2 spp.)
Sphaeriales incertae sedis
  *Oceanitis (1 sp.)
  Ophiobolus (1 marine sp.)
  *Savoryella (1 sp.)
  *Torpedospora (2 spp.)
Loculoascomycetes
  Dothideales
  Herpotrichiellaceae
    Herpotrichiella (1 marine sp.)
  Mycoporaceae
    Leiophloea (1 marine sp.)
  Mycosphaerellaceae
    Didymella (3 marine spp.)
    Mycosphaerella (7 marine spp.)
  Patellariaceae
    Banhegyia (1 facultative marine sp.)
    *Kymadiscus (1 sp.)
  Pleosporaceae
    Didymosphaeria (4 marine spp.)
    *Halotthia (1 sp.)
    *Helicascus? (1 sp.)
    Keissleriella (1 marine sp.)
    Leptosphaeria (13 marine spp.)
    *Manglicola? (1 sp.)
    Massarina (1 marine sp.)
    Microthelia (1 marine sp.)
    *Paraliomyces (1 sp.)
    Phaeosphaeria (2 marine spp.)
    Pleospora (7 marine spp.)
    *Pontoporeia (1 sp.)
    *Thalassoascus? (1 sp.)
    Trematosphaeria (2 marine spp.)
  Ascomycotina incertae sedis
    *Crinigera (1 sp.)

TABLE XVIII *(Continued)*

Ascomycotina (continued)

    *Orcadia (1 sp.)
    Sphaerulina (2 marine spp.)

Basidiomycotina

Gasteromycetes
  Melanogastrales
    Melanogastraceae
      *Nia (1 sp.)
Hymenomycetes
  Aphyllophorales
    Corticiaceae
      *Digitatispora (1 sp.)
    Aphyllophorales incertae sedis
      *Halocyphina (1 sp.)
Teliomycetes
  Ustilaginales
    Tilletiaceae
      Melanotaenium (1 marine sp.)

Deuteromycotina

Hyphomycetes
  Agonomycetales
    Agonomycetaceae
      Papulaspora (1 marine sp.)
  Hyphomycetales
    Moniliaceae
      Blodgettia (1 marine sp.)
      *Botryophialophora (1 sp.)
      Clavatospora (1 marine sp.)
      *Varicosporina (1 sp.)
    Dematiaceae
      Alternaria (doubtful marine isolates)
      *Asteromyces (1 sp.)
      *Cirrenalia (5 spp.)
      Cladosporium (1 marine sp.; also doubtful marine isolates)
      Clavariopsis (1 marine sp.)
      *Cremasteria (1 sp.)
      Dendryphiella (2 marine spp.)
      Dictyosporium (1 marine sp.)
      Drechslera (1 marine sp.)
      Humicola (1 marine sp.)
      Monodictys (1 marine sp.)
      *Orbimyces (1 sp.)
      Periconia (2 marine spp.)
      Sporidesmium (1 marine sp.)
      Stemphylium (1 marine sp.; also doubtful marine isolates)
      Trichocladium (1 marine sp.)
      Zalerion (2 marine spp.) ·

(Continued)

TABLE XVIII *(Continued)*

Deuteromycotina (continued)
Tuberculariales
  Tuberculariaceae
    Allescheriella (1 marine sp.)
    Epicoccum (doubtful marine isolates)
    Tubercularia (1 marine sp.)
Coelomycetes
  Sphaeropsidales
    Sphaerioidaceae
      Ascochyta (1 marine sp.)
      Ascochytula (1 marine sp.)
      Camarosporium (3 marine spp.)
      Coniothyrium (1 marine sp.)
      Cytospora (1 marine sp.)
      Diplodia (1 marine sp.)
      Macrophoma (probably several spp.)
      *Phialophorophoma (1 sp.)
      Phoma (3 marine spp. named; probably others occurring)
      Rhabdospora (1 marine sp.)
      Robillarda (1 marine sp.)
      Septoria (2 marine spp.)
      Stagonospora (2 marine spp.)
    Excipulaceae
      Dinemasporium (1 marine sp.)
  Melanconiales
    Melanconiaceae
      Sphaceloma (1 marine sp.)

Each of the remaining orders has only one genus with marine members. Laboulbeniales are rare in marine habitats because they are host specific and their hosts, predominantly insects, seldom occur submerged in salt-water habitats. The one representative of the Phyllachorales, *Buergenerula spartinae,* is restricted to salt-marsh cordgrass and most probably evolved from terrestrial ancestors. Finally, the Spathulo-sporales, with the single genus *Spathulospora,* is believed to be an original marine group that descended from marine ancestors (see Chapter 22). The number of species in this order may be limited because of their restricted parasitic way of life on the genus *Ballia* of the Rhodophyta.

Filamentous Basidiomycetes as a group are extremely rare in marine habitats. Only four species are known from marine waters (Table XVIII), and this low number indicates that Basidiomycetes did not originate in the ocean and were also less well equipped to invade the marine environment secondarily. Tresner and Hayes (1971) determined that Basidiomycetes were the least tolerant when they compared NaCl tolerance of 975

TABLE XIX

Taxonomic Distribution of Filamentous Marine Ascomycetes

Classes and orders with marine representatives	Genera (total)	Monotypic genera	Genera with exclusively marine species	Species
Discomycetes				
Helotiales	1	0	0	1
Plectomycetes				
Eurotiales	2	2	2	2
Pyrenomycetes				
Laboulbeniales	1	0	0	1
Phyllachorales	1	0	0	1
Spathulosporales	1	0	1	5
Sphaeriales	33	13	25	84
Loculoascomycetes				
Dothideales	20	7	7	51
Ascomycotina incertae sedis	3	2	2	4
Total	62	24	37	149

TABLE XX

Taxonomic Distribution of Filamentous Marine Deuteromycetes

Suprageneric taxa with marine representatives	Genera (total)	Monotypic genera	Genera with exclusively marine species	Species
Hyphomycetes				
Agonomycetales				
Agonomycetaceae	1	0	0	1
Hyphomycetales				
Moniliaceae	4	2	2	4
Dematiaceae	17	3	4	26
Tuberculariales				
Tuberculariaceae	3	0	0	3
Coelomycetes				
Sphaeropsidales				
Sphaerioidaceae	13	1	1	20
Excipulaceae	1	0	0	1
Melanconiales				
Melanconiaceae	1	0	0	1
Total	40	6	7	56

species of terrestrial fungi. The highest salt tolerance was found in the smuts (Ustilaginales), which might explain the frequent marine occurrence of yeasts with heterobasidiomycetous life cycles similar to those of the smuts (Fell, 1970) and the marine occurrence of the smut *Melanotaenium ruppiae*, a parasite of a saline phanerogam. The family Corticiaceae, to which *Digitatispora marina* belongs, may be another group with a high tolerance to NaCl (Kohlmeyer, 1974a).

The distribution of marine Deuteromycotina among suprageneric taxa (Tables XVIII and XX), in contrast to the foregoing groups, does not give clues to the terrestrial or marine origin of the species as long as a natural system of this subdivision is missing. The detection of further perfect–imperfect connections (Table XXI) will probably give some

---

TABLE XXI

Marine Ascomycetes with Imperfect or Spermatial States

Buergenerula spartinae: Conidial or spermatial state (Kohlmeyer and Gessner, *Can. J. Bot.* **54**, 1764, 1976)

Ceriosporopsis circumvestita: chlamydospores in culture (Kohlmeyer and Kohlmeyer, *Nova Hedwigia Z. Kryptogamenkd.* **12**, 195, 1966)

Ceriosporopsis halima: chlamydospores in culture (Kohlmeyer and Kohlmeyer, *Icones Fungorum Maris,* Table 31, 1964–1969)

Chadefaudia marina: possibly a pycnidial or spermogonial state (Kohlmeyer, *Bot. Mar.,* **16**, 204, 1973)

Corollospora pulchella: conidial state *Clavariopsis bulbosa* (Shearer and Crane, *Mycologia* **63**, 253, 1971)

Didymella fucicola: pycnidial or spermogonial state (Kohlmeyer, *Phytopathol. Z.* **63**, 342, 1968; Kohlmeyer and Kohlmeyer, *Icones Fungorum Maris,* Table 82, 1964–1969)

Didymella gloiopeltidis: pycnidial or spermogonial state (Kohlmeyer and Kohlmeyer, unpublished; see p. 382)

Didymella magnei: possibly a pycnidial or spermogonial state (Kohlmeyer and Kohlmeyer, unpublished; see p. 384)

Didymosphaeria danica: pycnidial or spermogonial state *Phoma marina* (Lind, *in:* "Danish fungi as represented in the herbarium of E. Rostrup," Nordisk Forlag, Copenhagen, p. 214, 1913)

Haloguignardia decidua: pycnidial or spermogonial state (Kohlmeyer and Kohlmeyer, unpublished; see p. 330)

Haloguignardia irritans: pycnidial or spermogonial state (Estee, *Univ. Calif. Publ. Bot.* **4**, 311, 1913)

Haloguignardia oceanica: pycnidial or spermogonial state (Kohlmeyer, *Mar. Biol.* **8**, 344, 1971)

Haloguignardia tumefaciens: pycnidial or spermogonial state (Kohlmeyer, *Mar. Biol.* **8**, 347–348, 1971)

Halonectria milfordensis: pycnidial or spermogonial state (Kohlmeyer and Kohlmeyer, *Icones Fungorum Maris,* Table 74, 1964–1969)

Halosphaeria cucullata: conidial state *Periconia prolifica* (Kohlmeyer, *Can. J. Bot.* **47**, 1478, 1969)

Halosphaeria mediosetigera: conidial state *Trichocladium achrasporum* (Shearer and Crane, *Mycologia* **69**, 1218, 1977)

25. Classification

TABLE XXI (Continued)

Halotthia posidoniae: pycnidial or spermogonial state (Kohlmeyer, *Nova Hedwigia Z. Kryptogamenkd.* **4,** 396, 1962)

Heleococcum japonense: conidial state in culture (Tubaki, *Trans. Mycol. Soc. Jpn.* **8,** 5, 1967)

Leiophloea pelvetiae: pycnidial or spermogonial state (Kohlmeyer, *Bot. Mar.* **16,** 206–209, 1973)

Leptosphaeria albopunctata: pycnidial or spermogonial state in pure culture (Kohlmeyer and Kohlmeyer, *Icones Fungorum Maris,* Table 34, 1964–1969)

Leptosphaeria avicenniae: possible pycnidial or spermogonial state *Rhabdospora avicenniae* (Kohlmeyer and Kohlmeyer, *Mycologia* **63,** 853, 1971)

Leptosphaeria obiones: pycnidial or spermogonial state in pure culture (Wagner, *Nova Hedwigia Z. Kryptogamenkd.* **9,** 45, 1965)

Leptosphaeria typhicola: pycnidial or spermogonial state in pure culture (Lucas and Webster, *Trans. Br. Mycol. Soc.* **50,** 118, 1967)

Massarina cystophorae: pycnidial or spermogonial state (Kohlmeyer and Kohlmeyer, unpublished; see p. 427)

Mycosphaerella ascophylli: probable spermogonial or pycnidial state *Septoria ascophylli* (Kohlmeyer, *Phytopathol. Z.* **63,** 345, 1968; Kohlmeyer and Kohlmeyer, *Icones Fungorum Maris,* Table 76, 1964–1969)

Mycosphaerella staticicola: pycnidial or spermogonial state (Kohlmeyer and Kohlmeyer, unpublished; see p. 391)

Mycosphaerella suaedae-australis: pycnidial state *Septoria suaedae-australis* (Hansford, *Proc. Linnean Soc. N. S. W.* **79,** 123, 1954)

Phaeosphaeria ammophilae: possible imperfect state *Tiarospora perforans* (von Höhnel, *Hedwigia* **60,** 141, 1918)

Phaeosphaeria typharum: pycnidial state *Hendersonia typhae* (Webster, *Trans. Br. Mycol. Soc.* **38,** 405, 1955)

Pharcidia balani: pycnidial or spermogonial state (Kohlmeyer and Kohlmeyer, *Icones Fungorum Maris,* Table 66, 1964–1969)

Pharcidia laminariicola: pycnidial or spermogonial state (Kohlmeyer, *Bot. Mar.* **16,** 209, 1973)

Pharcidia rhachiana: pycnidial or spermogonial state (Kohlmeyer, *Bot. Mar.* **16,** 210, 1973)

Phycomelaina laminariae: pycnidial or spermogonial state (Kohlmeyer, *Phytopathol. Z.* **63,** 350–356, 1968)

Pleospora triglochinicola: conidial state *Stemphylium triglochinicola* (Webster, *Trans. Br. Mycol. Soc.* **53,** 481, 1969)

Pleospora vagans var. *spartinae:* pycnidial state in pure culture (Webster and Lucas, *Trans. Br. Mycol. Soc.* **44,** 429, 1961)

Thalassoascus tregoubovii: pycnidial or spermogonial state (Kohlmeyer, *Nova Hedwigia Z. Kryptogamenkd.* **6,** 134, 1963)

Turgidosculum complicatulum: pycnidial or spermogonial state (Nylander, *Flora* **67,** 212, 1884)

Turgidosculum ulvae: pycnidial or spermogonial state (Kohlmeyer and Kohlmeyer, *Bot. Jahrb. Syst. Pflanzengesch. Pflanzengeogr.* **92,** 429, 1972)

Zopfiella marina: conidial state in culture (Furuya and Udagawa, *J. Jpn. Bot.* **50,** 249, 1975)

---

answers, and conidial ontogeny needs to be examined (see Chapter 20). The greatest number of deuteromycetous species occurs in the Dematiaceae (26 spp.), namely, Hyphomycetes with dark hyphae and conidia, and in Sphaerioidaceae (20 spp.), that is, Coelomycetes forming pycnidia.

Suprageneric taxa are not characterized in the descriptive part of this treatise because descriptions of classes, orders, and families have been published recently in Ainsworth, Sparrow, and Sussman's treatise *The Fungi* (1973) and such characterizations would be repetitive and beyond the scope and function of this book. Furthermore, the marine members of the suprageneric groups are often not the most characteristic examples of such higher taxa. Therefore, we describe in Chapter 26 only those higher groups that include marine representatives exclusively, namely, Spathulosporales, Spathulosporaceae, and Halosphaeriaceae, and list the other orders and families without descriptions. All genera and species are fully described.

26. Taxonomy and Descriptions of Filamentous Fungi

I. Ascomycotina
   A. Discomycetes
      1. Helotiales
         a. Orbiliaceae

Orbilia Fries

Summa Veg. Scand., p. 357, 1849

=*Hyalinia* Boudier, *Bull. Soc. Mycol. Fr.* **1,** 114, 1885
=*Orbiliopsis* Sydow, *Ann. Mycol.* **22,** 308, 1924 (Saccardo, *Syll. Fung.* **18,** 139, 1906, as subgenus)
=*Pteromyces* Bommer, Rousseau et Saccardo, *Ann. Mycol.* **3,** 507, 1905; Saccardo, *Syll. Fung.* **22,** 725, 1913

LITERATURE: Clements and Shear (1931); Dennis (1968); Korf (1973). ASCOCARPS small, discoid (apothecialike), superficial, subsessile, waxy, light-colored, often translucent when fresh. EXCIPULA composed of large, subglobose or angular, thin-walled cells. PARAPHYSES abruptly swollen at the tips, surpassing the asci and cohering with them firmly to produce the waxy, translucent texture of the hymenium. ASCI eight-spored, small, usually cylindrical–clavate, with a forked base and truncate tip, not blued with iodine. ASCOSPORES small, one-celled, hyaline. MODE OF LIFE: Saprobic, mostly on wood, bark, or decaying fungi. TYPE SPECIES: *Orbilia leucostigma* Fr. NOTE: There is one species occurring on hosts of marine origin besides about 30 terrestrial taxa; synonyms after Clements and Shear (1931), description after Dennis (1968) and Korf (1973). ETYMOLOGY: From the Latin, *orbis* = disk, circle, in reference to the disklike ascocarps.

Orbilia marina Phillips ex Boyd *in* Smith

Trans. Br. Mycol. Soc. **3**, 113, 1908 [1909] [*Calloria marina* Phillips, *in* Smith (1908) = unpublished manuscript name]

LITERATURE: Boyd (1901, 1909, 1911, 1916); O. Eriksson (1973).

Fig. 37

ASCOCARPS 250–425 μm in diameter, at first concave, becoming convex and discoid with a circular outline at maturity, apothecialike, erumpent, then superficial, sessile, rooted in the substrate with scleroplectenchymatous hyphae, waxy, initially light orange-colored, later darker, solitary or gregarious. PARAPHYSES filamentous, branched, septate, abruptly swollen at the tips, surpassing the asci and cohering with them firmly; apices 3–6 μm in diameter, enclosed in gelatinous sheaths; gelatinous material forming a continuous, transparent layer over the hymenium; the latter about 50 μm high. ASCI about 40–75 × 5–7.5 μm, eight-spored, cylindrical–clavate, tapering at the base, apically rounded, unitunicate, thin-walled, with apical ring-shaped apparatuses, 1.5 μm in

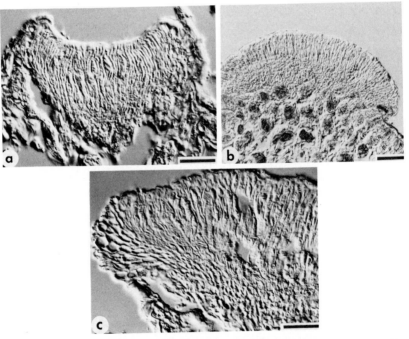

Fig. 37. *Orbilia marina* from *Ascophyllum nodosum.* (a) Young ascocarp (6-μm section). (b) Mature ascocarp (4-μm section). (c) Edge of mature ascocarp (6-μm section). (Collection of Boyd from K; all bars = 25 μm; in Nomarski interference contrast.)

diameter, not bluing in iodine. ASCOSPORES 7.5–10 × 4–5 μm, obliquely one-seriate, ellipsoidal, one-celled, smooth-walled, hyaline, with or without a thin gelatinous sheath. MODE OF LIFE: Saprobic. HOSTS: Decaying thalli of *Ascophyllum nodosum, Fucus serratus, F. vesiculosus, Fucus* sp., and *Halidrys siliquosa*. RANGE: Atlantic Ocean—Great Britain (Scotland), Norway, Sweden; Baltic Sea—Denmark (Sjaelland), Sweden. NOTE: Description based on Boyd (in Smith, 1908), O. Eriksson (1973), and on our examination of type material from Herb. K; although the fungus is not marine in the strict sense because it is usually found growing above the high-tide line, we have included *O. marina* because the substrates are all of marine origin. ETYMOLOGY: From the Latin, *marinus* = marine, in reference to the habitat of the species.

B. Plectomycetes
1. Eurotiales
a. Eurotiaceae

Amylocarpus Currey

Proc. R. Soc. London **9**, 119–123, 1857–1859
=*Plectolitus* Kohlm., *Nova Hedwigia* **2**, 328, 1960

ASCOCARPS solitary or gregarious, ellipsoidal to subglobose, erumpent, mostly superficial, nonostiolate, coriaceous, light-colored, dissolving irregularly above. PARAPHYSES absent. ASCI eight-spored, clavate or ellipsoidal, apiculate and pedunculate, unitunicate, thin-walled with thickened apical wall, early deliquescing; ascogenous hyphae and asci scattered irregularly throughout the ascocarp center between sterile hyphae. ASCOSPORES subglobose to ovoidal, one-celled, hyaline, staining blue with iodine, provided with numerous awl-like, radiating appendages. SUBSTRATE: Dead wood. TYPE SPECIES: *Amylocarpus encephaloides* Currey. The genus is monotypic. ETYMOLOGY: From the Greek, *amylon* = starch, and *carpos* = fruit, in reference to the amyloid reaction of the ascocarp contents upon treatment with iodine.

Amylocarpus encephaloides Currey

Proc. R. Soc. London **9**, 119–123, 1857–1859

=*Plectolitus acanthosporum* Kohlm., *Nova Hedwigia* **2**, 329, 1960

LITERATURE: Bommer and Rousseau (1891); Cavaliere (1966a); Henningsson (1974, 1976a,b); G. C. Hughes (1968, 1969); G. C. Hughes and

Chamut (1971); T. W. Johnson (1963c); T. W. Johnson and Sparrow (1961); Koch (1974); Kohlmeyer (1961a, 1968d, 1971b); Kohlmeyer and Kohlmeyer (1964–1969); Lind (1913); Lindau (1898, 1899); Rostrup (1884, 1888).

Fig. 38

ASCOCARPS 0.3–1.4 mm high, 0.3–3 mm in diameter, ellipsoidal or sub-globose, erumpent and finally superficial, rarely partly immersed, seated in the substrate with broad subicles, nonostiolate, cleistothecial, coriaceous or cartilaginous, cream-colored, yellow or reddish-yellow, amber-colored and corny when dried, solitary or gregarious, rarely two grown together; wall dissolving irregularly above at maturity. PERIDIUM 32–80 μm thick, two-layered; pseudoparenchymatous on the outside, prosenchymatous on the inside. PARAPHYSES absent. ASCI 27–43.5 × 18–26.5 μm in the sporogenous part, 8–16 × 3–5.5 μm in the peduncle, eight-spored, broadly clavate or ellipsoidal, apiculate, pedunculate, unitunicate, thin-walled with thickened apical wall, aphysoclastic, without apical apparatuses, deliquescing at ascospore maturity; ascogenous hyphae and asci scattered irregularly throughout the ascocarp center between capillitiumlike, sterile hyphae. ASCOSPORES 8–16 μm in diameter (excluding appendages), subglobose to ovoidal, one-celled, hyaline, staining blue in iodine, appendaged, containing one large oil globule; 10–25 awl-shaped appendages, 5.5–10 × 1 μm, slender, acuminate, rigid, distributed more or less evenly over the ascospore surface. MODE OF LIFE: Saprobic.

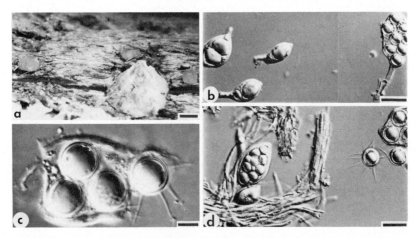

Fig. 38. *Amylocarpus encephaloides.* (a) Ascocarps on the wood surface near a barnacle (bar = 1.5 mm). (b) Immature and mature asci (bar = 20 μm). (c) Ruptured ascus, showing one ascospore with expanded appendages (bar = 5 μm). (d) Asci between capillitiumlike hyphae, ascospores on the left side (bar = 15 μm). (Herb. J.K. 3844; b–d in Nomarski interference contrast.)

SUBSTRATES: Intertidal wood, usually at or above the high-water mark (e.g., *Acer* sp., *Aesculus hippocastanum, Alnus* sp., *Arbutus menziesii, Betula pubescens, Betula* sp., *Fagus* sp., *Fraxinus* sp., *Populus* sp., *Quercus* sp., and *Salix* sp.). RANGE: Atlantic Ocean—Belgium, Canada (Newfoundland), Denmark, Great Britain [Scotland (J. Kohlmeyer, unpublished), Wales], United States (Maine, Massachusetts); Baltic Sea— Denmark (Sjaelland), Germany (Rügen, G.D.R.), Sweden; Pacific Ocean—Canada (British Columbia), Chile (Isla Hoste), United States (Washington). ETYMOLOGY: From the Greek, *encephalos* = brain, and *-oides* = -like, in reference to the ascocarps, appearing to the describer as having "a convoluted membrane."

Eiona Kohlm.

Ber. Dtsch. Bot. Ges. **81**, 58, 1968

ASCOCARPS gregarious, subglobose, finally cupulate by irregular dissolution of the upper wall, superficial, cleistothecial, without ostioles, coriaceous, brown, surface crinkled. PARAPHYSES absent; venter of ascocarp filled at random with ascogenous hyphae. ASCI eight-spored, clavate or ellipsoidal, unitunicate, thin-walled, early deliquescing, developing scattered throughout the ascocarp cavity. ASCOSPORES ellipsoidal, one-celled, hyaline, at both ends with subgelatinous, irregular appendages. SUBSTRATE: Dead wood. TYPE SPECIES: *Eiona tunicata* Kohlm. The genus is monotypic. ETYMOLOGY: From the Greek, *eion* = coast, shore, or beach, referring to the habitat of the fungus.

Eiona tunicata Kohlm.

Ber. Dtsch. Bot. Ges. **81**, 58, 1968

LITERATURE: Koch (1974); Kohlmeyer and Kohlmeyer (1964–1969).

ASCOCARPS 140–530 μm high, 170–600 μm in diameter, subglobose, finally cupulate by irregular dissolution of the upper wall, superficial, cleistothecial, without ostioles, coriaceous, brown, surface crinkled, gregarious. PERIDIUM two-layered; outer wall 17–75 μm thick, prosenchymatous, hyaline, crumpled and full of clefts; cells with small lumina; inner wall 10–15(–20) μm thick, pseudoparenchymatous, yellowish-brown to olivaceous, composed of isodiametric or elongate cells with thick walls and large lumina. PARAPHYSES absent; venter of ascocarp filled at random with hyaline, filiform, septate, ramose ascogenous hyphae, 1–1.5 μm in diameter, early deliquescing. ASCI 20–22.5 × 10–12 μm, eight-spored, broadly clavate or ellipsoidal, unitunicate, thin-walled, aphysoclastic, early deliquescing, without apical apparatuses, developing scattered throughout the ascocarp cavity. ASCOSPORES 7.5–12 × 3.5–5 μm, ellipsoidal, one-celled, hyaline, at both ends appendaged; appendages at first caplike and terminal, then subterminal, irregularly radiate, delicate, subgelatinous, ca. 3.5 μm long, developing from the exospore. MODE OF LIFE: Saprobic. SUBSTRATE: Dead, immersed wood. RANGE: Atlantic Ocean—Denmark (Jutland). ETYMOLOGY: Specific name from the Latin, *tunicatus* = clothed in a tunic, referring to the appendages covering the ascospores.

C. Pyrenomycetes
1. Laboulbeniales
a. Laboulbeniaceae

Laboulbenia Montagne et Robin *in* Robin

Histoire Naturelle des Végétaux Parasites qui Croissent sur l'Homme et sur les Animaux Vivants. Baillière et Fils, Paris, p. 622, 1853.
=*Ceraiomyces* Thaxter, *Proc. Am. Acad. Arts Sci.* **36**, 410, 1901
=*Eumisgomyces* Spegazzini, *An. Mus. Nac. Hist. Nat. Buenos Aires* **23**, 176, 1912
=*Laboulbeniella* Spegazzini, *An. Mus. Nac. Hist. Nat. Buenos Aires* **23**, 188, 1912
=*Schizolaboulbenia* Middlehoek, *Fungus* **27**, 73, 1957
=*Thaxteria* Giard, *C. R. Séances Soc. Biol. Ses Fil., Sér. 9* **4**, 156, 1892

RECEPTACLE proper consists of a basal cell (I) that forms a dark foot, the attachment organ to the host, and a subbasal cell (II) that subtends two cells (III and VI), placed side by side; cell III subtends two smaller cells (IV and V) that support the appendages; cell VI subtends cell VII bearing the perithecium. PERITHECIA solitary, sessile or stalked, compressed, elongate–ellipsoidal, ostiolate, consisting of three basal cells that subtend usually five tiers of cells composed of four vertically oriented rows of cells; presumably arising following fertilization of an archicarp bearing a filamentous, simple, or branched trichogyne. APPENDAGE typically consisting of two basal cells, giving rise to series of branches; the fertile ones bearing antheridia. ANTHERIDIA simple, flask-shaped, producing spermatia endogenously. ASCI four-spored, ellipsoidal, unitunicate, thin-walled, early deliquescing; developing at the base of the ascocarp venter. ASCOSPORES cylindrical, one-septate in the lower third, hyaline, surrounded by a mucilaginous sheath. ANTHERIDIA and perithecia borne on the same individual (monoecious), rarely on separate individuals (dioecious). HOSTS: Chitinous membranes of insects (Hexapoda) and mites (Arachnida). TYPE SPECIES: *Laboulbenia rougetii* Mont. et Robin *in* Robin. NOTE: Only one marine species is known among over 500 taxa; description after Thaxter (1896) and Benjamin (1971, 1973); synonyms following Benjamin (1971). ETYMOLOGY: Commemorating A. Laboulbène (1825–1898), French entomologist.

Laboulbenia marina Picard

C. R. Séances Soc. Biol. Ses Fil. **65**, 484–485, 1908

Fig. 39

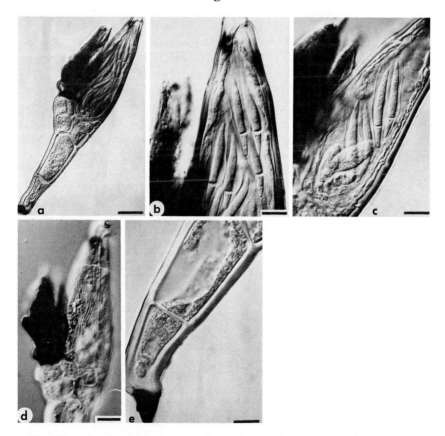

Fig. 39. *Laboulbenia marina* from the marine beetle, *Aepus robini.* (a) Thallus (bar = 25 μm). (b) Tip of ascocarp with two-celled ascospores surrounded by a mucilaginous sheath (bar = 10 μm). (c) Asci (arrow) and ascospores inside the ascocarp (bar = 10 μm). (d) Dark appendages, wavy ornamentations on surface of ascocarp (bar = 15 μm). (e) Basal cells of thallus with black, footlike attachment organ (bar = 10 μm). (Slide R. K. Benjamin No. 2537; in Nomarski interference contrast.)

THALLUS length 150–230 μm (including perithecium), surface with indistinct wavy lines. RECEPTACLES 105–112 μm long, 48–53 μm in diameter, consisting of an elongate basal cell (I) that forms a blackish foot, the attachment organ to the host, and of a longer subbasal cell (II) that subtends two short cells (III and VI), placed side by side; cell III subtends two smaller cells (IV and V) that support the appendages; cell VI subtends cells VII and VIII bearing the perithecium. PERITHECIA 76–118 μm high, 38–44 μm in diameter, elongate–ellipsoidal, wider at the base, conical and attenuate at the apex, sessile, ostiolate, hyaline to light brown, finally with a darker zone around the ostiole, solitary. APPENDAGES originating from cells IV and V that are topped by a black cell bearing two unequal cells of lighter color; from each of these arise up to four simple appendages; appendages 30–38 μm long, 8–10 μm in diameter at the base, 4 μm at the tip, broad and dark at the base, tapering toward the light-colored apex, three- or four-celled, constricted at the septa. ANTHERIDIA not observed. ASCI four-spored, ellipsoidal to clavate, unitunicate, thin-walled, aphysoclastic, without apical apparatuses, early deliquescing; developing at the base of the ascocarp venter. ASCOSPORES 26–35 × 4 μm (excluding sheath), elongate fusiform, pointed at the apex, rounded at the base, slightly curved, one-septate in the lower third, hyaline, surrounded by a mucilaginous sheath, leaving the perithecium and attaching to the host in pairs. MODE OF LIFE: Parasitic. HOSTS: On the base of elytra or on hairs of *Aepus robini*, a beetle living in the *Laminaria* zone. RANGE: Atlantic Ocean—France. Description in part after J. Kohlmeyer (unpublished), based on slide No. 2537 of R. K. Benjamin. ETYMOLOGY: Name from the Latin, *marinus* = marine, referring to the habitat.

2. Phyllachorales

a. Physosporellaceae

Buergenerula Sydow

Ann. Mycol. **34**, 392, 1936

=*Yukonia* Sprague, *Res. Stud. Wash. State Univ.* **30**(2), 45, 1962

LITERATURE: Barr (1976b, 1977); Dennis (1968); Müller (1950).

HYPHAE septate, with or without hyphopodia. ASCOCARPS subglobose, obpyriform, or ellipsoidal, immersed, ostiolate, papillate, periphysate, coriaceous, brown. PERIDIUM composed of several layers of cells, forming a textura angularis. PSEUDOPARAPHYSES (or apical paraphyses?) filamentous, attached at the top to a pseudoparenchyma of small, subglobose, thin-walled, hyaline cells. Asci eight-spored, cylindrical, fusiform or subclavate, unitunicate, persistent, thick-walled, with a nonamyloid apical apparatus that stains faintly blue in ink; developing at the base of the ascocarp venter. Ascospores biseriate, elliptical–fusiform, or clavate, often curved, two- or three-septate, hyaline. Hosts: Terrestrial and aquatic plants. Type Species: *Buergenerula biseptata* (Rostrup) Sydow. Note: One marine species is known besides two terrestrial taxa. Etymology: Named in honor of Prof. Oskar Bürgener (1876–1966), collector of the type specimen, teacher in Stralsund (German Democratic Republic).

Buergenerula spartinae Kohlmeyer et Gessner

Can. J. Bot. **54**, 1764, 1976

Literature (mostly as *Sphaerulina pedicellata* Johnson): Gessner (1976); Gessner and Goos (1973a,b); Gessner *et al.* (1972); Gessner and Kohlmeyer (1976); Goos and Gessner (1975); E. B. G. Jones (1962a); Schultz and Quinn (1973); Sieburth *et al.* (1974); E. E. Webber (1970).

Fig. 40

Hyphae 2.5–8 μm in diameter, branching and anastomosing, septate, thick-walled, light- or dark-colored, forming olive-brown to blackish, lobed hyphopodia, 14–37 μm in diameter. Ascocarps 300–475 μm high (excluding the necks), 200–450 μm in diameter, subglobose to obpyriform, ostiolate, papillate, coriaceous, light or dark brown, solitary, immersed, often developing in air chambers of the host tissue or between culm and leaf sheath. Peridium 30–60 μm thick, two-layered; outer layer 8–16 μm thick, composed of three to five layers of flat, brown cells with narrow lumina; inner layer 20–30 μm thick, composed of eight to ten layers of ellipsoidal, thin-walled, hyaline to light brown cells with large lumina; cells of both layers forming a textura angularis, filled with oil globules. Papillae or necks 100–280 μm high, 120–180 μm in diameter, apical or slightly lateral, cylindrical or conical; peridium composed of elongate cells with small lumina, merging toward the ostiolar canal into periphyses; periphyses filiform, wavy, with a narrow, refractive core, surrounded by a gelatinous outer layer. Pseudoparaphyses (or apical paraphyses?) 8–10 μm in diameter, filamentous, simple or branched, thin-walled, septate, slightly constricted at the septa, filled with oil globules; developing before the asci, at first attached at the tip to a pseudoparenchyma of small, subglobose, thin-walled, hyaline cells. Asci 120–190 × 18–20 μm, eight-spored, cylindrical to subfusiform, short pedunculate or nonpedunculate, unitunicate, thick-walled, with an ellipsoidal, ringlike apical apparatus, about 4 μm in diameter; developing at the base of the ascocarp venter and growing up between the pseudoparaphyses. ASCOSPORES 37–66 × 9.5–14 μm, biseriate, clavate, thick at the apex, tapering at the base, mostly curved, three (rarely four)-septate, not or slightly constricted at the septa, hyaline. CONIDIOPHORES (or spermatiophores?) formed laterally on hyphae in culture, stalked, branched, up to 20 μm long, bearing a small, phialidelike projection. CONIDIA (or spermatia?) curved, about 2 × 5 μm. MODE OF LIFE: Parasitic (?) and saprobic. HOSTS: *Spartina alterniflora* and *Spartina* sp. RANGE: Atlantic Ocean—Argentina (Buenos Aires), Canada (New Brunswick), Great Britain (England), United States (Connecticut, Florida, Maine, New Hampshire, New Jersey, North Carolina, Rhode Island, Virginia). ETYMOLOGY: Named after the host genus, *Spartina* (Gramineae).

3. Spathulosporales Kohlmeyer

Mycologia **65**, 615, 1973

Parasites in marine algae; hyphae absent; thalli composed of irregular, thick-walled cells, forming crusts or intracellular stromata; producing stalked or sessile antheridia with spermatia, and ascogonia with trichogynes; ascocarps superficial, ostiolate, periphysate, subiculate, eparaphysate, with a thick peridium, hairy or glabrous; ascogenous cells pseudoparenchymatous, not hyphoid; asci thin-walled, unitunicate, aphysoclastic, early deliquescing, without apical apparatuses; ascospores one-celled, appendaged.

a. Spathulosporaceae Kohlmeyer

Mycologia **65**, 615, 1973

Diagnosis as in the order. TYPE GENUS: *Spathulospora* Cavaliere et Johnson.

Fig. 40. *Buergenerula spartinae.* (a) Mature ascocarp (16-μm longitudinal section; holotype Herb. J.K. 3498; bar = 50 μm). (b) Section (16μm) through ostiole with periphyses (holotype Herb. J.K. 3498; bar = 25 μm). (c) Immature ascocarp (12-μm longitudinal section) with pseudoparaphyses; the neck will develop from the apical meristematic region (paratype Herb. J.K. 3503; bar = 25 μm). (d) Hyphae and hyphopodia (holotype Herb. J.K. 3498; bar = 15 μm). (e) Ascospore stained with ink (Herb. Gessner; bar = 5 μm). (f) Ascus with apical apparatus (holotype Herb. J.K. 3498; bar = 20 μm). (g) Ascospores and immature ascus (paratype Herb. J.K. 3706; bar = 15 μm). (h) Tip of immature ascus and ascospore, wall pulled away from cytoplasm, apical apparatus in between (Herb. Gessner; bar = 15 μm). (d and e in bright field, all others in Nomarski interference contrast.)

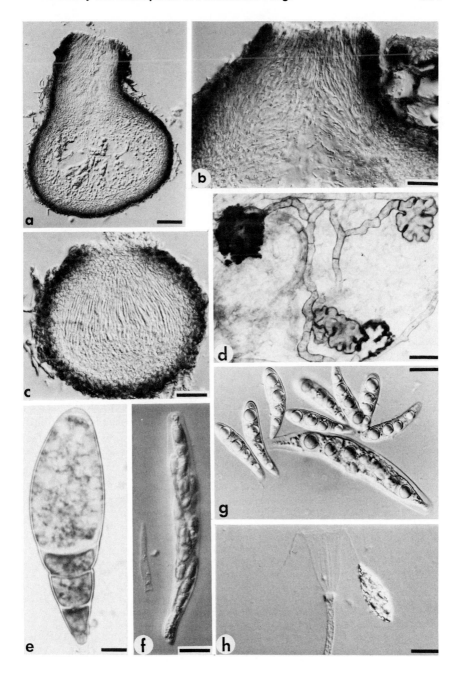

Spathulospora Cavaliere et Johnson

Mycologia **57**, 927, 1965

THALLUS crustose, surrounding host cells, bearing phialidelike antheridia and filamentous trichogynes; vegetative hyphae absent; assimilative cells forming a hypostroma in the host cells. ASCOCARPS solitary or geminate, subglobose to pyriform, seated with a subiculum on the thallus, ostiolate, papillate or epapillate, coriaceous, dark-colored, with or without hairs. PARAPHYSES lacking. ASCI eight-spored, clavate, unitunicate, thin-walled, early deliquescing, attached in the lower part of the locule. ASCOSPORES fusiform, cylindrical or ellipsoidal, one-celled, hyaline, each end with an appendage. HOSTS: Marine algae (Rhodophyta). TYPE SPECIES: *Spathulospora phycophila* Cavaliere et Johnson. NOTE: All members of the genus are obligate marine species. ETYMOLOGY: From the Latin, *spathulo-* = spathulate, and *spora* = spore, in reference to the spathulate ascospore tips of the type species, *S. phycophila*.

Spathulospora adelpha Kohlm.

Mycologia **65**, 615–617, 1973

LITERATURE: Kohlmeyer (1974a,b); Kohlmeyer and Kohlmeyer (1975).
Fig. 41
THALLUS crustose, surrounding the host cells, bearing sterile and fertile hairs, and trichogynes; vegetative hyphae absent; algal cells perforated by peglike cells that merge into intracellular assimilative cells, forming a hypostroma. ASCOCARPS 290–500 μm high, 150–310 μm in diameter, ovoidal, superficial, solitary or geminate, ostiolate, fuscous, leathery, subiculate, woolly. PERIDIUM 35–80 μm thick. STERILE HAIRS enclosing the perithecium and antheridia, curved, cylindrical, tapering, fuscous, 500–650 μm long, 8–15 μm in diameter at the base, 4–8 μm at the apex, septate, thick-walled, immersed in the wall of the perithecium. ANTHERIDIAL HAIRS shaped like a candelabrum; stalks 70–80 × 12–16 μm, cylin-

---

Fig. 41. *Spathulospora adelpha* from *Ballia callitricha.* (a) Hairy ascocarp on host filament (Herb J. K. 2981; bar = 100 μm). (b) Immature ascocarp (14-μm longitudinal section); host cell (arrow) in cross section (Herb. J.K. 3007; bar = 50 μm). (c) Mature ascocarp (8-μm longitudinal section) filled with ascospores (Herb. J.K. 2981; bar = 50 μm). From Kohlmeyer, *Mycologia* **65**, 616 (1973), Fig. 3. Reprinted with permission. (d) Young thallus with sterile hairs enclosing antheridia (Herb. J.K. 2983; bar = 50 μm). (e) Candelabrumlike arrangement of antheridia (Herb. J.K. 3003; bar = 25 μm). (f) Antheridia, the left one immature with closed tip (Herb. J.K. 2986; bar = 10 μm). From Kohlmeyer, *Mycologia* **65**, 616 (1973), Fig. 6. Reprinted with permission. (g) Tip of

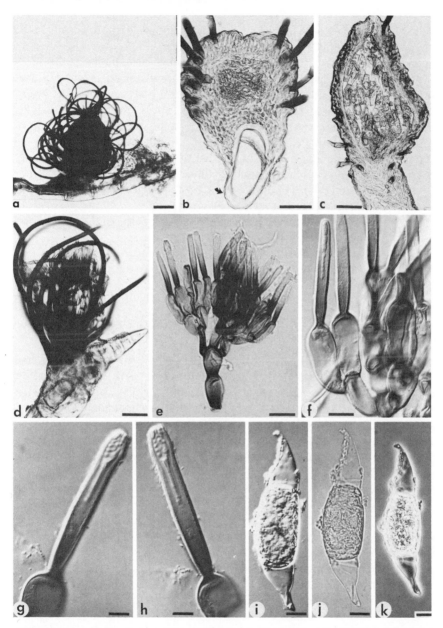

antheridium, enclosing spermatium with straight main body and coiled apical filament (Herb. J.K. 2983; bar = 5 μm). (h) Same as g. (i) Ascospore with a conical appendage at each end (Herb. J.K. 2981; bar = 10 μm). (j) Same ascospore as in i. (k) Same ascospore as in i. (All from type material; a–d and j in bright field, k in phase contrast, the others in Nomarski interference contrast.)

drical, two or three-septate, apically ramose and bearing 15–22 phialidelike antheridia. ANTHERIDIA lageniform, 55–72 μm long; venters cylindrical, 19–30 × 10–13 μm; necks obconical or cylindrical, 24–45 × 4–8 μm, forming spermatia; spermatia composed of an apical filiform, spirally coiled appendage and a basal cylindrical body, ca. 14–15 × 1 μm. TRICHOGYNES simple, cylindrical, septate. PAPILLAE absent; ostioles periphysate. PARAPHYSES lacking. ASCI eight-spored, clavate, thin-walled, aphysoclastic, early deliquescing. ASCOSPORES 70–104 × 16–23 μm (including appendages), fusiform, slightly curved, one-celled, hyaline; at each end with a conical appendage, filled with mucus. MODE OF LIFE: Parasitic. HOST: *Ballia callitricha*. RANGE: Pacific Ocean—Australia (Southeast Australia). ETYMOLOGY: From the Greek, *adelphe* = sister, in reference to the similarity between the lanose ascocarps of *S. adelpha* and those of *S. phycophila*.

Spathulospora antarctica Kohlm.

Mycologia **65**, 619–620, 1973

LITERATURE: Kohlmeyer (1974a); Kohlmeyer and Kohlmeyer (1975).
Fig. 42
THALLUS crustose, surrounding the host cells, bearing antheridia and trichogynes; vegetative hyphae absent; algal cells perforated by peglike cells that merge into intracellular assimilative cells, forming a hypo-stroma. ASCOCARPS 310–600 μm high, 270–500 μm in diameter, pyriform, superficial, solitary or geminate, papillate, ostiolate, fuscous to black, coriaceous, subiculate, glabrous. PERIDIUM 40–70 μm thick. ANTHERIDIA-BEARING CELLS short, 6.5 μm in diameter, with two or three phialidelike antheridia. ANTHERIDIA lageniform, 28–29 μm long; venters obconical, 6.5 μm in diameter; necks cylindrical to conical, 10–18 × 2.5–3 μm; spermatia not seen, probably simple and not coiled. TRICHOGYNES 30–160 × 6–8 μm, simple, cylindrical, septate, hyaline. PAPILLAE short; ostioles periphysate, about 16–18 μm in diameter. PARAPHYSES lacking. ASCI eight-spored, clavate, thin-walled, aphysoclas-tic, early deliquescing. ASCOSPORES 40–92 × 4–12 μm (including append-ages), elongate–ellipsoidal to cylindrical, one-celled, hyaline; at each end with an appendage; appendages conical, enclosing mucus; at first

---

Fig. 42. *Spathulospora antarctica* from *Ballia callitricha*. (a) Young thalli (arrows) on branches of the host (Herb. J.K. 2972; bar = 150 μm). (b) Immature ascocarp (6-μm longitudinal section), the base surrounding or attached to host cells (Herb. J.K. 3017; bar = 50 μm). (c) Mature ascocarp (10-μm section) filled with ascospores (Herb. J.K. 2989; bar = 25 μm). (d) Surface of young thallus with protruding trichogynes (Herb. J.K.

3018; bar = 50 μm). From Kohlmeyer, *Mycologia* **65,** 621 (1973), Fig. 25. Reprinted with permission. (e) Cross section (6 μm) through subiculum surrounding algal cell; penetration pegs perforate the algal wall and give rise to an intracellular stroma of assimilating cells, which fill the host cell completely (Herb. J.K. 3017; bar = 25 μm). (f) Phialidelike antheridia (Herb. J.K. 3005; bar = 10 μm). From Kohlmeyer, *Mycologia* **65,** 621 (1973), Fig. 20. Reprinted with permission. (g) Ascospore with mucilage oozing from both apical appendages (Herb. J.K. 3017; bar = 10 μm). (All from type material; a–d in bright field, the others in Nomarski interference contrast.)

closed, later opening at the apices and releasing the mucilage. MODE OF LIFE: Parasitic. HOST: *Ballia callitricha*. RANGE: Atlantic Ocean— Argentina, Chile (Tierra del Fuego, Santa Cruz), Falkland Islands; Indian Ocean—Kerguelen Islands; Pacific Ocean—Macquarie Island. ETYMOLOGY: From the Latin, *antarcticus,* in reference to the geographic distribution of the species, which is restricted to Antarctica and adjacent islands.

Spathulospora calva Kohlm.

Mycologia **65,** 622–623, 1973

LITERATURE: Kohlmeyer (1974a,b); Kohlmeyer and Kohlmeyer (1975).

Fig. 43

THALLUS crustose, surrounding the host cells, bearing a few sterile hairs around the ostiole, with antheridia and trichogynes; vegetative hyphae absent; algal cells perforated by peglike cells that merge into intracellular assimilative cells, forming a hypostroma; fungus inducing in the host a wild growth of hairs around the ascocarps. ASCOCARPS 300–400 μm high, 220–340 μm in diameter, enclosed within a hairy gall of the host, subglobose, superficial, solitary or geminate, ostiolate, fuscous, coriaceous, subiculate, bald except for one to three (rarely up to seven) hairs around the ostiole. PERIDIUM 22–65 μm thick. STERILE HAIRS slightly curved, cylindrical, tapering, fuscous, 50–75 × 8–10 μm, septate, thick-walled, immersed in the walls of the perithecium. ANTHERIDIAL HAIRS shaped like a candelabrum; stalks 40 × 5.5 μm, cylindrical, one-septate, apically ramose and bearing several phialidelike antheridia. ANTHERIDIA lageniform; venters cylindrical, 22 × 6.5–12.5 μm; necks cylindrical, 40–50 × 3.5–6.5 μm, forming spermatia; spermatia composed of an apical

---

Fig. 43. *Spathulospora calva* from *Ballia callitricha.* (a) Ascocarps inducing wild growth of hairs in the host (bar = 150 μm). (b) Longitudinal section (10 μm) through ostiole with periphyses, sterile hair near the orifice, ascospores in the venter (bar = 25 μm). From Kohlmeyer, *Mycologia* **65,** 624 (1973), Fig. 32. Reprinted with permission. (c) Ascospores with apical appendages enclosing a pitted "cushion" (bar = 10 μm). From Kohlmeyer, *Mycologia* **65,** 626 (1973), Fig. 42. Reprinted with permission. (d) Mature ascocarp (16-μm longitudinal section; bar = 50 μm). (e) Tip of antheridium enclosing undeveloped spermatium, which is connected at the base to cytoplasm of antheridial cavity (bar = 10 μm). (f) Apex of antheridium with mature spermatium, tip of coiled appendage passing into the ringlike reinforcement of the opening (bar = 5 μm). From Kohlmeyer, *Mycologia* **65,** 626 (1973), Fig. 39. Reprinted with permission. (g) Ascospores (bar = 10 μm). (h) Apices of antheridia, spermatial appendage coiled in one tip, uncoiled and extended through the orifice in another (bar = 5 μm). (i) Young thallus with antheridia on the host (bar = 10 μm). (j) Upper part of antheridium with enclosed spermatium (bar = 10 μm). From Kohlmeyer, *Mycologia* **65,** 626 (1973), Fig. 38. Re-

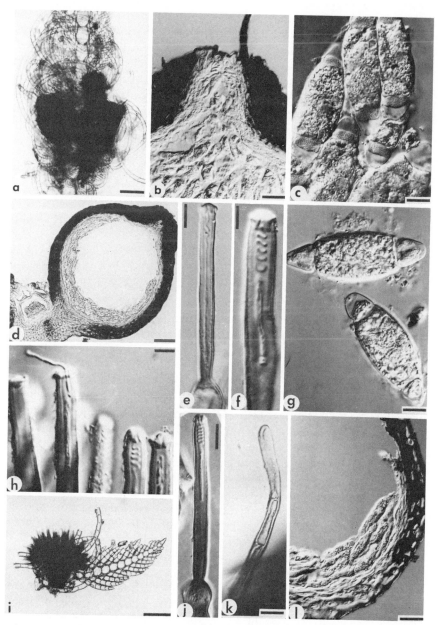

printed with permission. (k) Tip of septate trichogyne (bar = 15 μm). (l) Section (8 μm) through peridium with loose ascospores (bar = 25 μm). (All from type material; a and i Herb. J.K. 2976, the others Herb. J.K. 3010; a, d, i, and k in bright field, the others in Nomarski interference contrast.)

filiform, spirally coiled appendage, 23 × 1 μm, and a basal cylindrical body, 16–24 × 1.5–2 μm. TRICHOGYNES 175 × 6.5–7.5 μm, simple, cylindrical, septate, apically hyaline and thin-walled, basally light brown. Without papillae; ostioles periphysate, about 45 μm in diameter. PARAPHYSES lacking. ASCI 95–100 × 32–48 μm, eight-spored, clavate, thin-walled, aphysoclastic, early deliquescing. ASCOSPORES 45–56 × 14–19.5 μm (including appendages), ellipsoidal, one-celled, hyaline; at each end with a semiglobose appendage, enclosing a foveolate cushion; slightly constricted around the bases of the appendages. MODE OF LIFE: Parasitic. HOST: *Ballia callitricha*. RANGE: Pacific Ocean—Australia (Victoria), New Zealand. ETYMOLOGY: From the Latin, *calvus* = bald, in reference to the bald ascocarps.

Spathulospora lanata Kohlm.

Mycologia **65**, 625–626, 1973

LITERATURE: Kohlmeyer (1974a); Kohlmeyer and Kohlmeyer (1975).
Fig. 44

THALLUS crustose, composed of seven to ten layers of radiating dark brown cells, surrounding the host cells, bearing trichogynes and many long hairs with antheridia; vegetative hyphae absent; algal cells perforated by peglike cells that merge into intracellular assimilative cells, forming a hypostroma. ASCOCARPS 230–370 μm high, 160–290 μm in diameter, ovoidal, superficial, solitary or geminate, ostiolate, fuscous, leathery, subiculate, woolly. PERIDIUM 20–54 μm thick, two-layered; cells of the outer layer large, irregularly pyramidal, subglobose, truncate or ellipsoidal, brown, giving rise to the hairs; cells of the inner layer flattened, hyaline. ANTHERIDIAL HAIRS enclosing the perithecium, curved, cylindrical, fuscous, 200–340 μm long, 13–14 μm in diameter at the base, 8–12 μm at the apex, four- or five-septate, thick-walled, immersed in the wall of the ascocarp, apically branching dichotomously and bearing more than 20 phialidelike antheridia. ANTHERIDIA lageniform, 16–32 μm long; venters ellipsoidal, 8–16 × 5–8 μm; necks obconical or cylindrical, 10–16 × 2–3 μm, forming spermatia; spermatia hyaline, nonseptate, rod-shaped, 8–10 × 1.4–1.6 μm, cupulate at the base, apically covered by a mucilaginous cap, 1.3–2 × 0.5–1 μm. TRICHOGYNES 270–320 μm long, four-septate, simple, cylindrical, hyaline. PAPILLAE short or lacking; ostioles periphysate. PSEUDOPARENCHYMA of thin-walled cells filling venter of young ascocarps. ASCI ca. 53 × 28 μm, eight-spored, cylindrical to clavate, thin-walled, aphysoclastic, early deliquescing. ASCOSPORES 39–62(–74) × 8–11.5(–16) μm (including appendages), fusiform, straight or slightly

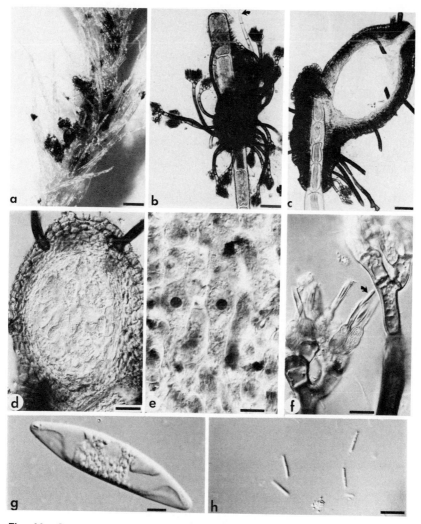

Fig. 44. *Spathulospora lanata* on *Ballia hirsuta*. (a) Ascocarps on algal filaments (Herb. J.K. 3488; bar = 400 μm). (b) Young thallus with antheridia-bearing hairs and trichogynes (arrow) (Herb. J.K. 3349; bar = 50 μm). (c) Mature ascocarp (16-μm longitudinal section) and thallus surrounding algal filament (Herb. J.K. 3349; bar = 50 μm). (d) Ascocarp (12-μm section) with immature asci (Herb. J.K. 3360; bar = 25 μm). (e) 10-μm section through central part of ascocarp, immature asci with large nuclei stained in acetocarmine (Herb. J.K. 3360; bar = 10 μm). (f) Immature (right) and mature (left) antheridial branches, spermatium released from neck of one antheridium (arrow) (Herb. J.K. 3360; bar = 10 μm). (g) Ascospore with a lenticular appressed appendage below each apex (Herb. J.K. 3349; bar = 5 μm). (h) Spermatia (Herb. J.K. 3360; bar = 10 μm). (a–c in bright field, the others in Nomarski interference contrast.)

curved, apically spathulate, one-celled, hyaline; below both ends covered with a lenticular appendage; germ tube forming a lobed appressorium. MODE OF LIFE: Parasitic. HOSTS: *Ballia hirsuta* and *B. scoparia*. RANGE: Pacific Ocean—New Zealand (North and South Island, Antipodes Island). ETYMOLOGY: From the Latin, *lanatus* = woolly, covered with long curled hairs, resembling wool, in reference to the lanate ascocarps.

Spathulospora phycophila Cavaliere et Johnson

Mycologia **57**, 927–928, 1965

LITERATURE: Kohlmeyer (1969e, 1973a, 1974a); Kohlmeyer and Kohlmeyer (1975).

Fig. 45

THALLUS crustose, surrounding the host cells, bearing sterile and fertile hairs, and trichogynes; vegetative hyphae absent; algal cells perforated by peglike cells that merge into intracellular assimilative cells, forming a hypostroma; fungus sometimes inducing in the host a wild growth of hairs

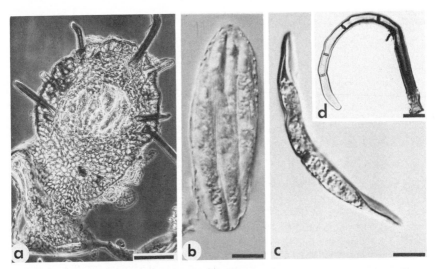

Fig. 45. *Spathulospora phycophila* from *Ballia* spp. (a) Immature ascocarp (8-μm longitudinal section; type BPI, bar = 50 μm). (b) Ascus (Herb. J.K. 3501 on *B. scoparia*; bar = 15 μm). (c) Ascospore with an appressed appendage below each end (type BPI; bar = 10 μm). From Kohlmeyer, *Mycologia* **65**, 628 (1973), Fig. 48. Reprinted with permission. (d) Sterile hair from ascocarp (type BPI; bar = 35 μm). (a in phase contrast, b and c in Nomarski interference contrast, d in bright field.)

around the ascocarps. ASCOCARPS (110–)157–320 μm in diameter, sub-globose, superficial, solitary, ostiolate, fuscous, leathery, subiculate, woolly. PERIDIUM 12–40 μm thick, two-layered; cells of the outer layer irregularly pyramidal or subglobose, brown, giving rise to the hairs; cells of the inner layer flattened, hyaline. HAIRS of two kinds enclosing the ascocarp. ANTHERIDIAL HAIRS curved, cylindrical, fuscous, 150–160 × 10–12 μm, two- to four-septate, thick-walled, immersed in the wall of the ascocarp, apically branching dichotomously and bearing less than ten phialidelike antheridia. STERILE HAIRS similar to the fertile ones, but more frequent and stronger curved to form half or full circles, 120–280 × 10–18.5 μm, 2–9(–13) septa. ANTHERIDIA lageniform, 20–22 μm long; venters ellipsoidal, 9.5–14 × 5.5–9 μm; necks obconical or cylindrical, 4–9.5 × 2.2–3.3 μm, forming spermatia; spermatia hyaline, nonseptate, filiform, 8.5 × 0.5 μm. TRICHOGYNES 27 × 8.5–9 μm, simple, cylindrical, hyaline. PAPILLAE short or lacking. PARAPHYSES lacking. ASCI 80–125 × 30–40 μm, eight-spored, cylindrical to clavate, thin-walled, aphysoclastic, early deliquescing. ASCOSPORES 80–110 × 10–13 μm (including appendages), fusiform, straight or curved, apically spathulate, one-celled, hyaline, covered by a thin gelatinous sheath that merges into a thick appendagelike cover around the tips. MODE OF LIFE: Parasitic. HOSTS: *Ballia callitricha* and *B. scoparia*. RANGE: Pacific Ocean—Australia (Victoria, South Australia), New Zealand (North Island). ETYMOLOGY: From the Greek, *phykos* = alga and *philus* = -loving, in reference to the algae-inhabiting mode of life of the fungus.

Key to the Species of *Spathulospora*

1. Ascocarps without any hairs; fungus on *Ballia callitricha* from antarctic or subantarctic
     waters . . . . *S. antarctica*
1'. Ascocarps enclosed by hairs, or at least bearing a few hairs around the ostiole; fungi on
     B. callitricha, B. hirsuta, or *B. scoparia* in Australia or New Zealand 2
        2(1') Ascospores longer than 65 μm 3
        2'(1') Ascospores usually shorter than 65 μm 4
3(2) Antheridia, when present, few, on long hairs that are similar to the sterile hairs;
     antheridial necks shorter than 10 μm. Ascospores less than 14 μm in diameter, spathulate
     to spoon-shaped at the apices, covered by a thin gelatinous sheath, merging into a thick
     cover around the tips . . . . *S. phycophila*
3'(2) Many antheridia on short stalks, different from the sterile hairs; antheridial necks
     longer than 20 μm. Ascospores more than 14 μm in diameter, with a conical, pointed
     appendage at each end . . . . *S. adelpha*
        4(2') Ascocarps completely surrounded by long, brown hairs, most of which bear
           antheridia at their apices; ascospores with spathulate tips and subterminal
           appendages . . . . *S. lanata*
        4'(2') Ascocarps with a few short, sterile hairs around the ostiole; ascospore tips not
           spathulate, appendages terminal . . . . *S. calva*

4. Sphaeriales
a. Diaporthaceae

Gnomonia Cesati et DeNotaris

Sferiacei Italici **1**, 57, 1863

=*Gnomoniopsis* Berlese, *Icones Fungorum* **1**, 93, 1894
=*Monopelta* Kirschstein, *Ann. Mycol.* **37**, 113, 1939
=*Rostrocoronophora* Munk, *Dan. Bot. Ark.* **15**(2), 98, 1953

LITERATURE: J. F. Morgan-Jones (1953, 1958, 1959); Müller and von Arx (1962, 1973); Munk (1957).

ASCOCARPS solitary, globose or depressed, ostiolate, usually with a long cylindrical, apical neck, periphysate, dark. PERIDIUM often two-layered, composed of several layers of more or less flattened cells; dark on the outside, lighter colored toward the center. PARAPHYSES absent. ASCI four- to eight-spored, fusiform or subclavate, thin-walled, unitunicate, with an apical ring; early detached from the ascogenous tissue and floating in the center of the ascocarp. ASCOSPORES elongate, often fusiform and provided with pointed apical appendages, one-septate in or near the center, rarely multiseptate, hyaline. MODE OF LIFE: Saprobic or parasitic. TYPE SPECIES: *Gnomonia setacea* (Pers. ex Fr.) Cesati et DeNotaris. NOTE: Synonyms and description after Müller and von Arx (1962); three marine species are recognized among about 50 terrestrial taxa. ETYMOLOGY: From the Greek, *gnomon* = index of a sundial, in reference to the long necks of ascocarps that usually extend above the surface of the substrate.

Gnomonia longirostris Cribb et Cribb

Univ. Queensl. Pap., Dep. Bot. **3**, 101, 1956

LITERATURE: T. W. Johnson and Sparrow (1961); Kohlmeyer and Kohlmeyer (1977); Nair (1970); Poole and Price (1972); Tubaki (1968, 1969); Tubaki and Ito (1973).

Figs. 46a and 46b

ASCOCARPS 100–220 µm in diameter, subglobose or bottle-shaped, immersed or partly immersed, ostiolate, with long necks, membranaceous, brown or almost hyaline, solitary or gregarious. NECKS 80–1125 × 13–40 µm, cylindrical or slightly pointed at the apex, often curved and with a variable diameter, rarely branching, projecting over the surface of the substrate. PARAPHYSES absent. ASCI 50–80 × 17–21 µm, eight-spored, cylindrical–clavate, ellipsoidal–clavate or subfusiform, short pedunculate, unitunicate, at first thick-walled apically, becoming thin-walled, with an apical pore; wall irregularly deliquescing. ASCOSPORES 13–19.5 × 4–7.5 μm, irregularly biseriate, elongate ellipsoidal or irregularly ellipsoidal, one-septate at the middle, slightly constricted at or near the septum, hyaline. MODE OF LIFE: Saprobic. SUBSTRATE: Dead branch of *Avicennia germinans, A. marina* var. *resinifera,* drift coconut, driftwood, and test panels (e.g., *Fagus sylvatica,* bamboo). RANGE: Atlantic Ocean— Bermuda, Great Britain (England); Indian Ocean—India (Kerala); Pacific Ocean—Australia (Queensland), Japan. NOTE: Description based on Cribb and Cribb (1956) and on our examination of the type (slide 32A), kindly provided by Dr. A. B. Cribb; we found *G. longirostris* different from *Lignincola laevis,* a similar species with catenophyses and asci without apical pore. ETYMOLOGY: From the Latin, *longus* = long and *-rostris* = beaked, in reference to the long necks of the ascocarps.

Gnomonia marina Cribb et Cribb

Univ. Queensl. Pap., Dep. Bot. **3**, 100, 1956

LITERATURE: T. W. Johnson and Sparrow 1961.
Figs. 46c–46e
ASCOCARPS 140–300 μm in diameter, bottle-shaped, mostly immersed, ostiolate, papillate, membranaceous, surrounded by brown hyphae 3–5 μm thick, subhyaline, light brown or fuscous, solitary or gregarious. PAPILLAE 100–560 × 42–140 μm, lighter colored than the ascocarps, cylindrical, simple or rarely bifurcate, immersed or projecting over the

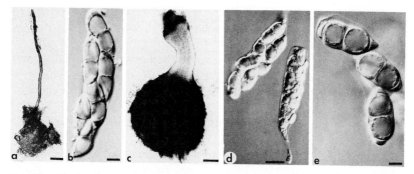

Fig. 46. Marine *Gnomonia* spp. (a) *G. longirostris*, ascocarp with long neck (bar = 75 μm). (b) *G. longirostris*, ascus (bar = 5 μm). (c) *G. marina*, ascocarp (bar = 50 μm). (d) *G. marina*, mature and immature ascus (bar = 15 μm). (e) *G. marina*, tip of ascus, ascospores with apical caps (bar = 5 μm). (All from type slides Herb. A.B. Cribb; a and c in bright field, the others in Nomarski interference contrast.)

surface of the substrate; ostiolar canal periphysate. PARAPHYSES absent. Asci 95–132 × 18–28 μm, eight-spored, clavate to subcylindrical, apically somewhat truncate, short pedunculate, unitunicate, thin-walled, but thickened at the apex and provided with a pore, persistent, physoclastic. ASCOSPORES 18–23(–26) × 9–12 μm (excluding appendages), cylindrical–ellipsoidal, apically rounded, one-septate at the middle, not or slightly constricted at the septum, hyaline, with a gelatinous appendage at each apex; appendages 1.5–3 μm long, apical or rarely subapical, caplike, round or apiculate, sometimes deciduous or deliquescing, subglobose in immature, nonseptate ascospores. MODE OF LIFE: Saprobic. SUBSTRATE: In dead roots of *Avicennia marina* var. *resinifera, Rhizophora mangle, R. racemosa,* and unidentified mangrove roots (the last three: J. Kohlmeyer, unpublished). RANGE: Atlantic Ocean—Liberia (Herb. J.K. 1823), United States (Florida; Herb. J.K. 1711); Pacific Ocean—American Samoa (Pago Pago; Herb. J.K. 1688), Australia (Queensland). NOTE: Description based on Cribb and Cribb (1956), our examination of the type (slide No. 80), kindly sent on loan by Dr. A.B. Cribb, and on our own collections; a putative record of *G. marina* by Cavaliere (1966c) is probably a different species having possibly stromatic ascocarps and ascospores without appendages; *G. marina* appears to be restricted to tropical or subtropical regions; our collections from tropical and subtropical areas (Brazil, Florida, Hawaii, Liberia, South Africa) indicate that there are other, undescribed species with appendaged ascospores, similar to *G. marina,* but differing in dimensions and other characters. ETYMOLOGY: From the Latin, *marinus* = marine, in reference to the habitat of the species.

Gnomonia salina E. B. G. Jones

Trans. Br. Mycol. Soc. **45**, 107, 1962

ASCOCARPS 490–700 μm in diameter, subglobose, superficial or partly immersed, ostiolate, papillate, coriaceous, dark brown to grayish-black, solitary. PAPILLAE 120–650 μm long, 140 μm in diameter, cylindrical or tapering at the apex. PARAPHYSES absent. ASCI 210–370 × 35–50 μm, eight-spored, clavate, pedunculate, unitunicate, thin-walled, but thickened at the apex, persistent, physoclastic. ASCOSPORES 36–72 × 20–32 μm (excluding appendages), cylindrical–ellipsoidal, apically rounded, one-septate at the middle, not or slightly constricted at the septum, hyaline, with a gelatinous appendage at each apex; appendages apical or subapical, caplike, round, deliquescing. MODE OF LIFE: Saprobic. SUBSTRATE: Driftwood, drift *Spartina townsendii,* and *Spartina* sp. RANGE: Atlantic Ocean—Great Britain (England); known only from the original description. NOTE: Description based on E. B. G. Jones (1962a) and our examination of the type (three slides, Herb. IMI Nos. 80274–80276). ETYMOLOGY: From the Latin, *salinus* = saline, in reference to the marine habitat of the species.

Key to the Marine Species of *Gnomonia*

1. Ascospores without apical caplike appendages, spore diameter less than 8 μm
. . . . *G. longirostris*
1'. Ascospores with apical appendages, spore diameter more than 8 μm 2
   2(1') Ascospore length more than 30 μm . . . . *G. salina*
   2'(1') Ascospore length less than 30 μm . . . . *G. marina*

b. Halosphaeriaceae Müller et von Arx ex Kohlmeyer

Can. J. Bot. **50**, 1951–1952, 1972

Marine saprobes, rarely parasites or symbionts; in algae, emersed or submersed phanerogams, wood, bark, leaves, or other cellulosic plant material; ascocarps of some species developing on hard substrates, such as grains of sand or calcareous shell fragments. ASCOCARPS subglobose, cylindrical or pyriform, light- or dark-colored, occasionally subiculate, rarely stromatic, superficial or immersed. Ostioles papillate or long cylindrical, rarely absent; ostiolar canal periphysate or with a thin-walled pseudoparenchyma. PERIDIUM soft or subcarbonaceous to carbonaceous, composed of flattened or irregularly polygonal, thick- or thin-walled cells, forming a textura angularis. PARAPHYSES absent; center of immature ascocarps consisting of polygonal, thin-walled, pseudoparenchymatous cells, sometimes with pit connections, at maturity separating to form catenophyses, or compressed by the growing asci and dissolving. ASCI fusiform, clavate or rarely subglobose, without apical apparatuses, rarely with a simple pore, unitunicate, thin-walled or rarely thick-walled below the apex, aphysoclastic, deliquescing at or before ascospore maturity, rarely persistent; hymenial layer at the base of the ascocarp venter, flat or convex. ASCOSPORES hyaline or light brown, one- or multicelled, rarely muriform, mostly provided with characteristic ornamentations or gelatinous sheaths, or both; mature ascospores filling the ascocarp venter, released singly through the ostiole or rarely within the ascus, which swells and dissolves after dispersal. TYPE GENUS: *Halosphaeria* Linder. NOTE: Barr (1976a) includes Halosphaeriaceae in the order Sordariales, but we prefer to keep the family in the Sphaeriales at this time. Scanning-electron microscope studies by E.B.G. Jones and Moss (*Mar. Biol.* **49**, 11–26, 1978) add important details to the knowledge of appendage formation in marine Ascomycetes. These authors state that ". . . ascocarp structure is of significance at the familial level only . . ." and continue "the separation of genera must be on the basis of ascospore structure and development of the appendages." Future investigations have to show if taxonomic conclusions on the generic level based solely on appendage formation are justified.

Key to the Genera of Halosphaeriaceae

1. Ascospores filiform (at least 10 times longer than broad) 2
1'. Ascospores not filiform 5
    2(1) Ascospores straight or slightly curved, without septa or apical chambers
    . . . . ***Bathyascus***
    2'(1) Ascospores folded or strongly curved in the ascus, S- or α-shaped after release,
    septate or with apical chambers 3
3(2') Ascospores nonseptate; pointed apical chambers filled with mucilage . . ***Lulworthia***
3'(2') Ascospores septate; apical chambers absent 4
    4(3') Ascospores broad at one end, filamentous at the other . . . . ***Trailia***
    4'(3') Ascospore diameter more or less equal throughout . . . . ***Lindra***
5(1') Ascospores one-celled, with a tuft of hairs at each apex and four tufts around the
equator . . . . ***Nautosphaeria***
5'(1') Ascospores without tufts, or if tufts present, spores septate 6
    6(5') Ascospores without appendages or gelatinous sheaths, one-septate; pseudo-
    parenchyma of ascocarpial center forming catenophyses 7
    6'(5') Ascospores with appendages, or gelatinous sheaths, or both; ascocarps with or
    without catenophyses 9
7(6) Ascus thick-walled below the apex, with a simple pore at the flat tip, persistent and
remaining attached after ascospore release; ascospores thick-walled . . . ***Aniptodera***
7'(6) Ascus thin-walled, without pore, dissolving at ascospore maturity or becoming detached; ascospores thin-walled 8
    8(7') Ascospores narrow, discharged from ascocarp while enclosed in intact asci;
    central portion of ascus swelling in water . . . . ***Lignincola***
    8'(7') Ascospores broad, released from dissolving asci within the ascocarp . . ***Nais***
9(6') Ascocarps carbonaceous or subcarbonaceous, often without ostioles or with short
papillae only; often attached to grains of sand or calcareous particles, superficial .10
9'(6') Ascocarps leathery, with distinct ostioles, necks often long; without affinity to
grains of sand or calcareous particles, superficial or immersed 11
    10(9) Ascospores enclosed in striate, gelatinous sheaths; three-septate or
    muriform . . . . ***Carbosphaerella***
    10'(9) Ascospores with tubular or fibrous appendages; fibers developing by fragmenta-
    tion of the exospore; 0- to 13-septate . . . . ***Corollospora***
11(9') Ascospores one-celled; ascocarps in or on algae . . . . ***Chadefaudia***
11'(9') Ascospores with one or more septa; ascocarps in other substrates (wood, bark, and
other cellulose-containing materials) 12
    12(11') Ascospores 2- to 11-septate; appendages only apically attached . ***Haligena***
    12'(11') Ascospores typically one-septate; appendages apical, or apically and later-
    ally inserted 13
13(12') Asci persistent; ascospore appendages only apical, at first caplike, then becoming
detached and ladlelike, developing into a long sticky filament . . . . . ***Halosarpheia***
13'(12') Asci deliquescing; ascospore appendages different 14

14(13') Ascocarps stromatic and subcarbonaceous, or nonstromatic and leathery; immature or also mature ascospores enclosed in a thick exosporic sheath that is pierced by the developing apical appendages; additional three to four lateral, subcylindrical appendages may occur; they are provided with small terminal caps; the apical appendages may be tubular . . . . *Ceriosporopsis*

14'(13') Ascocarps always nonstromatic, leathery; ascospores without exosporic sheath; appendages different from the foregoing, typically developing by fragmentation of the exospore; appendages not tubular, but thornlike apical processes occur in some species . . . . *Halosphaeria*

Aniptodera Shearer et Miller

Mycologia **69**, 893, 1977

ASCOCARPS globose or subglobose, immersed or superficial, ostiolate, papillate, membranaceous, hyaline to light brown. NECKS cylindrical, periphysate, brown at the tip. PSEUDOPARENCHYMA of thin-walled cells filling venter of young ascocarps, breaking up into catenophyses. ASCI eight-spored, clavate, short pedunculate, unitunicate, thin-walled except for a thick-walled area below the apex, flattened and refractive at the tip, and provided with a simple pore, slightly constricted below the apex, relatively persistent even after ascospore release, developing at the base of the ascocarp venter. ASCOSPORES ellipsoidal, one-septate, hyaline, thick-walled, without appendages. MODE OF LIFE: Saprobic. TYPE SPECIES: *Aniptodera chesapeakensis* Shearer et Miller. NOTE: The genus is monotypic; it is separated from the closely related genera *Lignincola* and *Nais* by features of the ascus apex and thick-walled ascospores; future investigations on the ontogeny and ultrastructure have to prove if these characters are sufficient to keep *A. chesapeakensis, L. laevis,* and *N. inornata* in different genera; slight differences in ascocarp pigmentation between these genera are not significant enough to be used for generic separations, especially if the darker ascocarps of *A. chesapeakensis* in *Juncus* and *Spartina* are considered; most genera of Halosphaeriaceae are characterized by ascospore ornamentations, which are absent in *Aniptodera;* however, because of the thick ascospore wall, *A. chesapeakensis* could be interpreted as an intermediate form between genera with appendaged and nonappendaged ascospores; electron microscopical and cytochemical studies of ascospores of *A. chesapeakensis* should clarify if the walls are multilayered and comparable to walls of species with appendages developing from an outer wall layer. ETYMOLOGY: From the Greek, *aniptos* = unwashed, and *deire* = neck, probably in reference to the dark apex of the ascocarp neck in the type material.

Aniptodera chesapeakensis Shearer et Miller

Mycologia **69**, 894, 1977

ASCOCARPS 130–300 μm high, 170–325 μm in diameter, globose, sub-globose, or ellipsoidal, immersed or superficial, ostiolate, papillate, membranaceous, hyaline, light brown or grayish-brown, solitary. PERIDIUM 12–16 μm thick, composed of elongated, thin-walled cells with large lumina, forming a textura angularis, merging into the pseudoparenchyma of the venter. NECKS 81–326 μm long, 36–80 μm in diameter, cylindrical, periphysate, brown at or below the tip. PSEUDOPARENCHYMA of thin-walled, large, polygonal cells filling venter of young ascocarps, eventually breaking up into catenophyses. ASCI 64–116 × 14–38 μm, eight-spored, clavate, short pedunculate, unitunicate, thin-walled, except for a thick-walled area below the apex, flattened and refractive at the tip, and provided with a simple pore, slightly constricted below the apex, rather persistent, even after ascospore release, developing at the base of the ascocarp venter. ASCOSPORES 21–37 × 7–15 μm, ellipsoidal, one-septate, not constricted at septum, hyaline, thick-walled, without appendages. MODE OF LIFE: Saprobic. SUBSTRATES: Submerged test panels (balsa) and bases of dead, brown leaves of *Juncus roemerianus* and *Spartina alterniflora* (J. Kohlmeyer, unpublished). RANGE: Atlantic Ocean—United States (Maryland, North Carolina). NOTE: Description based on Shearer and Miller (1977) and our numerous collections of *J. roemerianus* and *S. alterniflora* from North Carolina (Herb. J.K. 3681); our material agrees completely with the type description, but the ascocarps are embedded in the host tissue and are often light brown, and darker in the upper part; Shearer and Miller collected the species in mesohaline and oligohaline waters (salinity, 1.3–17.8°/oo), whereas our material came from one site with 33.4°/oo salinity (inlet of Mullet Pond, Shackleford Banks, Carteret County, North Carolina) and has been collected almost every month since March 1975 in Broad Creek, Carteret County (salinity at this site fluctuating between 1 and 34°/oo); hence, *A. chesapeakensis* is capable of growing at high salinities also. ETYMOLOGY: Named after Chesapeake Bay, recipient of Patuxent River, Maryland, where the type material was collected.

Bathyascus Kohlm.

Rev. Mycol. **41**, 190–191, 1977

ASCOCARPS solitary, subglobose or ellipsoidal, partly or completely immersed, ostiolate, papillate, coriaceous, brownish. PERIDIUM thin, forming a textura angularis. PARAPHYSES absent; center of immature ascocarps filled with thin-walled pseudoparenchymatous cells. Asci eight-spored, fusiform to clavate, unitunicate, thin-walled, early deliquescing, ripening simultaneously on a small-celled ascogenous tissue at the base of the ascocarp venter. Ascospores filiform, one-celled, hyaline, without appendages. Mode of Life: Saprobic. Substrate: Wood. Range: Deep sea. Type Species: *Bathyascus vermisporus* Kohlm. Etymology: From the Greek, *bathys* = deep, deep sea, and *askos* = ascus, in reference to the origin of the genus.

Bathyascus vermisporus Kohlm.

Rev. Mycol. **41**, 191, 1977

Fig. 47a

Ascocarps 150–260 μm high (including papilla), 215–480 μm in diameter, subglobose or ellipsoidal, partly or totally immersed, ostiolate, papillate, coriaceous, dark brown above, light brown to almost hyaline at the base, solitary. Peridium 40–45 μm thick around the ostiole, 20–30 μm at the base and sides, composed of five or six layers of thin-walled, polygonal, elongated cells with large lumina, forming a textura angularis, merging into the pseudoparenchyma of the center. Papillae or necks 60–380 μm long, 60–70 μm in diameter, cylindrical or obtusely conical, dark brown; ostiolar canal about 15 μm in diameter, periphyses absent; walls of some necks extend into lumina of wood vessels and, thus, appear warty. Paraphyses absent; center of immature ascocarps filled with hyaline, thin-walled pseudoparenchymatous cells that become compressed by the developing asci. Asci about 70–80 × 12–14 μm, eight-spored, fusiform to clavate, unitunicate, thin-walled, early deliquescing, ripening simultaneously at the base of the ascocarp venter; ascogenous tissue composed of thin-walled, polygonal cells, 6–12 μm in diameter. Ascospores 50–72 × 4–5.5 μm, filiform or rarely elongate fusoid, straight or slightly curved, often apically attenuate and thick-walled, one-celled, hyaline, without appendages; nucleus about 2 μm in diameter. Mode of Life: Saprobic. Substrate: Submerged wood. Range: Pacific Ocean—deep sea (off California at 1615–1720 m). Etymology: From the Latin, *vermis* = worm and *spora* = spore, in reference to the vermiform ascospores.

Carbosphaerella I. Schmidt

Feddes Repert. **80**, 108, 1969

Ascocarps solitary, globose or subglobose, superficial, subiculate, ostiolate or nonostiolate, papillate or epapillate, carbonaceous, black.

PARAPHYSES absent. ASCI four- to eight-spored, obpyriform to sub-globose, short stipitate, unitunicate, thin-walled, early deliquescing. AS-COSPORES ellipsoidal or ovoid, triseptate or muriform, apical cells lighter colored than the others, surrounded with a gelatinous, striate sheath. SUBSTRATE: Dead wood. TYPE SPECIES: *Carbosphaerella pleosporoides* I. Schmidt. NOTE: Both members of the genus are obligate marine species. ETYMOLOGY: From the Latin, *carbo* = coal and *sphaerella* = little ball, in reference to the carbonaceous ascocarps.

Carbosphaerella leptosphaerioides I. Schmidt

Nat. Naturschutz Mecklenburg **7**, 9–10, 1969 (publ. 1971)

LITERATURE: Koch (1974); Kohlmeyer (1971b); Schmidt (1974).
Figs. 47b and 47c

ASCOCARPS 90–210 μm in diameter, globose to subglobose, superficial, sometimes seated with subicula on grains of sand, with or without os-tioles, papillate or epapillate, carbonaceous, black, solitary or gregarious. PERIDIUM 5–12 μm thick, composed of two to three layers of cells,

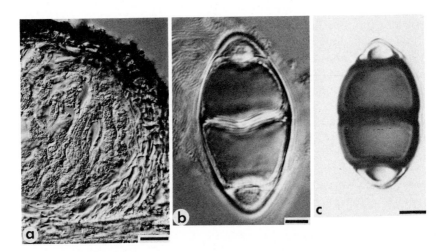

Fig. 47. (a) *Bathyascus vermisporus,* ascocarp (10-μm longitudinal section; Herb. J.K. 1947; bar = 25 μm). From Kohlmeyer, *Rev. Mycol.* **41,** 192 (1977), Fig. 2. Reprinted with permission. (b) *Carbosphaerella leptosphaerioides,* ascospore with striate, mucilaginous sheath (Herb. J.K. 1926; bar = 5 μm). (c) *C. leptosphaerioides,* ascospore with dark inner and hyaline apical cells, sheath dissolved (Herb. J.K. 1926; bar = 5 μm). From Kohlmeyer, *Trans. Br. Mycol. Soc.* **57** (1971), Plate 38, Fig. 2. Reprinted with permission. (All in Nomarski interference contrast.)

subglobose at the outside, flattened toward the inside. PAPILLAE, when present, short, conical, near the basal subiculum and pointing downward. PARAPHYSES absent. ASCI 65–90 × 45–60 μm (excluding stipe), stipe 12–17 × 6.5 μm, eight-spored, ovoidal or obpyriform, short stipitate, unitunicate, thin-walled, aphysoclastic, without apical apparatuses, early deliquescing. ASCOSPORES (25–)27–42 × (12.5–)16–24 μm (exluding sheath), ellipsoidal, triseptate, not or slightly constricted at the septa, central large cells dark brown, apical small cells hyaline or light brown; septa with a central porus, ca. 1 μm in diameter; except for the apical cells, surrounded by a gelatinous, persistent, irregular sheath, 4–12.5 μm thick, hyaline or faintly brown, appearing striated or dotted with parallel fibers embedded in a matrix; germination from the apical cells only. MODE OF LIFE: Saprobic. SUBSTRATE: Intertidal wood (*Abies, Picea, Pinus,* and *Quercus* spp.). RANGE: Atlantic Ocean—Denmark (Jutland), United States (Maine, North Carolina); Baltic Sea—Denmark (Sjaelland), Germany (Rügen, G.D.R.). ETYMOLOGY: Name referring to the resemblance of the ascospores to those of the genus *Leptosphaeria.*

Carbosphaerella pleosporoides I. Schmidt

Feddes Repert. **80**, 108, 1969

LITERATURE: Schmidt (1974).

ASCOCARPS 100–190 μm in diameter, globose, superficial, sometimes seated with subicula on grains of sand, rarely immersed in the substrate, with or without ostioles, papillate or epapillate, carbonaceous, black, solitary. PAPILLAE, when present, short, conical, near the basal subiculum and pointing downward. PARAPHYSES absent. ASCI 26.5–40 × 23–39 μm (excluding stipe), stipe up to 8 μm long, four-spored, obpyriform to subglobose, short stipitate, unitunicate, thin-walled, aphysoclastic, without apical apparatuses, early deliquescing. ASCOSPORES 22–30 × (11.5–)13–20 μm (excluding sheath), ovoidal or ellipsoidal, muriform, constricted at the septa, central cells dark brown, apical cells light brown; at first surrounded completely by a gelatinous, persistent, irregular sheath that eventually retracts from the apical cells; sheath 2–5 μm thick around the center, hyaline, appearing striated or dotted with parallel fibers embedded in a matrix; germination from all cells. MODE OF LIFE: Saprobic. SUBSTRATE: Intertidal wood. RANGE: Baltic Sea—Germany (Rügen, G.D.R.; known only from the original collection). ETYMOLOGY: Name referring to the resemblance of the ascospores to those of the genus *Pleospora.*

Key to the Species of *Carbosphaerella*

1. Ascospores with three cross septa . . . . *C. leptosphaerioides*
1'. Ascospores muriform . . . . *C. pleosporoides*

Ceriosporopsis Linder *in* Barghoorn and Linder

Farlowia **1**, 408, 1944

=*Ceriosporella* Cavaliere, *Nova Hedwigia* **10**, 393, 1966 (non Berlese, 1902)
=*Marinospora* Cavaliere, *Nova Hedwigia* **11**, 548, 1966

ASCOCARPS solitary or gregarious, subglobose to cylindrical, immersed or becoming exposed, ostiolate, papillate, coriaceous or subcarbonaceous, light brown to black. PSEUDOPARENCHYMA of large, thin-walled cells filling venter of young ascocarps; deliquescing, or central core developing into pseudoparaphyses or catenophyses. ASCI eight-spored, clavate, pedunculate, unitunicate, thin-walled, early deliquescing. ASCOSPORES ellipsoidal, one-septate, hyaline, surrounded by an exosporic sheath that forms apical, mucus-containing tubes or is ruptured by outward growing apical or also lateral mucilaginous appendages that stain metachromatically in 0.1% toluidine blue. SUBSTRATES: Dead intertidal wood or other cellulose-containing materials. TYPE SPECIES: *Ceriosporopsis halima* Linder. NOTE: Reasons for reducing *Marinospora* to synonymy of *Ceriosporopsis* were discussed in detail by Kohlmeyer (1971b), and the distinction between *Ceriosporopsis* and *Halosphaeria* was defined later (Kohlmeyer, 1972b). The ascocarps of *C. calyptrata* and *C. cambrensis* appear to be stromatic; however, these species belong definitely to Sphaeriales (Ascohymeniales) and not to Pseudosphaeriales (Ascoloculares). ETYMOLOGY: From *Ceriospora* Niessl, a related genus, and the Greek, *opsis* = aspect, appearance, indicating the resemblance between the two genera.

Ceriosporopsis calyptrata Kohlm.

Nova Hedwigia **2**, 301, 1960

≡*Ceriosporella calyptrata* (Kohlm.) Cavaliere, *Nova Hedwigia* **10**, 394, 1966
≡*Marinospora calyptrata* (Kohlm.) Cavaliere, *Nova Hedwigia* **11**, 548, 1966
=*Ceriosporopsis longissima* Kohlm., *Nova Hedwigia* **4**, 398, 1962
≡*Ceriosporella longissima* (Kohlm.) Cavaliere, *Nova Hedwigia* **10**, 394, 1966
≡*Marinospora longissima* (Kohlm.) Cavaliere, *Nova Hedwigia* **11**, 548, 1966

LITERATURE: Byrne and Jones (1974); Cavaliere (1968); G. C. Hughes (1969); T. W. Johnson (1963d, 1968); T. W. Johnson and Sparrow (1961); E. B. G. Jones (1963b, 1968b); E. B. G. Jones and Irvine (1971); Kirk and Catalfomo (1970); Koch (1974); Kohlmeyer [1959 (sub *Halosphaeria appendiculata* Linder), 1962a, 1963d,e, 1964, 1971b, 1972b]; Kohlmeyer and Kohlmeyer (1964–1969); Malacalza and Martínez (1971); Schaumann (1968, 1969); Schneider (1977).

Fig. 48

ASCOCARPS 295–605 μm high, 480–770 μm in diameter, ellipsoidal, ovoidal or subglobose, immersed, ostiolate, papillate, subcarbonaceous or subcoriaceous, light brown to black, solitary or gregarious. PERIDIUM 12–80 μm thick, composed of 7–30 layers of thick-walled cells with large

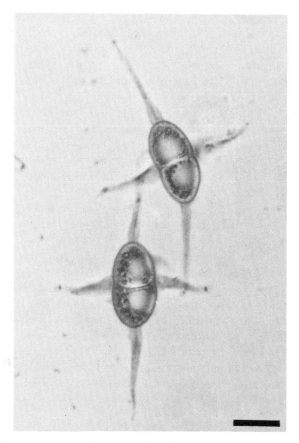

Fig. 48. *Ceriosporopsis calyptrata*, appendaged ascospores, stained in picro-anilin blue (Herb. J.K.; phase contrast; bar = 15 μm).

lumina; on the outside intimately adhering to decomposing wood particles; toward the inside merging into the thin-walled pseudoparenchyma of the venter. PAPILLAE or necks up to 1050 μm long, 55–120 μm in diameter, subconical or cylindrical, papilliform or elongated; ostiolar canal periphysate. PSEUDOPARENCHYMA of large, thin-walled, polygonal or subglobose cells filling venter of young ascocarps; eventually forming catenophyses, 30–40 × 3.2–13 μm, septate, ramose, persistent. ASCI 104–190 × 22–37 μm, eight-spored, clavate, somewhat apiculate, pedunculate, unitunicate, thin-walled, aphysoclastic, early deliquescing, without apical apparatuses, developing in the base of the ascocarp venter. ASCOSPORES 20–36 × (7–)9–19 μm (excluding appendages), broadly ellipsoidal, one-septate, constricted at the septum, hyaline, appendaged; provided at each end with a single, terminal appendage, obclavate or subcylindrical, tapering, 8.5–90(–187) × 2.5–7.5 μm; around the septum three or four similar, but often shorter (less than 4 μm), radiating appendages; ascospores at first surrounded by an exosporic sheath that is forced away from the spores by outward-growing appendages and fractures; small, cupuliform, thin caps cover the apices of the appendages; these caps, 2.3–4.9 μm long, 1.5–3.7 μm in diameter, may invert; a refractive area is seen in the tip of most appendages. MODE OF LIFE: Saprobic. SUBSTRATES: Intertidal wood and test panels (e.g., *Fagus sylvatica, Larix* sp., *Picea* sp., *Quercus* sp.). RANGE: Atlantic Ocean—Argentina (Buenos Aires), Belgium, Denmark, France, Germany (Helgoland and mainland, F.R.G.), Great Britain [England, Isle of Man, Scotland (J. Kohlmeyer, unpublished)], Iceland, Spain, United States (Connecticut, Massachusetts); Baltic Sea—G.F.R.; Mediterranean—France, Italy (Sicily); Pacific Ocean—Canada (British Columbia), United States (California, Washington). NOTE: *C. longissima* is reduced to synonymy because the main difference between this species and *C. calyptrata,* namely, length of ascospore appendages, is not considered sufficient anymore to separate both species; collections with intermediate lengths of appendages (Koch, 1974; Kohlmeyer, 1963d; Malacalza and Martínez, 1971; Schaumann, 1969) indicate that appendage length is variable and may depend on environmental factors; furthermore, T. W. Johnson (1963d) discovered an exosporic sheath in *C. calyptrata* that was described also in *C. longissima* (Kohlmeyer, 1962a). ETYMOLOGY: From the Latin, *calyptratus* = bearing a caplike covering, in reference to the small caps over the apices of ascospore appendages.

Ceriosporopsis cambrensis Wilson

Trans. Br. Mycol. Soc. **37**, 276, 1954

LITERATURE: Eaton and Dickinson (1976); Gold (1959); G. C. Hughes (1968, 1969); T. W. Johnson and Sparrow (1961); E. B. G. Jones (1962c, 1963c, 1965, 1968b, 1971a, 1972); Meyers (1957); Schaumann (1969); Wilson (1963).

ASCOCARPS 276–548 μm high, 577–862 μm long, 260–300 μm in diameter, ellipsoidal, immersed, later becoming exposed, ostiolate, papillate, coriaceous, pale brown to black, solitary or gregarious in rows. PERIDIUM composed of thick-walled cells on the outside; flattened, thin-walled cells on the inside, merge into the pseudoparenchyma of the venter; decomposing wood particles adhering to the outside of the wall. PAPILLAE or necks up to 880 μm long, cylindrical, centric or eccentric; ostiolar canal periphysate. PSEUDOPARENCHYMA of large, thin-walled cells surrounding a hyphoid, central core in young ascocarps; central tissue developing into pseudoparaphyses (or catenophyses?). ASCI 110–130 × 24–26 μm, eight (rarely seven)-spored, clavate, sessile or short pedunculate, bitunicate (?), aphysoclastic, early deliquescing, without apical apparatuses, developing in the base of the ascocarp venter. ASCOSPORES 29–31.5 × 10.5–14.5 μm (excluding appendages), ellipsoidal, one-septate, slightly constricted at the septum, hyaline, appendaged; provided at each end with a single terminal, subcylindrical, tapering, deliquescing appendage, 36–75 × 3–6 μm; ascospores at first surrounded by a thin exosporic sheath that encloses the curved-in appendage initials. MODE OF LIFE: Saprobic. SUBSTRATES: Intertidal wood and test panels (e.g., *Abies* sp., *Alnus glutinosa, Fagus sylvatica, Pinus sylvestris,* Douglas fir). RANGE: Atlantic Ocean—Canada (Newfoundland), Germany (Helgoland, F.R.G.), Great Britain (England, Isle of Man, Scotland, Wales), United States (North Carolina); Pacific Ocean—Canada (British Columbia). ETYMOLOGY: From the Latin, *cambrensis* = from Wales (*Cambria*), in reference to the origin of the type material.

Ceriosporopsis circumvestita (Kohlm.) Kohlm.

Can. J. Bot. **50**, 1953, 1972

≡*Halosphaeria circumvestita* Kohlm., *Nova Hedwigia* **2**, 307, 1960

LITERATURE: Byrne and Jones (1974); Cavaliere (1966b, 1968); Churchland and McClaren (1973); Eaton and Dickinson (1976); G. C. Hughes (1969); T. W. Johnson (1963a); E. B. G. Jones (1963a, 1965, 1968b, 1972); E.B.G. Jones and Irvine (1971); E.B.G. Jones and Le Campion-Alsumard (1970b); Kirk (1976); Kohlmeyer (1962a, 1963d); Kohlmeyer and Kohlmeyer (1964–1969, 1966).

ASCOCARPS 135–325 μm high, 150–450 μm in diameter, subglobose or ovoid, immersed or superficial, light or dark brown, sometimes reddish-brown or pale below, coriaceous, solitary or gregarious. PERIDIUM 13–16 μm thick, composed of four or five layers of thick-walled cells, roundish or ellipsoidal on the outside, elongate on the inside and merging into the pseudoparenchyma of the venter. PAPILLAE or necks 102–850 × 41–55 μm, conical or subcylindrical, sometimes absent, dark at the base, light-colored at the apex; ostiolar canal filled with a small-celled pseudoparenchyma. PSEUDOPARENCHYMA of thin-walled, polygonal cells filling venter of young ascocarps. ASCI eight-spored, clavate, peduncu-late, unitunicate, thin-walled, aphysoclastic, deliquescing before asco-spore maturity and rarely observed, without apical apparatuses. ASCO-SPORES 16.5–25 × 9–13 μm (excluding appendages), ellipsoidal, one-septate, slightly constricted at the septum, hyaline, surrounded by an irregular mucilaginous cover; covers subcylindrical at both ends, 4.5–11 μm long, 4–7.5 μm in diameter, forming irregular lobes (6.5–10.5 μm long, 3.5–8 μm in diameter) around the septum; outer cover encloses at each end a single cylindrical or almost filiform appendage (stain with hematoxylin!). CHLAMYDOSPORES from pure cultures 3.5–7.5 μm in diameter, reddish or grayish brown, catenulate; chains up to 150 μm long, up to 17-celled, intercalary or terminal, simple or ramose, straight or curved; single cells 6–12 μm long, globose, ellipsoidal or subcylindrical, deciduous. MODE OF LIFE: Saprobic. SUBSTRATES: Intertidal or drifting wood, or other plant remains, and test panels (e.g., *Arundo donax, Fagus sylvatica, Pinus sylvestris*). RANGE: Atlantic Ocean—France, Great Brit-ain (England, Isle of Man, Scotland, Wales), Iceland; Mediterranean—France; Pacific Ocean—Canada (British Columbia), United States (Wash-ington). ETYMOLOGY: From the Latin, *circumvestitus* = clothed all round, in reference to the appendage-covered ascospores.

Ceriosporopsis halima Linder *in* Barghoorn et Linder

Farlowia **1**, 409, 1944

=*Ceriosporopsis barbata* Höhnk, *Veroeff. Inst. Meeresforsch. Bremerha-ven* **3**, 210, 1955

LITERATURE: Anastasiou (1963b); Apinis and Chesters (1964); Barg-hoorn (1944); Biernacka (1965); Byrne and Eaton (1972); Byrne and Jones (1974, 1975b); Cavaliere (1966b, 1968); Churchland and McClaren (1973); Cribb and Cribb (1956); S.E.J. Furtado *et al.* (1977); Gold (1959); Henningsson (1974); Höhnk (1956); G. C. Hughes (1969, 1974); Irvine *et al.* (1972); T. W. Johnson (1958a, 1963d, 1968); T. W. Johnson and Sparrow (1961); E. B. G. Jones (1962a,c, 1963a,c, 1965, 1968b, 1972); E.

B. G. Jones *et al.* (1972); E. B. G. Jones and Oliver (1964); Kirk (1966, 1976); Kirk and Catalfomo (1970); Kirk *et al.* (1974); Koch (1974); Kohlmeyer (1958a,b, 1959, 1960, 1962a, 1963c,d, 1967, 1968d, 1969c, 1971b, 1972b, 1976); Kohlmeyer and Kohlmeyer (1964–1969); Lutley and Wilson (1972a); Malacalza and Martínez (1971); Meyers (1957); Meyers and Reynolds (1959a,c, 1960b); Meyers *et al.* (1960, 1963b); Neish (1970); Pisano *et al.* (1964); Poole and Price (1972); Raghukumar (1973); Schaumann (1968); Schmidt (1967, 1974); Schneider (1976); Shearer (1972b); Tubaki (1966, 1969); Tubaki and Ito (1973); Wilson (1951, 1954, 1956a, 1963, 1965).

Fig. 49a

ASCOCARPS 80–481 μm high, 130–503 μm in diameter or long, sub-globose, ellipsoidal or cylindrical, immersed or superficial, ostiolate, papillate, coriaceous, dark brown to black, reddish-brown when empty, solitary or gregarious. PERIDIUM 8–26 μm thick, composed of three to four (to ten) layers of thin-walled, elongate cells with large lumina, merging into the pseudoparenchyma of the venter. PAPILLAE or necks up to 706 μm long, 16–30(–56) μm in diameter, centric or eccentric, cylindrical; ostiolar canal at first filled with a small-celled pseudoparenchyma. PSEUDOPARENCHYMA of thin-walled, polygonal cells filling venter of young ascocarps. ASCI 56–89 × 14–22 μm, eight-spored, ellipsoidal, subclavate or subfusiform, short pedunculate, unitunicate, thin-walled, aphysoclastic, deliquescing before ascospore maturity, without apical apparatuses, developing at the base of the ascocarp venter. ASCOSPORES 18–27(–35) × 6–12 μm (excluding appendages), ellipsoidal to fusiform-ellipsoidal, one-septate, slightly or strongly constricted at the septum, hyaline, surrounded by a gelatinous, exosporic sheath that is pierced at each apex by an outward growing appendage; appendages 5–8 μm in diameter, of variable length, terminal, simple, subcylindrical, tapering, finally becoming viscous and filamentous. CHLAMYDOSPORES from pure cultures 6–17 μm in diameter, reddish-brown, catenulate; chains up to 90 μm long, up to 13-celled, terminal, simple or rarely ramose, curved, frequently ¼ or ¾ times coiled; single cells globose, ellipsoidal or sub-cylindrical, sometimes increasing in diameter from base to apex. MODE OF LIFE: Saprobic. SUBSTRATES: Intertidal or drifting wood, bark, rope, or other plant remains, test panels (*Abies firma, Abies* sp., *Alnus* sp., *Araucaria* sp., *Arundo donax, Betula* sp., *Dicorynia paraensis, Fagus crenata, F. sylvatica, Fagus* sp., *Laphira procera, Ocotea rodiaei, Pinus sylvestris, Pinus* sp., *Pseudotsuga* sp., *Quercus* sp., *Salix* sp., *Ulmus* sp., balsa, bamboo), *Spartina townsendii,* stone of *Prunus persica,* and cone of *Pseudotsuga taxifolia.* RANGE: Atlantic Ocean—Argentina (Buenos Aires), Canada (Nova Scotia), Denmark, France, Germany (F.R.G),

Great Britain (England, Isle of Man, Scotland, Wales), Iceland, Norway, Portugal, Spain (mainland and Tenerife), Sweden, United States (Connecticut, Florida, Maryland, Massachusetts, Mississippi, New Hampshire, New Jersey, North Carolina, Rhode Island, South Carolina, Texas, Virginia); Baltic Sea—Denmark (Sjaelland), Germany (F.R.G.), Poland, Sweden; Black Sea—Bulgaria, U.S.S.R. (J. Kohlmeyer, unpublished); Indian Ocean—India (Madras); Mediterranean—France, Greece, Italy (mainland and Sicily), Spain; Pacific Ocean—Australia (New South Wales, Queensland), Canada (British Columbia), Japan, New Zealand (North Island), Peru, United States [California, Hawaii (Hawaii, Kauai, Maui, Oahu), Washington]. NOTE: The species was also found on *Tamarix aphylla* in an inland salt lake, the Salton Sea, California (Anastasiou, 1963b). ETYMOLOGY: From the Greek, *halimos* = marine, referring to the habitat of the species.

Ceriosporopsis tubulifera (Kohlm.) Kirk *in* Kohlmeyer

Can. J. Bot. **50**, 1953, 1972

≡*Halosphaeria tubulifera* Kohlm., *Nova Hedwigia* **2**, 312, 1960

LITERATURE: Byrne and Jones (1974); Cavaliere (1966b, 1968); Eaton (1972); Henningsson (1974); G. C. Hughes (1968, 1969); T. W. Johnson (1963a); T. W. Johnson and Sparrow (1961); E. B. G. Jones (1972); Kirk (1966, 1976); Koch (1974); Kohlmeyer (1963d, 1971b); Kohlmeyer and Kohlmeyer (1964–1969); Neish (1970); Schaumann (1968); Schmidt (1974); Schneider (1976, 1977).

Figs. 49b and 49c

ASCOCARPS 165–324 μm high, 231–574 μm in diameter, subglobose, cylindrical or elongate–cylindrical, immersed or superficial, ostiolate, papillate, coriaceous, brown or black, sometimes almost hyaline below, solitary or gregarious. PERIDIUM 8–17.5(–35) μm thick, composed of 3 to 5 (to 12) layers of ellipsoidal, subglobose or polygonal, thick-walled cells; merging toward the inside into the pseudoparenchyma of the venter. PAPILLAE or necks 46.5–165 μm long, 33–73 μm in diameter, conical or cylindrical, centric or eccentric. PSEUDOPARENCHYMA of large, thin-walled, polygonal, deliquescing cells filling venter of young ascocarps; sometimes forming chains, but no distinct catenophyses. ASCI eight-spored, clavate, somewhat apiculate, pedunculate, unitunicate, thin-walled, aphysoclastic, deliquescing before ascospore maturity, without apical apparatuses. ASCOSPORES 14.5–23 × (7–)8.5–11 μm (excluding appendages), ellipsoidal, rarely subacuminate, one-septate, constricted at the septum, hyaline, appendaged; ascospores surrounded by an exosporic sheath that is drawn out at each apex into a cylindrical or broadly conical

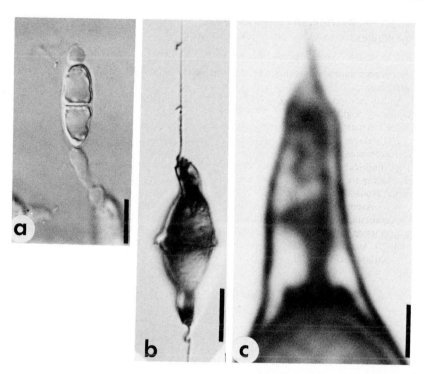

Fig. 49. *Ceriosporopsis* spp. (a) *C. halima,* ascospore with apical appendages (type slide No. 17, Barghoorn in FH; bar = 10 μm). (b) *C. tubulifera,* ascospore with mucilage exuding from the apical tubes and forming long threads (Herb. J.K. 1671; bar = 10 μm). From Kohlmeyer, *Trans. Br. Mycol. Soc.* **57** (1971), Plate 40, Fig. 27. Reprinted with permission. (c) *C. tubulifera,* ascospore apex with exosporic tube filled with mucilage (Herb. J.K. 1671; bar = 2 μm). (a and b in Nomarski interference contrast, c in bright field.)

tube, 10–14.5 μm long, 5.5–8.5 μm in diameter, containing a mucilaginous appendage; eventually, the internal appendages become viscous, oozing out of the orifices of the tubes and forming long threads; the exosporic sheath forms a 4.5–8.5 μm-thick annulus around the septum, finally becoming lobed by splitting and rupturing. MODE OF LIFE: Saprobic. SUBSTRATES: Intertidal and drifting wood and test panels (e.g., *Betula pubescens, Fagus sylvatica, Pinus sylvestris, Quercus* sp., *Salix* sp.). RANGE: Atlantic Ocean—Canada (Newfoundland, Nova Scotia), Denmark, Germany (F.R.G.), Great Britain (England, Isle of Man), Iceland, United States [Connecticut, Maine, Massachusetts, New Jersey (J. Kohlmeyer, unpublished)]; Baltic Sea—Denmark (Sjaelland), Germany (F.R.G.), Sweden; Pacific Ocean—Canada (British Columbia), United States (Washington). ETYMOLOGY: From the Latin, *tubulus* = small tube and *-fer* = -bearing, in reference to the ascospores adorned with tubelike appendages.

Key to the Species of Ceriosporopsis

1. Ascospore appendages consisting of a tube at each apex and a rupturing ring around the septum . . . . *C. tubulifera*
1'. Apical ascospore appendages solid, not tubular; equatorial appendages absent or not distinctly ringlike 2
   2(1') Ascospores with equatorial appendages 3
   2'(1') Ascospores without equatorial appendages 4
3(2) Ascospores surrounded by an irregular mucilaginous cover without sharp outlines, subcylindrical at both ends and irregularly lobed around the septum *C. circumvestita*
3'(2) Ascospores at each end with a distinct obclavate or subcylindrical, tapering appendage; around the septum three or four similar ones . . . . *C. calyptrata*
   4(2') Ascospores usually longer than 28 μm; wider than 10 μm; centrum of ascocarp containing persistent, wide filaments, similar to pseudoparaphyses or catenophyses . . . . *C. cambrensis*
   4'(2') Ascospores usually shorter than 28 μm; diameter 6–12 μm; centrum of ascocarp without persistent filaments . . . . *C. halima*

Chadefaudia Geneviève Feldmann

Rev. Gén. Bot. **64**, 150, 1957

=*Mycophycophila* Cribb et Cribb, *Univ. Queensl. Pap., Dep. Bot.* **4**, 42, 1960

LITERATURE: Kohlmeyer (1972b, 1973b); Müller and von Arx (1973).

ASCOCARPS solitary or gregarious, subglobose, hemispherical or depressed conical, superficial or immersed, ostiolate, papillate or epapillate, coriaceous to carbonaceous, dark-colored. PARAPHYSES absent; venter of young ascocarps filled with deliquescing pseudoparenchyma of thin-walled cells with large lumina. ASCI eight-spored, subglobose to clavate, unitunicate, thin-walled, early deliquescing. ASCOSPORES ellipsoidal, one-celled, hyaline, each end with a gelatinous, caplike appendage. HOSTS OR PHYCOBIONTS: Marine algae (Rhodophyta). TYPE SPECIES: *Chadefaudia marina* G. Feldmann. NOTE: All members of the genus are obligate marine species. ETYMOLOGY: Named after the French botanist M. Chadefaud.

Chadefaudia balliae Kohlm.

Mycologia **65**, 244–245, 1973

Figs. 50a and 50b

ASCOCARPS 230–600 μm high, 200–525 μm in diameter, subglobose or ovoid, immersed among the filaments in host stalks, ostiolate, papillate or epapillate, subcarbonaceous, black, smooth or verrucose, solitary or gregarious. PERIDIUM 40–50 μm (65–110 μm at the ostiole) thick, enclosing filaments of the host. PAPILLAE short or absent, 50–220 μm high, 80–180 μm in diameter, conical; ostioles 35–50 μm in diameter, periphysate. PSEUDOPARENCHYMA of thin-walled cells filling venter of young ascocarps. ASCI thin-walled, early deliquescing. ASCOSPORES 29–38 × 14.5–21.5 μm (including appendages), ellipsoidal, one-celled, hyaline, appendaged at both ends; appendages 2–5 μm thick, subgelatinous, each covering one-third of the spore like helmets or caps, irregular, thickest over indentations of the cell wall below the apex on opposed sides of the ascospore. MODE OF LIFE: Parasitic. HOST: *Ballia callitricha*. RANGE: Pacific Ocean—Australia (Victoria).

Chadefaudia corallinarum (Crouan et Crouan) Müller et von Arx *in* Ainsworth, Sparrow, et Sussman

The Fungi **4A,** 116, 1973

≡*Sphaeria corallinarum* Crouan et Crouan, *Florule du Finistère,* Klincksieck, Paris, and Brest, p. 24, 1867

≡*Physalospora corallinarum* (Crouan et Crouan) Saccardo, *Syll. Fung.* **1,** 448, 1882

≡*Mycophycophila corallinarum* (Crouan et Crouan) Kohlm., *Nova Hedwigia* **6,** 128, 1963

LITERATURE: Cribb and Cribb (1969); Kohlmeyer (1967, 1972b, 1973b, 1976); Kohlmeyer and Kohlmeyer (1964–1969).

Figs. 50c–50e

ASCOCARPS 75–180 μm high, 120–305 μm in diameter, subglobose, hemispherical or depressed conical, sometimes with a flat base and a thin margin, superficial on the surface of macroalgae and marine phanerogams among epiphytic microalgae, ostiolate, papillate or epapillate, subcarbonaceous, black, solitary or gregarious. PERIDIUM 15–30 μm thick. PAPILLAE short or absent, 36–102 μm high, 30–51 μm in diameter, conical; ostioles 11–15.5 μm in diameter, periphysate. PSEUDOPARENCHYMA of thin-walled cells filling venter of young ascocarps. ASCI 23–31 × 17–21.5 μm, eight-spored, subglobose to broadly clavate, with short stipe, unitunicate, thin-walled, aphysoclastic, early deliquescing, without apical apparatuses, developing in the base of the ascocarp venter. ASCOSPORES (9–)12–19 × (3.5–4–7.5(–8) μm (excluding appendages), irregularly triseriate, elongate–ellipsoidal, one-celled, hyaline, appendaged at both ends; appendages 1.2–1.8 μm high, 1.2–2.4 μm in diameter, subgelatinous, semiglobose, terminally attached, finally becoming viscous and stretching out. MODE OF LIFE: Symbiotic. HOSTS: Among crustaceous calcified microalgae (e.g., *Dermatolithon pustulatum, Dermatolithon* sp., *Epilithon membranaceum*), on macroalgae [e.g., *Ballia callitricha* (J. Kohlmeyer, unpublished), *Corallina mediterranea, C. officinalis, Euptilota formosissima* (J. Kohlmeyer, unpublished), *Halimeda opuntia, H. tuna, Jania adhaerens, J. corniculata, J. longifurca, J. rubens, Laminaria hyperborea, Lithophyllum tortuosum, Pseudolithophyllum expansum, Sargassum* sp., *Udotea petiolata*], and on marine phanerogams (e.g., *Halodule wrightii, Thalassia testudinum*). RANGE: Atlantic Ocean—the Bahamas, Canary Islands, Columbia, England, France, United States (Florida), West Indies (St. Thomas); Black Sea—U.S.S.R. (J. Kohlmeyer, unpublished); Indian Ocean—Mauritius; Mediterranean—Algeria, France, Greece, Italy, Yugoslavia; Pacific Ocean—Australia (Queensland, South Australia), New Zealand (J. Kohlmeyer, unpublished). NOTE: The associations between *C. corallinarum* and epiphytic marine algae are considered primitive lichenizations (Kohlmeyer, 1973b). ETYMOLOGY: Name referring to some of the substrates, namely, coralline algae.

Chadefaudia gymnogongri (J. Feldmann) Kohlm.

Bot. Mar. **16**, 202, 1973

[non *Chadefaudia gymnogongri* (J. Feldmann) von Arx et Müller *in* Farnham et Jones, *Trav. Stn. Biol. Roscoff* **20**, 8, 1973, illegitimate name]
≡*Macrophoma gymnogongri* J. Feldmann, *Soc. Hist. Nat. Afr. Nord* **31**, 167, 1940
≡*Mycophycophila gymnogongri* (J. Feldmann) Cribb et Cribb, *Univ. Queensl. Pap., Dep. Bot.* **4**, 43, 1960

LITERATURE: Cribb and Herbert (1954); Cribb and Cribb (1955); T. W. Johnson and Sparrow (1961).

Figs. 50f–50h

Fig. 50. *Chadefaudia* spp. (a) *C. balliae*, ascocarp (16-μm longitudinal section), basally attached to filaments of *Ballia callitricha* (holotype Herb. J.K. 2991; bar = 50 μm). (b) *C. balliae*, ascospores with subapical caplike appendages (Herb. J.K. 2991; bar = 15 μm). (c) *C. corallinarum*, ascospore with apical cap (type Herb. Crouan; bar = 3 μm). (d) *C. corallinarum*, subglobose ascocarp (8-μm longitudinal section) from *B. callitricha* (Herb. J.K. 3646; bar = 15 μm). (e) *C. corallinarum*, depressed conical ascocarp with flat base (8-μm longitudinal section) from *Laminaria hyperborea* (Herb. J.K. 2961; bar = 20 μm). (f) *C. gymnogongri*, longitudinal section of ascocarp immersed in *Ptilonia australasica* (slide A.B. Cribb; bar = 50 μm). (g) *C. gymnogongri*, habit of ascocarps on *Gymnogongrus norvegicus* (Herb. L, 941.95 . . . 360; bar = 1 mm). (h) *C. gymnogongri*, section through peridium and pseudoparenchyma of ascocarp from *P. australasica* (slide A.B. Cribb; bar = 10 μm). (a and e in bright field, the others in Nomarski interference contrast.)

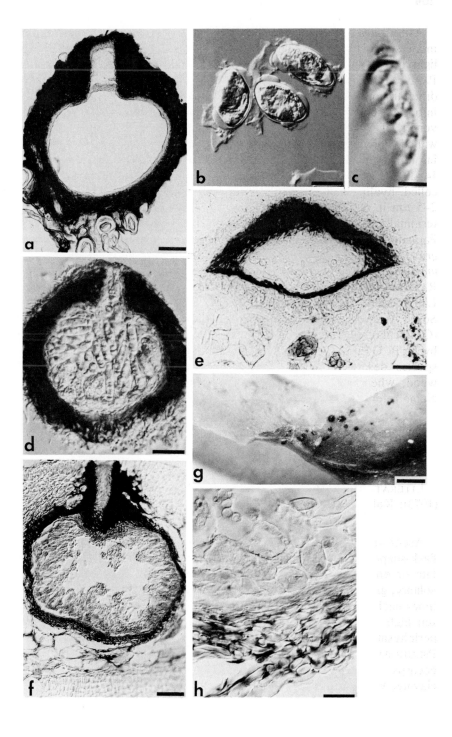

ASCOCARPS 105–370 μm high, 100–360 μm in diameter, subglobose, immersed, ostiolate, papillate, coriaceous, dark brown to blackish above, light brown below, solitary or gregarious. PERIDIUM 8–50 μm thick. PAPILLAE or necks 30–225 μm long, 35–80(–160) μm in diameter, cylindrical to conical, periphysate: ostiolar canal extending basally like a tube into the ascocarp cavity. PSEUDOPARENCHYMA of thin-walled cells filling venter of young ascocarps. ASCI broadly clavate, unitunicate, thin-walled, aphysoclastic, deliquescing before ascospore maturity, developing in a hymenium all around the inner perithecial venter. ASCOSPORES 14–20 × 3.5–6.5(–7.5) μm (excluding appendages), ellipsoidal, one-celled, hyaline, appendaged at both ends; appendages 1.5–2.5 μm high, 2–5 μm in diameter, subgelatinous, caplike, ellipsoidal to semiglobose, terminally attached, finally swelling and becoming soft. MODE OF LIFE: Parasitic. HOSTS: *Curdiea coriacea, Gigartina intermedia, Gymnogongrus norvegicus, Laurencia concinna, L. flexilis, L. heteroclada, L. pygmaea, L. succisa, L. tenera, Microcladia coulteri, Ptilonia australasica.* RANGE: Atlantic Ocean—France; Indian Ocean—Antarctica (Wilkes Land); Mediterranean—Algeria; Pacific Ocean—Australia (Queensland, South Australia, Tasmania), New Zealand (South Island), United States (California). NOTE: The fungus identified as *Macrophoma gymnogongri* by E. B. G. Jones (1962a) from Britain appears to be different from the species described above; Jones's fungus developed saprobically on wood, whereas *C. gymnogongri* is a parasite on algae. ETYMOLOGY: Name refers to one host genus, *Gymnogongrus.*

Chadefaudia marina Geneviève Feldmann

Rev. Gén. Bot. **64**, 150, 1957

LITERATURE: T. W. Johnson and Sparrow (1961); Müller and von Arx (1973); Kohlmeyer (1973b).

Figs. 51a–51c

ASCOCARPS 90–180 μm high, 90–200 μm in diameter, subglobose to flask-shaped, immersed in yellow–green spots of the host, ostiolate, papillate or epapillate, coriaceous, dark brown above, light brown below, solitary, gregarious or sometimes united. PERIDIUM 10–30 μm thick, sometimes enclosing single cells of the host. PAPILLAE short or absent, 25–80 μm high, 34–60 μm in diameter, cylindrical or conical; ostiolar canal periphysate, without tubelike extension into the ascocarp cavity. PSEUDOPARENCHYMA of thin-walled cells filling venter of young ascocarps. ASCI 23–34 × 12–16 μm, eight-spored, ellipsoidal to broadly clavate, with short stipe, unitunicate, thin-walled, aphysoclastic, early deliquescing, without apical apparatuses, forming a hymenium along the inner wall in the lower two-thirds of the ascocarp; stipe deliquescing before ascospore maturity, and asci finally floating in the ascocarp center. ASCOSPORES 12.5–16.5 × 4–5.5 μm (excluding appendages), at first fusiform, then ellipsoidal, one-celled, hyaline, appendaged at both ends; appendages 2–3 μm high, 4 μm in diameter, gelatinous, subglobose, terminally attached. MODE OF LIFE: Parasitic. HOST: *Rhodymenia palmata*. RANGE: Atlantic Ocean—France. NOTE: This species might have a pycnidial or spermogonial state (Kohlmeyer, 1973b); *C. marina* resembles *C. gymnogongri* but differs from the latter by generally smaller sizes of ascocarps and ascospores, by the absence of a tubelike ostiolar extension into the ascocarp cavity, and by the host. ETYMOLOGY: From the Latin, *marinus* = marine, in reference to the habitat of the species.

Chadefaudia polyporolithi (Bonar) Kohlm.

Bot. Mar. **16**, 205, 1973

≡*Mycophycophila polyporolithi* Bonar, *Mycologia* **57**, 379–380, 1965

Figs. 51d–51g

ASCOCARPS 120–315 μm high, 310–550 μm in diameter, hemispherical or depressed conical, with a flat base, superficial, ostiolate, epapillate, or rarely papillate, carbonaceous, black, solitary or gregarious, sometimes united. PERIDIUM 40–85 μm thick near the ostiole, 4–12 μm at the base. PAPILLAE, when present, short, cylindrical or conical, 40 μm high, 55 μm in diameter; ostiolar canal periphysate, 10–40 μm in diameter, extending basally like a tube into the ascocarp cavity. PSEUDOPARENCHYMA of thin-walled cells filling venter of young ascocarps. ASCI 46–72 × 12–21 μm, eight-spored, ventricose–clavate, unitunicate, thin-walled, aphysoclastic, early deliquescing, without apical apparatuses, forming a hymenium at the base of the ascocarp venter. ASCOSPORES 16–24 × (3.5–)5–8 μm (excluding appendages), two- or three-seriate, elongate–ellipsoidal, one-celled, hyaline, appendaged at both ends; appendages 2–3 μm high, 3–4.5 μm in diameter, gelatinous, subglobose, terminally attached, sometimes deciduous. MODE OF LIFE: Parasitic. HOSTS: *Polyporolithon conchatum* on *Calliarthron* and *Corallina* sp., *P. reclinatum,* unidentified Rhodophyta. RANGE: Pacific Ocean—United States (California). NOTE: *C. polyporolithi* resembles *C. corallinarum* but differs from the latter by generally larger sizes of ascocarps, asci, and ascospores; by thicker walls; by the presence of a tubelike ostiolar extension into the ascocarp cavity; and by the mode of life. ETYMOLOGY: Name referring to a host genus, *Polyporolithon*.

Key to the Species of *Chadefaudia*

1. Ascospores longer than 25 μm, diameter over 10 μm . . . . **C. balliae**
1'. Ascospores shorter than 25 μm, diameter below 10 μm 2
   2(1') Ascocarps superficial, but often surrounded by epiphytic, microscopic algae; base mostly flat 3
   2'(1') Ascocarps immersed in algal tissue; base rounded 4
3(2) Symbiotic fungus; ascocarp diameter generally below 300 μm; thickness of upper wall less than 40 μm; ostiolar projection into ascocarp cavity absent . . . **C. corallinarum**
3'(2) Parasitic fungus; ascocarp diameter generally over 300 μm; thickness of upper wall usually over 40 μm; ostiolar canal forming a tubelike projection into the ascocarp cavity . . . . **C. polyporolithi**
   4(2') Maximal ascospore length 16.5 μm (excluding appendages); ostiolar projection into ascocarp cavity absent; host *Rhodymenia* . . . . **C. marina**
   4'(2') Maximal ascospore length 20 μm (excluding appendages); ostiolar canal forming a tubelike projection into the ascocarp cavity; different host genera **C. gymnogongri**

Corollospora Werdermann

Notizbl. Bot. Gart. Berlin **8**, 248, 1922

=*Arenariomyces* Höhnk, *Veroeff. Inst. Meeresforsch. Bremerhaven* **3**, 28, 1954
=*Peritrichospora* Linder *in* Barghoorn et Linder, *Farlowia* **1**, 414, 1944

LITERATURE: Kohlmeyer (1962b, 1972b).

ASCOCARPS solitary or gregarious, globose to ellipsoidal, superficial or rarely immersed, with or without subicula, ostiolate or nonostiolate, papillate or epapillate, carbonaceous or rarely coriaceous, black, often seated on grains of sand or calcareous shells of marine animals. PSEUDOPARENCHYMA of large, thin-walled, deliquescing cells filling venter of young ascocarps; walls in most species with pitlike thickenings, connecting plasmatic strands of neighboring cells. ASCI eight-spored, fusiform or clavate, unitunicate, thin-walled, early deliquescing. ASCO-SPORES ellipsoidal to fusiform, amerosporous or phragmosporous, hyaline, or central cells light brown, appendaged; setalike appendages developing by fragmentation of the exospore, forming tufts or rows on the surface; some species with rigid, apical thorns or tubes. SUBSTRATE: Dead wood, bark, or other cellulose-containing substrates, decaying algae; arenicolous. TYPE SPECIES: *Corollospora maritima* Werdermann. NOTE: The genus was originally described as a representative of Sphaeropsidales (Fungi Imperfecti). All members of the genus are obligate marine species. ETYMOLOGY: From the Latin, *corolla* = little crown and *spora* = spore, in reference to the appendaged ascospores of *C. maritima*.

Fig. 51. *Chadefaudia* spp. (a) Longitudinal section (8 μm) through mature ascocarp of *C. marina* in *Rhodymenia palmata* (Herb. J.K. 2979; bar = 20 μm). (b) Same as a, section (8 μm) through immature ascocarp (bar = 20 μm). From Kohlmeyer, *Bot.*

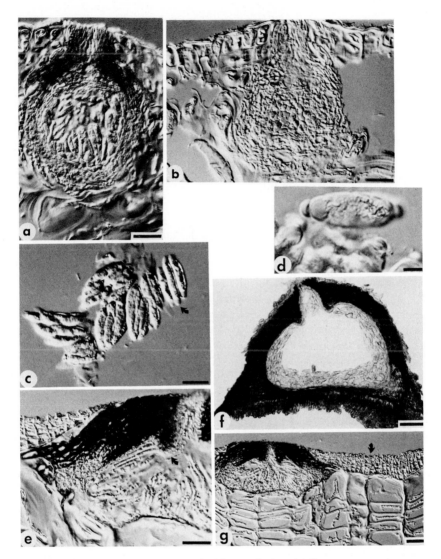

Mar. **16,** 205 (1973), Fig. 8. Reprinted with permission. (c) Same as a, mature ascospores with apical caps (arrow), immature spores enclosed in asci (bar = 15 μm). (d)
Ascospores of *C. polyporolithi* with apical caps (type Herb. UC 1272162; bar = 5 μm).
(e) Same as d, ascocarp in longitudinal section (6 μm), ostiole forming tubelike extension (arrow) into the ascocarp cavity (Herb. J.K. 2612; bar = 25 μm). (f) Same as d,
longitudinal section (16 μm) through ascocarp on *Polyporolithon conchatum* (type
Herb. UC 1272162; bar = 50 μm). (g) Same as d, ascocarp (6-μm section) associated
with crustose red alga (arrow) (Herb. J.K. 2612; bar = 50 μm). (f in bright field, the
others in Nomarski interference contrast.)

Corollospora comata (Kohlm.) Kohlm.

Ber. Dtsch. Bot. Ges. **75**, 126, 1962

≡*Peritrichospora comata* Kohlm., *Nova Hedwigia* **2**, 323, 1960

LITERATURE: Byrne and Jones (1974); Cavaliere (1966b); Churchland and McClaren (1973); Eaton and Dickinson (1976); Henningsson (1974); G. C. Hughes (1969); T. W. Johnson (1963b); Koch (1974); Kohlmeyer (1963e, 1966a,b, 1968d, 1971b, 1972b); Kohlmeyer and Kohlmeyer (1964–1969); Neish (1970); Schmidt (1974); Schneider (1976, 1977).

Figs. 52a and 52b

ASCOCARPS 251–670 μm high, 267–820 μm in diameter, subglobose or ellipsoidal, immersed or partly immersed, with or without ostioles, papillate or epapillate, carbonaceous, black, solitary or gregarious. PERIDIUM 16–40 μm thick, composed of dark, thick-walled cells, roundish at the outside, oblong toward the center, forming a textura angularis. PAPILLAE short or absent. PSEUDOPARENCHYMA of thin-walled cells filling venter of young ascocarps; deliquescing at ascospore maturity. ASCI 119–163 × 28.5–47.5 μm, eight-spored, broadly fusiform or clavate, pedunculate, unitunicate, thin-walled, aphysoclastic, without apical apparatuses, early deliquescing. ASCOSPORES (31.5–)35–54 × 12–17 μm (excluding appendages), bi- or triseriate, ellipsoidal or broadly fusiform, predominantly five-septate (rarely three-, four-, or six-septate), constricted at the septa, central cells fuscous, apical cells hyaline, appendaged; appendages developing by fragmentation of the exospore, setalike, flexible, attached in a tuft to each apex and in four or five tufts around the central septum; apical setae 8.5–14.5 μm long, lateral setae 12–20 μm long. MODE OF LIFE: Saprobic. SUBSTRATE: Intertidal and drifting wood (e.g., *Abies* sp., *Fagus sylvatica, Larix* sp., *Picea* sp., *Pinus sylvestris, Pinus* sp.) and bark; ascospores accumulate in foam along the shore. RANGE: Atlantic Ocean—Canada (Nova Scotia), Denmark, Great Britain (Isle of Man), United States (Maine, North Carolina); Baltic Sea—Denmark, Germany (F.R.G.), Sweden; Pacific Ocean—Canada (British Columbia), United States (California, Washington). ETYMOLOGY: From the Latin, *comatus* = provided with hair or a crest, in reference to the tufts of setae of the ascospores.

Corollospora cristata (Kohlm.) Kohlm.

Ber. Dtsch. Bot. Ges. **75**, 126, 1962

≡*Peritrichospora cristata* Kohlm., *Nova Hedwigia* **2**, 324, 1960

LITERATURE: Cavaliere [1966b, 1968 (sub *C. comata*)]; Henningsson (1974); G. C. Hughes (1968, 1969); G. C. Hughes and Chamut (1971); T. W. Johnson (1963b); E. B. G. Jones (1972); E. B. G. Jones and Irvine (1971, 1972); Koch (1974); Kohlmeyer (1962a, 1963d,e, 1968d, 1971b, 1972b); Kohlmeyer and Kohlmeyer (1964–1969); Neish (1970); Schaumann (1968, 1969); Schmidt (1974); Schneider (1976, 1977).

Fig. 52c

ASCOCARPS 201–680 μm high, 241–910 μm in diameter, subglobose or ellipsoidal, immersed or partly immersed, with or without ostioles, papillate or epapillate, carbonaceous, black, solitary or gregarious. PERIDIUM 19–29 μm thick, composed of dark, thick-walled cells with small lumina; roundish at the outside, oblong and compressed toward the center, merging into the pseudoparenchyma; forming a textura angularis. PAPILLAE 86–470 μm long, 59–120 μm in diameter, conical or subcylindrical, centric or eccentric. PSEUDOPARENCHYMA of thin-walled cells filling venter of young ascocarps; deliquescing at ascospore maturity. ASCI 64–89.5 × 29–32 μm, eight-spored, broadly fusiform or clavate, pedunculate, unitunicate, thin-walled, aphysoclastic, without apical apparatuses, early deliquescing. ASCOSPORES 24–38.5(–41) × 8.5–16.5 μm (excluding ap-

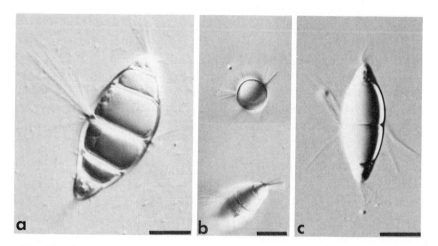

Fig. 52. *Corollospora* spp. (a) Five-septate ascospore of *C. comata* with tufts of setalike appendages at the apices and around the equator (Herb. J.K. 1919; bar = 10 μm). (b) Same as a, ascospores showing attachment of appendages (Herb. J.K. 1919; bar = 15 μm). (c) Three-septate ascospore of *C. cristata* with apical and lateral appendages (Herb. J.K. 2454; bar = 10 μm). From Kohlmeyer, *Trans. Br. Mycol. Soc.* **57** (1971), Plate 38, Fig. 10. Reprinted with permission. (All in Nomarski interference contrast.)

pendages), bi- or triseriate, ellipsoidal, predominantly three-septate (rarely two-, four-, or five-septate), constricted at the septa, central cells fuscous, apical cells hyaline, appendaged; appendages developing by fragmentation of the exospore, setalike, flexible, attached in a tuft to each apex and in several tufts around the central septum; apical setae 4.5–8 μm long, lateral setae 7.5–9 μm long. MODE OF LIFE: Saprobic. SUBSTRATE: Intertidal and drifting wood and wooden panels (e.g., *Dicorynia paraensis, Laphira procera, Ocotea rodiaei*); ascospores accumulate in foam along the shore. RANGE: Atlantic Ocean—Canada (Newfoundland, Nova Scotia), Denmark, Germany (Helgoland and mainland, F.R.G.), Great Britain (Scotland; Kohlmeyer, unpublished), Iceland, Sweden, United States (Maine); Baltic Sea—Germany (F.R.G., G.D.R.), Sweden; Mediterranean—France; Pacific Ocean—Canada (British Columbia), Chile (Isla Hoste), United States (California, Washington). NOTE: Cavaliere (1966b) considered *C. cristata* a synonym of *C. comata,* a suggestion that most later authors did not accept. ETYMOLOGY: From the Latin, *cristatus* = crested, in reference to the tufts of setae of the ascospores.

Corollospora intermedia I. Schmidt

Nat. Naturschutz Mecklenburg **7**, 6, 1969 (publ. 1971)

LITERATURE: Kohlmeyer (1972b); Schaumann (1972); Schmidt (1974).

ASCOCARPS 100–310(–575) μm in diameter, subglobose or ovoid, immersed or superficial, sometimes seated with subicula on grains of sand or cases of bryozoa, ostiolate, papillate, carbonaceous, black, solitary. PERIDIUM 10–20 μm thick, two-layered; outer layer composed of roundish cells with large lumina, inner layer consisting of elongate cells with narrow lumina. PAPILLAE 40–90 μm long, 40–60 μm in diameter, conical, near the basal subiculum and pointing downward; ostiolar canal at first filled with pseudoparenchymatous, small cells. PSEUDOPARENCHYMA of thin-walled, polygonal cells filling venter of young ascocarps; deliquescing at ascospore maturity; cell walls with pitlike thickenings, connecting plasmatic strands of neighboring cells. ASCI early deliquescing. ASCOSPORES 25–34(–35.5) × 7–11.5 μm (excluding appendages), fusiform, straight or slightly curved, three-septate (rarely more or less septa), constricted at the septa, hyaline, appendaged; at both ends with a single, terminal appendage, 9.5–14 × 1.5–2 μm, thornlike, slender, attenuate, rigid, straight or somewhat curved, at the tip with a refractive body and bearing a small cap or fibers that develop by peeling off of the exospore; peritrichous around the central septum with 10–18 flexible setae, 10–16

μm long, which develop by fragmentation of the exospore; setae attached to a narrow equatorial, beltlike thickening of the wall. MODE OF LIFE: Saprobic. SUBSTRATE: Rotting *Fucus vesiculosus*; also isolated from the sediment along a beach; ascospores accumulate in foam along the shore. RANGE: Atlantic Ocean—France, Germany (Helgoland, F.R.G.); Baltic Sea—Germany (G.D.R.). ETYMOLOGY: From the Latin, *intermedius* = intermediate, in reference to the intermediate position of the species between the related *C. maritima* and *C. lacera*.

Corollospora lacera (Linder *in* Barghoorn et Linder) Kohlm.

Ber. Dtsch. Bot. Ges. **75**, 126, 1962

≡*Peritrichospora lacera* Linder *in* Barghoorn et Linder, *Farlowia* **1**, 415, 1944

LITERATURE: Catalfomo *et al.* (1972–1973); Kirk *et al.* (1974); Koch (1974); Kohlmeyer (1960, 1966a,b, 1968d, 1971b, 1972b).

Fig. 53

ASCOCARPS 307–468 μm in diameter, subglobose to ovoid, superficial or partly immersed, sometimes seated with subicula on grains of sand, ostiolate, papillate or epapillate, carbonaceous, black, solitary or scattered. PERIDIUM 24–35 μm thick, composed of dark, thick-walled cells, roundish at the outside, oblong toward the center, forming a textura angularis; subicula prosenchymatous. PAPILLAE 40 μm high, 70 μm in diameter, conical, near the basal subiculum and pointing downward; ostiolar canal 40 μm in diameter, at first filled with pseudoparenchymatous, hyaline cells, ca. 6 μm in diameter. PSEUDOPARENCHYMA of thin-walled, polygonal, or rounded cells, about 25 μm in diameter, filling venter of young ascocarps; deliquescing at ascospore maturity; cell walls with pitlike thickenings, connecting plasmatic strands of neighboring cells. ASCI 112–150 × 18–28 μm, eight-spored, fusiform or clavate, unitunicate, thin-walled, aphysoclastic, deliquescing before ascospore maturity; originating from an ascogenous tissue of polygonal cells, 5–6 μm in diameter, in the upper part of the ascocarp venter; directed downward toward the ostiole. ASCOSPORES 39–60(–63) × 10–16(–19) μm (excluding apical thorns and setae), fusiform, straight or slightly curved, five-septate (rarely four- or six-septate), constricted at the septa, hyaline, appendaged; at both ends with a single, terminal appendage, 9.5–24 (–28.5) μm long, 2.2–4.5 μm in diameter at the base, 1.5–1.8 μm at the apex, thornlike, attenuate, rigid, straight or somewhat curved, at the tip with a refractive body and bearing a tube that develops by peeling off of the exospore, 19.5–33(–39) × 2–6 μm; peritrichous around the central septum with flexible setae, 12–16.5 μm long, which develop by fragmenta-

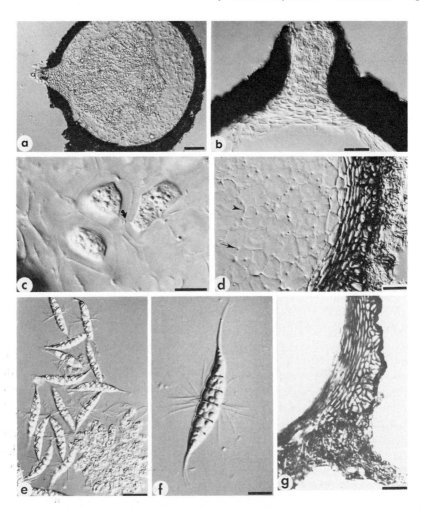

Fig. 53. *Corollospora lacera* from pure cultures. (a) Longitudinal section (4 μm) through ascocarp with lateral ostiole (bar = 50 μm). (b) Section (16 μm) through ostiole, canal filled with hyaline pseudoparenchyma before release of ascospores (bar = 20 μm). (c) Section (20 μm) through central pseudoparenchyma of ascocarp; pitlike thickenings (arrow) in the cell walls connect plasmatic strands of neighboring cells (bar = 10 μm). (d) 4-μm section through peridium and central pseudoparenchyma with pits in top (arrow) and side view (arrowhead) (bar = 15 μm). (e) Ascospores and immature asci (bar = 25 μm). (f) Ascospore with apical thorns, lateral and apical setae (bar = 10 μm). (g) 4-μm section through peridium and subiculum (bar = 20 μm). (e and f Herb. J.K. 3807, the others Herb. Brooks; g in bright field, the others in Nomarski interference contrast.)

tion of the exospore. MODE OF LIFE: Saprobic. SUBSTRATE: Intertidal wood (e.g., *Abies* sp., *Picea* sp., *Pinus* sp., *Populus* sp.), on incubated leaves of *Thalassia testudinum;* ascospores accumulate in foam along the shore. RANGE: Atlantic Ocean—Bahamas (Great Abaco) and Bermuda (J. Kohlmeyer, unpublished), Denmark, Sierra Leone (A. A. Aleem, personal communication), United States (Georgia, Massachusetts, North Carolina, Rhode Island); Pacific Ocean—United States (California). Description in part after J. Kohlmeyer (unpublished), based on material of R. D. Brooks. ETYMOLOGY: From the Latin, *lacer* = torn, mangled, in reference to the fragmenting exospore of the ascospores.

Corollospora maritima Werdermann

Notizbl. Bot. Gart. Berlin **8**, 248, 1922

=*Arenariomyces cinctus* Höhnk, *Veroeff. Inst. Meeresforsch. Bremerhaven* **3**, 28, 1954
=*Peritrichospora integra* Linder *in* Barghoorn et Linder, *Farlowia* **1**, 414, 1944

LITERATURE: Anastasiou (1963b); Anastasiou and Churchland (1969); Apinis and Chesters (1964); Barghoorn (1944); Block *et al.* (1973); Byrne and Eaton (1972); Byrne and Jones (1974, 1975a,b); Catalfomo *et al.* (1972–1973); Cavaliere (1966b, 1968); Cavaliere and Johnson (1966b); Eaton (1972); Fize (1960); Gessner and Goos (1973a,b); Henningsson (1974); Höhnk (1954b, 1956); G. C. Hughes (1969, 1974); Irvine *et al.* (1972); T. W. Johnson (1956b, 1963b, 1968); T. W. Johnson *et al.* (1959); T. W. Johnson and Sparrow (1961); E. B. G. Jones (1962a,c, 1963c, 1968b, 1971a, 1972, 1973); E. B. G. Jones and Irvine (1971); E. B. G. Jones and Jennings (1964, 1965); E. B. G. Jones and LeCampion-Alsumard (1970a,b); E. B. G. Jones *et al.* (1972); Kirk (1966, 1976); Kirk and Catalfomo (1970); Kirk *et al.* (1974); Koch (1974); Kohlmeyer (1958a,b, 1959, 1960, 1962a,b, 1963d,e, 1964, 1966a,b, 1967, 1968c,d, 1969c, 1971b, 1972b, 1976); Kohlmeyer and Kohlmeyer (1964–1969, 1971a, 1977); Le Campion-Alsumard (1970); Lutley and Wilson (1972b); Meyers (1957, 1968b, 1971a); Meyers and Reynolds (1959a,c, 1960b); Meyers and Scott (1967); Meyers *et al.* (1960, 1963b); Neish (1970); Peters *et al.* (1975); Raghukumar (1973); Schafer and Lane (1957); Schaumann (1968, 1969, 1974b); Schmidt (1967, 1974); Schneider (1976, 1977); Shearer (1972b); Tubaki (1966, 1969); Tubaki and Ito (1973); Wagner-Merner (1972); Wilson (1951).

Fig. 54a

ASCOCARPS 98–400 μm in diameter, globose or subglobose, superficial

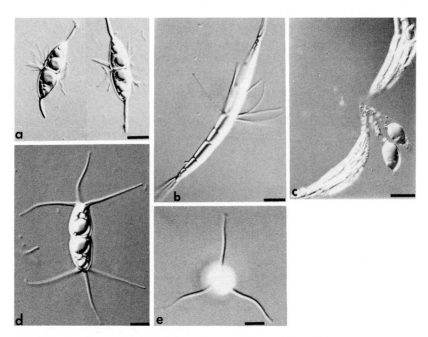

Fig. 54. *Corollospora* spp. (a) Ascospores of *C. maritima* with apical thorns and equatorial setae (Herb. Brooks; bar = 10 μm). (b) Ascospore of *C. pulchella* (Herb. J.K. 2630; bar = 10 μm). (c) Same as b, immature and mature asci (Herb. J.K. 2630; bar = 25 μm). (d) Ascospore of *C. trifurcata* with three appendages at each apex (Herb. Brooks; bar = 10 μm). (e) Same as d, apical view of ascospore (Herb. Brooks; bar = 10 μm). (All in Nomarski interference contrast.)

or partly immersed, often seated with subicula on grains of sand or other hard surfaces (shells of barnacles or mollusks; exoskeletons of bryozoa), with or without ostioles, papillate or epapillate, carbonaceous, metallic black, solitary or gregarious. PERIDIUM 12–40 μm thick, composed of dark, thick-walled cells, roundish or polygonal at the outside, oblong toward the center, forming a textura angularis; subicula prosenchymatous. PAPILLAE, when present, 17–72 μm long, 21–55 μm in diameter, conical or subcylindrical, apically, laterally, or near the basal subiculum and pointing downward. PSEUDOPARENCHYMA of thin-walled, polygonal or rounded cells filling venter of young ascocarps, deliquescing at ascospore maturity; cell walls with pitlike thickenings, connecting plasmatic strands of neighboring cells. ASCI 72–140 × 17–50 μm, eight-spored, fusiform or subclavate, sometimes apiculate, short stipitate, unitunicate, thin-walled, aphysoclastic, without apical apparatuses, early deliquescing. ASCOSPORES 20–34(–53) × (4–)6–11(–14) μm (excluding apical thorns and setae), fusiform or subellipsoidal, one-septate, constricted at the septum, hyaline, appendaged; at both ends with a single, terminal appendage, 7–17.5(–23) × 1–1.5 μm, spine- or thornlike, slender, attenuate, rigid, straight or somewhat curved, at the tip with a refractive body and bearing a small cap or fibers, 8–10 μm long, that develop by peeling off of the exospore; peritrichous around the septum with 8 or more flexible, ribbon-shaped setae, 5–16(–20) × 1 μm, which develop by fragmentation of the exospore; setae attached to a narrow equatorial, beltlike thickening of the wall; ascospores germinate usually subterminally. MODE OF LIFE: Saprobic. SUBSTRATES: Rotting algae (*Ceramium* sp., *Fucus vesiculosus*, *Macrocystis* sp., *Sargassum* sp.); marine phanerogams (*Spartina alterniflora*, *Thalassia testudinum*, *Zostera marina*); intertidal bark, cones of conifers (*Larix* sp., *Pinus* sp.); intertidal wood and driftwood, test blocks, pilings (*Abies firma*, *Betula pubescens*, *Casuarina* sp., *Dicorynia paraensis*, *Fagus crenata*, *F. sylvatica*, *Fagus* sp., *Laphira procera*, *Picea* sp., *Pinus sylvestris*, *Pinus* sp., *Populus* sp., *Pseudotsuga menziesii*, *Quercus* sp., *Tamarix aphylla*, balsa); submerged polyurethane foam; ascospores accumulate in foam along the shore. RANGE: Atlantic Ocean—Argentina, Bahamas (Great Abaco; J. Kohlmeyer, unpublished), Bermuda, Brazil, Canada (Nova Scotia), Colombia, Denmark, France, Germany (Büsum, Helgoland and mainland, F.R.G.), Great Britain [England, Scotland (J. Kohlmeyer, unpublished), Wales], Iceland, Mexico (Yucatan), Sierra Leone (A. A. Aleem, personal communication), Spain (Tenerife and mainland), Sweden, Tobago and Trinidad (J. Kohlmeyer, unpublished), United States (Connecticut, Florida, Georgia, Maine, Maryland, Massachusetts, New Jersey, North Carolina, Rhode Island, Virginia); Baltic Sea—Denmark (Sjaelland), Germany (Rügen and mainland, G.D.R.; F.R.G.), Sweden; Black Sea—Bulgaria; Indian Ocean—India (Madras), SriLanka (Ceylon); Mediterranean—France, Italy (Sicily and mainland), Yugoslavia; Pacific Ocean—Canada (British Columbia), Japan, Mexico (Oaxaca), New Zealand (North Island), Peru (Ica), United States [California, Hawaii (Hawaii, Kauai, Maui, Oahu), Washington]. ETYMOLOGY: From the Latin, *maritimus* = marine, referring to the habitat of the species.

Corollospora pulchella Kohlm., Schmidt et Nair

Ber. Dtsch. Bot. Ges. **80**, 98–99, 1967

Conidial state: *Clavariopsis bulbosa* Anastasiou, *Mycologia* **53**, 11, 1961

LITERATURE: Anastasiou (1961, 1963b); G. C. Hughes (1974); Kohlmeyer (1968c, 1969c, 1972b); Kohlmeyer and Kohlmeyer (1964–

1969, 1971a); Nair (1970); S. Nilsson (1964); Raghukumar (1973); Schmidt (1974); Shearer (1972a,b); Shearer and Crane (1971); Tubaki (1968, 1969); Tubaki and Ito (1973).

Figs. 54b and 54c

ASCOCARPS 80–295 μm in diameter, globose to subglobose, rarely subovoid, superficial on hard surfaces or immersed in soft substrates, sometimes seated with subicula on grains of sand or shells of marine mollusks, ostiolate, papillate, carbonaceous, black, tuberculate, solitary or gregarious. PERIDIUM 10–20 μm thick, dark brown to black, two-layered; outer layer composed of roundish cells with large lumina; inner layer consisting of elongate cells with narrow lumina. PAPILLAE 24–70 μm long, 20–60 μm in diameter, cylindrical or conical, black, tuberculate; predominantly near the basal subiculum and pointing downward; ostiolar canal at first filled with pseudoparenchymatous, small cells. PSEUDOPARENCHYMA of thin-walled, polygonal or rounded cells filling venter of young ascocarps; deliquescing at ascospore maturity; cell walls with pitlike thickenings, 1.8–2.2 μm in diameter, connecting plasmatic strands of neighboring cells. ASCI 100–140 × 20–30 μm, eight-spored, cylindrical to fusiform, short pedunculate, unitunicate, thin-walled, aphysoclastic, without apical apparatuses, early deliquescing; originating from an ascogenous tissue of polygonal cells, 3–5 × 2.5 μm in diameter, in the upper or lateral part of the ascocarp venter; directed downward toward the ostiole. ASCOSPORES 52.5–102.5(–112.5) × 7–12(–16) μm (excluding appendages), fusiform, slightly curved, 7-septate (rarely 9- to 13-septate), constricted at the septa, hyaline, appendaged; appendages developing by fragmentation of the exospore, setalike, flexible, attached in a tuft to a conical papilla at each apex and peritrichous around the central septum; about 7 apical appendages, 12.5–20(–24) × 1 μm; about 15 lateral appendages, 15–29 × 1 μm.

CONIDIOPHORES 10–300 × 2–5 μm, cylindrical, septate, simple or branched, hyaline. CONIDIA tetraradiate, septate, slightly constricted at the septa, hyaline to light brown, developing by transformation of the inflated apex of the conidiophore; basal arm one-septate; proximal cell 10–20 μm high, 5–11.5 μm in diameter, ellipsoidal or ovoid, truncate at the base, light brown; distal cell 6.5–13 × 6.5–13 μm, cylindrical or shortly three-branched, fuscous; three divergent arms arising simultaneously from the inflated distal cell of the basal arm, (4–)20–70 × 4–6(–8.5) μm, cylindrical, one- to seven-septate, light brown. CHLAMYDOSPORES 9–18 × 4–14 μm, catenulate, 2–20 cells per chain, strongly constricted at the septa, rarely branched, formed in acropetal succession, deep olive-brown to black. MODE OF LIFE: Saprobic. SUBSTRATES: Submerged wood (*Tamarix aphylla,* test panels of *Mangifera indica, Pinus echinata,* and balsa), mangrove roots, rotting algae (*Fucus vesiculosus*); ascospores accumulate in foam along the shore. RANGE: Atlantic Ocean—Liberia, Sierra Leone (A.A. Aleem, personal communication; conidia in foam), Trinidad (J. Kohlmeyer, unpublished), United States (Maryland, North Carolina); Baltic Sea—Germany (G.D.R.); Indian Ocean—India (Madras, Kerala State), South Africa; Pacific Ocean—Japan, Mexico (Oaxaca), Pago Pago, United States [California, Hawaii (Kauai, Oahu)]. ETYMOL-OGY: From the Latin, *pulchellus* (diminutive of *pulcher*) = very pretty, in reference to the ornamented ascospores.

Corollospora trifurcata (Höhnk) Kohlm.

Ber. Dtsch. Bot. Ges. **75**, 126, 1962

≡*Arenariomyces trifurcatus* Höhnk, *Veroeff. Inst. Meeresforsch. Bremerhaven* **3**, 30 1954
≡*Halosphaeria trifurcata* (Höhnk) Cribb et Cribb, *Univ. Queensl. Pap.*, *Dep. Bot.* **3**, 99, 1956
≡*Peritrichospora trifurcata* (Höhnk) Kohlm., *Nova Hedwigia* **3**, 89, 1961

LITERATURE: Byrne and Jones (1974); Catalfomo *et al.* (1972–1973); Cavaliere (1966b); Fize (1960); Henningsson (1974); Höhnk (1954b, 1956); G. C. Hughes (1968, 1969, 1974); T. W. Johnson (1956b, 1963b); T. W. Johnson and Sparrow (1961); Kirk and Catalfomo (1970); Kirk *et al.* (1974); Koch (1974); Kohlmeyer (1960, 1962a, 1963d, 1966a, 1967, 1968c, 1969c, 1971b, 1972b, 1976); Kohlmeyer and Kohlmeyer (1964–1969, 1971a, 1977); Meyers (1957); Raghukumar (1973); Schmidt (1974); Schneider (1976); Shearer (1972b); Tubaki (1968, 1969).

Figs. 54d and 54e

ASCOCARPS 100–325 μm in diameter, globose or subglobose, immersed or superficial, often flattened at the base and seated with subicula on grains of sand or calcareous animal shells, with or without ostioles, papillate or epapillate, carbonaceous or rarely coriaceous, black or dark brown, sometimes covered by short, brown hyphae, solitary or rarely gregarious. PERIDIUM 5–18 μm thick, composed of dark, thick-walled cells, roundish at the outside, oblong toward the center, forming a textura angularis; subicula prosenchymatous. PAPILLAE up to 30 μm long, conical or subcylindrical. PSEUDOPARENCHYMA of thin-walled polygonal or rounded cells, filling venter of young ascocarps, deliquescing at ascospore maturity; cell walls with inconspicuous pitlike thickenings, connecting plasmatic strands of neighboring cells. ASCI 60–80 × 19–23 μm, eight-spored, fusiform or subclavate, short stipitate, unitunicate, thin-walled, aphysoclastic, early deliquescing; originating from an ascogenous tissue of polygonal cells in the base of the ascocarp venter. ASCOSPORES (19–)

24–38.5 × 7–16.5 μm (excluding appendages), fusiform, ellipsoidal or oblong, one-septate, with or without slight constriction at the septum, hyaline, appendaged; at both ends with three (rarely four to seven) terminal or subterminal appendages, which develop by fragmentation of the exospore; appendages 15–39 × 1.5–2 μm, slender, attenuate, rigid, curved, sometimes with an apical thickening. MODE OF LIFE: Saprobic. SUBSTRATES: Intertidal and drifting wood (*Arbutus menziesii, Picea* sp., *Pinus* sp., *Quercus* sp., *Tamarix* sp.), test panels (balsa, *Dicorynia paraensis, Fagus sylvatica, Laphira procera*), bark, cones of conifers; ascospores accumulate in foam along the shore. RANGE: Atlantic Ocean—Argentina, Bahamas (Great Abaco; J. Kohlmeyer, unpublished), Bermuda, Canada (Newfoundland), Denmark, France, Germany (F.R.G.), Great Britain [Isle of Man, Scotland (J. Kohlmeyer, unpublished)], Mexico (Yucatan), Sierra Leone (A. A. Aleem, personal communication), Spain (Tenerife and mainland), Tobago and Trinidad (J. Kohlmeyer, unpublished), United States (Florida, Georgia, Maryland, Massachusetts, New Jersey, North Carolina, South Carolina, Virginia); Baltic Sea—Germany (F.R.G. and G.D.R.), Sweden; Black Sea—Bulgaria (J. Kohlmeyer, unpublished); Indian Ocean—India (Madras); Mediterranean—France, Yugoslavia; Pacific Ocean—Australia (Queensland), Canada (British Columbia), Japan, Mexico (Oaxaca), Peru (Ica), United States [California, Hawaii [Hawaii, Kauai, Oahu), Washington]. ETYMOLOGY: From the Latin, *trifurcatus* = three-forked, in reference to the three appendages at each end of the ascospore.

Corollospora tubulata Kohlm.

Ber. Dtsch. Bot. Ges. **81,** 53–54, 1968

LITERATURE: G. C. Hughes (1969); Koch (1974); Kohlmeyer (1972b); Kohlmeyer and Kohlmeyer (1964–1969).

Fig. 55

ASCOCARPS 430–880 μm high, 440–913 μm in diameter, subglobose, superficial, subiculate, ostiolate, papillate, subcarbonaceous, black, solitary, often covered by grains of sand. PERIDIUM 28–144 μm thick, black or dark brown, three-layered; outer layer (up to 95 μm thick) composed of large, more or less rectangular cells with large lumina; middle layer composed of small, irregular, flat cells with small lumina; inner layer composed of large elongate cells with large lumina, merging into the hyaline pseudoparenchyma of the venter. PAPILLAE 55–200 μm high, 70–130 μm in diameter, conical, predominantly near the basal subiculum and pointing downward; ostiolar canal at first filled with pseudo-

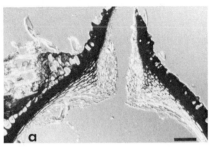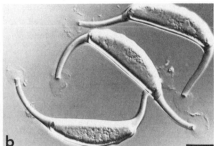

Fig. 55. *Corollospora tubulata*, from isotype at IMS. (a) Longitudinal section through ostiole showing three layers of peridium (bar = 50 μm). (b) Ascospores with apical tubes releasing mucus (bar = 25 μm). (Both in Nomarski interference contrast.)

parenchymatous, hyaline, small cells. PSEUDOPARENCHYMA of thin-walled, polygonal cells filling venter of young ascocarps, deliquescing at ascospore maturity; cell walls with pitlike thickenings, connecting plasmatic strands of neighboring cells. ASCI 135–192 × 24–32 μm, eight-spored, fusiform to clavate, unitunicate, thin-walled, aphysoclastic, deliquescing before ascospore maturity. ASCOSPORES 137–152 × 17.5–18.5 μm (including apical tubes), fusiform, curved, frequently S- or C-shaped, repand on one side, smooth on the other, one-celled, thick-walled, hyaline; both ends with a tubelike, mucus-filled appendage, separated from the cell proper by a wall; tubes 38–47 × 4–4.5 μm, curved, rigid, slightly tapering, with an apical porus covered by a lidlike membrane, rupturing at maturity and releasing mucus that forms a persistent gelatinous globule at the mouth of the tube. MODE OF LIFE: Saprobic. SUBSTRATE: Intertidal wood (e.g., *Picea* sp., *Pinus* sp.). RANGE: Atlantic Ocean—Denmark; Pacific Ocean—Canada (British Columbia). ETYMOLOGY: From the Latin, *tubulatus* = provided with tubes, referring to the ascospores.

Key to the Species of *Corollospora*

1. Ascospores one-celled, C- or S-shaped, with a long, mucus-filled tube at each end
. . . . *C. tubulata*

1'. Ascospores with one septum or more, straight or only slightly bent 2

   2(1') Ascospores without setae around the central septum; three rigid, thornlike appendages at each end . . . . *C. trifurcata*

   2'(1') Ascospores with setae around the central septum; apical appendages different 3

3(2') Ascospores at each end with one thornlike, rigid appendage without cytoplasm; outer wall layer of these thorns may peel off to form thin tubes or setae 4

3'(2') Ascospores with tufts of delicate setae on the tips of apical cells that contain cytoplasm and oil globules 6

   4(3) Ascospores one-septate . . . . *C. maritima*

4'(3) Ascospores with three or more septa 5
5(4') Ascospores three-septate, shorter than 36 μm (appendages excluded) *C. intermedia*
5'(4') Ascospores typically five-septate, longer than 36 μm (appendages excluded)
. . . . *C. lacera*
6(3') Ascospores with seven or more septa, hyaline throughout . . . *C. pulchella*
6'(3') Ascospores with six or less septa, central cells fuscous 7
7(6') Mature ascospores typically three-septate; asci less than 100 μm long
. . . . *C. cristata*
7'(6') Mature ascospores typically five-septate; asci longer than 100 μm . . *C. comata*

Haligena Kohlm.

Nova Hedwigia **3**, 87, 1961

ASCOCARPS solitary or gregarious, globose to ellipsoidal, immersed or becoming exposed, ostiolate, papillate or rarely epapillate, coriaceous to subcarbonaceous, light-colored to black. PSEUDOPARENCHYMA of large, thin-walled, deliquescing cells filling venter of young ascocarps. ASCI eight (rarely two or four)-spored, clavate or ellipsoidal, unitunicate, thin-walled, early deliquescing. ASCOSPORES ellipsoidal to fusiform or cylindrical, phragmosporous, hyaline, with gelatinous sheaths or appendages; appendages apical or subapical, small and caplike or at first covering the ascospore completely, often becoming viscous and forming long threads. SUBSTRATES: Dead wood, bark, or decaying phanerogams of salt marshes, polyurethane. TYPE SPECIES: *Haligena elaterophora* Kohlm. NOTE: All members of the genus are obligate marine or brackish-water species. ETYMOLOGY: From the Greek, *haligenes* = borne in the sea, referring to the habitat of the fungus.

Haligena amicta (Kohlmeyer) J. Kohlmeyer et E. Kohlmeyer comb. nov.

≡*Sphaerulina amicta* Kohlm., *Nova Hedwigia* **4**, 414, 1962 (basionym)

LITERATURE: Koch (1974); Kohlmeyer (1963d); Schmidt (1974).

ASCOCARPS 187–360 μm high, 205–340 μm in diameter, globose, subglobose, or depressed ellipsoidal, immersed, ostiolate, papillate, coriaceous, light to reddish-brown, solitary. PERIDIUM 16–19 μm thick, composed of four or five layers of elongate, thick-walled cells, forming a textura angularis, intimately bound to the rotten wood. PAPILLAE 47–220 μm long, 36–44 μm in diameter, cylindrical; ostiolar canal at first filled with a hyaline, thin-walled pseudoparenchyma, finally with indistinct, papilliform periphyses. PSEUDOPARENCHYMA of thin-walled cells filling venter of young ascocarps; deliquescing at ascospore maturity. ASCI

48–72 × 15–28 μm, eight (rarely two)-spored, clavate, pedunculate, uni-tunicate, thin-walled, aphysoclastic, without apical apparatuses, early deliquescing; developing at the base of the ascocarp venter. ASCOSPORES (15–)18–25.5(–27) × (7–)8–11.5 μm (excluding the sheath), cylindrical or ellipsoidal, three-septate, slightly constricted at the septa, hyaline, en-closed by an indistinct gelatinous, 3- to 6-μm-thick sheath (use India ink!). MODE OF LIFE: Saprobic. SUBSTRATE: Rotten driftwood (e.g., *Picea* sp., *Populus* sp.). RANGE: Atlantic Ocean—Denmark, France; Baltic Sea—Germany (G.D.R.); Mediterranean—France, Yugoslavia; Pacific Ocean—United States (Washington). NOTE: Description of asci and as-cospores partly based on Koch (1974); *H. amicta* had to be removed from *Sphaerulina*, a genus of bitunicate Ascomycetes (von Arx and Müller, 1975); instead, *H. amicta* belongs in Sphaeriales, Halosphaeriaceae and can best be accommodated in *Haligena*; the ascospore sheath of *H. amicta* corresponds to the envelope of immature ascospores of *H. elaterophora*. ETYMOLOGY: From the Latin, *amictus* = clothed, covered, in reference to the ascospores enclosed in a gelatinous sheath.

Haligena elaterophora Kohlm.

Nova Hedwigia **3**, 87–88, 1961

LITERATURE: Apinis and Chesters (1964); Catalfomo *et al.* (1972–1973); Eaton and Irvine (1972); Eaton and Jones (1971a,b); Henningsson (1974); Kirk *et al.* (1974); Koch (1974); Kohlmeyer [1960 (sub cf. *Ceriosporopsis* sp.), 1963d, 1967, 1968d, 1971b, 1972b]; Kohlmeyer and Kohlmeyer (1964–1969); Peters *et al.* (1975); Schmidt (1974).

Fig. 56

ASCOCARPS 205–580 μm in diameter, globose or ovoid, immersed or superficial, ostiolate, papillate or epapillate, subcoriaceous or subcar-bonaceous, black, solitary or gregarious. PERIDIUM 13–29 μm thick, composed on the outside of thick-walled irregular cells, forming a textura epidermoidea; cells on the inside elongate, merging into the pseudoparenchyma of the venter. PAPILLAE short or absent, conical; ostiolar canal periphysate. PSEUDOPARENCHYMA of thin-walled cells filling venter of young ascocarps; eventually breaking up into cateno-physes. ASCI 82–184 × 22.5–49.5 μm, eight (rarely four)-spored, clavate and somewhat apiculate, pedunculate, unitunicate, thin-walled, aphysoclas-tic, without apical apparatuses, early deliquescing; developing at the base of the ascocarp venter on a small-celled ascogenous tissue. ASCOSPORES 24–54.5 × (9–)10–17.5(–19.5) μm (excluding appendages), biseriate, ob-long ellipsoidal, three (rarely up to five)-septate, constricted at the septa,

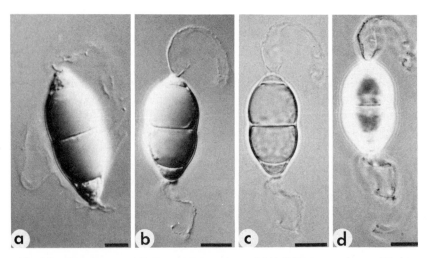

Fig. 56. Ascospores of *Haligena elaterophora*. (a) Veil-like appendages still covering the ascospore (bar = 5 μm). (b) Apical appendages expanded (bar = 10 μm). (c) Same ascospore as in b (bar = 10 μm). (d) Same ascospore as in b (bar = 10 μm). (Herb. J.K. 1922; a and b in Nomarski interference contrast, c in bright field, d in phase contrast.)

hyaline, appendaged; immature ascospores surrounded by a gelatinous sheath that expands at maturity, remaining attached with a cone-shaped or semiglobose protrusion to both apices; expanded elaterlike appendages, curved, attenuate, channeled, spoon-shaped at the base, 64–104 × 5–11 μm. MODE OF LIFE: Saprobic. SUBSTRATES: Intertidal and drifting wood, test panels (e.g., *Fagus sylvatica, Picea* sp., *Pinus sylvestris, Pinus* sp.), submerged bark, *Spartina townsendii*. RANGE: Atlantic Ocean—Denmark, Germany (F.R.G.), Great Britain (England, Wales), Spain (Tenerife and mainland), United States (Maine); Baltic Sea—Denmark, Germany (G.D.R.), Sweden; Pacific Ocean—United States (Washington). ETYMOLOGY: From the Greek, *elater* = driver and *-phorus* = -bearing, in reference to the ascospore appendages, reminiscent of elaters of Equisetaceae.

Haligena spartinae E. B. G. Jones

Trans. Br. Mycol. Soc. **45**, 245, 1962

LITERATURE: Apinis and Chesters (1964); Gessner and Goos (1973a,b); Henningsson (1974); G. C. Hughes (1969); Kohlmeyer (1968d, 1972b);

Schmidt (1974); Schneider (1977); Shearer (1972b); Shearer and Crane (1971).

ASCOCARPS 190–400 μm high, 230–390 μm in diameter, globose to subglobose, immersed, ostiolate, periphysate, papillate, dark brown to black, thick-walled, solitary or gregarious. ASCI 120–215 × 30–60 μm, eight-spored, clavate, pedunculate, unitunicate, thin-walled, aphysoclastic, early deliquescing. ASCOSPORES 40–75.5(–90) × 14–21.5 μm (excluding appendages), ellipsoidal to fusiform, 4–7(–9) septa, predominantly 5 septa, with or without constrictions at the septa, appendaged; at each end with a terminal or subterminal, small appendage, 7–18 × 0.5–1.5 μm, at first spoon-shaped, attached to the spore wall, mucilaginous, eventually forming a viscous thread. MODE OF LIFE: Saprobic. SUBSTRATES: *Agropyron junceiforme* and *A. pungens* (litter and fermentation layers), *Spartina alterniflora, S. townsendii, Typha* sp., driftwood, test panels (balsa, *Betula pubescens*). RANGE: Atlantic Ocean—Denmark, Great Britain (England, Wales), United States (Maryland, Rhode Island); Baltic Sea—Germany (G.D.R.), Sweden; Pacific Ocean—Canada (British Columbia). ETYMOLOGY: Name referring to the host genus, *Spartina* (Gramineae).

Haligena unicaudata E. B. G. Jones et Le Campion-Alsumard

Nova Hedwigia **19**, 574, 1970

LITERATURE: Byrne and Jones (1974); E. B. G. Jones and Le Campion-Alsumard (1970a); Kohlmeyer (1972b).

ASCOCARPS 400–500(–700) μm in diameter, globose, immersed or becoming exposed, ostiolate, papillate, coriaceous, brown to black, solitary, staining the substrate. PAPILLAE or necks 200–400 × 50–100 μm, ostiolar canal periphysate. PARAPHYSES absent. ASCI 60–75 × 15–20 μm, eight-spored, clavate, thin-walled, aphysoclastic, early deliquescing; developing at the base of the ascocarp venter on an ascogenous tissue. ASCOSPORES 36–60 × 2.5–4(–5) μm (excluding appendages); cylindrical or fusiform, three (rarely four or five)-septate, not constricted at the septa, hyaline, appendaged; appendages 15–36 μm long, terminal or subterminal, more or less lateral, forming an irregular sheath around the upper tip or along the upper side of the ascospore. MODE OF LIFE: Saprobic. SUBSTRATES: Submerged polyurethane, test panels (*Fagus sylvatica*). RANGE: Atlantic Ocean—Great Britain (England); Mediterranean—France. ETYMOLOGY: From the Latin, *unicaudatus* = provided with one tail, in reference to the ascospores bearing a single apical appendage.

Haligena viscidula J. Kohlmeyer et E. Kohlmeyer

Nova Hedwigia **9**, 92, 1965

LITERATURE: E. B. G. Jones *et al.* (1972); Kohlmeyer (1968c, 1972b); Kohlmeyer and Kohlmeyer (1964–1969); Shearer (1972b); Shearer and Crane (1971); Tubaki and Ito (1973).

ASCOCARPS 120–240 μm high, 100–252 μm in diameter, globose or subglobose, ellipsoidal or depressed, immersed, ostiolate, papillate, coriaceous, cream-colored, yellowish or brownish, solitary; occasionally staining the substrate cardinal to maroon red. PERIDIUM 13–25 μm thick, hyaline or pale cream-colored, composed of four to six layers of thick-walled, roundish or elongate cells, merging toward the center into the pseudoparenchyma. PAPILLAE or necks 160–250 × 25–36 μm, cylindrical; ostiolar canal with short periphyses. PSEUDOPARENCHYMA of thin-walled cells filling venter of young ascocarps; deliquescing at ascospore maturity. ASCI eight-spored, clavate or ellipsoidal, unitunicate, thin-walled, aphysoclastic, without apical apparatuses, deliquescing before ascospore maturity; developing at the base of the ascocarp venter. ASCOSPORES (37.5–)44–79(–89) × (3–)4–6.5 μm (excluding appendages), arranged in parallel in the ascus, long cylindrical or fusiform, straight or slightly curved, 5–11(–16) septa, predominantly 11 septa, not or slightly constricted at the septa, hyaline, appendaged; at both ends a subterminal, caplike, mucilaginous appendage, 6–10 μm long, eventually becoming viscous and forming a long thread. MODE OF LIFE: Saprobic. SUBSTRATES: Intertidal wood and test panels (*Pinus sylvestris, Pinus* sp., balsa, bamboo), bark of mangrove root (*Rhizophora racemosa*). RANGE: Atlantic Ocean—Ivory Coast, Liberia, United States (Maryland, North Carolina); Pacific Ocean—Japan. ETYMOLOGY: From the Latin, *viscidulus* = somewhat viscid, in reference to the viscous, sticky ascospore appendages.

Key to the Species of *Haligena*

1. Ascospores covered by a gelatinous sheath; at maturity without appendages

   . . . . *H. amicta*

1′. Ascospores at maturity with apical appendages 2

   2(1′) Ascospores with a single appendage at the upper end . . . . . *H. unicaudata*

   2′(1′) Ascospores with appendages at both ends 3

3(2′) Ascospores three (rarely up to five)-septate; appendages long, at first enclosing the ascospore completely . . . . *H. elaterophora*

3′(2′) Ascospores more than three-septate; appendages short, forming small caps at the ascospore apices 4

   4(3′) Ascospores over 10 μm in diameter, predominantly five-septate

   . . . . *H. spartinae*

   4′(3′) Ascospores less than 10 μm in diameter, predominantly 11-septate

   . . . . *H. viscidula*

Halosarpheia J. Kohlmeyer et E. Kohlmeyer

Trans. Br. Mycol. Soc. **68**, 208, 1977

ASCOCARPS solitary or gregarious, obpyriform to subglobose, immersed or partly immersed, ostiolate, papillate, coriaceous, brown to black. PSEUDOPARENCHYMA of large, thin-walled cells filling the venter of young ascocarps, breaking up to form catenophyses. ASCI eight-spored, clavate, pedunculate, unitunicate, thick-walled below the apex, thin-walled in the peduncle, persistent, developing at the base of the ascocarp venter; mature asci breaking off at the base from the ascogenous tissue. ASCOSPORES ellipsoidal, one-septate, hyaline, with apical appendages; appendages are exosporic remnants, at first caplike, stiff and homogenous, attached to the ascospore apices, at maturity becoming soft and transforming into fibers. MODE OF LIFE: Saprobic. SUBSTRATES: Immersed wood and roots. TYPE SPECIES: *Halosarpheia fibrosa* Kohlm. et Kohlm. The genus is monotypic. ETYMOLOGY: Anagram of the closely related genus *Halosphaeria* Linder (Halosphaeriaceae).

Halosarpheia fibrosa J. Kohlmeyer et E. Kohlmeyer

Trans. Br. Mycol. Soc. **68**, 208, 1977

Fig. 57

ASCOCARPS 380–440 μm high, 340–450 μm in diameter, obpyriform to subglobose, immersed or partly immersed, ostiolate, papillate, coriaceous, brown to black, immersed part lighter colored than the exposed neck and top. PERIDIUM 40–60 μm thick, two-layered, forming a textura angularis; outer stratum composed of two or three layers of polygonal or subglobose dark or light brown cells with large lumina; inner stratum composed of seven to ten layers of elongated, hyaline cells with narrow lumina, merging with the pseudoparenchyma of the venter; ascocarps sometimes covered by brown hyphae. NECKS 140–530 μm long, 130–200 μm in diameter at the base, 90–110 μm in diameter at the apex, subconical to cylindrical; ostiolar canal periphysate; periphyses 15–35 × 2–3 μm. PSEUDOPARENCHYMA of thin-walled, large, polygonal cells filling the venter of young ascocarps before the development of ascogenous cells, breaking up to form catenophyses; the latter merging apically with periphyses. ASCI 160–220 × 34–46 μm, eight-spored, clavate, pedunculate, unitunicate, thick-walled below the apex, thin-walled in the peduncle, persistent, without apical apparatuses, developing at the base of the ascocarp venter on a small-celled ascogenous tissue that is separated from the peridium by thin-walled, polygonal cells of the sterile pseudoparenchyma; asci mature successively, breaking off basally from the ascogenous tissue without releasing the ascospores; ascospore dispersal has not been observed thus far. Ascospores 32–44 × 18–24 μm (excluding appendages), broad ellipsoidal, one-septate, not constricted at the septum, hyaline, with apical appendages; appendages are remnants of the exospore; one caplike, stiff, and homogeneous appendage is attached to each ascospore apex, 3–5 μm thick, 6–8(–11) μm in diameter, at maturity becoming soft and scooplike, eventually transforming into a coil of delicate fibers that uncoil and form long, sticky filaments, 0.5–1 μm in diameter. Mode of Life: Saprobic. Substrates: Immersed wood of mangroves (*Avicennia germinans, Rhizophora mangle*), other shoreline trees (e.g., *Tamarix gallica*), and submerged rotten stalk of palm leaf. Range: Atlantic Ocean—Bermuda, Brazil, United States (Florida). Etymology: From the Latin, *fibrosus* = fibrous, composed of separable threads or fibers, in reference to the ascospore appendages forming coils of filaments.

Halosphaeria Linder *in* Barghoorn et Linder

Farlowia **1**, 412, 1944

=*Antennospora* Meyers, *Mycologia* **49**, 501, 1957
=*Halosphaeriopsis* Johnson, *J. Elisha Mitchell Sci. Soc.* **74**, 44, 1958
=*Palomyces* Höhnk, *Veroeff. Inst. Meeresforsch. Bremerhaven* **3**, 212–213, 1955
=*Remispora* Linder *in* Barghoorn et Linder, *Farlowia* **1**, 409, 1944

Literature: Kohlmeyer (1972b).
Ascocarps solitary or gregarious, globose or subglobose, sometimes ovoid or ellipsoidal, immersed or becoming exposed, ostiolate, papillate, coriaceous or membranaceous, rarely subcarbonaceous, hyaline, cream-colored, brown or black. Pseudoparenchyma of large, thin-walled cells filling venter of young ascocarps; deliquescing or breaking up into catenophyses. Asci eight-spored, clavate or subfusiform, pedunculate, unitunicate, thin-walled, early deliquescing, developing at the base of the ascocarp venter. Ascospores ellipsoidal, rarely rhomboid or cylindrical, one-septate, hyaline, appendaged; appendages episporic, apically or also laterally attached to the ascospore wall, subcylindrical, obclavate, subglobose or pleomorphic, persistent or gelatinizing. Substrate: Dead wood, culms, or other cellulose-containing material. Type Species:

Fig. 57. *Halosarpheia fibrosa.* (a) Longitudinal section (16 μm) through mature ascocarp (Herb. J.K. 3726; bar = 50 μm). (b) Section (16 μm) through ostiole with periphyses, ascospores in the canal and being released (Herb. J.K. 3623; bar = 25 μm). (c) Asci in different stages of development (Herb. J.K. 3739; bar = 25 μm). (d) Section

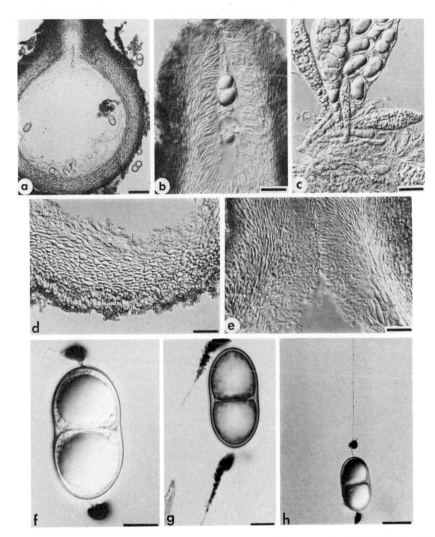

(6 μm) through basal part of peridium showing two different strata of cells (Herb. J.K. 3726; bar = 25 μm). (e) Section (16 μm) through base of ostiolar canal with periphyses (Herb. J.K. 3764; bar = 25 μm). (f) Ascospore with apical appendages becoming partly detached from the wall, stained in hematoxylin (Herb. J.K. 3740; bar = 10 μm). (g) Ascospore appendages being transformed into fibers, in hematoxylin (Herb. J.K. 3744; bar = 10 μm). From Kohlmeyer and Kohlmeyer, *Trans. Br. Mycol. Soc.* **68**, 213 (1977), Fig. 9. Reprinted with permission. (h) Apical appendage stalked and stretching out into a long filament, in hematoxylin (Herb. J.K. 3745; bar = 20 μm). From Kohlmeyer and Kohlmeyer, *Trans. Br. Mycol. Soc.* **68**, 213 (1977), Fig. 10. Reprinted with permission. (All from type material IMS and NY; a and g in bright field, the others in Nomarski interference contrast.)

Halosphaeria appendiculata Linder. NOTE: All members of the genus are obligate marine species. ETYMOLOGY: From the Greek, *hals* = sea, salt, and *Sphaeria* = a genus of Pyrenomycetes (*sphaera* = globe), in reference to the marine occurrence of the species and to the globose ascocarps.

Halosphaeria appendiculata Linder *in* Barghoorn et Linder

Farlowia **1**, 412, 1944

=*Remispora ornata* Johnson et Cavaliere, *Nova Hedwigia* **6**, 188, 1963

LITERATURE: Byrne and Jones (1974, 1975b); Catalfomo *et al.* (1972–1973); Cavaliere (1966b, 1968); Cribb and Cribb (1956); Eaton (1972); Fazzani and Jones (1977); Henningsson (1974); G. C. Hughes (1968, 1969, 1974); G. C. Hughes and Chamut (1971); Irvine *et al.* (1972); T. W. Johnson (1958b, 1968); T. W. Johnson and Sparrow (1961); E. B. G. Jones (1962a,b, 1963c, 1965, 1968b, 1971a, 1972); E. B. G. Jones and Irvine (1971); E. B. G. Jones *et al.* (1972); Kirk (1966, 1976); Kirk and Catalfomo (1970); Kirk *et al.* (1974); Koch (1974); Kohlmeyer (1960, 1962a, 1963d,e, 1967, 1968d, 1971b, 1972b); Kohlmeyer and Kohlmeyer (1964–1969); Lutley and Wilson (1972b); Malacalza and Martínez (1971); Meyers (1957); Neish (1970); Peters *et al.* (1975); Poole and Price (1972); Rheinheimer (1971); Schaumann (1968, 1969); Schmidt (1974); Schneider (1976); Shearer (1972b); Tubaki (1968, 1969); Tubaki and Ito (1973).

Fig. 58a

ASCOCARPS 140–330 µm high, 140–500 µm in diameter, subglobose, ovoid, ellipsoidal or depressed, immersed or partly immersed, ostiolate, papillate, coriaceous, dark brown or black, often almost hyaline below, solitary or gregarious. PERIDIUM 10–27.5(–47) µm thick, composed of four to five (rarely up to twelve) layers of thick-walled cells with large lumina, roundish at the outside, oblong toward the center, forming a textura angularis, merging into the pseudoparenchyma of the venter. PAPILLAE or necks up to 782 µm long, 34–69(–90) µm in diameter, cylindrical or conical, centric or eccentric, dark- or light-colored; ostiolar canal at first filled with a small-celled pseudoparenchyma. PSEUDOPARENCHYMA of thin-walled, large, polygonal, deliquescing cells filling venter of young ascocarps. ASCI 50–108 × 10–24 µm, eight (rarely four)-spored, clavate or subfusoid, pedunculate, unitunicate, thin-walled, aphysoclastic, without apical apparatuses, early deliquescing; developing at the base of the ascocarp venter on a small-celled ascogenous tissue. ASCOSPORES (16–)18–29 × (6–)8–12 µm (excluding appendages), ellipsoidal, one-septate, not or slightly constricted at the septum, hyaline, append-

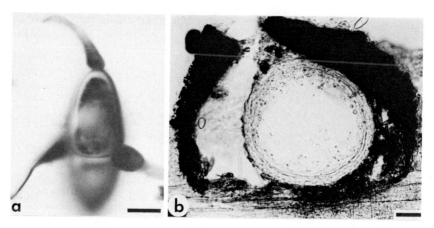

Fig. 58. *Halosphaeria* spp. (a) Ascospore of *H. appendiculata* with apical and equatorial appendages stained in hematoxylin (Herb. J.K. 1646; bar = 5 μm). (b) Longitudinal section through ascocarp of *H. hamata,* developed as a hypersaprobe inside an old fruiting body of *Leptosphaeria oraemaris* (Herb. J.K. 516; bar = 30 μm). (a in Nomarski interference contrast, b in bright field.)

aged; provided at each end with a single, terminal, somewhat membranous, episporic appendage, obclavate, attenuate, curved, flattened, spoon-shaped at the base, 7.5–18(−25) × 1–2.7 μm; around the septum four (rarely three or five) similar, radiating appendages. MODE OF LIFE: Saprobic. SUBSTRATES: Intertidal, submerged, and drifting wood, bark, test panels (e.g., *Arbutus menziesii, Betula pubescens, Cryptomeria japonica, Dicorynia paraensis, Fagus sylvatica, Laphira procera, Ocotea rodiaei, Phyllostachys pubescens, Pinus densiflora, P. sylvestris, Platanus occidentalis, Populus* sp., *Quercus* sp., *Tilia* sp., balsa), submerged rhizome of *Arundo donax,* trunk of *Phoenix* sp. RANGE: Atlantic Ocean—Argentina (Buenos Aires), Canada (Newfoundland, Nova Scotia), Denmark, France, Germany (Helgoland and mainland, F.R.G.), Great Britain (England, Isle of Man, Scotland, Wales), Iceland, Spain (Tenerife and mainland), Sweden, United States (Connecticut, Maine, Maryland, Massachusetts, New Jersey, North Carolina, Rhode Island, South Carolina, Texas); Baltic Sea—Denmark (Sjaelland), Germany (F.R.G., G.D.R.); Black Sea—U.S.S.R. (J. Kohlmeyer, unpublished); Mediterranean—France; Pacific Ocean—Australia (Queensland), Canada (British Columbia), Chile (Isla Hoste), Japan, United States (California, Washington). NOTE: The circumscription of the species had to be amended because Linder's (1944) description was based on poorly preserved material (Kohlmeyer, 1971b); the ascospore structure was examined particularly by Kirk (1966) and Lutley and Wilson (1972b). ETYMOL-

OGY: From the Latin, *appendiculatus* = appendiculate, in reference to the appendaged ascospores.

Halosphaeria cucullata (Kohlm.) Kohlm.

Can. J. Bot. **50**, 1956, 1972

=*Remispora cucullata* Kohlm., *Mycologia* **56**, 770, 1964

CONIDIAL STATE: *Periconia prolifica* Anastasiou, *Nova Hedwigia* **6**, 260, 1963

LITERATURE: G. C. Hughes (1974); Kohlmeyer (1969b,c); Kohlmeyer and Kohlmeyer (1964–1969, 1971a, 1977); Newell (1976); Raghukumar (1973); Shearer (1972b); Tubaki (1969); Tubaki and Ito (1973); Wagner-Merner (1972).

ASCOCARPS 150–245 μm high, 150–250 μm in diameter, subglobose or ampulliform, immersed, ostiolate, papillate, coriaceous, brownish-black or brownish-red, solitary or gregarious. PERIDIUM 7.5–11.5 μm thick, composed of four to six layers of thick-walled, mostly ellipsoidal cells, forming a textura angularis, merging toward the center into the pseudoparenchyma. PAPILLAE or necks 55–100 μm long, 28–75 μm in diameter, cylindrical or conical; ostiolar canal at first filled with a small-celled pseudoparenchyma. PSEUDOPARENCHYMA of thin-walled, poly-gonal or rounded cells filling venter of young ascocarps; eventually break-ing up into catenophyses; cells 9–23 μm long, 6–15 μm in diameter. ASCI 45–70 × 15–30 μm, eight-spored, clavate, short pedunculate, unitunicate, thin-walled, aphysoclastic, without apical apparatuses, early deliquesc-ing, developing in the base of the ascocarp venter on a small-celled ascogenous tissue. ASCOSPORES 20–68.5 × 6–11.5 μm (excluding append-ages), cylindrical or rarely ellipsoidal, one-septate, upon germination often becoming two- to four-septate, slightly or not constricted at the septa, hyaline, with a caplike, subglobose, terminal, deciduous appendage at one end, 5–8.5 μm in diameter.

CONIDIOPHORES 5–200 × 2.5 μm, cylindrical, septate, simple or branched, hyaline, often forming pustules on the surface of the substrate; conidiogenous cell ellipsoidal or ovoid, hyaline, produced acrogenously. CONIDIA 6–13(–20) μm in diameter, one-celled, subglobose or ovoid, smooth, thick-walled, light brown with a reddish tint or dark brown, developing basipetally, catenulate, cells finally separating. MODE OF LIFE: Saprobic. SUBSTRATES: Intertidal and drifting wood, test panels (e.g., *Betula* sp., *Mangifera indica, Pinus taeda, Pinus* sp., balsa), man-grove roots, and seedlings of *Rhizophora mangle,* submerged branches of *Pluchea* × *fosbergii,* endocarp of *Cocos nucifera*; sometimes in the calcareous lining of teredinid tubes. RANGE: Atlantic Ocean—Bahamas (Great Abaco; J. Kohlmeyer, unpublished), Brazil (São Paulo), United States [Florida, Maryland, North Carolina, Virginia (P. W. Kirk, unpublished)]; Indian Ocean—India (Kerala, Madras), South Africa; Pacific Ocean—Australia (according to unpublished records of G. C. Hughes, 1974), Guatemala, Japan, Mexico (Oaxaca), United States [Hawaii (Kauai, Oahu)]. NOTE: Anastasiou (1963b) reported the imperfect state also from inland salt lakes, namely, the Salton Sea (California) on *Tamarix aphylla* and the Salt Lake on Oahu (Hawaii) on acacia wood; the species was collected in its perfect and imperfect states only in tropical and subtropical areas; one collection from antarctic waters tentatively assigned to *P. prolifica* by G. C. Hughes and Chamut (1971) probably belongs to a different species. ETYMOLOGY: From the Latin, *cucullatus* = hooded, in reference to the ascospore adorned with a caplike appendage.

Halosphaeria galerita (Tubaki) I. Schmidt *in* J. Kohlmeyer and E. Kohlmeyer, comb. nov.

≡*Remispora galerita* Tubaki, *Publ. Seto Mar. Biol. Lab.* **15**, 362, 1968 (basionym)

LITERATURE: Tubaki (1969); the following sub *Lentescospora submarina* Linder: Koch (1974); Kohlmeyer (1962a, 1963d, 1969c, 1971b); Kohlmeyer and Kohlmeyer (1964–1969, 1971a); Schmidt (1974).

ASCOCARPS 100–280 μm in diameter, globose, subglobose, or pyriform, immersed or becoming exposed, ostiolate, papillate, coriaceous, hyaline, yellowish or light brown, solitary. PERIDIUM 8–15 μm thick, composed of three or four layers of oblong, roundish or rectangular, thick-walled cells with large lumina, forming a textura angularis, merging into the pseudoparenchyma of the venter. PAPILLAE or necks 120–440(–500) × 20–50 μm, cylindrical; ostiolar canal periphysate. PSEUDOPARENCHYMA of thin-walled, large cells filling venter of young ascocarps, early deliquescing. ASCI 56–96 × 20–45 μm, eight-spored, clavate, ellipsoidal or ovoid, short pedunculate, unitunicate, thin-walled, aphysoclastic, without apical apparatuses, early deliquescing; developing at the base of the ascocarp venter on a small-celled ascogenous tissue. ASCOSPORES (15.5–) 20–28(–35) × 7–12.5(–15) μm (excluding appendages), ellipsoidal, one-septate (rarely with additional, abnormal septa), not or slightly constricted at the septum, hyaline, thick-walled at the apices, appendaged; at first a gelatinous sheath covers ascospores completely; later only a subglobose, faintly striate, subgelatinous appendage (stain with hematoxylin, violamin, or trypan blue!), 3.5–7.5 μm thick, covers each apex. MODE OF

LIFE: Saprobic. SUBSTRATES: Intertidal, drifting, and submerged wood, test panels (e.g., balsa). RANGE: Atlantic Ocean—Denmark, Mexico (Yucatan), Spain, United States (Maine, New Hampshire, North Carolina); Baltic Sea—Germany (G.D.R.); Mediterranean—France, Yugoslavia; Pacific Ocean—Japan, United States [Hawaii (Maui)]. NOTE: The new combination, proposed by I. Schmidt in an unpublished thesis (personal communication), is necessary because we regard *Remispora* as a synonym of *Halosphaeria* (Kohlmeyer, 1972b), and *H. galerita* fits well with the other species of *Halosphaeria*. It is possible that the fungi described as *Lentescospora submarina* and *Remispora galerita* are identical [as was surmised earlier (e.g., Kohlmeyer, 1969c)], but this cannot be proven with certainty because the original description of *L. submarina* is insufficient and the type material does not permit recognition of critical features (Kohlmeyer, 1969c); therefore, *L. submarina* must be declared a *nomen dubium*. Our earlier collections, as well as those of Koch (1974) and Schmidt (1974), are included under *H. galerita* because they clearly belong to this species; the following publications are not considered in the foregoing description because the identity of the species identified in these papers as *Lentescospora submarina* is not certain: Apinis (1964); Cavaliere (1968); T. W. Johnson (1958b); T. W. Johnson and Sparrow (1961); Poole and Price (1972); Henningsson (1974). ETYMOLOGY: From the Latin, *galeritus* = wearing a hood or skullcap, in reference to the caplike appendages of the ascospores.

Halosphaeria hamata (Höhnk) Kohlm.

Can. J. Bot. **50**, 1956, 1972

≡*Ceriosporopsis hamata* Höhnk, *Veroeff. Inst. Meeresforsch. Bremerhaven* **3**, 211, 1955
≡*Remispora hamata* (Höhnk) Kohlm., *Ber. Dtsch. Bot. Ges.* **74**, 305, 1961

LITERATURE: Apinis and Chesters (1964); Byrne and Eaton (1972); Byrne and Jones (1974, 1975b); Catalfomo *et al.* (1972–1973); Cavaliere (1966b, 1968); Gessner and Goos (1973a,b); Henningsson (1974); Höhnk (1956); G. C. Hughes (1969, 1974); T. W. Johnson (1958a); T. W. Johnson and Cavaliere (1963); T. W. Johnson and Sparrow (1961); E. B. G. Jones (1962a,b, 1963c, 1968b, 1972); E. B. G. Jones and Irvine (1971); Kirk *et al.* (1974); Koch (1974); Kohlmeyer (1960, 1962a, 1963d,e); Kohlmeyer and Kohlmeyer (1964–1969); Meyers (1957); Neish (1970); Peters *et al.* (1975); Schaumann (1968, 1969); Schmidt (1974); Schneider (1976, 1977); Shearer (1972b).

Fig. 58b

Ascocarps 165–250 μm in diameter, globose or subglobose, immersed or partly immersed, ostiolate, papillate, coriaceous, cream-colored, grayish or light brown, solitary or gregarious. Peridium 8.5–13 μm thick, composed of about four layers of elongate, thick-walled cells, forming a textura angularis, merging into the pseudoparenchyma of the venter. Necks 245–535(–890) μm long, up to 59 μm in diameter, centric or eccentric, straight or curved, cylindrical, one or rarely several on an ascocarp; ostiolar canal periphysate. Pseudoparenchyma of large, thin-walled, polygonal, subglobose, or ellipsoidal cells filling venter of young ascocarps; eventually breaking up into catenophyses. Asci 64–93 × 17–22.5 μm, eight-spored, clavate or subfusiform, short pedunculate, unitunicate, thin-walled, aphysoclastic, early deliquescing, without apical apparatuses, developing in the base of the ascocarp venter. Ascospores 16.5–29.5 × 6.5–10.5 μm (excluding appendages), ellipsoidal, one-septate, slightly or not constricted at the septum, hyaline, appendaged; at each end provided with a cap- or hooklike, thin appendage, which is subapically attached to the spore wall, probably developing from the exospore; appendages eventually becoming viscous, stretching and dissolving. Mode of Life: Saprobic or hypersaprobic (ascocarps developing in decaying fruiting bodies of other fungi). Substrates: Intertidal, submerged, and drifting wood, test panels (e.g., *Betula pubescens, Dicorynia paraensis, Fagus sylvatica, Laphira procera, Pinus sylvestris,* balsa), *Agropyron pungens, Spartina alterniflora, S. townsendii, Typha* sp. Range: Atlantic Ocean—Canada (Nova Scotia), Denmark, Germany (Helgoland and mainland, F.R.G.), Great Britain (England, Isle of Man, Scotland, Wales), Iceland, Spain, United States (Maryland, North Carolina, Rhode Island, Texas); Baltic Sea—Germany (F.R.G., G.D.R.), Sweden; Mediterranean—France; Pacific Ocean—Canada (British Columbia), United States (California, Washington). Note: *Halosphaeria hamata* has been confused with other species bearing superficially similar ascospore appendages, for example, *Ceriosporopsis cambrensis* (T. W. Johnson, 1958a; T. W. Johnson and Sparrow, 1961); Australian collections of G. C. Hughes (1974) differ from the typical *H. hamata* from temperate regions; in addition, a collection from Brazil (J. Kohlmeyer, unpublished) included in *R. hamata* in our iconography (Kohlmeyer and Kohlmeyer, 1964–1969) does not belong to this species as indicated by a reexamination. Furthermore, material identified by T. W. Johnson and Cavaliere (1963) as *Remispora hamata* has two-layered ascocarp walls, in contrast to a simple wall in our collections (e.g., Kohlmeyer, 1961b) on which the present description is based. The type material of *H. hamata* has not become available for comparison. ETYMOLOGY: From the Latin, *hamatus* = barbed, hooked at the tip, in reference to the hooklike ascospore appendages.

Halosphaeria maritima (Linder) Kohlm.

Can. J. Bot. **50**, 1956, 1972

≡*Remispora maritima* Linder *in* Barghoorn et Linder, *Farlowia* **1**, 410, 1944
=*Remispora lobata* Höhnk, *Veroeff. Inst. Meeresforsch. Bremerhaven* **3**, 206, 1955

LITERATURE: Biernacka (1965); Byrne and Jones (1974); Cavaliere (1966b, 1968); Eaton (1972); Eaton and Dickinson (1976); Eaton and Irvine (1972); Eaton and Jones (1971a,b); Henningsson (1974); Höhnk (1956); G. C. Hughes (1968, 1969); G. C. Hughes and Chamut (1971); Irvine *et al.* (1972); T. W. Johnson (1956b); T. W. Johnson and Cavaliere (1963); T. W. Johnson and Sparrow (1961); E. B. G. Jones (1972); E. B. G. Jones and Irvine (1971); E. B. G. Jones *et al.* (1972); Koch (1974); Kohlmeyer (1960, 1962a, 1963d, 1969c, 1971b); Kohlmeyer and Kohlmeyer (1964–1969); Malacalza and Martínez (1971); Meyers (1957); Rheinheimer (1971); Schaumann (1968, 1969); Schmidt (1974); Schneider (1976, 1977); Tubaki (1968, 1969); Wilson (1951).

Figs. 59a and 59b

ASCOCARPS 180–570(–670) μm in diameter, globose, subglobose, or ovoid, immersed or superficial, ostiolate, papillate, coriaceous, almost hyaline, cream-colored or smoke-gray, sometimes the upper part darker than the lower, solitary or gregarious. PERIDIUM 8–32 μm thick, composed of six to nine layers of thick-walled, more or less elongated cells with large lumina, forming a textura angularis, merging into the pseudoparenchyma of the venter. NECKS 100–390(–600) × 20–60 μm, subcylindrical or truncate–conical, sometimes forked or two on one ascocarp, centric or eccentric. PSEUDOPARENCHYMA of thin-walled, large, ellipsoidal or polygonal cells filling venter of young ascocarps; eventually breaking up into catenophyses, 7–9.5 μm in diameter. ASCI 72–148 × 21.5–39 μm, eight-spored, clavate or broadly fusoid, somewhat apiculate, pedunculate, unitunicate, thin-walled, aphysoclastic, without apical apparatuses, early deliquescing, developing at the base of the ascocarp venter. ASCOSPORES 18–30(–31.5) × 8–13 μm (excluding appendages), ellipsoidal, ovoid or broadly ellipsoidal, one-septate, not constricted at the septum, hyaline, appendaged; at first surrounded by a subgelatinous, exosporic sheath that unfolds, remaining attached at both ends of the

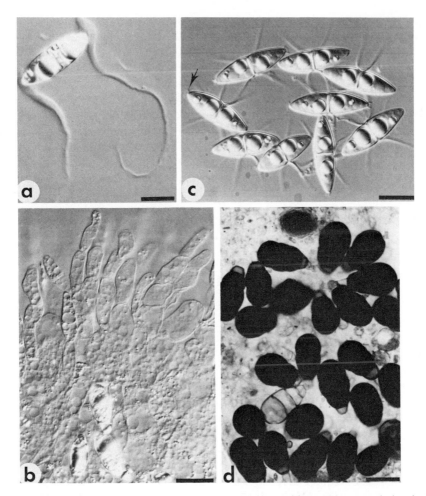

Fig. 59. *Halosphaeria* spp. (a) Ascospore of *H. maritima* with expanded apical appendages (Herb. J.K. 3317; bar = 10 μm). (b) Catenophyses of *H. maritima* (Herb. J.K. 3317; bar = 20 μm). (c) Ascospores of *H. mediosetigera* with equatorial crescent-shaped appendages and small apical caps (arrow) (Herb. I. Schmidt; bar = 15 μm). (d) Conidia of *Trichocladium achrasporum,* the imperfect state of *H. mediosetigera* (Herb. J.K. 2629; bar = 15 μm). (d in bright field, the others in Nomarski interference contrast.)

ascospore; appendages pleomorphic, mostly geminate and yoke-shaped, apices attenuate and irregularly stretched out. MODE OF LIFE: Saprobic. SUBSTRATES: Intertidal and drifting wood, test panels (e.g., *Abies* sp., *Fagus sylvatica, Picea* sp., *Pinus sylvestris, Pinus* sp., balsa). RANGE: Atlantic Ocean—Argentina (Buenos Aires), Canada [New Brunswick (J. Kohlmeyer, unpublished), Newfoundland], Denmark, France, Germany

(Helgoland, Wangerooge, and mainland, F.R.G.), Great Britain (England, Isle of Man, Wales), Iceland, Norway (J. Kohlmeyer, unpublished), Portugal, Sweden, United States (Maine, Massachusetts, North Carolina, South Carolina); Baltic Sea—Denmark, Germany (F.R.G., G.D.R.), Poland, Sweden; Mediterranean—France; Pacific Ocean—Canada (British Columbia), Chile (Isla Hoste), Japan, United States [California, Hawaii (Maui), Washington]. NOTE: Ascospore sizes from a North Carolina collection by T. W. Johnson (1956b) are excluded from the description because they are outside of the size range of all other collections; either the author included the appendages or he was dealing with a different species. ETYMOLOGY: From the Latin, *maritimus* = marine, referring to the habitat of the species.

Halosphaeria mediosetigera Cribb et Cribb

Univ. Queensl. Pap., Dep. Bot. **3,** 100, 1956

≡*Halosphaeriopsis mediosetigera* (Cribb et Cribb) Johnson, *J. Elisha Mitchell Sci. Soc.* **74,** 44, 1958
=*Halosphaeria mediosetigera* var. *grandispora* Kohlm., *Nova Hedwigia* **2,** 310, 1960

CONIDIAL STATE: *Trichocladium achrasporum* (Meyers et Moore) Dixon *in* Shearer et Crane, *Mycologia* **63,** 244, 1971.

LITERATURE: Anastasiou (1963b); Catalfomo *et al.* (1972–1973); Cavaliere (1966b, 1968); Eaton (1972); Eaton and Irvine (1972); Eaton and Jones (1971a); S. E. J. Furtado *et al.* (1977); Henningsson (1974, 1976a); G. C. Hughes (1969, 1974); Irvine *et al.* (1972); Jensen and Sguros (1971a,b); T. W. Johnson (1963a, 1968); T. W. Johnson and Sparrow (1961); E. B. G. Jones (1963b, 1968b, 1972); E. B. G. Jones and Irvine (1971); Kirk (1966, 1970, 1976); Kirk *et al.* (1974); Koch (1974); Kohlmeyer (1959, 1960, 1962a, 1963d,e, 1971b); Kohlmeyer and Kohlmeyer (1964–1969, 1966); Meyers (1968a,b, 1971a); Meyers and Hopper (1966, 1967); Meyers and Scott (1968); Meyers and Simms (1967); Meyers *et al.* (1963a,b, 1964a); Moss and Jones (1977); Peters *et al.* (1975); Pisano *et al.* (1964); Schaumann (1968); Schmidt (1967, 1974); Schneider (1976); Sguros and Simms (1963a,b, 1964); Sguros *et al.* (1971, 1973); Shearer (1972b); Shearer and Crane (1977); Tubaki and Ito (1973); Tyndall and Kirk (1973).

Figs. 59c and 59d

ASCOCARPS 138–590 μm high, 197–830 μm in diameter, subglobose or ellipsoidal, immersed or partly immersed, ostiolate, papillate, subcarbonaceous, black, solitary or gregarious. PERIDIUM 12–32 μm thick, composed of 5 to 8 (to 20) layers of irregular, polygonal, thick-walled cells, forming a textura angularis, merging into the pseudoparenchyma of the venter. PAPILLAE or necks 35–320 μm long, 35–80(–140) μm in diameter, conical or subcylindrical, centric or eccentric, dark; ostiolar canal filled with a small-celled pseudoparenchyma. PSEUDOPARENCHYMA of thin-walled, large cells filling venter of young ascocarps, early deliquescing. ASCI 62.5–168(–200) × 20–48 μm, eight-spored, clavate, pedunculate, unitunicate, thin-walled, aphysoclastic, without apical apparatuses, early deliquescing; developing at the base of the ascocarp venter on a small-celled ascogenous tissue. ASCOSPORES 24–44.5 × 8–17(–20) μm (excluding appendages), ellipsoidal or subfusiform, one-septate, not or slightly constricted at the septum, hyaline, appendaged; at both ends with a small cap that may invert; around the septum three (rarely four) crescent-shaped, rigid, attenuate appendages, (12–)24.5–40 μm long, up to 1.3 μm in diameter at the base, obliquely attached to the septum, developing by fragmentation of the exospore. SPORODOCHIA-like structures occasionally found on wood in the natural habitat, superficial, compact, fuscous to black. CONIDIOPHORES absent or short, zero- to four-septate, simple, formed laterally on hyphae, hyaline to light brown or fuscous. CONIDIA (blastoconidia) (15–)20–34(–45) × (8–)10–24 μm, clavate, ovoid or obpyriform, two- to five-septate, constricted at the septa, straight or slightly curved, increasing in diameter from base to apex, formed singly on the conidiophores; apical cells subglobose, dark brown; basal cells conical or subcylindrical, subhyaline to light brown or fuscous. CHLAMYDOSPORES 6–16 μm in diameter, catenulate, dark brown; chains up to 500 μm long, up to 35-celled, terminal, rarely intercalary, simple; cells increasing in diameter from base to apex, 7.5–19 μm long, globose or ellipsoidal. MODE OF LIFE: Saprobic. SUBSTRATES: Intertidal and drifting wood and bark, test panels (e.g., *Dicorynia paraensis, Fagus sylvatica, Laphira procera, Ocotea rodiaei, Pinus monticola, P. sylvestris, Pinus* sp., *Platanus occidentalis, Populus* sp., *Tilia americana,* balsa), dead, exposed roots of *Liquidambar styraciflua* (J. Kohlmeyer, unpublished) and of *Tamarix* sp., submerged pine cone, wood of pneumatophores of *Avicennia germinans,* bark of prop roots of *Rhizophora racemosa,* filamentous algae on beechwood panels, straw. RANGE: Atlantic Ocean—Canada (Newfoundland), Denmark, France, Germany (F.R.G.), Great Britain [England, Scotland (J. Kohlmeyer, unpublished), Wales], Iceland, Liberia, Mexico, Portugal, Sénégal, Spain, United States [Connecticut, Florida, Maryland, New Jersey (J. Kohlmeyer, unpublished), North Carolina, Rhode Island (J. Kohlmeyer, unpublished), Texas, Virginia]; Baltic Sea—Denmark (Sjaelland), Germany (F.R.G., G.D.R.), Sweden; Black Sea—Bulgaria, U.S.S.R. (J. Kohlmeyer, unpublished);

Indian Ocean—South Africa; Mediterranean—France; Pacific Ocean—Australia (Queensland), Canada (British Columbia), Japan, United States (California, Washington). NOTE: The species was also found on *Tamarix aphylla* in an inland salt lake, the Salton Sea, California (Anastasiou, 1963b). While earlier collections appeared to indicate the existence of a large-spored variety, *H. m.* var. *grandispora,* subsequent collections have been made with ascospore diameters overlapping both var. *mediosetigera* and var. *grandispora* (Cavaliere, 1968; Henningsson, 1974; T. W. Johnson, 1968; Kohlmeyer, 1962a; Schaumann, 1968; Schmidt, 1967); therefore, the reduction to synonymy of var. *grandispora* by T. W. Johnson and Sparrow (1961) is recognized at the present time. However, collections by Cribb and Cribb (1956), T. W. Johnson (1958b), Koch (1974), and Tubaki and Ito (1973), containing ascospores with small diameters, show the need for a thorough investigation of ascospore variability, possibly using pure cultures of different geographical origin. Three-septate ascospores (Cribb and Cribb, 1956; T. W. Johnson, 1958b) have never been observed in our collections. Shearer and Crane (1977) discovered the connection between the perfect and imperfect states. ETYMOLOGY: From the Latin, *medius* = middle and *setiger* = bristly, in reference to the bristlelike appendages that are attached to the equator of the ascospores.

Halosphaeria pilleata (Kohlm.) Kohlm.

Can. J. Bot. **50**, 1957, 1972

≡*Remispora pilleata* Kohlm., *Nova Hedwigia* **6**, 319, 1963

LITERATURE: Cavaliere (1968); Henningsson (1974); G. C. Hughes (1968, 1969); Koch (1974); Kohlmeyer (1968d, 1971b); Kohlmeyer and Kohlmeyer (1964–1969); Schaumann (1968); Schmidt (1974).

ASCOCARPS 231–364 µm high, 209–364 µm in diameter, globose to subglobose, immersed or partly immersed, ostiolate, papillate, subcoriaceous, almost black above, grayish below, solitary or gregarious. PERIDIUM 19–30 µm thick, composed of five or six layers of rectangular, thick-walled cells with large lumina, forming a textura angularis, merging into the pseudoparenchyma of the venter. PAPILLAE or necks 178–507 µm long, 75–103 µm in diameter at the base, 31–55 µm at the apex, elongate–subconical, centric; ostiolar canal periphysate. PSEUDO-PARENCHYMA of thin-walled, large, polygonal, deliquescing cells filling venter of young ascocarps. ASCI 96–129 × 24–49 µm, eight-spored, clavate, basally attenuate, pedunculate, unitunicate, thin-walled, aphysoclastic, without apical apparatuses, early deliquescing; developing at the base of the ascocarp venter on a small-celled ascogenous tissue. Ascospores 24–34(–36.5) × 12.5–19(–20.5) μm (excluding appendages), rhomboid or rarely subellipsoidal, one-septate, not constricted at the septum, thick-walled, hyaline, appendaged; at first surrounded by an exosporic, subgelatinous cover that unfolds, remaining attached at the apices, finally stretching out and becoming veil-like, pleomorphic; inconspicuous striation in the appendages close to the ascospore wall becomes evident by stains (e.g., hematoxylin). MODE OF LIFE: Saprobic. SUBSTRATES: Intertidal and drifting wood, test panels (e.g., *Abies* sp., *Picea* sp., *Pinus sylvestris, Pinus* sp.). RANGE: Atlantic Ocean—Canada (Newfoundland), Denmark, Germany (F.R.G.), Great Britain [Scotland (J. Kohlmeyer, unpublished)], Iceland, United States (Maine); Baltic Sea—Denmark (Sjaelland), Germany (G.D.R.), Sweden; Pacific Ocean—Canada (British Columbia). ETYMOLOGY: From the Latin, *pilleatus* = capped or wearing a cap, in reference to the appendaged ascospores.

Halosphaeria quadricornuta Cribb et Cribb

Univ. Queensl. Pap., Dep. Bot. **3**, 99, 1956

≡*Antennospora quadricornuta* (Cribb et Cribb) Johnson, *J. Elisha Mitchell Sci. Soc.* **74**, 46, 1958
=*Antennospora caribbea* Meyers, *Mycologia* **49**, 503, 1957

LITERATURE: Becker and Kohlmeyer (1958a,b); Cribb and Cribb (1969); G. C. Hughes (1974); T. W. Johnson and Sparrow (1961); E. B. G. Jones (1968a,b); E. B. G. Jones *et al.* (1972); Kohlmeyer (1958b, 1968c, 1969b,c, 1976); Kohlmeyer and Kohlmeyer (1964–1969, 1977); Meyers [1953 and 1954 (as "Form 2"), 1968b, 1971a]; Meyers and Reynolds (1959a, 1960b); Meyers *et al.* (1960); Nair (1970); Raghukumar (1973); Ritchie (1954); Tubaki (1968, 1969).

Fig. 60a

ASCOCARPS 130–260(–514) μm high, 140–285 μm in diameter, subglobose or ellipsoidal, immersed or becoming exposed, ostiolate, papillate, coriaceous or subcarbonaceous, dark brown to black, often surrounded by dark brown hyphae, solitary or gregarious. PERIDIUM 9–12.5 μm thick, composed of three or four layers of small, irregular, polygonal, thick-walled cells, forming a textura angularis, merging more or less abruptly into the large-celled pseudoparenchyma of the venter. PAPILLAE or necks 70–560 μm long, 20–70(–93) μm in diameter, subconical or cylindrical, centric or eccentric; ostiolar canal indistinctly periphysate. PSEUDOPARENCHYMA of thin-walled, large, polygonal or ellipsoidal, thin-walled cells filling venter of young ascocarps. ASCI eight-spored,

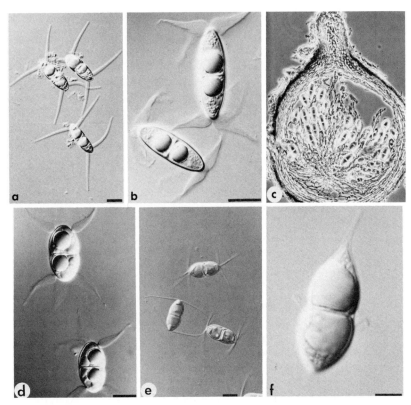

Fig. 60. *Halosphaeria* spp. (a) Ascospores of *H. quadricornuta* with two appendages below each apex (Herb. J.K. 3744; bar = 10 μm). (b) Ascospores of *H. quadriremis* with striate apical appendages (Herb. Brooks; bar = 10 μm). (c) Longitudinal section (12 μm) through ascocarp of *H. quadriremis*, catenophyses between ascospores (Herb. Malacalza: bar = 50 μm). (d) Ascospores of *H. salina* with striate apical appendages (Herb. J.K. 2737; bar = 10 μm). (e) Ascospores with slender type of appendages of *H. salina* (Herb. J.K. 2583; bar = 10 μm). (f) Ascospore of *H. torquata* with apical thorn and lateral annulus (Herb. J.K. 1673; bar = 5 μm). (c in phase contrast, the others in Nomarski interference contrast.)

clavate, pedunculate, unitunicate, thin-walled, aphysoclastic, without apical apparatuses, deliquescing before ascospore maturity; developing at the base of the ascocarp venter on a small-celled ascogenous tissue. ASCOSPORES 20–35 × (6–)8–12 μm (excluding appendages), ellipsoidal, one-septate, not or slightly constricted at the septum, hyaline, appendaged; at each end two subterminal, cylindrical, attenuate, stiff appendages, 20–37 μm long, 1–2 μm in diameter; pairs of appendages at right angles to one another. MODE OF LIFE: Saprobic. SUBSTRATES: Intertidal and drifting wood and bark, catamarans, test panels (e.g., *Bombax malabaricum, Dipterocarpus* sp., *Fagus sylvatica, Pinus sylvestris, Pseudotsuga* sp.), dead roots of mangroves (*Avicennia marina* var. *resinifera, Rhizophora mangle*), immersed dead branches of *Conocarpus erecta* and *Hibiscus tiliaceus,* drifting endocarp of *Cocos nucifera,* mantle of tunicates; ascocarps often in or under calcareous linings of empty shipworm tubes. RANGE: Atlantic Ocean—Antigua, Bahamas (Abaco, Bimini, Eleuthera, Exuma), Belize (British Honduras), Bermuda, Colombia, Cuba, Haiti, Liberia, Martinique, Mexico (Campeche), Panama, Puerto Rico, Trinidad, United States (Florida, Texas), Venezuela, Virgin Islands (St. Croix); Indian Ocean—Aden, India (Kerala, Madras, Maharashtra), Malaysia, Mozambique, Singapore; Pacific Ocean—Australia (Queensland), Fiji Islands, Japan (Tane Island), Mexico (Oaxaca), Philippines, Tahiti, United States [Hawaii (Hawaii, Oahu)]. NOTE: Data of T. W. Johnson (1958b), T. W. Johnson *et al.* (1959), and T. W. Johnson and Sparrow (1961) from North Carolina are omitted in the list of distribution because the presence of paraphyses and a gelatinous ascospore sheath are mentioned for *H. quadricornuta,* indicating that these authors have dealt with a different species; also, collections from California (E. B. G. Jones *et al.*, 1972) and from Germany (E. B. G. Jones, 1968b) are deleted because *H. quadricornuta* appears to be restricted to tropical waters (sensu G. C. Hughes, 1974). ETYMOLOGY: From the Latin, *quadri-* = four- and *cornutus* = horned, in reference to the ascospores, adorned by four spurlike appendages.

Halosphaeria quadriremis (Höhnk) Kohlm.

Can. J. Bot. **50**, 1957, 1972

≡*Palomyces quadriremis* Höhnk, *Veroeff. Inst. Meeresforsch. Bremerhaven* **3**, 213, 1955
≡*Arenariomyces quadriremis* (Höhnk) Meyers, *Mycologia* **49**, 505, 1957
≡*Remispora quadriremis* (Höhnk) Kohlm., *Nova Hedwigia* **2**, 332, 1960

LITERATURE: Cavaliere (1966b); Eaton and Dickinson (1976); Henningsson (1974, 1976b); Höhnk (1956); G. C. Hughes (1969, 1974); T. W. Johnson and Cavaliere (1963); T. W. Johnson and Sparrow (1961); E. B. G. Jones and Jennings (1964); E. B. G. Jones *et al.* (1972); Kohlmeyer (1958a,b, 1959, 1962a, 1963d,e, 1967, 1971b); Kohlmeyer and Kohlmeyer (1964–1969); Malacalza and Martínez (1971); Schaumann (1968); Schmidt (1974); Schneider (1977); Shearer (1972b); Tubaki (1968, 1969); Tubaki and Ito (1973).

Figs. 60b and 60c

ASCOCARPS 135–400 µm in diameter, globose or subglobose, immersed or superficial, ostiolate, papillate, membranaceous or coriaceous, cream-colored, grayish or brown, sometimes the top darker than the base, solitary or gregarious. PERIDIUM 12–22 µm thick, composed of cells with large lumina, merging into the pseudoparenchyma of the venter. PAPIL-LAE or necks up to 600 µm long, 24–58 µm in diameter, cylindrical, centric or eccentric, one or several occurring on one ascocarp. PSEUDOPARENCHYMA of thin-walled polygonal to ellipsoidal cells filling venter of young ascocarps; eventually breaking up into catenophyses. ASCI 70–100 × 20–40 µm, eight-spored, clavate or subfusiform, short pedunculate, unitunicate, thin-walled, aphysoclastic, without apical apparatuses, early deliquescing; developing at the base of the ascocarp venter. ASCOSPORES 18–30(–34) × 8–12(–15.5) µm (excluding appendages), ellipsoidal, one-septate, not or slightly constricted at the septum, hyaline, appendaged; at each end four radiating appendages, developing by fragmentation of the exospore; appendages 12–21.5 µm long, 2.5–4 µm in diameter at the base, terminal, obclavate, curved, attenuate, semirigid, inconspicuously striate by fiberlike elements embedded in the subgelatinous matrix. MODE OF LIFE: Saprobic. SUBSTRATES: Intertidal and drifting wood, test panels (e.g., *Cryptomeria japonica, Pinus sylvestris, Pinus* sp., balsa). RANGE: Atlantic Ocean—Argentina (Buenos Aires), France, Germany [Helgoland (J. Kohlmeyer, unpublished), Amrum and mainland, F.R.G.], Great Britain [England; Scotland (J. Kohlmeyer, unpublished)], Portugal, Spain (Tenerife and mainland), Sweden, United States (Connecticut, Maryland, North Carolina, Rhode Island); Baltic Sea—Denmark, Germany (G.D.R.), Sweden; Mediterranean—Italy (Sicily and mainland); Pacific Ocean—Canada (British Columbia), Japan, United States (Washington). ETYMOLOGY: From the Latin, *quadri-* = four- and *remus* = an oar, in reference to the ascospores adorned with several stretched-out appendages.

Halosphaeria salina (Meyers) Kohlm.

Can. J. Bot. **50**, 1957, 1972

≡*Arenariomyces salina* Meyers, *Mycologia* **49**, 505, 1957
≡*Remispora salina* (Meyers) Kohlm., *Mycologia* **60**, 262, 1968

LITERATURE: T. W. Johnson *et al.* (1959); Kohlmeyer (1969b,c, 1976); Kohlmeyer and Kohlmeyer (1964–1969, 1971a); Meyers and Reynolds (1959a,c); Meyers *et al.* (1963b).

Figs. 60d and 60e

ASCOCARPS 100–425 µm in diameter, globose or subcylindrical, immersed, partly immersed, or becoming exposed, ostiolate, papillate, membranaceous or coriaceous, dark brown to black, surrounded by brown hyphae, solitary or gregarious. PERIDIUM 8–12 μm thick, composed of three to five layers of flat, thick-walled cells with large lumina, forming a textura angularis, merging into the pseudoparenchyma of the venter. PAPILLAE 68–165 μm long, 28–52 μm in diameter, cylindrical or conical; ostiolar canal in the upper part filled with a thin-walled pseudoparenchyma, in the lower part with thick-walled, hyaline cells. PSEUDOPARENCHYMA of thin-walled, large, polygonal to ellipsoidal cells filling venter of young ascocarps. ASCI 60–80 × 29–41 μm, eight-spored, clavate or ellipsoidal, short pedunculate, unitunicate, thin-walled, aphysoclastic, without apical apparatuses, early deliquescing; developing at the base of the ascocarp venter on a small-celled ascogenous tissue. ASCOSPORES 19–28 × 8–13.5 μm (excluding appendages), ellipsoidal, one-septate, not or slightly constricted at the septum, hyaline, appendaged; at each end three or four (rarely five) radiating appendages, developing by fragmentation of the exospore; appendages 12–19 μm long, 1.5–2.5 μm in diameter at the base, subterminal, obclavate, curved, attenuate, semirigid, indistinctly spoon-shaped at the base, inconspicuously striate by fiberlike elements embedded in the subgelatinous matrix. MODE OF LIFE: Saprobic. SUBSTRATES: Intertidal and drifting wood, test panels (e.g., *Pinus* sp.), mesocarp of drifting *Cocos nucifera,* sometimes in the calcareous lining of teredinid tubes. RANGE: Atlantic Ocean—Bahamas (Bimini), Belize (British Honduras), Colombia, Liberia, United States (Florida, North Carolina); Indian Ocean—South Africa; Pacific Ocean—Mexico (Chiapas, Oaxaca), United States [Hawaii (Hawaii, Kauai)]. NOTE: *Halosphaeria salina* is the tropical counterpart of the similar species from temperate waters, *H. quadriremis* (Kohlmeyer, 1968c); pit connections similar to those of *Corollospora* spp. have been observed in the pseudoparenchyma of *H. salina* (J. Kohlmeyer, unpublished). ETYMOLOGY: From the Latin, *salinus* = saline, in reference to the marine habitat of the species.

Halosphaeria stellata (Kohlm.) Kohlm.

Can. J. Bot. **50,** 1957, 1972

≡*Remispora stellata* Kohlm., *Nova Hedwigia* **2,** 334, 1960

LITERATURE: Henningsson (1974, 1976b); G. C. Hughes (1969); Irvine *et al.* (1972); T. W. Johnson and Cavaliere (1963); T. W. Johnson and Sparrow (1961); E. B. G. Jones (1972); E. B. G. Jones and Irvine (1971); Koch (1974); Kohlmeyer (1972b); Kohlmeyer and Kohlmeyer (1964–1969); Schaumann (1968); Schmidt (1974); Schneider (1976, 1977).

ASCOCARPS 172–471 μm high, 226–471 μm in diameter, subglobose or ovoid, immersed or becoming exposed, ostiolate, papillate, membranaceous or coriaceous, cream-colored, yellowish or brownish, solitary or gregarious. PERIDIUM 12–25 μm thick, composed of about six layers of roundish or ellipsoidal, thick-walled cells, forming a textura angularis, merging into the large-celled pseudoparenchyma of the venter. PAPILLAE or necks up to 622 μm long, 41–70 μm in diameter, conical, centric. PSEUDOPARENCHYMA of thin-walled, polygonal or ellipsoidal cells filling venter of young ascocarps; eventually breaking up into catenophyses, 8–14 μm in diameter. ASCI 72–80 × 19–28 μm, eight-spored, clavate or subfusiform, pedunculate, unitunicate, thin-walled, aphysoclastic, without apical apparatuses, early deliquescing. ASCOSPORES 24–30.5 × 8.5–12.5 μm (excluding appendages), ellipsoidal, one-septate, not or slightly constricted at the septum, hyaline, appendaged; at each end generally six (rarely more or fewer) radiating appendages, developing by fragmentation of the exospore; appendages 13–24.5 μm long, 2.5–6.5(–8) μm in diameter at the base, terminal, subclavate, curved, attenuate, semirigid, slightly channeled on the inner side, inconspicuously striate by fiberlike elements embedded in the subgelatinous matrix. MODE OF LIFE: Saprobic. SUBSTRATES: Intertidal and drifting wood and bark, test panels (e.g., *Abies* sp., *Alnus* sp., *Fraxinus* sp., *Ocotea rodiaei*, *Picea* sp., *Pinus sylvestris*, *Pinus* sp., *Quercus* sp., *Salix* sp.), submerged cones of *Pseudotsuga* sp. RANGE: Atlantic Ocean—Denmark, Germany (F.R.G.), Great Britain (England); Baltic Sea—Denmark (Sjaelland), Germany (F.R.G., G.D.R.), Sweden; Pacific Ocean—Canada (British Columbia), United States (Washington). ETYMOLOGY: From the Latin, *stellatus* = starry, starred, in reference to the starlike, radiating ascospore appendages.

Halosphaeria torquata Kohlm.

Nova Hedwigia **2**, 311, 1960

LITERATURE: Cavaliere (1966b, 1968); G. C. Hughes (1969); G. C. Hughes and Chamut (1971); T. W. Johnson (1963a); E. B. G. Jones (1963b, 1968b, 1972); E. B. G. Jones and Irvine (1971); Kohlmeyer (1971b, 1972b); Kohlmeyer and Kohlmeyer (1964–1969).

Fig. 60f

ASCOCARPS 225–336 μm high, 251–363 μm in diameter, elongate–cylindrical or subglobose, immersed, sometimes partly immersed or superficial, ostiolate, papillate, membranaceous, almost hyaline, cream-colored or pale brown, solitary or gregarious. PERIDIUM 12–26 μm thick, composed of four to eight layers of irregular or elongate, thin-walled cells with large lumina, forming a textura angularis, merging into the pseudoparenchyma of the venter. PAPILLAE or necks 198–396 μm long, 59–99 μm in diameter, short cylindrical or subconical, centric or eccentric, rarely absent. PSEUDOPARENCHYMA of thin-walled, large, polygonal or ovoid cells filling venter of young ascocarps; eventually breaking up into catenophyses, 6.5–11 μm in diameter. ASCI 80–150 × 20–32.5 μm, eight-spored, clavate or subfusiform, somewhat apiculate, pedunculate, unitunicate, thin-walled, aphysoclastic, without apical apparatuses, early deliquescing; developing at the base of the ascocarp venter. ASCOSPORES 20–30.5 × 10–16 μm (excluding appendages), broadly ellipsoidal or rarely subrhomboid, one-septate, not or slightly constricted at the septum, hyaline, appendaged; provided at each end with a single, terminal, sub-cylindrical, rigid appendage, 8–14.5 μm long, 1.3–3.3 μm in diameter at the base, tapering to 0.7–1 μm in diameter at the apex; around the septum covered by a tubular annulus, 4–6 μm thick. MODE OF LIFE: Saprobic. SUBSTRATES: Intertidal and drifting wood, test panels (e.g., *Fagus sylvatica*, *Tsuga heterophylla*. RANGE: Atlantic Ocean—Great Britain (Scotland), Iceland, United States (Connecticut); Pacific Ocean—Canada (British Columbia), Chile (Isla Hoste), United States (Washington). ETYMOLOGY: From the Latin, *torquatus* = bearing a collar, in reference to the ascospore adorned with a collarlike appendage around the septum.

Halosphaeria trullifera (Kohlm.) Kohlm.

Can. J. Bot. **50**, 1957, 1972

≡*Remispora trullifera* Kohlm., *Nova Hedwigia* **6**, 321, 1963 (ut *"trullifer* Kohlm."*)

LITERATURE: Koch (1974); Kohlmeyer (1971b); Kohlmeyer and Kohlmeyer (1964–1969); Schaumann (1968, 1969).

ASCOCARPS 200–745(–820) μm in diameter, globose, immersed or becoming exposed, ostiolate, papillate, coriaceous, cream-colored or brown, rarely dark brown or black, sometimes light-colored below, darker above; surrounded by brown hyphae. PERIDIUM 27.5–41 μm thick, consisting on the outside of five or six layers of yellowish, polygonal, thick-walled cells, forming a textura angularis, merging into the thin-walled pseudoparenchyma of the venter. NECKS 240–680 μm long, 61–172 μm in diameter, subconical to cylindrical, dark near the venter, lighter colored at the apex; ostiolar canal periphysate. PSEUDOPARENCHYMA of thin-walled, large, polygonal cells filling venter of young ascocarps; eventually breaking up into catenophyses. Asci 118–182 × 29–40.5 μm, eight-spored, clavate, basally attenuate and pedunculate, unitunicate, thin-walled, aphysoclastic, without apical apparatuses, early deliquescing, developing at the base of the ascocarp venter on a small-celled ascogenous tissue. Ascospores (23–)24–32.5(–35.5) × 14–17.5(–20) μm (excluding appendages), ellipsoidal or oblong ellipsoidal, one-septate, not or slightly constricted at the septum, hyaline (or pale grayish-brown?), appendaged; at each end a scoop-shaped, terminal appendage, 3.5–4.5 μm high, 6–10 μm in diameter, developing from the exospore, finally becoming viscous and forming long threads. Mode of Life: Saprobic. Substrate: Intertidal wood. Range: Atlantic Ocean—Denmark (?), Germany (Helgoland and mainland, F.R.G.), Spain, United States (Connecticut); Pacific Ocean—United States (Washington). Note: A record from Denmark (Koch, 1974) is accepted with reservation because of ascospore lengths [33–44(–76) μm] exceeding dimensions of the type and other collections. Etymology: From the Latin, *trulla* = small ladle and *-fer* = -carrying, in reference to the scoop-shaped ascospore appendages.

Key to the Species of *Halosphaeria*

1. Extended appendages of mature ascospores apically or subapically attached only
. . . . 2
1'. Apical and lateral ascospore appendages present 11
  2(1) One subglobose cap at one end of the ascospore only . . . . . . *H. cucullata*
  2'(1) Appendages at both ends of the ascospore 3
3(2') Two spurlike, stiff appendages at each end of the ascospore; these pairs at right angles to one another . . . . *H. quadricornuta*
3'(2') Appendages not in such pairs 4
  4(3') Appendages starlike, composed of three or more distinct arms, radiating from the ascospore apices 5
  4'(3') Appendages not composed of radiating, equal, apical arms 7
5(4) Generally three or four, subterminally attached appendages at each ascospore apex; species of tropical waters . . . . *H. salina*
5'(4) Four or more, terminally attached appendages at each ascospore apex; species of temperate waters 6
  6(5') Generally four radiating appendages at each ascospore apex
. . . . *H. quadriremis*
  6'(5') Generally six appendages at each ascospore apex . . . . *H. stellata*
7(4') Young ascospores completely surrounded by an exosporic sheath that unfolds and develops into pleomorphic, long appendages 8
7'(4') Appendages covering only part of the ascospore near the apices; or, if young ascospores are enclosed by a sheath, it disappears at maturity, leaving subglobose apical appendages 9
  8(7) Ascospores thick-walled, rhomboid, up to 20.5 μm in diameter; appendages usually simple, not yoke-shaped . . . . *H. pilleata*
  8'(7) Ascospores thin-walled, ellipsoidal, up to 13 μm in diameter; appendages mostly geminate and yoke-shaped . . . . *H. maritima*
9(7') Ascospore appendages subapical, hooklike; eventually gelatinizing, stretching and dissolving . . . . *H. hamata*

9'(7') Ascospore appendages apical, persistent or gelatinizing 10
    10(9') Ascospore diameter generally less than 13 μm; appendage a persistent, sub-
    globose cap with faint radiating striae (stain!) . . . . **H. galerita**
    10'(9') Ascospore diameter more than 13 μm; appendage at first scoop-shaped, then
    gelatinizing, stretching and dissolving . . . . **H. trullifera**
11(1') Lateral ascospore appendages crescent-shaped; at each apex a small, inconspicuous
cap . . . . **H. mediosetigera**
11'(1') Lateral ascospore appendages not crescent-shaped; apical appendages long, subcylindrical or obclavate 12
    12(11') Lateral and apical ascospore appendages identical, obclavate
    . . . . **H. appendiculata**
    12'(11') Ascospores with a tubular annulus around the septum; at each end with a
    subcylindrical thorn . . . . **H. torquata**

Lignincola Höhnk

Veroeff. Inst. Meeresforsch. Bremerhaven **3**, 216, 1955

ASCOCARPS solitary or gregarious, subglobose or ellipsoidal, immersed or superficial, ostiolate, papillate, coriaceous, light-colored to blackish. PSEUDOPARENCHYMA of thin-walled cells with large lumina filling venter of young ascocarps; eventually breaking up into catenophyses. ASCI eight-spored, clavate or subfusiform, pedunculate, unitunicate, thin-walled, aphysoclastic, persistent; developing at the base of the ascocarp venter; breaking off at maturity, and released through the ostiole. ASCO-SPORES ellipsoidal, one-septate, hyaline, without appendages. SUB-STRATE: Dead wood and bark. TYPE SPECIES: *Lignincola laevis* Höhnk. The genus is monotypic. ETYMOLOGY: From the Latin, *lignum* = wood and *incola* = inhabitant, in reference to the substrate of the fungus.

Lignincola laevis Höhnk

Veroeff. Inst. Meeresforsch. Bremerhaven **3**, 216, 1955

LITERATURE: Apinis and Chesters (1964); Byrne and Eaton (1972); Byrne and Jones (1974); Catalfomo *et al.* (1972–1973); Cavaliere (1966b, 1968); Henningsson (1974); Höhnk (1968); G. C. Hughes (1969); T. W. Johnson *et al.* (1959); T. W. Johnson and Sparrow (1961); Kirk (1976); Kirk *et al.* (1974); Kohlmeyer (1960, 1963d, 1966a, 1969c, 1971b, 1972b); Kohlmeyer and Kohlmeyer (1964–1969, 1971a); Meyers (1957); Meyers and Reynolds (1959a, 1960b); Meyers *et al.* (1960, 1963b); Peters *et al.* (1975); Raghukumar (1973); Schaumann (1968, 1969, 1974b); Schmidt (1974); Shearer (1972b); Tubaki (1968, 1969); Tubaki and Ito (1973).

Fig. 61

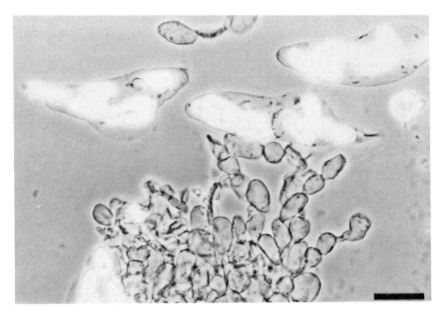

Fig. 61. Free asci and catenophyses of *Lignincola laevis* (Herb. J.K. 2691; bar = 25 μm; phase contrast).

ASCOCARPS 125–250(–386) μm in diameter, subglobose or ellipsoidal, immersed or superficial, ostiolate, papillate, coriaceous, hyaline, light brown, fuscous or blackish, solitary or gregarious. PERIDIUM 13–16 μm thick, composed of two to five layers of elongate, thick-walled cells with large lumina, forming a textura angularis; merging into the pseudoparenchyma of the venter. PAPILLAE or necks up to 4 mm long, 25–40 μm in diameter, cylindrical, centric or eccentric; ostiolar canal at first filled with a small-celled pseudoparenchyma. PSEUDOPARENCHYMA of thin-walled cells with large lumina filling venter of young ascocarps; eventually breaking up into catenophyses. ASCI 49–69 × 15–20 μm, eight-spored, clavate or subfusiform, short pedunculate, unitunicate, thin-walled, sometimes slightly thickened at the apex, aphysoclastic, without apical apparatuses, persistent; developing at the base of the ascocarp venter on a tissue of small ascogenous cells; stipe becoming detached at ascospore maturity, and asci released through the ostiole into the water while enclosing the ascospores; central part of ascus swelling in water. ASCOSPORES (12.5–)16–24 × (5–)6–8 μm, irregularly biseriate, ellipsoidal, one-septate, slightly constricted at the septum, hyaline, without appendages. MODE OF LIFE: Saprobic. SUBSTRATES: Intertidal wood, bark, roots of shoreline trees, test panels (e.g., balsa, bamboo, *Fagus sylvatica, Hibiscus tiliaceus, Liquidambar styraciflua, Pachira aquatica, Pluchea* × *fosbergii, Quercus* sp., *Rhizophora mangle,* possibly also *Avicennia germinans*), litter of *Agropyron pungens* and *Spartina townsendii.* RANGE: Atlantic Ocean—Brazil, Cayman Islands, Germany (Helgoland and mainland, F.R.G.), Great Britain (England), Iceland, Spain, Sweden, United States (Alabama, Connecticut, Florida, Maryland, Massachusetts, North Carolina, Texas); Baltic Sea—Germany (G.D.R.), Sweden; Indian Ocean—India (Madras), South Africa; Pacific Ocean—Canada (British Columbia), Guatemala, Japan, United States [California, Hawaii (Maui, Oahu), Washington]. Data in part after J. Kohlmeyer (unpublished). NOTE: A similar species, *Gnomonia longirostris,* is distinguished from *L. laevis* by the absence of catenophyses and the presence of an apical pore in the ascus (Kohlmeyer, 1972b). ETYMOLOGY: From the Latin, *laevis* = smooth, free from hairs, in reference to the unadorned ascospores.

Lindra Wilson

Trans. Br. Mycol. Soc. **39,** 411, 1956

ASCOCARPS solitary or gregarious, semiglobose or ellipsoidal, immersed or becoming exposed, ostiolate, papillate or epapillate, carbonaceous or coriaceous, dark-colored. PSEUDOPARENCHYMA of large, thin-walled, deliquescing cells filling venter of young ascocarps. ASCI eight-spored, cylindrical to clavate, unitunicate, thin-walled, early deliquescing. ASCOSPORES filiform, multiseptate, hyaline, tips inflated. SUBSTRATES: Dead wood, marine phanerogams, or algae. TYPE SPECIES: *Lindra inflata* Wilson. NOTE: All members of the genus are obligate marine species. ETYMOLOGY: Probably named after the American mycologist David H. Linder (1899–1946).

Lindra inflata Wilson

Trans. Br. Mycol. Soc. **39,** 411, 1956

LITERATURE: Cavaliere (1966a); G. C. Hughes (1968, 1969); T. W. Johnson and Sparrow (1961); E. B. G. Jones *et al.* (1972); Koch (1974); Kohlmeyer (1968d); Neish (1970).

ASCOCARPS 250–397 μm high, 308–611 μm in diameter, semiglobose or ellipsoidal, immersed or becoming exposed, ostiolate, papillate or epapillate, carbonaceous to subcarbonaceous, black, solitary or gregarious, sometimes several fused together. PERIDIUM 40–60 μm thick, two-layered; outer layer composed of irregular, thick-walled, dark-colored cells, sometimes forming lateral outgrowths; inner layer composed of elongate to polygonal, thick-walled, hyaline cells. PAPILLAE centric or eccentric, short, conoid, pointed. PSEUDOPARENCHYMA of large, thin-walled, deliquescing cells filling venter of young ascocarps. ASCI 231–463 × 22–38 μm, eight-spored, cylindrical to clavate, unitunicate, thin-walled, aphysoclastic, without apical apparatuses, early deliquescing. ASCOSPORES 210–415 × 4–6 μm, filiform, 30- to 50-septate (predominantly 36-septate; stain with Congo red or acid Fuchsin lactophenol or mount in glycerine jelly!), not constricted at the septa, hyaline; at each end provided with a globose or semiglobose, gelatinous appendage, 7.5–18.5 μm in diameter. MODE OF LIFE: Saprobic. SUBSTRATES: Intertidal and drifting wood, test panels (e.g., *Picea* sp., *Pinus sylvestris,* mahogany). RANGE: Atlantic Ocean—Canada (Newfoundland, Nova Scotia), Denmark, Great Britain (Wales), Norway, United States (North Carolina); Pacific Ocean—Canada (British Columbia). ETYMOLOGY: From the Latin, *inflatus* = inflated, in reference to the ascospore apices.

Lindra marinera Meyers

Mycologia **61**, 488, 1969

LITERATURE: Kohlmeyer and Kohlmeyer (1971a); Meyers (1971a); Meyers and Scott [1968 (sub *Lindra* sp.)]; Wagner-Merner (1972).

ASCOCARPS (100–)110–135 × (100–)115–150 μm, subglobose, immersed or erumpent, ostiolate, papillate, coriaceous, scattered. PERIDIUM 10–15 μm thick, two-layered; outer layer composed of irregular, elongate, dark cells; inner layer composed of rhomboid, thick-walled, hyaline cells. PAPILLAE or necks 15–150 × 25–35 μm, short or long cylindrical, straight or irregular. PSEUDOPARENCHYMA of thin-walled, polygonal, deliquescing cells filling venter of young ascocarps. ASCI cylindrical, elongate, thin-walled, early deliquescing. ASCOSPORES 152–230 μm long, 3.5–7 μm in diameter in the middle, 2.5–4.5 μm at the apices, filiform, tapering toward both ends, curved (S-, U-, or α-shaped), 6–17 septa, not or barely constricted at the septa, hyaline; tips slightly inflated. MODE OF LIFE: Perthophytic (?) and saprobic. SUBSTRATE: Leaves of *Thalassia testudinum*. RANGE: Atlantic Ocean—Bahamas (Great Abaco; J. Kohlmeyer, unpublished), Mexico (Yucatan), Tobago (J. Kohlmeyer, unpublished), United States (Florida). NOTE: Ascospores may accumulate in foam along the seashore near beds of *T. testudinum. Lindra marinera* varies little from *L. thalassiae* and could be considered a short-spored variety of the latter.

ETYMOLOGY: From the Spanish, *marinero* = seaworthy, the seaman, referring to the marine habitat of the species.

Lindra thalassiae Orpurt, Meyers, Boral, et Simms

Bull. Mar. Sci. Gulf. Caribb. **14**, 406, 1964

LITERATURE: E. B. G. Jones *et al.* (1971); Kohlmeyer (1971a, 1972c); Kohlmeyer and Kohlmeyer (1964–1969, 1977); Meyers (1968b, 1969, 1971b); Meyers and Hopper (1967); Meyers and Simms (1965, 1967); Meyers *et al.* (1965); Tubaki (1969).

Figs. 62a and 62b

ASCOCARPS 115–200 μm high (excluding papillae), 127–200 μm in diameter, globose, subglobose, or depressed, immersed, ostiolate, papillate, coriaceous, dark brown to black, gregarious. PERIDIUM 7.5–10 μm thick, composed of four or five layers of irregular, mostly elongate, thick-walled cells, merging into the pseudoparenchyma of the venter; outer layer of cells forming a textura epidermoidea. PAPILLAE 22.5–84 μm high (in culture up to 500 μm long), 27.5–37.5(–50) μm in diameter, conical or cylindrical; ostiolar canal at first filled with a pseudoparenchyma of small, hyaline, thin-walled cells. PSEUDOPARENCHYMA of large, thin-walled, polygonal or roundish, deliquescing cells filling venter of young ascocarps. ASCI eight-spored, cylindrical to subfusiform or subclavate, curved, unitunicate, thin-walled, aphysoclastic, without apical apparatuses, early deliquescing. ASCOSPORES (220–)230–390 × 3–6 μm, filiform, tapering toward both apices; 1–1.5 μm in diameter at the apex; curved (S-, U-, or α-shaped), 14–26 septa, not or barely constricted at the septa, hyaline; tips slightly inflated. MODE OF LIFE: Perthophytic and saprobic. SUBSTRATES: Leaves of *Thalassia testudinum*, vesicles of *Sargassum* sp. RANGE: Atlantic Ocean—Bahamas (Great Abaco; J. Kohlmeyer, unpublished), Bermuda, Sargasso Sea, United States (Florida); Pacific Ocean—Japan. ETYMOLOGY: Name referring to the host genus from which the species was isolated initially.

Key to the Species of *Lindra*
1. Ascospores not tapering, not strongly curved; on wood . . . . *L. inflata*
1'. Ascospores tapering, strongly curved; on algae or leaves of marine phanerogams . 2
    2(1') Ascospores predominantly shorter than 230 μm . . . . *L. marinera*
    2'(1') Ascospores predominantly longer than 230 μm . . . . *L. thalassiae*

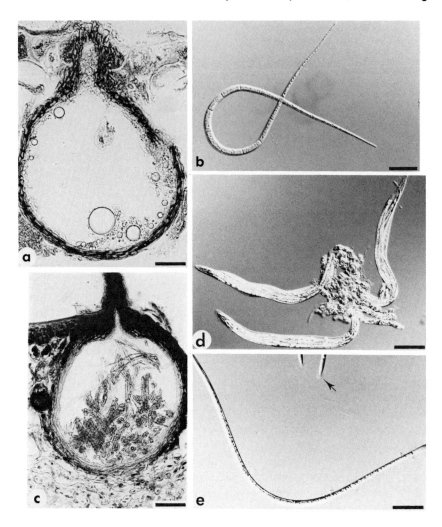

Fig. 62. Scolecosporous Ascomycetes. (a) Longitudinal section (4 μm) through ascocarp of *Lindra thalassiae* from air vesicle of *Sargassum* sp. (Herb. J.K. 2729; bar = 25 μm). (b) Septate ascospore of *L. thalassiae* (Herb. J.K. 2728; bar = 25 μm). (c) Longitudinal section (16 μm) through ascocarp of *Lulworthia* sp. in *Laminaria digitata* (Herb. J.K. 3155; bar = 50 μm). (d) Asci of *Lulworthia* sp. attached to ascogenous cells (Herb. J.K. 2939; bar = 50 μm). (e) Ascospore of *Lulworthia* sp., apical chambers releasing mucus (arrow) (Herb. J.K. 2939; bar = 25 μm). (a and c in bright field, the others in Nomarski interference contrast.)

Lulworthia Sutherland

Trans. Br. Mycol. Soc. **5**, 259, 1916

=*Halophiobolus* Linder *in* Barghoorn and Linder, *Farlowia* **1**, 415, 1944

LITERATURE: Cavaliere and Johnson (1966a); Cribb and Cribb (1955); G. C. Hughes (1969); T. W. Johnson and Sparrow (1961); Kohlmeyer (1968c); Kohlmeyer and Kohlmeyer (1964, 1971b); Meyers *et al.* (1964b); Petrak (1952).

Figs. 62c–62e

ASCOCARPS solitary or gregarious, globose to cylindrical, immersed or superficial, ostiolate, papillate or with a long neck, coriaceous to subcarbonaceous, brown to black. PARAPHYSES absent. ASCI eight-spored, cylindrical, clavate or fusiform, unitunicate, thin-walled, early deliquescing. ASCOSPORES filamentous, curved, hyaline, cylindrical, but tapering at each end into a conical process or apical chamber; this chamber is filled with mucus that is released through an apical pore. MODE OF LIFE: Parasitic or saprobic. TYPE SPECIES: *Lulworthia fucicola* Sutherland. NOTE: The genus is in critical need of a revision; we do not agree with Cavaliere and Johnson (1966a), who made all described 13 species and 2 varieties of the genus synonyms of one species, namely, *L. medusa* (Ellis et Everh.) Cribb et Cribb; besides three species accepted by us there are more species, or possibly varieties or races that cannot be distinctly circumscribed at the present time (Kohlmeyer, 1972b). Only work with pure cultures derived from a variety of substrates and geographical areas can elucidate morphological and physiological differences between these taxa; therefore, the doubtful species are lumped together as *Lulworthia* spp. in the following. ETYMOLOGY: From Lulworth (Dorset, England), the collecting site of the type species.

Lulworthia fucicola Sutherland

Trans. Br. Mycol. Soc. **5**, 259, 1916

=*Halophiobolus cylindricus* Linder *in* Barghoorn and Linder, *Farlowia* **1**, 416, 1944

≡*Lulworthia cylindrica* (Linder) Cribb et Cribb, *Univ. Queensl. Pap.*, *Dep. Bot.* **3**, 79, 1955

LITERATURE: Barghoorn (1944); G. C. Hughes (1969); T. W. Johnson and Sparrow (1961); E. B. G. Jones (1963a, 1968b, 1971a, 1972); E. B. G. Jones and Irvine (1971); Kirk (1966); Koch (1974); Kohlmeyer (1963d, 1972b); Meyers (1957); Schmidt (1974).

ASCOCARPS 440–762 μm long, 123–155 μm in diameter when elongate;

350–486 μm in diameter when roundish; globose, cylindrical or subcylindrical, immersed or superficial, ostiolate, papillate or with a long neck, carbonaceous, black; when immersed with a subhyaline, soft base. PAPILLAE or necks up to 500 μm long, 22–27 μm in diameter, black, sometimes with a bulbous base. PARAPHYSES absent. ASCI 90–126 × 14–20 μm, eight-spored, clavate or fusiform, unitunicate, thin-walled, early deliquescing. ASCOSPORES 70–110(–126) × (3.6–)4.5–5.8 μm (including apical chambers), filamentous, curved, hyaline, tapering at each end into an elongate, conical process or apical chamber; processes 7.2–11.5 × 2.2–3 μm, filled with mucus that is released through an apical pore. MODE OF LIFE: Perthophytic (?) and saprobic. SUBSTRATES: In the base of *Fucus vesiculosus,* driftwood, intertidal wood, and test panels (e.g., *Fagus sylvatica, Pinus sylvestris, Quercus* sp.). RANGE: Atlantic Ocean—Denmark, Great Britain (England, Scotland, Wales), United States (Massachusetts); Baltic Sea—Germany (F.R.G., G.D.R.); Pacific Ocean—Canada (British Columbia), United States (Washington). NOTE: Description combined from authors listed above. ETYMOLOGY: From *Fucus* (Phaeophyta) and the Latin suffix, -*cola* = -dweller, in reference to the host of the type collection.

Lulworthia grandispora Meyers

Mycologia **49**, 513, 1957

=*L. grandispora* var. *apiculata* Johnson, *Mycologia* **50**, 159, 1958

LITERATURE: Cavaliere and Johnson (1966a,b); Gold (1959); T. W. Johnson *et al.* (1959); T. W. Johnson and Sparrow (1961); Kirk (1966); Kohlmeyer and Kohlmeyer (1971a, 1977); Meyers and Reynolds (1959a).

ASCOCARPS 180 × 306 μm in diameter, globose or subglobose to pyriform, immersed or superficial, ostiolate, with a long neck, brown to black, solitary or gregarious. NECKS 75–1400 × 15–33 μm, cylindrical, straight or curved, sometimes two on one ascocarp. PARAPHYSES absent. ASCI eight-spored, elongate-fusiform or cylindrical, unitunicate, thin-walled, early deliquescing. ASCOSPORES 500–756 × 3–5 μm (including apical chambers), filamentous, curved, hyaline, tapering at each end into an elongate, conical process or apical chamber; processes 3.6–7 μm long, acute or rounded, filled with mucus that is released through an apical pore. MODE OF LIFE: Saprobic. SUBSTRATES: Test panels (*Acer* sp., *Pinus* sp.), prop roots and submerged branches of *Rhizophora mangle* and *R. racemosa,* submerged fruits of *Casuarina* sp. (J. Kohlmeyer, unpublished), roots of *Pachira aquatica.* RANGE: Atlantic Ocean—Bahamas (Great Abaco), Bermuda, Brazil (São Paulo), Liberia, United States (Florida, North Carolina); Pacific Ocean—Guatemala, Peru (Tumbes; J.

Kohlmeyer, unpublished). NOTE: Description after Meyers (1957), T. W. Johnson and Sparrow (1961), and Kohlmeyer and Kohlmeyer (1971a); the species appears to be restricted to tropical and subtropical waters. ETYMOLOGY: From the Latin, *grandis* = large, great, big, and *spora* = spore, in reference to the long ascospores.

Lulworthia kniepii Kohlm.

Nova Hedwigia **6**, 140–141, 1963

[Nomina nuda: *Ophiobolus kniepii* Ade et Bauch *in* Bauch, *Pubbl. Stn. Zool. Napoli* **15**, 389–390, 1936; *Lulworthia kniepii* (Ade et Bauch) Petrak, *Sydowia* **10**, 297, 1956; names invalid according to the International Code of Botanical Nomenclature because not provided with a Latin diagnosis]

LITERATURE: T. W. Johnson and Sparrow (1961); Kohlmeyer (1967, 1969b,c); Meyers (1957).

ASCOCARPS 300–410 μm in diameter, globose, immersed in discolored areas of the host, ostiolate, papillate or with a long neck, membranaceous to subcarbonaceous, blackish-brown, solitary or gregarious. NECKS 140–300 × 50 μm, cylindrical, dark-colored, emerging above the surface of the host. PERIDIUM composed of pseudoparenchymatous cells, forming a textura angularis. PARAPHYSES absent. ASCI about 270 × 7–10 μm, eight-spored, elongate–fusiform, unitunicate, thin-walled, early deliquescing. ASCOSPORES 200–270 × 2.5–6 μm (including apical chambers), filamentous, curved, hyaline, tapering at each end into an elongate, conical process or apical chamber; processes 4.5–6 μm long, acute or rounded, filled with mucus that is released through an apical pore. MODE OF LIFE: Parasitic. HOSTS: Calcified Rhodophyta, for example, *Clathromorphum compactum* (J. Kohlmeyer, unpublished), *Lithophyllum tortuosum, Lithophyllum* sp., *Porolithon onkodes, Pseudolithophyllum expansum;* possibly also *L. incrustans* and *L. racemus.* RANGE: Atlantic Ocean—Canada (Newfoundland; J. Kohlmeyer, unpublished), Spain (Canary Islands); Mediterranean—Italy, France; Pacific Ocean—United States [Hawaii (Oahu)]. NOTE: Description after Bauch (1936) and Kohlmeyer (1963b, 1969c). ETYMOLOGY: Named in honor of H. Kniep (1881–1930), German botanist.

Lulworthia spp.*

NOTE: As explained in the generic description, those species of *Lulworthia* that cannot be distinctly circumscribed are summarily dealt with in

* Named species not recognized by us.

the following. It should be pointed out, however, that we do not consider this group as one single species.

Lulworthia attenuata Johnson, *Mycologia* **50,** p. 157, 1958. SUBSTRATE: test panel (*Liriodendron* sp.). RANGE: Atlantic Ocean—United States (North Carolina). LITERATURE: Cavaliere and Johnson (1966a); T. W. Johnson and Sparrow (1961).

Lulworthia conica Johnson, *Mycologia* **50,** p. 156, 1958. SUBSTRATE: test panel (*Picea* sp.). RANGE: Atlantic Ocean—United States (North Carolina). LITERATURE: Cavaliere and Johnson (1966a); T. W. Johnson and Sparrow (1961).

Lulworthia cylindrica (Linder) Cribb et Cribb, *Univ. Queensl. Pap., Dep. Bot.* **3,** p. 79, 1955 (≡*Halophiobolus cylindricus* Linder, *in* Barghoorn et Linder, *Farlowia* **1,** p. 416, 1944). SUBSTRATE: wood. RANGE: Atlantic Ocean—United States (Massachusetts). LITERATURE: Barghoorn (1944); Cavaliere and Johnson (1966a).

Lulworthia floridana Meyers, *Mycologia* **49,** p. 515, 1957. SUBSTRATES: driftwood and test panels (e.g., *Araucaria* sp., balsa, *Fagus sylvatica*, *Phoenix* sp., *Pinus monticola*, *P. sylvestris*, *Pseudotsuga douglasii*, *P. menziesii*), cellulose tape, manila cordage, rhizomes of *Ruppia maritima* and *Spartina alterniflora*. RANGE: Atlantic Ocean—Cameroun, Cayman Islands, Cuba, France, Great Britain (England, Scotland, Wales), Martinique, Norway, Puerto Rico, United States (Florida, Massachusetts, North Carolina, Rhode Island); Indian Ocean—India, South Yemen; Mediterranean—France, Italy; Pacific Ocean—Australia, Canada (British Columbia), New Zealand, Singapore, United States (California, Washington). LITERATURE: Cavaliere and Johnson (1966a); Churchland and McClaren (1973); Eaton (1972); Gessner and Goos (1973b); Gold (1959); G. C. Hughes (1969); T. W. Johnson (1958a); T. W. Johnson and Sparrow (1961); E. B. G. Jones (1962a,b, 1963c, 1968a,b, 1971a, 1972); E. B. G. Jones *et al.* (1971); E. B. G. Jones and Jennings (1964); E. B. G. Jones *et al.* (1972); E. B. G. Jones and Oliver (1964); Kohlmeyer (1960, 1962a, 1963b,c); Meyers (1966, 1968a,b, 1971a); Meyers and Reynolds (1959a,c, 1960b); Meyers and Scott (1968); Meyers and Simms (1967); Meyers *et al.* (1960, 1963a, 1964a); Orpurt *et al.* (1964); Pisano *et al.* (1964).

Lulworthia halima (Diehl et Mounce) Cribb et Cribb, *Univ. Queensl. Pap., Dep. Bot.* **3,** p. 80, 1955 [≡*Ophiobolus halimus* Diehl et Mounce *in* Mounce et Diehl, *Can. J. Res.* **11,** p. 246, 1934; ≡*Halophiobolus halimus* (Diehl et Mounce) Linder *in* Barghoorn et Linder, *Farlowia* **1,** p. 419, 1944; ≡*Linocarpon halimum* (Diehl et Mounce) Petrak, *Sydowia* **6,** p. 388, 1952]. SUBSTRATE: leaves and rhizomes of *Zostera marina*. RANGE: Atlantic Ocean—Canada (New Brunswick, Nova Scotia, Québec), Denmark, Great Britain (England, Ireland), United States (Maine);

Mediterranean—France. LITERATURE: Cavaliere and Johnson (1966a); Conners (1967); T. W. Johnson and Sparrow (1961); Kohlmeyer (1963d); Meyers (1957); H. E. Petersen (1935); Renn (1936); Shoemaker (1976); Tutin (1934).

Lulworthia longirostris (Linder) Cribb et Cribb, *Univ. Queensl. Pap., Dep. Bot.* **3,** p. 80, 1955 (≡*Halophiobolus longirostris* Linder *in* Barghoorn et Linder, *Farlowia* **1,** p. 418, 1944). SUBSTRATE: intertidal wood. RANGE: Atlantic Ocean—Great Britain (Ireland), United States (Maine). LITERATURE: Cavaliere and Johnson (1966a); Petrak (1952); Wilson (1956b).

Lulworthia longispora Cribb et Cribb, *Univ. Queensl. Pap., Dep. Bot.* **3,** p. 80, 1955. SUBSTRATES: roots of *Avicennia marina* var. *resiniferae,* intertidal wood and bark. RANGE: Pacific Ocean—Australia (Queensland). LITERATURE: Cavaliere and Johnson (1966a); Lloyd and Wilson (1962); Meyers (1957).

Lulworthia medusa (Ellis et Everh.) Cribb et Cribb, *Univ. Queensl. Pap., Dep. Bot.* **3,** p. 80, 1955 [≡*Ophiobolus medusa* Ellis et Everh., *J. Mycol.* **1,** p. 150, 1885; ≡*Halophiobolus medusa* (Ellis et Everh.) Linder *in* Barghoorn et Linder, *Farlowia* **1,** p. 419, 1944; ≡*Linocarpon medusa* (Ellis et Everh.) Petrak, *Sydowia* **6,** p. 388, 1952]. SUBSTRATES: *Agropyron pungens, Spartina alterniflora, S. cynosuroides, S. townsendii, Zostera marina,* submerged leaves and twigs of mangroves and *Tamarix aphylla,* driftwood and intertidal wood, test panels (e.g., *Betula pubescens, Fagus* sp., *Pinus sylvestris, Pinus* sp., cedar, fir, sycamore). RANGE: Atlantic Ocean—Argentina (Buenos Aires), Canada (Newfoundland, Nova Scotia), Great Britain (England, Wales), Iceland, Sweden, United States (Florida, New Jersey, North Carolina, South Carolina, Texas, Virginia); Baltic Sea—Sweden; Indian Ocean—India (Madras); Mediterranean—France; Pacific Ocean—Canada (British Columbia), Chile, United States [California (Salton Sea), Hawaii (Oahu)]. LITERATURE: Anastasiou (1963b); Anastasiou and Churchland (1969); Apinis and Chesters (1964); Cavaliere (1968); Cavaliere and Johnson (1966a); Davidson (1973, 1974a); Ellis and Everhart (1892); Goodman (1959); Henningsson (1974, 1976a,b); G. C. Hughes (1968, 1969); G. C. Hughes and Chamut (1971); Irvine *et al.* (1972); T. W. Johnson (1956b); T. W. Johnson and Sparrow (1961); E. B. G. Jones (1963a); E. B. G. Jones and Irvine (1971); E. B. G. Jones and Jennings (1964, 1965); Kirk (1966); Kohlmeyer (1963d); Lloyd and Wilson (1962); Malacalza and Martínez (1971); Meyers (1957); Mounce and Diehl (1934); Neish (1970); Parguey-Leduc (1967); Raghukumar (1973); Shoemaker (1976).

Lulworthia medusa var. *biscaynia* Meyers, *Mycologia* **49,** pp. 516–517, 1957. SUBSTRATE: wood panels (*Pinus* sp.). RANGE: Atlantic Ocean—

Bahamas, Bermuda, Puerto Rico, United States (Florida, North Carolina). LITERATURE: Cavaliere and Johnson (1966a); T. W. Johnson (1958a); T. W. Johnson and Sparrow (1961); Meyers and Reynolds (1959a); Ritchie (1957).

Lulworthia opaca (Linder) Cribb et Cribb, *Univ. Queensl. Pap., Dep. Bot.* **3**, p. 79, 1955 (≡*Halophiobolus opacus* Linder *in* Barghoorn et Linder, *Farlowia* **1**, p. 417, 1944). SUBSTRATES: intertidal rope and wood, driftwood and test panels (e.g., *Fagus* sp., *Pinus* sp., *Quercus* sp., *Tamarix aphylla*), litter of *Agropyron pungens* and *Spartina townsendii, Laminaria saccharina,* sediment. RANGE: Atlantic Ocean—Germany, Great Britain (England, Scotland, Wales), Sweden, United States (Florida, Massachusetts, North Carolina, Texas, Virginia); Baltic Sea—Germany (F.R.G.); Pacific Ocean—Canada (British Columbia), United States [California (also Salton Sea), Washington]. LITERATURE: Anastasiou (1963b); Apinis and Chesters (1964); Barghoorn (1944); Cavaliere and Johnson (1966a); Gustafsson and Fries (1956); Höhnk (1954b, 1955, 1956); G. C. Hughes (1969); T. W. Johnson and Sparrow (1961); E. B. G. Jones (1963a, 1964); E. B. G. Jones and Jennings (1964); Kohlmeyer (1960); E. E. Webber (1970); Wilson (1951, 1954, 1956b).

Lulworthia purpurea (Wilson) Johnson, *Mycologia* **50**, p. 154, 1958 (≡*Halophiobolus purpureus* Wilson, *Trans. Br. Mycol. Soc.* **39**, p. 403, 1956). SUBSTRATES: driftwood, intertidal rope, wood and test panels (e.g., *Fagus sylvatica, Pinus sylvestris*), polyurethane foam. RANGE: Atlantic Ocean—Great Britain (England, Scotland, Wales), Ivory Coast, Norway, United States (North Carolina); Indian Ocean—India (Kerala, Maharashtra), South Africa, South Yemen; Mediterranean—France; Pacific Ocean—Australia (New South Wales), Malaysia, New Zealand. LITERATURE: Cavaliere and Johnson (1966a); S. E. J. Furtado *et al.* (1977); T. W. Johnson and Sparrow (1961); E. B. G. Jones (1962a,b, 1963c, 1964, 1965, 1968a,b, 1971a, 1972); E. B. G. Jones and Irvine (1971); E. B. G. Jones and Le Campion-Alsumard (1970a,b); E. B. G. Jones and Oliver (1964); E. B. G. Jones *et al.* (1972).

Lulworthia rotunda Johnson, *Mycologia* **50**, p. 154, 1958. SUBSTRATE: test panel (*Tilia* sp.). RANGE: Atlantic Ocean—United States (North Carolina). LITERATURE: T. W. Johnson and Sparrow (1961).

Lulworthia rufa (Wilson) Johnson, *Mycologia* **50**, p. 154, 1958 (≡*Halophiobolus rufus* Wilson, *Trans. Br. Mycol. Soc.* **39**, p. 405, 1956). SUBSTRATES: driftwood, test panels (*Fagus sylvatica, Pinus sylvestris,* yellow poplar). RANGE: Atlantic Ocean—France, Great Britain (England, Scotland, Wales), United States (North Carolina); Pacific Ocean—United States (Washington). LITERATURE: Cavaliere and Johnson (1966a); T. W. Johnson and Sparrow (1961); E. B. G. Jones (1962a, 1963c, 1965, 1968b,

1971a, 1972); E. B. G. Jones and Oliver (1964); E. B. G. Jones *et al.* (1972); Kohlmeyer (1960).

Lulworthia salina (Linder) Cribb et Cribb, *Univ. Queensl. Pap., Dep. Bot.* **3**, p. 80, 1955 (≡*Halophiobolus salinus* Linder *in* Barghoorn et Linder, *Farlowia* **1**, p. 419, 1944). SUBSTRATES: intertidal manila rope, bark and wood, test panels (e.g., *Liriodendron tulipifera, Pinus* sp., *Tilia* sp.), stumps of *Fucus vesiculosus*. RANGE: Atlantic Ocean—Great Britain (Wales), United States (Connecticut, Massachusetts, New Hampshire, North Carolina); Pacific Ocean—Canada (British Columbia), United States (California, Washington). LITERATURE: Barghoorn (1944); Cavaliere and Johnson (1966a); G. C. Hughes (1969); T. W. Johnson (1956b, 1958a); T. W. Johnson *et al.* (1959); Kohlmeyer (1960); Meyers (1957); Wilson (1951, 1960).

Lulworthia submersa Johnson, *Mycologia* **50**, p. 156. 1958. SUBSTRATE: wooden test panels (*Acer* sp., *Pinus sylvestris*). RANGE: Atlantic Ocean—Great Britain (England, Wales), United States (North Carolina and South Carolina). LITERATURE: Cavaliere and Johnson (1966a); Gold (1959); T. W. Johnson *et al.* (1959); T. W. Johnson and Sparrow (1961); E. B. G. Jones (1962a, 1963c).

Lulworthia sp. (unidentified species). SUBSTRATES: intertidal cordage, corn cob (*Zea mays*), bark, wood, test panels (e.g., balsa, *Dicorynia paraensis, Fagus sylvatica, Fagus* sp., *Laphira procera, Pinus sylvestris, Tilia* sp.), submerged parts of mangroves [e.g., bark and wood of roots and branches, seedlings, leaves of *Avicennia africana, A. germinans, Conocarpus erecta* (J. Kohlmeyer, unpublished), *Hibiscus tiliaceus, Rhizophora mangle, R. racemosa*], deteriorating algae (*Laminaria hyperborea, L. saccharina, Saccorhiza polyschides*; J. Kohlmeyer, unpublished), rhizomes of *Cymodocea manatorum, Juncus roemerianus* (J. Kohlmeyer, unpublished), and *Thalassia testudinum*, polychaete tubes (*Chaetopterus variopedatus*), polyurethane foam. RANGE: Atlantic Ocean—Bahamas (Great Abaco; J. Kohlmeyer, unpublished), Bermuda, Brazil, Canada (New Brunswick, Nova Scotia), Denmark, Germany (Helgoland and mainland, F.R.G.), Great Britain (England), Iceland, Liberia, Mexico, Sargasso Sea, Senegal, Spain (Canary Islands), Tobago and Trinidad (J. Kohlmeyer, unpublished), United States (Connecticut, Florida, Maine, Maryland, Massachusetts, North Carolina, Rhode Island), Venezuela; Baltic Sea—Germany (F.R.G., G.D.R.), Poland; Black Sea—U.S.S.R. (J. Kohlmeyer, unpublished); Indian Ocean—India (Kerala); Mediterranean—France, Italy (mainland and Sicily); Pacific Ocean— Japan, United States [Alaska, Hawaii (Oahu)]. LITERATURE: Biernacka (1965); Byrne and Eaton (1972); Byrne and Jones (1974, 1975b); Fell and Master (1973); Gunn and Dennis (1971); Irvine *et al.*

(1972); T. W. Johnson (1968); E. B. G. Jones (1972); E. B. G. Jones and Irvine (1971, 1972); Kirk (1976); Koch (1974); Kohlmeyer (1958a,b, 1959, 1963d, 1964, 1967, 1968c, 1969c,d, 1971b, 1972a, 1976); Kohlmeyer and Kohlmeyer (1971a, 1977); Le Campion-Alsumard (1970); Lee and Baker (1973); Meyers (1954, 1971a); Meyers and Hopper (1967); Meyers and Reynolds (1957, 1958, 1960a, 1963); Meyers and Simms (1967); Meyers *et al.* (1964a,b); Nair (1970); Pisano *et al.* (1964); Rheinheimer (1971); Schaumann (1968, 1969, 1974b,c); Schmidt (1974); Schneider (1977); Shearer (1972b); Tubaki and Ito (1973).

Key to the Species of *Lulworthia*
1. Ascospores longer than 500 μm; species on mangroves and wood in tropical and subtropical regions . . . . *L. grandispora*
1'. Ascospores shorter than 500 μm 2
    2(1') Ascospore length usually 110 μm or less; species of temperate regions . . . . *L. fucicola*
    2'(1') Ascospores longer than 110 μm; in temperate or tropical regions 3
3(2') Parasitic in calcified Rhodophyta . . . . *L. kniepii*
3'(2') Saprobic in other substrates . . . . *Lulworthia* sp.

Nais Kohlm.

Nova Hedwigia **4**, 409, 1962

ASCOCARPS gregarious, subglobose or depressed, immersed or superficial, ostiolate, papillate, coriaceous, blackish. PSEUDOPARENCHYMA of large, thin-walled cells filling venter of young ascocarps, eventually breaking up into catenophyses. ASCI eight-spored, clavate, unitunicate, thin-walled, early deliquescing. ASCOSPORES ellipsoidal, one-septate, hyaline, without appendages. SUBSTRATES: Dead wood and culms. TYPE SPECIES: *Nais inornata* Kohlm. The genus is monotypic. ETYMOLOGY: From the Greek, *nais* = a water nymph, in reference to the aquatic habitat of the fungus.

Nais inornata Kohlm.

Nova Hedwigia **4**, 409, 1962

LITERATURE: Byrne and Eaton (1972); Byrne and Jones (1974); Catalfomo *et al.* (1972–1973); Churchland and McClaren (1973); Davidson (1974b); Eaton and Irvine (1972); Eaton and Jones (1971a,b); Gessner and Goos (1973a,b); Irvine *et al.* (1972); T. W. Johnson and Sparrow [1961 (sub *Melanopsamma* sp.)]; E. B. G. Jones and Steward (1972); Kirk *et al.*

(1974); Kohlmeyer (1972b); Kohlmeyer and Kohlmeyer (1964–1969); Peters *et al.* (1975); Schneider (1976); Schmidt (1974); Shearer (1972b).

ASCOCARPS 135–250 μm high, 200–370 μm in diameter, subglobose or depressed, immersed or superficial, ostiolate, papillate, coriaceous, dark brown to black, gregarious. PERIDIUM 12–15 μm thick, composed of two or three layers of thick-walled, oblong or ellipsoidal cells, merging into the pseudoparenchyma of the venter. PAPILLAE or necks 100–620 × 30–44 μm, cylindrical; ostiolar canal periphysate. PSEUDOPARENCHYMA of thin-walled cells filling venter of young ascocarps; eventually breaking up into catenophyses, 55–240 × 5.5–36 μm, including up to 15 cells per chain. ASCI 85–160 × 25–33 μm, eight-spored, clavate, short pedunculate, unitunicate, thin-walled, aphysoclastic, without apical apparatuses, early deliquescing; developing at the base of the ascocarp venter. ASCO-SPORES 22–30 × 11.5–15.5 μm, broadly ellipsoidal, one-septate, slightly or not constricted at the septum, hyaline, with one large oil globule in each cell and many small ones near the septum and the apices; without appendages. MODE OF LIFE: Saprobic. SUBSTRATES: Intertidal and drifting wood and bark, test panels (e.g., *Dicorynia paraensis, Fagus sylvatica, Laphira procera, Ocotea rodiaei, Pinus sylvestris, Pinus* sp., balsa), *Spartina alterniflora*. RANGE: Atlantic Ocean—Great Britain (England, Wales), United States (Maryland, Mississippi, North Carolina, Rhode Island); Baltic Sea—Germany (F.R.G., G.D.R.); Mediterranean—France; Pacific Ocean—Canada (British Columbia). NOTE: The species was also found on *Juncus* sp. in a saline lake in Wyoming (Davidson, 1974b). ETYMOLOGY: From the Latin, *inornatus* = unadorned, in reference to the simple ascospores, lacking appendages.

Nautosphaeria E. B. G. Jones

Trans. Br. Mycol. Soc. **47**, 97, 1964

ASCOCARPS solitary or gregarious, globose to subglobose, immersed, ostiolate, papillate, coriaceous, light-colored. PSEUDOPARENCHYMA of thin-walled, deliquescing cells filling venter of young ascocarps. ASCI eight-spored, clavate to ellipsoidal, unitunicate, thin-walled, early deliquescing. ASCOSPORES ellipsoidal, one-celled, hyaline, with tufts of bristlelike appendages. SUBSTRATE: Dead wood. TYPE SPECIES: *Nautosphaeria cristaminuta* E. B. G. Jones. The genus is monotypic. ETYMOLOGY: From the Latin, *nauta* = a sailor and *sphaera* = globe, in reference to the spherelike ascocarps and the marine habitat of the fungus.

Nautosphaeria cristaminuta E. B. G. Jones
Trans. Br. Mycol. Soc. **47**, 97, 1964

LITERATURE: Byrne and Jones (1974); E. B. G. Jones (1968b, 1971a, 1972); E. B. G. Jones and Irvine (1971); Koch (1974); Kohlmeyer (1963d, 1967, 1972b); Kohlmeyer and Kohlmeyer (1964–1969).

ASCOCARPS (144–)180–288 μm high, 82–198 μm in diameter, globose or subglobose, immersed, ostiolate, papillate, coriaceous, hyaline or cream-colored, solitary or gregarious, anchored with brownish hyphae in the substrate. PERIDIUM 8–10 μm thick, composed of four or five layers of hyaline, thick-walled cells, ellipsoidal at the outside, compressed toward the center, and merging into the pseudoparenchyma. PAPILLAE or necks 36–396 × 20–30 μm, cylindrical; sometimes wartlike papillae covering the outside; ostiolar canal at first filled with small, thin-walled pseudoparenchymatous cells. PSEUDOPARENCHYMA of thin-walled cells filling venter of young ascocarps; deliquescing at ascospore maturity. ASCI ca. 35 × 20 μm, eight-spored, broadly clavate or ellipsoidal, pedunculate, unitunicate, thin-walled, aphysoclastic, without apical apparatuses, early deliquescing; developing at the base of the ascocarp venter. ASCOSPORES 13–18 × 7–11 μm (excluding appendages), ellipsoidal, one-celled, hyaline (possibly becoming grayish or fuscous), with a tuft of bristlelike appendages at each end and four tufts around the equator; appendages 4–11.5 μm long, inserted in the wall. MODE OF LIFE: Saprobic. SUBSTRATES: Intertidal and drifting wood, test panels (e.g., *Fagus sylvatica*). RANGE: Atlantic Ocean—Denmark, Great Britain (Isle of Man), Spain (Tenerife and mainland); Mediterranean—France. ETYMOLOGY: From the Latin, *crista* = crest or terminal tuft and *minutus* = very small, in reference to the tuftlike appendages of the ascospores.

Trailia Sutherland
Trans. Br. Mycol. Soc. **5**, 149, 1915

LITERATURE: Rogerson (1970).

ASCOCARPS gregarious or scattered, subglobose, completely immersed, ostiolate, with a long neck, coriaceous, light-colored. PARAPHYSES absent. ASCI eight-spored, cylindrical, thin-walled, developing at the base of the ascocarp venter. ASCOSPORES filiform, bent double and coiled in the ascus, tapering from a broad to a narrow end, one- to four-septate, hyaline. HOSTS: Marine algae (Phaeophyta). TYPE SPECIES: *Trailia ascophylli* Sutherland. The genus is monotypic. NOTE: The genus was placed in Hyponectrieae by Sutherland (1915c), but it should be transferred as a doubtful genus to Sphaeriales, Halosphaeriaceae, because asci do not line the sides and base of the ascocarp cavity, and apical paraphyses are absent. ETYMOLOGY: Commemorating J. W. H. Trail (1851–1919), Scottish mycologist.

Trailia ascophylli Sutherland

Trans. Br. Mycol. Soc. **5**, 149, 1915

LITERATURE: Kohlmeyer (1968e); Wilson (1951, 1960).

Fig. 63

HYPHAE 1.5–3 µm in diameter, septate, hyaline, forming a network in the host tissue. ASCOCARPS 50–150 µm in diameter, subglobose, completely immersed, ostiolate, with a long neck, coriaceous, hyaline, thin-walled, asci and ascospores visible through the wall; gregarious or scattered in blackened areas of the host. PERIDIUM 13–19 µm thick, composed of hyaline, irregular cells with large lumina; cells on the outside usually smaller than those adjoining the center. NECKS 140–450 × 7.5–19 µm, cylindrical, straight or curved, tapering, hyaline, wall appearing hyphoid. PARAPHYSES absent. ASCI 45–55 × 9–12 µm, eight-spored, cylindrical to clavate, curved, slightly constricted above the broad lower end, unitunicate, thin-walled, aphysoclastic, without apical apparatuses, early deliquescing, developing at the base of the ascocarp venter, maturing successively. ASCOSPORES 90–110 µm long, 3–4 µm in diameter at one end, 1 µm at the other, filamentous, tapering, thick-walled in the broad apex, one- to four-septate in the broad end, not constricted at the septa, hyaline, bent double and coiled in the ascus, both ends pointing toward the ascus apex; after release remaining curved or coiled. MODE OF LIFE: Parasitic or perthophytic. HOSTS: *Ascophyllum nodosum, Fucus* sp. RANGE: Atlantic Ocean—Great Britain (Scotland, Wales), United States (Maine). NOTE: Description in part after J. Kohlmeyer (unpublished), based on material on *A. nodosum* from Wales of I. M. Wilson; the species occurs often in the company of *Orcadia ascophylli*; ascospores may accumulate in foam along the seashore near stands of *A. nodosum* and *Fucus* spp. ETYMOLOGY: Named after a host plant, *Ascophyllum*.

c. Hypocreaceae

Halonectria E. B. G. Jones

Trans. Br. Mycol. Soc. **48**, 287, 1965

LITERATURE: Rogerson (1970).

ASCOCARPS solitary or gregarious, globose or subglobose, partly or totally immersed, ostiolate, papillate, coriaceous, orange-colored or brownish. PARAPHYSES absent. ASCI eight-spored, clavate, pedunculate, unitunicate, thin-walled, early deliquescing, developing along the inner wall of the ascocarp, up to the ostiolar canal. ASCOSPORES elongate, fusiform or cylindrical, one-celled, hyaline. PYCNIDIA (or spermogonia?) solitary or gregarious, obpyriform or cylindrical, partly immersed or superficial, ostiolate, rostrate, coriaceous, reddish or brownish. CONID-IOPHORES elongate, simple, septate. CONIDIA (or spermatia?) filiform, one-celled, hyaline. SUBSTRATE: Dead wood. TYPE SPECIES: *Halonectria milfordensis* E. B. G. Jones. The genus is monotypic. ETYMOLOGY: From the Greek, *hals* = sea, salt, and *Nectria* = a related genus, in reference to the marine occurrence of the fungus.

Halonectria milfordensis E. B. G. Jones

Trans. Br. Mycol. Soc. **48**, 287, 1965

LITERATURE: Eaton and Dickinson (1976); G. C. Hughes (1969); E. B. G. Jones (1968b, 1972); E. B. G. Jones and Irvine (1971); Kohlmeyer (1971b); Kohlmeyer and Kohlmeyer (1964–1969).

ASCOCARPS 130–250 μm high, 105–180 μm in diameter, globose or subglobose, partly or totally immersed, or becoming exposed, ostiolate, papillate, coriaceous, bright orange-colored to pale reddish-brown, somewhat dark above, lighter at the base, solitary or gregarious. PERIDIUM 10–17.5 μm thick, composed of thick-walled, irregularly elon-gate or roundish cells with large lumina. PAPILLAE or necks 108–345 μm long, 30–100 μm in diameter, cylindrical, orange-colored to reddish-brown, protruding from the surface of the substrate; ostiolar canal at first filled with elongate, hyaline, thin-walled cells, possibly forming peri-physes. PARAPHYSES absent. ASCI 21.5–34 × 4–8.5 μm, eight-spored, cla-vate, short pedunculate, unitunicate, thin-walled, aphysoclastic, without apical apparatuses, early deliquescing, developing along the inner wall of the ascocarp, up to the ostiolar canal. ASCOSPORES 16.5–29(–30) × 2–3.5 μm, elongate fusiform, rarely cylindrical or ellipsoidal, slightly curved, one-celled, one- to three-guttulate, hyaline. PYCNIDIA (or spermogonia?) 140–170 μm high, 45–55 μm in diameter, obpyriform or cylindrical, partly immersed or superficial, ostiolate, rostrate, coriaceous, pale reddish-brown, solitary or gregarious. CONIDIOPHORES elongate, simple, at-tenuate, septate, producing conidia singly at the apex. CONIDIA (or sper-

---

Fig. 63. *Trailia ascophylli* from *Ascophyllum nodosum* (Herb. Wilson). (a) Whole mount of thin-walled ascocarp showing the contents (bar = 20 μm). (b) Ascospore with broad apex and tapering base (bar = 10 μm). (c) Asci showing the bent condition of the enclosed septate ascospores (bar = 10 μm). (All in Nomarski interference contrast.)

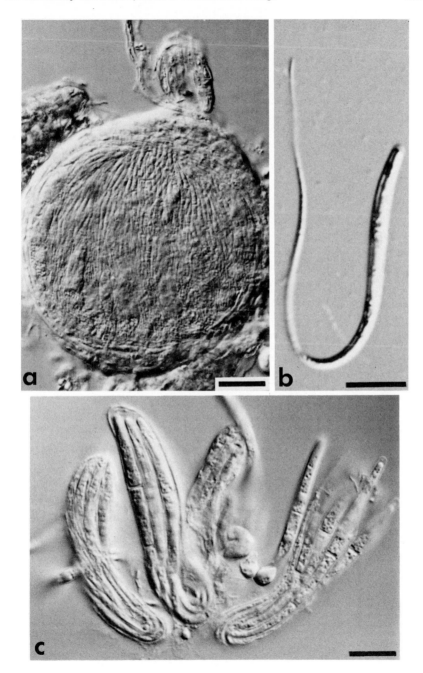

matia?) 10–16 × 0.5 μm, filiform, straight or curved, one-celled, hyaline. MODE OF LIFE: Saprobic. SUBSTRATES: Intertidal wood, test panels (e.g., *Abies* sp., *Fagus sylvatica, Pinus sylvestris, Quercus* sp.). RANGE: Atlantic Ocean—Great Britain (England, Scotland, Wales), United States (Maine, Massachusetts); Pacific Ocean—Canada (British Columbia), United States (Washington). ETYMOLOGY: Name refers to the type locality in Milford Haven (Pembrokeshire, South Wales, Great Britain).

Heleococcum Jørgensen

Bot. Tidsskr. **37,** 417, 1922

LITERATURE: Müller and von Arx (1973); Rogerson (1970).

ASCOCARPS globose, superficial, nonostiolate, membranaceous, white to orange-colored. PERIDIUM thin, transparent, pseudoparenchymatous, smooth, dissolving irregularly at maturity. PARAPHYSES absent. ASCI eight-spored, globose, sessile, unitunicate, thin-walled, scattered irregularly throughout the center of the ascocarp. ASCOSPORES ellipsoidal, one-septate, smooth, thick-walled, hyaline to orange-colored. SUBSTRATE: Wood or soil. TYPE SPECIES: *Heleococcum aurantiacum* Jørgensen. NOTE: One marine species is known besides the terrestrial type species. ETYMOLOGY: From the Greek, *helos* = marsh and *kokkos* = berry, seed, in reference to the habitat and globular ascocarps of the type species.

Heleococcum japonense Tubaki

Trans. Mycol. Soc. Jpn. **8,** 5, 1967

ASCOCARPS 250–400 μm in diameter, globose, superficial, nonostiolate, cleistothecial, membranaceous, white, later light orange-colored or pale brown, gregarious. PERIDIUM transparent, pseudoparenchymatous, composed of several layers of angular or rounded cells, dissolving at maturity. PARAPHYSES absent. ASCI 44–50 × 30–40 μm, eight-spored, globose or subglobose, sessile, unitunicate, thin-walled, aphysoclastic, without apical apparatuses, deliquescing at ascospore maturity, scattered irregularly throughout the center of the ascocarp. ASCOSPORES 18–21 × 10–13 μm, broad ellipsoidal or ovoidal, one-septate in the middle, not constricted at the septum, smooth or slightly rough, thick-walled, hyaline. CONID-IOPHORES 20–28 × 3–4 μm, mostly unbranched, erect. developing from aerial hyphae. CONIDIA 8–12 × 3–4 μm, ellipsoidal, hyaline, originating as blown-out ends of the conidiophores and accumulating in loose clusters; imperfect state of the *Trichothecium* type. MODE OF LIFE: Saprobic.

SUBSTRATE: Wooden test panels (*Abies firma* and balsa). RANGE: Pacific Ocean—Japan (known only from the original collection). NOTE: Description based on Tubaki (1967). ETYMOLOGY: From the Latin, *japonensis* = Japanese, in reference to the origin of the type material.

Hydronectria Kirschstein

Verh. Bot. Ver. Mark Brandenburg **67**, 87, 1925

LITERATURE: Müller and von Arx (1962, 1973); Rogerson (1970).

ASCOCARPS solitary or gregarious, subglobose, immersed in a crust composed of hyaline hyphae or in wood, ostiolate, periphysate, usually without papillae, fleshy–leathery, orange-brownish to orange-yellowish; peridium thick, especially around the ostiole, two-layered. APICAL PARAPHYSES absent in mature ascocarps, or present and finally deliquescing. ASCI eight-spored, clavate, ventricose or subcylindrical, unitunicate, thin-walled, deliquescing; developing all along the wall of the ascocarp venter or in the lower part of the venter. ASCOSPORES ellipsoidal, wide fusiform or ovoid, one-septate, constricted at the septum, hyaline. SUBSTRATES: On rocks in freshwater, on marine wood. TYPE SPECIES: *Hydronectria kriegeriana* Kirschstein. NOTE: The genus contains one freshwater species (*H. kriegeriana*) and one marine species. ETYMOLOGY: From the Greek, *hydor* = water, and *Nectria* Fries = a related genus, in reference to the aquatic occurrence of the species.

Hydronectria tethys J. Kohlmeyer et E. Kohlmeyer

Nova Hedwigia **9**, 95, 1965

LITERATURE: Kohlmeyer (1969c); Kohlmeyer and Kohlmeyer (1964–1969, 1971a, 1977); Lee and Baker (1973).

ASCOCARPS 275–380 μm high, 315–460 μm in diameter, subglobose or depressed-ellipsoidal, partly or rarely completely immersed, ostiolate, epapillate or clypeoid thickened around the ostiole, fleshy–leathery, orange-yellowish, gregarious or frequently confluent; the color results from orange-colored oil droplets concentrated in the apical part of the peridium and the paraphyses. PERIDIUM above and at the sides 70–75 μm thick, at the base 20–30 μm, two-layered; on the outside composed of irregularly polygonal, thick-walled cells, on the inside of elongate, flattened cells that merge into paraphyses. OSTIOLAR CANAL 17.5–32.5 μm in diameter, conical, lined with periphyses 1–2 μm in diameter. APICAL PARAPHYSES 2 μm in diameter, septate, in the dome of the venter attached to large isodiametric cells with large lumina, finally deliquescing. Asci 90–105 × 15–23 μm, eight-spored, clavate or subcylindrical, short pedunculate, unitunicate, thin-walled at maturity, aphysoclastic, without apical apparatuses, developing at the base of the ascocarp venter. Ascospores 17–26 × 8.5–13 μm, one- or biseriate, ellipsoidal or ovoid, one-septate somewhat below the center, slightly constricted at the septum, hyaline; with about six longitudinal ridges, 0.5–0.8 μm in diameter, running around the spore from one pole to the other. Mode of Life: Saprobic. Substrates: Driftwood, roots of mangroves (*Avicennia germinans, Rhizophora mangle*), wood and bark of submersed branches of *Conocarpus erecta* (J. Kohlmeyer, unpublished), *Hibiscus tiliaceus,* and *Tamarix gallica.* Range: Atlantic Ocean—Bahamas [Great Abaco (J. Kohlmeyer, unpublished)], Bermuda, United States (Florida); Indian Ocean—South Africa; Pacific Ocean—United States [Hawaii (Oahu)]. Etymology: From the Greek, *Tethys* = a sea goddess, or the sea, in reference to the marine habitat of the species.

Nectriella Nitschke *in* Fuckel

Jahrb. Nassau. Ver. Naturk. **23/24,** 175, 1869–1870 (non *Nectriella* Saccardo 1877)

=*Charonectria* Saccardo, *Michelia* **2,** 72, 1880
=*Cryptonectriella* (v. Höhnel) Weese, *Sitzungsber. Kais. Akad. Wiss. Wien, Math.-Naturwiss. Kl.* **128**(1), 715, 1919
=*Pronectria* Clements *in* Clements and Shear, *The Genera of Fungi.* Wilson, New York, p. 282, 1931

Literature: Arnold (1967); Müller and von Arx (1962); Rogerson (1970).

Ascocarps solitary or gregarious, globose, without stromata, immersed, ostiolate, papillate, light-colored. Peridium mostly thin, light-colored, fleshy, composed of concentric layers of flattened, thin-walled, hyaline or yellowish cells. Papillae semiglobose, breaking through the surface of the substrate; ostiolar canal periphysate. Paraphyses filamentous, often early deliquescing. Asci eight-spored, cylindrical, thin-walled, unitunicate, with a simple apical apparatus that does not blue in iodine. Ascospores ellipsoidal, ovoid or fusiform, one-septate near the middle, hyaline. Mode of Life: Saprobic. Type Species: *Nectriella fuckelii* Nitschke *in* Fuckel. Note: One marine species is known besides about 20 terrestrial taxa; description and synonyms after Müller and von Arx (1962); Arnold (1967) discussed the difficulties in selecting the type species of *Nectriella.* Etymology: From *Nectria,* a related genus, and the diminutive suffix *-ellus.*

Nectriella laminariae O. Eriksson

Sven. Bot. Tidsk. **58**, 233, 1964

ASCOCARPS 250–320 μm high, 200–330 μm in diameter, bottle-shaped or subglobose, immersed or erumpent, ostiolate, epapillate, fleshy, light brown or red, darker around the apex, gregarious or rarely solitary. PERIDIUM 20–30 μm thick, composed of three layers of cells; cells of outer layer thick-walled, hyaline, forming a disk around the ostiole; cells of the broad central layer elongated, colored; cells of the inner layer thin-walled, flattened, hyaline. OSTIOLAR CANAL 25–35 μm in diameter, periphysate. PARAPHYSES filamentous. ASCI 65–80 × 7–10 μm, eight-spored, cylindrical, nonpedunculate, more or less rounded at the apex, thin-walled, without apical apparatuses. ASCOSPORES 13–20 × 7–9 μm, uni- to biseriate, ellipsoidal, one-septate at the middle, constricted at the septum, smooth, hyaline. MODE OF LIFE: Saprobic. HOST: Dead stipes of *Laminaria* sp. RANGE: Atlantic Ocean—Norway (known only from the original collections). NOTE: Description after O. Eriksson (1964); we examined isotype material, kindly provided by Dr. Eriksson. ETYMOLOGY: From the host genus, *Laminaria* (Phaeophyta).

d. Polystigmataceae

Haloguignardia Cribb et Cribb

Univ. Queensl. Pap., Dep. Bot. **3**, 97, 1956

GALLS grapelike, often with fingerlike, radiating processes. ASCOCARPS subglobose or ellipsoidal, immersed in the gall processes, ostiolate, shortly papillate, coriaceous, hyaline. PARAPHYSES absent. ASCI eight-spored, ellipsoidal, clavate or cylindrical, unitunicate, thin-walled, early deliquescing. ASCOSPORES ellipsoidal to fusiform, one-celled, hyaline, with a conical, acute, chambered or striate appendage covering each end like caps; appendages persistent or deciduous. SPERMOGONIA subglobose to ellipsoidal, immersed in the gall processes, ostiolate, papillate, coriaceous, hyaline. SPERMATIOPHORES septate, filiform, lining the interior walls and lobes of the spermogonia. SPERMATIA ellipsoidal, one-celled, hyaline. HOSTS: Marine algae (Phaeophyta). TYPE SPECIES: *Haloguignardia decidua* Cribb et Cribb [not *H. longispora* Cribb et Cribb, as designated later by Cribb and Cribb (1960a)]. NOTE: All members of the genus are obligate marine species. ETYMOLOGY: From the Greek, *hals* = sea, salt, and *Guignardia* = a superficially similar genus, in reference to the marine occurrence of the fungus.

Haloguignardia decidua Cribb et Cribb

Univ. Queensl. Pap., Dep. Bot. **3**, 97, 1956

LITERATURE: Cribb and Cribb (1960b); T. W. Johnson and Sparrow (1961); Kohlmeyer (1971a, 1972c).

Fig. 64

GALLS irregular, knotty, composed of fingerlike, clavate or irregularly rounded processes; on stems and blades of the host. ASCOCARPS 170–420 μm in diameter, subglobose, immersed singly in the tissue of a gall process, ostiolate, papillate or epapillate, subcarbonaceous (?). PERIDIUM 14–40 μm thick, pseudoparenchymatous, black. PAPILLAE short, cylindrical, up to 110 μm long, up to 86 μm in diameter. PARAPHYSES absent. ASCI 52–70 × 20–32.5 μm, eight-spored, ellipsoidal or clavate, short stipitate, unitunicate, thin-walled, aphysoclastic, early deliquescing, without apical apparatuses. ASCOSPORES 38.5–48 × 5.5–9 μm (including appendages; 23–43 μm long excluding appendages), fusiform, straight or rarely curved, one-celled, tapering at the appendaged apices, after loss of appendages truncate or rounded, hyaline; appendages 3–11 μm long, deciduous, acute or rounded, conical, with delicate striae perpendicular to the long axis. SPERMOGONIA 250–320 μm high, 240–280 μm in diameter, subglobose, immersed singly in the tissue of fingerlike projections of separate composite galls, ostiolate, periphysate, shortly papillate; walls hyaline; spermatiophores lining the walls of the locule. MODE OF LIFE: Parasitic. HOSTS: *Sargassum daemelii, Sargassum* sp. RANGE: Pacific Ocean—Australia (Queensland, New South Wales). NOTE: Description partly based on our examination of the type slide (''Queensland Marine Fungi No. 110'') kindly provided by Dr. A. B. Cribb and of material from Herb. L (Herb. Lugd. Bat. No. 937, 82 . . . 28). ETYMOLOGY: From the Latin, *deciduus* = deciduous, in reference to the ephemeral ascospore appendages.

Haloguignardia irritans (Setchell et Estee *in* Estee) Cribb et Cribb

Univ. Queensl. Pap., Dep. Bot. **3**, 98, 1956

≡*Guignardia irritans* Setchell et Estee *in* Estee, *Univ. Calif., Berkeley, Publ. Bot.* **4**, 311, 1913

LITERATURE: T. W. Johnson and Sparrow (1961); Kohlmeyer (1971a, 1972c, 1974b).

Figs. 65 and 66

GALLS 3–20 mm in diameter, grapelike, containing ascocarps or spermogonia; composed of many fingerlike processes, 1–5 mm long, 0.5–1.5

Fig. 64. *Haloguignardia decidua* (type slide A.B. Cribb). (a) Thin-walled ascus with immature ascospores (bar = 10 μm). (b) Immature ascospore, appendages not developed as yet (bar = 5 μm). (c) Ascospore with striate apical appendages (bar = 5 μm). (d) Ascospore appendages dissolving (bar = 5 μm). (e) Group of eight ascospores from a dissolved ascus, showing the central nucleus in each cell (bar = 10 μm). (All in Nomarski interference contrast.)

mm in diameter, radiating from a center; on stems, blades, and vesicles of the host. ASCOCARPS 385–860 μm high, 320–900 μm in diameter, subglobose or ellipsoidal, immersed singly in the apical tissue of a gall process, ostiolate, papillate, coriaceous, hyaline. PERIDIUM 16–32 μm thick, composed of hyaline, flattened cells. PAPILLAE 115–180 μm high, 80–185 μm in diameter, cylindrical, periphysate; ostiole closed with a gelatinous substance; ostiolar canal projecting into the ascocarp cavity with a persistent pseudoparenchymatous tube composed of thin-walled cells with large lumina. PARAPHYSES absent; centers of young ascocarps filled with pseudoparenchymatous, hyaline, thin-walled, deliquescing

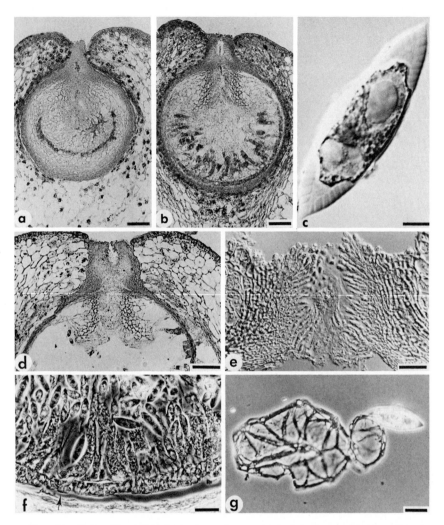

Fig. 65. *Haloguignardia irritans* from *Cystoseira osmundacea.* (a) 16-μm longitudinal section through immature ascocarp immersed in the gall tissue; centrum filled with pseudoparenchyma; the zone of ascogenous cells appears as a dark crescent (Herb. J.K. 2606; bar = 50 μm). (b) Immature ascocarp (16-μm section); young asci originate on the ascogenous tissue; a tubular extension of the ostiole projects into the ascocarp cavity (Herb. J.K. 2607; bar = 50 μm). (c) Ascospore with caplike, chambered appendages (Herb. J.K. 2606; bar = 5 μm). (d) Section (8 μm) through upper part of ascocarp with ostiole, ostiolar tube extending into ascocarp cavity (Herb. J.K. 2607; bar = 50 μm). (e) 8-μm section through ostiole showing periphyses (Herb. J.K. 2646; bar = 25 μm). (f) 16-μm section through base of mature ascocarp with peridium at the bottom; ascogenous tissue (arrow) giving rise to oblong asci; mature ascospores (above) fill venter of ascocarp (Herb. J.K. 2607; bar = 25 μm). (g) Pseudoparenchymatous cells and ascospore from ascocarp center (Herb. J.K. 2606; bar = 15 μm). (c and e in Nomarski interference contrast, f and g in phase contrast, the others in bright field.)

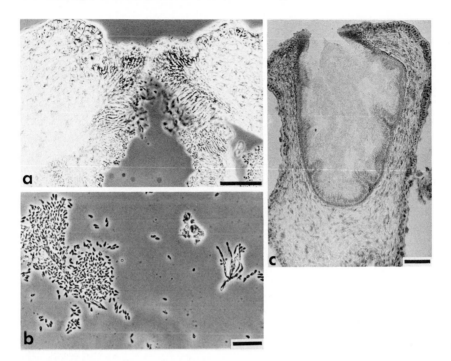

Fig. 66. *Haloguignardia irritans* from *Cystoseira osmundacea.* (a) Longitudinal section (8 μm) through ostiole of spermogonium with periphyses (Herb. J.K. 2640; bar = 50 μm). (b) Spermatiophores and spermatia (Herb. J.K. 2639; bar = 25 μm). (c) Section (6 μm) through spermogonium, showing internal lobes bearing spermatiophores (Herb. J.K. 2640; bar = 50 μm). (a and b in phase contrast, c in bright field.)

cells. Asci 130–160 × 26–32 μm, eight-spored, clavate or oblong–ventricose, unitunicate, thin-walled, aphysoclastic, early deliquescing, without apical apparatuses, developing along the inner wall of the ascocarp on an ascogenous tissue, almost up to the ostiolar tube. Ascospores (30–)34–50 × (10–)12–15 μm (including appendages), oblong–ellipsoidal to fusiform, one-celled, tapering at the appendaged apices, hyaline; appendages persistent, covering the spore apices like conical caps, divided into five to seven annular chambers, filled with mucus; distal chamber biturbinate, with an apical porus, releasing mucus. Spermogonia 420–740 μm high, 350–615 μm in diameter, ellipsoidal, immersed singly in the apices of fingerlike processes of ♂ galls, ostiolate, papillate, coriaceous, hyaline. Peridium 10–18 μm thick. Papillae 50–70 μm high, 110–200 μm in diameter, short cylindrical, periphysate; ostiole closed with a gelatinous substance. Spermatiophores 20–44 × 2 μm, simple or ramose, septate, filiform, lining the interior walls and pseudoparenchymatous lobes of the spermogonia. SPERMATIA 2.5–3.5 × 1.5–2 μm, ellipsoidal or subglobose, one-celled, with gelatinous cylindrical, basal appendages, often catenulate, hyaline. MODE OF LIFE: Parasitic. HOSTS: *Cystoseira osmundacea* and *Halidrys dioica*. RANGE: Pacific Ocean—United States (California). NOTE: Galls caused by *H. irritans* are often infected by the hyperparasitic Deuteromycete *Sphaceloma cecidii*, causing a black discoloration of the tissues. Description in part after J. Kohlmeyer (unpublished). ETYMOLOGY: From the Latin, *irritare* = stimulate, incite, provoke, in reference to the gall-producing activity of the fungus.

Haloguignardia oceanica (Ferdinandsen et Winge) Kohlm.

Mar. Biol. **8,** 344, 1971

≡*Phyllachorella oceanica* Ferdinandsen et Winge, *Mycologia* **12**, 103, 1920

LITERATURE: T. W. Johnson and Sparrow (1961); Kohlmeyer (1972c); Winge (1923).

Fig. 67

GALLS 2–5.4 mm in diameter, 3.3–8.7 mm long, grapelike, containing ascocarps or spermogonia, little constricted around the fruiting bodies; on stems of the host. ASCOCARPS 350–650 μm high, 275–510 μm in diameter, subglobose or flask-shaped, immersed in the surface of the gall, ostiolate, papillate, fleshy–leathery, with dark brown clypei, hyaline at the bases, forming a pseudostroma, gregarious. PERIDIUM 14–26 μm thick, composed of flattened cells, merging to the outside into the pseudostroma. PAPILLAE 125–280 μm high, 100–200 μm in diameter; ostiolar canal 44–60 μm in diameter, with hyaline periphysoid cells. PARAPHYSES absent; centers of young ascocarps filled with pseudoparenchymatous, hyaline, thin-walled, deliquescing cells. ASCI 50–80 × 22–32 μm, eight-spored, clavate, pedunculate, unitunicate, thin-walled, aphysoclastic, early deliquescing, without apical apparatuses, developing along the inner wall of the ascocarp on an ascogenous tissue, almost up to the ostiolar canal. ASCOSPORES 19.5–32(–36) × 9–14 μm (excluding appendages), ellipsoidal, one-celled, thick-walled, hyaline; at first fusiform, with an apiculate appendage at each end; appendages deciduous, covering the spore apices like conical caps, with delicate striae or chambers. SPERMOGONIA 210–260 μm in diameter, subglobose, immersed, wall distinctly separated from the algal tissue, ostiolate, apapillate, coriaceous, hyaline to light brown, gregarious. PERIDIUM 10–20 μm thick. OSTIOLES round, periphysate. SPERMATIOPHORES 25–47.5 × 2–4.5 μm, simple, septate, filiform, at-

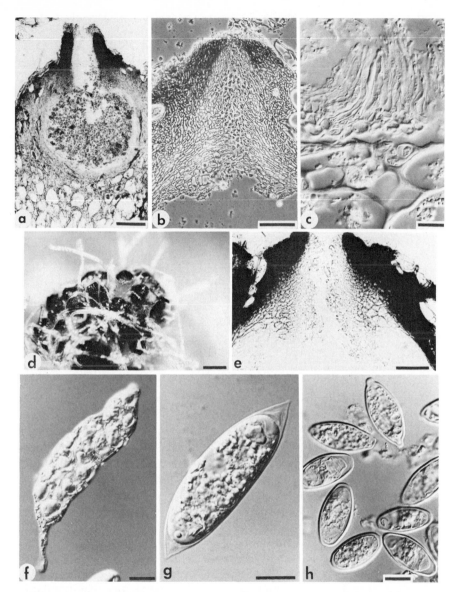

Fig. 67. *Haloguignardia oceanica* from *Sargassum* spp. (type material C). (a) Longitudinal section (16 μm) through ascocarp (bar = 100 μm). (b) Section (6 μm) through ostiole, development of ostiolar canal by dissolution of central cells (bar = 50 μm). (c) Section (8 μm) through peridium of spermogonium, spermatiophores above (bar = 10 μm). (d) Gall with black ascocarps (bar = 500 μm). (e) Section (16 μm) through ostiole of mature ascocarp (bar = 50 μm). (f) Immature ascus (bar = 10 μm). (g) Ascospore with apical appendages (bar = 5 μm). (h) Ascospores that have lost their appendages (bar = 10 μm). (a, d, and e in bright field, b in phase contrast, the others in Nomarski interference contrast.)

tenuate, lining the interior walls of the spermogonia. SPERMATIA 3–4 ×
1.5–2 μm, ellipsoidal, one-celled, hyaline. MODE OF LIFE: Parasitic.
HOSTS: *Sargassum fluitans, S. natans.* RANGE: Atlantic Ocean—
Bahamas [washed up on Great Abaco (J. Kohlmeyer, unpublished)],
Sargasso Sea, United States (washed up in North Carolina). NOTE: Galls
caused by *H. oceanica* are often infected by the hyperparasite
Sphaceloma cecidii, causing a black discoloration of the tissues; descrip-
tion in part after J. Kohlmeyer (unpublished). ETYMOLOGY: From the
Latin, *oceanicus* = pertaining to the ocean, in reference to the marine
occurrence of the species.

Haloguignardia tumefaciens (Cribb et Herbert) Cribb et Cribb

Univ. Queensl. Pap., Dep. Bot. **3**, 98, 1956

≡*Guignardia tumefaciens* Cribb et Herbert, *Univ. Queensl. Pap., Dep.
  Bot.* **3**, 9–10, 1954
=*Haloguignardia longispora* Cribb et Cribb, *Univ. Queensl. Pap., Dep.
  Bot.* **3**, 98, 1956

LITERATURE: Cribb and Cribb (1960b); T. W. Johnson and Sparrow
(1961); Kohlmeyer (1971a, 1972c, 1974b).

Fig. 68

GALLS 23 mm long, 5 mm broad, and up to 4 mm high, grapelike,
containing both ascocarps and spermogonia; composed of many fingerlike, clavate processes, 1.5–4 mm long, 0.4–1 mm in diameter, radiating
from a center; on stems and blades of the host. ASCOCARPS 400–600 μm
high, 240–500(–700) μm in diameter, subglobose to ellipsoidal, immersed
singly (or rarely as twins) in the gall processes, ostiolate, papillate,
coriaceous, hyaline. PERIDIUM 10–20 μm, near ostioles 30–40 μm thick,
composed of hyaline, flattened cells. PAPILLAE 60–130 μm high, 60–160
μm in diameter, conical; ostiolar canal 60–100 μm in diameter, filled with
periphyses, projecting slightly into the ascocarp cavity with a persistent
pseudoparenchymatous tube composed of thin-walled cells with large
lumina. PERIPHYSES 30–100 μm long, 3–6 μm in diameter, cylindrical,
thickened at the apex, sometimes extending above the ostiole. PARA-
PHYSES absent; centers of young ascocarps filled with pseudoparenchymatous, hyaline, thin-walled, deliquescing cells. ASCI 140 × 60 μm, eight-
spored, cylindrical, clavate or oblong–ventricose, unitunicate, thin-
walled, aphysoclastic, early deliquescing, without apical apparatuses.
ASCOSPORES (30–)35–70 × (10–)15–24(–26) μm (including appendages),
ellipsoidal to fusiform, one-celled, tapering at the appendaged apices,

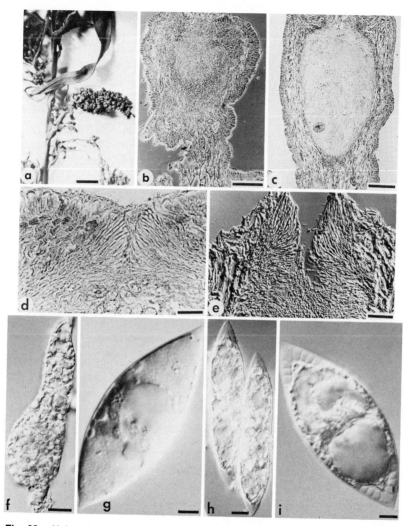

Fig. 68. *Haloguignardia tumefaciens* on *Sargassum* spp. (a) Gall (bar = 4 mm). (b) Longitudinal section (6 μm) through immature ascocarp (bar = 100 μm). (c) Section (6 μm) through spermogonium (bar = 100 μm). (d) Section (6 μm) through ostiole of ascocarp with periphyses (bar = 25 μm). (e) Section (6 μm) through ostiole of spermogonium with periphyses (bar = 25 μm). (f) Immature ascus (bar = 10μm). (g) Immature ascospore; apical appendages not formed as yet (bar = 5 μm). (h) Mature ascospores (bar = 5 μm). (i) Ascospore with caplike, chambered appendages (bar = 5 μm). (a–e from Herb. Agardhiorum 2066 and 2095 in LD; f, g, and i from type slide A.B. Cribb; h from type slide of *H. longispora*, A.B. Cribb; b in phase contrast; a, c, and d in bright field, the others in Nomarski interference contrast.)

hyaline; appendages persistent or rarely deciduous, covering the spore apices like conical caps, divided into four or five annular chambers; distal chamber biturbinate. SPERMOGONIA 340–660 μm high, 160–290 μm in diameter, elongate ellipsoidal, immersed singly in the apices of fingerlike processes of the galls, ostiolate, papillate, coriaceous, hyaline. PERIDIUM 12–16 μm thick. PAPILLAE 60–80 μm high, 80–120 μm in diameter, conical or short cylindrical, ostiolar canal 40–50 μm in diameter, filled with periphyses. PERIPHYSES 40–60 × 3–4 μm, cylindrical, apically thickened. SPERMATIOPHORES lining the interior walls of the spermogonia. SPERMATIA 3–4 × 1.5–2 μm, ellipsoidal, one-celled, hyaline. MODE OF LIFE: Parasitic. HOSTS: *Sargassum decipiens, S. fallax, S. globulariaefolium, S. sinclairii, Sargassum* sp. RANGE: Atlantic Ocean—Sargasso Sea; Pacific Ocean—Australia (New South Wales, South Australia, Tasmania), New Zealand (West Coast). NOTE: Galls caused by *H. tumefaciens* are sometimes infected by the hyperparasitic Deuteromycete *Sphaceloma cecidii,* causing a black discoloration of the tissues; description in part after J. Kohlmeyer (unpublished); we have examined the type slide of *Haloguignardia longispora,* the description of which is based on one specimen only; it cannot be distinguished from *H. tumefaciens* and is regarded as a synonym. ETYMOLOGY: From the Latin, *tumor* = tumor, swelling, and *faciens* = producing, in reference to the gall-producing ability of the species.

Key to the Species of *Haloguignardia*

1. Host *Cystoseira* or *Halidrys* . . . . **H. irritans**
1′. Host *Sargassum* spp. 2
   2(1′) Ascospore appendages persistent . . . . **H. tumefaciens**
   2′(1′) Ascospore appendages deciduous 3
3(2′) Galls with fingerlike projections; ascospores including appendages longer than 35 μm, up to 9 μm in diameter; Pacific Ocean . . . . **H. decidua**
3′(2′) Galls without fingerlike projections; ascospores including appendages commonly shorter than 35 μm, 9–14 μm in diameter; Atlantic Ocean . . . . . . **H. oceanica**

Phycomelaina Kohlm.

Phytopathol. Z. **63,** 350, 1968

PSEUDOSTROMATA forming black spots in the stems of algae. ASCOCARPS gregarious, associated with spermogonia, subglobose or ampulliform, immersed in the pseudostroma, ostiolate, epapillate, fleshy–leathery, hyaline, clypeus dark-colored. PARAPHYSES present. ASCI eight-spored, rarely six-spored, cylindrical or clavate, stipitate, unitunicate, thin-walled. ASCOSPORES ellipsoidal or cylindrical, one-septate, hyaline, with a subglobose appendage at each end. SPERMOGONIA gregarious, lentiform, immersed, ostiolate, epapillate, fleshy–leathery, hyaline, clypeus dark-colored. SPERMATIOPHORES cylindrical to conical, simple or ramose. SPERMATIA ellipsoidal, one-celled, hyaline, enteroblastic, catenate, appendaged. HOSTS: Marine algae (Phaeophyta). TYPE SPECIES: *Phycomelaina laminariae* (Rostrup) Kohlm. The genus is monotypic. ETYMOLOGY: From the Greek, *phykos* = alga and *melainein* = to blacken, in reference to the formation of blackish spots on the alga.

Phycomelaina laminariae (Rostrup) Kohlm.

Phytopathol. Z. **63**, 350, 1968

≡*Dothidella laminariae* Rostrup, *Bot. Tidssk.* **19**, 213, 1894/1895
≡*Endodothella laminariae* (Rostrup) Theissen et Sydow, *Ann. Mycol.* **13**, 582, 1915
≡*Placostroma laminariae* (Rostrup) Meyers, *Mycologia* **49**, 480, 1957
=*Hypoderma laminariae* Sutherland, *New Phytol.* **14**, 190, 1915

LITERATURE: T. W. Johnson and Sparrow (1961); Kohlmeyer (1971b); Kohlmeyer and Kohlmeyer (1964–1969); Ostenfeld-Hansen (1897); Vestergren (1900); E. E. Webber (1970).

Fig. 69

PSEUDOSTROMATA forming black, ellipsoidal spots in the cortex of algal stems, enclosing cells of the hosts. ASCOCARPS 150–295 μm high, 130–220 μm in diameter, subglobose or ampulliform, immersed in the pseudostroma, ostiolate, epapillate, fleshy–leathery, hyaline below and at the sides, with a blackish-brown clypeus, gregarious, associated with spermogonia. PERIDIUM 12–37 μm thick, composed of polygonal, hyaline, thin-walled cells, merging with the pseudostroma. PAPILLAE absent; ostioles 20–33 μm in diameter, round, with some periphysoid hyaline cells, 2.5 μm in diameter, around the opening. PARAPHYSES 2–2.5 μm in diameter, septate, early deliquescing, in young ascocarps forming a central pseudotissue that is pressed aside by developing asci. ASCI 60–90 × (10–)12–15 μm, eight-spored, rarely six-spored, cylindrical or clavate, stipitate, unitunicate, young thick-walled, mature thin-walled, aphysoclastic, without apical apparatuses; developing in the base and side of the ascocarp venter. ASCOSPORES (17.5–)20–25(–27.5) × 5.5–8 μm (excluding appendages), uni- to biseriate, ellipsoidal or cylindrical with rounded apices, one-septate, slightly or distinctly constricted at the septum, hyaline, appendaged at each end; appendages 5.5–12.5 μm in diameter, gelatinous, hyaline, at first subglobose, caplike, terminal, later becoming

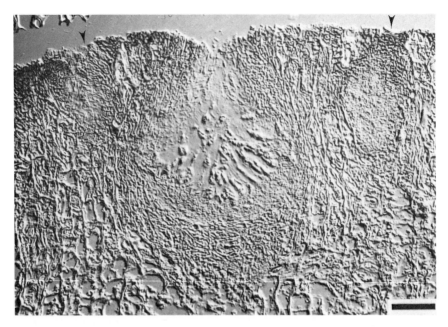

Fig. 69. Longitudinal section (6 μm) through ascocarp and spermogonia (arrowheads) of *Phycomelaina laminariae* in *Laminaria* sp. (Herb. J.K. 2505; bar = 50 μm; in Nomarski interference contrast).

soft, elongate, attenuate, up to 31 μm long. SPERMOGONIA 100–220 μm high, 52–76(–100) μm thick, 80–140 μm long, lentiform, immersed in the pseudostroma among ascocarps, with one or several locules, opening with a slitlike porus 75–150 μm long, 10–12(–17) μm wide, epapillate, fleshy-leathery, hyaline below and at the sides, with a blackish-brown clypeus, gregarious. SPERMATIOPHORES 7.5–12(–15) × 2–3 μm, cylindrical to conical, simple or ramose, lining the walls of the locule. SPERMATIA 3.5–4.5 × 1.7–2 μm, ellipsoidal, truncate or rounded below, one-celled, hyaline, enteroblastic, catenate, with a basal, short cylindrical, gelatinous appendage. MODE OF LIFE: Parasitic. HOSTS: *Laminaria agardhii, L. digitata, L. longicruris, L. saccharina;* possibly *Alaria esculenta.* RANGE: Atlantic Ocean—France, Great Britain (Ireland, Orkney Islands), Greenland, Iceland, Norway (Jan-Mayen and mainland), United States (Maine, Massachusetts, New Hampshire, Rhode Island). NOTE: Sizes given by E. E. Webber (1970) for *P. laminariae* are not included because they are about twice as high as those in all collections seen by us. ETYMOLOGY: From the host genus, *Laminaria* (Phaeophyta).

e. Sordariaceae

Biconiosporella Schaumann

Veroeff. Inst. Meeresforsch. Bremerhaven **14**, 24, 1972

ASCOCARPS solitary or gregarious, ovoid, pyriform or flask-shaped, superficial or immersed, ostiolate, periphysate, papillate, coriaceous, brown to black. PARAPHYSES absent; venter of young ascocarps filled with a pseudoparenchyma of thin-walled cells with large lumina, forming catenophyses. ASCI eight-spored, cylindrical, short stipitate, unitunicate, thin-walled, physoclastic, with an apical plate, persistent; developing at the base of the ascocarp venter. ASCOSPORES unequally biconical, triseptate, central cells dark brown to black, apical cells lighter, with a germ pore at each end, tuberculate around the equator. SUBSTRATE: Dead wood. TYPE SPECIES: *Biconiosporella corniculata* Schaumann. NOTE: The genus is monotypic and has affinities to Sordariaceae as well as Sphaeriaceae. ETYMOLOGY: From the Latin *bi-* = two, *conus* = cone, and *sporella* = little spore, in reference to the biconic ascospores.

Biconiosporella corniculata Schaumann

Veroeff. Inst. Meeresforsch. Bremerhaven **14**, 24–25, 1972
Fig. 70

ASCOCARPS 350–532 μm high (including necks), 155–297 μm in diameter, ovoid, pyriform or flask-shaped, superficial or immersed, ostiolate, papillate, coriaceous, brown to black, solitary or gregarious, with or

Fig. 70. Immature hyaline and mature dark ascospores of *Biconiosporella corniculata;* the upper left spore in view from the tip, showing the tubercules around the equator (Herb. J.K. 3117; bar = 15 μm; in Nomarski interference contrast).

without hyphae on the surface. PERIDIUM 8–17 μm thick, three-layered; outer layer prosenchymatous, appearing as textura epidermoidea in surface view; the other two layers pseudoparenchymatous; the central one with thick-walled, dark cells; the inner one with thin-walled, hyaline cells, merging into the central pseudoparenchyma. PAPILLAE or necks 71–187 μm high, 53–84 μm in diameter, obtuse–conical to cylindrical; ostiolar canal periphysate, toward the ascocarp cavity merging into pseudoparenchyma. PARAPHYSES absent; venter of young ascocarps filled with a pseudoparenchyma of thin-walled cells with large lumina, eventually breaking up into catenophyses. ASCI (155–)173–205 × (18.5–)20.5–22 μm, eight-spored, cylindrical, short stipitate, unitunicate, thin-walled, physoclastic, with an apical plate, persistent; developing at the base of the ascocarp venter. ASCOSPORES 26.5–43 × 16–28.5 μm, obliquely uniseriate, unequally biconical, unequally triseptate, thick-walled, not or slightly constricted at the septa, apices rounded or obtuse; the largest middle cell obtuse–biconical, dark brown to black, with 7–8(–9) tubercules around the equator; the second middle cell smaller, light to dark brown; apical cells usually lighter colored than the central ones, provided with a germ porus at each apex, 2–2.6 μm in diameter; cross septa with a central porus, 0.6–1(–1.9) μm in diameter; a thin, deliquescing sheath covering ascospores within the ascus. MODE OF LIFE: Saprobic. SUBSTRATE: Intertidal wood. RANGE: North Sea—Germany (Helgoland, F.R.G.); Pacific Ocean—United States (California). Description in part after J. Kohlmeyer (unpublished). ETYMOLOGY: From the Latin, *corniculatus* = with small horns, in reference to the equatorial protuberances of the ascospores.

Zopfiella Winter *in* Rabenhorst

Kryptogamenflora 1(Abt. 2), 56, 1884

LITERATURE: von Arx (1973); Huang (1973); Malloch and Cain (1971); Udagawa and Furuya (1972, 1974); Udagawa and Horie (1974); Udagawa and Takada (1974).

ASCOCARPS solitary or gregarious, globose to subglobose, superficial or rarely immersed, nonostiolate, thin-walled, irregularly dehiscing, covered with hairs. PERIDIUM pseudoparenchymatous, membranaceous, cephalothecoid in some species, cells forming a textura angularis. PARAPHYSES early deliquescing, indistinct or absent. ASCI four- to eight-spored, clavate to cylindrical or rarely subglobose, pedunculate, unitunicate, deliquescing, in some species with an apical ring, fasciculate or irregularly arranged. Ascospores uni-, bi-, or triseriate, at first one-celled, hyaline, becoming one-septate in the lower third, forming a large ellipsoidal, dark upper cell and a small, mostly cylindrical, hyaline, often collapsing basal cell; the upper cell may become divided by a horizontal septum in some species; with an apical or subapical germ pore. Mode of Life: Saprobic. Type Species: *Zopfiella tabulata* Winter. Note: Two out of twelve species are known from marine habitats. Etymology: Named in honor of F. W. Zopf (1846–1909), German mycologist, and the Latin diminutive suffix *-ellus*.

Zopfiella latipes (Lundqvist) Malloch et Cain

Can. J. Bot. **49**, 876, 1971

≡*Tripterospora latipes* Lundqvist, Bot. Not. **122**, 592, 1969

Literature: Furuya and Udagawa (1973); Lundqvist (1972); Shearer (1972b); Tubaki and Ito (1973).

Fig. 71a

Ascocarps 120–700 μm in diameter, globose or subglobose, superficial or immersed, nonostiolate, coriaceous, irregularly dehiscing, dark brown to black, covered with hyaline to grayish- or yellowish-brown, septate, branched hairs, 1.5–4 μm in diameter; solitary. Peridium 40–50 μm thick, semitransparent, composed of three or four layers of irregular or angular, thin-walled cells of 5–12 μm in diameter, forming a textura angularis. Paraphyses up to 12 μm in diameter, composed of vesicular cells, early deliquescing. Asci 80–120 × 12–18 μm, eight-spored, clavate, broadest in the middle, short pedunculate, apically truncate, unitunicate, deliquescing, with a simple apical ring, 2.1 μm wide; fasciculate. Ascospores biseriate, ellipsoidal, becoming one-septate in the lower third; slightly constricted at the septum; large upper cell 16–22(–25) × 10–13 (–15) μm, ellipsoidal, apex conical or umbonate, base truncate, olivaceous to brown, thin-walled, smooth, with a subapical germ pore, 1 μm in diameter; small lower cell (pedicel) 4–8(–9) μm long, 3.5–7(–8) μm in diameter, broadly cylindrical, apex truncate, base broadly rounded, hyaline, at maturity without cytoplasm; the base and one side of the pedicel thin-walled, collapsing, and giving it a cuplike shape; collapsed pedicel appearing triangular in lateral view. Mode of Life: Saprobic. Substrates: Submerged wood panels (e.g., balsa wood) and driftwood under mangroves. Range: Atlantic Ocean—United States (Maryland); Indian Ocean—India [Andhra Pradesh and Kerala State (J. Kohlmeyer, unpublished)], South Africa (J. Kohlmeyer, unpublished); Pacific

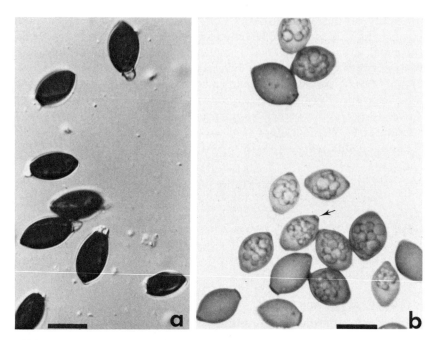

Fig. 71. Ascospores of *Zopfiella* spp. (a) *Z. latipes*, basal hyaline cells mostly collapsed (Herb. J.K. 2631). (b) *Z. marina*, basal cells collapsed, arrow pointing at germ pore (type culture NHL 2731). (Bars = 15 μm; a in Nomarski interference contrast, b in bright field.)

Ocean—Japan. NOTE: The foregoing description is partly based on examination of own material (J. Kohlmeyer, unpublished); Malloch and Cain (1971) have transferred all *Tripterospora* species with a broad, thick-walled, late differentiated basal ascospore cell, including *T. latipes*, to *Zopfiella*; Lundqvist (1972) believes that *Tripterospora* and *Zopfiella* might perhaps better be united, but he chooses to separate them in the traditional way, namely, on the basis of septation of the apical ascospore cell, even though he considers this a dubious generic character; in addition to marine substrates *Z. latipes* was isolated from terrestrial habitats in Denmark, Jamaica, and Japan (Furuya and Udagawa, 1973; Lundqvist, 1969, 1972) and, therefore, is accepted as a facultative marine species; Lundqvist (1969) found a *Humicola* imperfect state in pure cultures of *Z. latipes*, whereas Furuya and Udagawa (1973) described a different imperfect form in cultures of their isolates. ETYMOLOGY: From the Latin, *latus* = broad and *pes* = foot, in reference to the broad basal ascospore cell.

Zopfiella marina Furuya et Udagawa

J. Jpn. Bot. **50**, 249, 1975

Fig. 71b

ASCOCARPS 180–450 μm in diameter, globose to subglobose, superficial or immersed, nonostiolate, coriaceous, irregularly dehiscing, black, almost glabrous or loosely covered by some curved, hyaline, septate, simple hyphae, 2–2.5 μm in diameter at the base; solitary. PERIDIUM 15–28 μm thick, semitransparent, composed of about four layers of angular, thin-walled cells of 5–14 μm in diameter, forming a textura angularis. PARAPHYSES composed of vesicular cells, early deliquescing. ASCI 75–90 × 14–20 μm, eight-spored, clavate, broadest in the middle, short pedunculate, unitunicate, deliquescing, with an indistinct apical ring. ASCOSPORES biseriate, clavate, becoming one-septate in the lower third; slightly constricted at the septum; large upper cell (14–)15–20 × 10–13(–14) μm, ellipsoidal, apex slightly umbonate, base truncate, olivaceous to dark brown, thin-walled, smooth, with an apical germ pore, about 1 μm in diameter; small lower cell (pedicel) 8–10 μm long, 3–4 μm in diameter, elongate cylindrical, apex truncate, base rounded, straight or slightly curved, collapsing at age; collapsed pedicel appearing more or less truncate in lateral view. CONIDIA in culture 4–6 × 2.5–3 μm, ovate to elongate, one-celled, hyaline, smooth, solitary, monoblastic. MODE OF LIFE: Saprobic. SUBSTRATE: Isolated from marine sediment at depth of 120 m. RANGE: Pacific Ocean—East China Sea (known only from the type collection). NOTE: The description is based on Furuya and Udagawa (1975) and on our examination of the type culture, NHL 2731, kindly supplied by Dr. Udagawa. ETYMOLOGY: From the Latin, *marinus* = marine, in reference to the origin of the type isolate.

Key to the Marine Species of *Zopfiella*

1. Ascospore with subapical germ pore; basal cell broadly cylindrical, up to 7–8 μm in diameter . . . . **Z. latipes**
1'. Ascospore with apical germ pore; basal cell elongate cylindrical, up to 4 μm in diameter . . . . **Z. marina**

f. Sphaeriaceae

Abyssomyces Kohlm.

Ber. Dtsch. Bot. Ges. **83**, 505, 1970 (publ. 1971)

ASCOCARPS solitary, subglobose or pyriform, superficial, ostiolate, periphysate, papillate, coriaceous, brown, setose. PARAPHYSES absent. ASCI eight-spored, subcylindrical or fusiform, stipitate, unitunicate, thin-walled, persistent; developing at the base of the ascocarp venter. ASCOSPORES subcylindrical, triseptate, hyaline, with a semiglobose appendage at each end. SUBSTRATE: Marine invertebrates (hydrozoa). TYPE SPECIES: *Abyssomyces hydrozoicus* Kohlm. The genus is monotypic. ETYMOLOGY: From the Greek, *abyssos* = the abyss and *mykes* = fungus, in reference to the deep-sea occurrence of the fungus.

Abyssomyces hydrozoicus Kohlm.

Ber. Dtsch. Bot. Ges. **83**, 505–506, 1970 (publ. 1971)
Fig. 72

ASCOCARPS 130–155 μm high (excluding necks), 135–155 μm in diameter, subglobose or pyriform, superficial, ostiolate, papillate, coriaceous, light brown, solitary, setose. PERIDIUM 12–20 μm thick, three-layered; with a crustose, light brown, noncellular film on the outside, hyaline and cellular on the inside; outer cell layer composed of irregular, truncate-ellipsoidal cells, giving rise to rigid, brown setae; setae 8–60 μm long,

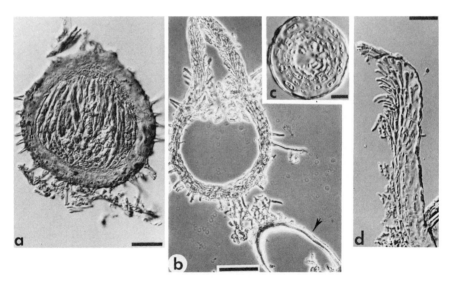

Fig. 72. *Abyssomyces hydrozoicus* from hydrozoan exoskeleton (isotype Herb. J.K. 2754, IMS). (a) Longitudinal section (8 μm) through ascocarp (bar = 25 μm). (b) 6-μm section through empty ascocarp showing attachment of hyphae to hydrozoan tube (arrow) (bar = 50 μm). (c) Cross section (6 μm) through neck of ascocarp (bar = 10 μm). (d) Longitudinal section (6 μm) through neck with periphyses (bar = 15 μm). (b in phase contrast, the others in Nomarski interference contrast.)

3.3–4.8 μm in diameter at the base, 2–2.2 μm at the apex, awl-shaped, simple or one-septate; cells of the inner wall layer flattened. PAPILLAE or necks 95–120 μm high, 40–57 μm in diameter, subconical or cylindrical; ostioles periphysate. PARAPHYSES absent. ASCI 60–88 × 6–9 μm, eight-spored, subcylindrical or fusiform, short stipitate, unitunicate, thin-walled, aphysoclastic, without apical apparatuses, persistent; developing at the base of the ascocarp venter. ASCOSPORES 18–20 × 3.5–4 μm (excluding appendages), biseriate, subcylindrical, straight or slightly curved, triseptate, not constricted at the septa, hyaline, appendaged at each end, sometimes surrounded by a mucous cover; appendages 0.5–1 μm in diameter, hyaline, semiglobose, caplike, terminal. MODE OF LIFE: Parasitic (or saprobic?). SUBSTRATES: Chitinous hydrorhiza and hydrocaulon of hydrozoa. RANGE: Atlantic Ocean—Open sea (near South Orkney Islands; known only from the original collection). ETYMOLOGY: From the Greek, *hydra* = polyp and *zoon* = animal, and the suffix -*icus* = belonging to, in reference to the substrate, namely, hydrozoa.

Chaetosphaeria L. R. Tulasne

Selecta Fung. Carp. **2**, 252, 1863

=*Chaetolentomita* Maubl. apud Maubl. et Rang., *Bol. Agric. São Paulo* **16**, 313, 1915
=*Didymopsamma* Petrak, *Ann. Mycol.* **23**, 80, 1925
=*Lentomita* Niessl. *Verh. Naturf. Ver. Brünn* **14**, 44, 1876
=*Melanopsamma* Niessl emend. Sacc., *Michelia* **1**, 347, 1878
=*Montemartinia* Curzi, *Atti Ist. Bot. Univ. Pavia, Ser. 3*, **3**, 84, 1927
=*Urnularia* Karst. teste Sacc., *Syll. Fung.* **15**, 439, 1901

LITERATURE: Booth (1957, 1958); Müller and von Arx (1962).
ASCOCARPS solitary or gregarious, globose, superficial, sometimes on a thin stroma or anchored basally in the substrate, often subiculate, ostiolate, periphysate, papillate, carbonaceous, black, usually developing among conidial fructifications. PARAPHYSES filamentous, apically free and often deliquescing at maturity. ASCI eight-spored, clavate, unitunicate, thin-walled, wall thickened at the apex and enclosing an apical plate, persistent, developing laterally and basally along the wall in the ascocarp venter. ASCOSPORES cylindrical to broadly fusiform, one-septate to phragmosporous, hyaline. SUBSTRATE: Old wood and bark. TYPE SPECIES: *Chaetosphaeria innumera* L. R. Tulasne. NOTE: Description and synonyms after Müller and von Arx (1962); only one marine species is known among about 20 species. ETYMOLOGY: From the Greek, *chaite* = long hair, bristle, and *Sphaeria* = a genus of Pyrenomycetes (*sphaera* =

globe), in reference to the globose ascocarps seated amidst bristlelike conidiophores.

Chaetosphaeria chaetosa Kohlm.

Nova Hedwigia **6,** 307–308, 1963

LITERATURE: Koch (1974); Kohlmeyer (1967, 1971b); Kohlmeyer and Kohlmeyer (1964–1969); Schaumann (1969).

ASCOCARPS 170–275 μm high, up to 275 μm in diameter, subglobose or pyriform, immersed or partly immersed, ostiolate, papillate, subcoriaceous, dark brown or black, anchored with brown hyphae in the substrate, solitary or gregarious. PERIDIUM 11–30 μm thick, on the outside composed of irregular, thick-walled cells with dark incrustations, toward the interior merging into three or four layers of hyaline, flattened, thin-walled cells, forming a textura angularis. PAPILLAE or necks 55–385 μm long, 20–70 μm in diameter, subcylindrical; ostiolar canal periphysate. PARAPHYSES 1.5–4.5 μm in diameter, septate, ramose, apically free, cohering and forming a subgelatinous ball that encloses the asci. ASCI 92–135 × 12–18 μm, eight-spored, cylindrical to clavate, basally attenuate, pedunculate, unitunicate, thin-walled, thickened at the apex and enclosing a disk-shaped, perforated apical apparatus, persistent; developing at the base of the ascocarp venter. ASCOSPORES 24–36.5 × 6–11.5 μm, biseriate, fusiform or elongate–ellipsoidal, three-septate, constricted at the septa, hyaline, peritrichous around the central septum with 10–15 flexible setae, 6–11 μm long. MODE OF LIFE: Saprobic. SUBSTRATES: Intertidal and drifting wood, palm trunk. RANGE: Atlantic Ocean— Denmark, Germany (Helgoland, F.R.G.), Spain (Tenerife and mainland), United States (Massachusetts); Black Sea—Bulgaria. ETYMOLOGY: From the Greek, *chaite* = long hair, bristle, and the Latin suffix, *-osus* = indicating abundance or full or marked development, in reference to the setose ascospores.

Pontogeneia Kohlm.

Bot. Jahrb. **96,** 200–201, 1975

ASCOCARPS solitary or gregarious, subglobose or ovoid, superficial or partly immersed, ostiolate, papillate or epapillate, coriaceous, dark-colored. PARAPHYSES distinct, thick, septate, or indistinct. ASCI eight-spored, clavate or fusiform, unitunicate, thin-walled, early deliquescing, attached in the lower part of the locule. ASCOSPORES filiform, rarely fusoid–ellipsoidal or ellipsoidal, often curved, phragmosporous, hyaline. HOSTS: Marine algae (Chlorophyta and Phaeophyta). TYPE SPECIES: *Pontogeneia padinae* Kohlm. NOTE: All members of the genus are obligate marine species. ETYMOLOGY: From the Greek, *pontogeneia* = the ocean-borne one, in reference to the marine origin of the genus and its species.

Pontogeneia calospora (Patouillard) Kohlm.

Bot. Jahrb. **96,** 205, 1975

≡*Zignoella calospora* Patouillard, *J. Bot., Paris* **11,** 242, 1897

Fig. 73

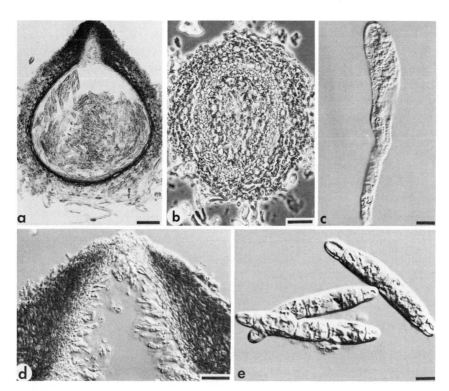

Fig. 73. *Pontogeneia calospora* from *Castagnea chordariaeformis* (Herb. J.K. 3073). (a) Longitudinal section (12 μm) through mature ascocarp filled with ascospores (bar = 50 μm). (b) Section (6 μm) through immature ascocarp (bar = 25 μm). (c) Immature ascus (bar = 10 μm). (d) Section (12 μm) through ostiole with periphyses (bar = 25 μm). (e) Ascospores (bar = 10 μm). (a in bright field, b in phase contrast, the others in Nomarski interference contrast.)

ASCOCARPS 330–480 μm high, 300–390 μm in diameter, subglobose, superficial or with immersed base, ostiolate, papillate, coriaceous, light to dark brown, solitary or gregarious. PERIDIUM 40–54 μm (60 μm at the ostiole) thick, two-layered, sometimes enclosing filaments of the host; inner layer composed of flat, elongate cells; outer layer composed of more or less loosely entwined hyphae; both layers easily separable. PAPILLAE short, conical; ostiolar canal 40–100 μm in diameter, conical, periphysate. PERIPHYSES 4.5–5 μm in diameter, at the top of the canal 2.5 μm in diameter. PARAPHYSES indistinct. ASCI 100–130 × 30–35 μm, eight-spored, clavate, rounded at the apex, tapering at the base, unitunicate, thin-walled, aphysoclastic, early deliquescing, without apical apparatuses, developing in the base of the ascocarp venter. ASCOSPORES (54–)62–85 × 8–12(–14) μm, cylindrical, straight or slightly curved, phragmosporous, 3–5(–6)-septate, slightly constricted at the septa, rounded at the apices, hyaline; diameter to length ratio 1:5–11. MODE OF LIFE: Parasitic. HOST: *Castagnea chordariaeformis* (Phaeophyta). RANGE: Atlantic Ocean—France, Spain. ETYMOLOGY: From the Greek, *calos* = beautiful and *spora* = spore, in reference to the ascospores.

Pontogeneia codiicola (Dawson) J. Kohlmeyer et E. Kohlmeyer comb. nov.

≡*Sphaerulina codiicola* Dawson, *Occas. Pap. Allan Hancock Found.* **8,** 20–21, 1949 (as "*S. codicola*"; see *Index of Fungi* **4,** 185, 1973), basionym.

Figs. 74a–74e

ASCOCARPS 300–435 μm high, 265–450 μm in diameter, pyriform, subglobose or ellipsoidal, immersed among utricles of the host, ostiolate, papillate, coriaceous, shiny, dark brown, solitary; surrounded by branched, septate, hyaline nutritional hyphae, 3–6 μm in diameter. PERIDIUM 16–26 μm thick, composed of four to six layers of brown, elongate cells with large lumina, forming a textura angularis. PAPILLAE 125–375 μm long, 60–120 μm in diameter, cylindrical, dark brown with a hyaline tip, projecting above the surface of the algae; ostiolar canal lined with periphyses, 3–4 μm in diameter, two- to three-septate. PARAPHYSES 8–10 μm in diameter, filamentous, septate, standing between and around the asci. ASCI 105–125 × 25–42 μm, eight-spored, at first ellipsoidal, then clavate, short pedunculate, unitunicate, thin-walled, aphysoclastic, deliquescing, without apical apparatuses; developing successively at the base of the ascocarp venter. ASCOGENOUS HYPHAE 3–6 μm in diameter, refractive, separated from the basal peridium by a small-celled tissue. ASCOSPORES 52–79 × 20–28 μm, elongate–ellipsoidal to cylindrical with

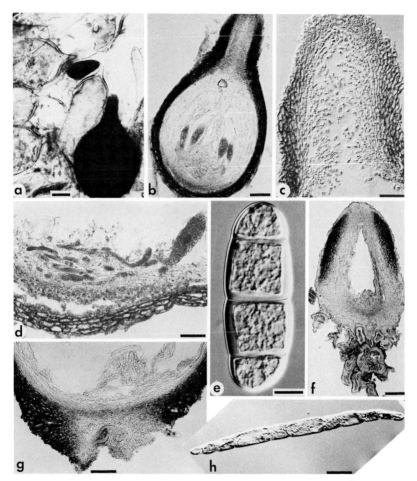

Fig. 74. *Pontogeneia* spp. (a) Ascocarp of *P. codiicola* between pointed utricles of *Codium fragile* (bar = 100 μm). (b) Longitudinal section (16 μm) through mature ascocarp of *P. codiicola (bar = 50 μm)*. (c) Section (4 μm) thrcugh ostiole of *P. codiicola* with periphyses (bar = 20 μm). (d) Section (4 μm) through base of ascocarp of *P. codiicola*; ascogenous hyphae and immature asci stained dark in violamin (bar = 25 μm). (e) Ascospore of *P. codiicola* (bar = 10 μm). (f) Longitudinal section (8 μm) through immature ascocarp of *P. cubensis* on *Halopteris scoparia* (bar = 50 μm). (g) Section (8 μm) through subiculate base of ascocarp of *P. cubensis* (bar = 50 μm). (h) Ascospore of *P. cubensis* (bar = 25 μm). (a–e holotype Dawson 4766b, AHFH; f–h type PC; c, e, and h in Nomarski interference contrast, the others in bright field.)

rounded ends, often curved, three-septate, not or slightly constricted at the septa, hyaline; diameter to length ratio 1:2–4. MODE OF LIFE: Parasitic. HOSTS: *Codium fragile* and *C. simulans* (Chlorophyta). RANGE: Pacific Ocean—Mexico (Baja California), United States (California). NOTE: Description based on our examination of the holotype from Herb. AHFH (Dawson 4766b) and other authentical material of Dawson (AHFH Nos. 36417, 36516, 57399); in the diagnosis, Dawson (1949) mentioned the absence of paraphyses, which, however, are definitely present in the material examined; annulations of ascospores described by Dawson are artifacts; *P. codiicola* had to be removed from *Sphaerulina*, a genus of bitunicate Ascomycetes (von Arx and Müller, 1975); instead, *P. codiicola* belongs in Sphaeriales, Sphaeriaceae and can best be accommodated in *Pontogeneia*. ETYMOLOGY: From the host genus, *Codium*, and the Latin suffix, *-cola* = -dweller, in reference to the substrate of the species.

Pontogeneia cubensis (Hariot et Patouillard) Kohlm.

Bot. Jahrb. **96**, 207, 1975

≡*Zignoella cubensis* Hariot et Patouillard, *Bull. Soc. Mycol. Fr.* **20**, 65, 1904

Figs. 74f–74h

ASCOCARPS 520–840 μm high, 400–520 μm in diameter, ovoid to obpyriform, superficial, subiculate, ostiolate, epapillate, coriaceous to subcarbonaceous, black, sometimes lighter colored at the base, solitary or gregarious. PERIDIUM 60–80 μm (at the ostiole 50 μm) thick, two-layered; on the outside composed of large brown, polygonal, subglobose or irregular cells, 8–14 μm in diameter, with large lumina, merging toward the interior into hyaline, compressed cells with narrow lumina. PARAPHYSES or PSEUDOPARENCHYMA not observed. ASCI deliquesce early and are not preserved in the type material. ASCOSPORES 215–325 × 13.5–17(–21) μm, 7–9 μm in diameter at the tips, filiform, curved, phragmosporous, (10–)12–13-septate, slightly constricted at the septa, tapering and rounded at the apices, hyaline; diameter to length ratio 1:(10–)13–24. MODE OF LIFE: Parasitic. HOST: *Halopteris scoparia* (Phaeophyta). RANGE: Atlantic Ocean—Cuba (known only from the original collection). ETYMOLOGY: From Cuba, and the Latin suffix *-ensis,* indicating the origin of the type material.

Pontogeneia enormis (Patouillard et Hariot) Kohlm.

Bot. Jahrb. **96**, 208, 1975

≡*Zignoella enormis* Patouillard et Hariot, *J. Bot., Paris* **17**, 228, 1903

Ascocarps 650–800 μm high, 400 μm in diameter, ovoid to obpyriform, superficial, ostiolate, epapillate, coriaceous, dark brown to black, solitary, easily detached. Peridium 50–80 μm thick, two-layered; on the outside composed of large, brown, irregular cells with large lumina, merging toward the interior into hyaline, compressed cells with narrow lumina. Paraphyses indistinct. Asci eight-spored, early deliquescing and not preserved in the type material, long clavate, at the apex obtusely rounded, tapering at the base. Ascospores 280–350 × 12–14 μm, filiform, more or less curved, phragmosporous, four- to five-septate, not constricted at the septa, tapering and rounded at the apices, hyaline; diameter to length ratio 1:20–29. Mode of Life: Parasitic. Host: *Halopteris scoparia* (Phaeophyta). Range: Atlantic Ocean—Spain (known only from the original collection). Etymology: From the Latin, *enormis* = very large, in reference to the long ascospores.

Pontogeneia padinae Kohlm.

Bot. Jahrb. **96**, 201, 1975
Fig. 75

Ascocarps 340–440 μm high, 320–450 μm in diameter, subglobose, superficial on the base of the algae, but surrounded and covered by the basal filaments of the host, ostiolate, papillate or rarely epapillate, coriaceous, black, sometimes with dark- or light-brown bases, solitary or rarely gregarious. Peridium 32–52 μm thick, composed of large thin-walled cells (8–44 × 4–12 μm) with large lumina, merging at the outside into hyphae, enclosing algal filaments, sometimes even in the neck. Papillae or necks 140–650 μm long, 80–130 μm in diameter, periphysate. Periphyses 4–5 μm in diameter. Paraphyses at first filling the ascocarp, later pushed aside by developing asci and standing predominantly along the wall of the venter; septate, constricted at the septa, simple or rarely forked, 7–15 μm in diameter, composed of cells 16–85 μm long. Asci 140–200 × 25–46 μm, eight-spored, fusiform to clavate, stipitate, unitunicate, thin-walled, aphysoclastic, early deliquescing, without apical apparatuses, developing in the base of the ascocarp venter. Ascogenous tissue composed of thin-walled, pseudoparenchymatous cells, 6–8 × 3–4 μm. Ascospores 99–167(–172) × 10–12 μm, filiform, straight or slightly curved, phragmosporous, 6–9(–10)-septate, not or slightly constricted at the septa, rounded at the apices, hyaline, containing one or two large oil globules in each cell; diameter to length ratio 1:8–17. Mode of Life: Parasitic (or symbiotic?). Host: *Padina durvillaei* (Phaeophyta). Range: Pacific Ocean—Mexico (Gulf of California; known only from the original collection). Etymology: Name refers to the host genus, *Padina*.

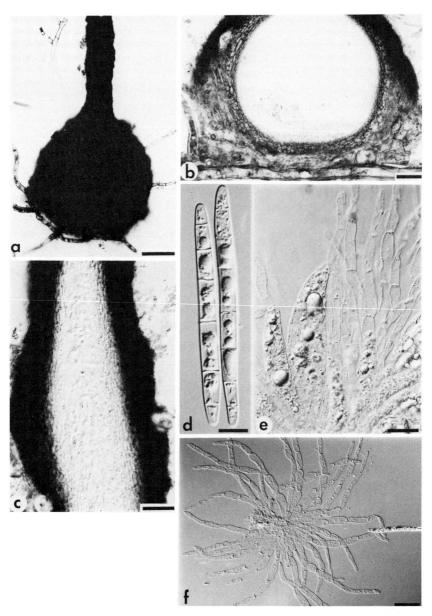

Fig. 75. *Pontogeneia padinae* on *Padina durvillaei* from type NY and IMS. (a) Whole mount of ascocarp between algal filaments (bar = 100 μm). (b) Longitudinal section (20 μm) through ascocarp attached to host (bar = 50 μm). (c) Section (20 μm) through neck of ascocarp with periphyses (bar = 25 μm). (d) Ascospores (bar = 15 μm). (e) Paraphyses and immature asci (bar = 25 μm). (f) Bundle of paraphyses, ascospore on the right (bar = 50 μm). (a and b in bright field, the others in Nomarski interference contrast.)

Pontogeneia valoniopsidis (Cribb et Cribb) Kohlm.

Bot. Jahrb. **96**, 209, 1975

≡*Zignoella valoniopsidis* Cribb et Cribb, *Univ. Queensl. Pap., Dep. Bot.* **4**, 41, 1960

Thin, black mycelial layer covering basal parts of filaments of the host. ASCOCARPS 730 μm in diameter, subglobose, superficial, subiculate, ostiolate, with a short papilla, subcarbonaceous, black, solitary. PARAPHYSES 4–7 μm in diameter, multiseptate. ASCI four-spored (?), ellipsoidal, early deliquescing. ASCOSPORES 72–100 × 21–30 μm, fusoid–ellipsoidal, straight or slightly curved, phragmosporous, two- to five-septate, not constricted at the septa, hyaline; end cells usually with homogenous contents, the others multivacuolate; diameter to length ratio 1:2.5–5. MODE OF LIFE: Parasitic. HOST: *Valoniopsis pachynema* (Chlorophyta). RANGE: Pacific Ocean—Australia (Queensland; known only from the original collection). NOTE: Description after Cribb and Cribb (1960a). ETYMOLOGY: Name refers to the host genus, *Valoniopsis*.

Key to the Species of *Pontogeneia*

1. Ascospore diameter 20 μm or more; hosts *Codium* or *Valoniopsis* 2
1'. Ascospore diameter below 20 μm; different host genera 3
  2(1) Ascospores up to 79 μm long, three-septate; on *Codium* . . . . ***P. codiicola***
  2'(1) Ascospores up to 100 μm long, two- to five-septate; on *Valoniopsis*
  . . . . ***P. valoniopsidis***
3(1') Ascospores longer than 200 μm; host *Halopteris* 4
3'(1') Ascospores shorter than 200 μm; different host genera 5
  4(3) Ascospores with ten or more septa . . . . ***P. cubensis***
  4'(3) Ascospores with five or less septa . . . . ***P. enormis***
5(3') Ascospores three- to six-septate, normally shorter than 85 μm; host *Castagnea* . . . . ***P. calospora***
5'(3') Ascospores normally six- to nine-septate, longer than 95 μm; host *Padina*
  . . . . ***P. padinae***

g. Verrucariaceae

Pharcidia Koerber

Parerga Lichenologica, Breslau, p. 469–470, 1865

=*Epicymatia* Fuckel, *Symb. Mycol.* p. 118, 1869

LITERATURE: Janex-Favre (1965); Müller and von Arx (1962).
ASCOCARPS solitary, small, globose to ellipsoidal, often apically flattened, ostiolate, periphysate, epapillate, dark. PERIDIUM particularly thick around the ostiole, composed of thick-walled, subglobose, or, near

the center, flattened cells. PARAPHYSES numerous, often hanging downward into the ascocarp cavity. ASCI four- to eight-spored, elongate–clavate, weakly pedunculate, thick-walled, unitunicate. ASCOSPORES ellipsoidal or elongate, one-septate in the center, hyaline. MODE OF LIFE: Mostly parasitic in lichens. TYPE SPECIES: *Pharcidia congesta*. NOTE: Description after Müller and von Arx (1962); three marine-occurring species are recognized besides about fifteen terrestrial taxa; *P. pelvetiae* Sutherland (1915a) is a doubtful species that has not been collected again since its original description, and type material is missing. ETYMOLOGY: From the Greek, *pharkis* = wrinkle, in reference to the wrinkly appearance of the apothecium of the host lichen, caused by the parasite, *P. congesta*.

Pharcidia balani (Winter) Bauch

Pubbl. Stn. Zool. Napoli **15**, 379, 1936

≡*Epicymatia balani* Winter *in* Hariot, *J. Bot., Paris* **1**, 233–234, 1887
≡*Didymella balani* (Winter) Feldmann, *Trav. Stn. Biol. Roscoff, Suppl.* **6**, 136, 1954
≡*Melanopsamma balani* (Winter) Meyers, *Mycologia* **49**, 485, 1957
=*Didymella conchae* Bonar, *Univ. Calif., Berkeley, Publ. Bot.* **19**, 188, 1936
=*Pharcidia marina* Bommer, *Bull. Soc. Belge Microsc.* **17**, 151, 1891
[In the lichenological literature treated as a lichen: *Arthopyrenia sublitoralis* (Leight.) Arnold, *Ber. Bayer. Bot. Ges.* **1**, Appendix, 1891.]

LITERATURE: Böttger (1967, 1969); J. Feldmann (1937); Henssen and Jahns [1974 (sub *Arthopyrenia halodytes*)]; T. W. Johnson and Sparrow (1961); von Keissler [1937 (sub *Thelidium litorale*)]; Klement and Doppelbaur (1952); Kohlmeyer (1967, 1969b); Kohlmeyer and Kohlmeyer (1964–1969); Santesson (1939); Schaumann (1969).

Fig. 76a

ASCOCARPS 130–200 μm high, 215–345 μm in diameter, subglobose, partly or completely immersed, ostiolate, epapillate, black and carbonaceous above and at the sides, pallid and coriaceous below, solitary or gregarious, rarely two of them joined; with or without algal phycobionts. PERIDIUM 35–95 μm thick in the involucrellum, 8–12 μm thick in the

Fig. 76. *Pharcidia* spp. (a) Fruiting bodies of *P. balani* in the test of *Chthamalus montagui* (Herb. J.K. 3837; bar = 500 μm). (b) Ascus of *P. laminariicola,* gelatinous caps (arrow) on ascospores (Herb. J.K. 3118; bar = 10 μm). (c) Longitudinal section (8 μm) through ascocarp of *P. laminariicola* on *Laminaria digitata* (Herb. J.K. 3118; bar = 25 μm). (d) Section (6 μm) through ascocarp of *P. rhachiana* on *L. digitata* (Herb. J.K. 3119; bar = 25 μm). (e) Ascospores of *P. rhachiana* (Herb. J.K. 3119; bar = 5 μm). From

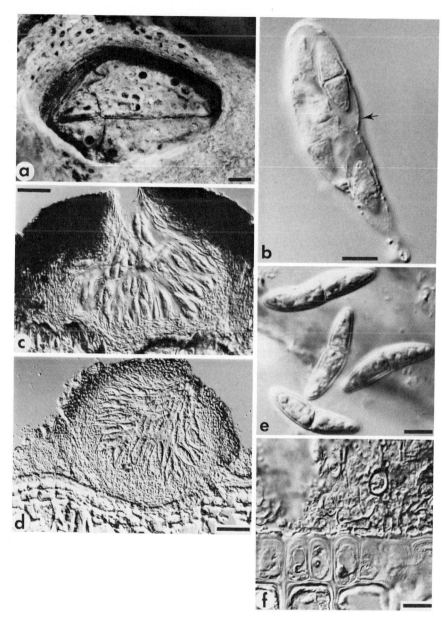

Kohlmeyer, *Bot. Mar.* **16,** 211 (1973), Fig. 31. Reprinted with permission. (f) Section (2 μm) through base of ascocarp of *P. rhachiana* showing the distinct separation of the fungus from the supporting host, *L. digitata* (Herb. J.K. 3119; bar = 10 μm). (b–f types NY and IMS; all in Nomarski interference contrast.)

basal excipulum, composed of irregular cells with small lumina. PARA-
PHYSES 1–2 μm in diameter, septate, ramose, reticulate, with gelatinous
walls, enclosing the asci firmly, projecting into the wide ostiole. ASCI
60–85(–93) × 10–25 μm, eight-spored, clavate to cylindrical, short
pedunculate, unitunicate, thick-walled and broadly rounded at the apex,
without apical apparatuses, persistent; developing at the base of the
ascocarp venter. ASCOSPORES 12–23(–27) × 5–9 μm, obliquely biseriate,
ellipsoidal or obovoid, one-septate near the center, not or slightly constricted at the septum, hyaline. PYCNIDIA (or spermogonia?) 85–140 μm
in diameter, globose or subglobose, partly or completely immersed, ostiolate, epapillate or shortly papillate, carbonaceous, black, gregarious.
CONIDIOPHORES (or spermatiophores?) 10–15 μm long, cylindrical, simple, lining the walls of the locule. CONIDIA (or spermatia?) 2.5–3.5 ×
1–1.5 μm, ellipsoidal, one-celled, hyaline, catenulate. MODE OF LIFE:
Saprobic or facultatively symbiotic with various species of microscopic
algae. HOSTS: Calcareous shells of marine animals, namely, mollusks and
cirripedes (*Acmaea, Balanus, Burnupena, Cerithium, Chthamalus, El-
minius, Fissurella, Helcioniscus, Lithotrys, Littorina, Mitella, Murex,
Nacella, Nerita, Nucella, Octomeris, Oxystele, Patella, Patinella,
Planaxis, Pollicipes, Risella, Tegula, Tetraclita*). RANGE: Arctic
Ocean—Norway, U.S.S.R.; Atlantic Ocean—Argentina (Buenos Aires,
Tierra del Fuego), Brazil (Rio de Janeiro, São Paulo), Canada [New
Brunswick, Newfoundland (J. Kohlmeyer, unpublished)], Denmark
(mainland and Faeroe Islands), Falkland Islands, France, Germany (main-
land and islands of Baltrum, För, Helgoland, and Juist, F.R.G.), Great
Britain [England, Ireland, Scotland (J. Kohlmeyer, unpublished), Wales],
Iceland, Norway, Portugal, South Africa, Spain (mainland and Canary
Islands), St. Helena, Sweden, United States [Maine (J. Kohlmeyer, un-
published), Massachusetts, Virginia]; Indian Ocean—Kenya, Mauritius,
Réunion, South Africa; Mediterranean—France [mainland (J.
Kohlmeyer, unpublished) and Corsica], Italy (Sardinia), Yugoslavia (J.
Kohlmeyer, unpublished); Pacific Ocean—Australia (New South Wales,
Tasmania), Canada (British Columbia), Chile (Chiloe, Guaitecas Island,
Tierra del Fuego), Ecuador (Galapagos Islands), Hong Kong, Indonesia
(Java, Sumatra, Timor), Japan, Malaysia (North Borneo), Mexico [So-
nora (J. Kohlmeyer, unpublished)], New Zealand (North Island), Peru (Ica
and Lima), United States (California, Washington). NOTE: The species
could be treated as a lichen, forming associations between the Ascomycete and blue–green algae; *Arthopyrenia sublitoralis* seems to be the valid
lichen name for the organism (Santesson, 1939); however, according to
Santesson, a revision of the genus *Arthopyrenia* is necessary; we prefer to

use the oldest valid fungal name, *Pharcidia balani,* until a revision of *Arthopyrenia* will be made and, possibly, an earlier valid lichen name will be found. ETYMOLOGY: Named after a host genus, *Balanus.*

Pharcidia laminariicola Kohlm.

Bot. Mar. **16**, 209, 1973 (as *"laminaricola"*)

Figs. 76b and 76c

ASCOCARPS 130–210 µm high, 180–300 µm in diameter, ellipsoidal, basally flattened, superficial, ostiolate, epapillate, subcarbonaceous and black above, pallid below, with a thin dark-brown or blue–green layer close to the surface of *Laminaria,* solitary, gregarious or several in stromalike groups, rarely associated with pycnidia. PERIDIUM clypeoid and 42–66 µm thick near the ostiole, composed of subglobose cells; laterally 22–36(–50) µm thick, merging with hyphae and surrounding filaments of *Ectocarpus*; basally 14–18(–26) µm thick, composed of small, hyaline cells. OSTIOLES 8–22 µm in diameter. PARAPHYSES forming a dense mass of vertical parallel hyphae; at maturity indistinct, transformed into a gelatinous substance in which fragments of the original filaments can still be seen, enclosing the asci firmly and also filling the ostiolar canal. ASCI 47–56 × 14–17 µm, eight-spored, clavate, short pedunculate, unitunicate, thick-walled, aphysoclastic, without apical apparatuses, not bluing in iodine; developing all along the inner wall of the ascocarp, maturing successively. ASCOSPORES 17.5–25 × 6.5–8 µm (excluding appendages), biseriate, rhomboid or biturbinate, almost pointed at the apices, one-septate, not or slightly constricted at the septum, hyaline; both ends covered by a caplike, subgelatinous, 3- to 5-µm-thick appendage. PYCNIDIA (or spermogonia?) 95–110 µm high, 55–110 µm in diameter, subglobose, superficial, ostiolate, epapillate, subcoriaceous, dark brown, solitary or gregarious. PERIDIUM 16 µm thick near the ostiole, 5 µm at the base. CONIDIOPHORES (or spermatiophores?) 12 × 2 µm, developing in a layer all along the inner wall of the fruiting body. CONIDIA (spermatia?) about 2 × 0.5 µm, ellipsoidal, one-celled, hyaline. MODE OF LIFE: Symbiotic in lichenoid association with epiphytic *Ectocarpus fasciculatus* (Phaeophyta). HOST: Stalk of *Laminaria digitata*. RANGE: Atlantic Ocean—Norway (known only from the original collection). ETYMOLOGY: From the host genus, *Laminaria,* and the Latin *-cola* = dweller, in reference to the substrate of the fungus.

Pharcidia rhachiana Kohlm.

Bot. Mar. **16,** 210, 1973

Figs. 76d–76f

ASCOCARPS 52–140 μm high, 70–155 μm in diameter, subglobose or ellipsoidal, basally flattened, superficial, ostiolate, epapillate, subcarbonaceous, dark brown above, hyaline below, solitary or gregarious. PERIDIUM 18–28 μm thick near the ostiole, 10–15 μm at the sides, 6–10 μm at the base. OSTIOLES 8–14 μm in diameter, periphysate. PARAPHYSES filling the center of young ascocarps, pushed aside by developing asci. ASCI 29–40 × 9–12.5 μm, eight-spored, clavate to fusiform, short pedunculate, unitunicate, thick-walled, aphysoclastic, without apical apparatuses, not bluing in iodine; developing all along the inner wall of the ascocarp. ASCOSPORES 12.5–20.5 × 4–5 μm, biseriate, ellipsoidal, straight or faintly curved, one-septate, not or slightly constricted at the septum, hyaline, without appendages. PYCNIDIA (or spermogonia?) 30–45 μm high, 55–60 μm in diameter, subglobose, superficial, ostiolate, epapillate, subcoriaceous, brown above, hyaline below, solitary. PERIDIUM 6–14 μm thick. CONIDIOPHORES (or spermatiophores?) about 6 × 1.5 μm, cylindrical, developing in a layer all along the inner wall of the fruiting body. CONIDIA (or spermatia?) about 1.5 × 0.5 μm, ellipsoidal, one-celled, hyaline. MODE OF LIFE: Symbiotic in lichenoid association with epiphytic blue–green and probably brown algae. HOST: Hapteres of *Laminaria digitata*. RANGE: Atlantic Ocean—Norway (known only from the original collection). ETYMOLOGY: From the Greek, *rhachia* = rocky shore, surge, and the suffix *-anus* = indicating the origin, in reference to the habitat of the species.

Key to the Marine Species of Pharcidia

1. In calcareous shells of marine mollusks and cirripeds . . . . **P. balani**
1'. On the surface of Phaeophyta 2
    2(1') Ascospores with apical caplike appendages; spore diameter more than 6 μm . . . . **P. laminariicola**
    2'(1') Ascospores without appendages; diameter less than 6 μm . . . **P. rhachiana**

Turgidosculum J. Kohlmeyer et E. Kohlmeyer

Bot. Jahrb. **92,** 429, 1972

MYCELIUM growing between layers or groups of algal cells. ASCOCARPS solitary or gregarious, subglobose, immersed or erumpent, ostiolate, epapillate, coriaceous, top and base dark brown, sides hyaline or dark; ostiolar canal periphysate and with a turgid pulvillus. PARAPHYSES absent;

centrum with or without a gelatinous matrix enclosing the asci. ASCI eight-spored, clavate, pedunculate, unitunicate, at first thick-walled, finally deliquescing, developing along the inner wall of the ascocarp up to the ostiolar canal. ASCOSPORES ellipsoidal to ovoid, one-celled, hyaline. SPERMOGONIA solitary, subglobose to lentiform, immersed, irregularly chambered, ostiolate, epapillate, coriaceous, top and base brown, sides hyaline. SPERMATIOPHORES conical or cylindrical, simple. SPERMATIA filiform or subglobose, one-celled, hyaline. HOSTS: Marine algae (Chlorophyta). TYPE SPECIES: *Turgidosculum ulvae* (Reed) Kohlm. et Kohlm. The genus embraces two marine species. ETYMOLOGY: From the Latin, *turgidus* = turgid, swollen, and *osculum* = little mouth, in reference to the gelatinous swelling cushion closing the ostiole of *T. ulvae*.

Turgidosculum complicatulum (Nylander) J. Kohlmeyer et E. Kohlmeyer comb. nov.

≡Leptogiopsis complicatula Nylander, *Flora (Jena)* **67**, 211–212, 1884 (basionym)

=*Guignardia alaskana* Reed, *Univ. Calif., Berkeley, Publ. Bot.* **1**, 161, 1902

≡*Laestadia alaskana* (Reed) P. A. Saccardo et D. Saccardo *in* Saccardo, *Syll. Fung.* **17**, 576, 1905

=*Laestadia prasiolae* Winter, *Hedwigia* **26**, 16–17, 1887

≡*Guignardia prasiolae* (Winter) Lemmermann, *Abh. Naturwiss. Ver. Bremen* **17**, 199, 1901 [non *Guignardia prasiolae* (Winter) Reed 1902, superfluous new combination]

=*Laestadia tessellata* Winter ex Hariot, Algues, p. 29 *in* Mission Scientifique du Cap Horn, 1882–1883, Vol. 5, Botanique, 1889 (*nomen nudum*)

=*Physalospora prasiolae* Hariot, *J. Bot., Paris* **1**, 233, 1887 (*nomen nudum*)

[In the lichenological literature often treated as a lichen: *Mastodia tessellata* (Hooker f. et Harvey) Hooker f. et Harvey ex Hooker, Flora Antarctica, Part II, Botany of Fuegia etc., p. 499 *in* The Botany, The Antarctic Voyage of H. M. Discovery Ships Erebus and Terror in the Years 1839–1843, London, 574 pp., 1847; also sometimes included under the algal partner, *Prasiola tessellata* (Hooker et Harvey) Kützing.]

LITERATURE: Ahmadjian (1967); Brodo (1977); Cotton (1909); Dodge (1973); Henssen and Jahns (1974); Hue (1909); T. W. Johnson and Sparrow (1961); Knebel (1936); Lamb (1948); Letrouit-Galinou (1969); Meyers (1957); Nagai (1940); Printz (1964); Wainio (1903); Weber and Shushan (1959); Zanefeld (1969).

Fig. 77

MYCELIUM forming a thallus similar to a textura intricata, enclosing algal cells that form tetrads or rows. ASCOCARPS 240–300 μm high, 200–450 μm in diameter, subglobose, immersed in the thallus, ostiolate, epapillate, coriaceous, top dark brown, sides and base also often with dark areas, solitary or gregarious. PERIDIUM 42–80 μm thick around the ostiole, 14–22 μm at the base and sides, two-layered; outer layer composed of flattened cells with narrow lumina, merging at the base and sides into the thallus, with brown incrustations; inner layer composed of polygonal, hyaline cells with large lumina, forming a textura angularis. PAPILLAE absent; ostiolar canal formed schizogenously, about 20 μm in diameter, periphysate, closed by a gelatinous, faintly striate, turgescent matrix; periphyses sparse, anastomosing at the base. PARAPHYSES absent in mature ascocarps, but centrum filled with a gelatinous, faintly striate matrix, surrounding the asci; short papillate hyphae (periphyses?) hang down from the dome of the ascocarp venter. ASCI (25–)30–46(–57) × (7–)9–15 μm, eight-spored, clavate to subcylindrical, short pedunculate, unitunicate, without apical apparatuses, at first thick-walled, finally deliquescing, developing along the inner wall of the ascocarp, up to the ostiolar canal. ASCOSPORES (8.5–)11–17.5(–18.5) × 3–5 μm, elongate–ellipsoidal to cylindrical or rarely fusiform, with rounded ends, one-celled, hyaline, at maturity released into the venter of the ascocarp. SPERMOGONIA 160–240 μm high, 170–280 μm in diameter, similar to ascocarps, subglobose to lentiform, immersed, irregularly chambered, ostiolate, epapillate, coriaceous, top and base brown, sides hyaline, solitary or gregarious. PERIDIUM 12–24 μm thick. SPERMATIOPHORES about 10 × 1.5 μm, simple, cylindrical, lining the walls and lobes of the spermogonial cavity. SPERMATIA 2 × 1 μm, subglobose or ellipsoidal, one-celled, hyaline. MODE OF LIFE: Symbiotic; forming mycophycobioses (or parasitic?). HOSTS: *Prasiola borealis* and *P. tessellata*. RANGE: Pacific Ocean— Canada (British Columbia: Queen Charlotte Islands), Chile (Tierra del Fuego), United States (Alaska, Kodiak and Unalaska Islands), U.S.S.R. (Siberia, Kuril Islands: Paramushir Island). NOTE: The nomenclature of the *Prasiola*-association is confused, to say the least. Since 1845 phycologists, mycologists, and lichenologists have assigned names to this composite, which was at some time included in the algae, fungi or lichens. The earliest name is *Ulva tessellata* Hooker f. et Harvey (Hooker and

Fig. 77. *Turgidosculum complicatulum* on *Prasiola* spp. (a) Habit of the association growing on rock (bar = 5 mm). (b) 4-μm section through thallus with spermogonium (bar = 50 μm). (c) Section (4 μm) through thallus showing the separation of groups of algal cells by fungal mycelium (bar = 20 μm). (d) Thallus in surface view (bar = 25 μm). (e) 2-μm section through immature ascocarp (bar = 50 μm). (f) 12-μm section through mature ascocarp (bar = 50 μm). (g) 4-μm section through ostiole; ostiolar canal closed by a gelatinous matrix covering the periphyses (bar = 15 μm). (h) 4-μm section through base of ascocarp showing immature asci and gelatinous matrix

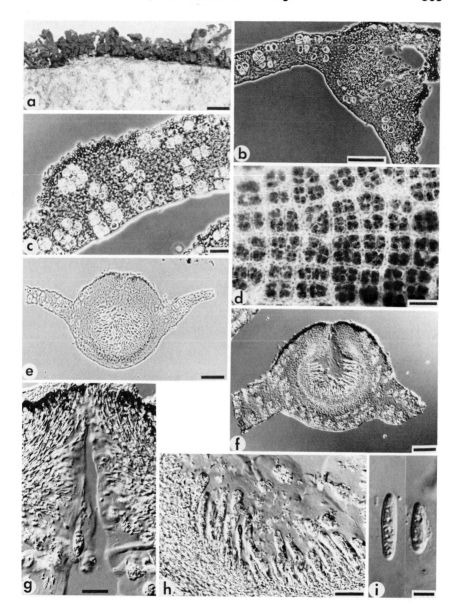

in the venter above them (bar = 20 μm). (i) Ascospores (bar = 5 μm). (a from holotype Nylander 41012 in H; b, c, e, f, and i from type of *Laestadia prasiolae* in F; d, g, and h from type of *Guignardia alaskana* in BPI and F; a and d in bright field; b, c, and e in phase contrast; the others in Nomarski interference contrast.)

Harvey, 1845), described in error (as *U. "tesellata"*) as a marine species
and later transferred by Hooker and Harvey (in Hooker, 1847) to *Mastodia*, as *M. tessellata.** These authors clearly characterized *M. tessellata*
as an alga from a freshwater stream and lake on Kerguelen's Land. We
agree with Brodo (1977), who states that the epithet *"tessellata"* does not
take priority over later epithets applied to the fungal component. Therefore, the lichen name *Leptogiopsis complicatula* Nylander (1884) from
Siberia is the earliest available name accompanied with the description of
the fungus. We regard the fungus as a symbiont of *Prasiola* and transfer it
to the genus *Turgidosculum* because it is closely related to the type
species, *T. ulvae*, which belongs to the Sphaeriales, probably in the
Verrucariaceae. The description of *T. complicatulum* is based on our
examinations of the holotype of *Leptogiopsis complicatula* Nylander
(H-NYL, Nos. 41010–41012†) and syntypes of *Guignardia alaskana* Reed
(Coll. Setchell, Nos. 4021 & 5138, BPI, FH, NY) and *Laestadia prasiolae*
Winter (Coll. Hariot, FH, K); we conclude that these taxa cannot be
separated on a morphological basis because gross characters and dimensions of the fungal component are identical and *Leptogiopsis complicatula* has the priority. *Dermatomeris georgica* Reinsch from South
Georgia (Reinsch, 1890, Pl. 19, Figs. 1–5) is possibly another synonym of
T. complicatulum, but type material is not available, according to curators
of herbaria B and ER. However, dimensions of ascocarps, asci, and
ascospores in *D. georgica* are smaller than those of *T. complicatulum*. In
the absence of a type specimen, we consider *D. georgica* to be a *nomen
dubium*. We have excluded collection sites and data cited by authors
other than Nylander (1884), Winter (1887), Reed (1902), Nagai (1940),
Weber and Shushan (1959), and Brodo (1977) because it is not clear in all
cases whether the specimens derived from marine, terrestrial, or freshwater habitats, and species other than *T. complicatulum* might be involved.
The host plants of different geographical areas appear not to differ greatly
from each other, although Knebel (1936, p. 12) separates *Prasiola borealis*
from *P. tessellata* by sharp-cornered or rounded cells, respectively.
ETYMOLOGY: From the Latin, *complicatum* = folded together, and the

---

* We have seen syntype material of *Ulva tessellata* Hooker f. et Harvey (=*Prasiola
tessellata*; Nos. 655 and 657, Herb. BM ex K) from Kerguelen Islands and found the
associated fungus similar to the fungal component in Nylander's *Leptogiopsis complicatula*
from Siberia.

† These specimens in the Nylander collection are not marked as types and Nylander's
protologue [*Flora (Jena)* **67**, 211–212, 1884] does not refer to a particular specimen, but the
labels read "Fret. Behringii, Konyambay. E. Almquist (Exped. Vega)." Therefore, we
consider this collection to be the holotype.

diminutive suffix *-ulum,* in reference to the many-folded thalli of the infested *Prasiola.*

Turgidosculum ulvae (Reed) J. Kohlmeyer et E. Kohlmeyer

Bot. Jahrb. **92,** 429, 1972

≡*Guignardia ulvae* Reed, *Univ. Calif., Berkeley, Publ. Bot.* **1,** 160, 1902

LITERATURE: Ahmadjian (1967); Doty (1947); T. W. Johnson and Sparrow (1961); Meyers (1957); Norris (1971); Norris and Abbott (1972); Kohlmeyer (1974b).

Fig. 78

MYCELIUM similar to a textura intricata, growing between the two layers of algal cells. ASCOCARPS 330–560 μm high, 330–600 μm in diameter, subglobose, immersed or erumpent, ostiolate, epapillate, coriaceous,

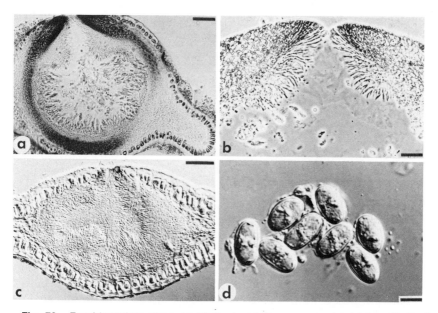

Fig. 78. *Turgidosculum ulvae* on *Blidingia minima* var. *vexata.* (a) Longitudinal section (16 μm) through ascocarp (Herb. Setzer 3750; bar = 50 μm). (b) Section (4 μm) through ostiole, showing gelatinous pulvillus closing ostiolar canal and covering the periphyses (Herb. Setzer 3750). (c) Longitudinal section (2 μm) through spermogonium between the upper and lower algal layers (type in F; bar = 25 μm). (d) Group of ascospores from a dissolved ascus (type in F; bar = 5 μm). (a in bright field, b in phase contrast, c and d in Nomarski interference contrast.)

top and base dark brown, sides hyaline, solitary or gregarious, rarely two of them joined. PERIDIUM 40–70 μm thick above, 20–40 μm at the base. PAPILLAE absent; ostiolar canal periphysate, in immersed plants closed by a gelatinous, turgescent swelling cushion. PARAPHYSES absent. ASCI 43–57 × 9–15.5 μm, eight-spored, clavate, pedunculate, unitunicate, without apical apparatuses, at first thick-walled, finally deliquescing, developing along the inner wall of the ascocarp up to the ostiolar canal. ASCOSPORES (8.5–)10–13(–14) × 3.5–7 μm, broadly ellipsoidal to ovoid, one-celled, hyaline, at maturity released into the venter of the ascocarp. SPERMOGONIA 100–200 μm high, 210–350 μm in diameter, lentiform, immersed, irregularly chambered, ostiolate, epapillate, coriaceous, top and base brown, sides hyaline. SPERMATIOPHORES 8–14 × 1.7–2.4 μm, conical, simple, lining the walls of the locule. SPERMATIA 8–11 × 0.6–1 μm, filiform, one-celled, hyaline. MODE OF LIFE: Symbiotic; forming mycophycobioses (or parasitic?). HOST: *Blidingia minima* var. *vexata* (≡*Ulva vexata*). RANGE: Pacific Ocean—Canada (British Columbia), United States (California, Oregon, Washington). NOTE: Description in part after J. Kohlmeyer (unpublished). ETYMOLOGY: Named after the host genus, *Ulva* (Chlorophyta).

Key to the Species of *Turgidosculum*

1. Ascocarp center with a gelatinous matrix enclosing asci; ascospores elongate ellipsoidal, up to 5 μm in diameter; in *Prasiola*. . . . . **T. complicatulum**
1'. Gelatinous matrix filling the ostiole only, asci free; ascospores broad ellipsoidal, up to 7 μm in diameter; in *Blidingia* . . . . **T. ulvae**

h. Sphaeriales incertae sedis

***Oceanitis* Kohlm.**

Rev. Mycol. **41**, 193–194, 1977

ASCOCARPS gregarious, subglobose to ellipsoidal, seated on the surface of a thin hypostroma, ostiolate, periphysate, epapillate, fleshy, brownish to dull orange-colored. PERIDIUM thick, forming a textura angularis. PARAPHYSES absent; center of immature ascocarps filled with thin-walled pseudoparenchymatous cells. ASCI eight-spored, clavate, unitunicate, thin-walled, ripening successively on an ascogenous tissue at the bottom of the ascocarp venter, becoming detached at maturity at the base and pushed upward by young asci, eventually dissolving in the upper part of the ascocarp venter. ASCOSPORES filamentous, one-septate, hyaline, with a filiform apical appendage. SUBSTRATE: Wood. RANGE: Deep sea. TYPE SPECIES: *Oceanitis scuticella* Kohlm. NOTE: The genus belongs to Sphaeriales, but the exact taxonomic position is uncertain; some characters (fleshy, light-colored ascocarp, hypostroma) point to the Hypocreaceae, others (pseudoparenchymatous ascocarp center, early deliquescing asci, ascospore appendages) to the Halosphaeriaceae. ETYMOLOGY: *Oceanitis,* a daughter of the marine god Oceanus.

Oceanitis scuticella Kohlm.

Rev. Mycol. **41,** 194, 1977

Fig. 79

ASCOCARPS 1.4–2.03 mm high, 1.4–1.78 mm in diameter, subglobose to ellipsoidal, round at the top, flat at the base, seated on the surface of a thin hypostroma, ostiolate, epapillate, fleshy, brown to dull orange-colored, gregarious. HYPOSTROMA 30–120 μm thick, light-colored, composed of small, polygonal to rounded cells with large lumina; masses of lilac-colored hyphae can be found in the wood vessels under the stromata. PERIDIUM 450–470 μm thick at the apex, 240–260 μm at the sides, composed of 17–35 layers of hyaline, thick-walled, polygonal, isodiametric cells with large lumina, forming a textura angularis; merging toward the center into flattened cells; outer 3–7 cells filled with small yellowish globules. OSTIOLES 25–30 μm in diameter; ostiolar canal periphysate. PARAPHYSES absent; center of immature ascocarps filled with hyaline, thin-walled, polygonal, isodiametric, pseudoparenchymatous cells, which are eventually compressed by the asci. ASCI 70–90 × 12–17 μm, eight-spored, clavate, unitunicate, thin-walled, without apical apparatuses, ripening successively on an ascogenous tissue at the bottom of the ascocarp venter, becoming detached at maturity at the base and pushed upward by young asci, eventually dissolving in the upper part of the ascocarp venter. ASCOSPORES 60–80 × 4–6 μm, filiform to elongate fusiform, one-septate, hyaline, with one apical appendage; appendages 32–50 μm long, 2(–3) μm in diameter at the base, filamentous, tapering toward the apex, wavy, whiplike, at first attached to the side of the ascospore, becoming detached from the wall, but adhering with its base to the apex of the ascospore. MODE OF LIFE: Saprobic. SUBSTRATE: Submerged wood. RANGE: Atlantic Ocean—Deep sea (Gulf of Angola in 3975 m; known only from the original collection). ETYMOLOGY: From the Latin, *scutica* = whip, and the diminutive suffix *-ellus,* indicating resemblance, in reference to the whiplike, appendaged ascospores.

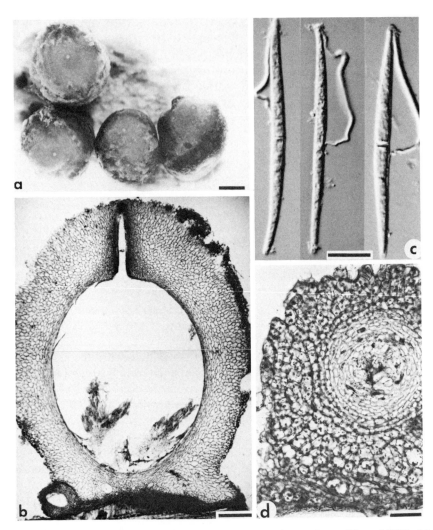

Fig. 79. *Oceanitis scuticella* from wood in the deep sea (type PC and IMS). (a) Group of ascocarps on the wood surface (bar = 0.5 mm). (b) Longitudinal section (16 μm) through mature ascocarp (bar = 200 μm). (c) Ascospores with whiplike appendages (bar = 15 μm). (d) Section (8 μm) through immature ascocarp stained in violamin (bar = 25 μm). b and c from Kohlmeyer, *Rev. Mycol.* **41**, 195 (1977), Figs. 9 and 12. Reprinted with permission. (c in Nomarski interference contrast, the others in bright field.)

Ophiobolus Riess

Hedwigia **1**, 27, 1854

NOTE: The generic description of *Ophiobolus* is omitted because the only recognized marine species, *O. australiensis*, does not belong to this genus; several other marine species assigned at one time to *Ophiobolus* are listed among the doubtful or rejected species, or in the genus *Lulworthia*.

Ophiobolus australiensis Johnson et Sparrow

Fungi in Oceans and Estuaries, Cramer, Weinheim, p. 419, 1961

≡*Ophiobolus littoralis* Cribb et Cribb, *Univ. Queensl. Pap., Dep. Bot.* **3**, 101, 1956 [non *Ophiobolus littoralis* (Crouan et Crouan) Saccardo 1883]

Fig. 80

ASCOCARPS 210–560 μm in diameter, subglobose or ellipsoidal, immersed or almost superficial, ostiolate, papillate, membranaceous or subcarbonaceous, black, solitary, surrounded by wide brown hyphae, especially at the base. PAPILLAE or necks up to 280 μm long, 22–26 μm in diameter, papilliform or subcylindrical, centric or eccentric. PARAPHYSES absent. ASCI 110–165 × 20–22.5 μm, eight-spored, fusiform to subclavate

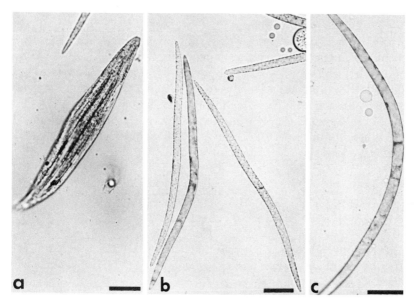

Fig. 80. *Ophiobolus littoralis* from type slide A.B. Cribb. (a) Ascus. (b) Ascospores. (c) Ascospore with a central septum. (All bars = 20 μm; all in bright field.)

or elongate–ellipsoidal, short pedunculate, immature asci with a small apiculus, unitunicate, thin-walled, without apical apparatuses, early deliquescing. ASCOSPORES 115–200 × (4.5–)5.5–7.5 μm, parallel and often slightly curved in the ascus, filiform, cylindrical throughout with rounded apices and somewhat tapering to the tips, non- or rarely one-septate in the center, not constricted at the septum, curved or rarely straight, hyaline, germinating apically; without appendages or end chambers. MODE OF LIFE: Saprobic. SUBSTRATE: Dead roots of *Avicennia marina* var. *resinifera*. RANGE: Pacific Ocean—Australia (Queensland; known only from the original collection). NOTE: Description based on Cribb and Cribb (1956) and on our examination of the type slide kindly provided by Dr. A. B. Cribb; septa were found in nongerminated ascospores; this sphaeriaceous species should be removed from *Ophiobolus* (sensu Holm, 1957), a genus of Pleosporales with bitunicate asci; however, no change is proposed at this time because new collections with early developmental stages of ascocarps are needed to assign the species properly. ETYMOLOGY: From the Latin, *australiensis* = from Australia, in reference to the origin of the type collection.

Savoryella E. B. G. Jones et Eaton

Trans. Br. Mycol. Soc. **52**, 161, 1969

ASCOCARPS solitary or gregarious, subglobose, immersed or partly immersed, ostiolate, periphysate, papillate, membranaceous, brown. PARAPHYSES absent. ASCI eight-spored, cylindrical or clavate, short pedunculate, unitunicate, thin-walled, persistent. ASCOSPORES ellipsoidal, triseptate, central cells brown, apical cells hyaline. SUBSTRATE: Dead wood. TYPE SPECIES: *Savoryella lignicola* E. B. G. Jones et Eaton. The genus is monotypic. ETYMOLOGY: Named in honor of J. G. Savory, British mycologist, with the Latin diminutive suffix *-ellus*.

Savoryella lignicola E. B. G. Jones et Eaton

Trans. Br. Mycol. Soc. **52**, 162, 1969

LITERATURE: Eaton (1972); Eaton and Irvine (1972); Eaton and Jones (1971a,b); E. B. G. Jones (1972); Raghukumar (1973); Tubaki and Ito (1973).

Fig. 81a

ASCOCARPS 177–350 μm high, 120–250 μm in diameter, subglobose or ellipsoidal, immersed or partly immersed, ostiolate, papillate, mem-

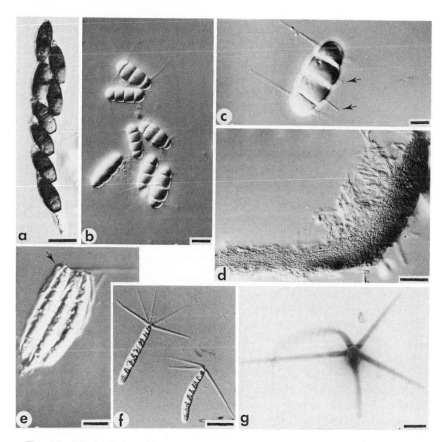

Fig. 81. Marine Sphaeriales of uncertain affinities. (a) Ascus of *Savoryella lignicola* (type IMI 129784; bar = 20 μm). (b) Appendaged ascospores of *Torpedospora ambispinosa* (isotype IMS Herb. J.K. 178; bar = 10 μm). (c) Ascospore of *T. ambispinosa* with unusual appendage formation (arrows) from the central cells (Herb. J.K. 178; bar = 5 μm). (d) Hand section through peridium and paraphyses of *T. ambispinosa* (Herb. J.K. 178; bar = 25 μm). (e) Ascus of *T. radiata;* most appendages (arrow) still attached to the ascospores (Herb. J.K. 1634; bar = 10 μm). (f) Ascospores of *T. radiata* with expanded appendages (Herb. J.K. 2737; bar = 15 μm). (g) Apical view of ascospore, appendages stained in hematoxylin (Herb. J.K. 1634; bar = 5 μm). (g in bright field, the others in Nomarski interference contrast.)

branaceous, subhyaline to dark brown. PAPILLAE 48–165 μm long, about 72 μm in diameter; ostiolar canal periphysate. PARAPHYSES absent. ASCI 100–180 × 16–24 μm, eight-spored, elongate–cylindrical or clavate, short pedunculate, unitunicate, thin-walled, persistent. ASCOSPORES 24–33.5 (−43) × 8.5–13 μm, uni- to biseriate, ellipsoidal, triseptate, constricted at the septa; central cells large, brown; apical cells small, hyaline. MODE OF

LIFE: Saprobic. SUBSTRATE: Driftwood and test panels (e.g., *Fagus sylvatica, Ocotea rodiaei, Pinus sylvestris,* balsa, bamboo). RANGE: Atlantic Ocean—Great Britain (England, Wales); Indian Ocean—India (Madras); Pacific Ocean—Japan. NOTE: Description based on E. B. G. Jones and Eaton (1969), Raghukumar (1973), and Tubaki and Ito (1973); the species is a facultative marine fungus that has been collected in freshwater as well as in seawater. ETYMOLOGY: From the Latin, *lignum* = wood and *-cola* = -dweller, in reference to the substrate of the species.

***Torpedospora* Meyers**

Mycologia **49**, 496, 1957

LITERATURE: Kohlmeyer (1969c, 1972b).

ASCOCARPS solitary, subglobose to ellipsoidal, immersed or superficial, ostiolate, papillate or epapillate, subcarbonaceous to subcoriaceous, dark-colored. PARAPHYSES ramose, growing irregularly through the ascocarp venter; persistent or deliquescing. ASCI eight-spored, clavate to ellipsoidal, unitunicate, thin-walled, aphysoclastic, early deliquescing. ASCOSPORES cylindrical to elongate–ellipsoidal, triseptate, hyaline, with several radiating appendages at one or both ends. SUBSTRATE: Dead wood or bark. TYPE SPECIES: *Torpedospora radiata* Meyers. NOTE: Both members of the genus are obligate marine fungi; the taxonomic position of *Torpedospora* is uncertain and ontogenetic studies are needed to observe early stages in the development of ascocarps and ascospores; *Torpedospora* cannot be included in Halosphaeriaceae because a central pseudoparenchyma is absent in the ascocarps, and asci appear to originate at random throughout the venter. ETYMOLOGY: From the Latin, *torpedo* = name of a fish (electric ray, causing stiffness or numbness = *torpedo* in Latin; later used for cigar-shaped, destructive submarine projectiles) and *spora* = spore, in reference to the "torpedo"-shaped ascospores of the type species.

***Torpedospora ambispinosa* Kohlm.**

Nova Hedwigia **2**, 336, 1960

LITERATURE: T. W. Johnson (1963c); T. W. Johnson and Sparrow (1961); Kohlmeyer and Kohlmeyer (1964–1969).

Figs. 81b–81d

ASCOCARPS 241–328 μm high, 276–380 μm in diameter, globose, subglobose or ellipsoidal, immersed or superficial, ostiolate, papillate or epapillate, subcarbonaceous or subcoriaceous, dark brown to black, solitary. PERIDIUM 20.5–37 μm thick, composed of irregular, more or less oblong, thick-walled cells with small lumina. PAPILLAE centric, eccentric or lateral, short, broadly conical or cylindrical. PARAPHYSES growing irregularly through the ascocarp venter, early deliquescing. ASCI eight-spored, broadly clavate or subglobose, unitunicate, thin-walled, aphysoclastic, deliquescing before ascospore maturity. ASCOSPORES 17–24.5 × 6.5–10 μm, cylindrical to elongate–ellipsoidal, triseptate, slightly constricted at the septa, hyaline, appendaged; appendages subterminal, radiating, four to seven at each end, 9–17.5 × 0.3–0.5 μm, rigid, attenuate, straight or slightly curved, possibly developing by outgrowth of the spore cytoplasm. MODE OF LIFE: Saprobic. SUBSTRATE: Intertidal wood. RANGE: Pacific Ocean—United States (California, Washington). ETYMOLOGY: From the Latin, *ambi* = on both sides and *spinosus* = spiny, in reference to the ascospores, bearing spinelike appendages at both ends.

Torpedospora radiata Meyers

Mycologia **49**, 496, 1957

LITERATURE: Byrne and Jones (1975a); S. E. J. Furtado *et al.* (1977); G. C. Hughes (1974); T. W. Johnson (1958b); T. W. Johnson *et al.* (1959); T. W. Johnson and Sparrow (1961); E. B. G. Jones (1973); E. B. G. Jones *et al.* (1972); Kirk (1976); Kohlmeyer (1959, 1963d, 1967, 1968c, 1969c); Kohlmeyer and Kohlmeyer (1964–1969, 1971a); Lutley and Wilson (1972b); Meyers (1968a,b, 1971a); Meyers and Reynolds (1959a,c, 1960b, 1963); Meyers *et al.* (1960, 1963b); Nair (1970); Pisano *et al.* (1964); Raghukumar (1973); Schaumann (1974b); Tubaki (1968, 1969); Tubaki and Ito (1973).

Figs. 81e–81g

ASCOCARPS (75–)100–340 μm high, 100–361 μm in diameter, subglobose to pyriform, immersed or superficial, ostiolate, papillate, subcoriaceous, fuscous or dark brown above, subhyaline, gray or brownish below, gregarious. PERIDIUM 15–25 μm thick, two-layered; outer layer composed of subglobose, thick-walled cells with small lumina; inner layer composed of elongate, thick-walled cells with larger lumina. PAPILLAE or necks up to 225 μm long, 18–32 μm in diameter, cylindrical, fuscous; ostiolar canal at first filled with small, thin-walled pseudoparenchymatous cells. PARAPHYSES 0.7–1.3 μm in diameter, septate, ramose, growing irregularly through the venter of the ascocarp. ASCI 103–181 × 12–16 μm, eight-spored, clavate or oblong–ellipsoidal, sessile or short pedunculate, unitunicate, thin-walled, aphysoclastic, without apical apparatuses, early deliquescing; developing along the inner wall of the lower half of the ascocarp. Ascospores 30.5–52 × 4–9 μm, cylindrical or clavate, broader at the apex, triseptate, rarely with more septa, not or slightly constricted at the septa, hyaline, appendaged; three or four, rarely five, radiating appendages on the lower end, 19–39.5 × 1.5–4.5 μm, semirigid, straight or slightly curved, with a thick base, tapering toward the apex; a skeleton of parallel fibrils visible upon electron-optical examination. Mode of Life: Saprobic. Substrates: Intertidal wood and bark, test panels (e.g., *Pinus sylvestris, Pinus* sp., *Tilia* sp., balsa, bamboo), pneumatophores of *Avicennia germinans,* bark of roots of *Rhizophora mangle,* endo- and exocarp of submerged *Cocos nucifera.* Range: Atlantic Ocean— Bahamas [Great Abaco (J. Kohlmeyer, unpublished)], Great Britain (Wales), Ivory Coast, Liberia, Mexico (Veracruz, Yucatan), Norway, Sierra Leone (A. A. Aleem, personal communication), Spain (Tenerife and mainland), Tobago (J. Kohlmeyer, unpublished), United States (Florida, North Carolina); Indian Ocean—India (Kerala State, Madras); Mediterranean—France, Italy (Sicily and mainland); Pacific Ocean— Japan, New Zealand (North Island), Samoa (Pago Pago), United States [Hawaii (Hawaii, Kauai, Maui)]. Etymology: From the Latin, *radiatus* = provided with spokes or rays, in reference to the radiating ascospore appendages.

Key to the Species of *Torpedospora*

1. Ascospores 30 μm or longer; appendages at one end only . . . . *T. radiata*
1′. Ascospores shorter than 30 μm; radiating appendages at both ends
. . . . *T. ambispinosa*

D. Loculoascomycetes
   1. Dothideales
      a. Herpotrichiellaceae

Herpotrichiella Petrak

Ann. Mycol. **12,** 472, 1914

=*Didymotrichiella* Munk, *Dan. Bot. Ark.* **15**(2), 131, 1953

Literature: Barr (1972); Müller and von Arx (1962); Munk (1957); von Arx and Müller (1975).

Ascocarps solitary, globose or conical, small, superficial, ostiolate, epapillate, thin-walled, brownish or grayish, setose or with a rough surface. Asci eight-spored, cylindrical or clavate, bitunicate. Ascospores fusiform or ellipsoidal, one- to several-septate, in some species with vertical septa, hyaline to gray or olive brown. SUBSTRATES: Rotting wood, leaves, or other fungi. TYPE SPECIES: *Herpotrichiella moravica* Petrak. NOTE: Description and synonym following the authors listed above; only one marine member is known among about six terrestrial species. ETYMOLOGY: From *Herpotrichia* Fuckel and the Latin diminutive suffix *-ella,* indicating a relationship to this genus.

Herpotrichiella ciliomaris Kohlm.

Nova Hedwigia **2,** 313, 1960

LITERATURE: T. W. Johnson (1963c); T. W. Johnson and Sparrow (1961); Kohlmeyer (1963d); Kohlmeyer and Kohlmeyer (1964–1969).

ASCOCARPS 77–226 μm in diameter, globose or ovoid, superficial or rarely partly embedded, ostiolate, epapillate, membranaceous, whitish, grayish, light or even dark blue or almost black, solitary or gregarious, hyaline or light brown hyphae at the base, setose around the ostiole. PERIDIUM 10–18 μm thick, composed of three or four layers of subglobose or ellipsoidal, thick-walled cells, 2–9 × 1.5–4.5 μm, with large lumina, forming a textura angularis. OSTIOLES 32–56 μm in diameter, closed by hyaline periphyses; 8–45 brown, awl-shaped, septate or nonseptate setae, developing from the outer peridial cells, surround the ostiole; setae 21–56 μm long, 4.5–9.5 μm in diameter at the base, 1.5–5 μm in diameter at the apex. PSEUDOPARAPHYSES (or apical paraphyses?) 4.5–23 × 0.7–2.5 μm, simple, covering the wall of the ascocarp venter, merging into the periphyses. ASCI (38.5–)58–80 × 14.5–21 μm, eight-spored, cylindrical or subclavate, pedunculate, slightly curved inside the ascocarp, bitunicate, apically thick-walled, physoclastic, without apical apparatuses; developing at the base of the ascocarp venter. ASCOSPORES (16.5–)20–28 × 7.5–10 μm (excluding appendages), bi- or triseriate, ellipsoidal or subovoid, one-septate below the center, constricted at the septum, hyaline, appendaged; at each end a subterminal crown of 20–30 radiating, filamentous, stiff appendages, about 8–8.5 μm long. MODE OF LIFE: Saprobic. SUBSTRATE: Intertidal or permanently submerged bark of conifers and deciduous trees. RANGE: Atlantic Ocean—Germany (F.R.G.), Spain; Pacific Ocean—United States (Washington). NOTE: According to Barr (1972) *Herpotrichiella* lacks apical pseudoparaphyses and has a locule of the *Dothidea* type. Therefore, developmental studies should ascertain the taxonomic position of *H. ciliomaris*. ETYMOLOGY: From the Latin, *cilium* = marginal hair, cilium, and *mare* = the sea, in reference to the appendaged ascospores and the marine occurrence of the species.

b. Mycoporaceae

Leiophloea S. F. Gray

Natural Arrangement of British Plants, London, Vol. I, p. 495, 1821
(sub *Lejophlea*), non *Leiophloea* Trevis.

=*Arthopyrenia* Mass., p.p., *Ric. Auton. Lich.* p. 165, 1852
=*Mycarthopyrenia* Keissler, p.p., *Ann. Naturhist. Mus. Wien* **34**, 71,
1921
=*Paraphysothele* Keissler, apud Rabenh., *Kryptogamenflora,* **9**, Pàrt 1,
No. 2, p. 210, 1937 (non Zschacke, l.c. **9**, Part 1, No. 1, p. 564, 1934)
=*Pseudarthopyrenia* Keissler, *Rev. Bryol. Lichénol,* **8**, 32, 1935
=*Thelidium* Keissler, p.p., apud Rabenh., *Kryptogamenflora,* **9**, Part
1, No. 2, p. 217, 1937 (non Mass., *Framment. Lich.* p. 15, 1855)

LITERATURE: Riedl (1961, 1962); von Arx and Müller (1975).

ASCOCARPS solitary or gregarious, subglobose, basally immersed in a smooth, flat, crustlike thallus, ostiolate; epapillate or short papillate, dark-colored above, sometimes also below. PERIDIUM thick, two-layered; outer layer thick and hard, inner layer membranaceous. PSEUDO-PARAPHYSES filiform, hyaline. ASCI eight-spored, bitunicate. ASCO-SPORES one-septate, hyaline. SUBSTRATE: Mainly dead plant material (e.g., bark or wood). TYPE SPECIES: *Leiophloea analepta* (Ach.) S. F. Gray. NOTE: Most species are lichenized with *Trentepohlia,* although in some species the lichenization is not distinct; only one marine species is known among about 50 terrestrial taxa; synonyms after Riedl (1962). ETYMOLOGY: From the Greek, *leios* = smooth, polished, and *phloios* = bark, rind, in reference to the smooth thallus of the lichen association.

Leiophloea pelvetiae (Sutherland) J. Kohlmeyer et E. Kohlmeyer comb. nov.

≡*Dothidella pelvetiae* Sutherland, *Trans. Br. Mycol. Soc.* **5**, 154, 1915
(basionym)
≡*Placostroma pelvetiae* (Sutherland) Meyers, *Mycologia* **49**, 480, 1957
=*Plowrightia pelvetiae* Fragoso, *Mem. Soc. Esp. Hist. Nat.* **11**, 110–
111, 1919

LITERATURE: Kohlmeyer (1973b).

Fig. 82

ASCOCARPS enclosed in stromata, single or rarely two or three united, 60–140 μm high, 120–340 μm in diameter, 310–590 μm long in bi- or

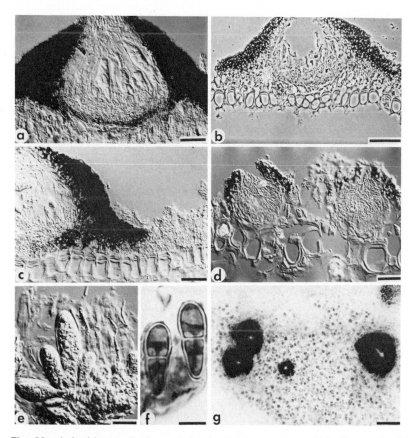

Fig. 82. *Leiophloea pelvetiae,* a lichenized Ascomycete on *Pelvetia canaliculata.* (a) Longitudinal section (10 μm) through ascocarp (bar = 25 μm). From Kohlmeyer, *Bot. Mar.* **16,** 206 (1973), Fig. 16. Reprinted with permission. (b) Section (2 μm) through ascocarp showing structure of involucrellum (bar = 50 μm). (c) Section (4 μm) through part of ascocarp showing fungus and associated one-celled blue–green algae on the surface of the large cells of *P. canaliculata* (bar = 25 μm). (d) Longitudinal section (2 μm) through spermogonia (bar = 20 μm). (e) Mature and immature asci (bar = 15 μm). From Kohlmeyer, *Bot. Mar.* **16,** 206 (1973), Fig. 18. Reprinted with permission. (f) Ascospores stained in violamin (bar = 5 μm). (g) Surface view of ascocarps and a small pycnidium between epiphytic blue–green algae (bar = 50 μm). (a, b, d, and f from type of *Plowrightia pelvetiae* in MA; c from Herb. J.K. 2959; e and g from Herb. J.K. 2962; b in phase contrast, f and g in bright field, the others in Nomarski interference contrast.)

trilocular stromata, depressed conoidal, outline almost circular with ir-regular edges, superficial, ostiolate, epapillate, carbonaceous, black, gre-garious. PERIDIUM of the top and sides (involucrellum) 20–40 μm thick, black, extending into the associated epiphytic algae, composed of irregu-lar, thick-walled cells with small lumina, usually enclosing some epiphytes; basis 8–12 μm thick, light brown, attached to the cortex of *Pelvetia* or to epiphytic algae, forming a textura angularis. LOCULES 45–120 μm high, 65–140 μm in diameter, subglobose to pyriform, eventu-ally truncate. OSTIOLES 14–40 μm in diameter, finally opening up to 60 μm, without periphyses, forming schizogenously. PSEUDOPARAPHYSES 1.1–2 μm in diameter, anastomosing, sparingly branching, composed of a central core that stains dark in violamin and of an outer gelatinous matrix; in mature ascocarps gelatinizing, enclosing the asci as a firm ball. ASCI 40–60 × 13–18 μm, eight-spored, clavate, short pedunculate, bitunicate, thick-walled, physoclastic; endoascus remaining protruding from ectoas-cus after ascospore discharge, filling the ostiolar canal; without apical apparatuses; developing at the base of the ascocarp venter. ASCOSPORES 11.5–16 × 5–6.5 μm, biseriate, elongate ovoidal, one-septate near the middle, not or slightly constricted at the septum, upper cell broader than lower one, hyaline. PYCNIDIA (or spermogonia?) 40–110 μm high, 44–68 μm in diameter, subglobose, outline circular or elongate with irregular edges, superficial, ostiolate, epapillate, subcarbonaceous, black, gregari-ous, near ascocarps and sometimes connected with them. PERIDIUM of the top 10–26 μm thick, clypeoid, brown; at the sides and base 2.4–3.2 μm thick, composed of two or three layers of cells with large lumina, hyaline, forming a textura angularis. OSTIOLES 4–16 μm, finally up to 50 μm in diameter. CONIDIOPHORES (or spermatiophores?) 7.2–8.8 × 1–1.6 μm, cylindrical to elongate–conical, slightly tapering toward the apex, simple, forming conidia singly at the tip, hyaline, lining the walls of the locule. CONIDIA (or spermatia?) 2–2.6 × 0.8–1.4 μm, bacilliform, hyaline, holoblastic. MODE OF LIFE: Symbiotic in lichenoid association with vari-ous species of epiphytic blue–green algae. HOST: Epiphytic on *Pelvetia canaliculata*. RANGE: Atlantic Ocean—France, Great Britain [England, Scotland (Orkney Island)], Spain. NOTE: We compared the fungus with the type species of *Dothidella* (Microthyriaceae), namely, *D. australis* (Herb. Spegazzini, La Plata) and concluded that they have nothing in common (see also Müller and von Arx, 1962). Instead, *L. pelvetiae* belongs to the Mycoporaceae and is transferred to *Leiophloea*, a genus containing lichenized species; Dr. E. Müller also examined material of the fungus and concurred with our view (E. Müller, personal communica-tion). ETYMOLOGY: Named after the host genus, *Pelvetia*.

c. Mycosphaerellaceae

Didymella Saccardo ex Saccardo

Michelia **2**, 57, 1880

≡*Didymella* Sacc., *Michelia* **1**, 377, 1878, non rite publ.
=*Arcangelia* Sacc., *Bull. Soc. Mycol. Fr.* **5**, 115, 1890
=*Didymolepta* Munk, *Dan. Bot. Arkiv* **15**(2), 110, 1953
=*Haplotheciella* von Höhnel, *Sitzungsber. Kais. Akad. Wiss. Wien. Math.-Naturwiss. Kl., Abt. 1* **128**, 614, 1919
=*Mycosphaerellopsis* von Höhnel, *Ann. Mycol.* **16**, 157, 1918

LITERATURE: Corbaz (1957); Holm (1953, 1975); Müller and von Arx (1962); von Arx and Müller (1975).

ASCOCARPS solitary, globose, depressed or flattened, without well-developed stroma, immersed or erumpent, ostiolate, papillate, dark; often associated with conidial states in the form-genera *Ascochyta, Diplodina,* and *Phoma.* PSEUDOPARAPHYSES filiform, simple, often connected to each other by a gelatinous matrix. ASCI eight-spored, cylindrical, weakly pedunculate, rounded and thickened at the apex, bitunicate. ASCOSPORES ovoid to ellipsoidal, one-septate in or below the center, rarely above the middle, mostly constricted at the septum, hyaline. MODE OF LIFE: Mostly parasitic on leaves of higher plants, but often maturing on the dead substrates. TYPE SPECIES: *Didymella exigua* (Niessl) Sacc. NOTE: Only three marine occurring species are known among about 50 terrestrial taxa; synonyms and taxonomic placement following von Arx and Müller (1975). ETYMOLOGY: From the Greek, *didymos* = in pairs or two-celled, and the Latin diminutive suffix *-ellus,* in reference to the two-celled ascospores.

Didymella fucicola (Sutherland) Kohlm.

Phytopathol. Z. **63**, 342, 1968

≡*Didymosphaeria fucicola* Sutherland, *New Phytol.* **14**, 188, 1915
=*Didymosphaeria pelvetiana* Sutherland, *New Phytol.* **14**, 185, 1915

LITERATURE: Kohlmeyer and Kohlmeyer (1964–1969); Wilson and Knoyle (1961).

Fig. 83

ASCOCARPS 120–186 μm high, 170–275 μm in diameter (including clypeus), ellipsoidal, depressed or pyriform, totally or almost completely immersed, ostiolate, epapillate, coriaceous, brownish-black above, hyaline below, clypeate, solitary or gregarious. PERIDIUM 78–112 μm

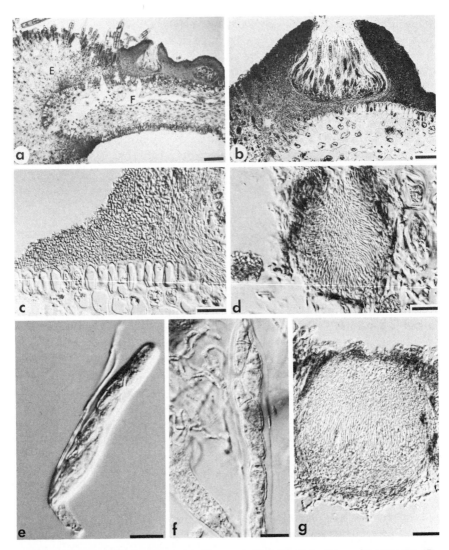

Fig. 83. *Didymella fucicola.* (a) 10-μm section through *Fucus vesiculosus* (F), partly covered by *Elachista fucicola* (E) and ascocarps of *D. fucicola* (bar = 100 μm). (b) Longitudinal section (6 μm) through mature ascocarp superficially seated on *F. vesiculosus* and adjoining thallus of *E. fucicola* to the left (bar = 50 μm). (c) 2-μm section through basis of ascocarp on *F. vesiculosus* (bar = 25 μm). (d) Longitudinal section (6 μm) through spermogonium in *Pelvetia canaliculata* (bar = 15 μm). (e) Ascus (bar = 10 μm). (f) Ascospores enclosed in thick-walled ascus (bar = 10 μm). (g) 10-μm section through immature ascocarp filled with pseudoparaphyses (bar = 25 μm). (a–c from Herb. J.K. 3120; d–g from Herb. Wilson on *P. canaliculata;* a and b in bright field, the others in Nomarski interference contrast.)

thick in the clypeus, composed of irregular, rounded, thick-walled cells with small lumina; laterally and basally 25–33 μm thick, composed of polygonal, thin-walled cells with large lumina, forming a textura angularis, enclosing cells of the host. OSTIOLAR CANAL 25–50(–70) μm in diameter, surrounded by polygonal, thin-walled, hyaline cells, about 3 μm in diameter. PSEUDOPARAPHYSES 1–2 μm in diameter, apically thickened, septate, simple or ramose, with gelatinous walls, enclosing the asci as a firm ball, projecting into the ostiole. ASCI 50–90(–115) × 10–22 μm, eight-spored, clavate or subcylindrical, short pedunculate, bitunicate, thick-walled, physoclastic, endoascus remaining protruding from ectoascus after ascospore discharge, without apical apparatuses; developing at the base of the ascocarp venter. ASCOSPORES 16–23 × 6–8 μm, biseriate, obovoidal or ellipsoidal, one-septate somewhat below the center, slightly constricted at the septum, hyaline, possibly becoming yellowish in age. SPERMOGONIA 58–85 μm high, 40–87 μm in diameter, subglobose, totally or almost completely immersed, ostiolate, epapillate, coriaceous, brownish-black above, hyaline below, clypeate, solitary or gregarious. PERIDIUM 19–21 μm thick in the clypeus, composed of irregular, thick-walled cells; laterally and basally 11–13 μm thick, composed of polygonal, thin-walled cells with large lumina. OSTIOLES circular. SPERMATIOPHORES 9–13.5 × 2–2.5 μm, phialidic, cylindrical to conical, simple, lining the walls of the locule. SPERMATIA 3.1–3.8 × 1.4–2 μm, ellipsoidal, one-celled, hyaline. MODE OF LIFE: Perthophytic. HOSTS: Midribs and air vesicles of lower branches of live *Fucus spiralis* and *F. vesiculosus,* vegetative thalli of *Pelvetia canaliculata;* often associated with bases of *Elachista clandestina* and *E. fucicola.* RANGE: Atlantic Ocean—Canada [Québec (J. Kohlmeyer, unpublished)], France (J. Kohlmeyer, unpublished), Great Britain [England, Scotland (Orkney Islands), Wales], Norway (J. Kohlmeyer, unpublished), United States (Maine, Massachusetts, Rhode Island). NOTE: *Didymosphaeria pelvetiana* is reduced to synonymy of *Didymella fucicola* because these fungi described in the same paper by Sutherland (1915b) from *Pelvetia* and *Fucus* are morphologically identical and can be separated only by the different host plants. Sutherland (1915b) already noted a striking resemblance between the two, but separated them on the basis of alleged size differences in ascocarps and asci and of supposedly different shapes of ascospores. No type material of Sutherland's species was available for reexamination (Kohlmeyer, 1968e, 1974b). However, new collections of *D. pelvetiana* made by Dr. Irene M. Wilson (Wilson and Knoyle, 1961), and kindly placed at our disposal, permitted a comparison between the fungus from *Pelvetia* with our ample collections of *D. fucicola* (Kohlmeyer, 1968e, also unpublished). Dimensions of ascocarps and asci are identical in both hosts, and ascospores are hyaline, ellipsoidal, and slightly constricted. We did not observe the slight yellowish coloration of ascospores mentioned by Sutherland (1915b). In the absence of type material of either species, we decided to retain *D. fucicola* and reduce *D. pelvetiana* to synonymy because a neotype for the former had been designated before (Kohlmeyer, 1968e). The retention of *D. pelvetiana* would have required the transfer to *Didymella* and, therefore, would have created a new binomial. ETYMOLOGY: From the host genus, *Fucus,* and the Latin ending *-cola* = -dweller.

Didymella gloiopeltidis (Miyabe et Tokida) J. Kohlmeyer et E. Kohlmeyer comb. nov.

≡*Guignardia gloiopeltidis* Miyabe et Tokida, *Bot. Mag.* **61**, 118, Figs. 1–6, 1948 (basionym)

LITERATURE: G. Feldmann (1957); T. W. Johnson and Sparrow (1961); Meyers (1957).

Fig. 84

ASCOCARPS 110–190 μm high, 80–190 μm in diameter, subglobose, completely immersed, ostiolate, epapillate or short papillate, coriaceous, dark brown with a lighter colored basis, solitary, seated on a network of thin, hyaline hyphae. PERIDIUM 8–15 μm thick in the base and sides, 14–24 μm around the ostiole, composed of hyphoid, loosely packed cells, forming a textura intricata with brown incrustations. OSTIOLAR CANAL about 15–20 μm in diameter in young ascocarps, opening to about 30 μm in mature ones by expanded endoasci of emptied asci, filled with periphyses; the round opening surrounded by irregular tufts of brown hyphae. PSEUDOPARAPHYSES filiform, numerous, apically merging with periphyses. ASCI 48–80 × 14–18.5 μm, eight-spored, oblong–clavate to ventricose, short or long pedunculate, bitunicate, thick-walled, physoclastic, wall with a canal in the tip, but without apical apparatus; developing in parallel at the base of the ascocarp venter with the curved stalks deeply embedded in the ascogenous tissue; endoascus protruding from ectoascus after ascospore release. ASCOSPORES (14–)17–22 × 4–5.5 μm, biseriate, obovoid, subclavate or subcylindrical, one-septate in the lower third (apiosporous), not constricted at the septum, hyaline; upper cell longer and wider than the lower one; rarely inverted in the ascus; usually germinating from the large apical cell. PYCNIDIA (or spermogonia?) 100–130 μm high, 100–120 μm in diameter, subglobose, immersed, opening widely with a circular ostiole, epapillate, coriaceous, brown to black. PERIDIUM 10–12 μm thick, forming a textura intricata with brown incrustations.

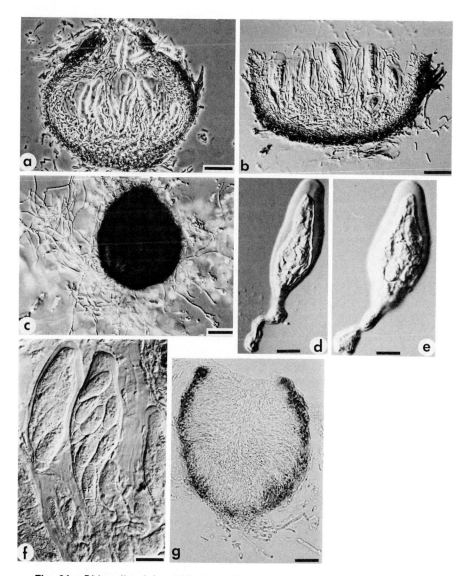

Fig. 84. *Didymella gloiopeltidis* from *Gloiopeltis furcata*, syntype in AHFH, No. 27570. (a) Longitudinal section (8 μm) through ascocarp (bar = 30 μm). (b) 6-μm section through ascocarp, separated from host and spreading out by swelling of asci and pseudoparaphyses (bar = 25 μm). (c) Whole mount of ascocarp surrounded by hyphae (bar = 35 μm). (d) Ascus showing apical canal (bar = 10 μm). (e) Ascus attached to ascogenous cell (bar = 10 μm). (f) Mature asci with ascospores (bar = 10 μm). (g) Longitudinal section (6 μm) through spermogonium (bar = 25 μm). (a in phase contrast, g in bright field, the others in Nomarski interference contrast.)

CONIDIOPHORES (or spermatiophores?) developing all along the wall, filamentous. CONIDIA (or spermatia?) about 6–10 × 0.7 μm, bacilliform, one-celled, hyaline. MODE OF LIFE: Parasitic. HOST: *Gloiopeltis furcata* (Rhodophyta); causing light-colored discolorations in the alga. RANGE: Pacific Ocean—Japan. NOTE: Description based on our examination of isotype material from Herb. AHFH (No. 27570) and specimens from the type locality (J.K. No. 2727) collected and kindly sent to us by Dr. Hiroshi Yabu; G. Feldmann (1957), Meyers (1957), and T. W. Johnson and Sparrow (1961) all state that *G. gloiopeltidis* does not belong to *Guignardia* without proposing any changes; G. Feldmann (1957) suggests a relationship to *Mycosphaerella;* however, our examination, showing distinctive pseudoparaphyses and parallel asci, rather indicates a kinship with *Didymella* (Mycosphaerellaceae, sensu von Arx and Müller 1975). ETYMOLOGY: Named after the host genus, *Gloiopeltis* (Rhodophyta).

Didymella magnei G. Feldmann

Rev. Gén. Bot. **65,** 414–415, 1958

LITERATURE: T. W. Johnson and Sparrow (1961).

Fig. 85

ASCOCARPS 70–120 μm high, 80–130 μm in diameter, subglobose, im-

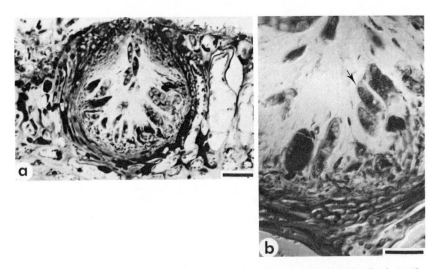

Fig. 85. *Didymella magnei* in *Rhodymenia palmata.* (a) Longitudinal section through ascocarp (bar = 20 μm). (b) Ascospores with inconspicuous sheath (arrow) in section of an ascocarp (bar = 10 μm). (Type slide R.4 of G. Feldmann; a in bright field, b in Nomarski interference contrast.)

mersed, ostiolate, epapillate, subcarbonaceous or subcoriaceous, blackish. PERIDIUM 19–28 μm thick in the dark, thickened, apical part, 6–13 μm thick at the hyaline, thinner sides and base; composed of ellipsoidal or elongate, thick-walled cells with large lumina, forming a textura angularis, enclosing some cells of the host. OSTIOLE circular, about 2 μm in diameter. PSEUDOPARAPHYSES (or paraphyses?) filamentous, simple or ramose, eventually deliquescing. ASCI 25–38 × 7.5–10 μm, eight-spored, ellipsoidal, nonpedunculate, unitunicate, thin-walled, without apical apparatuses; developing on an ascogenous tissue that lines the base and sides of the ascocarp venter. ASCOSPORES (8–)11.5–16.5 × (2–)3–4 μm, biseriate, elongate–ellipsoidal, one-septate, not constricted at the septum, hyaline, enclosed in an inconspicuous, gelatinous sheath. MODE OF LIFE: Parasitic. HOST: *Rhodymenia palmata*. RANGE: Atlantic Ocean—France (known only from the original collection). NOTE: Description in part after J. Kohlmeyer [unpublished; examinations of type slides (from Herb. G. Feldmann) stained in hematoxylin, and Feulgen plus Fast Green, respectively]; *D. magnei* appears to have a pycnidial or spermogonial state; the thin-walled, aphysoclastic asci indicate that this species should be removed from *Didymella*, but fresh material is required to obtain missing information on the ontogeny of the species. ETYMOLOGY: Named in honor of the collector, F. Magne.

Key to the Marine Species of *Didymella*
1. Asci shorter than 45 μm, ascospores usually shorter than 16 μm; on *Rhodymenia* . . . . *D. magnei*
1'. Asci longer than 45 μm, ascospores usually longer than 16 μm; on other hosts . . 2
  2(1') Ascospore septate in the lower third, diameter less than 6 μm; on Rhodophyta . . . . *D. gloiopeltidis*
  2'(1') Ascospore septate near the center, diameter 6 μm or more; on Phaeophyta . . . . *D. fucicola*

Mycosphaerella Johanson

Oefvers. Foerh. K. Sven. Vetensk.-Akad. **41**, 163, 1884

[Synonyms: See Barr (1972); von Arx and Müller (1975)]

LITERATURE: Barr (1972); Dennis (1968); Holm (1975); Luttrell (1973); Müller and von Arx (1962); von Arx (1949); von Arx and Müller (1975).

ASCOCARPS solitary or gregarious, globose, subglobose, conical or depressed, in some species surrounded by a hyphoid stroma or even enclosed in a compact stroma, immersed or erumpent, ostiolate, epapillate or with a short papilla, thin-walled to medium thick-walled. PERIDIUM composed of one to four layers of dark, polygonal cells. PSEUDO-

PARAPHYSES usually absent; young ascocarps filled with a pseudo-parenchyma of thin-walled, polygonal or rounded, deliquescing cells. ASCI eight-spored, elongate–cylindrical, ventricose, ovoid or rarely short clavate, bitunicate, arising in a fascicle or parallel to each other from a small-celled ascogenous tissue at the base of the ascocarp venter. ASCOSPORES biseriate or irregularly arranged, ellipsoidal to elongate, usually three times as long as wide, one-septate near the middle, hyaline, in some species becoming brownish in age or surrounded by a gelatinous sheath. MODE OF LIFE: Saprobic on dead substrates, rarely in spots on living leaves. TYPE SPECIES: *Mycosphaerella punctiformis* (Pers. ex Fr.) Starbäck ≡ *Sphaerella punctiformis* (Pers. ex Fr.) Rabenh. NOTE: Description after the above-named authors; five *Mycosphaerella* species have been described from marine habitats, whereas about 500 terrestrial species are known; conidial states belong to Hyphomycetes or Coelomycetes. ETYMOLOGY: From the Greek, *mykes* = fungus and *Sphaerella* = a genus of Pyrenomycetes (*sphaera* = globe and *-ellus* = a Latin diminutive suffix).

Mycosphaerella ascophylli Cotton

Trans. Br. Mycol. Soc. **3**, 96, 1907

≡*Sphaerella ascophylli* (Cotton) Saccardo et Trotter *in* Saccardo. *Syll. Fung.* **22**, 147, 1913
=*Mycosphaerella pelvetiae* Sutherland, *New Phytol.* **14**, 34–35, 1915
≡*Sphaerella pelvetiae* (Sutherland) Trotter *in* Saccardo, *Syll. Fung.* **24**, 849, 1928

PROBABLE SPERMOGONIAL STATE: *Septoria ascophylli* Melnik et Petrov, *Nov. Sist. Niz. Rast.* pp. 211–212, 1966.

LITERATURE: Barr (1972); Boyd (1911); Church (1919); David (1943); J. Feldmann (1954); T. W. Johnson and Howard (1968); T. W. Johnson and Sparrow (1961); Kingham and Evans (1977); Kohlmeyer (1968e, 1974b); Kohlmeyer and Kohlmeyer (1964–1969, 1972a); Kremer (1973); Mattick (1953); Meyers (1957); Smith and Ramsbottom (1915); F. C. Webber (1967); Wilson (1951, 1960).

Fig. 86

MYCELIUM systemic in the host plants; hyphae 1–1.5(–2) μm in diameter, septate, branched, intercellular. ASCOCARPS 65–130(–155) μm high, 55–130(–160) μm in diameter, ovoid or obpyriform, truncate above, with a bundle of interwoven, rootlike hyphae at the base, immersed in receptacles, rarely in stalks of the hosts, ostiolate, epapillate, coriaceous, dark brown above, pallid below, solitary or gregarious. PERIDIUM 15–17.5 μm thick in the upper part, 5–10 μm in the sides and base, composed of three

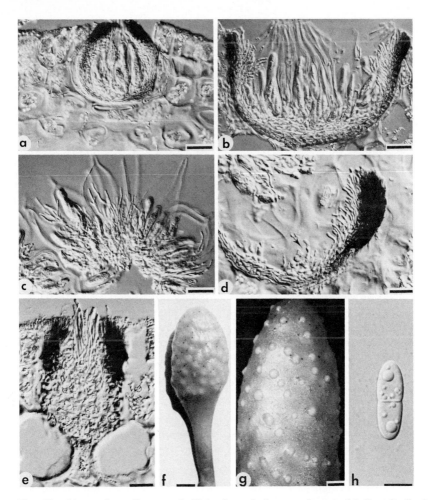

Fig. 86. *Mycosphaerella ascophylli* in *Ascophyllum nodosum.* (a) Longitudinal section through ascocarp in *A. n.* ecad *mackaii* (Herb. South 4725; bar = 25 μm). (b) 8-μm section through ascocarp, spread out to show expanded empty asci (Herb. J.K. 2481; bar = 25 μm). (c) Asci and pseudoparaphyses (Herb. South; bar = 25 μm). (d) Section through ascocarp showing ascogenous hyphae (arrow) and periphyses (Herb. South 4725; bar = 15 μm). (e) 8-μm section through immature ascocarp with trichogynes protruding from the ostiole (Herb. J.K. 2481; bar = 15 μm). (f) Algal receptacle dotted by black ascocarps (Herb. J.K. 2481; bar = 2 mm). (g) Receptacle of *A. nodosum* with black ascocarps, mucilage containing algal sexual products is oozing from the conceptacles (bar = 1 mm). (h) Ascospore (bar = 5 μm). (f and g close-up photographs, the others in Nomarski interference contrast.)

or four layers of thick-walled, elongate or ellipsoidal cells with large lumina, forming a textura angularis. OSTIOLAR CANAL 12–24(–36) μm in diameter, lined with periphyses; trichogynes emerging from ostioles in young ascocarps; ostioles prominent. PSEUDOPARAPHYSES 12–25 × 1–2.5 μm, simple or rarely ramose, parietal, indistinct, apically merging into periphyses. ASCI 37–64(–68) × 11–21.5 μm, eight-spored, ellipsoidal or clavate, pedunculate, bitunicate, thick-walled, physoclastic, without apical apparatuses; the thick-walled apex is pierced by a narrow canal; asci developing at the base of the ascocarp venter; tips of empty asci may project through the ostiole. ASCOSPORES 15–22.5(–25) × 4–5.5(–6.5) μm, bi- or triseriate, ellipsoidal, one-septate somewhat below the middle, not or slightly constricted at the septum, hyaline. SPERMOGONIA (probably identical with *Septoria ascophylli* from *Ascophyllum*) 60–100 μm high, 40–60 μm in diameter, flask-shaped, ostiolate, epapillate, thin-walled. SPERMATIOPHORES 7–10 × 2 μm, cylindrical to obclavate, simple, lining the base and sides of the locule. SPERMATIA 12.5–30 × 0.3–1 μm, filiform, straight or curved, 0–2-septate, hyaline. MODE OF LIFE: Symbiotic; forming mycophycobioses. HOSTS: *Ascophyllum nodosum, A. n.* forma *mackaii* and *Pelvetia canaliculata*. RANGE: Atlantic Ocean—Canada [Newfoundland, Nova Scotia, Québec (J. Kohlmeyer, unpublished)], Denmark (Faeroe Islands), France, Germany (Helgoland, F.R.G.), Great Britain (England, Ireland, Scotland, Wales), Iceland, Norway (J. Kohlmeyer, unpublished), Sweden, United States (Maine, Massachusetts). ETYMOLOGY: Named after a host genus, *Ascophyllum* (Phaeophyta).

Mycosphaerella pneumatophorae Kohlm.
Ber. Dtsch. Bot. Ges. **79**, 32, 1966

LITERATURE: Kohlmeyer (1968c, 1976); Kohlmeyer and Kohlmeyer (1964–1969, 1971a).

Fig. 87a

ASCOCARPS 90–170 μm high, 110–175 μm in diameter, subglobose or depressed ellipsoidal, eventually more or less cupuliform, immersed in the bark tissue, finally apically exposed, ostiolate, epapillate, coriaceous, brownish-black above, pallid below, solitary. PERIDIUM 15–20 μm thick, composed of irregular, polygonal, thick-walled, dark brown cells with large lumina in the upper part of the ascocarp, with similar, but light-colored cells with small lumina in the base forming a textura angularis, enclosing fragments of hostal cell walls. OSTIOLAR CANAL about 35 μm in diameter, finally enlarged by a crumbling-away of cells of the peridium, developing into an irregular opening of about 75 μm in diameter, exposing

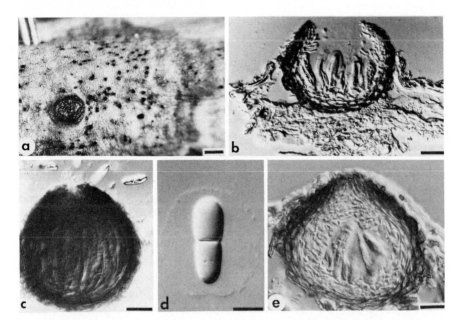

Fig. 87. *Mycosphaerella* spp. (a) Black ascocarps of *M. pneumatophorae* in bark of pneumatophore of *Avicennia germinans* (type IMS, Herb, J.K. 1737c; bar = 500 μm). (b) Longitudinal section (6 μm) through ascocarp of *M. salicorniae* in *Salicornia ambigua* (type material of *Sphaerella peruviana* in LPS, No. 2172; bar = 15 μm). (c) Whole mount of thin-walled ascocarp of *M. salicorniae* (Herb. J.K. 3520; bar = 15 μm). (d) Ascospore of *M. salicorniae* with gelatinous sheath (Herb. J.K. 3772; bar = 5 μm). (e) Longitudinal section through ascocarp of *M. staticicola* (type in FH; bar = 15 μm). (a = close-up photograph, the others in Nomarski interference contrast.)

the asci. PSEUDOPARAPHYSES forming an indistinct network in immature ascomata, absent in mature fruiting bodies. ASCI 37–60 × 15–21 μm, eight-spored, ellipsoidal or clavate, short pedunculate, bitunicate, thick-walled, physoclastic, without apical apparatuses; endoascus remaining protruding from ectoascus after ascospore discharge and filling the ostiolar opening of the ascocarps; developing at the base of the ascocarp venter. ASCOSPORES 14–18(–21) × (6–)7–8.5 μm, bi- or triseriate, obovoid, one-septate somewhat above the middle, slightly constricted at the septum; upper cell wider than lower one; lower cell tapering at the base; hyaline. MODE OF LIFE: Parasitic (or saprobic?). HOSTS: Bark of living pneumatophores of *Avicennia africana* and *A. germinans*. RANGE: Atlantic Ocean—Bahamas, Colombia, Liberia, Mexico (Veracruz), United States (Florida). ETYMOLOGY: From the Greek, *pneuma* = breath and *-phorus* = -carrying, in reference to the characteristic substrate of the fungus, namely, pneumatophores.

Mycosphaerella salicorniae (Auerswald) Petrak

Hedwigia **74**, 35, 1934

≡*Sphaeria salicorniae* Auerswald, Botanischer Tauschverein 1863 (fide Gonnermann and Rabenhorst 1869)
≡*Lizonia salicorniae* (Auerswald) Auerswald, Botanischer Tauschverein 1866 (fide Gonnermann and Rabenhorst 1869)
≡*Sphaerella salicorniae* (Auerswald) Auerswald *in* Gonnermann et Rabenhorst, Mycol. Eur., No. 5, Pyrenomycetes, p. 16, Dresden 1869
=*Sphaerella peruviana* Spegazzini, *An. Soc. Cient. Argent.* **12**, 115, 1881

LITERATURE: J. B. Ellis and Everhart (1892); Kohlmeyer and Kohlmeyer (1977); Săvulescu (1947); Winter (1887).

Figs. 87b–87d

ASCOCARPS 22–80 μm high, 24–80(–90) μm in diameter, globose, sub-globose or pyriform, immersed under the epidermis or erumpent, ostiolate, epapillate, coriaceous, light or dark brown, gregarious, often united by strands of brown cells or hyphae. PERIDIUM 6–10 μm thick, composed of two or three layers of elongate or ellipsoidal cells with large lumina, forming a textura angularis. OSTIOLAR CANAL about 6–10 μm in diameter, lined with periphyses. PSEUDOPARAPHYSES absent in mature ascocarps. ASCI 20–40(–48) × 10–18(–20) μm, eight-spored, at first cylindrical, later obclavate to ellipsoidal, bitunicate, thick-walled, physoclastic, without apical apparatuses; developing at the base of the ascocarp venter. ASCO-SPORES (8–)10–18 × (2–)3–6 μm, bi- or triseriate, clavate–oblong or elongate–ellipsoidal, one-septate near the middle, not or slightly constricted at the septum, upper cell somewhat larger than the lower one, hyaline, in fresh material surrounded by an inconspicuous gelatinous sheath, about 5 μm thick. MODE OF LIFE: Saprobic. HOSTS: Drying stalks and inflorescences of Chenopodiaceae, namely, *Salicornia ambigua* (=*S. peruviana*), *S.* cf. *fruticosa, S. herbacea, S. h.* var. *procumbens, S. subterminalis, S. virginica, Suaeda fruticosa.* RANGE: Atlantic Ocean— Argentina (Buenos Aires), Bahamas [Great Abaco (J. Kohlmeyer, unpublished)], Bermuda, Great Britain (England), Mexico (Yucatan), United States (Connecticut, North Carolina); Mediterranean—France; Pacific Ocean—Peru (Ica); in addition to these marine sites the species has been collected in inland saline or alkaline habitats in Argentina, Germany (G.D.R.), Rumania, and the United States (California). NOTE: The foregoing description is mostly based on our examination of 18 collections (J. Kohlmeyer, unpublished); among this material are type specimens of *Sphaerella salicorniae* (Herb. B) and *S. peruviana* (Herb. LPS) that are identical; the following collection in Herb. B is designated as lectotype: "Georg Winter, *Sphaerella salicorniae* Awd., Kötzschau b. Leipzig, leg. Auerswald"; Winter (1887, p. 373) notes that the description of *S. salicorniae* rendered by Auerswald, in particular the sizes, does not agree with the type material that is in his, Winter's, possession; it can be assumed that the specimen collected by Auerswald and provided with Winter's herbarium label is the type. *Mycosphaerella staticicola* from Plumbaginaceae is closely related to but not identical with *M. salicorniae*. ETYMOLOGY: Named after a host genus, *Salicornia*.

Mycosphaerella staticicola (Patouillard) S. Dias

Mem. Soc. Brot. **21**, 72, 1970–1971

≡*Sphaerella staticicola* Patouillard, *Catalogue Raisonné des Plantes Cellulaires de la Tunésie,* Paris, p. 104, 1897 (sub *S. staticecola*)

Fig. 87e

ASCOCARPS 60–80 μm high, 60–120 μm in diameter, ellipsoidal to subglobose, immersed under the epidermis, ostiolate, epapillate or rarely with a short papilla, membranous, light brown, more or less regularly scattered, connected to each other by thick-walled, branched, irregular hyphae, 4–10 μm in diameter. PERIDIUM 10–14 μm thick, composed of three layers of thin-walled, ellipsoidal or irregularly rounded cells, brown at the outside, hyaline toward the center, forming a textura angularis. OSTIOLES about 10 μm in diameter, usually under stomata; ostiolar canal lined with cylindrical, hyaline periphyses, 10–14 × 2 μm, merging downward into short paraphysislike cells (4–8 μm long, 2–3 μm in diameter) that cover the wall of the ascocarp cavity. ASCI 30–50 × 14–18 μm, eight-spored, ventricose, thick-walled, especially near the apex, without apical apparatuses, developing at the base of the ascocarp venter. ASCOSPORES 12–15 × 4–6 μm, multiseriate, ellipsoidal, one-septate near the middle, not or slightly constricted at the septum, upper cell larger than lower one, hyaline. PYCNIDIA (or spermogonia?) 60–75 μm high, 70–130 μm in diameter, ellipsoidal, immersed under the epidermis among ascocarps, ostiolate, short papillate, membranous, light brown, scattered. PERIDIUM 20 μm thick around the ostiole, 5 μm at the base. CONIDIOPHORES (spermatiophores?) lining wall of ascocarp venter, almost up to the ostiole. CONIDIA (spermatia?) 4 × 1.5 μm, ellipsoidal to cylindrical, hyaline. MODE OF LIFE: Saprobic. HOSTS: Drying inflorescences of Plumbaginaceae, namely, *Armeria pungens* and *Limonium* sp. (=*Statice*). RANGE: Atlantic Ocean—Portugal; Mediterranean—Tunisia. NOTE: The foregoing description is mostly based on our examination of the type material of *M. staticicola* in Herb. FH (Patouillard, collected in

Gabès, Tunisia, Feb. 1893); we found a similar species that may be identical with *M. staticicola* on *Limonium brasiliense* in Argentina (Buenos Aires, J. K. Nos. 3528, 3593); the Argentinian fungus is not fully matured and appears to differ from *M. staticicola* only in a pseudoparenchyma composed of polygonal cells, 6–10 × 4–6 μm, in the ascocarp venter; the well-preserved type material of *M. staticicola* contains more or less loosely attached short cylindrical cells lining the walls of the ascocarp center; it is possible that the Argentinian fungus is an early developmental state of *M. staticicola*. ETYMOLOGY: Named after a host genus, *Statice* (=*Limonium*).

Mycosphaerella suaedae-australis Hansford

Proc. Linn. Soc. N. S. W. **79**, 122–123, 1954

IMPERFECT STATE: *Septoria suaedae-australis* Hansford

ASCOCARPS about 150 μm in diameter, punctiform, immersed, ostiolate, papillate, membranous, black above, pale below, glabrous, scattered. PERIDIUM thin, composed of one or two layers of dark brown to paler colored, angular cells. PAPILLA blunt, reaching the surface of the epidermis. PSEUDOPARAPHYSES absent. ASCI 60 × 13 μm, eight-spored, subcylindrical, apex broadly rounded, basally attenuate into a short nodose peduncle, mostly curved, fasciculate. ASCOSPORES 18–20 × 3–3.5 μm, multiseriate, arranged in parallel in the ascus, narrowly ellipsoidal, ends rounded and slightly attenuate, one-septate, not constricted at the septum, smooth, hyaline. PYCNIDIA mixed between ascocarps, up to 150 μm in diameter, punctiform, immersed, ostiolate, papillate, membranous, black above, paler below, glabrous, loosely to closely scattered. PERIDIUM thin, composed of angular cells, dark brown near the ostiole, subhyaline at the base. PAPILLA with a wide pore, penetrating the epidermis. CONIDIA 45–58 × 3 μm, filiform, straight or curved, slightly attenuate at both rounded ends, three-septate, not constricted at the septa, smooth, hyaline. MODE OF LIFE: Saprobic. HOST: Dead stems of *Suaeda australis*. RANGE: Pacific Ocean—Australia (South Australia; known only from the original collection). NOTE: The foregoing description is based on the diagnosis given by Hansford (1954). ETYMOLOGY: Named after the host species.

Mycosphaerella sp. I

ASCOCARPS 100–140 μm high, 100–140 μm in diameter, pyriform, immersed under the epidermis or erumpent, ostiolate, epapillate, coriaceous, brown, blue–green to black around the ostiole, gregarious. PERIDIUM 30–40 μm thick around the ostiole, 10–16 μm thick in the base and sides, composed of large ellipsoidal cells. OSTIOLAR CANAL lined with periphyses, 10–15 × 3 μm. ASCI 80 × 32 μm, eight-spored, clavate to ellipsoidal, bitunicate, thick-walled, without apical apparatuses; developing at the base of the ascocarp venter. ASCOSPORES 23–28 × 8–10 μm, ellipsoidal to fusiform, slightly curved, one-septate near the middle, slightly constricted at the septum, thick-walled, hyaline. MODE OF LIFE: Saprobic. HOSTS: *Spartina alterniflora* and *S.* cf. *densiflora*. RANGE: Atlantic Ocean—Argentina (Buenos Aires). NOTE: This undescribed species appears to be common on decaying *Spartina* spp. in Argentina (collections Herb. J. K. Nos. 3509–3512, I.M.S.); it is not identical with *Mycosphaerella spartinae* (Ell. et Ev.) Tomílin (*Nov. Sist. Niz. Rast.*, p. 190, 1967; basionym *Sphaerella spartinae* Ell. et Ev., *J. Mycol.* **4**, 97, 1888, type from Farlow Herb. examined); *M. spartinae* is not a typical *Mycosphaerella* and not marine as it occurs on inland *Spartina* in Nebraska.

Mycosphaerella sp. II

ASCOCARPS 75–100 μm in diameter, subglobose, immersed, ostiolate, epapillate, coriaceous, brown, gregarious. ASCI 40 × 18 μm, eight-spored, ellipsoidal, thick-walled. ASCOSPORES 15–20 × 5–8 μm, ellipsoidal, straight or slightly curved, one-septate near the middle, not or slightly constricted at the septum, hyaline. MODE OF LIFE: Saprobic. HOSTS: *Spartina* cf. *pectinata* and *Spartina* sp. RANGE: Atlantic Ocean—Canada (New Brunswick, Nova Scotia). NOTE: This species was collected by us in Canada (Herb. J. K. Nos. 3380 and 3384); it is not identical with *M. spartinae* (see remarks under *Mycosphaerella* sp. I).

Key to the Marine Species of *Mycosphaerella*

1. Systemic endophyte in algae (*Ascophyllum, Pelvetia*) . . . . ***M. ascophylli***
1'. Saprobes in higher plants 2
  2(1') In bark of mangrove roots . . . . ***M. pneumatophorae***
  2'(1') In other hosts 3
3(2') In Plumbaginaceae . . . . ***M. staticicola***
3'(2') In other hosts 4
  4(3') In Chenopodiaceae 5
  4'(3') In *Spartina* spp. 6
5(4) Ascocarps more than 100 μm in diameter; ascospores 18 μm long or longer . . . . ***M. suaedae-australis***
5'(4) Ascocarps less than 100 μm in diameter; ascospores up to 18 μm long, with a gelatinous sheath . . . . ***M. salicorniae***
  6(4') Ascospore length 23 μm or more . . . . ***Mycosphaerella* sp. I**
  6'(4') Ascospore length 20 μm or less . . . . ***Mycosphaerella* sp. II**

d. Patellariaceae

Banhegyia Zeller et Tóth

Sydowia **14**, 326, 1960

ASCOCARPS solitary or gregarious, at first semiglobose to conical, later discoid, erumpent, finally superficial, sessile, often rooted in the substrate, fleshy–leathery, dark-colored. EXCIPULA laterally thick, at the margin and base thin, forming a textura angularis. HYPOTHECIA little developed, merging into the foot. EPITHECIA dark-colored, crumbling, finally exposing the hymenium. PSEUDOPARAPHYSES clavate, simple, hyaline with dark apex, surpassing the asci. ASCI eight-spored, clavate or ellipsoidal, pedunculate, bitunicate, apically thick-walled, reacting blue with iodine. ASCOSPORES ellipsoidal, one-septate, hyaline to light brown, with setalike appendages at both ends. SUBSTRATES: Dead wood and bark. TYPE SPECIES: *Banhegyia setispora* Zeller et Tóth. The genus is monotypic. ETYMOLOGY: Named after the Hungarian mycologist J. Bánhegyi.

Banhegyia setispora Zeller et Tóth

Sydowia **14**, 327–328, 1960

=*Celidium proximellum* (Nyl.) Karst. var. *uralensis* Naumoff, *Bull. Soc. Mycol. Fr.* **30**, 384, 1914 [*Banhegyia uralensis* (Naumoff) Kohlm., *Trans. Br. Mycol. Soc.* **50**, 138, 1967, is a nomenclaturally superfluous combination]

LITERATURE: Kohlmeyer and Kohlmeyer (1964–1969); Müller and von Arx (1962).

ASCOCARPS 120–460 μm in diameter, at first semiglobose or truncate–conical, later discoid, apothecialike, erumpent, finally superficial, sessile, often rooted in the substrate with a wedge-shaped foot; fleshy–leathery, dark brown, solitary or gregarious. EXCIPULA laterally thick, at the margin and base thin, light brown; cells large, isodiametrical or subglobose, forming a textura angularis. HYPOTHECIA little developed, scarcely differentiated from the hymenium, hyaline to light brown, merging at the base into the dark brown foot. EPITHECIA brownish, crumbling, finally exposing the hymenium. PSEUDOPARAPHYSES 70–86 × 1.5 μm, apically broadened to 3–7 μm in diameter, clavate, simple, septate, hyaline with brownish apex, surpassing the asci. ASCI 33–83 × 15–27 μm, eight-spored, clavate or ellipsoidal, short pedunculate, bitunicate, apically thick-walled, thinner toward the base, without apical apparatuses, reacting blue with iodine. ASCOSPORES 15–27 × 6–10 μm (excluding appendages), obliquely biseriate, ellipsoidal, one-septate, constricted at the septum, hyaline to light brown, appendaged; at both ends three to eight straight, terminal, radiating, bristlelike, deciduous setae, up to 14 μm long. MODE OF LIFE: Saprobic. SUBSTRATE: Intertidal conifer wood. RANGE: Atlantic Ocean—Spain (Tenerife). NOTE: The fungus is facultative marine and was also collected in terrestrial habitats at high altitudes on *Juniperus communis* (Bück Mountains, Hungary; Ural Mountains, U.S.S.R.; Alps, probably Switzerland). ETYMOLOGY: From the Latin, *seta* = bristle and *spora* = spore, in reference to the appendaged ascospores.

Kymadiscus J. Kohlmeyer and E. Kohlmeyer

Mycologia **63**, 837, 1971

ASCOCARPS solitary or gregarious, semiglobose or discoid, superficial, sessile, rooted in the substrate, fleshy–leathery, dark-colored. EXCIPULA composed of thick-walled cells with large lumina, forming a textura angularis. HYPOTHECIA composed of thin-walled cells, forming a textura angularis. EPITHECIA dark-colored, crumbling, finally exposing the hymenium. PSEUDOPARAPHYSES clavate, simple, hyaline, surpassing the asci. ASCI eight-spored, clavate, pedunculate, bitunicate, apically thick-walled, reacting blue with iodine. ASCOSPORES ellipsoidal or obovoid, one-septate in the lower third, brownish, indistinctly striate. SUBSTRATE: Dead wood. TYPE SPECIES: *Kymadiscus haliotrephus* (Kohlm. et Kohlm.) Kohlm. et Kohlm. The genus is monotypic. ETYMOLOGY: From the Greek, *kyma* = wave, surge or current, and *diskos* = disk, in reference to the habitat of the fungus and to the shape of the ascocarp.

Kymadiscus haliotrephus (J. Kohlmeyer et E. Kohlmeyer) J. Kohlmeyer et E. Kohlmeyer

Mycologia **63**, 837, 1971

≡*Buellia haliotrepha* J. Kohlm. et E. Kohlm., *Nova Hedwigia* **9**, 90, 1965

LITERATURE: Hafellner and Poelt (1976); Kohlmeyer (1969c); Kohlmeyer and Kohlmeyer (1964–1969, 1977).

Figs. 88a–88c

ASCOCARPS 200–320 μm high, 360–1000 μm in diameter, at first subglobose, becoming semiglobose or discoid, flat or convex, apothecialike,

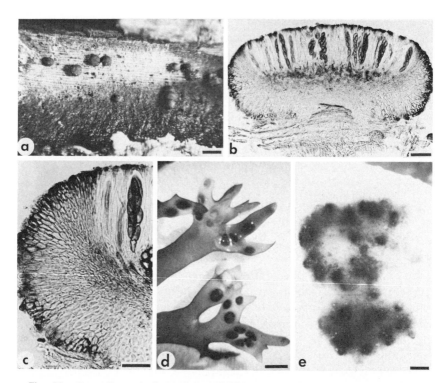

Fig. 88. *Kymadiscus haliotrephus* and *Didymosphaeria danica.* (a) Apothecialike ascocarps of *K. haliotrephus* on the surface of decorticated prop root (type IMS, Herb. J.K. 1744; bar = 1.5 mm). (b) Section (8 μm) through ascocarp of *K. haliotrephus (Herb. J.K. 2682; bar = 30 μm).* (c) Section (8 μm) through excipulum of *K. haliotrephus* (Herb. J.K. 2682; bar = 25 μm). (d) Cystocarps of *Chondrus crispus* blackened by fruiting bodies of *D. danica* (Herb. J.K. 1988; bar = 4 mm). (e) Cluster of ascocarps of *D. danica* in cystocarps of the host (Herb. J.K. 2919; bar = 400 μm). (b and c in bright field, the others close-up photographs.)

superficial, sessile, rooted in the substrate with an obconical foot, fleshy–leathery, dark reddish-brown, appearing almost black when dry, solitary or gregarious. EXCIPULA composed of thick-walled, subglobose to polygonal cells with large lumina, diverging from the center in parallel rows, merging into the sclerenchymatous foot, forming a textura angularis; barely arching over the hymenium. HYPOTHECIA composed of thin-walled cells, forming a textura angularis, claret-colored, merging basally into the hyaline foot. EPITHECIA brownish, crumbling, finally exposing the hymenium. PSEUDOPARAPHYSES 85–125 × 1.2–1.9 μm, at the apex 2.5–4 μm in diameter, clavate, simple, rarely branching, septate, hyaline, surpassing the asci. ASCI 70–100 × 17.5–20 μm, eight-spored, clavate, short pedunculate, bitunicate, apically thick-walled, thinner toward the base, without apical apparatuses; the ectoascus secretes a gelatinous sheath that reacts blue with iodine. ASCOSPORES (15–)18–27.5(–31.5) × 7.5–11.5(–14.5) μm, obliquely one- or biseriate, ellipsoidal or obovoid, one-septate in the lower third, constricted at the septum, at first grayish-green, becoming brownish; epispore with delicate, forked, longitudinal striations. MODE OF LIFE: Saprobic. SUBSTRATES: Intertidal wood and bark, roots of shoreline trees (*Avicennia germinans, Quercus virginiana, Rhizophora mangle;* possibly *Hibiscus tiliaceus*). RANGE: Atlantic Ocean—Bahamas, Bermuda, Trinidad (J. Kohlmeyer, unpublished), United States (Florida, Georgia); Pacific Ocean—United States [Hawaii (Oahu)]. ETYMOLOGY: From the Greek, *hals* = sea, salt, and *trepho* = to grow up, feed, in reference to the marine habitat of the species.

e. Pleosporaceae

Didymosphaeria Fuckel

Jahrb. Nassau. Ver. Naturkd. **23/24,** 140, 1869–1870

=*Cryptodidymosphaeria* (Rehm) von Höhnel, as subgenus: *Ann. Mycol.* **4,** 265, 1906; as genus: *Sitzungsber. Kais. Akad. Wiss. Wien, Math.-Naturwiss. Kl., Abt. 1* **126,** 359, 1917
=*Didymascina* von Höhnel, *Ann. Mycol.* **3,** 331, 1905
=*Didymosphaerella* Cooke, *Grevillea* **19,** 3, 1890
=*Haplovalsaria* von Höhnel, *Sitzungsber. Kais. Akad. Wiss. Wien, Math.-Naturwiss. Kl., Abt. 1* **128,** 583, 1919
=*Massariellopsis* Curzi, *Atti 1st. Bot. Univ. Pavia* **3,** 162, 1927
=*Phaeodothis* Sydow, *Ann. Mycol.* **2,** 166, 1904
=*Rhynchostomopsis* Petrak et Sydow, *Ann. Mycol.* **21,** 378, 1923

LITERATURE: Holm (1957); Müller and von Arx (1962); Scheinpflug (1958); von Arx and Müller (1975).

ASCOCARPS globose or flask-shaped, immersed or erumpent, often with a dark clypeus or hyphoid stroma, ostiolate, short papillate. OSTIOLAR CANAL in primitive species filled by a hyaline pseudoparenchyma, in evolved species by hyphoid cells or delicate pseudoparaphyses. PERIDIUM composed of several layers of roundish, isodiametric or slightly flattened, mostly small and thick-walled, brown cells. PSEUDO-PARAPHYSES in primitive species wide, septate and almost cellular, in more advanced species thin, filiform and branched. ASCI four- or eight-spored, cylindrical or claviform, short or long pedunculate, slightly thickened at the apex, bitunicate, developing at the base of the ascocarp venter. ASCOSPORES uni- or biseriate, elongate, one-septate near the middle, brown. MODE OF LIFE: Saprobic, hyperparasitic, or in lichens. TYPE SPECIES: *Didymosphaeria futilis* (Berk. et Broome) Rehm. NOTE: Description and synonyms combined from the authors listed above. Four marine-occurring species are known among about 100 terrestrial taxa; *D. pelvetiana* Sutherland (1915b) is a doubtful species. ETYMOLOGY: From the Greek, *didymos* = in pairs or two-celled, and *Sphaeria* = a genus of Pyrenomycetes (*sphaera* = globe), in reference to the two-celled ascospores and globose ascocarps.

Didymosphaeria danica (Berlese) Wilson et Knoyle

Trans. Br. Mycol. Soc. **44**, 57, 1961

≡*Leptosphaeria danica* Berlese, *Icones Fung.* **1**, 87, 1892
=*Leptosphaeria chondri* Rosenvinge, *Bot. Tidsskr.* **27**, XXXV, 1906
=*Leptosphaeria marina* Rostrup, *Bot. Tidsskr.* **17**, 234, 1889 (non J. B. Ellis et Everhart 1885)
≡*Didymosphaeria marina* (Rostrup) Lind, Danish fungi as represented in the herbarium of E. Rostrup, Nordisk Forlag, Copenhagen, p. 214, 1913 (imperfect or spermogonial state: *Phoma marina* Lind)
=*Sphaerella chondri* H. L. Jones, *Oberlin Coll. Lab.* **9**, 3, 1898
≡*Guignardia chondri* (Jones) Estee *in* Bauch, *Pubbl. Staz. Zool. Napoli* **15**, 378, 1936

IMPERFECT STATE: *Phoma marina* Lind, Danish fungi as represented in the herbarium of E. Rostrup, Nordisk Forlag, Copenhagen, p. 214, 1913.

LITERATURE: Cotton (1909); J. Feldmann (1954); Kohlmeyer (1964, 1971b); Prince and Kingsbury (1973); Wilson (1951).

Figs. 88d and 88e

ASCOCARPS 125–267 μm high, 110–300 μm in diameter, ampulliform to subglobose, immersed, confined to cystocarps and tetrasporic pustules of the host, ostiolate, papillate, coriaceous, pale, except for the black pseudoclypeus, gregarious. PERIDIUM composed of interwoven hyphae, small-celled, indistinct above, particularly in the ostiolar region. PAPILLAE elongate conical or cylindrical. PSEUDOPARAPHYSES thin, septate, simple or branched, attached at both ends. ASCI 70–94 × 10–15 μm, eight-spored, subclavate to subcylindrical, short pedunculate, bitunicate, thick-walled, physoclastic; developing at the base and lower side of the ascocarp venter. ASCOSPORES (25–)33–40(–44) × 5–7(–8) μm, irregularly biseriate, elongate fusiform, one-septate below the center; lower cell cylindrical, rounded; upper cell broad in the lower half and pointed apically; slightly constricted at the septum, hyaline, but also becoming pale yellowish in age. SPERMOGONIA 100–192 μm high, 63–113 μm in diameter, subglobose to ampulliform, immersed in cystocarps and tetrasporic pustules of the host before the ascocarps, ostiolate, papillate, coriaceous, pale, except for the black pseudoclypeus, gregarious. PERIDIUM composed of interwoven hyphae, small-celled, indistinct above, particularly in the ostiolar region. PAPILLAE elongate conical or cylindrical. SPERMATIOPHORES 16–18 μm long, subcylindrical or almost ampulliform, simple or rarely branched at the base, lining the base and side of the spermogonial venter. SPERMATIA about 4 × 1 μm, oblong to ellipsoidal, one-celled, hyaline, originating in mucilaginous, basipetal strings at the apices of the spermatiophores. MODE OF LIFE: Parasitic. HOST: *Chondrus crispus*. RANGE: Atlantic Ocean—Denmark, France, Great Britain (England, Scotland, Wales), United States (Massachusetts, Rhode Island). NOTE: The occurrence of *D. danica* on grasses and litter (*Agropyrum pungens, Puccinellia maritima*) as listed by Apinis and Chesters (1964) is unlikely, because the fungus is a parasite on a submerged alga; the light color of the ascospores indicates that *D. danica* may belong to another genus, possibly *Didymella*. ETYMOLOGY: From the Latin, *danicus* = Danish, in reference to the origin of the type material.

Didymosphaeria enalia Kohlm.

Ber. Dtsch. Bot. Ges. **79**, 28, 1966

LITERATURE: G. C. Hughes (1974); Kohlmeyer (1968c, 1969c,d); Kohlmeyer and Kohlmeyer (1964–1969, 1971a, 1977); Lee and Baker (1973).

Figs. 89a–89c

ASCOCARPS 295–480 μm high (including papilla), 140–520 μm in diameter, subglobose, ampulliform or depressed ellipsoidal, partly or completely immersed, ostiolate, papillate, clypeate, carbonaceous, black, solitary. PERIDIUM 12.5–17.5 μm thick, composed of about six or more layers of irregular roundish or elongate, thick-walled cells with usually small lumina, forming a textura angularis; merging on the outside into thick, brown hyphae or stromatic structures. PAPILLAE 80–145 μm long, 140–300 μm in diameter (including clypeus), conical, surrounded by a blackish-brown clypeus; ostiolar canal obturbinate, filled with long, delicate, hyaline periphyses; the porus is closed by somewhat thicker, shorter, hyaline hyphae. PSEUDOPARAPHYSES 1.5–2 μm in diameter, septate, rarely branched, attached at both ends and reaching into the ostiolar canal; connected to each other by a gelatinous outer layer and thereby forming a compact ball that firmly encloses the asci. ASCI 117–135 × 12.5–15.5 μm, eight-spored, cylindrical, pedunculate, bitunicate, thick-walled, physoclastic, without apical apparatuses; developing at the base of the ascocarp venter. ASCOSPORES (15.5–)16.5–23 × (6.5–)7.5–10(–11) μm, obliquely uniseriate, ellipsoidal, one-septate, constricted at the septum, dark brown, verrucose to verruculose; sometimes with a distinct small, hyaline tubercle at each apex, probably a germ pore. MODE OF LIFE: Saprobic. SUBSTRATES: Intertidal wood, dead roots, prop roots, and pneumatophores of trees along the shore (*Avicennia africana, A. germinans, Mangifera indica, Quercus virginiana, Rhizophora mangle, R. racemosa*). RANGE: Atlantic Ocean—Bahamas, Bermuda, Liberia, Mexico (Veracruz, Yucatan), Trinidad (J. Kohlmeyer, unpublished), United States (Florida, Georgia, North Carolina); Pacific Ocean—Australia (Queensland), Eniwetok (J. Kohlmeyer, unpublished), Guatemala, United States [Hawaii (Oahu)]. ETYMOLOGY: From the Greek, *enalios* = in, at, or of, the sea, in reference to the marine habitat of the species.

Didymosphaeria maritima (Crouan et Crouan) Saccardo

Syll. Fung. **1**, 703, 1882

≡*Sphaeria maritima* Crouan et Crouan, *Florule du Finistère*, Paris, p. 27, 1867 (non *Sphaeria maritima* Cooke et Plowright, *Grevillea* **5**, 120, 1877)

Figs. 89d–89g

ASCOCARPS 270–420 μm high (including papilla), 300–450 μm in diameter, subglobose to ellipsoidal, superficial, partly or sometimes totally immersed and erumpent, ostiolate, papillate, subcarbonaceous, blackish-brown, gregarious. PERIDIUM 30–70 μm thick, prosenchymatous; surface verruculose by emerging thick-walled cells; peridium dark brown at the outside, hyaline near the venter and merging into pseudoparaphyses; irregular cells, occluded by dark brown deposits, forming a textura epidermoidea; enclosing cells of the host. PAPILLAE

Fig. 89. *Didymosphaeria* spp. (a) Ascus of *D. enalia* with verrucose ascospores (Herb. J.K. 1956; bar = 15 μm). (b) Same as a, ectoascus ruptured, but still adhering to apex (Herb. J.K. 2682; bar = 20 μm). (c) Necks of ascocarps of *D. enalia* breaking through the wood surface (type IMS, Herb. J.K. 1744; bar = 500 μm). (d) 4-μm section through ostiole of *D. maritima* (type Herb. Crouan; bar = 25 μm). (e) Longitudinal section (4 μm) through ascocarp of *D. maritima* (type Herb. Crouan; bar = 50 μm). (f) Ascus of *D. maritima* (type Herb. Crouan; bar = 15 μm). (g) Ascospores of *D. maritima* (type Herb. Crouan; bar = 10 μm). (h) Ascus tip of *D. rhizophorae* with apical apparatus (type IMS, Herb. J.K. 2390; bar = 10 μm). (i) Ascospores of *D. rhizophorae* with delicate wavy surface structure (type IMS, Herb. J.K. 2390; bar = 10 μm). (c close-up photograph, d in phase contrast, e and g in bright field, the others in Nomarski interference contrast.)

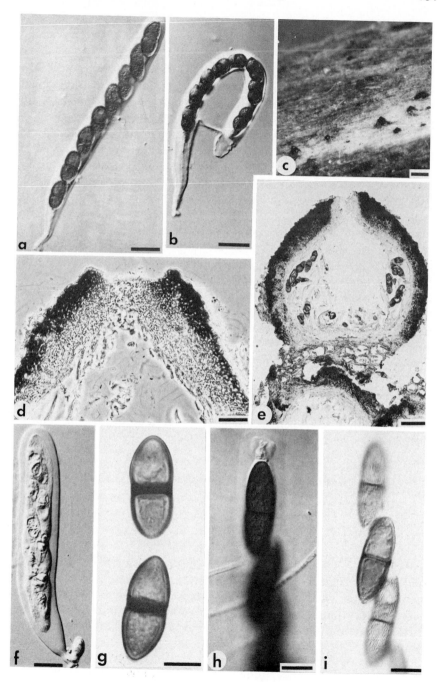

absent or short, about 40 μm high, 140 μm in diameter; ostiolar canal 50–90 μm high, 45–70 μm in diameter, cylindrical, filled with hyaline periphyses, 0.5 μm in diameter. PSEUDOPARAPHYSES 1–2 μm in diameter, filiform, simple or branched, numerous, attached at both ends. ASCI 130–145 × 20–24 μm, eight-spored, cylindrical, short pedunculate, bitunicate, thick-walled, physoclastic, without apical apparatuses; developing in a layer at the base of the ascocarp venter. ASCOSPORES 24–32 × 12–15 μm, uniseriate, ellipsoidal, one-septate in the middle, constricted at the septum, smooth, thick-walled, reddish-brown, darkest around the septum. MODE OF LIFE: Saprobic. SUBSTRATE: Dead stems of *Halimione portulacoides*. RANGE: Atlantic Ocean—France (known only from the type collection). NOTE: The description is based on examination of the type material deposited (sub *Sphaeria maritima* on "*Obione portulacoides*") in the Crouan collection of the Laboratoire de Biologie Marine, Collège de France, Concarneau (J. Kohlmeyer, unpublished); the paper capsule containing the type specimen has the following inscription: "*Sphaeria maritima* Crouan mscr., sur l'*Obione portulacoides* Moq." (slides in Herb. IMS, J.K. No. 3097); the species can be considered marine because the host plant had been submersed as shown by the remains of small red algae on the stem fragments of the type collection; gelatinous apical or lateral bodies seen on some ascospores could indicate that the spores are appendaged, but these bodies could also be mere remnants of ascus cytoplasm, and fresh collections have to be made to clarify this question; *D. maritima* agrees with the modern concept of *Didymosphaeria* as described by Holm (1957), Müller and von Arx (1962), and Scheinpflug (1958). ETYMOLOGY: From the Latin, *maritimus* = marine, referring to the habitat of the species.

Didymosphaeria rhizophorae J. Kohlmeyer et E. Kohlmeyer

Icones Fungorum Maris **1**, (4 and 5), Tabula 62, Cramer, Weinheim, 1967

LITERATURE Kohlmeyer (1976).

Figs. 89h and 89i

ASCOCARPS 300–490 μm high, 200–360 μm in diameter, obpyriform, partly or completely embedded, ostiolate, papillate, subcarbonaceous to subcoriaceous, blackish-brown, gregarious, sometimes two grown together. PERIDIUM 37–45 μm thick, two-layered; the outer, thick stratum pseudostromatic, composed of irregular or roundish, dark brown cells, on the outside with a more or less recognizable hyphal structure, enclosing some decaying cells of the host; inner stratum thin, composed of four or five layers of hyaline, polygonal, elongate, thin-walled cells with large lumina, merging into the pseudoparaphyses. PAPILLAE or necks 65–95 μm in diameter at the apex, conical, emerging in immersed ascocarps; ostiolar canal periphysate; periphyses 1.5–2.5 μm in diameter, septate, simple. PSEUDOPARAPHYSES 1–1.5 μm in diameter, septate, simple or rarely branched, connected to each other by a gelatinous outer layer, persistent. ASCI 150–175 × 14–17.5 μm, eight-spored, cylindrical, short pedunculate, bitunicate, thick-walled, physoclastic, with refractive apical apparatus ("manchon périapical"), without reaction in iodine; developing at the base of the ascocarp venter. ASCOSPORES 23–32(–33) × 9–12 μm, straight or obliquely uniseriate in the ascus, ellipsoidal, one-septate in the center, not or slightly constricted at the septum, dark brown, striate by delicate costae that run parallel or in a slight angle to the longitudinal axis of the ascospore; wall slightly thickened at the apex, without germ pores, germinating apically. MODE OF LIFE: Saprobic (or perthophytic?). SUBSTRATES: Dead intertidal wood, bark, and wood of roots of *Rhizophora mangle*. RANGE: Atlantic Ocean—Bahamas [Great Abaco (J. Kohlmeyer, unpublished)], Colombia, United States (Florida), Venezuela. ETYMOLOGY: Named after the host plant, *Rhizophora* (Rhizophoraceae).

Key to the Marine Species of *Didymosphaeria*

1. On *Chondrus*; Ascospores hyaline, becoming pale yellowish in age   . . . ***D. danica***
1'. On phanerogams and their remains; ascospores dark brown   2
    2(1') A north-temperate species on *Halimione*; ascospores smooth   . . ***D. maritima***
    2'(1') Tropical–subtropical species on mangroves and wood; ascospores ornamented   3
3(2') Ascospores verrucose or verruculose; length 23 μm or less   . . . . . . ***D. enalia***
3'(2') Ascospores delicately striate; length 23 μm or more   . . . . . . ***D. rhizophorae***

Halotthia Kohlm.

Nova Hedwigia **6**, 9, 1963

STROMATA black, subepidermal. ASCOCARPS solitary, gregarious or confluent, conical or semiglobose, at first immersed, then erumpent, ostiolate, epapillate, carbonaceous, thick-walled, black. PSEUDOPARAPHYSES numerous, persistent, ramose. ASCI eight-spored, cylindrical, stipitate, bitunicate, thick-walled, persistent, developing in the angle between covering and basal wall of the ascocarp. ASCOSPORES ellipsoidal, one-septate, brown, thick-walled, septum with a central porus. SPERMOGONIA associated with ascocarps, gregarious or confluent, scutellate, immersed, then erumpent, ostiolate, epapillate, carbonaceous, black, irregularly chambered. SPERMATIOPHORES obclaviform, simple. SPERMATIA subglobose to ellipsoidal, one-celled, hyaline. HOSTS: Marine phanerogams (*Posidonia*). TYPE SPECIES: *Halotthia posidoniae* (Dur. et

Mont. *in* Mont.) Kohlm. The genus is monotypic. ETYMOLOGY: From the Greek, *hals, halos,* f. = the sea, and *Otthia* = a closely related genus.

Halotthia posidoniae (Durieu et Montagne *in* Montagne) Kohlm.

Nova Hedwigia **6**, 9, 1963

≡*Sphaeria posidoniae* Durieu et Montagne *in* Montagne, *Sylloge Generum Specierumque Cryptogamarum,* p. 229, Baillière et Fils, Paris, 1856*
≡*Amphisphaeria posidoniae* (Durieu et Montagne *in* Montagne) Cesati et DeNotaris, *Comment. Soc. Crittogam. Ital.* **1**, 224, 1863

LITERATURE: Desmazières (1850)*; Durieu and Montagne (1869)*; Ollivier (1928, 1930); Kohlmeyer (1959, 1962a, 1964); Kohlmeyer and Kohlmeyer (1964–1969).

Fig. 90

STROMATA 55–110 µm thick, black, subepidermal in cortex of defoliated rhizomes of the host. ASCOCARPS 770–1075 µm high, 1.5–2.1 mm in diameter, broadly conical or semiglobose, flattened at the base, enclosed in the stroma, at first immersed, later erumpent, ostiolate, epapillate, carbonaceous, thick-walled, black, solitary, gregarious or confluent. PERIDIUM in the top 230–285 µm thick, in the sides 165–275 µm, plectenchymatous, enclosing cells of the host. PAPILLAE absent; ostioles 110–140 µm in diameter toward the outside; canal obconical, 90–110 µm in diameter near the cavity, filled with tips of entwined pseudoparaphyses. PSEUDOPARAPHYSES 1.5–2 µm in diameter, septate, ramose, persistent. ASCI 275–290 × 25–35 µm, eight-spored, cylindrical, attenuate at the base, short stipitate, bitunicate, thick-walled, physoclastic, persistent, with a canal in the tip, developing in the angle between covering and basal wall of the ascocarp, directed toward the ostiole. ASCOSPORES 37–60.5 × 16.5–26 µm, uniseriate, ellipsoidal, subcylindrical or obtuse–fusiform, one-septate, constricted at the septum, dark brown, with a dark band around the septum; walls thickened at both ends; septum with a central porus, 0.5 µm in diameter. SPERMOGONIA 250–350 µm high, 1.5 mm long, 500 µm wide, scutellate, subepidermal, in the cortex near ascocarps, immersed, finally erumpent, irregularly chambered, opening with a porus,

---

* The fungus was possibly first described in Montagne 1856 [see Hawksworth and Booth (1974) on *Sphaeria biturbinata,* a similar case]. However, Desmazières distributed *Sphaeria posidoniae* as No. 2062 in fascicle 42 of his *Plantes Cryptogames de France,* published in 1850, citing "Flore d' Algérie" as place of publication without mentioning the year. If Desmazières had seen in print the description of *S. posidoniae* in 1850 or before, "*Exploration Scientifique de l'Algérie, Bot.* **1**, 502–503" must be considered the first place of publication, but the exact year of issue of this volume appears to be in doubt (Hawksworth and Booth, 1974).

Fig. 90. Ascocarps of *Halotthia posidoniae* on rhizome of *Posidonia oceanica* (Herb. J.K. 477; bar = 3 mm).

5 μm in diameter, epapillate, carbonaceous, black, gregarious or confluent. SPERMATIOPHORES obclaviform, simple, lining the walls of the locule. SPERMATIA 2–3.5 μm in diameter, subglobose, ovoid or ellipsoidal, one-celled, hyaline. MODE OF LIFE: Perthophytic (or parasitic?). HOST: Living rhizomes of *Posidonia oceanica*. RANGE: Mediterranean—Algeria, France, Greece, Italy, Yugoslavia, Libya, Spain. NOTE: The fungus reported by Crouan and Crouan (1867) from rhizomes of "*Zostera oceanica*" as *Sphaeria posidoniae* is not identical with the species from *Posidonia, H. posidoniae*, as shown by examinations of material from Herb. Crouan at Concarneau (J. Kohlmeyer, unpublished); T. W. Johnson and Sparrow (1961) list a collection of *Amphisphaeria posidoniae* at Cape Horn, allegedly made by Hariot (1889); this record is erroneous and probably caused by a mistake in translation. ETYMOLOGY: Named after the host plant, *Posidonia* (Potamogetonaceae).

Helicascus Kohlm.

Can. J. Bot. **47**, 1471, 1969

STROMATA with several locules, black, immersed in dead roots. ASCOCARPS depressed ampulliform, horizontally arranged under a black pseudoclypeus, united radially into valsoid groups; ostioles connected into a common, central, periphysate porus. PSEUDOPARAPHYSES numerous, persistent, delicate threads. ASCI eight-spored, subcylindrical to oblong clavate, stipitate, bitunicate, thick-walled, physoclastic, with apical apparatuses; ectoascus forming a third layer at the base; endoascus swelling, stretching and coiling at maturity. ASCOSPORES obovoidal, one-septate, constricted at the septum, brown; walls two-layered; at first covered by a gelatinous sheath. HOSTS: Mangrove trees (*Rhizophora*). TYPE SPECIES: *Helicascus kanaloanus* Kohlm. The genus is monotypic. ETYMOLOGY: From the Greek, *helix* = coil, spiral, and *askos* = ascus, in reference to the coiled endoascus of mature asci.

Helicascus kanaloanus Kohlm.

Can. J. Bot. **47**, 1471, 1969

STROMATA 0.60–0.78 mm high, 1.25–2.75 mm wide, lenticular, immersed, black, carbonaceous, enclosing 3–4(–5) loculi covered by a pseudoclypeus; stromatic tissue around the loculi composed of more or less isodiametric cells, at the periphery built of vertically arranged, longitudinal cells; pseudoclypeus composed of host cells enclosed in black stromatic fungus material. ASCOCARPS 235–370 μm high, 440–800 μm long, depressed ampulliform, united radially into valsoid groups, horizontally arranged under a black pseudoclypeus; ostioles 70–170 μm in diameter, converging at the center, uniting into one common, central porus. PERIDIUM absent; partitions between loculi formed of brown, isodiametric or elongated cells of the stroma. PAPILLAE 270–435 μm high, 255–300 μm in diameter, slightly rising over the surface of the pseudoclypeus, subconical; canal filled with thick, bright orange-colored to yellowish periphyses. PSEUDOPARAPHYSES numerous, persistent, delicate threads that mix with periphyses in the ostiolum. ASCI 250–335 × 25–30 μm, eight-spored, subcylindrical, finally oblong clavate (400–480 μm long), stipitate, bitunicate, thick-walled, physoclastic, apically multilayered and annulate; ectoascus forming a third, thin permeable outer layer around the base; at maturity, endoascus swelling in water and becoming coiled, finally stretching and pushing the ascus into the ostiolar canal. ASCOSPORES 36.5–48.5 × 18–22.5 μm, uniseriate, obovoidal, unequally two-celled, upper cell usually large; distinctly constricted at the septum, dark brown, sometimes at one or both ends apiculate; walls two-layered; endospore 1.7–3.4 μm thick, brown, at each end provided with a germ porus; septum with a central porus, about 4 μm in diameter; exospore thin; at first enclosed in a gelatinous, dissolving sheath, 2.7–5.4 μm thick, with funnel-shaped, apical indentations. MODE OF LIFE: Saprobic. HOST: Dead roots of *Rhizophora mangle*. RANGE: Pacific Ocean—United States [Hawaii (Oahu); known only from the original collection]. ETYMOLOGY: From Kanaloa, a Hawaiian sea god, and the Latin suffix -*anus,* indicating connection or relation, in reference to the origin of the type material, namely, Hawaii.

Keissleriella von Höhnel

Sitzungsber. Kais. Akad. Wiss. Wien, Math.-Naturwiss. Kl., Abt 1 **128**, 592, 1919

=*Coenosphaeria* Munk, *Dan. Bot. Ark.* **15**, 133, 1953
=*Trichometasphaeria* Munk, *Dan. Bot. Ark.* **15**, 135, 1953
=*Zopfinula* Kirschstein, *Ann. Mycol.* **37**, 98, 1939

LITERATURE: Bose (1961); Dennis (1968); Koponen and Mäkelä (1975); Luttrell (1973); Müller and von Arx (1962).

ASCOCARPS solitary or gregarious, globose or semiglobose, immersed or rarely erumpent, ostiolate, porus lined with unicellular, dark or light-colored setae, nonpapillate or rarely papillate, thick-walled above, thin-walled at the sides and base, brown, often clypeate. PSEUDOPARAPHYSES thin, initially attached at both ends, later becoming free at the apex. ASCI four- to eight-spored, cylindrical to clavate or rarely saccate, bitunicate, often thin-walled and wall thickened at the apex. ASCOSPORES fusiform, ellipsoidal or sometimes clavate, one- or multiseptate, hyaline, rarely light brown in age, often surrounded by a mucous sheath or provided with apical appendages. SUBSTRATE: Living or dead tissues of higher plants or hyperparasitic in ascocarps of other Ascomycetes. TYPE SPECIES: *Keissleriella cladophila* (Niessl) Corbaz. NOTE: Description and synonyms after Müller and von Arx (1962); only one marine-occurring species is known among about 14 terrestrial species. ETYMOLOGY: Commemorating K. von Keissler (1872–1965), Austrian mycologist.

Keissleriella blepharospora J. Kohlmeyer et E. Kohlmeyer

Nova Hedwigia **9**, 97, 1965

LITERATURE: Gunn and Dennis (1971); Kohlmeyer (1968c, 1969c,d, 1976); Kohlmeyer and Kohlmeyer (1964–1969, 1971a, 1977).

Fig. 91

ASCOCARPS 125–230 μm high, 225–425 μm in diameter (including dark clypei), depressed ellipsoidal, subglobose or pyriform, immersed in the bark, ostiolate, papillate, coriaceous, subhyaline, clypeate, solitary or gregarious. PERIDIUM 10–15 μm thick, hyaline to light brown, composed of several layers of subglobose, ellipsoidal or elongate, thick-walled cells, forming a textura prismatica; at the edge intimately connected to decomposing bark cells. PAPILLAE short, subconical, surrounded by a brownish-black, 30- to 37.5-μm-thick pseudoclypeus that encloses bark cells of the host; ostiolar canal 12.5–27.5 μm in diameter, periphysate. PSEUDOPARAPHYSES 1.5–2.5 μm in diameter, sparsely sep-

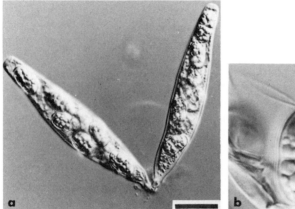
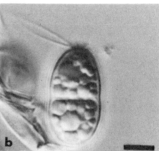

Fig. 91. *Keissleriella blepharospora* from *Rhizophora mangle*. (a) Asci (Herb. J.K. 2736; bar = 20 μm). (b) Ascospore with setalike appendages at one end (Herb. J.K. 2736; bar = 5 μm). From Kohlmeyer and Kohlmeyer, *Mycologia* **63**, 834 (1971), Fig. 4. Reprinted with permission. (Both in Nomarski interference contrast.)

tate, rarely branched, sometimes connected by mucilage to one another, attached at the base and apex. Asci 62–75 × 11–15 μm, eight-spored, cylindrical or oblong–ventricose, apically truncate, short pedunculate, indistinctly bitunicate, thin-walled, thick-walled at the apex and provided with an apical plate; developing at the base of the ascocarp venter. Ascospores 12–21 × 6–8 μm (excluding appendages), one- or biseriate, ellipsoidal, one-septate, not or slightly constricted at the septum, thin-walled, hyaline, appendaged at one end; four to seven radiating, terminal setae, deciduous, about 13 μm long. Mode of Life: Parasitic (or saprobic?). Host: Bark of seedlings and roots of *Rhizophora mangle*. Range: Atlantic Ocean—Bahamas, Bermuda, Colombia, Mexico (Veracruz), Trinidad (J. Kohlmeyer, unpublished), United States [Florida, North Carolina (on drifting seedlings of *R. mangle*)]; Pacific Ocean—Peru [Tumbes (J. Kohlmeyer, unpublished)], United States [Hawaii (Oahu)]. Note: Superficially, *K. blepharospora* may resemble species of *Microthyrium* of the Hemisphaeriales (see J. P. Ellis, 1976), in particular by the ciliate ascospores. However, haustoria are absent in *K. blepharospora*, ascocarps are immersed and often clypeate, and the persistent pseudoparaphyses stand between the basally produced asci. A microthyriaceous Ascomycete, possibly of terrestrial origin, with superficial flat ascocarps occurs sometimes on the cuticle of *Rhizophora* seedlings; it may be confused with *K. blepharospora* since both have hyaline, two-celled ascospores with apical setae, but ascospores of *K. blepharospora* are wider than those of the cuticle-inhabiting species. ETYMOLOGY: From the Greek, *blepharis* = bristle and *spora* = spore, in reference to the setae-bearing ascospores.

Leptosphaeria Cesati et DeNotaris

Comment. Soc. Crittogam. Ital. **1**, 234, 1863 (nom. cons. prop.)

=*Ampullina* Quélet, *Mém. Soc. Emul. Montbéliard,* **3**, 1876
=*Astrotheca* Hino, *Bull. Miyazaki Coll. Agric. For.* **10**, 57, 1938
=*Baumiella* P. Hennings *in* H. Baum, *Ber. Kunene-Exped., Berlin,* p. 105, 1902
=*Bilimbiospora* Auerswald *in* Rabenhorst, *Fungi Europaei* No. 261, 1861, nom. rej. prop.
=*Dothideopsella* von Höhnel, *Sitzungsber. Kais. Akad. Wiss. Wien, Math.-Naturwiss. Kl., Abt. I* **124**, 70, 1915
=*Exilispora* Tehon et Daniels, *Mycologia* **19**, 112, 1927
=*Macrobasis* Starbäck, *Bih. K. Sven. Vetenskaps-Akad. Handl., Afd. 3* **19**, No. 2, 97, 1894
=*Metasphaeria* Saccardo, *Syll. Fung.* **2**, 156, 1883
=*Phaeoderris* von Höhnel, *Sitzungsber. Kais. Akad. Wiss. Wien, Math.-Naturwiss. Kl., Abt. I* **120**, 462, 1911
=*Pocosphaeria* Saccardo, *Syll. Fung.* **2**, 32, 1883
=*Sclerodothis* von Höhnel, *Ann. Mycol.* **16**, 69, 1918
=*Syncarpella* Theissen et Sydow, *Ann. Mycol.* **13**, 631, 1915

LITERATURE: Holm (1957, 1975); Koponen and Mäkelä (1975); Luttrell (1973); Müller (1950); von Arx and Müller (1975).

ASCOCARPS solitary or gregarious, usually broadly conical, often flattened, mostly smooth and always without stiff hairs, usually without stroma, immersed, ostiolate, generally with a small neck; tendency to form dothideaceous fructifications. PERIDIUM generally thick, sides often strongly thickened toward the base, in most species scleroplectenchymatous, composed of large, globose, thick-walled cells. PSEUDO-PARAPHYSES numerous, filamentous. ASCI generally eight-spored, numerous, clavate to subcylindrical, short pedunculate, in some species with a refractive apical apparatus, bitunicate. ASCOSPORES ellipsoidal, fusiform or cylindrical, rarely scolecosporous, with three or more transverse septa, mostly guttulate, yellowish-brown to almost hyaline. MODE OF LIFE: Saprobic or sometimes parasitic, mostly on stalks of dicotyledons. TYPE SPECIES: *Leptosphaeria doliolum* (Pers. ex Fr.) Ces. et DeNot. NOTE: Thirteen marine species are known among 50 (to 200?) taxa; description mainly after Holm (1957); synonyms following von Arx and Müller (1975); conidial states belong to Coelomycetes in the form-genera *Camarosporium, Hendersonia, Phoma, Plenodomus, Rhabdospora,* and *Stagonospora.* ETYMOLOGY: From the Greek, *leptos* = slender, thin, and

Sphaeria = a genus of Pyrenomycetes (*sphaera* = globe), probably in reference to the mostly small and roundish ascocarps.

Leptosphaeria albopunctata (Westendorp) Saccardo

Syll. Fung. **2**, 72, 1883

≡*Sphaeria albopunctata* Westend., *Bull. Acad. R. Belge,* Sér. 2, **7**, 87, 1859

≡*Heptameria (Leptosphaeria) albopunctata* (Westend.) Cooke, *Grevillea* **17**, 32, 1889

=*Leptosphaeria spartinae* Ellis et Everhart, *J. Mycol.* **1**, 43, 1885

=*Leptosphaeria sticta* Ellis et Everhart, *J. Mycol.* **1**, 43, 1885

=*Sphaeria incarcerata* Berkeley et Cooke *in* Berkeley, *Grevillea* **4**, 152, 1876 (non *Sphaeria incarcerata* Desmaz. 1846)

≡*Heptameria (Leptosphaeria) incarcerata* (Berk. et Cooke) Cooke, *Grevillea* **17**, 33, 1889

≡*Leptosphaeria incarcerata* (Berk. et Cooke) Saccardo, *Syll. Fung.* **2**, 86, 1883

LITERATURE: Cavaliere (1966c, 1968); Dennis (1968); J. B. Ellis and Everhart (1892); Gessner and Goos (1973b); Gold (1959); G. C. Hughes (1974); T. W. Johnson (1956a); T. W. Johnson and Sparrow (1961); E. B. G. Jones (1962a); Kohlmeyer and Kohlmeyer (1964–1969); Lucas and Webster (1967); Müller (1950); Neish (1970); Raghukumar (1973); Schmidt (1974); Sutton and Pirozynski (1965); Szaniszlo and Mitchell (1971); Szaniszlo *et al.* (1968); Wagner-Merner (1972).

ASCOCARPS 161–400 μm high (including papillae), 217–450 μm in diameter, subglobose, subovoid or rarely depressed, completely immersed, ostiolate, short papillate, carbonaceous, black, gregarious, arranged in rows; hyphae blackening the epidermis of the host. PERIDIUM near the ostiole 40–45(–70) μm thick, at the sides and base 17.5–27.5 μm, composed of roundish or elongate, thick-walled cells, forming a textura angularis. PAPILLAE about 60 μm high, 80–110 μm in diameter, short, conical, perforating the epidermis; ostiolar canal lined with periphyses (2.5–3.5 μm in diameter). PSEUDOPARAPHYSES 2–2.5 μm in diameter, septate, at the septa thickened to 2.5–3.5 μm, ramose. ASCI (120–)134–210(–293) × 10–24(–30) μm, eight-spored, cylindrical or subclavate, short pedunculate, bitunicate, thick-walled, physoclastic, without apical apparatuses; developing at the base of the ascocarp venter. ASCOSPORES 26–55 × (8–)10–15 μm, obliquely one-seriate or biseriate, fusiform, broadly fusiform or ellipsoidal, rarely slightly curved, five (to seven) -septate, constricted at the septa, third cell often the largest, pale yellowish to light yellow–brown. PYCNIDIA (or spermogonia?) from pure culture

120–500 μm in diameter, globose to ellipsoidal, ostiolate, papillate, sub-carbonaceous, black. PERIDIUM up to 30 μm thick. CONIDIA (or spermatia?) 4–6(–8.5) × 2–3(–4) μm, cylindrical to ellipsoidal, one-celled, subhyaline or light yellowish-brown, appearing reddish-brown in masses. MODE OF LIFE: Saprobic. HOSTS: Dead culms of *Juncus maritimus, Phragmites communis, Spartina alterniflora, S. townsendii, Spartina* sp., and driftwood. RANGE: Atlantic Ocean—Argentina [Buenos Aires (J. Kohlmeyer, unpublished)], Belgium, Canada [New Brunswick (J. Kohlmeyer, unpublished), Nova Scotia], Great Britain (England, Wales), Iceland, United States (Florida, Mississippi, New Jersey, North Carolina, Rhode Island, South Carolina, Virginia); Baltic Sea—Germany (G.D.R.); Indian Ocean—India (Madras). NOTE: The synonymy follows Johnson (1956a); Lucas and Webster (1967) surmise that Westendorp's fungus from *Phragmites* is different from the *Spartina* fungus, and they prefer to use the name *L. spartinae* for the latter; until the question about the identity is solved we are using the name *L. albopunctata*, which has become accepted in marine mycological literature. ETYMOLOGY: From the Latin, *albus* = white and *punctatus* = dotted, in reference to the periphysate ostioles that appear as white dots in the blackened epidermis of the host.

Leptosphaeria australiensis (Cribb et Cribb) G. C. Hughes

Syesis **2**, 132, 1969

≡*Metasphaeria australiensis* Cribb et Cribb, *Univ. Queensl. Pap., Dep. Bot.* **3**, 79, 1955

LITERATURE: Cribb and Cribb (1969); Gold (1959); G. C. Hughes (1974); T. W. Johnson and Sparrow (1961); Kohlmeyer (1966a, 1969c,d, 1976); Kohlmeyer and Kohlmeyer (1964–1969, 1971a, 1977); Meyers (1957).

Figs. 92a–92d

ASCOCARPS 100–200 μm high, 120–195 μm in diameter, obpyriform, immersed, ostiolate, papillate, coriaceous, fuscous or light brown, lighter colored at the base, gregarious or solitary. PERIDIUM 10–25 μm thick, two-layered, forming a textura angularis; outer layer composed of small, irregularly rounded, thick-walled, pigmented cells; inner layer composed of trapezoid, thin-walled, light-colored cells with large lumina. PAPILLAE or necks 70–225 μm long, 30–50 μm in diameter at the tip, 70–75 μm at the base, conical or subcylindrical, fuscous or almost hyaline; ostioles lined with periphyses, 1.5–2.2 μm in diameter. PSEUDOPARAPHYSES 1–2.2 μm in diameter, septate, somewhat branched, with gelatinous walls, persistent. Asci 70–105 × 10–15 μm, eight-spored, clavate–fusiform, pedunculate, indistinctly bitunicate, thick-walled when young, later relatively thin-walled, with apical apparatuses (pulvillus and annulus); developing at the base of the ascocarp venter. Ascospores 19–27 × 5.5–9 μm, biseriate, ellipsoidal, fusiform or clavate–fusiform, three-septate, constricted at the septa, hyaline. Mode of Life: Saprobic in bark and wood of dead roots or seedlings of *Avicennia germinans, A. marina* var. *resinifera, Mangifera indica, Rhizophora mangle, R. racemosa*, submerged branches of *Casuarina* sp. and *Conocarpus erecta* (Kohlmeyer, unpublished), *Hibiscus tiliaceus, Tamarix gallica*, and intertidal wood. Range: Atlantic Ocean—Bahamas (Great Abaco), Bermuda, Colombia, Liberia, Trinidad (Kohlmeyer, unpublished), United States (Florida, North Carolina), Venezuela; Pacific Ocean—Australia (Queensland), United States [Hawaii (Hawaii, Oahu)]. Note: Collections from temperate and antarctic waters identified as *L. australiensis* (G. C. Hughes, 1969; G. C. Hughes and Chamut, 1971), or as being close to this species (Tubaki and Ito, 1973), do not belong to this warm-water fungus (see also G. C. Hughes, 1974). Etymology: Specific name referring to the country of origin of the type collection.

Leptosphaeria avicenniae J. Kohlmeyer et E. Kohlmeyer

Nova Hedwigia **9**, 98, 1965

Literature: Kohlmeyer (1968c, 1969d); Kohlmeyer and Kohlmeyer (1964–1969, 1971a, 1977).

Figs. 92e and 92f

Ascocarps 340–420 μm high (including papillae), 260–300 μm in diameter, pyriform, half immersed, ostiolate, papillate, carbonaceous, black, gregarious, developing in light-colored spots on pneumatophores. Peridium 28–36 μm thick, composed of irregular, thick-walled cells, forming a textura angularis, merging externally into a stromatic tissue

---

Fig. 92. Marine *Leptosphaeria* spp. (a) Ascospores and tip of ascus (arrow) of *L. australiensis* (Herb. J.K. 2733; bar = 10 μm). (b) Same as a (bar = 10 μm). (c) Same as a, ascospore (bar = 5 μm). (d) Same as a, tips of asci and pseudoparaphyses (Herb. J.K. 2680; bar = 15 μm). (e) Ascus of *L. avicenniae* (Herb. J.K. 2735; bar = 10 μm). From Kohlmeyer and Kohlmeyer, *Mycologia* **63**, 834 (1971), Fig. 9. Reprinted with permission. (f) Large ascocarps of *L. avicenniae* (arrow) and smaller pycnidia of *Rhabdospora avicenniae* in light-colored spots of pneumatophore of *Avicennia germinans* (Herb. J.K. 3715; bar = 1 mm). (g) Ascospores of *L. marina* (from type of *L. macrosporidium* in IMI, No. 80279; bar = 15 μm). (h) Longitudinal section (16 μm) through ascocarp of *L. marina* in *Spartina* sp. (Herb. J.K. 3378; bar = 50 μm). (f close-up photograph, h in bright field, the others in Nomarski interference contrast.)

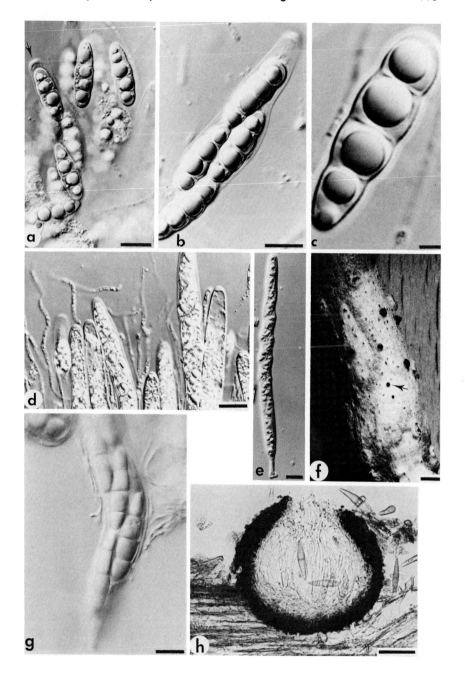

covered with brownish hyphae. PAPILLAE conical, short; ostiolar canal 70 μm in diameter, periphysate. PSEUDOPARAPHYSES septate, simple or ramose. ASCI 112–158 × 7.5–10 μm, eight-spored, cylindrical, short pedunculate, bitunicate, thick-walled, physoclastic, without apical apparatuses; developing at the base of the ascocarp venter. ASCOSPORES 17.5–25 × 5.5–8 μm, uniseriate, ellipsoidal, three-septate, slightly constricted at the septa, hyaline, covered by a gelatinous sheath, 2.5–5 μm thick. MODE OF LIFE: Parasitic (or perthophytic?) HOSTS: Pneumatophores, bark and wood of submerged branches and tree trunks of *Avicennia africana* and *A. germinans.* RANGE: Atlantic Ocean—Bahamas [Great Abaco (J. Kohlmeyer, unpublished), North Bimini], Bermuda, Liberia, Mexico (Veracruz), United States (Florida). NOTE: Black pycnidia of *Rhabdospora avicenniae* developing next to ascocarps of *L. avicenniae* may represent an imperfect (or spermogonial?) state (Kohlmeyer and Kohlmeyer, 1971a). ETYMOLOGY: Named after the host genus, *Avicennia.*

Leptosphaeria contecta Kohlm.

Nova Hedwigia **6**, 314, 1963

LITERATURE: Henningsson (1974, 1976a,b); Kohlmeyer [1960 (sub *Leptosphaeria* sp. form I), 1971a]; Kohlmeyer and Kohlmeyer (1964–1969); Schaumann (1968, 1969).

ASCOCARPS 176–352 μm high, 231–440 μm in diameter, globose or ellipsoidal, immersed or half immersed, ostiolate, papillate, subcarbonaceous, black above, brownish-gray below, gregarious or rarely confluent, often surrounded by thick hyphae. PERIDIUM 16.5–25 μm thick, composed of polygonal, thick-walled cells with large lumina, forming a textura angularis. PAPILLAE 69–137 μm long, 69–151 μm in diameter, conical or subcylindrical, black; ostiolar canal filled with periphysoid hyphae. PSEUDOPARAPHYSES 2.5–3 μm in diameter, septate, simple or ramose. ASCI 102–162 × 16.5–22 μm, eight-spored, cylindrical or subclavate, short pedunculate, bitunicate, thick-walled, physoclastic, with an apical canal but without apical apparatuses; developing at the base of the ascocarp venter. ASCOSPORES 32–44 × 8.5–12 μm, one-seriate in the base, biseriate in the top, fusiform, slightly curved, three (rarely four) -septate, constricted at the septa, hyaline, covered by a hyaline sheath, 9–11 μm thick; the second cell from the top is usually the largest. MODE OF LIFE: Saprobic. SUBSTRATES: Intertidal or submerged wood (e.g., *Betula pubescens* panels, lobster pots). RANGE: Atlantic Ocean—Germany (Helgoland and mainland, G.F.R.), Sweden, United States

(Maine, Massachusetts); Baltic Sea—Sweden; Pacific Ocean—United States (Washington). ETYMOLOGY: From the Latin, *contectus* = covered, in reference to the ascospore sheath.

Leptosphaeria halima Johnson

Mycologia **48**, 502, 1956

LITERATURE: Borut and Johnson (1962); T. W. Johnson and Sparrow (1961); Kohlmeyer and Kohlmeyer (1964–1969).

ASCOCARPS 90–252 μm high, 100–216 μm in diameter, subglobose to pyriform, immersed or superficial, ostiolate, papillate, subcarbonaceous or subcoriaceous, black, solitary or gregarious. PERIDIUM 10–12.5 μm thick, composed of four or five layers of ellipsoidal or irregularly rounded, thick-walled cells with large lumina, forming a textura angularis. PAPIL-LAE about 36–40 μm in diameter, subconical, short; ostiolar canal periphysate. PSEUDOPARAPHYSES 2–2.5 μm in diameter, septate, ramose. ASCI 64–108 × 8–14 μm, eight-spored, subclavate to subcylindrical, short pedunculate, bitunicate, thick-walled, physoclastic, without apical ap-paratuses; developing at the base of the ascocarp venter. ASCOSPORES 12–18 × 5–7(–8) μm, biseriate, oblong–ellipsoidal, subcylindrical or sub-fusiform, three-septate, constricted at the septa, yellow–brown; sur-rounded by a gelatinous, indistinct sheath that may be demonstrated by using India ink; the second cell from the top is largest. MODE OF LIFE: Saprobic. HOSTS: Dead culms of *Spartina alterniflora,* test panels (e.g., *Liriodendron tulipifera*), and driftwood. RANGE: Atlantic Ocean—United States (Florida, North Carolina, Virginia). ETYMOLOGY: From the Greek, *halimos* = marine, referring to the habitat of the species.

Leptosphaeria marina Ellis et Everhart

J. Mycol. **1**, 43, 1885 (also in *Hedwigia* **25**, 109, 1886; non *Leptosphaeria marina* Rostrup 1889)

≡*Heptameria marina* (Ell. et Everh.) Cooke, *Grevillea* **18**, 32, 1889
≡*Metasphaeria marina* (Ell. et Everh.) Berlese, *Icon. Fung.* **1**, 140, 1894
≡*Leptosphaeria treatiana* Saccardo, *Syll. Fung.* **10**, 923, 1892 (superflu-ous name)
=*Leptosphaeria macrosporidium* E. B. G. Jones, *Trans. Br. Mycol. Soc.* **45**, 103, 1962

LITERATURE: Apinis and Chesters (1964); Dennis (1968); J. B. Ellis and Everhart (1892); T. W. Johnson (1956a); T. W. Johnson and Sparrow (1961); Lucas (1963); Wagner-Merner (1972).

Figs. 92g and 92h

Ascocarps 105–400 μm high, 187–550 μm in diameter, globose, sub-globose, globose–conical or subellipsoidal, immersed in culms and covered by the blackened epidermis, later erumpent, or superficial on wood, ostiolate, papillate, carbonaceous, black, solitary. Peridium 20–28 μm thick, black, thickest around ostiole, composed of compressed cells. Papilla short, perforating the epidermis, deciduous and leaving a large opening. Pseudoparaphyses filiform, simple or branched. Asci 112–220 × 18–35 μm, eight-spored, clavate or cylindrical, pedunculate, bitunicate, thick-walled, walls thinner at maturity with an apical chamber, but without apical apparatuses. Ascospores 35–68(–72) × 8–12(–14) μm, triseriate above, uniseriate below, fusiform or clavate–fusiform, straight or slightly curved, one- to three-septate, slightly constricted at the septa, especially at the medium septum, with obtuse apices, subhyaline, becoming yellowish at maturity. Mode of Life: Saprobic. Hosts: Decaying culms of *Juncus roemerianus, Spartina alterniflora, S. townsendii, Spartina* sp. and driftwood. Range: Atlantic Ocean—Canada [New Brunswick, Nova Scotia (J. Kohlmeyer, unpublished)], Great Britain (England and Wales), Portugal, United States [Delaware, Florida, Maine (J. Kohlmeyer, unpublished), New Jersey, North Carolina, South Carolina, Virginia]. Note: Berlese (1894) illustrates one- to five-septate ascospores, based on type material of *L. marina; Juncus maritimus,* allegedly found as host of *L. marina* in North Carolina (T. W. Johnson, 1956a), is omitted because this rush does not occur in the Carolinas (Radford *et al.*, 1968). Etymology: From the Latin, *marinus* = marine, referring to the habitat of the species.

Leptosphaeria neomaritima Gessner et Kohlmeyer

Can. J. Bot. **54**, 2032, 1976

≡*Sphaeria maritima* Cooke et Plowright *in* Cooke, *Grevillea* **5**, 120, 1877; later homonym of *Sphaeria maritima* Crouan et Crouan, *Florule du Finistère,* Paris, p. 27, 1867

≡*Leptosphaeria maritima* (Cooke et Plowright) Saccardo, *Syll. Fung.* **2**, 73, 1883; non *Leptosphaeria maritima* Hollós, *Ann. Mus. Nat. Hung.* **5**, 46, 1907

Literature: Berlese (1894); Dennis (1968); T. W. Johnson (1956a); T. W. Johnson and Sparrow (1961).

Figs. 93a and 93b

Ascocarps 100–185 μm high, 115–250 μm in diameter, globose, sub-globose or ellipsoidal, immersed, later erumpent, ostiolate, papillate,

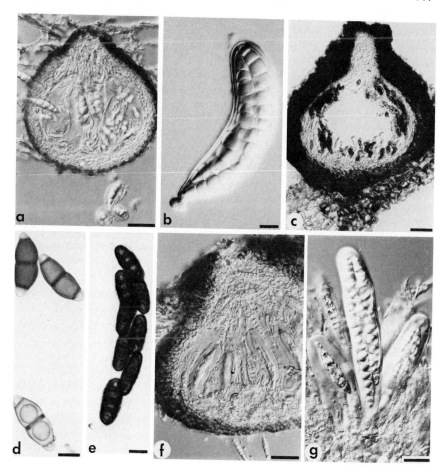

Fig. 93. Marine *Leptosphaeria* spp. (a) Longitudinal section (12 μm) through as-cocarp of *L. neomaritima* from *Spartina* sp.; ascospores with gelatinous sheaths at the bottom (Herb. J.K. 3384; bar = 25 μm). (b) Ascus of *L. neomaritima* (Herb. J.K. 3384; bar = 10 μm). (c) Longitudinal section (16 μm) through ascocarp of *L. obiones* on *Halimione portulacoides* (type Herb. Crouan; bar = 50 μm). (d) Ascospores of *L. obiones* from *Spartina* sp. (Herb. J.K. 3508; bar = 10 μm). (e) Ascus of *L. peruviana* from *Salicornia ambigua* (type Herb. Spegazzini 2172, LPS; bar = 5 μm). (f) Longitudinal section (16 μm) through ascocarp of *L. typhicola* from *Spartina* sp. (Herb. J.K. 3508; bar = 25 μm). (g) Immature and mature asci of *L. typhicola* (Herb. J.K. 3508; bar = 15 μm). (d and e in bright field, the others in Nomarski interference contrast.)

carbonaceous or subcarbonaceous, brown to black, solitary. PERIDIUM about 15 μm thick, mostly composed of five layers of cells. PAPILLAE short, black, perforating the epidermis. PSEUDOPARAPHYSES filiform, simple, filling the ostiolar canal. ASCI 90–145 × (9–)14–30 μm, eight-spored, clavate or cylindrical, pedunculate, bitunicate, thick-walled. AS-COSPORES (30–)32–45 × (6–)8–14 μm, biseriate, fusiform to elongate-ellipsoidal, straight or mostly slightly curved, three- to five-septate, constricted at the septa, especially at the central septum, the third cell from the top usually the largest, at first hyaline, becoming yellowish to yellow–brown at maturity, covered by a gelatinous sheath, 3–4 μm thick. MODE OF LIFE: Saprobic. HOSTS: Decaying culms of *Juncus maritimus, J. roemerianus, Spartina alterniflora, S. townsendii,* and *Spartina* sp. RANGE: Atlantic Ocean—Canada (Nova Scotia), Great Britain (England), United States (Florida, North Carolina, Virginia). NOTE: Description based mainly on T. W. Johnson and Sparrow (1961); two records, published under *L. maritima,* have been deleted because these collections are probably not identical with the species circumscribed above; one fungus with narrower ascospores (3.5–5.5 μm in diameter) occurred in a montane habitat of the Punjab (India) on *Elymus dasystachys* (Wehmeyer, 1963); the other, listed by Munk (1957) from Denmark on *Juncus,* is possibly marine but differs from *L. neomaritima* by 10- to 13-septate ascospores; Munk's fungus is perhaps identical with *Leptosphaeria typhicola;* we found a senescent *Leptosphaeria* with ascospores resembling *L. neomaritima* on *S. alterniflora* in Argentina. ETYMOLOGY: From the Latin, *neo-* = new- and *maritima* = marine (original specific epithet), indicating the relationship of the new to the old name that was pre-occupied by an earlier homonym.

Leptosphaeria obiones (Crouan et Crouan) Saccardo

Syll. Fung. **2,** 24, 1883

≡*Pleospora obiones* Crouan et Crouan, *Florule du Finistère,* Paris, p. 22, 1867 (sub *P. obionei*)
=*Didymosphaeria spartinae* Grove, *J. Bot. (London)* **71,** 259, 1933
=*Leptosphaeria discors* Saccardo et Ellis *in* Saccardo, *Michelia* **2,** 567, 1882
≡*Metasphaeria discors* (Sacc. et Ellis *in* Sacc.) Sacc., *Syll. Fung.* **2,** 173, 1883
≡*Passeriniella discors* (Sacc. et Ellis *in* Sacc.) Apinis et Chesters, *Trans. Br. Mycol. Soc.* **47,** 432, 1964
=*Passeriniella incarcerata* Berlese, *Icon. Fung.* **1,** 51, 1892 [as *P. incarcerata* (Berkeley et Cooke) Berlese; Berlese's description and illus-

tration in Tab. XXXVIII, Fig. 1, are apparently based on *L. obiones* and not on *Sphaeria incarcerata* Berkeley et Cooke (syn. of *L. albopunctata*), which has five-septate ascospores and occurs on the same host]

LITERATURE: Cavaliere (1968); Dennis (1968); Gessner and Goos (1973a,b); Gessner and Kohlmeyer (1976); Gold (1959); Goodman (1959); Henningsson (1974); G. C. Hughes (1969); T. W. Johnson (1956a,c); T. W. Johnson and Sparrow (1961); E. B. G. Jones (1962a); Kohlmeyer (1966a, 1971b); Kohlmeyer and Kohlmeyer (1964–1969); Lucas and Webster (1967); Poole and Price (1972); Raghukumar (1973); Schaumann (1968); Schmidt (1974); Szaniszlo and Mitchell (1971); Tubaki (1968, 1969); Wagner (1965, 1969); E. E. Webber (1970).

Figs. 93c and 93d

ASCOCARPS 105–295 μm high, 200–440 μm in diameter, subglobose or ellipsoidal, immersed or erumpent, ostiolate, epapillate or perforating the epidermis with a short papilla, carbonaceous or subcarbonaceous, black above, brown or light brown below, sometimes clypeate, usually covered by brown hyphae, gregarious. PERIDIUM in the upper part 36–47 μm thick, in the sides and base 15–25 μm thick, composed of irregular or compressed, thick-walled cells, forming a textura angularis, enclosing cells of the host. PSEUDOPARAPHYSES septate, ramose, filling the ostiolar canal. ASCI 141–291 × 15–24 μm, eight-spored, clavate or subcylindrical, short pedunculate, bitunicate, thick-walled, physoclastic, without apical apparatuses; developing at the base of the ascocarp venter. ASCOSPORES 25–36(–38) × 9–14 μm, one-seriate in the base, biseriate in the top, ellipsoidal or broadly fusoid, slightly curved, three-septate, constricted at the septa; central cells brown or yellowish-brown, large; apical cells subhyaline, small. SPERMOGONIA globose or subglobose, immersed or superficial, ostiolate, light to dark brown. PERIDIUM two-layered; outer layer dark brown, consisting of several layers of pseudoparenchymatous cells; inner layer hyaline or pale brown, composed of thin-walled, oblong to ovoid–polygonal cells. SPERMATIOPHORES subcylindrical, simple, unbranched, lining the walls of the spermogonium, forming spermatia apically by abstriction. SPERMATIA 1 × 2 μm, one-celled, subglobose, hyaline. MODE OF LIFE: Saprobic. HOSTS: Dead culms of *Agropyron junceiforme*, *Halimione portulacoides*, *Spartina alterniflora*, *S. cynosuroides*, *S. townsendii*, *Spartina* sp., intertidal and drifting wood (e.g., *Quercus* sp.), bark and bamboo, test panels (e.g., *Betula pubescens*, *Fagus sylvatica*). RANGE: Atlantic Ocean—Argentina (Buenos Aires), France, Germany (F.R.G.), Great Britain (England, Wales), Iceland, United States [Alabama (J. Kohlmeyer, unpublished), Connecticut,

Florida, Georgia (J. Kohlmeyer, unpublished), Maine, Maryland, Massachusetts, Mississippi, New Jersey, North Carolina, Rhode Island, South Carolina, Texas, Virginia]; Baltic Sea—Germany (G.D.R.), Sweden; Indian Ocean—India (Madras); Pacific Ocean—Canada (British Columbia), Japan. NOTE: Description based on type material from Herb. Crouan at Concarneau and own collections; description of spermogonia after Wagner (1965); a fungus from mangrove areas in tropical waters, resembling *L. obiones* but having distinctly smaller ascospores (Cribb and Cribb, 1960a, from Australia; our own unpublished collections from Liberia and Mexico), is omitted in the compilation because it appears to be a new species in need of thorough examination; abnormally high dimensions of *L. obiones* reported by Cavaliere (1968), E. B. G. Jones (1962a), and E. E. Webber (1970) are not included; species with similar ascospores, for example, *L. paucispora* or *Savoryella lignicola,* may be mistaken for *L. obiones;* also, *Juncus maritimus,* allegedly found as host of *L. obiones* in North Carolina (T. W. Johnson, 1956c), is left out because this rush does not occur in the Carolinas (Radford *et al.,* 1968). ETYMOLOGY: Named after a host genus, *Obione.*

Leptosphaeria oraemaris Linder *in* Barghoorn et Linder

Farlowia **1,** 413, 1944

LITERATURE: Anastasiou (1963b); Brooks *et al.* (1972); Catalfomo *et al.* (1972–1973); Cavaliere (1966c, 1968); Churchland and McClaren (1973); Dennis (1968); Henningsson (1974); T. W. Johnson (1956a); T. W. Johnson and Sparrow (1961); E. B. G. Jones (1962a, 1963c, 1965, 1968b, 1971a); Kirk *et al.* (1974); Koch (1974); Kohlmeyer (1959, 1960, 1961b, 1962a, 1963d, 1967, 1971b); Kohlmeyer and Kohlmeyer (1964–1969); Neish (1970); Peters *et al.* (1975); Poole and Price (1972); Schaumann (1968, 1969); Schmidt (1974); Shearer (1972b); Sieburth *et al.* (1974); Szaniszlo and Mitchell (1971); Tubaki (1968, 1969); Tubaki and Ito (1973); E. E. Webber (1966).

ASCOCARPS 125–240 μm high, 135–310(−372) μm in diameter, subglobose, ellipsoidal or broadly ovoid, partly or totally immersed, ostiolate, short papillate or epapillate, carbonaceous or subcarbonaceous, black, base often lighter colored, solitary or gregarious. PERIDIUM 20–25 μm thick at the top, 15–20 μm at the sides and base, composed of six to eight layers of irregular or elongate polygonal, thick-walled cells with large lumina, forming a textura angularis. OSTIOLAR CANAL periphysate. PSEUDOPARAPHYSES 1–1.5 μm in diameter, septate, ramose, anastomosing. Asci (80–)96–141 × 8–12 μm, eight-spored, cylindrical or subclavate, short pedunculate, bitunicate, thick-walled, physoclastic, without apical apparatuses; developing at the base of the ascocarp venter. Ascospores 17–29(–32) × (4–)5–8 μm, biseriate, fusiform or rarely oblong–ellipsoidal or subclavate, straight or slightly curved, one- or three-septate, strongly constricted at the central septum, less so at the others, pale brown, becoming darker in age. Mode of Life: Saprobic. Hosts: On decaying culms of *Arundo donax, Glaux maritima, Spartina alterniflora, S. townsendii,* intertidal wood, driftwood, test panels (e.g., *Fagus sylvatica, Phyllostachys pubescens, Pinus monticola, P. sylvestris, Populus* sp., *Quercus* sp., balsa). Range: Atlantic Ocean—Belgium, Canada (Nova Scotia), Denmark, Germany (Helgoland and mainland, F.R.G.), Great Britain (England, Scotland, Wales), Iceland, Spain (Canary Islands), Sweden, United States [Connecticut, Florida, Maine, Maryland, Massachusetts, New Hampshire (J. Kohlmeyer, unpublished), North Carolina, Rhode Island, Texas]; Baltic Sea—Germany (G.D.R.), Sweden; Mediterranean—France, Italy (Sicily); Pacific Ocean—Japan, United States (California, Washington). Note: The species was reported from immersed *Tamarix aphylla* in an inland salt lake, namely, the Californian Salton Sea (Anastasiou, 1963b); however, ascospores of this fungus were verruculose and surrounded by a thin gelatinous sheath, features not found in marine collections of *L. oraemaris.* Etymology: From the Latin, *ora* = coast and *mare* = the sea, in reference to the marine habitat of the species.

Leptosphaeria paucispora A. B. Cribb et J. W. Cribb

Univ. Queensl. Pap., Dep. Bot. **4**, 41, 1960

Ascocarps 84–140 μm in diameter, flask-shaped, immersed, ostiolate, papillate, membranaceous, cream-colored to brown, solitary. Papillae 70–100 μm long, 40–55 μm in diameter, stout, pallid. Pseudoparaphyses up to 1 μm in diameter, scarce, simple or ramose. Asci 82–106 × 18–23 μm, two-spored, cylindrical–clavate, thin-walled at maturity. Ascospores 36–50 × 13–16.5 μm, fusoid–ellipsoidal, three-septate, slightly constricted at the septa, central cells brown, apical cells hyaline. Mode of Life: Saprobic. Substrate: Submerged wood near mangroves. Range: Pacific Ocean—Australia (Queensland; known only from the type collection). Note: Description based on Cribb and Cribb (1960a); the species may not belong to *Leptosphaeria* because these authors described perithecia, paraphyses, and thin-walled asci; type material has not become available; therefore, the final disposition of *L. paucispora* has to be postponed. ETYMOLOGY: From the Latin, *pauci-* = few- and *spora* = spore, in reference to the asci containing but two ascospores.

Leptosphaeria pelagica E. B. G. Jones

Trans. Br. Mycol. Soc. **45**, 105, 1962

LITERATURE: Apinis and Chesters (1964); Dennis (1968); Gessner and Goos (1973a,b); Kohlmeyer [1960 (sub *Leptosphaeria* sp. Form II)]; Neish (1970); Poole and Price (1972).

ASCOCARPS 115–201 μm high, 152–280 μm in diameter, globose, subglobose or ellipsoidal, superficial or immersed, ostiolate, papillate, subcarbonaceous, dark brown to black, sometimes with lighter colored bases, solitary. PAPILLAE 62–103 μm high, 49–74 μm in diameter, cylindrical, base dark, apex light-colored. PSEUDOPARAPHYSES 1.6–2.4 μm in diameter, filiform, simple or sparingly ramose. ASCI 100–149 × 14.5–24 μm, eight-spored, subclavate to long cylindrical, bitunicate, thick-walled. ASCOSPORES 28–36(–40) × 8–12 μm, fusiform to subclavate, straight or curved, three-septate, constricted at the septa, apically rounded, hyaline. MODE OF LIFE: Saprobic. HOSTS: Decaying culms of *Agropyron junceiforme*, *A. pungens*, *Puccinellia maritima*, *Spartina alterniflora*, *S. townsendii*, *Spartina* sp., intertidal wood, driftwood, and test panels (e.g., *Fagus sylvatica*). RANGE: Atlantic Ocean—Canada [New Brunswick (J. Kohlmeyer, unpublished), Nova Scotia], Great Britain (England, Wales), United States (Rhode Island); Pacific Ocean—United States (Washington). NOTE: Description after E. B. G. Jones (1962a) and Kohlmeyer (1960; sub *Leptosphaeria* sp. Form II); comparison of isotype material of *L. pelagica* (Herb. I.M.I. No. 87372) with our collection from the Pacific Northwest of the United States (Herb. J.K. 204) showed them to be identical. ETYMOLOGY: From the Latin, *pelagicus* = pertaining to the sea, in reference to the marine habitat of the species.

Leptosphaeria peruviana Spegazzini

An. Soc. Cient. Argent. **12**, 179, 1881

≡*Leptosphaeria promontorii* Saccardo, *Syll. Fung.* **2**, 22–23, 1883 (superfluous name)

Fig. 93e

ASCOCARPS 130–150 μm in diameter, lenticular–subglobose, immersed, mostly covered by the epidermis and perforating the cortex with a small ostiole, thin-walled, black. PSEUDOPARAPHYSES absent. ASCI 48–60 × 10–15 μm, eight-spored, cylindrical–clavate, at the apex obtusely rounded, at the base tapering into a short peduncle, bitunicate, thick-walled. ASCOSPORES 12–16 × 4–5.5 μm, biseriate, cylindrical–ellipsoidal, at both ends obtusely rounded, three-septate, constricted at the septa, second cell from the top largest, dirty olive-brown. MODE OF LIFE: Saprobic. HOST: Decaying stems of *Salicornia ambigua* (=*S. peruviana*). RANGE: Atlantic Ocean—Argentina (Buenos Aires; known only from the original collection). NOTE: The description is mostly based on Spegazzini's diagnosis and on our examination of scarce type material in Herb. LPS; the species is poorly circumscribed and is in need of a thorough reexamination of fresh material. ETYMOLOGY: Named after the original epithet of the host plant.

Leptosphaeria typhicola Karsten

Mycol. Fenn. **2**, p. 100, 1873, *in* Bidrag till Kännedom af Finlands Natur och Folk, *Fin. Vetenskaps-Soc.* **23**, 1873

LITERATURE: Berlese (1894); Lucas and Webster (1967).

Figs. 93f and 93g

ASCOCARPS 180–220 μm high, 200–430 μm in diameter, subglobose to depressed–ellipsoidal, immersed or erumpent, ostiolate, epapillate or with short papillae perforating the epidermis, subcoriaceous, dark brown, solitary or gregarious. PERIDIUM 10–20(–32) μm thick, composed of three to five layers of large, thin-walled cells with large lumina, forming a textura angularis. PSEUDOPARAPHYSES 2–3 μm in diameter, septate, simple, merging apically with pseudoparenchyma, which fills the ostiolar canal before maturation of the ascocarp. ASCI 90–140(–170) × 18–26(–34) μm, eight-spored, cylindrical to clavate, short pedunculate, bitunicate, thick-walled; developing at the base of the ascocarp venter. ASCOSPORES 34–52(–62) × 7–10(–13) μm, bi- or triseriate in the upper part of the ascus, uniseriate below, fusiform, seven- to eleven-septate, slightly constricted at the septa, particularly around the thickest cell (fourth or fifth from the top), straight or curved, at first hyaline, becoming light brown (or golden brown) and verrucose in age, surrounded by a gelatinous, 2- to 4-μm-thick sheath. PYCNIDIA 320–400 μm in diameter, globose, black. PERIDIUM 32–48 μm thick, covered by brown hyphae. CONIDIA 3–4 × 2.5–3 μm, globose, ovoid or ellipsoidal, hyaline. MODE OF LIFE: Saprobic. HOSTS: Decaying culms of *Juncus roemerianus* (J. Kohlmeyer, unpublished) and *Spartina* sp. RANGE: Atlantic Ocean—Argentina (Buenos Aires), United States (North Carolina). NOTE: The description is based on

Lucas and Webster's (1967) and our own notes; the fungus listed by Munk (1957, p. 356) as *Leptosphaeria maritima* on *Juncus atricapillus* from Denmark belongs most probably to *L. typhicola*; the species also occurs in freshwater locations in England and Finland on *Phragmites communis* and *Typha latifolia*; therefore, *L. typhicola* can be considered a facultative marine fungus. ETYMOLOGY: From a host genus, *Typha*, and the Latin, -*cola* = -dweller.

Key to the Marine Species of *Leptosphaeria*

1. Ascospores three-septate, both central cells brown, large; apical cells almost hyaline, small 2
1'. Ascospores with three or more septa; concolorous throughout 3
  2(1) Asci two-spored; ascospores longer than 36 μm; diameter 13–16.5 μm . . . . *L. paucispora*
  2'(1) Asci eight-spored; ascospores usually shorter than 36 μm; diameter 9–14 μm . . . . *L. obiones*
3(1') Ascospores with seven to eleven septa . . . . *L. typhicola*
3'(1') Ascospores with less septa 4
  4(3') Ascospores surrounded by a gelatinous sheath (use India ink or phase contrast!) 5
  4'(3') Ascospores without gelatinous sheath 8
5(4) Ascospores longer than 30 μm 6
5'(4) Ascospores shorter than 30 μm 7
  6(5) Ascospores three- to five-septate, becoming yellowish-brown in age . . . . *L. neomaritima*
  6'(5) Ascospores three (rarely four)-septate, always hyaline . . . . . . *L. contecta*
7(5') Ascospores usually longer than 18 μm, hyaline; on living pneumatophores of mangroves (*Avicennia* spp.) . . . . *L. avicenniae*
7'(5') Ascospores usually shorter than 18 μm, yellow–brown; on other substrates . . . . *L. halima*
  8(4') Ascospores five (rarely seven)-septate . . . . *L. albopunctata*
  8'(4') Ascospores with less septa 9
9(8') Ascospores shorter than 17 μm, olive-brown . . . . *L. peruviana*
9'(8') Ascospores 17 μm long or longer, hyaline, yellowish or brown 10
  10(9') Ascospores always hyaline, strictly three-septate 11
  10'(9') Ascospores yellowish or brown in age, one- to three-septate 12
11(10) Ascospores longer than 27 μm; in temperate regions . . . . *L. pelagica*
11'(10) Ascospores 27 μm or shorter; in tropical and subtropical regions *L. australiensis*
  12(10') Ascospores longer than 33 μm; diameter 8 μm or more; subhyaline to yellowish . . . . *L. marina*
  12'(10') Ascospores shorter than 33 μm; diameter (4–)5–8 μm; pale brown or darker . . . . *L. oraemaris*

Manglicola J. Kohlmeyer et E. Kohlmeyer

Mycologia **63**, 840–841, 1971

ASCOCARPS solitary, obtuse clavate to fusiform, stipitate, superficial, seated in the substrate with a hypostroma, ostiolate, epapillate, coriaceous, olive-brown, periphysate. PSEUDOPARAPHYSES numerous, septate, simple or reticulate. ASCI eight-spored, cylindrical, bitunicate, thick-walled, developing at the base of the ascocarp venter between the pseudoparaphyses. ASCOSPORES fusiform, apiculate, one-septate; apical cell large, dark brown; basal cell small, light brown; deliquescing appendages cover both apices. HOST: Mangrove trees (*Rhizophora*). TYPE SPECIES: *Manglicola guatemalensis* Kohlm. et Kohlm. The genus is monotypic. ETYMOLOGY: From the Spanish, *mangle* = mangrove, and the Latin, *-cola* = -dweller, in reference to the habitat and host.

Manglicola guatemalensis J. Kohlmeyer et E. Kohlmeyer

Mycologia **63**, 841, 1971

Fig. 94

ASCOCARPS 835–1275 μm high, 185–387 μm in diameter around the center, 95–165 μm in diameter around the base, 100–185 μm in diameter around the apex, obtusely clavate to obtusely fusiform, stipitate, superficial; seated in the substrate with a hypostroma, composed of pseudoparenchymatous cells and brown hyphae, 10–20 μm in diameter; ostiolate, epapillate, coriaceous, olive-brown, solitary; stipe composed of a cortex of polygonal, brown cells, and of a central core (50–80 μm in diameter) of hyaline, interwoven hyphae of 5–8 μm in diameter. PERIDIUM 36–50 μm thick, composed of three to five layers of cells. PAPILLAE absent; ostioles surrounded by hyaline, clavate hyphae, 4–6 μm in diameter; periphyses 2–3 μm in diameter, simple or ramose. PSEUDOPARAPHYSES 2–4 μm in diameter, numerous, septate, simple or reticulate. ASCI 275–326 × 24–28 μm, eight-spored, cylindrical, bitunicate, thick-walled, developing at the base of the ascocarp venter between the pseudoparaphyses. ASCOSPORES 80–109 × 18–34 μm, uniseriate, fusiform, apiculate, unequally one-septate, constricted at the septum; apical cell large, chestnut-brown; basal cell 18–27 μm long, 13–15.5 μm in diameter, turbinate, light brown; deliquescing appendages cover both apices; apical appendage cylindrical, basal appendage subglobose. MODE OF LIFE: Saprobic. HOST: Dead roots of *Rhizophora mangle*. RANGE: Pacific Ocean—Guatemala (known only from the original collection). ETYMOLOGY: Specific name referring to the country of origin.

Massarina Saccardo

Syll. Fung. **2**, 153, 1883

=*Abaphospora* Kirschstein, *Ann. Mycol.* **37**, 97, 1939
=*Amphididymella* Petrak, *Engler's Bot. Jahrb.* **62**, 94, 1928

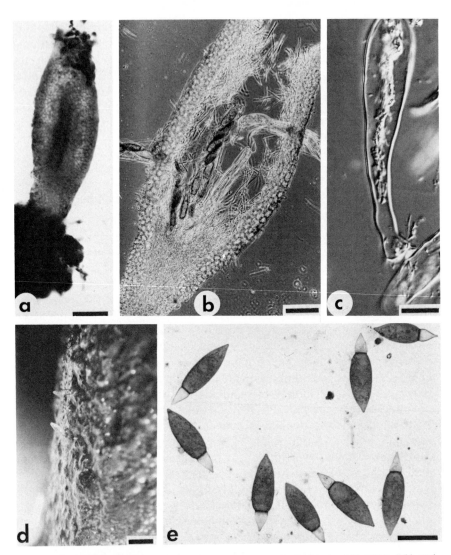

Fig. 94. *Manglicola guatemalensis* from type material Herb. J.K. 2700 (NY and IMS). (a) Whole mount of ascocarp (bar = 100 μm). From Kohlmeyer and Kohlmeyer, *Mycologia* **63,** 842 (1971), Fig. 17. Reprinted with permission. (b) Longitudinal section (28 μm) through ascocarp (bar = 50 μm). (c) Immature ascus (bar = 15 μm). (d) Two ascocarps on the wood surface (bar = 1 mm). (e) Ascospores (bar = 50 μm). (a and e in bright field, b in phase contrast, c in Nomarski interference contrast, d close-up photograph.)

=*Clypeothecium* Petrak *in* Sydow et Petrak, *Ann. Mycol.* **20**, 182–183, 1922

=*Holstiella* P. Hennings, *Pilze Ostafrikas*, p. 33, *in* A. Engler "Die Pflanzenwelt Ostafrikas und der Nachbargebiete," Part C, Reimer, Berlin 1895

=*Massarinula* Géneau de Lamarlière, *Rev. Gén. Bot.* **6**, 321, 1894

=*Oraniella* Spegazzini, *An. Mus. Nac. Hist. Nat. Buenos Aires* **19**, 378, 1909

=*Parasphaeria* Sydow, *Ann. Mycol.* **22**, 297, 1924

=*Phragmosperma* Theissen et Sydow, *Ann. Mycol.* **14**, 450, 1916

=*Pseudodiaporthe* Spegazzini, *An. Mus. Nac. Hist. Nat. Buenos Aires* **19**, 358–359, 1909

=*Trematostoma* (Saccardo) Shear, *Mycologia* **34**, 272, 1942

LITERATURE: Bose (1961); Holm (1957); Luttrell (1973); Müller and von Arx (1962); Munk (1956); von Arx and Müller (1975); Webster (1965).

ASCOCARPS solitary, globose or depressed, immersed or rarely erumpent, or covered by a more or less distinctly developed stromatic clypeus, ostiolate, mostly epapillate, dark; sometimes associated with conidial states in the form-genera *Coniothyrium*, *Ceratophoma*, *Microsphaeropsis*, and others. PERIDIUM mostly thick, composed of thick-walled, flattened cells, forming concentric layers. OSTIOLAR CANAL surrounded by small, thick-walled, brown cells; at first filled with a hyaline, small-celled pseudoparenchyma, the lower cells of which are attached to the apices of pseudoparaphyses. PSEUDOPARAPHYSES filiform, septate. ASCI four- to eight-spored, ventricose or cylindrico-clavate, subsessile or weakly pedunculate, wall thickened at the apex, bitunicate. ASCOSPORES ellipsoidal, fusiform or clavate, straight or curved, rounded at the apices, one- or multiseptate, hyaline, rarely becoming light brown, often covered by a gelatinous sheath. MODE OF LIFE: Saprobic or hypersaprobic. TYPE SPECIES: *Massarina eburnea* (Tul.) Sacc. NOTE: Only 1 marine species is known besides 1 limnic and about 35 terrestrial taxa; synonyms and description after Bose (1961) and Müller and von Arx (1962). ETYMOLOGY: Diminutive of a related genus, *Massaria*.

Massarina cystophorae (Cribb et Herbert) J. Kohlmeyer et E. Kohlmeyer comb. nov.

≡*Otthiella cystophorae* Cribb et Herbert, *Univ. Queensl. Pap., Dep. Bot.* **3**, 10, 1954 (basionym)

≡*Melanopsamma cystophorae* (Cribb et Herbert) Meyers, *Mycologia* **49**, 485, 1957

LITERATURE: T. W. Johnson and Sparrow (1961).

Fig. 95

GALLS up to 1.5 cm in diameter, induced by the fungus in old parts of

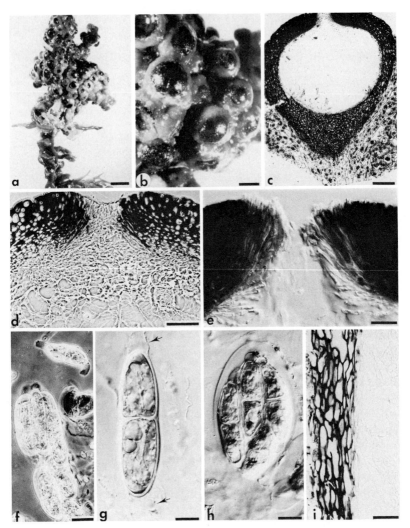

Fig. 95. *Massarina cystophorae* in *Cystophora* spp. (a) Gall enclosing black as-cocarps and spermogonia (bar = 2 mm). (b) Projections of gall with black fungal fruiting bodies (bar = 500 μm). (c) Longitudinal section (16 μm) through ascocarp (bar = 150 μm). (d) Section through ostiole of immature ascocarp (bar = 50 μm). (e) Section (16 μm) through ostiole of mature ascocarp with periphyses (bar = 25 μm). (f) Asco-spores with apical appendages (bar = 25 μm). (g) Ascospore with apical caps (arrow) (bar = 10 μm). (h) Ascus, ascospore appendages already developed (bar = 15 μm). (i) Section through peridium, ascocarp centrum to the right (bar = 25 μm). (d and f–i from type slide A.B. Cribb on *C. retroflexa;* the others from Herb. Parsons A625 on *C. subfarcinata*; a and b close-up photographs, c and i in bright field, d and f in phase contrast, the others in Nomarski interference contrast.)

the stipe of the host, enclosing ascocarps and spermogonia in more or less pronounced projections. ASCOCARPS 650–950 μm high, 650–1000 μm in diameter, subglobose to turbinate, half or three-quarters immersed in the tissue of the gall, ostiolate, epapillate, subcarbonaceous, black, gregarious. PERIDIUM 105–130 μm thick around the ostiole, 30–120 μm at the sides, composed of ellipsoidal to subglobose, thick-walled, dark brown cells with large lumina, forming a textura angularis, not enclosing cells of the host, extending into a wedgelike base. OSTIOLAR CANAL about 80 μm in diameter, surrounded by subglobose, thick-walled, brown cells of the peridium; these merge into elongate brown cells and the latter into hyaline periphyses. PSEUDOPARAPHYSES 3–8 μm in diameter, septate, simple, numerous, enclosed in a gelatinous matrix. ASCI 105–150 × 50 –64 μm, eight-spored, ellipsoidal to subclavate, short pedunculate, bitunicate, thick-walled, physoclastic, without apical apparatuses; developing at the base and sides of the ascocarp venter. ASCOSPORES 50–65(–73) × (15–) 16–23(–25) μm, irregularly triseriate, elongate–ellipsoidal or cylindrico-ellipsoidal with broadly rounded apices, one-septate at the center, not or slightly constricted at the septum, hyaline, thick-walled; at each end provided with a subglobose or cylindrical, caplike, gelatinous appendage. SPERMOGONIA about 600 μm in diameter, subglobose, three-quarters immersed in the tissue of the gall, ostiolate, epapillate, subcarbonaceous, black, gregarious, mixed with ascocarps in the same gall. PERIDIUM 40–90 μm thick, similar to the peridium of ascocarps. SPERMATIOPHORES about 20 μm long, cylindrical, lining the inner walls and the partitions in the cavities of the spermogonia. SPERMATIA about 3–4 × 2 μm, ellipsoidal, hyaline. MODE OF LIFE: Parasitic. HOSTS: *Cystophora retroflexa* and *C. subfarcinata* (Phaeophyta). RANGE: Pacific Ocean—Australia (Tasmania, Western Australia). NOTE: Description based on Cribb and Herbert (1954) and on our examination of type slides and of material from Esperance (Herb. M. J. Parsons, A625), kindly provided by Drs. A. B. Cribb and M. J. Parsons, respectively; the species cannot remain in *Otthiella,* a synonym of *Otthia* according to Müller and von Arx 1962, nor can it be assigned to *Melanopsamma* (Sphaeriales); however, it agrees well with the circumscription of *Massarina* as given by Bose (1961) and Müller and von Arx (1962). ETYMOLOGY: Named after the host plant, *Cystophora* (Phaeophyta).

Microthelia Koerber

Systema Lichenum Germaniae, Breslau, p. 372, 1855

=*Astrosphaeriella* Sydow, *Ann. Mycol.* **11,** 260, 1913
=*Jahnula* Kirschstein, Ann. Mycol. **34,** 196, 1936

=*Kirschsteiniella* Petrak, *Ann. Mycol.* **21**, 331, 1923
=*Microtheliomyces* Ciferri et Tomaselli, *Atti Ist. Bot. Univ. Lab. Crittogam., Pavia*, Ser. 5, **10**, 32 and 59, 1953

LITERATURE: Henssen and Jahns (1974); Lamb (1963); Müller and von
Arx (1962); Scheinpflug (1958).

ASCOCARPS solitary or gregarious, conical, or sometimes semiglobose
or depressed, seated superficially on the substrate with a broad base, or
developing under the epidermis or periderm and erumpent; rarely surrounded by a pseudostromatous tissue; without hypostroma; ostiolate,
epapillate or rarely short papillate, thick-walled. PERIDIUM above and at
the sides thick, composed of thick-walled, elongate or isodiametric cells
that merge at the base into elongate cells with thinner walls, arranged in
regular layers. OSTIOLAR CANAL filled with pseudoparaphyses or hyphoid
cells. PSEUDOPARAPHYSES filiform, septate, often branched, numerous.
ASCI four- to eight-spored, cylindrical, clavate or ventricose, bitunicate,
wall thickened at the apex. ASCOSPORES fusiform, rarely clavate or ovoid,
one-septate near the middle, brown, grayish-brown or dirty green. SUB-
STRATE: Dead plant material, mainly bark and wood. TYPE SPECIES:
Microthelia biformis (Leight.) Massee. NOTE: Description after Müller
and von Arx (1962), synonyms after Lamb (1963) and Müller and von Arx
(1962); the genus embraces lichenized and nonlichenized species; only 1
marine species is known among about 70 terrestrial species; *M. ver-
ruculosa* (Anastasiou 1963b) was described from an inland saline lake,
namely, the Salton Sea, California. ETYMOLOGY: From the Greek, *micros*
= small and *thele* = nipple, probably in reference to the shape of the
ascocarps.

Microthelia linderi Kohlm.

Trans. Br. Mycol. Soc. **57**, 483, 1971

≡*Amphisphaeria maritima* Linder *in* Barghoorn et Linder, *Farlowia* **1**,
411, 1944
≡*Microthelia maritima* (Linder) Kohlm., *Nova Hedwigia* **2**, 322, 1960
(non *Microthelia maritima* Bouly de Lesdain, *Recherches sur les Lichens
des Environs de Dunkerque*, Michel, Dunkerque, pp. 254–255, 1910)

LITERATURE: Barghoorn (1944); Byrne and Eaton (1972); Byrne and
Jones (1974, 1975b); Eaton (1972); Eaton and Dickinson (1976); G. C.
Hughes (1969); T. W. Johnson (1968); T. W. Johnson and Sparrow (1961);
E. B. G. Jones (1968b, 1972); E. B. G. Jones *et al.* (1972); Kohlmeyer
(1959, 1962a, 1967); Kohlmeyer and Kohlmeyer (1964–1969); Schmidt
(1974); Schneider (1976); Wilson (1951).

ASCOCARPS 57–128 μm high, 104–268 μm in diameter, subconical, semiglobose, rarely subglobose, with a flattened base, occasionally membranous at the edge, superficial or partly immersed, ostiolate, short papillate or epapillate, carbonaceous, black, gregarious. PERIDIUM 30–35 μm thick above and at the sides, two-layered, outer layer brownish-black, inner layer hyaline; composed of subglobose, thick-walled cells with usually large lumina, forming a textura angularis; basis of peridium crustlike, 2.5–5 μm thick. OSTIOLAR CANAL 20–30 μm in diameter, periphysate. PSEUDOPARAPHYSES septate, ramose, attached at both ends. ASCI 37–60(–74) × 8–14 μm, eight-spored, clavate or elongate–ellipsoidal, pedunculate, bitunicate, thick-walled at the apex, physoclastic, without apical apparatuses; developing at the base of the ascocarp venter. ASCOSPORES 13.5–21 × 5–7.5 μm, bi- or rarely triseriate, ellipsoidal to subfusiform, one-septate somewhat below the center, constricted at the septum, brown. MODE OF LIFE: Saprobic. SUBSTRATES: Driftwood and drifting bark, intertidal and submerged wood, test panels (e.g., *Pinus sylvestris,* bamboo); predominantly on resin-coated conifer wood. RANGE: Atlantic Ocean—Great Britain (England, Scotland, Wales), Iceland, Norway, Spain (Tenerife), United States (Connecticut, Maine, Massachusetts, South Carolina); Baltic Sea—Denmark, Germany (F.R.G., G.D.R.); Mediterranean—France; Pacific Ocean—Canada (British Columbia), United States (Washington). NOTE: Collections with hysterotheciumlike, or membranous, ascocarps and those with elongated necks (Cavaliere, 1966c) most probably belong to a different species. ETYMOLOGY: Named after the American mycologist David H. Linder (1899–1946).

Paraliomyces Kohlm.

Nova Hedwigia **1**, 81, 1959

STROMATA black, immersed. ASCOCARPS solitary, subglobose to pyriform, subiculate or nonsubiculate, immersed or erumpent, ostiolate, papillate or epapillate, carbonaceous, thick-walled, black. PSEUDOPARAPHYSES filamentous, numerous, persistent. ASCI eight-spored, cylindrical, pedunculate, bitunicate, thick-walled. ASCOSPORES ellipsoidal to subfusiform, one-septate, hyaline, surrounded by a gelatinous sheath that contracts to form a lateral, flat appendage. SUBSTRATE: Wood and roots of mangrove trees. TYPE SPECIES: *Paraliomyces lentiferus* Kohlm. The genus is monotypic. ETYMOLOGY: From the Greek, *paralios* = coastal and *mykes* = fungus, referring to the habitat of the fungus.

Paraliomyces lentiferus Kohlm.

Nova Hedwigia **1**, 81–82, 1959

LITERATURE: T. W. Johnson and Sparrow (1961); Kohlmeyer (1968c); Kohlmeyer and Kohlmeyer (1964–1969); Müller and von Arx (1962).

Fig. 96

STROMATA black, immersed, penetrating into the substrate with dark brown hyphae. ASCOCARPS up to 680 μm high, 540 μm in diameter, subglobose to pyriform, subiculate or nonsubiculate, immersed or erumpent, ostiolate, papillate or epapillate, carbonaceous, thick-walled, black, solitary. PAPILLAE short or absent; ostiolar canal ca. 70 μm in diameter, periphysate. PSEUDOPARAPHYSES 140–160 × 0.5–1.2 μm, filamentous, numerous, persistent. ASCI 85–115 × 13–17 μm, eight-spored, cylindrical, short pedunculate, bitunicate, thick-walled, without apical apparatuses; developing in a hymenium in the lower half of the ascocarp. ASCOSPORES 17–26.5(–27.5) × 8–12 μm, ellipsoidal to subfusiform, one-septate below the middle, constricted at the septum, hyaline; at first

Fig. 96. Longitudinal section through ascocarp of *Paraliomyces lentiferus* in wood, stained in picro-anilin blue (Herb. J.K. 144, holotype B; bar = 150 μm; in bright field).

surrounded by a gelatinous sheath that contracts to form a lateral, lentiform, viscous appendage over the septum, 7.5–12.5 μm in diameter, 1–3 μm thick. MODE OF LIFE: Saprobic. SUBSTRATES: Submersed wood (fishing boats), dead roots and submerged branches of mangrove trees (*Avicennia germinans, Rhizophora mangle*). RANGE: Atlantic Ocean—Mexico (Veracruz), United States (Florida); Indian Ocean—India (Madras). ETYMOLOGY: From the Latin, *lens* = lens and *-fer* = -carrying, in reference to the ascospores, bearing a lentiform appendage.

Phaeosphaeria Miyake

Bot. Mag. **23** (266), 93, 1909 (in Japanese)

J. Coll. Agric., Tokyo Imp. Univ. **2**(4), 246, 1910 (in German)

=*Leptosphaeria* Ces. et DeNot. subgen. *Leptosphaerella* Saccardo, *Syll. Fung.* **2**, 47, 1883

LITERATURE: Eriksson (1967a); Hedjaroude (1968); Holm (1957); Koponen and Mäkelä (1975); von Arx and Müller (1975).

ASCOCARPS solitary or gregarious, subglobose to pyriform, in some species with a tendency to form dothideaceous stromata, immersed or subepidermal, ostiolate, papillate or with a long neck, generally thinwalled, smooth or covered by hyphae. PERIDIUM pseudoparenchymatous, composed of small rounded or flattened cells with rather thin walls. PSEUDOPARAPHYSES generally numerous, filamentous, hyaline and distinctly septate. ASCI eight-spored, clavate to cylindrical, short pedunculate, bitunicate, in some species with apical apparatuses. ASCOSPORES bi- to triseriate, fusoid to ellipsoidal, three- to multiseptate, often with one distinctly inflated cell, generally yellowish-brown, rarely hyaline. MODE OF LIFE: Saprobic, rarely parasitic; almost always on monocotyledons, rarely on dicotyledons or cryptogams. TYPE SPECIES: *Phaeosphaeria oryzae* Miyake. NOTE: Only 2 of about 30 species are known from estuarine habitats (*P. typharum*) or habitats exposed to saltwater spray (*P. ammophilae*); description following the authors listed above. ETYMOLOGY: From the Greek, *phaeo-* = dark, and *Sphaeria* = a genus of Pyrenomycetes (*sphaera* = globe), in reference to the dark ascospores and the globose ascocarps.

Phaeosphaeria ammophilae (Lasch) J. Kohlmeyer et E. Kohlmeyer

Icones Fungorum Maris **1**(3), Tabula 55, Cramer, Weinheim 1965

≡*Sphaeria ammophilae* Lasch, *Flora (Jena)* **8**, 282, 1850; *Bot. Ztg.* **8**,

438–440, 1850; Klotzsch-Rabenhorst, *Herbarium Mycologicum* **1**, No. 1340, 1850

≡*Leptosphaeria ammophilae* (Lasch) Cesati et DeNotaris, *Comment. Soc. Crittogam. Ital.* **1**, 236, 1863

=*Leptosphaeria littoralis* Saccardo, *Michelia* **1**, 38, 1877

≡*Phaeosphaeria littoralis* (Saccardo) Holm, *Symb. Bot. Ups.* **14**(3), 121, 1957

=*Sphaeria sabuletorum* Berkeley et Broome, *Ann. Mag. Nat. Hist.,* Ser. 2 **9**, 382, 1852

≡*Leptosphaeria sabuletorum* (Berk. et Br.) von Höhnel, *Hedwigia* **60**, 141, 1918

≡*Metasphaeria sabuletorum* (Berk. et Br.) Saccardo, *Syll. Fung.* **2**, 180, 1883

Possible IMPERFECT STATE: *Tiarospora perforans* (Roberge) von Höhnel (*Hedwigia* **60**, 141, 1918; for a list of synonyms see Grove, 1935; description in Pirozynski and Shoemaker, 1971).

LITERATURE: Apinis and Chesters (1964); Berlese (1894); Dennis (1964, 1968); Eriksson (1967a,b); Hedjaroude (1968); Holm (1952); Jaap (1907a,b); Larsen (1952).

ASCOCARPS 300–400 μm high (including papillae), 200–430 μm in diameter, subglobose, ellipsoidal or pyriform, immersed, ostiolate, papillate, carbonaceous, black, solitary or gregarious. PERIDIUM 15–25 μm thick, composed of three to five layers of polygonal or ellipsoidal, thin-walled cells with large lumina, forming a textura angularis. PAPILLAE 100 μm high, 100–150 μm in diameter, cylindrical or conical, often surrounded by a more or less developed clypeus; ostiolar canal 40–100 μm in diameter, conical, initially filled with a pseudoparenchyma of small, thin-walled, hyaline cells. PSEUDOPARAPHYSES about 4 μm in diameter, septate, simple, rarely ramose, with a gelatinous, hyaline outer layer. ASCI (125–)135–195 × (25–)27.5–36.5 μm, eight-spored, clavate or subcylindrical, short pedunculate, bitunicate, thick-walled, without apical apparatuses; developing at the base of the ascocarp venter. ASCOSPORES (35–)38.5–51(–55) × 12–15(–16.5) μm (excluding the sheath), biseriate, ellipsoidal or fusiform, 5–7(–8)-septate, usually 6-septate, slightly constricted at the septa, central cell broadest, yellowish to yellow–brown, with a gelatinous sheath that is constricted around the central septum and provided with a trumpet-shaped canal at the apices, reaching down to the spore wall. MODE OF LIFE: Saprobic. HOST: *Ammophila arenaria.* RANGE: Atlantic Ocean—Belgium, Denmark (Isle of Rømø), Germany (F.R.G.; mainland, isles of Amrum and Langeoog). Great Britain [England, Scotland (Isle of Rhum)], Netherlands, Norway, Sweden; Baltic Sea—Denmark, Finland, Germany [F.R.G.; G.D.R. (Isle of Rügen)], Sweden (mainland, isles of Gotland and Öland), U.S.S.R. (Latvian S.S.R.); Mediterranean—Italy. NOTE: Records from hosts other than *A. arenaria* have to be reconfirmed (Holm, 1957): *Agropyrum pungens, Ammophila arenaria* × *Calamagrostis epigejos, Elytrigia juncea, Scirpus* sp. ETYMOLOGY: Named after the host genus, *Ammophila.*

Phaeosphaeria typharum (Desmaz.) Holm

Symb. Bot. Ups. **14**(3), 126, 1957

≡*Sphaeria scirpicola* D. C. var. *typharum* Desmazières, *Plantes Crypto-games de France,* 2nd ed., No. 1778, 1849
≡*Leptosphaeria typharum* (Desmaz.) Karsten, *Mycol. Fenn.* **2**, p. 100, 1873, *in* Bidrag till Kännedom af Finlands Natur och Folk, *Fin. Vetenskaps-Soc.* **23**, 1873
≡*Pleospora typharum* (Desmaz.) Fuckel, *Jahrb. Nassau. Ver. Naturkd.* **23/24**, 137, 1869–1870
≡*Sphaeria typharum* (Desmaz.) Rabenh., *Herbarium Mycologicum,* 2nd ed., No. 731, 1858
=*Leptosphaeria kunzeana* Berlese, *Icones Fung.* **1**, 66, 1892

IMPERFECT STATE: *Hendersonia typhae* Oudemans, *Arch. Néerl. Sci.* **8**, 19, 1873.

LITERATURE: Apinis and Chesters (1964); Gessner and Goos (1973b); Goodman (1959); Hedjaroude (1968); E. B. G. Jones (1962a); Parguey-Leduc (1966); Webster (1955).

Fig. 97

ASCOCARPS 100–160 μm high, 75–160 μm in diameter, subglobose to pyriform, immersed, ostiolate, short papillate or epapillate, coriaceous, light brown, darker around the ostiole, solitary or gregarious. PERIDIUM 8–10(–15) μm thick, composed of two to four layers of thin-walled cells with large lumina, yellowish-brown on the outside, hyaline near the centrum, forming a textura angularis; mature ascospores visible through the peridium. PAPILLAE short or absent; ostiolar canal lined with hyaline, cylindrical periphyses, 8–10 × 2.5–3.5 μm. PSEUDOPARAPHYSES 3–4.5 μm in diameter, septate, simple. ASCI (70–)80–100 × (20–)22–30 μm, eight-spored, obclavate, ellipsoidal or cylindrical, short pedunculate, bitunicate, thick-walled, without apical apparatuses; developing at the base of the central cavity on a small lenticular tissue of ascogenous cells. ASCOSPORES (21–)24–34(–42) × (8–)9–12(–16) μm, bi- or triseriate or irregularly arranged in the ascus, ellipsoidal, three-septate, constricted at the septa, second cell broadest, straight or slightly curved, yellowish-brown to dark brown, thick-walled. PYCNIDIA up to 350 μm in diameter, subglobose or of irregular shape, immersed, ostiolate, simple or composed of several loculi. PERIDIUM composed on the outside of about five

Fig. 97. *Phaeosphaeria typharum* from *Spartina* sp. (Herb. J.K. 3377). (a) Whole mount of thin-walled ascocarp (bar = 25 μm). (b) Asci in different stages of development (bar = 15 μm). (a in bright field, b in Nomarski interference contrast.)

layers of olive-brown, flattened cells, on the inside of three to five layers of hyaline, prismatic cells, forming a textura angularis. CONIDIA 30–90 × 6–8 μm, obclavate, three- to eight-septate, constricted at the septa, yellow to brown, thick-walled, with a tapering, 4-μm-long, hyaline appendage at the apex; developing on the prismatic cells of the inner pycnidial wall. MODE OF LIFE: Saprobic. HOSTS: *Spartina alterniflora, S. townsendii, Spartina* spp. RANGE: Atlantic Ocean—Argentina (Buenos Aires), Canada (New Brunswick, Nova Scotia, Québec), Great Britain (England), United States (Maine, North Carolina, Rhode Island). NOTE: The description of the perfect state is mostly based on own collections (J. Kohlmeyer, unpublished); the description of the imperfect state is based on Webster's (1955) data; the species has also been collected in freshwater locations on *Typha angustifolia* and *T. latifolia* in Belgium, Denmark, Finland, France, Germany, Great Britain, Sweden, Switzerland, and the United States (California, Colorado, Delaware, Montana, North Dakota) [see Berlese (1894); Dennis (1968); Ellis and Everhart (1892); Hedjaroude (1968); Holm (1957); Karsten (1873); Larsen (1952); Müller (1950); Munk (1957); Pugh and Mulder (1971); Webster (1955)]; therefore, *P. typharum* can be considered a facultative marine fungus; *Leptosphaeria typharum* reported by Wehmeyer (1952) on a number of terrestrial gramineae from

Mount Rainier National Park (Washington) differs in many respects from the marine and freshwater fungus and is probably a different species. ETYMOLOGY: Named after a host genus, *Typha*.

Key to the Marine Species of *Phaeosphaeria*

1. Ascospores 3-septate, usually shorter than 35 μm . . . . *P. typharum*
1'. Ascospores 5–7(–8)-septate, longer than 35 μm . . . . *P. ammophilae*

Pleospora Rabenhorst ex Cesati et DeNotaris

Comment. Soc. Crittogam. Ital. **1**, 217, 1863

=Chaetoplea Clements, *The Genera of Fungi, 1st ed. Minneapolis, 1909 (as subgen. in Saccardo, Syll. Fung.* **2**, 279, 1883)
=*Cleistotheca* Zukal, *Oesterr. Bot. Z.* **43**, 163, 1893
=*Cleistothecopsis* Stevens et True, *Ill., Agric. Exp. Stn., Bull.* **220**, 530, 1919
=*Graphyllium* Clements, *Rep. Bot. Surv. Nebr.* **5**, 6, 1901
=*Montagnula* Berlese, *Icon, Fung.* **2**, 68, 1896
=*Pseudopleospora* Petrak, *Ann. Mycol.* **17**, 84, 1919

LITERATURE: Corlett (1975); Deighton (1965); Donk (1962); Müller (1951); von Arx and Müller (1975); Wehmeyer (1961).

ASCOCARPS solitary or gregarious, small or of medium size, usually globose and smooth, mostly without stroma, clypeus or superficial hyphae, immersed, ostiolate, rarely without distinct ostiole, papillate. PERIDIUM usually thinner than 50 μm, pseudoparenchymatous. PSEUDOPARAPHYSES numerous, filamentous or forming a paraphysoid tissue. ASCI eight-spored, numerous or, in small ascocarps, less numerous, cylindrical, claviform, or sometimes wide ovoidal, mostly distinctly pedunculate, thick-walled, bitunicate. ASCOSPORES ellipsoidal, claviform or even fusiform, with transverse and longitudinal septa (muriform), often constricted in the middle, round in cross section, yellowish or brownish, sometimes with gelatinous sheaths or appendages. MODE OF LIFE: Mostly saprobic, sometimes parasitic, on higher plants. TYPE SPECIES: *Pleospora herbarum* (Pers. ex Fr.) Rabenh. NOTE: Six marine species are recognized among about 200 taxa; additional species may be found in marine habitats, but only those are included that can be positively identified thus far; description and synonyms after von Arx and Müller (1975) and Müller (1951); conidial states belong to the form-genera *Alternaria, Dendryphion, Phoma,* and *Stemphylium; Pleospora purpurascens* Santesson *in* Gustafsson and Fries (1956) is a *nomen nudum.* ETYMOLOGY: From the Greek, *pleon* = more and *spora* = spore, in reference to the multichambered ascospores.

Pleospora gaudefroyi Patouillard

Tabulae Analyticae Fungorum, Paris, Deuxième Sér., p. 40, No. 602,
1886

=*Pleospora lignicola* Webster et Lucas, *Trans. Br. Mycol. Soc.* **44,** 431,
1961
=*Pleospora salicorniae* Jaap. *Verh. Bot. Ver. Prov. Brandenburg* **49,** 16,
1907 (non *Pleospora salicorniae* Dangeard 1888)
≡*Pleospora herbarum* (Fr.) Rabenh. var. *salicorniae* (Jaap) Jaap, *Ann.
Mycol.* **14,** 17, 1916 (non *Pleospora herbarum* f. *salicorniae* Auerswald
in Rabenhorst, *Fungi Europaei Exsiccati, Cent.* **2,** No. 145, 1860, invalid
name)
=*Pleospora salsolae* Fuckel var. *schoberiae* Saccardo, *Michelia* **2,** 69,
1880
≡*Pleospora schoberiae* (Sacc.) Berlese, *Icon. Fung.* **2,** 23, 1895

LITERATURE: Berlese (1888); Kohlmeyer (1962c); Kohlmeyer and
Kohlmeyer (1964–1969).

Figs. 98a–98c

ASCOCARPS 170–250 μm high, 180–350 μm in diameter, subglobose to
ellipsoidal, immersed or erumpent, ostiolate, short papillate or epapillate,
carbonaceous, black, gregarious. PERIDIUM 16–25 μm thick, composed of
three to five layers of roundish or ellipsoidal, thick-walled cells with large
lumina, forming a textura angularis. PAPILLAE short or absent; ostiolar
canal 45–50 μm in diameter. PSEUDOPARAPHYSES 2–3.5 μm in diameter,
septate, ramose. ASCI 125–200 × (22–)25–36 μm, eight-spored, broadly
clavate or elongate–ellipsoidal, often bent parallel to the peridial curva-
ture, short pedunculate, bitunicate, thick-walled, physoclastic, without
apical apparatuses; developing at the base of the ascocarp venter. ASCO-
SPORES (20–)30–46.5(–55.5) × (8.5–)13.5–20.5(–25) μm, biseriate, ellipsoidal or elongate–ovoidal, muriform, with seven (to nine) transverse and
one to three longitudinal septa, strongly constricted around the center,
slightly at the other septa, upper part of the spore usually wider than the
lower part, yellowish-brown to golden-brown, covered by a 9- to 12-μm-
thick gelatinous sheath that is slightly constricted around the equator; two
(to three) subconical, gelatinous appendages, 33–48(–75) × 12–21 μm,
extend through the sheath at each apex. MODE OF LIFE : Saprobic.
HOSTS: Dead culms of *Halimione portulacoides* and *Salicornia ambigua*

Fig. 98. Marine *Pleospora* spp. (a) Longitudinal section (16 μm) through ascocarp
of *P. gaudefroyi* in *Salicornia* sp. (Herb. J.K. 3520; bar = 50 μm). (b) Ascus of *P.
gaudefroyi* with immature ascospores surrounded by gelatinous sheaths (Herb. J.K.
2904; bar = 15 μm). (c) Ascospores of *P. gaudefroyi* with gelatinous sheaths and
subconical appendages (Herb. J.K. 3520; bar = 25 μm). (d) Ascus of *P. pelagica* from

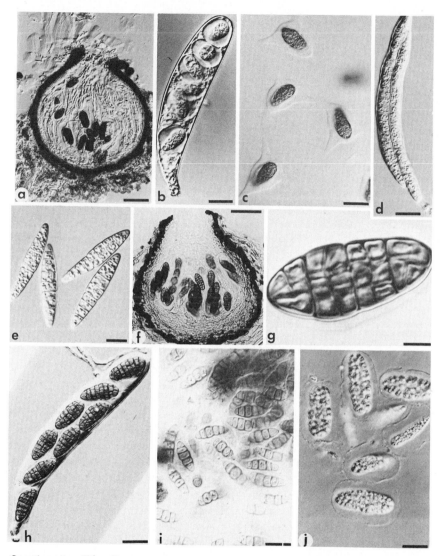

Spartina alterniflora (Herb. J.K. 3680; bar = 20 μm). (e) Same as d, ascospores (bar = 10 μm). (f) Longitudinal section through ascocarp of *P. pelvetiae* from *Laminaria* sp. (neotype, Herb. Desmazières 1772, PC; bar = 50 μm). (g) Same as f, ascospore (bar = 5 μm). (h) Ascus of *P. pelvetiae* from *Laminaria* sp. (in collection of *Cladosporium algarum* in K; bar = 25 μm). (i) Ascospores of *P. spartinae* (paratype slide Sheffield 2061; bar = 20 μm). (j) Mature and immature ascospores of *Pleospora* sp. I from *Salicornia virginica* (Herb. J.K. 3777; bar = 20 μm). (f in bright field, the others in Nomarski interference contrast.)

(both J. Kohlmeyer, unpublished), *S. fruticosa, S. herbacea, Salicornia* sp., *Suaeda maritima,* intertidal wood and driftwood (*Quercus* sp.). RANGE: Atlantic Ocean—Argentina [Buenos Aires (J. Kohlmeyer, unpublished)], France, Germany (Isle of Sylt, F.R.G.), Great Britain (England); Mediterranean—France, Yugoslavia. NOTE: We have examined the apparent type specimen from Herb. PC, deriving from the herbarium of E. Gaudefroy (as *Pleospora salsolae* Fuckel f. *schoberiae* Sacc.); it was collected by O. Hariot in the "Marais de la Pointe de Touquet, près Etaples (Pas de Calais), 15 Août 79"; this material contains ascospores with the characteristic appendages described and illustrated by Patouillard (1886); *Pleospora* spp. on *Salicornia* spp. from inland habitats are often wrongly identified as *P. salicorniae* Jaap (e.g., Săvulescu, 1947); the latter species is a synonym of *P. gaudefroyi,* but the inland *Pleospora* spp. lack the typical ascospore appendages of *P. gaudefroyi* (Kohlmeyer, 1962c) and, therefore, belong to a different species. ETYMOLOGY: Named after the supplier of the type specimen, E. Gaudefroy.

Pleospora pelagica Johnson

Mycologia **48**, 504, 1956

LITERATURE: T. W. Johnson and Sparrow (1961); Kohlmeyer (1966a); Kohlmeyer and Kohlmeyer (1964–1969); Wehmeyer (1961).
Figs. 98d and 98e.

ASCOCARPS 160–350 μm high, 185–390 μm in diameter, subglobose to ellipsoidal, immersed, ostiolate, short papillate, subcarbonaceous, black, solitary or gregarious; hyphae blackening the epidermis of the host. PERIDIUM 25–35 μm thick, two-layered, forming a textura angularis; outer layer composed of about five rows of irregular, thick-walled cells with small lumina; inner layer composed of three to six rows of polygonal, thin-walled cells with large lumina, merging into pseudoparaphyses. PAPILLAE 40–60 μm high, 52–68 μm in diameter, subconical; ostiolar canal filled with periphysoid hyphae. PSEUDOPARAPHYSES persistent, septate, simple or rarely ramose, apically cylindrical or clavate. ASCI 151–218 × 16–28 μm, eight-spored, clavate or subcylindrical, short pedunculate, bitunicate, rather thin-walled, physoclastic, without apical apparatuses; developing at the base of the ascocarp venter. ASCOSPORES 35–52 × 9.5–15 μm, partly or completely biseriate, fusiform, fusoid ellipsoidal or clavate ellipsoidal, muriform, with seven to nine transverse septa and one longitudinal septum in one to eight of the central segments;

apical cells without longitudinal septa; constricted at the septa, particularly below the middle, yellow–brown. MODE OF LIFE: Saprobic. HOST: Dead culms of *Spartina alterniflora*. RANGE: Atlantic Ocean—United States (Florida, North Carolina). NOTE: An immature fungus similar to *P. pelagica* was reported from wood from the Atlantic coast of France (Kohlmeyer, 1962a). ETYMOLOGY: From the Latin, *pelagicus* = pertaining to the sea, in reference to the habitat of the species.

Pleospora pelvetiae Sutherland

New Phytol. **14**, 41–42, 1915

=*Sphaeria herbarum* Fr. var. *fucicola* Roberge *in* Desmazières, *Plantes Cryptogames de France,* Ed. I, Fasc. 42, No. 2072, Lille, 1850

LITERATURE: T. W. Johnson and Sparrow (1961); Kohlmeyer (1973b).
Figs. 98f–98h

ASCOCARPS 180–340 μm high, 160–340 μm in diameter, subglobose, superficial or immersed, ostiolate, epapillate or papillate, leathery to subcarbonaceous, dark brown, to black, solitary or gregarious. PERIDIUM 24–50 μm thick, two-layered, forming a textura angularis; thin outer layer with brown incrustations; inner layer composed of about five rows of polygonal, thick-walled, hyaline cells with large lumina, becoming thin-walled and flattened near the center and merging into pseudoparaphyses. PAPILLAE short, conical; ostiolar canal 25–50 μm in diameter, formed schizogenously by crumbling away of thin-walled, subglobose, hyaline cells at the apex of the ascocarp. PSEUDOPARAPHYSES 2–2.5 μm in diameter, cylindrical, septate, attached at the base and apex, forming an interascicular tissue. ASCI 130–180 × 22–34 μm, eight-spored, cylindrical, short pedunculate, bitunicate, thick-walled, physoclastic, without apical apparatuses; developing at the base of the ascocarp venter. ASCOSPORES (25–)26.5–35 × 11.5–17 μm, partly or completely biseriate, ellipsoidal, muriform, with (six to) seven transverse septa and several longitudinal septa in each segment; broadest above the first-formed, central septum and strongest constricted around it, less so around the other septa, yellowish-brown to brown, surrounded by a gelatinous, deliquescing, 2- to 4-μm-thick sheath. MODE OF LIFE: Saprobic or perthophytic. HOSTS: Damaged or decaying red or brown algae (*Ceramium* sp., *Chondrus crispus, Furcellaria* sp., *Laminaria* sp., *Pelvetia canaliculata*). RANGE: Atlantic Ocean—France, Great Britain (Scotland), Norway. NOTE: Description mostly after Kohlmeyer (1973b); a Hyphomycete associated with ascocarps of *P. pelvetiae* was reported by Sutherland (1915a) and

Kohlmeyer (1973b); a possible synonym is *P. laminariana* Sutherland (1916b), a doubtful species. ETYMOLOGY: Named after a host genus, namely, *Pelvetia* (Phaeophyta).

Pleospora spartinae (Webster et Lucas) Apinis et Chesters

Trans. Br. Mycol. Soc. **47**, 432, 1964

≡*Pleospora vagans* Niessl var. *spartinae* Webster et Lucas, *Trans. Br. Mycol. Soc.* **44**, 427, 1961

Fig. 98i

ASCOCARPS 350–375 μm in diameter, globose, immersed, ostiolate, papillate, subcarbonaceous, black, solitary, covered at the sides and base by light brown hyphae; causing a black discoloration of the host tissue. PERIDIUM thick, composed of dark brown, polygonal cells. PAPILLAE conical, black, shining. PSEUDOPARAPHYSES filiform, long. ASCI 150–180 × 14–18 μm, eight-spored, cylindrical, apically rounded, pedunculate, bitunicate, thick-walled. ASCOSPORES 24–38 × 10–13 μm, uniseriate, ellipsoidal, boat-shaped or somewhat inequilateral, thick-walled, muriform, with usually five transverse septa and one longitudinal septum in the third or fourth cell from the top, sometimes without longitudinal septa, third cell the largest, constricted at the septa, particularly around the middle, yellowish-brown. PYCNIDIA 160–300 μm in diameter, globose or depressed–globose, dark brown to black, subepidermal. CONIDIA 36–72 × 3–6 μm, cylindrical, tapering at the apices, five- to seven-septate, slightly constricted at the septa, straight or somewhat curved, yellow to pale brown, with several small guttules near the septa. MODE OF LIFE: Saprobic. HOST: *Spartina townsendii*. RANGE: Atlantic Ocean—Great Britain (England). NOTE: Description after Webster and Lucas (1961); we have seen a slide of the paratype (Herb. Sheffield No. 2061). ETYMOLOGY: Named after the host genus, namely, *Spartina*.

Pleospora triglochinicola Webster

Trans. Br. Mycol. Soc. **53**, 481, 1969

≡*Pleospora maritima* Rehm, *Hedwigia* **35**, 149, 1896 (non *P. maritima* Bommer, Rousseau et Saccardo, *Syll. Fung.* **9**, 893, 1891)

CONIDIAL STATE: *Stemphylium triglochinicola* Sutton et Pirozynski, *Trans. Br. Mycol. Soc.* **46**, 519, 1963.

LITERATURE: Kohlmeyer and Kohlmeyer (1964–1969); Munk (1957); Riedl (1968); Simmons (1969); Wehmeyer (1948).

ASCOCARPS 400–500 μm in diameter, subglobose to depressed globose, immersed or erumpent, ostiolate, papillate, black, smooth, solitary or gregarious. PERIDIUM composed of dark brown, polygonal cells, 5–10 μm in diameter. PAPILLAE short, conical. PSEUDOPARAPHYSES about 4 μm in diameter, filamentous, deliquescing. ASCI 155–180 × 40–45 μm, eight-spored, broadly cylindrical to oblong–clavate, tapering abruptly into a knoblike base, bitunicate, thick-walled. ASCOSPORES 45–65 × 16.5–25 μm, biseriate, ellipsoidal to fusiform, somewhat broadened at the upper end, with seven transverse septa and one to two longitudinal septa in most segments, apical segments usually with a single longitudinal septum; end cells obtuse or bluntly pointed, constricted at the central transverse septum, slightly at the others; golden-brown to almost olive-brown, surrounded by a mucilaginous sheath. CONIDIOPHORES 50–175(–350) μm long, 6–12 μm in diameter, cylindrical, swollen at the base and apex (up to 15 μm), often also along the length, smooth, three- to ten-septate, simple or rarely branched, straight or slightly curved, light brown to yellow–brown, single or in groups of two to six, forming conidia singly after one or two proliferations. CONIDIA developing singly through a pore at the apex of the conidiophore, 40–82 × 18–46 μm, irregularly obclavate to ob-pyriform, with a broad base, tapering to the rounded apical beak, at the base with a short lateral beak and a dark abscission scar (10–12 μm in diameter), muriform, up to twelve transverse septa, each segment divided by one to four longitudinal or oblique septa, constricted at the primary transverse septa, less so at the secondary septa, greenish-brown to almost black, smooth or rarely inconspicuously roughened. MODE OF LIFE: Saprobic. HOST: Old leaves and inflorescences of *Triglochin maritima*. RANGE: Atlantic Ocean—Denmark (Sjaelland, Island of Samsø), Great Britain (England); Arctic Ocean—Norway. NOTE: Description in part after Simmons (1969) and Webster (1969) and our examination of type material (Rehm: Ascomyceten No. 1188 in Herb. B; *Stemphylium triglochinicola* in Herb. IMI No. 95209) and a collection of Professor A. Munk; the fungus from wood identified as *P. maritima* Rehm by Cavaliere (1966c) has smaller ascospores and probably represents a different species. ETYMOLOGY: From the host genus, *Triglochin*, and the Latin, -*cola* = -dweller.

Pleospora sp. I

Fig. 98j

ASCOCARPS 190–250 μm high, 275–355 μm in diameter, ellipsoidal, immersed or erumpent, ostiolate, short papillate, coriaceous, dark brown to black, solitary or gregarious. PERIDIUM 14–26 μm thick, composed of about five layers of large, polygonal cells with large lumina, forming a textura angularis. PAPILLAE 80–100 μm high, 90–110 μm in diameter, cylindrical, centric or eccentric. PSEUDOPARAPHYSES 4–6 μm in diameter, simple, septate, persistent. ASCI 130–185 × 30–35 μm, eight-spored, cylindrical to broadly clavate, short pedunculate, bitunicate, thick-walled, physoclastic, without apical apparatus, but with an apical chamber, tip covered with an inconspicuous gelatinous, caplike layer. ASCOSPORES 40–52 × (14–)16–19 μm, biseriate, cylindrical with obtuse apices or ellipsoidal, muriform, with about eight to ten transverse septa and several longitudinal septa per segment, septal pattern difficult to discern, with three major constrictions, yellow–brown, surrounded by a gelatinous, 5- to 12.5-μm-thick sheath that swells in water. MODE OF LIFE: Saprobic. HOST: Dead culms of *Salicornia virginica*. RANGE: Atlantic Ocean—Bermuda, United States (New Jersey, North Carolina). NOTE: Description based on collections Herb. J.K. 1774, 1982, 1994, 3718, 3772, 3775, 3777, and 3781; the fungus is similar to *Pleospora valesiaca* on *Carex* spp. in inland habitats, namely, the Alps, but differs from this species by the ascospores, which are oblong–clavate and tapering at the basis in *P. valesiaca* (Müller, 1951), but cylindrical and nontapering in the marine fungus; both have a gelatinous ascospore sheath; Dr. E. Müller kindly examined the *Pleospora* from *Salicornia* and was not convinced that it was identical with *P. valesiaca;* it also appears uncertain if the *Pleospora* from *Juncus maritimus* from a marine habitat in England and identified as *P. valesiaca* (Lucas and Webster, 1964) belongs to this species; we include the *Pleospora* from *Salicornia* because it is common on this host, but we refrain from naming it at this time since its difference from similar *Pleospora* spp. from other shore or salt-marsh plants (*Batis, Cakile, Crithmum, Limonium, Salicornia* spp.) cannot be established with certainty; pure cultures are needed to compare and separate the marine members of this *Pleospora* complex; *Pleospora* sp. I is clearly different from *P. salsolae* Fuckel, of which we have seen type and authentical material from *Salsola kali* (Fuckel, Fungi rhenani 814, in B; Herb. Barbey-Boissier No. 466, in BPI); the name *P. salsolae* is often wrongly used in collections of *Pleospora* on *Salicornia*; we have been unable to locate type specimens of *P. salicorniae* Dangeard, a species described from *Salicornia herbacea* in France, and agree with Wehmeyer (1961) that this name should be discarded.

Key to the Marine Species of *Pleospora*

1. Ascospores with a gelatinous sheath and a pair of subconical appendages extending through the sheath at each end (use India ink or phase contrast!)   . . . **P. gaudefroyi**
1'. Ascospores without sheaths; or, if a sheath present, without appendages   . . .   2

Pontoporeia Kohlm.

Nova Hedwigia **6**, 5–6, 1963

ASCOCARPS solitary or gregarious, globose, seated in the surface of the substrate with weakly developed hypostromata, superficial, nonostiolate, carbonaceous, black, smooth. PSEUDOPARAPHYSES filiform, simple or branched, developing along the inside of the peridium. ASCI eight-spored, broadly clavate, ovoid or subellipsoidal, long pedunculate, bitunicate, thick-walled; developing at the periphery of a hemispherical pulvinus, which is composed of radiating hyphae and which fills the base and center of the ascocarp venter. ASCOSPORES biturbinate or subellipsoidal, apically papillate and provided with germ pores, one-septate, dark brown; walls two-layered, composed of a light-colored, thick endospore and a dark, thin exospore. SUBSTRATE: Rhizomes of sea grasses (*Posidonia*). TYPE SPECIES: *Pontoporeia biturbinata* (Durieu et Montagne) Kohlm. The genus is monotypic. NOTE: *Pontoporeia* was synonymized with *Zopfia* Rabenhorst by Malloch and Cain (1972), a decision accepted by Hawksworth and Booth (1974). The following features distinguish *P. biturbinata* from *Zopfia* as characterized by Hawksworth and Booth (1974) and appear to justify the retention of *P. biturbinata* in a separate genus. First, the asci originate at the periphery of a subglobose pulvinus that is composed of delicate hyphae radiating from the base of the as-cocarp venter (this pulvinus is not mentioned by the above-cited authors); second, there is not indication that the central portion of the ascocarps is originally occupied by hyaline pseudoparenchymatous cells (a family character of Zopfiaceae sensu Hawksworth and Booth), but, instead, the center is made up of filaments; third, the peridial structure is irregular and can be termed cephalothecoid only with a stretch of the imagination; fourth, the ascospore walls are two-layered, composed of a thin, dark brown exospore and a thick, light-colored endospore provided with a germ pore at each end (Kohlmeyer, 1963a); Hawksworth and Booth (1974) mention only hyaline apical papillae as being characteristic for *P.*

biturbinata, omitting the two-layered nature of the ascospore wall; and fifth, the marine occurrence of *P. biturbinata*, as opposed to the terrestrial distribution of the twelve species of *Zopfia*, indicates different physiological requirements in addition to the morphological differences; as we do not agree with Malloch and Cain's placement of the species in *Zopfia*, we do not assign it to Zopfiaceae, but prefer to keep *Pontoporeia* in the Pleosporaceae. ETYMOLOGY: From the Greek, *Pontoporeia* = one of the Nereids (*pontoporeuo* = to cross the ocean), in reference to the marine habitat of the fungus.

Pontoporeia biturbinata (Durieu et Montagne) Kohlm.

Nova Hedwigia **6,** 6, 1963

≡*Sphaeria biturbinata* Dur. et Mont. *in* Montagne, *Syll. Gen. Sp. Crypt.* p. 228, 1856*

≡*Amphisphaeria biturbinata* (Dur. et Mont.) Saccardo, *Syll. Fung.* **1,** 729, 1882

≡*Zopfia biturbinata* (Dur. et Mont.) Malloch et Cain, *Can. J. Bot.* **50,** 67, 1972

LITERATURE: Kohlmeyer and Kohlmeyer (1964–1969).

Fig. 99

ASCOCARPS 805–1120 μm high, 805–1375 μm in diameter, globose, seated in the surface of the substrate with weakly developed hypostromata, superficial, nonostiolate, carbonaceous, black, smooth, solitary or gregarious. PERIDIUM 110–137.5 μm thick, on the outside composed of small, irregular, thick-walled cells, occluded by irregular, dark brown deposits; toward the center composed of slightly larger, less pigmented cells with larger lumina, giving rise to pseudoparaphyses; forming a textura angularis. PSEUDOPARAPHYSES 1–1.5 μm in diameter, septate, filiform, simple or branched, developing along the inside of the peridium. ASCI 225–360 × 72–111 μm, eight-spored (rarely with less spores), broadly clavate, ovoid or subellipsoidal, long pedunculate, bitunicate, thick-walled (ectoascus thin-walled, endoascus thick-walled), without apical apparatuses, surrounded by pseudoparaphyses; developing successively at the periphery of a hemispherical pulvinus, which is composed of radiating hyphae and which fills the base and center of the ascocarp venter. ASCOSPORES 66–90 × 31.5–43.5 μm, bi- or triseriate, biturbinate or subellipsoidal, apically papillate and provided with germ pores, one-septate, constricted at the septum, dark brown to blackish-brown, walls

---

* See the footnote on p. 404, under *Halotthia*.

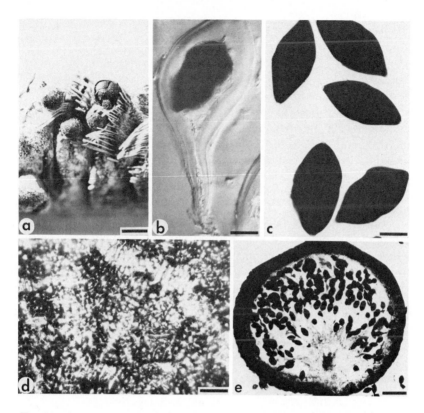

Fig. 99. *Pontoporeia biturbinata* from rhizomes of *Posidonia oceanica* (Herb. J.K. 819 in B). (a) Superficial ascocarps between fibers of decomposed leaves (bar = 1 mm). (b) Thick-walled ascus with one ascospore (bar = 20 μm). (c) Ascospores with hyaline apical germ pores (bar = 20 μm). (d) Tangential section (4 μm) through peridium, showing arrangement of cells (bar = 25 μm). (e) Longitudinal section through ascocarp, asci originating at the periphery of a hemispherical basal pulvinus (bar = 150 μm). (b in Nomarski interference contrast, the others in bright field.)

two-layered, composed of a hyaline to light brown, thick endospore (2–4 μm) and a dark brown, thin exospore; endospore sometimes pseudoseptate. MODE OF LIFE: Perthophytic (or parasitic?). HOST: Defoliated parts of living rhizomes of *Posidonia oceanica*. RANGE: Mediterranean—Algeria, Spain. NOTE: T. W. Johnson and Sparrow (1961) list a collection at Cape Horn, allegedly made by Hariot (1889); this record is erroneous and was probably caused by a mistake in translation. ETYMOLOGY: From the Latin, *bi-* = two- and *turbinatus* = turbinate, obconical, in reference to the shape of the ascospores.

Thalassoascus Ollivier

C. R. Hebd. Séances Acad. Sci. **182**, 1348–1349, 1926

ASCOCARPS solitary or gregarious, subglobose, with or without a stalk, superficial, subiculate, ostiolate, periphysate, epapillate, subcarbonaceous, black. PSEUDOPARAPHYSES numerous, ramose, persistent. ASCI eight-spored, clavate, pedunculate, bitunicate, thick-walled, developing along the inner wall of the lower two-thirds of the locule. ASCOSPORES ellipsoidal, one-septate, hyaline, thick-walled. SPERMOGONIA solitary or gregarious, subglobose, superficial, subiculate, ostiolate, epapillate, subcarbonaceous, black. SPERMATIOPHORES simple, cylindrical. SPERMATIA subglobose to ovoid, one-celled, hyaline. HOSTS: Marine algae (Phaeophyta). TYPE SPECIES: *Thalassoascus tregoubovii* Ollivier. The genus is monotypic. NOTE:von Arx and Müller (1975) place *Thalassoascus* in the Zopfiaceae; however, we prefer to include it tentatively in the Pleosporaceae because the ascospores are hyaline, and the ascocarps become distinctly ostiolate and periphysate, although in a late state of development (Kohlmeyer, 1967). ETYMOLOGY: From the Greek, *thalassa* = sea, ocean, and *askos* = ascus (skin, hose), in reference to the habitat of the Ascomycete.

Thalassoascus tregoubovii Olliver

C. R. Hebd. Séances Acad. Sci. **182**, 1348–1349, 1926 (sub *T.*
tregoubovi)

≡*Melanopsamma tregoubovii* (Ollivier) Ollivier, *Bull. Inst. Océanogr.* **522**,
　3, 1928
≡*Melanopsamma tregoubovii* (Ollivier) Ollivier var. *cutleriae* Ollivier,
　Ann. Inst. Océanogr. (Paris) (N.S.) **7**, 172, 1930
≡*Melanopsamma tregoubovii* (Ollivier) Ollivier var. *cystoseirae* Ollivier,
　Ann. Inst. Océanogr. (Paris) (N.S.) **7**, 172, 1930

LITERATURE: Aleem (1952); J. Feldmann (1931, 1932); T. W. Johnson and Sparrow (1961); Kohlmeyer (1963b, 1964, 1967); Kohlmeyer and Kohlmeyer (1964–1969); Meyers (1957); von Arx and Müller (1975).

Fig. 100

HYPHAE 4.5–7.5 μm in diameter, dark brown, septate, ramose, on and in the substrate, also forming stromalike crusts. ASCOCARPS on *Cystoseira* spp. clavate, verrucose, 1.1–1.75 mm high, 495–820 μm in diameter in the fertile, subglobose head, 200–550 μm in diameter in the sterile stalk (on species of *Aglaozonia*, *Dilophus*, and *Zanardinia* without stalk, smooth, 400–750 μm in diameter); superficial, subiculate, ostiolate, epapillate,

Fig. 100. Ascocarps of *Thalassoascus tregoubovii* on the base of *Cystoseira fimbriata* (Herb. J.K. 835; bar = 3mm).

subcarbonaceous, black, solitary or gregarious. PERIDIUM up to 130 μm thick, two-layered, composed of thick-walled, irregular cells, subglobose with large lumina on the outside, elongate and with smaller lumina on the inside; outer part black, inner part hyaline. PAPILLAE absent; ostiolar canal with periphyses. PSEUDOPARAPHYSES 2–3.5 μm in diameter, septate, ramose, persistent. ASCI 165–190 × 47–55 μm, eight-spored, broadly clavate, pedunculate, bitunicate, thick-walled, without apical apparatuses, developing along the inner wall of the lower two-thirds of the locule. ASCOSPORES 29–48 × 18.5–33.5 μm, biseriate, broadly ellipsoidal, one-septate, strongly constricted at the septum, hyaline, rarely light grayish, thick-walled. SPERMOGONIA 135–180(–350) μm in diameter, subglobose, superficial, subiculate, ostiolate, epapillate, verrucose, subcarbonaceous, black, solitary or gregarious; sometimes associated with ascocarps. SPERMATIOPHORES 25 × 2–3 μm, simple, cylindrical, lining the interior walls of the spermogonia. SPERMATIA 3–5 × 2–3.5 μm, subglobose, ellipsoidal or ovoid, one-celled, hyaline. MODE OF LIFE: Parasitic. HOSTS: *Aglaozonia chilosa, A. parvula, Cystoseira abiesmarina, C. discors, C. fimbriata, Zanardinia collaris.* RANGE: Atlantic Ocean—Spain (Tenerife); Mediterranean—Algeria, France, Tunisia. NOTE: Collections of the fungus on *Cystoseira* spp. (Kohlmeyer, 1963b, 1967) appear almost identical with the fungi on *Aglaozonia* and *Zanardinia* (Ollivier, 1930), except for the smooth ascocarps without stipes in the latter; until new material becomes available showing distinct specific or varietal differences between the fungi on the different hosts, they are treated as one species; the fungus on *Dilophus fasciola*, identified as *Melanopsamma tregoubovii* by Aleem (1952), may be a small-spored variety of the species. ETYMOLOGY: Named in honor of G. Tregouboff, subdirector of the marine station at Villefranche-sur-Mer, France.

Trematosphaeria Fuckel

Jahrb. Nassau. Ver. Naturkd. **23/24**, 161, 1869–1870

LITERATURE: Eriksson (1967a); Holm (1957).

ASCOCARPS solitary or gregarious, pyriform or broadly conical, somewhat or deeply immersed, ostiolate, papillate or epapillate, thick-walled. PERIDIUM composed of numerous layers of cells of variable size and shape, generally with thin walls but strongly and irregularly colored. PSEUDOPARAPHYSES filamentous, numerous. ASCI eight-spored, clavate or cylindrical, bitunicate, with a small refractive apical apparatus. ASCOSPORES ellipsoidal to fusiform, with one or more septa, generally more or less cinnamon-colored, frequently lighter at the apices. MODE OF LIFE: Mostly saprobic on wood; some species are aquatic. TYPE SPECIES: *Trematosphaeria pertusa* Fuckel. NOTE: Description after Holm (1957); 1 obligate and 1 facultative marine species are known among about 25 terrestrial species; *T. pertusa* was found by Eaton and Irvine (1972) and Eaton and Jones (1971a,b) on wood in cooling towers soaked with fresh and brackish water, however, not in natural marine habitats; the following names are *nomina nuda*: *T. oraemaris*, *T. purpurascens*, and *T. thalassica*. ETYMOLOGY: From the Greek, *trema* = hole, and *Sphaeria* = a genus of Pyrenomycetes (*sphaera* = globe), in reference to the ostiolate ascocarps.

Trematosphaeria britzelmayriana (Rehm) Saccardo

Syll. Fung. **2**, 120, 1883

≡*Melanomma megalosporum* Rehm var. *britzelmayrianum* Rehm, Ascomyceten Fasc. XII, No. 588, *Hedwigia* **20**, 51, 1881
=*Melanomma megalospora* (DeNot.) Rehm, Ascomyceten Fasc. 1-XI, No. 536a, *Ber. Naturhist. Ver. Augsburg* **26**, 126, 1881
=*Trematosphaeria megalospora* Saccardo, *Syll. Fung.* **2**, 120, 1883 (non *Sphaeria megalospora* DeNot., nec non *Trematosphaeria megalospora* Fabre)

LITERATURE: Eriksson (1967a); Henningsson (1974); Holm (1957).

ASCOCARPS up to 300 μm high, 200–400 μm in diameter, conical, pyriform or often somewhat elongate, partly immersed, ostiolate, short papillate, carbonaceous, black, solitary or gregarious. PERIDIUM 30–50 μm thick, composed of numerous layers of rounded or flattened, thin- or thick-walled cells, up to 15 μm in diameter; strong but irregularly colored. ASCI 100–140 × 16–24 μm, eight-spored, clavate, subpedunculate, thick-walled. ASCOSPORES 27–51 × (6.5–)8.5–13.5 μm, biseriate in the upper part, uniseriate in the lower part of the ascus, fusiform, obtusely rounded at the apices, four- to ten-septate (usually eight-septate), strongly constricted at the septa, smooth, chestnut-colored; the fourth cell usually largest. MODE OF LIFE: Saprobic. SUBSTRATE: Dead, submerged wood. RANGE: Baltic Sea. NOTE: This species was known from freshwater in Germany and Switzerland (Holm, 1957), but Henningsson (1974) collected it on *Betula pubescens* and *Pinus sylvestris* along the Swedish coast of the Baltic Sea in low salinities (between approximately 2 and 7°/₀₀); consequently, *T. britzelmayriana* is considered a facultative marine species; the foregoing description is mostly based on Holm's (1957) work. ETYMOLOGY: Named after the collector, the German mycologist M. Britzelmayr (1839–1909).

Trematosphaeria mangrovis Kohlm.

Mycopathol. Mycol. Appl. **34,** 1–2, 1968

LITERATURE: Kohlmeyer and Kohlmeyer (1964–1969).

ASCOCARPS 380–750 μm high (including papillae), 450–800 μm in diameter, conical or semiglobose, flattened at the base, seated partly in the substrate with subicula that consist of dark, plectenchymatous layers and downward of brown hyphae; ostiolate, short papillate, carbonaceous, black, gregarious, sometimes confluent. PERIDIUM 64–88 μm thick, two-layered, enclosing cells of the host; outer black layer 56–84 μm thick, scleroplectenchymatous, composed of thick-walled, brown cells with irregular, small lumina; inner hyaline layer 8–12 μm thick, composed of elongate, thin-walled cells with large lumina at the sides and in the angle between cover and basis of the ascocarps, merging into periphyses and pseudoparaphyses; cells of inner layer isodiametric under the asci. PAPILLAE 160–220 μm high, 160–180 μm in diameter, conical; ostiolar canal at first closed at the apex by polygonal, light brown cells with small lumina, 3–4.5 μm in diameter; filled with delicate, periphysoid hyphae. PSEUDOPARAPHYSES 1.6–2.2 μm in diameter, septate, simple or rarely reticulate. ASCI 190–220 × 20–22 μm, eight-spored, cylindrical, pedunculate, bitunicate, thick-walled, without apical apparatuses; developing at the base of the ascocarp venter. ASCOSPORES 30–41 × 10–13(–16.5) μm, obliquely uni- or biseriate, broad fusiform or ellipsoidal, sometimes slightly apiculate, somewhat curved, three-septate, slightly constricted at the septa, fuscous, sometimes lighter colored at the extreme apices. MODE OF LIFE: Perthophytic. HOST: Bark and wood of submerged roots of *Rhizophora racemosa.* RANGE: Atlantic Ocean—Liberia [known positively only from the type collections, but possibly occurring in Florida according to Wagner-Merner (1972), who found similar ascospores in sea foam]. ETYMOLOGY: From the Spanish, *mangle* = mangrove tree and the English, grove, in reference to the habitat of the species.

Key to the Marine Species of Trematosphaeria

1. Ascospores three-septate; tropical species . . . . **T. mangrovis**
1'. Ascospores four- to ten-septate; species of temperate zones . . . **T. britzelmayriana**

E. Ascomycotina incertae sedis

Crinigera I. Schmidt

Nat. Naturschutz Mecklenburg **7,** 11, 1969 (publ. 1971)

ASCOCARPS solitary, globose, superficial or immersed, subiculate, without ostioles, epapillate, coriaceous, whitish to light brown, with small appendages. PARAPHYSES absent. ASCI eight-spored, clavate, stipitate, unitunicate, apically thick-walled, later thin-walled, finally deliquescing, developing in a bunch. ASCOSPORES ellipsoidal, one-septate, hyaline, with subterminal appendages. SUBSTRATES: Bark, dead algae, and wood. TYPE SPECIES: *Crinigera maritima* I. Schmidt. NOTE: The genus is monotypic and cannot be placed properly because young developmental stages are unknown; similarities with Plectascales and Erysiphales (Schmidt, 1969b) are only superficial and do not indicate a relationship to these orders. ETYMOLOGY: From the Latin, *crinis* = the hair and *-ger* = bearing, in reference to the ascospores, covered by long, hairlike appendages.

Crinigera maritima I. Schmidt

Nat. Naturschutz Mecklenburg **7,** 11, 1969 (publ. 1971)

LITERATURE: Koch (1974); Schmidt (1974).

ASCOCARPS 150–250 μm in diameter, globose, superficial on hard surfaces or immersed in soft substrates, subiculate, without ostioles, epapillate, opening by irregular dissolution of the wall, coriaceous, whitish to light brown, with small, short appendages, solitary. PERIDIUM 17–39 μm thick, two-layered; outer wall consisting of several layers of isodiametric, brownish cells with small lumina, producing the appendages; inner wall

10–29 μm thick, composed of several layers of flattened, yellowish cells with larger lumina; appendages composed of a short stem and an irregular warty or branched head. PARAPHYSES absent. ASCI 26–27.5 × 13–20 μm (excluding stipe), eight-spored, clavate, stipitate, unitunicate, apically thick-walled, later thin-walled, without apical apparatuses, finally deliquescing, developing in a bunch. ASCOSPORES (7–)10–13 × 3.5–5(–6.5) μm, ellipsoidal, one-septate, slightly constricted at the septum, hyaline, with subterminal, delicate, hairlike appendages that develop by spirally unfolding of an outer layer of the wall. MODE OF LIFE: Saprobic. SUBSTRATES: Rotting *Fucus vesiculosus,* submerged bark, and dead intertidal wood (*Quercus* sp.). RANGE: Baltic Sea—Germany (G.D.R.); Pacific Ocean—United States (Washington). NOTE: The fungus described by Koch (1974) as *Crinigera* sp. on *Larix* sp., *Picea* sp., and *Pinus* sp. from the Atlantic Ocean (Jutland, Denmark) seems to differ from *C. maritima* by thicker ascus walls and by an irregular sheath covering the ascospores; most probably the Danish collections represent *C. maritima* in an early stage of development. Description in part after J. Kohlmeyer (unpublished). ETYMOLOGY: From the Latin, *maritimus* = marine, referring to the habitat of the species.

Orcadia Sutherland

Trans. Br. Mycol. Soc. **5**, 151, 1915

LITERATURE: Petch (1938); Rogerson (1970).

ASCOCARPS gregarious, subglobose or pyriform, completely immersed, ostiolate, papillate, coriaceous, light-colored. PSEUDOPARAPHYSES (or apical paraphyses?) simple or ramose, early deliquescing. ASCI eight-spored, clavate to subcylindrical, pedunculate, unitunicate, thin-walled with thickened apical wall, opening with an operculus; developing at the base of the ascocarp venter. ASCOSPORES cylindrical, triseptate, hyaline. SUBSTRATE: Marine algae (Phaeophyta). TYPE SPECIES: *Orcadia ascophylli* Sutherland. The genus is monotypic. ETYMOLOGY: Name refers to the origin of the type material; from the Latin, *Orcades* = Orkney (Scotland).

Orcadia ascophylli Sutherland

Trans. Br. Mycol. Soc. **5**, 151, 1915

=*Orcadia pelvetiana* Sutherland, *New Phytol.* **14**, 183, 1915

LITERATURE: T. W. Johnson and Sparrow (1961); Kohlmeyer (1968e,

1974a); Kohlmeyer and Kohlmeyer (1964–1969); Sutherland (1916b); Wilson (1951).

Fig. 101

Ascocarps (80–)100–230 μm high, (80–)130–180(–200) μm in diameter, subglobose or pyriform, completely immersed, ostiolate, papillate, coriaceous, subhyaline, gregarious. Peridium 15–25 μm thick, hyaline, two-layered; outer layer composed of subglobose to elongate, thick-walled, small cells with large lumina; inner layer composed of elongate to polygonal, thin-walled, larger cells. Papillae or necks 80–180 × 20–35

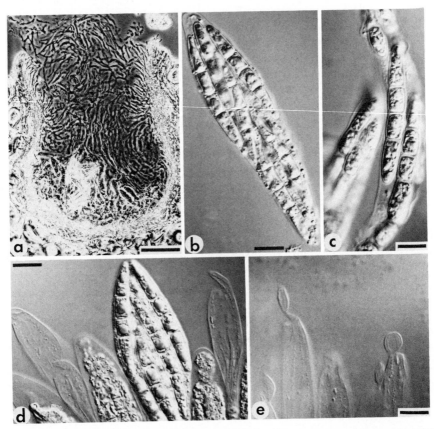

Fig. 101. *Orcadia ascophylli* in *Pelvetia canaliculata.* (a) Longitudinal section (10 μm) through ascocarp filled with pseudoparaphyses (Herb. J.K. 2951; bar = 50 μm). (b) Ascus (Herb. J.K. 2965; bar = 10 μm). (c) Ascospores (Herb. J.K. 2962; bar = 10 μm). (d) Immature, mature and empty asci, the right one showing operculus (Herb. J.K. 2965; bar = 10 μm). (e) Tips of empty asci with operculi (Herb. J.K. 2962; bar = 10 μm). (a in phase contrast, the others in Nomarski interference contrast.)

μm, subconical or cylindrical; ostiolar canal periphysate. PSEUDO-PARAPHYSES (or apical paraphyses?) septate, simple or ramose, early deliquescing. ASCI 50–110 × 10–20 μm, eight-spored, clavate, sub-cylindrical or subfusiform, short pedunculate, unitunicate, thin-walled, with thickened apical wall, without apical apparatuses, opening with an operculus; developing at the base of the ascocarp venter. ASCOSPORES 34.5–56 × 4–7 μm, bi- or triseriate, cylindrical, slightly curved, triseptate, constricted at the septa, hyaline (possibly yellowish at maturity), mul-tiguttulate. MODE OF LIFE: Perthophytic. SUBSTRATES: Damaged Phaeophyta, for example, *Ascophyllum nodosum, Fucus vesiculosus, Fucus* sp., *Pelvetia canaliculata*. RANGE: Atlantic Ocean—France (J. Kohlmeyer, unpublished), Great Britain [England (J. Kohlmeyer, unpub-lished), Scotland, Wales], Norway (J. Kohlmeyer, unpublished). ETYMOLOGY: Named after a host genus, *Ascophyllum*.

Sphaerulina Saccardo

Michelia **1**, 399, 1878

NOTE: The generic description of *Sphaerulina* is omitted because the only remaining species collected in marine habitats, *S. albispiculata* and *S. oraemaris,* are doubtfully assigned to this genus; *S. longirostris* San-tesson *in* Gustafsson et Fries (1956) is a *nomen nudum*.

Sphaerulina albispiculata Tubaki

Publ. Seto Mar. Biol. Lab. **15**, 366–367, 1968

LITERATURE: Tubaki (1969).

ASCOCARPS 200–340 μm high, 150–220 μm in diameter, globose, sub-globose or ovoidal, immersed, ostiolate, papillate, membranaceous, hyaline to light brown. PAPILLAE or necks 230–400 μm high, 100–220 μm in diameter, broadly ellipsoidal, cylindrical or obclavate, centric, covered with white bristlelike hairs; projecting above the substrate. PARAPHYSES present. ASCI 120–200 × 8–10 μm, eight-spored, clavate or cylindrical. ASCOSPORES 25–30 × 5–6 μm, partially or completely biseriate, fusiform or ellipsoidal, five- or six-septate, not constricted at the septa, straight or slightly curved, hyaline. MODE OF LIFE: Saprobic. SUBSTRATE: Driftwood. RANGE: Pacific Ocean—Japan. NOTE: We have examined type material (Herb. IFO, H-11592) kindly sent on loan by Dr. Tubaki, but we were unable to find intact asci or paraphyses; the material was senes-cent and many ascospores had germinated; in the absence of detailed information on the structure of asci we cannot clarify the taxonomic position of the species; it is possible that *S. albispiculata* had not developed in a marine habitat, because the wood contained tunnels filled with feces of insects, but showed no signs of marine life, such as borers or fouling organisms. ETYMOLOGY: From the Latin, *albus* = white and *spiculatus* = covered with fine points, in reference to the white hairy necks of ascocarps.

Sphaerulina oraemaris Linder *in* Barghoorn et Linder

Farlowia **1**, 413, 1944 (as *"orae-maris"*)

LITERATURE: Apinis and Chesters (1964); Cavaliere (1966c, 1968); Gold (1959); Gustafsson and Fries (1956); T. W. Johnson and Sparrow (1961); E. B. G. Jones (1963a); Neish (1970); Poole and Price (1972); Schmidt (1974).

Fig. 102

ASCOCARPS 190–220 μm high, 150–250 μm in diameter, subglobose, pyriform or ellipsoidal, partly or completely immersed, ostiolate, epapillate or short papillate, coriaceous, brown to blackish above, subhyaline to light fuscous below, solitary or gregarious. PERIDIUM 20–30 μm thick, often up to 40 μm and clypeoid around the ostiole, composed of small ellipsoidal to subglobose cells with small lumina, forming a textura angularis. PAPILLAE absent or, if present, 50–80 μm high, 60–100 μm in diameter; ostiolar canal 10–15 μm in diameter, conical, periphysate. PARAPHYSES 1.5–2.5 μm in diameter, filiform, simple, septate. ASCI 70–80 × 10–14 μm, eight-spored, clavate, apically rounded or obtuse, short pedunculate, unitunicate, apically thick-walled, without apical apparatuses, persistent; developing at the base of the ascocarp venter. ASCOSPORES 26–32 × (5–)6–8 μm, bi- or triseriate, subfusiform with rounded ends or elongate–ellipsoidal, three-septate, not or slightly constricted at the septa, smooth, hyaline. MODE OF LIFE: Saprobic. SUBSTRATES: Intertidal and drifting wood, test panels (e.g., *Fagus sylvatica, Fagus* sp.), in humus layer under *Ammophila arenaria*. RANGE: Atlantic Ocean—Canada (Nova Scotia), Great Britain (England), Iceland, United States (Florida, North Carolina); Baltic Sea—Germany (G.D.R.); Pacific Ocean—United States (California). NOTE: Description based on Linder (1944) and on our examination of type material from Herb. FH; *S. oraemaris* may easily be confused with other, *Leptosphaeria*-like Ascomycetes; fungi identified in the past as *S. oraemaris* may actually belong to other species because Linder's (1944) fungus was considered

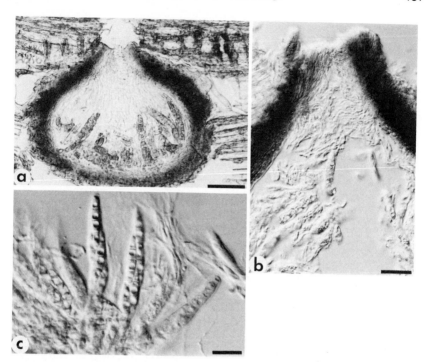

Fig. 102. *Sphaerulina oraemaris* from wood (type FH). (a) Longitudinal section (16 μm) through ascocarp with paraphyses and periphyses (bar = 50 μm). (b) 16-μm section through ostiole with periphyses (bar = 25 μm). (c) Asci (bar = 15 μm). (a in bright field, b and c in Nomarski interference contrast.)

until now as aparaphysate; however, paraphyses and periphyses are distinct in the type material of *S. oraemaris*; this unitunicate species does not belong to the bitunicate genus *Sphaerulina* (von Arx and Müller, 1975); a final taxonomic disposition of *S. oraemaris* cannot be made because asci in the type specimen are generally not well preserved; the possibility cannot be excluded that *S. oraemaris* is not of marine origin, because the type material, deriving from driftwood, consists of a small piece of wood (21 × 12 mm) without traces of marine life, and it is not known how long this substrate had been immersed in seawater. ETYMOLOGY: From the Latin, *ora* = coast and *mare* = the sea, in reference to the marine collecting site of the species.

Key to the Marine Species of *Sphaerulina*

1. Ascospores three-septate . . . . *S. oraemaris*
1'. Ascospores five- or six-septate . . . . *S. albispiculata*

II. Basidiomycotina
  A. Gasteromycetes
    1. Melanogastrales
      a. Melanogastraceae

Nia Moore et Meyers

Mycologia **51**, 874, 1959

LITERATURE: Doguet (1967); Dring (1973).

HYPHAE septate, ramose, with clamp connections, hyaline. BASIDIOCARPS solitary or gregarious, subglobose, superficial, pedicellate, light-colored, soft, thin-walled, villous or smooth, containing a homogenous gleba; protocysts preceding formation of basidia; opening by irregular rupture of the peridium. BASIDIA four- to eight-spored, consisting of a filament with a terminal, subglobose inflation; hyaline, without sterigmata. BASIDIOSPORES ovoid or ellipsoidal, one-celled, hyaline, with several slender appendages. SUBSTRATE: Dead wood. TYPE SPECIES: *Nia vibrissa* Moore et Meyers. The genus is monotypic. NOTE: Originally, *Nia* was described as a member of Deuteromycetes, but Doguet (1967) demonstrated its affinity with Gasteromycetes; Dring (1973) defined Doguet's (1967) protocysts as "primary basidia," without demonstrating that these produce basidiospores, and placed *Nia* in the Torrendiaceae. ETYMOLOGY: From Andean–Indian, *Ni* = personification of the sea, referring to the habitat of the fungus.

Nia vibrissa Moore et Meyers

Mycologia **51**, 874, 1959

LITERATURE: Brooks (1975); Byrne and Jones (1974); Doguet (1967, 1968, 1969); Fazzani and Jones (1977); Ingold (1972); T. W. Johnson (1968); E. B. G. Jones and Irvine (1971); Koch (1974); Kohlmeyer (1963d, 1964, 1971b); Kohlmeyer and Kohlmeyer (1964–1969, 1971a); Tubaki (1969).

Fig. 103

HYPHAE 2–3 µm in diameter, septate, ramose, with clamp connections, hyaline. BASIDIOCARPS 1–3 mm in diameter, subglobose, superficial, anchored in the substrate with an inconspicuous, cylindrical pedicel (50 µm long, 80 µm in diameter), whitish, yellowish, pinkish, and finally orange-colored, soft, thin-walled, villous (or smooth in culture), containing a homogenous gleba; opening by irregular rupture of the peridium; solitary or gregarious. PERIDIUM 10–15 µm thick, composed of hyphae with their

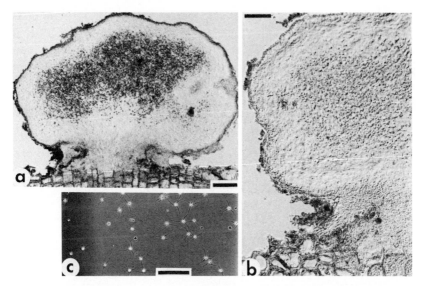

Fig. 103. *Nia vibrissa,* a marine Gasteromycete. (a) Longitudinal section (16 µm) through basidiocarp on prop root of *Rhizophora mangle;* basidiospores fill the center (Herb. J.K. 1732; bar = 100 µm). (b) Section through side of basidiocarp, some host cells filled with fungal hyphae (Herb. J.K. 1732; bar = 50 µm). (c) Appendaged basidiospores (bar = 100 µm). (a in bright field, b in Nomarski interference contrast, c in phase contrast.)

thick walls united, merging toward the center into a 20-µm-thick layer of separate, tangential hyphae, and bearing on the outside long hairs; external hyphae up to 275 µm long, 4–7 µm in diameter, thick-walled, straight or curved, sometimes slightly curled apically and uncinate. GLEBA at first composed of hyphae with apical inflated tips of 15–25 µm in diameter (protocysts), later replaced by basidia developing at random throughout the venter. BASIDIA 35–50 µm long, 2 µm in diameter, with an apical, subglobose inflation, 10–12 µm in diameter, bearing at the round apex four to eight basidiospores; hyaline, without sterigmata. BASIDIOSPORES (7.5–) 9–15 µm long, 6.5–11 µm in diameter, ovoid or ellipsoidal, one-celled, hyaline, appendaged; at the apex provided with a single, slender, flexible, attenuate, hyaline appendage, 20–47 µm long, less than 1 µm in diameter, terminally slightly inflated; four (rarely three or five) similar, subterminal, radiating appendages around the base, 20–32 µm long; at the point of attachment to the basidium with a short cylindrical projection. MODE OF LIFE: Saprobic. SUBSTRATES: Intertidal and drifting wood, test panels (e.g., *Fagus crenata, F. sylvatica, Picea* sp., *Pinus* sp., *Salix* sp., balsa), roots of *Rhizophora mangle,* stems and rhizomes of *Spartina alterniflora.*

RANGE: Atlantic Ocean—Bahamas [Great Abaco (J. Kohlmeyer, unpublished)], Brazil, Denmark, France, Germany (Helgoland, F. R. G.), Great Britain (England, Isle of Man), Iceland, Spain, United States (Florida, Massachusetts, North Carolina, Rhode Island); Mediterranean—France; Pacific Ocean—Japan, United States (California). NOTE: Basidiospores may accumulate in foam along the seashore. ETYMOLOGY: From the Latin, *vibrissa* = hair of the nostrils, in reference to the hairlike appendages of the basidiospores.

B. Hymenomycetes
1. Aphyllophorales
a. Corticiaceae

Digitatispora Doguet

C. R. Hebd. Séances Acad. Sci. **254**, 4338, 1962

LITERATURE: Donk (1964); Talbot (1973).

HYPHAE septate, ramose, with clamp connections, hyaline. BASIDIOCARPS of irregular shape, resupinate, light-colored, soft. HYMENIUM pulvinate, without cystidia. BASIDIA four-spored, cylindrical to clavate, hyaline, without sterigmata. BASIDIOSPORES tetraradiate, hyaline, deciduous; composed of a cylindrical basal cell and three radiating cylindrical apical cells. SUBSTRATE: Dead intertidal wood. TYPE SPECIES: *Digitatispora marina* Doguet. The genus is monotypic. ETYMOLOGY: From the Latin, *digitatus* = fingered and *spora* = spore, in reference to the fingerlike, radiating cells of the basidiospore.

Digitatispora marina Doguet

C. R. Hebd. Séances Acad. Sci. **254**, 4338, 1962

LITERATURE: Brooks (1975); Byrne and Jones (1974); Doguet (1962b, 1963, 1964); Donk (1964); Henningsson (1974); G. C. Hughes (1968, 1969); Koch (1974); Kohlmeyer (1963d); Kohlmeyer and Kohlmeyer (1964–1969).

Fig. 104

HYPHAE 3 µm in diameter, septate, ramose, with clamp connections, hyaline. BASIDIOCARPS 150–200 µm high, 1 mm wide, 1–4 mm long, outline elliptical or irregular, resupinate, whitish or grayish, soft; dried orange-colored, horny. HYMENIUM 100–120 µm high, pulvinate, without cystidia. BASIDIA 50 × 5–6.5 µm, four-spored, cylindrical or subclavate,

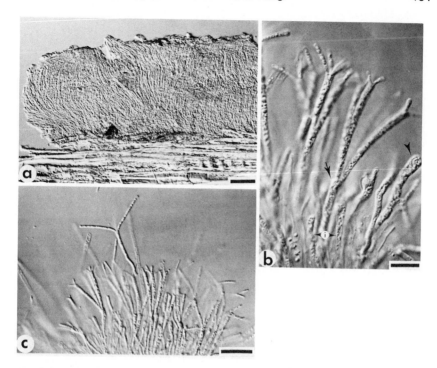

Fig. 104. *Digitatispora marina,* a marine Hymenomycete from wood. (a) Longitudinal section through basidiocarp seated on the wood surface (Herb. J.K. 940; bar = 50 μm). (b) Immature basidia (i), basidium bearing immature basidiospores (arrow), and basidium having shed its basidiospores (arrowhead) (Herb. J.K. 905; bar = 10 μm). (c) Hymenium with basidia bearing immature basidiospores, a mature spore above (Herb. J.K. 905; bar = 25 μm). (All in Nomarski interference contrast.)

erect or slightly curved, hyaline, without sterigmata. BASIDIOSPORES tetraradiate (rarely five- to seven-radiate), hyaline, deciduous; composed of a cylindrical or subclavate basal arm, 30–46(–56) × 3–4.5 μm, and three (rarely two or four to six) radiating, cylindrical, erect or slightly curved (rarely forked) apical arms, 26.5–41(–45) × 2–3.5 μm. MODE OF LIFE: Saprobic. SUBSTRATES: Intertidal and drifting wood, test panels (e.g., *Betula pubescens, Fagus sylvatica, Pinus sylvestris, Quercus* sp.). RANGE: Atlantic Ocean—Canada (Newfoundland), Denmark, France, Great Britain (England, Isle of Man), Sweden, United States (Rhode Island); Baltic Sea—Germany (F.R.G.), Sweden; Pacific Ocean—Canada (British Columbia), United States (California, Washington). NOTE: Donk (1964) questioned the nature of the sessile basidiospores and proposed the hypothesis that they "are in reality sterigmata acting as diaspores." ETYMOLOGY: From the Latin, *marinus* = marine, referring to the habitat.

b. Aphyllophorales incertae sedis

Halocyphina J. Kohlmeyer et E. Kohlmeyer

Nova Hedwigia **9**, 100, 1965

LITERATURE: Ginns and Malloch (1977). BASIDIOCARPS cyphelloid, solitary or gregarious, turbinate or clavate, eventually funnel-shaped or cupulate, pedunculate, superficial, light-colored, soft, thin-walled, tomentose. HYMENIUM lining the inner wall of the funnel. BASIDIA four-spored, clavate or cylindrical with a narrow base, nonseptate, hyaline, with (two to) four evanescent sterigmata. BASIDIOSPORES subglobose, one-celled, smooth, hyaline, nonamyloid, accumulating at maturity in the opening of the basidiocarp. SUBSTRATE: Dead wood. TYPE SPECIES: *Halocyphina villosa* Kohlm. et Kohlm. NOTE: Originally, *Halocyphina* was described as a member of Deuteromycetes, but Ginns and Malloch (1977) demonstrated its affinity with Aphyllophorales. ETYMOLOGY: From the Greek, *hals* = the sea, salt, and *Cyphina,* a genus of Deuteromycetes with similar fruiting bodies, in reference to the marine occurrence of the fungus.

Halocyphina villosa J. Kohlmeyer et E. Kohlmeyer

Nova Hedwigia **9**, 100, 1965

LITERATURE: Ginns and Malloch (1977); G. C. Hughes (1975); Kohlmeyer (1969c); Kohlmeyer and Kohlmeyer (1964–1969, 1977); Newell (1976).

Fig. 105

BASIDIOCARPS cyphelloid, 350–500 μm high, 310–440 μm in diameter in the apex, 100–150 μm in diameter in the stalk, turbinate or clavate, eventually funnel-shaped or cupulate, pedunculate, superficial in pro-tected parts of the wood, whitish or yellowish, soft, thin-walled, tomen-tose, solitary or gregarious. PERIDIUM 25–40 μm thick, composed of closely packed hyaline, rarely branched hyphae with thin walls and rare clamp connections, (1.2–)3–4(–6) μm in diameter; external hairs up to 100 μm long, 4–5 μm in diameter at the base, 1.5 μm at the apex, simple or dichotomously branched near the tip, hyaline, dissolving in 10% KOH. HYMENIUM lining the inner wall of the funnel-shaped basidiocarp; in immature fruiting bodies covered by a network of hyphae through which at maturity basidiospores are pushed to the outside. BASIDIA 13–22 × 5–9.5 μm, four-spored, clavate or cylindrical with a narrow base, nonsep-tate, hyaline, with (two to) four evanescent sterigmata, about 3.5 μm long. BASIDIOSPORES 8–10.5 × (7–)8–9.5(–10.5) μm, subglobose, one-celled, smooth, hyaline, nonamyloid, accumulating at maturity in the opening of

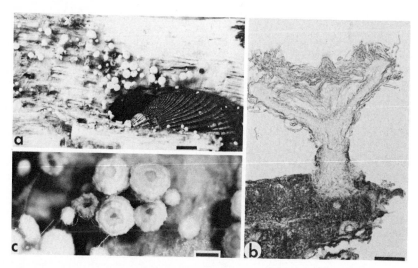

Fig. 105. *Halocyphina villosa.* (a) Basidiocarps on *Tamarix gallica* near a mussel (Herb. J.K. 3722; bar = 1 mm). (b) Longitudinal section (12 μm) through basidiocarp on wood of mangrove root (Herb. J.K. 1833; bar = 50 μm). (c) Basidiocarps in view from the top (Herb. J.K. 3722; bar = 200 μm). (a and c close-up photographs, b in bright field.)

the basidiocarp. MODE OF LIFE: Saprobic. SUBSTRATES: Dead roots of *Pandanus candelabrum, Rhizophora mangle,* and *R. racemosa,* submerged dead branches of *Conocarpus erecta, R. mangle,* and *Tamarix gallica,* culms of *Spartina* sp. RANGE: Atlantic Ocean—Bermuda, Liberia, United States (Florida, North Carolina); Pacific Ocean—United States [Hawaii (Oahu)]. NOTE: Description based in part on Ginns and Malloch (1977); the fungus is definitely marine because basidiocarps develop on intertidal wood next to marine organisms, namely, algae, shipworms, or barnacles. ETYMOLOGY: From the Latin, *villosus* = hairy, in reference to the hairy basidiocarps.

C. Teliomycetes
1. Ustilaginales
a. Tilletiaceae

Melanotaenium De Bary

Bot. Ztg. **32,** 105, 1874

LITERATURE: Beer (1920); Durán (1973).
SORI dark, in vegetative parts (stem or leaf tissues) of the hosts, at first

covered by the host epidermis, rupturing at maturity, lacking a sterile stromatic tissue; sporogenous mycelium converted into spores. SPORES uniformly dark pigmented, without germ pores, single at maturity. MODE OF LIFE: Parasitic. HOSTS: Pteridophyta and Spermatophyta. TYPE SPECIES: *Melanotaenium endogenum* (Ung.) De Bary. NOTE: Only 1 species among about 18 is known to occur on a halophytic host. ETYMOLOGY: From the Greek, *melas* = black and *tainia* = band, in reference to the dark-colored, bandlike sori of the fungi.

Melanotaenium ruppiae G. Feldmann

Rev. Gén. Bot. **66**, 36, 1959

LITERATURE: T. W. Johnson and Sparrow (1961).

HYPHAE 2–3 μm in diameter, ramose, septate. SORI developing in air chambers of the inner cortex of the stem and in air chambers on both sides of the median nerve in the base of leaves, without causing malformations in the host. SPORES (chlamydospores) 12–25 μm long, 12–13 μm in diameter, globose or ovoid, often with an irregular outline, dark fuscous, appearing blackish in masses, filled with small lipid globules; wall up to 2 μm thick, composed of a brown, smooth exospore and a hyaline endospore; often with one or two small blunt tips, the former points of attachment of the spores to the mycelium; rarely provided with a slightly pointed tip and two blunt protuberances at opposed ends. MODE OF LIFE: Parasitic. HOST: *Ruppia maritima*. RANGE: Mediterranean—France (known only from the type collection). ETYMOLOGY: Named after the host plant, *Ruppia* (Ruppiaceae).

III. Deuteromycotina
　A. Hyphomycetes
　　1. Agonomycetales
　　　a. Agonomycetaceae

Papulaspora Preuss

Linnaea **24**, 112, 1851

LITERATURE: S. J. Hughes (1958); Weresub (1974); Weresub and LeClair (1971).

HYPHAE septate, hyaline or rarely brown. CONIDIOPHORES and CONIDIA absent. PAPULASPORES composed of a core of cells enclosed in a cellular sheath, at maturity red or brown, scattered or gregarious, superficial; propagules originating as a hyphal coil, with sheathing cells produced from hyphae coiled around prospective central cells; central cells subglobose or angular with thick, pigmented walls; sheath composed of branching hyphae, forming a single layer of subglobose or elongated, thin-walled, light-colored cells. MODE OF LIFE: Saprobic or mycoparasitic. SUBSTRATES: Dead wood and other plant material. TYPE SPECIES: *Papulaspora sepedonioides* Preuss. NOTE: Description after Weresub (1974) and Weresub and LeClair (1971); there is one marine representative besides about ten terrestrial species. ETYMOLOGY: From the Latin, *papula* = pustule and *spora* = spore, in reference to the bulbous propagules.

Papulaspora halima Anastasiou

Nova Hedwigia **6,** 266, 1963

LITERATURE: Anastasiou and Churchland (1968, 1969); Churchland and McClaren (1973); Corte (1975); Kohlmeyer and Kohlmeyer (1964–1969); Neish (1970); Newell (1976); Shearer (1972b); Tubaki (1966, 1969); Weresub and LeClair (1971).

HYPHAE septate, ramose, anastomosing, hyaline or brown, possibly forming clamp connections; two types of hyphae—thick, brown ones, 3–5 μm in diameter; thin, hyaline ones, 1.5–2.2 μm in diameter, with brown, ringlike thickenings around the septa, eventually becoming brown; producing a brown pigment in the substrate. PAPULASPORES 35–870 μm in diameter, subglobose, black, superficial or immersed, originating from catenulate primordial cells; central cells 5–18 × 4–14 μm, subglobose to ellipsoidal, thin-walled, coalescent, brown. MODE OF LIFE: Saprobic. SUBSTRATES: Submerged wood and leaves of *Arbutus menziesii*, seedlings of *Rhizophora mangle*, driftwood and test panels (e.g., *Abies alba, A. firma, Betula papyrifera, Fagus sylvatica, Larix decidua, Ochroma lagopus, Olea europaea, Pinus pinaster, Populus alba, Quercus* sp., balsa). RANGE: Atlantic Ocean—Canada (Nova Scotia), United States (Florida, Maryland, Virginia); Mediterranean—Italy; Pacific Ocean—Canada (British Columbia), Japan, United States (California). NOTE: The species was also collected on submerged wood of *Tamarix aphylla* in an inland salt lake [Salton Sea, California (Anastasiou, 1963b)]; *P. halima* is doubtfully assigned to *Papulaspora* (Weresub and LeClair, 1971), because the "papulaspores" of this species do not show the characteristic sheath and central cells (see sectioned bulbil in Kohlmeyer and Kohlmeyer, 1964–1969, Table 89, Fig. 4); no taxonomical change is proposed because the position of this member of Mycelia Sterilia remains doubtful until its sexual state will be known. ETYMOLOGY: From the Greek, *halimos* = marine, referring to the habitat of the species.

2. Hyphomycetales
a. Moniliaceae

Blodgettia Wright

Trans. R. Ir. Acad. **28**, 25, 1881 (non *Blodgettia* Harvey, *Smithson. Contrib. Knowl.* **10**, 46–47, 1858)

≡*Blodgettiomyces* J. Feldmann, *Rev. Bryol. Lichénol.* **11**, 157, 1938 (superfluous name)

HYPHAE septate, branched, anastomosing, growing within the cell walls of algae. CHLAMYDOSPORES catenulate; chains terminal or intercalary, constricted at the septa, hyaline to brownish; cells subglobose to doliiform. MODE OF LIFE: Symbiotic. TYPE SPECIES: *Blodgettia bornetii* Wright. The genus contains a second species, *B. indica,* growing on unidentified stalks submerged in freshwater. NOTE: Harvey (1858) described *Blodgettia* as a genus of green algae, but the description was based on two discordant elements, namely, an alga and a fungus; therefore, the name must be rejected according to Article 70 of the International Code of Botanical Nomenclature: ". . . unless it is possible to select one of the elements as a satisfactory type." Wright (1881) followed the latter procedure, applying the name *Blodgettia* to the fungal partner; as *Blodgettia* sensu Harvey (1858) is not used in the algae, *Blodgettia* Wright is legitimate and *Blodgettiomyces* J. Feldmann (1938) is superfluous. ETYMOLOGY: Named in honor of the physician Dr. J. L. Blodgett (1809–1853) of Key West, Florida, an amateur botanist, and the Greek, *mykes* = fungus.

Blodgettia bornetii Wright

Trans. R. Ir. Acad. **28**, 25–26, 1881 (*Blodgettia confervoides* Harvey, *Smithson. Contrib. Knowl.* **10**, 48, 1858, based on two discordant elements: International Code of Botanical Nomenclature, Article 70)

≡*Blodgettiomyces bornetii* (Wright) J. Feldmann, *Rev. Bryol. Lichénol.* **11**, 157, 1938 (as *B. borneti;* superfluous combination)

LITERATURE: Børgesen (1913); Cribb and Cribb (1969); Harvey (1858); Howe (1905); T. W. Johnson and Sparrow (1961); Setchell (1926); Tubaki (1961); Verona and Benedek (1965); Wright (1876, 1880); Yamada (1934).

Fig. 106

MYCELIUM between the inner and outer wall layer of the host, pushing these layers apart. HYPHAE 2–4 μm in diameter, septate, branched, anastomosing, forming strands in more or less regular distances from each other, arranged parallel to the long axis of the plant, equally distributed in the alga but never penetrating into the cells, the cross walls or the very tips of the host. CHLAMYDOSPORES 35–150 μm long, 8–33 μm in diameter, forming two- to eight-celled terminal or intercalary chains, constricted at the septa; cells sometimes separated from each other by strong segmentations, depending on the position subglobose, ellipsoidal or doliiform, hyaline, light yellowish or yellow–brown, multiguttulate; originating on lateral branches between the hyphal strands. MODE OF LIFE: Symbiotic; forming mycophycobioses. HOSTS: *Cladophora caespitosa, C. fuliginosa.* RANGE: Atlantic Ocean—Bahamas [Great Abaco (J. Kohlmeyer, unpublished)], Bermuda, Guadeloupe, Puerto Rico, United States, (Florida), Virgin Islands (St. Croix, St. Thomas); Pacific Ocean—Australia (Queensland), Japan (Ryukyu Islands), Taiwan (Formosa), United States (Hawaii). NOTE: We examined material of *C. fuliginosa* from the algal

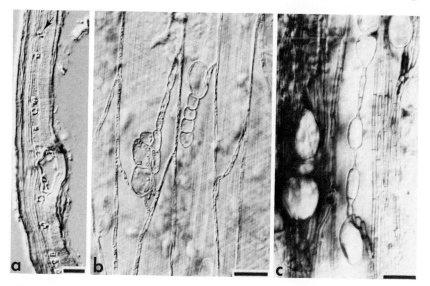

Fig. 106. *Blodgettia bornetii* in *Cladophora fuliginosa* (PC and Herb. J.K. 3078). (a) Cross section through the host wall with hyphae (arrow) embedded between the inner and outer layers of the wall; the large hole represents a section through a chlamydospore (bar = 10 μm). (b) Terminal chlamydospores and network of hyphae; fine striations in the background are components of the algal wall (bar = 25 μm). (c) Hyphae and intercalary chlamydospores in the wall of the host (bar = 25 μm). (All in Nomarski interference contrast.)

collection of Herb. PC (Farlow, Anderson, and Eaton, Algae Exsicc. Am. Bor. No. 44, Key West, leg. F. W. Hooper); Feldmann (1938) advanced the thought that this association could be considered a lichen; we prefer to classify it as a mycophycobiosis (Kohlmeyer and Kohlmeyer, 1972a); a fungus isolated from Antarctic soils and algae and identified as *Blodgettia borneti* (Tubaki, 1961) is possibly a different species because it is not growing symbiotically inside the wall of *Cladophora* and because of its Antarctic occurrence, which is outside the distribution range of *B. bornetii*. ETYMOLOGY: Named in honor of E. Bornet, French botanist.

Botryophialophora Linder *in* Barghoorn et Linder

Farlowia **1**, 403, 1944

HYPHAE septate, simple or branched, hyaline or light-colored. CONID-IOPHORES subglobose or irregular, lateral or apical on the main hyphae, bearing clusters of conidiogenous cells. CONIDIOGENOUS CELLS phialidic, flask-shaped, apex narrow, trumpet-shaped, simple, hyaline, on co-nidiophores or rarely produced singly on a hypha. CONIDIA (enteroblastic–phialidic) globose or subglobose, one-celled, hyaline. MODE OF LIFE: Saprobic. TYPE SPECIES: *Botryophialophora marina* Linder. The genus is monotypic. NOTE: von Arx (1970) lists the genus as a synonym of *Myrioconium* Sydow without indicating if type material of *B. marina* has been examined. ETYMOLOGY: From the Greek, *botrys* = bunch, *phiale* = urne, phialide, and *-phorus* = -bearing, in reference to the phialides arranged in bunches on the hyphae.

Botryophialophora marina Linder *in* Barghoorn et Linder

Farlowia **1**, 404, 1944

LITERATURE: T. W. Johnson and Sparrow (1961); Steele (1967); Wilson (1951).

Fig. 107

HYPHAE 4–10 μm in diameter, septate, simple or branched, hyaline or light-colored. CONIDIOPHORES subglobose or irregularly polygonal, lateral or apical on the main hyphae, bearing clusters of up to 20 or more

Fig. 107. *Botryophialophora marina* from type slide No. 8 of Barghoorn in FH. (a) Cluster of conidiogenous cells among hyphae. (b) Conidiophores (arrow) originating laterally on the main hyphae, globose conidia to the left. (Both bars = 10 μm; both in Nomarski interference contrast.)

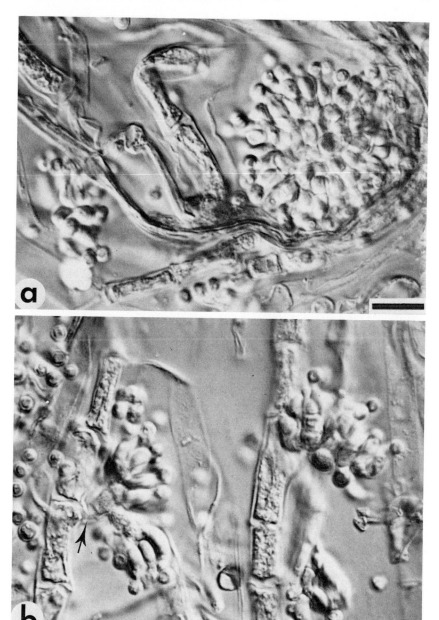

conidiogenous cells. CONIDIOGENOUS CELLS 3.5–8.5 μm long, 1.5–3.3 μm in diameter, phialidic, flask-shaped, apex narrow, trumpet-shaped, simple, hyaline, on conidiophores or rarely produced singly on a hypha. CONIDIA (enteroblastic–phialidic) 2–3.3 μm in diameter, globose or subglobose, one-celled, hyaline, containing a large vacuole. MODE OF LIFE: Saprobic. SUBSTRATES: Intertidal wood, test panels, coastal sand. RANGE: Atlantic Ocean—Great Britain (Wales), United States (Massachusetts); Pacific Ocean—United States (Hawaiian Islands). NOTE: Description based on Linder (1944) and our examination of the type slide (Herb. F); a putative isolation from soil of the Sonoran Desert (Ranzoni, 1968, sub *B. maritima*) may or may not be a different species; as the fungus has been merely isolated from marine habitats (Linder, 1944; Steele, 1967; Wilson, 1951) but has not been observed growing there, it is questionable if *B. marina* should remain on the list of marine fungi. ETYMOLOGY: From the Latin, *marinus* = marine, in reference to the origin of the type isolate.

Clavatospora S. Nilsson ex Marvanová et S. Nilsson

Trans. Br. Mycol. Soc. **57**, 531, 1971

≡*Clavatospora* S. Nilsson, *Symb. Bot. Ups.* **18**, 88, 1964 (without designation of a type species)

LITERATURE: Ingold (1975b).
HYPHAE septate, branched, hyaline, without sporodochia. CONIDIOPHORES simple or sparsely branched, hyaline. CONIDIOGENOUS CELLS phialidic, hyaline. CONIDIA (enteroblastic–phialidic) one-celled, tetraradiate, consisting of a clavate, straight, basal main axis, from which three narrow cylindrical or conical, elongate branches arise at the apex; hyaline. MODE OF LIFE: Saprobic. TYPE SPECIES: *Clavatospora longibrachiata* (Ingold) S. Nilsson ex Marvanová et S. Nilsson. NOTE: Description mostly based on S. Nilsson (1964); there are four freshwater and one marine species. ETYMOLOGY: From the Latin, *clavatus* = clavate and *spora* = spore, in reference to the clavate conidia.

Clavatospora stellatacula Kirk ex Marvanová et S. Nilsson

Trans. Br. Mycol. Soc. **57**, 531, 1971

≡*Clavatospora stellatacula* Kirk, *Mycologia* **61**, 178, 1969 (invalid name: Article 43 of the International Code of Botanical Nomenclature)

LITERATURE: Kirk and Catalfomo (1970).

HYPHAE 1–5 μm in diameter, septate, ramose, sometimes forming spirals, hyaline. CONIDIOPHORES 8–24 × 3–6 μm, phialidic, lageniform, simple or ramose, hyaline, clustered in groups of up to 30 on hyphae; the small openings collared, trumpet-shaped, light brown. CONIDIA 7–9 μm long, 6–7 μm in diameter (processes included), enteroblastic–phialidic, stellate, one-celled, hyaline; main axis clavate with (three or) four subconical, whorled, subapical processes. SUBSTRATE: Test panels (*Pinus ponderosa*). RANGE: Atlantic Ocean—United States (Virginia; known only from the type collection). NOTE: We have examined a subculture of the type isolate, kindly provided by Dr. Kirk. ETYMOLOGY: From the Latin, *stellatus* = stellate, starry, and the diminutive suffix -*culus,* in reference to the stellate, small conidia.

Varicosporina Meyers et Kohlmeyer

Can. J. Bot **43**, 916, 1965

HYPHAE septate, branched, hyaline to light brown. CONIDIOPHORES cylindrical, multiseptate, simple or rarely branched, hyaline. CONIDIOGENOUS CELLS monoblastic, integrated, terminal or lateral, determinate. CONIDIA acrogenous, solitary, branched (irregularly staurosporous), septate, hyaline, consisting of a system of axes; a main axis with two, rarely three, side branches, each side branch arising from the previously developed branch. MODE OF LIFE: Saprobic. TYPE SPECIES: *Varicosporina ramulosa* Meyers et Kohlmeyer. The genus is monotypic. ETYMOLOGY: From *Varicosporium* Kegel (Hyphomycetes) and the Latin suffix -*inus,* indicating a resemblance.

Varicosporina ramulosa Meyers et Kohlmeyer

Can. J. Bot. **43**, 916, 1965

LITERATURE: G. C. Hughes (1974, 1975); Kohlmeyer (1966a, 1969c); Kohlmeyer and Kohlmeyer (1964–1969, 1971a); Meyers (1968a,b, 1971a); Meyers and Hoyo (1966); Tubaki (1969); Wagner-Merner (1972).

Fig. 108

HYPHAE 2.5–5(–10) μm in diameter, septate, branched, hyaline to light brown. CONIDIOPHORES 20–200 × 2.5–4.5 μm, cylindrical, multiseptate, simple or rarely branched, geniculate, hyaline; single cells 9.5–25 μm long. CONIDIOGENOUS CELLS monoblastic, integrated, terminal or lateral, determinate. CONIDIA acrogenous, solitary, branched (irregularly

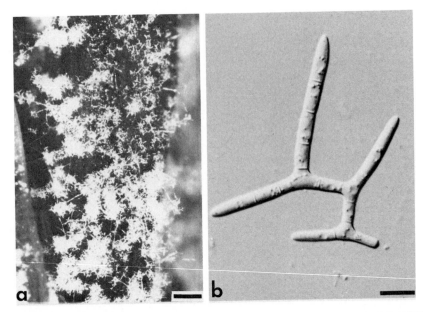

Fig. 108. *Varicosporina ramulosa* from leaves of *Zostera marina* (Herb. J.K. 3907). (a) Habit of sporulating fungal colonies on the leaf surface (bar = 200 µm). (b) Asco-spore (bar = 10 µm). (a = close-up photograph, b in Nomarski interference contrast.)

staurosporous), septate, slightly constricted at the septa, hyaline, mul-tiguttulate, consisting of a system of axes; main axis 27.5–45.5 µm long, 2–3.5 µm in diameter at the base, 3–5 µm at the apex, (1–)2–3(–7)-septate, slightly curved; first side branch 25–40 × 3.5–5.5 µm, 1–3(–4)-septate, almost at a right angle to the main axis; second side branch 19.5–34.5 × 4.5–5.5 µm, 1–3-septate, arising from the central cell of the first side branch; rarely a third side branch is formed from the central cell of the second side arm, 34.5 × 5.5 µm, four-celled. SCLEROTIA formed in pure cultures are black and usually attached to the walls of the culture vessels. MODE OF LIFE: Saprobic. SUBSTRATES: Isolated from drift and washed-up algae (*Hypnea charoides*, *Sargassum* sp.), sea grasses (*Thalassia testudinum*, *Zostera marina*), and synthetic sponge; conidia accumulate in foam along the shore in tropical and subtropical waters. RANGE: Atlantic Ocean—Mexico (Yucatan), Tobago (J. Kohlmeyer, unpublished), United States (Florida, North Carolina); Pacific Ocean—Japan, United States [Hawaii (Hawaii, Kauai, Oahu)]. ETYMOLOGY: From the Latin, *ramulosus* = bearing branchlets, in reference to the branched conidia.

b. Dematiaceae

Alternaria Nees von Esenbeck ex Fries

Das System der Pilze und Schwämme, Würzburg, p. 70, 1817

=?*Dictyocephala* Medeiros, *Publ. Inst. Mic. Univ. Recife* **372**, 13, 1962
=*Macrosporium* Fries, *Systema Mycologicum, Gryphiswaldiae* **3**, p. 373, 1832
=*Prathoda* Subramanian, *J. Indian Bot. Soc.* **35**, 73, 1956
=*Rhopalidium* Montagne, *Ann. Sci. Nat., Bot. Biol. Vég. Sér.* 2 **6**, 30, 1846

LITERATURE: Barron (1968); M. B. Ellis (1971, 1976); Joly (1964); Kendrick and Carmichael (1973); Neergaard (1945); Simmons (1967).

CONIDIOPHORES cylindrical, septate, simple or irregularly branched, straight or curved, basal cell occasionally swollen, smooth, bearing conidia at the uniperforate apex, yellowish to brown, singly or fascicled. CONIDIA (enteroblastic–tretic) ovoid, obclavate, obpyriform or ellipsoidal, with a basal pore, tapering or not into an apical beak, muriform, constricted at the septa, smooth or rough, olivaceous to brown, produced singly or in acropetal succession, forming simple or branched chains. MODE OF LIFE: Parasitic or saprobic. TYPE SPECIES: *Alternaria alternata* (Fries) Keissler. NOTE: Description and synonyms based on the authors listed above; there are about 50 species in the genus; the following species have been isolated from marine habitats: *A. fasciculata, A. geophila, A. humicola, A. tenuis* (= *A. alternata*), and *A. tenuissima;* we consider *A. maritima* Sutherland a dubious species and a *nomen rejiciendum* since no type material exists and the identity of the species cannot be established; in spite of numerous collections of *Alternaria* spp. from the marine environment no author has proven by direct observation that species of the genus are actually developing under natural marine conditions; as representatives of *Alternaria* are ubiquitous in all areas of the world the possibility cannot be excluded that marine isolates are not obligate or facultative marine species but are instead merely contaminants; until the opposite is proven we list isolates from marine habitats as *Alternaria* spp. ETYMOLOGY: From the Latin, *alternus* = alternate and the suffix *-arius,* indicating a connection or possession, in reference to the diameter of conidial chains, alternating between the thick bases and slender apices of the conidia.

Alternaria spp.

LITERATURE: Anastasiou and Churchland (1969); Biernacka (1965); Byrne and Jones (1974); Dickinson (1965); Dickinson and Pugh (1965b);

Gessner and Goos (1973a,b); Goodman (1959); Henningsson (1974); Höhnk (1955, 1956, 1967); G. C. Hughes (1969); T. W. Johnson (1956b); T. W. Johnson et al. (1959); T. W. Johnson and Sparrow (1961); Joly (1964); E. B. G. Jones (1962a, 1963c, 1968b, 1972); E. B. G. Jones and Irvine (1971); E. B. G. Jones and Oliver (1964); Kohlmeyer (1962a, 1966a, 1971b); Linder (1944); Meyers and Reynolds (1959b, 1960b); Meyers et al. (1960, 1964a, 1967b); Moreau and Moreau (1941); Neish (1970); Nicot (1958a,c); Picci (1966); Pugh (1966); Pugh et al. (1963); Pugh and Williams (1968); Ristanović and Miller (1969); Roth et al. (1964); Schaumann (1968, 1969); Schlichting (1971); Schmidt (1974); Siepmann (1959a,b); Siepmann and Johnson (1960); Sivanesan and Manners (1970); Sparrow (1937); Steele (1967); Sutherland (1916a); Wagner-Merner (1972).

SUBSTRATES: Isolated from seawater, sediment, driftwood, intertidal wood, wood panels (e.g., *Betula* sp., *Fagus sylvatica, Pinus sylvestris*), *Halimione portulacoides, Juncus roemerianus, Laminaria* sp., *Salsola kali, Spartina alterniflora, S. townsendii;* conidia observed in sea foam. RANGE: Atlantic Ocean—Canada (New Brunswick, Nova Scotia), France, Germany (Helgoland, Wangerooge, and mainland, F.R.G.), Great Britain (England, Scotland, Wales, Orkney Islands), Sweden, United States (Florida, Maine, Massachusetts, North Carolina, Rhode Island, Texas, Virginia); Baltic Sea—Germany (G.D.R.), Poland; Black Sea—Rumania; Indian Ocean—India (Madras); Mediterranean—France, Italy, Yugoslavia; Pacific Ocean—Canada (British Columbia), United States (Hawaii).

Asteromyces Moreau et Moreau ex Hennebert

Can. J. Bot. **40**, 1211–1213, 1962

LITERATURE: Barron (1968); M. B. Ellis (1971); Kendrick and Carmichael (1973); Moreau and Moreau (1941).

HYPHAE septate, branched, hyaline to brown. CONIDIOGENOUS CELLS developing singly, rarely intercalary, laterally on the hyphae, sessile or with a short stalk, at first cylindrical and hyaline, becoming subglobose or subclavate and fuscous, bearing several conidia. CONIDIA (blastoconidia) originating successively on denticles; the first conidium apical, the following ones in a lateral whorl; ovoid to obpyriform, one-celled, rarely septate, brown, seceding by rupture of the denticle or released in aggregates attached to the conidiogenous cell. MODE OF LIFE: Saprobic. SUBSTRATES: Algae and cellulosic substrates. TYPE SPECIES: *Asteromyces cruciatus* Moreau et Moreau ex Hennebert. The genus is monotypic. ETYMOLOGY: From the Greek, *aster* = star and *mykes* = fungus, in reference to the starlike appearance of the conidial aggregate.

Asteromyces cruciatus Moreau et Moreau ex Hennebert

Can. J. Bot. **40**, 1213, 1962

LITERATURE: Barron (1968); Brown (1958a,b); Byrne and Jones (1974, 1975a,b); Cole (1976); Colon (1940); M. B. Ellis (1971); E. B. G. Jones (1968b, 1972); E. B. G. Jones and Irvine (1971, 1972); E. B. G. Jones and Jennings (1964, 1965); E. B. G. Jones and Ward (1973); Kohlmeyer (1971b); Kohlmeyer and Kohlmeyer (1964–1969); Moreau and Moreau (1941); Nicot (1958a,c); Nolan (1972); Schaumann (1969, 1974b,c, 1975b); Schmidt (1974).

Fig. 109

HYPHAE 2–8 μm in diameter, septate, branched, often anastomosing, smooth, hyaline to brown. CONIDIOGENOUS CELLS 8–15 μm long, 1.5–4.5 μm in diameter, at first cylindrical and hyaline, becoming subglobose or subclavate and fuscous, bearing up to 13 conidia; developing singly, rarely intercalary, laterally on the hyphae, sessile or with a cylindrical one- to three-celled stalk; stalk up to 20 μm long, 1.5–2.5 μm in diameter; subterminal cells of the stalk may also become sporogenous. CONIDIA (blastoconidia) 9–20 × 4–9 μm, ovoid to obpyriform, one-celled, rarely one-septate, thin-walled, wall thickened at the apex, brown, originating

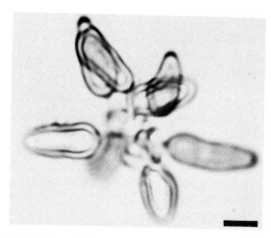

Fig. 109. *Asteromyces cruciatus*, aggregate of ascospores connected to the conidiogenous cell, as found in sea foam (Herb. J.K. 1943; bar = 5 μm; in bright field).

successively on denticles on the conidiogenous cell; the first conidium is apical, the following ones are in one or more lateral whorls of usually four conidia; denticles 2.5–9.5 × 0.5–2 μm, cylindrical; conidia seceding by rupture of the denticle, but usually released in aggregates attached to the conidiogenous cell. MODE OF LIFE: Saprobic. SUBSTRATES: Isolated from drift algae (e.g., *Cystoseira osmundacea, Egregia menziesii*) and cellulosic substrates (*Spartina* sp.) washed ashore, test panels (*Fagus sylvatica*), beach and dune sand under *Agropyrum* sp. and *Ammophila arenaria*, roots of *A. arenaria;* conidia accumulate in foam along the shore in temperate waters. RANGE: Atlantic Ocean—France, Germany (Helgoland), Great Britain (England, Wales), United States (Massachusetts, New Jersey); Baltic Sea—Germany (G.D.R.); Pacific Ocean—United States (California). NOTE: Description after Hennebert (1962) and Kohlmeyer and Kohlmeyer (1964–1969); since the "conidia" are usually distributed in nature as aggregates attached to the conidiogenous cell, the question may be raised if these aggregates should be considered multicelled conidia, comparable to conidia of *Clavariopsis bulbosa* and *Orbimyces spectabilis*. ETYMOLOGY: From the Latin, *cruciatus* = crossshaped, in reference to the crosslike arrangement of the conidia in each whorl.

Cirrenalia Meyers et Moore

Am. J. Bot. **47**, 346, 1960; emend. J. Kohlmeyer, *Ber. Dtsch. Bot. Ges.* **79**, 35, 1966

LITERATURE: M. B. Ellis (1976); Sutton (1973a).

CONIDIOPHORES present or obsolete, cylindrical, septate or nonseptate, acrogenous or laterally on the hyphae, hyaline or light brown. CONIDIOGENOUS CELLS monoblastic, integrated, terminal, determinate. CONIDIA acrogenous, solitary, helicoid, septate, constricted or not constricted at the septa, brownish; diameter and pigmentation increasing from base to apex; cells·dissimilar. MODE OF LIFE: Saprobic. TYPE SPECIES: *Cirrenalia macrocephala* (Kohlmeyer) Meyers et Moore. NOTE: Five of the six species of *Cirrenalia* are marine; only *C. donnae* Sutton occurs in a terrestrial habitat. ETYMOLOGY: From the Latin, *cirrus* = curl, lock, and the Greek, *enalios* = in or of the sea, in reference to the curved conidia and the marine occurrence.

Cirrenalia fusca I. Schmidt

Feddes Repert. **80**, 110, 1969

Fig. 110a

HYPHAE (1–)2.5–3.3 μm in diameter, septate, ramose, superficial or immersed in the substrate, yellow or brown. CONIDIOPHORES up to 23 μm long, ca. 2 μm in diameter, cylindrical, one- or two-septate, simple, sometimes obsolete, yellow or brown. CONIDIOGENOUS CELLS monoblastic, integrated, terminal, determinate. CONIDIA acrogenous, solitary, regularly or irregularly helicoid, conidia semicontorted to 1½ times contorted, (2–)3(–4)septate, constricted at the septa, brown to dark brown (never reddish); cells increasing in diameter and pigmentation from base to apex, distinctly dissimilar; spirals 20–36 μm high, 20–30 μm in diameter; terminal cell 13–33 μm high, 11–22.5 μm in diameter, reniform or ellipsoidal, fuscous; basal cells obtusely conical; central cells irregularly ellipsoidal or subglobose. MODE OF LIFE: Saprobic. SUBSTRATES: Dead rhizomes of *Phragmites communis* and unidentified reeds. RANGE: Baltic Sea—Germany (Hiddensee and mainland, G.D.R.); Mediterranean—France (J. Kohlmeyer, unpublished). NOTE: We examined paratype material kindly sent to us by Dr. I. Schmidt (Stralsund) and our own collection Herb. J. K. 549 on culms of a reed from an étang in southern France (IMS); the species differs from *C. pseudomacrocephala* by the number of septa and the shape of apical cells. ETYMOLOGY: From the Latin, *fuscus* = fuscous, a somber brown, in reference to the color of the conidia.

Cirrenalia macrocephala (Kohlmeyer) Meyers et Moore

Am. J. Bot. **47,** 347, 1960

≡*Helicoma macrocephala* Kohlmeyer, *Ber. Dtsch. Bot. Ges.* **71,** 99, 1958

LITERATURE: Apinis (1964); Becker (1961); Brooks *et al.* (1972); Byrne and Eaton (1972); Byrne and Jones (1974); Churchland and McClaren (1973); Eaton (1972); Eaton and Dickinson (1976); Eaton and Jones (1971a,b); M. B. Ellis (1976); Gessner and Goos (1973b); Henningsson (1974); G. C. Hughes (1969, 1974); Irvine *et al.* (1972); T. W. Johnson (1968); T. W. Johnson *et al.* (1959); T. W. Johnson and Gold (1959); T. W. Johnson and Sparrow (1961); E. B. G. Jones (1962a, 1963c, 1968a,b, 1971a, 1972); E. B. G. Jones and Irvine (1971); E. B. G. Jones *et al.* (1972); Kohlmeyer (1959, 1960, 1962a, 1963c–e, 1967, 1971b); Kohlmeyer *et al.* (1959); Kohlmeyer and Kohlmeyer (1964–1969); Meyers (1968a, 1971a); Schaumann (1968, 1969, 1974c, 1975a–c); Schmidt (1967, 1974); Schneider (1976, 1977); Shearer (1972b); Sieburth *et al.* (1974); Tubaki (1966, 1969).

Fig. 110b

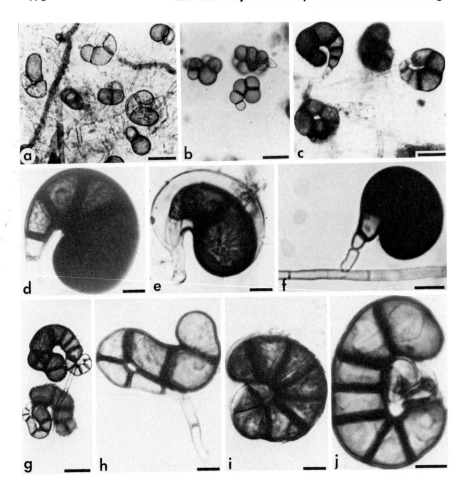

Fig. 110. Conidia of *Cirrenalia* spp. (a) *C. fusca* from *Phragmites communis* (paratype IMS; bar = 20 μm). (b) *C. macrocephala* from wood (Herb. J.K. 3912; bar = 15 μm). (c) *C. pseudomacrocephala* from wood (isotype IMS, Herb. J.K. 2447; bar = 20 μm). (d) *C. pygmea* from wood of prop root of *Rhizophora racemosa* (Herb. J.K. 1877; bar = 5 μm). (e) Same as d, conidium with internal proliferation (paratype IMS, Herb. J.K. 1841a; bar = 5 μm). (f) Same as d, conidium attached to conidiogenous cell on hypha (paratype IMS, Herb. J.K. 1815a; bar = 10 μm). (g) *C. tropicalis* from wood (isotype IMS, Herb. J.K. 1888c; bar = 15 μm). (h–j) Same as g (bar = 5 μm). (i in Nomarski interference contrast, the others in bright field.)

HYPHAE 1.5–2 μm in diameter, septate, ramose, hyaline to dark brown. CONIDIOPHORES 3.5–25 μm long, 2–5 μm in diameter, cylindrical, occasionally apically somewhat inflated, zero- to three-septate, simple, straight or curved, hyaline to yellowish. CONIDIOGENOUS CELLS monoblastic, integrated, terminal, determinate. CONIDIA acrogenous, solitary, helicoid, rarely straight, ¼ to 1 time contorted, two- to seven-septate, strongly constricted at the septa, multiguttulate, reddish fuscous; cells increasing in diameter and pigmentation from base to apex, distinctly dissimilar; spirals 12–31.5(–35) μm high, 12–23.5 μm in diameter; terminal cell 5.5–13.5 μm high, 6.5–14(–17) μm in diameter, subglobose, basally flattened, reddish fuscous, largest and darkest of all cells; basal cells 1.5–6(–7) μm high, 2.5–7 μm in diameter, semiglobose or obtusely conical, hyaline; central cells subglobose, cylindrical or doliiform, brownish. MODE OF LIFE: Saprobic. SUBSTRATES: Driftwood and intertidal wood and bark, test panels (e.g., *Abies firma,* balsa, *Cryptomeria japonica, Fagus sylvatica, Ocotea rodiaei, Picea* sp., *Pinus sylvestris, Populus* sp., *Pseudotsuga douglasii, P. taxifolia*), *Spartina alterniflora;* isolated from debris of *Ammophila arenaria;* frequently in tunnels and fecal pellets of wood-boring crustacea (e.g., *Limnoria* and *Sphaeroma,* Isopoda). RANGE: Atlantic Ocean—France, Germany (Amrum, Helgoland, and mainland, F.R.G.), Great Britain (England, Scotland, Wales), Iceland, Norway, Portugal, Spain (Canary Islands and mainland), Sweden, United States (Connecticut, Florida, Maryland, Massachusetts, New Jersey, North Carolina, Rhode Island, Texas, Virginia); Baltic Sea—Denmark, Germany (Hiddensee, Rügen, and mainland, G.D.R.; F.R.G.); Mediterranean—France, Greece, Italy (Sicily and mainland); Pacific Ocean—Australia (New South Wales), Canada (British Columbia), Japan, New Zealand, United States (California, Washington). NOTE: Other *Cirrenalia* species appear to replace *C. macrocephala* in tropical waters; also G. C. Hughes (1974) considers this species less common in tropical regions than in temperate regions; therefore, putative collections of *C. macrocephala* from tropical areas [Cameroun, Ivory Coast, Malaysia, South Africa (Jones *et al.*, 1972)] need to be reconfirmed. ETYMOLOGY: From the Greek, *makros* = large, long, and *kephale* = head, top, in reference to the conidia ending in a large apical cell.

Cirrenalia pseudomacrocephala Kohlmeyer

Mycologia **60**, 266, 1968

LITERATURE: M. B. Ellis (1976); Kohlmeyer and Kohlmeyer (1964–1969, 1977).

Fig. 110c

HYPHAE 2.5–5 μm in diameter, septate, superficial or immersed, brown. CONIDIOPHORES 23–30 μm long, 3–5 μm in diameter, cylindrical, zero- to two-septate, simple, straight or curved, acrogenous or laterally on hyphae, often remaining connected with detached conidia, sometimes obsolete, light brown. CONIDIOGENOUS CELLS monoblastic, integrated, terminal, determinate. CONIDIA acrogenous, solitary, helicoid, semicontorted to 1¼ times contorted, 3–5(–6)-septate, slightly or strongly constricted at the septa, fuscous to grayish-brown (upper three to five cells dark, lower one to three cells light-colored); cells increasing in diameter and pigmentation from base to apex, distinctly dissimilar; spirals 27.5–38.5 μm in diameter; terminal cell 16.5–18.5 μm high, 16.5–20 μm in diameter, subglobose to ellipsoidal, basally flattened, mostly largest and darkest of all cells; basal cells 5.5–10 μm high, 4.5–5.5 μm in diameter, obtusely conical or cylindrical, light brown; central cells subglobose or obtusely conical. MODE OF LIFE: Saprobic. SUBSTRATE: Driftwood under mangroves. RANGE: Atlantic Ocean—Bermuda, Mexico (Veracruz). NOTE: Differs from *C. macrocephala* by the color and diameter of conidia and apical cells. ETYMOLOGY: From the Greek, *pseudo-* = false-, and the specific epithet *macrocephala,* indicating a close resemblance between the two species.

Cirrenalia pygmea Kohlmeyer

Ber. Dtsch. Bot. Ges. **79**, 35, 1966

LITERATURE: Kohlmeyer (1968c, 1969b); Kohlmeyer and Kohlmeyer (1964–1969).

Figs. 110d–110f

HYPHAE 2.2–4.5 μm in diameter, septate, ramose, fuscous. CONIDIOPHORES obsolete. CONIDIA acrogenous, solitary, helicoid, contorted ½ or 1 time, three- or four-septate, not or slightly constricted at the septa, fist-shaped or reniform, black or fuscous, fulgent (upper three cells dark, lower two or three cells light-colored); cells increasing in diameter from base to apex, distinctly dissimilar; spirals 25.5–31 × 28.5–34 μm; terminal cell 16.5–23 μm in diameter, subglobose to reniform, basally flattened; basal cells 3.5–5.5 μm in diameter; central cells irregularly conical or almost wedge-shaped. MODE OF LIFE: Saprobic. SUBSTRATES: Bark and wood of dead roots of *Mangifera indica, Rhizophora mangle,* and *R. racemosa,* driftwood; also sporulating on calcareous linings of empty shipworm tubes. RANGE: Atlantic Ocean—Liberia, Trinidad (J. Kohlmeyer, unpublished); Pacific Ocean—Mexico (Chiapas). ETYMOLOGY: From the Greek, *pygme* = fist, in reference to the fist-shaped conidia.

Cirrenalia tropicalis Kohlmeyer

Mycologia **60**, 267, 1968

LITERATURE: M. B. Ellis (1976); Kohlmeyer and Kohlmeyer (1964–1969); Sutton (1973a).

Figs. 110g–110j

HYPHAE 2.5–5 µm in diameter, septate, superficial or immersed, light brown. CONIDIOPHORES 25–42 µm long, 2.5–5 µm in diameter, cylindrical, zero- to four-septate, simple, acrogenous or laterally on hyphae, often remaining connected with detached conidia, sometimes obsolete, straight or curved, light brown. CONIDIOGENOUS CELLS monoblastic, integrated, terminal, determinate. CONIDIA acrogenous, solitary, regularly or irregularly helicoid, mostly 1 to 1½ times contorted, rarely semicontorted, six- to twelve-septate, not or slightly constricted at the septa, umber to reddish-brown; cells increasing in diameter from base to apex, distinctly dissimilar; spirals 20–38.5 µm in diameter; terminal cell 9–15 µm high, 10–20 µm in diameter, subglobose to ellipsoidal, basally flattened; basal cells 5.5–10 µm high, 3–5 µm in diameter, cylindrical; central cells subglobose, obtusely conical or doliiform. MODE OF LIFE: Saprobic. SUBSTRATES: Wood of prop roots of *Rhizophora racemosa,* intertidal wood; also sporulating on calcareous linings of empty shipworm tubes. RANGE: Atlantic Ocean—Liberia. NOTE: Differs from *C. pseudomacrocephala* by the number of septa, less marked constrictions, smaller apical cells, and color of conidia. ETYMOLOGY: From the Latin, *tropicalis* = tropical, in reference to the origin of the type collection, namely, Liberia.

Key to the Marine Species of *Cirrenalia*
1. Mature conidia black, shiny, fist-shaped, not constricted at the obscured septa
. . . . *C. pygmea*
1'. Mature conidia brown, all septa distinctly visible, constricted at the septa . . 2
  2(1') Conidia six- to twelve-septate, slightly constricted at the septa . . *C. tropicalis*
  2'(1') Conidia mostly with less than six septa, distinctly constricted at the septa . 3
3(2') Conidia reddish-brown; height of apical cell less than 13.5 µm . . *C. macrocephala*
3'(2') Conidia without reddish tint; height of apical cell more than 13 µm 4
  4(3') Conidia 3–5(–6)-septate, apical cell subglobose to ellipsoidal; a tropical
  species . . . . *C. pseudomacrocephala*
  4'(3') Conidia three-septate, rarely two- or four-septate, apical cell often sausage-
  shaped; a temperate species . . . . *C. fusca*

Cladosporium Link ex Fries

Link, *Mag. Ges. Naturf. Freunde, Berlin* **7**, 37–38, 1815; Fries,
Systema Mycologicum, Gryphiswaldiae **1**, p. XLVI, 1821

≡*Sporocladium* Chevallier, *Flore Gén. Environs Paris, Paris* **1**, p. 36, 1826
=*Didymotrichum* Bonorden, *Handbuch der allgemeinen Mycologie*, Stuttgart, p. 89, 1851
=*Heterosporium* Klotzsch *in* Cooke, *Grevillea* **5**, 123, 1877
=*Myxocladium* Corda, *Icon. Fung.* **1**, 12, 1837

LITERATURE: Barron (1968); M. B. Ellis (1971, 1976); S. J. Hughes (1958); Kendrick and Carmichael (1973); de Vries (1952).

CONIDIOPHORES macronematous, semimacronematous or micronematous, septate, simple or only branched apically, straight or curved, smooth or verrucose, olive-brown or brown. RAMO-CONIDIA frequently occurring. CONIDIOGENOUS CELLS polyblastic, mostly integrated, terminal, intercalary or rarely discrete, sympodial, cylindrical, cicatrized with prominent scars. CONIDIA simple, cylindrical, doliiform, fusiform, ellipsoidal or subglobose, with a protuberant basal scar or with scars at each end, zero to three (or more)-septate, smooth, verruculose or echinulate, olive-brown or brown, usually catenate, also solitary, chains often branched, acropleurogenous. MODE OF LIFE: Saprobic or parasitic. TYPE SPECIES: *Cladosporium herbarum* (Pers.) Link ex S. F. Gray. NOTE: Description based on M. B. Ellis (1971), synonyms after S. J. Hughes (1958); there are about 500 species in the genus; *C. algarum* is found growing in marine habitats; other, doubtfully marine species of *Cladosporium* are often isolated by culture methods. ETYMOLOGY: From the Greek, *klados* = branch and *sporos* = seed, spore, in reference to the branched chains of conidia of the type species.

Cladosporium algarum Cooke et Massee *in* Cooke

Grevillea **16**, 80, 1888

≡*Heterosporium algarum* (Cooke et Massee) Cooke et Massee *in* Cooke, *Grevillea* **18**, 74, 1890

LITERATURE: T. W. Johnson and Sparrow (1961); Sutherland (1916a).
Fig. 111
CONIDIOPHORES 140–175 μm long, 8–13 μm in diameter at the base, 4 μm at the tip, macronematous, cylindrical, septate, simple, straight or somewhat curved, smooth, brown. RAMO-CONIDIA up to 34 μm long, 6 μm in diameter, three- or four-septate, cylindrical or elongate–ellipsoidal, bearing up to four conidia on small protuberances, smooth, brown. CONIDIOGENOUS CELLS polyblastic, integrated, terminal or intercalary, more or less cylindrical, cicatrized with prominent scars. CONIDIA 8–24 × 4.5–8 μm, ellipsoidal, with a protuberant basal scar or with scars at each end, 0–2(–3)-septate, not or slightly constricted at the septa, smooth,

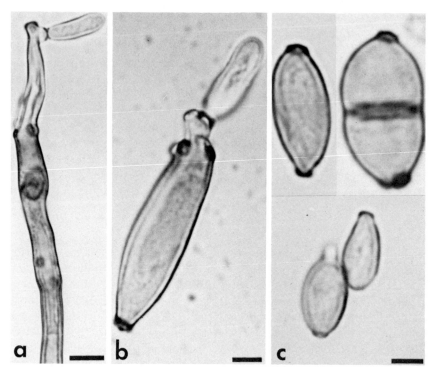

Fig. 111. *Cladosporium algarum* from *Laminaria flexicaulis* (lectotype K). (a) Conidiogenous cell with prominent scars, immature conidium attached to the tip (bar = 5 μm). (b) Ramo-conidium with scars and young conidium at the tip (bar = 3 μm). (c) Conidia with scars at the apices (bar = 3 μm). (All in bright field.)

thick-walled, olive-brown, solitary, rarely in short chains, acropleurogenous. MODE OF LIFE: Saprobic. SUBSTRATES: Decaying algae (e.g., *Laminaria flexicaulis* = *L. digitata*) washed up on the beach, dead rhizomes of *Zostera marina*, dead leaves of *Spergularia salina* and *Suaeda maritima* (J. Kohlmeyer, unpublished; J. K. Nos. 2910, 3168, 3171). RANGE: Atlantic Ocean—France, Germany [Helgoland, F.R.G. (J. Kohlmeyer, unpublished)], Great Britain (England). NOTE: The description is based on our examination of type material from Herbaria K and NY, identically labeled "*Cladosporium algarum* C. et Mass., on *Laminaria flexicaulis* LeJol., Baxhill nr. Hastings, 12/87, E. M. Holmes"; we have designated the specimen in Herb. K as the lectotype; the material from K is mounted on the same sheet with another collection on *Laminaria* sp. from west Kilbride, Ayrshire (Jan. 1890, leg. Boyd); this last collection was evidently the basis for the transfer of the species to

Heterosporium by Cooke and Massee (in Cooke, 1890); however, the Kilbride material is distinctly different from the original Hastings collection; conidia of the type specimen are smooth, whereas those of the later collection are distinctly echinulate and have a larger diameter; this wall ornamentation is evident also in immature conidia; therefore, the type material is not an immature state of the Kilbride collection; conidia of *C. algarum* have to be considered smooth, a fact that Sutherland (1916a) also recognized in his collections; but Sutherland pointed out that *C. algarum* may be associated with other Hyphomycetes having verrucose conidia. ETYMOLOGY: From the Latin, *algarum* = of the algae, in reference to the origin of the type material.

Cladosporium spp.

Representatives of this genus are frequently reported from marine collections; because the marine origin of these species cannot be established unless they are found growing *in situ,* we simply list the references dealing with *Cladosporium* spp. obtained by incubation methods or those identified to genus only: Cribb and Cribb (1969); Höhnk (1962a); T. W. Johnson and Sparrow (1961); Meyers *et al.* (1964a, 1967b); Nicot (1958a,c); Picci (1966); Quinta (1968); Reichenbach-Klinke (1956); Roth *et al.* (1964); Schaumann (1975a,b); Siepmann (1959a,b); Siepmann and Johnson (1960); Sparrow (1937); Steele (1967); Tubaki (1969).

Clavariopsis De Wildeman

Ann. Soc. Belge Microsc. **19**, 197, 1895 (non *Clavariopsis* Holtermann 1898)

LITERATURE: Ingold (1975b); S. Nilsson (1964); R. H. Petersen (1962, 1963); von Arx (1970).

CONIDIOPHORES simple. CONIDIA (thalloconidia) terminal, hyaline, branched; spore axis clavate, two-celled, straight; apical cell simultaneously forming three narrow or conical projections at the blunt apex. MODE OF LIFE: Saprobic. SUBSTRATES: Decaying leaves and wood. TYPE SPECIES: *Clavariopsis aquatica* De Wildeman. NOTE: Description based on Ingold (1975b) and R. H. Petersen (1962); there are two freshwater and one marine species. ETYMOLOGY: From the genus *Clavaria* Vaill. ex Fr. (Clavariaceae) and the Greek, *opsis* = appearance, in reference to the fancied resemblance of the clavate conidia to the basidiocarps.

Clavariopsis bulbosa Anastasiou

Mycologia **53**, 11, 1961

CONIDIAL STATE of *Corollospora pulchella* Kohlm., Schmidt et Nair. LITERATURE: Anastasiou (1963b); Kohlmeyer (1968c, 1969c); Kohlmeyer and Kohlmeyer (1964–1969); S. Nilsson (1964); Raghukumar (1973); Shearer (1972b); Shearer and Crane (1971); Tubaki (1968, 1969); Tubaki and Ito (1973).

See description of *Clavariopsis bulbosa* under *Corollospora pulchella* (p. 276). NOTE: S. Nilsson (1964) suggested that *C. bulbosa* could be removed from the genus *Clavariopsis* because of differences in morphology and habitat between the marine and the freshwater species; indeed, *C. bulbosa* appears closer related to *Orbimyces spectabilis* than to the other members of *Clavariopsis;* however, a transfer does not appear to be warranted because there is no perfect state of *O. spectabilis* known thus far. ETYMOLOGY: From the Latin, *bulbosus* = bulbous, in reference to the inflated basal cell of the conidia.

Cremasteria Meyers et Moore

Am. J. Bot. **47**, 348, 1960

HYPHAE septate, branched, hyaline. CONIDIOPHORES *sensu stricto* absent. CONIDIA (or chlamydospores?) subglobose to ellipsoidal, one-celled, brown, catenulate, rarely sessile, apical or on lateral hyphae. MODE OF LIFE: Saprobic. SUBSTRATE: Wood. TYPE SPECIES: *Cremasteria cymatilis* Meyers et Moore. The genus is monotypic. NOTE: Reservations concerning the identity of the genus are discussed under *C. cymatilis* (below). ETYMOLOGY: From the Greek, *kremasterion* = necklace bead [according to Meyers and Moore (1960); but *kremastos* = hanging, and *erion* = wool], in reference to the conidial appearance.

Cremasteria cymatilis Meyers et Moore

Am. J. Bot. **47**, 348, 1960

LITERATURE: Churchland and McClaren (1973); G. C. Hughes (1969); T. W. Johnson (1968); T. W. Johnson and Sparrow (1961); E. B. G. Jones (1971a); E. B. G. Jones and Jennings (1964); E. B. G. Jones *et al.* (1971, 1972); Kohlmeyer and Kohlmeyer (1964, 1964–1969); Meyers and

Reynolds (1960b); Meyers *et al.* (1960); Pisano *et al.* (1964); Tubaki (1969).

HYPHAE 1.5–4 μm in diameter, septate, branched, occasionally anastomosing, hyaline. CONIDIOPHORES *sensu stricto* absent. CONIDIA (or chlamydospores?) 7.5–22(–27) μm long, 7.5–13.5(–16) μm in diameter, subglobose to ellipsoidal, rarely pyriform, one-celled, rusty (light brown with some mixture of red), catenulate, rarely sessile, apical or on short lateral hyphae, basipetal; catenulae 65–184 μm long, 2- to 21-celled, distinctly constricted at the septa, increasing in diameter from base to apex, simple, rarely ramose, straight or curved; intercalary chlamydospores also occurring. MODE OF LIFE: Saprobic. SUBSTRATES: Intertidal wood and test panels (e.g., *Pinus palustris, P. sylvestris*). RANGE: Atlantic Ocean—France, Great Britain (England), Iceland, Norway; Pacific Ocean—Australia (New South Wales), Canada (British Columbia), United States (Alaska). NOTE: Description based on our examination of an Isotype slide (I: 263), kindly provided by Dr. R. T. Moore; T. W. Johnson and Sparrow (1961) and T. W. Johnson (1968) have questioned the validity of the species, proposing that it could represent nothing but chlamydospores of an Ascomycete; in addition, Kohlmeyer and Kohlmeyer (1964) pointed out that catenulate chlamydospores developing in pure cultures of *Halosphaeria circumvestita* and *H. mediosetigera* (Kohlmeyer and Kohlmeyer, 1966) are very similar to those of *C. cymatilis;* therefore, it appears doubtful that any collections from natural habitats can be matched with *C. cymatilis;* this doubt is strengthened by the fact that descriptions of *C. cymatilis* are based solely on a pure culture; the growth habit of the species could be entirely different in nature, as shown, for instance, in a comparison between conidia of *Humicola alopallonella* from cultures and those from nature (Kohlmeyer, 1969c). ETYMOLOGY: From the Greek, *kymatilos* = of the waves [according to Meyers and Moore (1960); but *kyma* = wave], in reference to the marine habitat.

Dendryphiella Bubák et Ranojević *in* Ranojević

Ann. Mycol. **12**, 417, 1914

LITERATURE: Barron (1968); M. B. Ellis (1971); S. J. Hughes (1958); Pugh and Nicot (1964); Reisinger (1968); Reisinger and Guedenet (1968).

CONIDIOPHORES macronematous, septate, simple or irregularly branched near the apex, smooth or verruculose, bearing conidia at the swollen tip and at intercalary swellings, light to dark or reddish brown, solitary, showing indefinite growth. CONIDIOGENOUS CELLS polytretic, integrated, terminal but becoming intercalary, sympodial, subglobose to clavate, cicatrized on the nodose swellings with scars close to each other. CONIDIA simple, cylindrical or oblong with rounded ends, or ellipsoidal, zero- to three-septate, smooth or verruculose, light brown to olive-brown, catenate or solitary, acropleurogenous. MODE OF LIFE: Saprobic. TYPE SPECIES: *Dendryphiella vinosa* (Berk. et Curt.) Reisinger. NOTE: Description based on Ellis (1971) and Reisinger (1968); S. J. Hughes (1958) considers this genus a synonym of *Dendryphion* Wallroth, a decision accepted by Barron (1968) but not by the other authors cited above; besides four terrestrial species there are two marine species in the genus. ETYMOLOGY: From the genus *Dendryphion* (Hyphomycetes) and the Latin diminutive suffix *-ellus*, indicating a relationship to this genus.

Dendryphiella arenaria Nicot

Rev. Mycol. **23**, 93, 1958

LITERATURE: Chadefaud (1969); M. B. Ellis (1976); Gessner and Goos (1973a,b); Hopper and Meyers (1966); T. W. Johnson and Sparrow (1961); E. B. G. Jones and Jennings (1964); Kohlmeyer (1966a); Kohlmeyer and Kohlmeyer (1964–1969); Meyers (1968b); Meyers and Hopper (1966, 1967); Meyers *et al.* (1963a, 1964a, 1965); Nicot (1958a,c); Pugh and Nicot (1964); Roth *et al.* (1964); Tubaki (1969); Wagner-Merner (1972).

Fig. 112a

HYPHAE 2.5–5(–7.5) μm in diameter, septate, subhyaline to light brown. CONIDIOPHORES 15–90 μm long, 2.5–3.5 μm in diameter, macronematous, cylindrical, zero- to two-septate, simple or rarely branched,

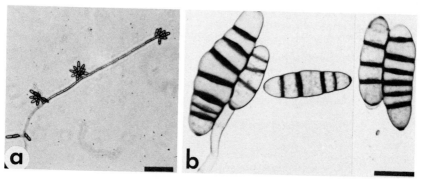

Fig. 112. Marine *Dendryphiella* spp. (a) Hypha of *D. arenaria* with loose heads of conidia, from incubated leaves and rhizomes of *Thalassia testudinum* (Herb. J.K. 2435; bar = 50 μm). (b) Conidia of *D. salina* from rotting *Laminaria* sp. (Herb. J.K. 2953; bar = 15 μm). (Both in bright field.)

straight or curved, also geniculate, apically somewhat swollen (up to 4–5 μm in diameter), pleurogenous on the mycelium, subhyaline to light brown. CONIDIA 9–20 × 3.5–6.5 μm, ovoid, ellipsoidal or cylindrical, with a distinct dark basal scar, one- to three-septate, slightly constricted at the septa, smooth or somewhat echinulate, brown, solitary, developing at the apex and on successive swellings of the conidiophore, forming small loose heads of two to four (rarely up to nine) conidia. MODE OF LIFE: Saprobic. SUBSTRATES: Decaying marine or estuarine plants (e.g., *Sargassum* sp., *Spartina alterniflora, Thalassia testudinum*); in beach sands under *Ammophila arenaria;* conidia occur in sea foam. RANGE: Atlantic Ocean—France, United States (Florida, North Carolina, Rhode Island); Black Sea—U.S.S.R. (J. Kohlmeyer, unpublished). NOTE: M. B. Ellis (1976), without giving explanations, transfers *D. arenaria* to *Scolecobasidium arenarium,* a proposition that we do not accept at this time; in the key of his earlier book, M. B. Ellis (1971) distinguishes *Scolecobasidium* from *Dendryphiella* by holoblastic conidiogenous cells in the former and enteroblastic cells in the latter; in the generic description (M. B. Ellis, 1971), *Scolecobasidium* is characterized by polyblastic denticulate conidiogenous cells, with long, narrow–cylindrical, threadlike denticles that often break across the middle and leave part attached to the conidium and part to the point of origin; *Dendryphiella,* however, has polytretic, cicatrized conidiogenous cells without denticles; Nicot (1958b) does not mention or illustrate any denticles in her description of the type material of *D. arenaria,* and conidia in our collections of this species are borne on the conidiogenous cell without such denticles; however, M. B. Ellis (1976) describes and illustrates both, conidiogenous cells and conidia, with distinct pegs; the ultrastructure of conidiogenesis has not been examined in *D. arenaria* and it is not known with certainty if the walls of the conidiogenous cell contribute to the formation of the conidia; in our opinion, conidia of *D. arenaria* have an enteroblastic origin and lack denticles, and, therefore, we prefer to keep the species in *Dendryphiella.* ETYMOLOGY: From the Latin, *arenarius* = growing on sand, in reference to the origin of the type material from beach sand.

Dendryphiella salina (Sutherland) Pugh et Nicot

Trans. Br. Mycol. Soc. **47,** 266, 1964

≡*Cercospora salina* Sutherland, *New Phytol.* **15,** 43, 1916

LITERATURE: Allaway and Jennings (1970a,b, 1971); Barron (1968); Byrne and Eaton (1972); Byrne and Jones (1974, 1975a,b); Chesters and Bull (1963a,b); Chupp (1953); Dickinson (1965); Dickinson and Pugh

(1965a–c); M. B. Ellis (1976); Gessner and Goos (1973a,b); Henningsson (1974); Holligan and Jennings (1972a–c); Jennings and Aynsley (1971); T. W. Johnson and Sparrow (1961); E. B. G. Jones (1962a, 1963c, 1968b, 1971a, 1972, 1973); E. B. G. Jones and Irvine (1971, 1972); E. B. G. Jones and Jennings (1964, 1965); E. B. G. Jones and Oliver (1964); E. B. G. Jones et al. (1971, 1972); Kohlmeyer (1971b); Kohlmeyer and Kohlmeyer (1964–1969); Lee and Baker (1972b); Meyers (1968a); Meyers and Scott (1968); Meyers et al. (1972); Poole and Price (1972); Pugh (1960, 1962a,b, 1974); Schaumann (1974b, 1975b); Schmidt (1974); Schultz and Quinn (1973); Tubaki and Asano (1965); Wilson (1951); Wilson and Knoyle (1961); Withers et al. (1975).

Fig. 112b

HYPHAE 2–8 µm in diameter, septate, hyaline, light brown or light olivaceous. CONIDIOPHORES 15–60 µm long, 4–4.5 µm in diameter, macronematous, cylindrical, septate, simple or rarely branched, straight or curved, apically or subapically somewhat swollen, hyaline, light olivaceous or pale brown. CONIDIA 14.5–75 × 5.5–10.5 µm, cylindrical or subellipsoidal, with a basal scar, 1–9(–11)-septate, predominantly 3- to 5-septate, slightly or not constricted at the septa, straight or somewhat curved, smooth, subhyaline, pale brown or light olivaceous, solitary or rarely in short chains, developing apically and laterally on small swellings of the conidiophore, forming small loose heads of three to six conidia; deciduous. MODE OF LIFE: Saprobic. SUBSTRATES: Decaying marine or estuarine plants [e.g., *Chondrus crispus, Furcellaria fastigiata, Laminaria hyperborea* (J. Kohlmeyer, unpublished), *Laminaria* sp., *Sargassum muticum*], driftwood and test panels (e.g., *Betula pubescens, Fagus sylvatica, Pinus sylvestris*), in the rhizosphere, on roots and seeds of salt-marsh plants (e.g., *Halimione portulacoides, Puccinellia maritima, Rhizophora mangle, Salicornia stricta, Spartina alterniflora, S. townsendii, Suaeda maritima*); conidia occur in sea foam. Range: Atlantic Ocean—France, Germany (Helgoland), Great Britain (England, Orkney Islands, Scotland, Wales), Norway, Sweden, United States (Massachusetts, Rhode Island); Baltic Sea—Germany (G.D.R.); Indian Ocean—Antarctica; Mediterranean—Greece, Italy; Pacific Ocean—Antarctica, United States [Hawaii (Oahu)]. NOTE: We made collections in England, France, Helgoland, Scotland, and New England on decaying algae and in foam; the species was recently transferred to *Scolecobasidium* by M. B. Ellis (1976); for the time being we chose to keep *D. salina* in the genus *Dendryphiella* because the conidial origin appears to be enteroblastic and denticles are absent in our collections (see Fig. 112b), although M. B. Ellis (1976) describes and illustrates such pegs on conidiogenous cells and conidia (see also discussion of *D. arenaria*, p. 488); Sutherland (1916a), in the type description, and Pugh and Nicot (1964) also do not mention conidial pegs; a submicroscopical study should clarify the conidiogenesis of *D. salina*. ETYMOLOGY: From the Latin, *salinus* = saline, in reference to the marine habitat of the species.

Key to the Marine Species of *Dendryphiella*

1. Conidia one- to three-septate; never longer than 20 μm . . . . **D. arenaria**
1'. Conidia usually three- to five-septate, rarely with less septa or up to eleven-septate; usually longer than 20 μm, up to 75 μm . . . . **D. salina**

Dictyosporium Corda

in Weitenweber's *Beiträge zur Gesammten Natur und Heilwissenschaft,* Prag **1**, p. 87, 1836

=*Speira* Corda, *Icon. Fung.* **1**, p. 9, 1837
=*Cattanea* Garovaglio, *Rend., R. Ist. Lomb. Sci. Lett.,* Ser. 2 **8**, 125, 1875

LITERATURE: Barron (1968); Damon (1952); M. B. Ellis (1971); S. J. Hughes (1958).

CONIDIOPHORES micronematous, mononematous, septate, irregularly branched, smooth, hyaline to brown. CONIDIOGENOUS CELLS monoblastic, integrated, terminal or rarely intercalary, determinate, cylindrical, doliiform, globose or subglobose. CONIDIA cheiroid, branched, branches arising from a single basal cell, mostly flattened in one plane, multiseptate, mostly constricted at the septa, smooth, olive to brown, solitary, a-crogenous or rarely pleurogenous. MODE OF LIFE: Saprobic. TYPE SPECIES: *Dictyosporium elegans* Corda. NOTE: Description and synonyms based on Damon (1952) and M. B. Ellis (1971); there is one marine representative among the ten species of the genus. ETYMOLOGY: From the Greek, *dictyon* = net and *sporos* = seed, spore, in reference to the netlike appearance of the multichambered conidia.

Dictyosporium pelagicum (Linder) G. C. Hughes ex Johnson et Sparrow

Fungi in Oceans and Estuaries, Cramer, Weinheim, p. 391, 1961 [as "*D. pelagica*"; non *D. pelagica* (Linder) G. C. Hughes ex E. B. G Jones, *Trans. Br. Mycol. Soc.* **46**, 137, 1963, superfluous new combination]

≡*Speira pelagica* Linder *in* Barghoorn et Linder, *Farlowia* **1**, 407, 1944 [as "*S. (Cattanea) pelagica*"]
=*Speira litoralis* Höhnk, *Veroeff. Inst. Meeresforsch. Bremerhaven* **3**, 221, 1955

LITERATURE: Anastasiou and Churchland (1969); Brooks *et al.* (1972); Byrne and Eaton (1972); Byrne and Jones (1974); Damon (1952); Eaton (1972); Eaton and Dickinson (1976); Eaton and Jones (1971a); Gessner and Goos (1973a,b); Höhnk (1956, 1958); G. C. Hughes (1969); G. C. Hughes and Chamut (1971); Irvine *et al.* (1972); T. W. Johnson (1956b); T. W. Johnson and Sparrow (1961); E. B. G. Jones (1963a, 1968b, 1971a, 1972); E. B. G. Jones and Irvine (1971); E. B. G. Jones and Oliver (1964); E. B. G. Jones *et al.* (1972); Kohlmeyer (1958a,b, 1960, 1962a, 1963d, 1971b); Kohlmeyer and Kohlmeyer (1964–1969); Poole and Price (1972); Schaumann (1968, 1969, 1974c, 1975a,b); Schmidt (1974); Shearer (1972b); Sieburth *et al.* (1974); Wilson (1951).

HYPHAE 1.3–3.1 μm in diameter, septate, ramose, hyaline to fuscous. CONIDIOPHORES 10–35 μm long, 1.5–2.5 μm in diameter, simple or rarely septate, sometimes swollen at the apex, hyaline to light brown. CONIDIA 12–66 × 9–28.5(–36) μm, ellipsoidal, ovoid, cylindrical, clavate or cheiroid, sometimes branched, multiseptate, constricted at the septa; rows of cells arising from a single basal cell; cells initially in three, rarely two, rows, then forming three to eight irregular rows; basal cells 2.5–7.5 μm high, 1.5–9.5 μm in diameter; apical cells 4–10 μm high, 3.5–8.5 μm in diameter; smooth, dark brown to almost black. MODE OF LIFE: Saprobic. SUBSTRATES: Submerged leaves, driftwood and drifting bark, intertidal bark and wood, test panels (e.g., *Fagus sylvatica, Liriodendron tulipifera, Ocotea rodiaei, Pinus sylvestris, Quercus* sp., balsa), *Spartina alterniflora*. RANGE: Atlantic Ocean—Cameroun, France, Germany (Helgoland and mainland, F.R.G.), Great Britain (England, Scotland, Wales), Ivory Coast, Spain, United States (Connecticut, Florida, Maryland, Massachusetts, North Carolina, Rhode Island, South Carolina, Texas); Baltic Sea—Denmark, Germany (F.R.G. and G.D.R.); Mediterranean—Italy (Sicily and mainland); Pacific Ocean—Canada (British Columbia), Chile, New Zealand, United States (Washington). NOTE: Damon (1952) and G. C. Hughes (1969) state that *Speira pelagica* does not belong in *Dictyosporium*; G. C. Hughes (1969) concludes that an affinity with *Monodictys* is indicated. ETYMOLOGY: From the Latin, *pelagicus* = pertaining to the sea, in reference to the marine occurrence of the species.

Drechslera Ito

Proc. Imp. Acad. Jpn. **6,** 355, 1930

=*Bipolaris* Shoemaker, *Can. J. Bot.* **37,** 882, 1959

LITERATURE: Barron (1968); Chidambaram *et al.* (1973); M. B. Ellis

(1971, 1976); Shoemaker (1962); Subramanian and Jain (1966); Sutton (1976); von Arx (1970).

STROMATA occur in some species. SCLEROTIA may be formed in culture. CONIDIOPHORES macronematous, mononematous, occasionally in tufts, septate, often geniculate, simple or sparsely branched in some species, straight or curved, mostly smooth, brown. CONIDIOGENOUS CELLS polytretic, integrated, terminal, often becoming intercalary, sympodial, cylindrical, cicatrized. CONIDIA simple, clavate or obclavate, cylindrical with rounded ends, ellipsoidal or fusiform, straight or curved, some species with a protuberant basal scar, pseudoseptate, smooth or rarely verruculose, thick-walled, with a pigmented exospore and a hyaline endospore, straw-colored, light, dark, or olivaceous brown, sometimes apical cells lighter colored than central ones, solitary, in some species catenate or with secondary conidiophores bearing conidia, acropleurogenous. MODE OF LIFE: Parasitic or saprobic. TYPE SPECIES: *Drechslera tritici-vulgaris* (Nisikado) Ito ex S. J. Hughes [imperfect state of *Pyrenophora tritici-repentis* (Died.) Drechsl.]. NOTE: Description based on M. B. Ellis (1971) and Shoemaker (1962); *Bipolaris* is a synonym of *Drechslera* fide Subramanian and Jain (1966); there is 1 marine representative among about 60 species of the genus. ETYMOLOGY: Named in honor of C. Drechsler (1892–), American mycologist.

Drechslera halodes (Drechsler) Subramanian et Jain

Curr. Sci. **35,** 354, 1966

=*Helminthosporium halodes* Drechsler, *J. Agric. Res.* **24,** 709, 1923
=*Bipolaris halodes* (Drechsler) Shoemaker, *Can. J. Bot.* **37,** 883, 1959
=*Exserohilum halodes* (Drechsler) Leonard et Suggs, *Mycologia* **66,** 290, 1974

LITERATURE: Chidambaram *et al.* (1973); M. B. Ellis (1971); Gessner and Kohlmeyer (1976); Leonard (1976); Subramanian (1971); Subramanian and Raghukumar (1974).

CONIDIOPHORES 60–150 μm long, 4–8 μm in diameter, single or paired, cylindrical, septate, often geniculate in the upper part, simple, straight or curved, smooth, brown. CONIDIA 20–105 × 10–20 μm, simple, cylindrical to ellipsoidal, slightly tapering toward the base, never rostrate, straight or somewhat curved, with a protuberant basal scar, with six to twelve pseudosepta, not constricted at the septa, smooth, thick-walled, brown; apical cells hyaline to light brown, separated from adjoining cells by distinct thick, dark-colored septa; central cells golden brown. MODE OF LIFE: Parasitic or saprobic. HOSTS: *Distichlis spicata, Spartina alter-*

niflora; also isolated from marine soil. RANGE: Atlantic Ocean—United States (Florida, New York, North Carolina, Rhode Island); Indian Ocean—India (Madras). NOTE: Description mostly after Drechsler (1923) and M. B. Ellis (1971); Chidambaram *et al.* (1973) give the following sizes of 13- to 18-septate conidia: 13.6–191.4 × 10.2–22.1 μm; Leonard (1976) considers *D. halodes* to be synonymous with *Exserohilum rostratum;* however, M. B. Ellis (1976) includes *Exserohilum* as a synonym of *Drechslera;* we use the name under which the marine collections have been reported, namely, *D. halodes,* until the taxonomic questions will be clarified by specialists of this group; *D. halodes* causes seedling and foot-rot diseases in a number of cultivated terrestrial plants (references in Chidambaram *et al.,* 1973); M. B. Ellis (1971) lists countries, mostly tropical or subtropical, in which *D. halodes* has been isolated from terrestrial sources; we include *D. halodes* as a possible facultative marine species. ETYMOLOGY: From the Greek, *hals* = salt, and the suffix *-odes,* indicating a resemblance (sic!), apparently in reference to the type habitat of the species, namely, a salt marsh.

Humicola Traaen emend. Fassatiová

Traaen: *Nyt Mag. Naturvidenskab.* **52,** 31–34, 1914; Fassatiová: *Česká Mykol.* **21,** 79, 1967

=*Melanogone* Wollenweber et Richter, *Zentralbl. Bakteriol., Parasitenkd., Infektionskr. Hyg., Abt. 2* **89,** 74, 1934

LITERATURE: Barron (1968); de Bertoldi *et al.* (1972, 1973); M. B. Ellis (1971); Gambogi (1969); Lepidi *et al.* (1972); Moorhouse and de Bertoldi (1975); Nicoli and Russo (1974); White and Downing (1953); von Arx (1970).

HYPHAE septate, ramose, hyaline or pigmented. Intercalary CHLA-MYDOSPORES occur in some species. CONIDIOPHORES micronematous or semimacronematous, cylindrical, sometimes oblong, simple or rarely irregularly branched, straight or curved, smooth, hyaline to light golden brown, sometimes indistinct. CONIDIOGENOUS CELLS monoblastic, integrated, terminal, determinate, cylindrical, doliiform, pyriform to funnel-shaped. CONIDIA simple, globose, rarely obovoid or pyriform, one-celled, smooth, light to golden or dark brown, solitary, acrogenous. PHIALIDIC state present in some species. MODE OF LIFE: Saprobic, mostly soil-inhabiting. TYPE SPECIES: *Humicola fuscoatra* Traaen. NOTE: Description mostly based on M. B. Ellis (1971) and Nicoli and Russo (1974); there is one marine representative among about nine species and three varieties of the genus; Moorhouse and de Bertoldi (1975) suggest the use of enzymatic characters to separate different taxa of *Humicola* be-

cause morphological features are highly variable in this genus. ETYMOL-
OGY: From the Latin, *humus* = earth, soil, and *-cola* = -dweller, in
reference to the habitat of the type species, namely, soil.

Humicola alopallonella Meyers et Moore

Am. J. Bot. **47**, 346, 1960

LITERATURE: Anastasiou and Churchland (1969); de Bertoldi *et al.*
(1972); Byrne and Eaton (1972); Byrne and Jones (1974); Churchland and
McClaren (1973); Eaton (1972); Eaton and Irvine (1972); Eaton and Jones
(1971a,b); Fassatiová (1967); S. E. J. Furtado *et al.* (1977); Henningsson
(1974, 1976a,b); G. C. Hughes (1968, 1969, 1974); Irvine *et al.* (1972); T.
W. Johnson and Sparrow (1961); E. B. G. Jones (1963a, 1965, 1968a,b,
1971a, 1972, 1973); E. B. G. Jones and Irvine (1971); E. B. G. Jones and
Stewart (1972); E. B. G. Jones *et al.* (1972); Kohlmeyer (1961a, 1963d,e,
1968c, 1969b,c); Kohlmeyer and Kohlmeyer (1964–1969, 1977); Lepidi *et
al.* (1972); Meyers (1968a, 1971a); Meyers and Reynolds (1959b, 1960b);
Meyers *et al.* (1960); Moorhouse and de Bertoldi (1975); Nicoli and Russo
(1974); Pisano *et al.* (1964); Schaumann (1968, 1974b, 1975a–c); Schmidt
(1967, 1974); Schneider (1976); Sguros and Simms (1963a); Sguros *et al.*
(1971, 1973); Shearer (1972b); Tubaki (1968, 1969); Tubaki and Ito (1973).

Fig. 113

HYPHAE septate, subhyaline to light brown. CONIDIOPHORES 3.5–6.5 ×
3.5–6 μm, macronematous, simple, one- to two-celled, smooth, subhyaline to light brown, lateral, short, sometimes indistinct. CONIDIA
10–22.5(–37.5) μm long, (8.5–)10–18 μm in diameter, obpyriform, ovoidal
or subglobose, one- or two-celled, fuscous; when two-celled, apical cell
large (8.5–15.5 × 7–12 μm), ovoid, fuscous, basal cell smaller, obconical
to cylindrical, light brown; conidiogenous cell usually remaining connected with the conidium. MODE OF LIFE: Saprobic. SUBSTRATES: Intertidal wood, driftwood, test panels (e.g., *Betula pubescens*, *Dicorynia
paraensis*, *Fagus sylvatica*, *Ocotea rodiaei*, *Pinus sylvestris*, *Tilia
americana*, balsa), submerged bone, leaves, bark of *Betula* sp., endocarp
of *Cocos nucifera* (J. Kohlmeyer, unpublished), branches of *Pluchea* ×
fosbergii and *Rhizophora mangle*, on the calcareous lining of empty
teredinid tubes. RANGE: Atlantic Ocean—Bahamas [Great Abaco (J.
Kohlmeyer, unpublished)], Barents Sea, Bermuda, Cameroun, Canada
(Newfoundland), France, Germany (Helgoland and mainland, F.R.G.),
Great Britain (England, Scotland, Wales), Ivory Coast, Liberia, Norway,
United States (Maryland, North Carolina, Texas); Baltic Sea—Germany
(F.R.G. and G.D.R.); Indian Ocean—Aden; Mediterranean—Italy;
Pacific Ocean—American Samoa [Pago Pago (J. Kohlmeyer, unpub-

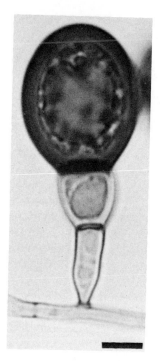

Fig. 113. Conidium of *Humicola alopallonella* attached to hypha (pure culture Ko 20; bar = 5 μm; in bright field).

lished, J.K. 1690)], Australia (New South Wales), Canada (British Columbia), Eniwetok (J. Kohlmeyer, unpublished, Herb. J.K. 3026), Japan, United States [Alaska, Hawaii (Hawaii, Oahu), Washington]. NOTE: We have examined the holotype (Herb. FH); conidia from pure cultures are very variable in shape and size and usually larger than conidia collected on the natural substrate; Lepidi *et al.* (1972) question the generic placement of *H. alopallonella,* whereas other authors cited above accept it. Putative records of *H. alopallonella* from freshwater habitats (Eaton, 1972; Eaton and Irvine, 1972; Eaton and Jones, 1971a,b; Schaumann, 1968) have to be reconfirmed because these identifications have not been supported by data or illustrations. ETYMOLOGY: From the Greek, *hals* = sea, salt, the Italian, *pallone* = ball, balloon, and the Latin diminutive suffix *-ellus,* in reference to the marine habitat and the balloon-shaped conidia of the species.*

* The epithet should actually read *"halopallonella"* because in Greek the first letter (alpha) bears a *Spiritus asper* that should have been transcribed in Latin as the letter *h,* but a correction of this orthographic error is not advisable because the first letter of the name is concerned (Article 73, Note 4, of the International Code of Botanical Nomenclature).

Monodictys S. J. Hughes

Can. J. Bot. **36,** 785, 1958

LITERATURE: M. B. Ellis (1971, 1976).

HYPHAE septate, branched, subhyaline or dark brown, immersed, partly immersed or superficial. CONIDIOPHORES micronematous or semimacronematous, mononematous, subcylindrical, cells occasionally swollen, septate, simple or irregularly branched, straight or curved, smooth, hyaline to brown, single or sometimes forming pustules. CO-NIDIOGENOUS CELLS monoblastic, integrated, terminal, determinate, cylindrical, doliiform or subglobose. CONIDIA oblong with rounded ends, subpyriform, clavate, ellipsoidal, subglobose or irregular, occasionally spirally twisted, muriform, sometimes constricted at the septa, smooth or verrucose, brown to black, acrogenous, solitary; basal cell sometimes inflated and lighter colored than the others. MODE OF LIFE: Saprobic. TYPE SPECIES: *Monodictys putredinis* (Wallr.) S. J. Hughes. NOTE: Description based on Ellis (1971) and S. J. Hughes (1958); von Arx (1970) considers *Monodictys* a synonym of *Pithomyces* Berk. et Br.; there are ten terrestrial species and one marine representative; *M. austrina* Tubaki (Tubaki and Asano, 1965) was isolated from soil and unidentified algae in Antarctica, and its marine occurrence has not been demonstrated. ETYMOLOGY: From the Greek, *monos* = single and *dictyon* = net, in reference to the netlike appearance of the singly produced, muriform conidia.

Monodictys pelagica (Johnson) E. B. G. Jones

Trans. Br. Mycol. Soc. **46,** 138, 1963

≡*Piricauda pelagica* Johnson, *J. Elisha Mitchell Sci. Soc.* **74,** 42, 1958
=*Piricauda arcticoceanorum* Moore, *Rhodora* **61,** 95, 1959

LITERATURE: Anastasiou and Churchland (1969); Becker (1961); Borut and Johnson (1962); Brooks *et al.* (1972); Byrne and Eaton (1972); Byrne and Jones (1974); Churchland and McClaren (1973, 1976); Curran (1975); Eaton and Dickinson (1976); S. E. J. Furtado *et al.* (1977); Gessner and Goos (1973a,b); Gold (1959); Henningsson (1974, 1976a,b); G. C. Hughes (1968, 1969, 1974); G. C. Hughes and Chamut (1971); Irvine and Jones (1975); T. W. Johnson (1968); T. W. Johnson and Sparrow (1961); E. B. G. Jones (1968b, 1971a, 1972); E. B. G. Jones and Jennings (1964); E. B. G. Jones *et al.* (1972); Koch (1974); Kohlmeyer (1959, 1960, 1962a, 1963d, 1971b); Kohlmeyer and Kohlmeyer (1964–1969); Meyers and Reynolds

(1959b, 1960b); Meyers *et al.* (1960); Neish (1970); Raghukumar (1973); Schaumann (1968, 1969, 1974b, 1975a–c); Schmidt (1967, 1974); Schneider (1976); Shearer (1972b); Tubaki (1968, 1969).

HYPHAE septate, rarely branched, dark brown or fuscous. CONID-IOPHORES short or lacking, cylindrical, zero- to two-septate, simple, lateral, dark brown. CONIDIA 15–41(–44) μm long, 12.5–37 μm in diameter, obpyriform, ovoid or rarely subglobose, muriform, not or slightly constricted at the septa, smooth, black, with one to three yellowish or light brown basal cells, solitary; the septation is only visible in immature conidia. CHLAMYDOSPORES 48–155 μm long, 6.5–21 μm in diameter, dark brown, intercalary, catenulate, composed of 4–25 cells; chains simple, rarely muriform. MODE OF LIFE: Saprobic. SUBSTRATES: Driftwood (e.g., *Quercus* sp.), intertidal wood and bark, test panels (e.g., *Betula pubescens, Dicorynia paraensis, Fagus sylvatica, Laphira procera, Ocotea rodiaei, Pinus monticola, P. sylvestris, Pseudotsuga menziesii,* balsa), *Spartina alterniflora,* stems of *Typha* sp., reeds and submerged leaves. RANGE: Atlantic Ocean—Canada (Newfoundland, Nova Scotia), Denmark, France, Gambia, Germany (Helgoland and mainland, F.R.G.), Great Britain (England, Ireland, Scotland, Wales), Iceland, Ivory Coast, Spain (Canary Islands and mainland), Sweden, United States (Maine, Maryland, Massachusetts, New Hampshire, North Carolina, Rhode Island, Virginia); Baltic Sea—Denmark, Germany (F.R.G., G.D.R.), Sweden; Black Sea—U.S.S.R. (J. Kohlmeyer, unpublished, Herb J. K. 3665); Indian Ocean—India (Madras); Mediterranean—France, Italy (Sicily); Pacific Ocean—Canada (British Columbia), Chile, Japan, New Zealand, Panama, United States (Alaska, California, Washington). NOTE: Davidson (1974b) found *M. pelagica* on *Juncus* sp. and wood in an inland saline lake [United States (Wyoming)]. ETYMOLOGY: From the Latin, *pelagicus* = pertaining to the sea, in reference to the marine occurrence of the species.

Orbimyces Linder *in* Barghoorn et Linder

Farlowia **1**, 404, 1944

CONIDIOPHORES obsolete; conidia developing singly, pleurogenously on short, truncate, hyaline sporogenous cells on the mycelium. CONIDIA consisting of a large, dark, subglobose cell with one or two crowns of septate, light-colored appendages at the distal end. MODE OF LIFE: Saprobic. TYPE SPECIES: *Orbimyces spectabilis* Linder. The genus is monotypic. ETYMOLOGY: From the Latin, *orbis* = orb, and the Greek, *mykes* = fungus, in reference to the resemblance of the conidia to the orb of the English royal regalia.

Orbimyces spectabilis Linder *in* Barghoorn et Linder

Farlowia **1,** 404–405, 1944

LITERATURE: Henningsson (1974); G. C. Hughes (1969); T. W. Johnson and Sparrow (1961); E. B. G. Jones (1973); E. B. G. Jones *et al.* (1972); Kohlmeyer (1960); Kohlmeyer and Kohlmeyer (1964–1969); Meyers (1968b, 1971a); Meyers and Hoyo (1966); Pisano *et al.* (1964); Schmidt (1974); Sguros and Simms (1963a); Tubaki (1969); Tyndall and Kirk (1973).

Fig. 114

CONIDIOPHORES obsolete; conidia developing singly, pleurogenously on short, truncate, hyaline sporogenous cells on the mycelium. CONIDIA consisting of a large dark cell with one or two crowns of septate, light-colored appendages at the distal end; basal cell 24–41.5 × 18–37 μm, subglobose or ovoid, thick-walled, shining black, smooth, with a scar or the sporogenous cell at the base; radiating appendages septate, consisting of a central cell and (3–)4–6(–7) arms; central cell with three to seven short projections, fuscous; arms 13–50 × 4–6.5 μm, cylindrical, (0–) 1–2(–4)-septate, slightly constricted at the septa, light brown at the base, subhyaline at the apex. MODE OF LIFE: Saprobic. SUBSTRATES: Driftwood, intertidal wood, test panels (e.g., *Pinus sylvestris*). RANGE: Atlantic Ocean—Ivory Coast, United States (Massachusetts, Rhode Island); Baltic Sea—Germany (G.D.R.), Sweden; Pacific Ocean—Canada (British Columbia), United States (Washington). NOTE: Description based on Linder (1944) and on our examination of pure culture R.D.B. 94, isolated from wood in Rhode Island and kindly made available to us by Dr. R. D. Brooks. ETYMOLOGY: From the Latin, *spectabilis* = notable, remarkable, in reference to the unique conidia.

Periconia Tode ex Fries

Tode, *Fungi Mecklenburgenses Selecti*, Lüneburg **2,** pp. 2–3, 1791; Fries, *Systema Mycologicum*, Gryphiswaldiae **1,** p. XLVII, 1821

=*Berkeleyna* O. Kuntze, *Revisio Generum Plantarum*, Leipzig **3,** p. 447, 1898
=*Harpocephalum* Atkinson, *N. Y. Agric. Exp. Stn., Ithaca Bull.* **3,** 41, 1897
=*Pachytrichum* Sydow, *Ann. Mycol.* **23,** 420, 1925
=*Sporocybe* Fries, *Systema Orbis Vegetabilis*, Lundae p. 170, 1825

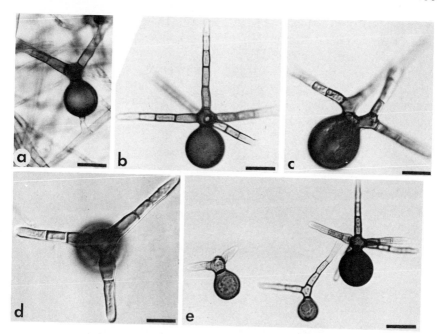

Fig. 114. *Orbimyces spectabilis* from pure culture (Brooks No. 94). (a) Conidium attached to conidiogenous cell on hypha. (b) Conidium with one crown of appendages. (c) Conidium with two crowns of appendages. (d) Conidial appendages in apical view. (e) Two immature conidia, a mature one to the right. (Bar in e = 20 µm, the others = 15 µm; c and d in Nomarski interference contrast, the others in bright field.)

> =*Sporodum* Corda, *Icon. Fung.* **1**, p. 18, 1837
> =*Trichocephalum* Costantin, *Les Mucédinées Simples,* Klincksieck, Paris, p. 106, 1888

LITERATURE: Barron (1968); M. B. Ellis (1971, 1976); Kendrick and Carmichael (1973); von Arx (1970).

STROMA often present, brown, pseudoparenchymatous. CONID-IOPHORES macronematous, occasionally micronematous, mononematous; macronematous ones stalked with a globose head, with or without branches, stipe straight or curved, brown or appearing black, smooth, rarely verrucose; the apex may be sterile and setiform. CONIDIOGENOUS CELLS monoblastic or polyblastic, discrete on the stipe and branches, determinate, ellipsoidal, globose or subglobose. CONIDIA globose, subglobose or rarely ellipsoidal, oblong or broad cylindrical, one-celled, verruculose, echinulate or rarely smooth, light to dark brown, catenate, often in branched chains, developing acropetally at one or several points on the surface of the conidiogenous cell. MODE OF LIFE: Saprobic or parasitic. Hosts: Mostly higher plants. Type Species: *Periconia lichenoides* Tode ex Mérat. Note: Description based on M. B. Ellis (1971); synonyms following M. B. Ellis (1971) and Kendrick and Carmichael (1973); there are 2 marine representatives among about 30 species. Etymology: From the Greek, *peri-* = about, around, surrounding, and *konia* = dust, ashes, probably in reference to the numerous dustlike conidia formed.

Periconia abyssa Kohlmeyer

Rev. Mycol. **41,** 202, 1977
Fig. 115

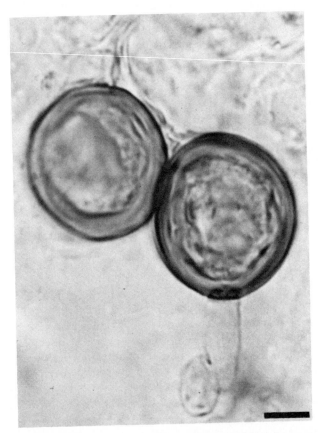

Fig. 115. Conidia of *Periconia abyssa* on conidiogenous cells, from wood in 5315 m (isotype IMS, Herb. J.K. 3811; bar = 5 μm, in bright field).

HYPHAE 3–6 μm in diameter, septate, light brown, superficial; some unbranched hyphae perpendicular to the surface may occur. CONIDIOGENOUS CELLS 5 μm in diameter, monoblastic, cylindrical to clavate, semimacronematous, hyaline to light brown. CONIDIA 16–20 μm in diameter, globose to subglobose, single or catenate, thin-walled, smooth, light brown, with pores between adjoining cells; often forming clumps on the surface of the substrate. MODE OF LIFE: Saprobic. SUBSTRATE: Submerged wood. RANGE: Atlantic Ocean—Gulf of Angola and Iberian deep sea (3975–5315 m). ETYMOLOGY: From the Greek, *abyssos* = the abyss or bottomless, in reference to the deep-sea occurrence of the species.

Periconia prolifica Anastasiou

Nova Hedwigia **6**, 260, 1963

CONIDIAL STATE of *Halosphaeria cucullata* (Kohlm.) Kohlm.
LITERATURE: Kohlmeyer (1969b,c); Kohlmeyer and Kohlmeyer (1964–1969, 1971a, 1977); Newell (1976); Raghukumar (1973); Shearer (1972b); Tubaki (1969); Tubaki and Ito (1973); Wagner-Merner (1972).
See description of *Periconia prolifica* under *Halosphaeria cucullata* (p. 290). ETYMOLOGY: From the Latin, *prolificus* = producing offsets, in reference to the abundant production of conidia on the conidiogenous cells.

Key to the Marine Species of *Periconia*
1. Conidia 16 μm in diameter or more; deep-sea species . . . . *P. abyssa*
1'. Conidia generally 13 μm in diameter or less; littoral species . . . . . . *P. prolifica*

Sporidesmium Link ex Fries

Link, *Mag. Ges. Naturf. Freunde, Berlin* **3**, 41, 1809; Fries, *Systema Mycologicum, Gryphiswaldiae* **1**, p. XL, 1821

=*Podoconis* Boedijn, *Bull. Jard. Bot. Buitenz.* Sér. 3 **13**, 133, 1933

LITERATURE: M. B. Ellis (1971, 1976); S. J. Hughes (1958); Moore (1958); von Arx (1970).
CONIDIOPHORES macronematous, mononematous, occasionally caespitose, simple, straight or curved, light to dark brown. CONIDIOGENOUS CELLS monoblastic, integrated, terminal, determinate or percurrent, cylindrical, doliiform or lageniform. CONIDIA acrogenous, solitary, cylindrical, fusiform, obclavate, obpyriform or obturbinate, occasionally rostrate, straight, curved or rarely sigmoid, transversely septate or pseudoseptate, smooth or verruculose, subhyaline, straw-colored, light to

dark brown, olivaceous or reddish brown. MODE OF LIFE: Saprobic. SUBSTRATES: Higher plants. TYPE SPECIES: *Sporidesmium atrum* Link. NOTE: Description based on M. B. Ellis (1971); there is 1 marine representative among about 70 terrestrial species. ETYMOLOGY: From the Greek, *sporos* = seed, spore, and *desma* = band, bandage, in reference to the elongate, bandlike conidia.

Sporidesmium salinum E. B. G. Jones

Trans. Br. Mycol. Soc. **46**, 135, 1963

LITERATURE: Byrne and Jones (1974); M. B. Ellis (1976); E. B. G. Jones (1968b, 1972); E. B. G. Jones and Irvine (1971); E. B. G. Jones and Jennings (1964, 1965); E. B. G. Jones *et al.* (1971); Kohlmeyer and Kohlmeyer (1964–1969).

HYPHAE 2.5–6 μm in diameter, septate, ramose, brown or dark brown, superficial or immersed. CONIDIOPHORES 40–61.5(–101.5) μm long, 4.5–7.5 μm in diameter, cylindrical, two- to six-septate, simple or rarely branched, straight or slightly curved, solitary, acrogenous or laterally on hyphae, brown. CONIDIA 120–296 × 11–34 μm, clavate or rarely cylindrical, two- to eight-septate, not or slightly constricted at the septa, straight or curved, smooth, brown; basal cells cylindrical, identical in diameter with the conidiophore; apical cells 61–257 × 11–34 μm, clavate; conidia sometimes with proliferating hyphae inside the apical cell. MODE OF LIFE: Saprobic. SUBSTRATE: Wood (test panels, e.g., *Fagus sylvatica* and *Pinus sylvestris*). RANGE: Atlantic Ocean—Great Britain (England, Scotland). NOTE: Description based on E. B. G. Jones (1963a) and on our examination of paratype material in Herb. B. ETYMOLOGY: From the Latin, *salinus* = saline, in reference to the marine habitat of the species.

Stemphylium Wallroth

Flora Cryptogamica Germaniae, Norimbergae Pars II p. 300, 1833

=*Epochniella* Saccardo, *Michelia* **2**, 127, 1880
=*Fusicladiopsis* Maire, *Bull. Soc. Bot. Fr.* **53**, 187, 1906
=*Soreymatosporium* Camara, *Acad. Sci. Lisboa, la Cl.* p. 18, 1930
=*Thyrodochium* Werdermann, *Ann. Mycol.* **22**, 188, 1924
=*Thyrospora* Tehon et Daniels, *Phytopathology* **15**, 718, 1925

LITERATURE: Carroll and Carroll (1973); Corlett (1973); M. B. Ellis (1971, 1976); Kendrick and Carmichael (1973); Simmons (1967, 1969); von Arx (1970); Wiltshire (1938).

STROMA occasionally present. CONIDIOPHORES macronematous,

mononematous, septate, nodose with dark apical and intercalary swellings, simple or sometimes branched, straight or curved, smooth or verruculose, light brown, brown or olivaceous, solitary or in groups. CoNIDIOGENOUS CELLS monoblastic, integrated, terminal, percurrent, initially clavate or subglobose, thin-walled at the apex, later becoming cuplike. CONIDIA oblong with rounded ends, ellipsoidal, obclavate or subglobose, some species with apical or lateral protrusions, muriform, with or without constrictions at the septa, smooth, verrucose or echinulate, cicatrized at the base, light brown, brown or olivaceous, solitary, acrogenous. MODE OF LIFE: Saprobic or parasitic. HOSTS: Mostly higher plants. TYPE SPECIES: *Stemphylium botryosum* Wallroth [*Stemphylium* state of *Pleospora herbarum* (Pers. ex Fr.) Rabenh.]. NOTE: Description based on M. B. Ellis (1971); synonyms following Kendrick and Carmichael (1973); M. B. Ellis (1971) and Simmons (1969) list nine species of *Stemphylium* of which *S. triglochinicola* occurs in a marine habitat; unidentified species of *Stemphylium* are frequently reported from the marine environment; *S. codii* Zeller and *S. maritimum* Johnson are dubious species and *nomina rejicienda* since no type material exists and the identity of the species cannot be established. ETYMOLOGY: Probably from the Greek, *staphyle* = grape, cluster of grapes, in reference to the ellipsoidal conidia.

Stemphylium triglochinicola Sutton et Pirozynski

Trans. Br. Mycol. Soc. **46**, 519, 1963

CONIDIAL STATE of *Pleospora triglochinicola* Webster.
LITERATURE: M. B. Ellis (1976); Kohlmeyer and Kohlmeyer (1964–1969); Simmons (1969); Webster (1969).
See description of *Stemphylium triglochinicola* under *Pleospora triglochinicola* (p. 443).

Stemphylium spp.

The numerous isolations of *Stemphylium* spp. from marine substrata do not prove that members of the genus are actually developing under natural marine conditions; marine isolates may be mere contaminants deriving from terrestrial habitats (see also *Alternaria* spp., p. 473).
LITERATURE: Curran (1975); Höhnk (1956); T. W. Johnson (1956b); T. W. Johnson and Sparrow (1961); E. B. G. Jones (1962a, 1963c, 1968b, 1972); E. B. G. Jones and Irvine (1971); E. B. G. Jones and Oliver (1964); E. B. G. Jones *et al.* (1972); Kohlmeyer (1971b); Pugh and Williams

(1968); Quinta (1968); Schaumann (1975a–c); Schmidt (1974); Siepmann and Johnson (1960); Steele (1967); Wagner-Merner (1972); Withers *et al.* (1975).

SUBSTRATES: Isolated from seawater, sediments, driftwood, intertidal wood, test panels (e.g., *Fagus sylvatica, Pinus sylvestris, Pinus* sp.), rope, *Salsola kali, Sargassum muticum, Spartina* sp.; conidia observed in sea foam. RANGE: Atlantic Ocean—Germany (F.R.G., Helgoland), Great Britain (England, Ireland, Wales), Portugal, United States (Florida, Maine, North Carolina); Baltic Sea—Germany (G.D.R.); Mediterranean—Italy; Pacific Ocean—United States (Hawaii).

Trichocladium Harz

Bull. Soc. Imp. Nat. Moscou **44**, 125–127, 1871

=*Culcitalna* Meyers et Moore, *Am. J. Bot.* **47**, 348, 1960

LITERATURE: Barron (1968); Dixon (1968); M. B. Ellis (1971); Hennebert (1971); S. J. Hughes (1952, 1969); Kendrick and Bhatt (1966); von Arx (1970).

CONIDIOPHORES mostly inconspicuous, poorly differentiated, forming short pedicels, zero- to three-septate, simple or rarely branched, straight or curved, smooth, bearing conidia at the apex, hyaline or light brown. CONIDIOGENOUS CELLS monoblastic or polyblastic, integrated, terminal or intercalary, determinate, cylindrical to doliiform. CONIDIA acrogenous or acropleurogenous, solitary, clavate, obovoid, pyriform or cylindrical, uni- to multiseptate, mostly thick-walled, smooth or verrucose, light to dark brown; cells may be unequally pigmented; germ pores occur in some or all cells. PHIALIDES may be formed in some pleomorphic species, producing small hyaline, one-celled conidia. MODE OF LIFE: Saprobic. TYPE SPECIES: *Trichocladium asperum* Harz. NOTE: Description based on Barron (1968), M. B. Ellis (1971), and Kendrick and Bhatt (1966); there are six terrestrial and one marine species; two additional *Trichocladium* species from marine habitats (Schmidt, 1974) are *nomina nuda*. ETYMOLOGY: From the Greek, *thrix* = hair and *klados* = branch, in reference to the branched, hairlike hyphae of the colonies.

Trichocladium achrasporum (Meyers et Moore) Dixon *in* Shearer et Crane, *Mycologia* **63**, 244, 1971

≡*Culcitalna achraspora* Meyers et Moore, *Am. J. Bot.* **47**, 349, 1960
=*Trichocladium achraspora* Dixon, *Trans. Br. Mycol. Soc.* **51**, 163, 1968
(*nomen nudum*)

CONIDIAL STATE of *Halosphaeria mediosetigera* Cribb et Cribb (see Shearer and Crane, 1977).

LITERATURE: Byrne and Eaton (1972); Byrne and Jones (1974); Chesters and Bull (1963a); Churchland and McClaren (1973); Cribb and Cribb (1969); Eaton and Dickinson (1976); M. B. Ellis (1976); S. E. J. Furtado *et al.* (1977); Henningsson (1974, 1976a,b); G. C. Hughes (1974); S. J. Hughes (1969); Jensen and Sguros (1971a,b); T. W. Johnson (1968); T. W. Johnson and Sparrow (1961); E. B. G. Jones (1963b, 1968b, 1972); E. B. G. Jones *et al.* (1972); Kirk and Catalfomo (1970); Kirk *et al.* (1974); Kohlmeyer (1963d, 1968c); Kohlmeyer and Kohlmeyer (1964–1969, 1971a); Meyers (1968a,b); Meyers and Reynolds (1959b, 1960b); Meyers *et al.* (1960, 1963a, 1964a); Peters *et al.* (1975); Pisano *et al.* (1964); Schaumann (1975c); Schmidt (1974); Sguros and Simms (1963a, 1964); Sguros *et al.* (1962, 1971, 1973); Shearer (1972b); Shearer and Crane (1977); Tubaki (1966, 1969); Tubaki and Ito (1973); Tyndall and Kirk (1973); Wagner-Merner (1972).

NOTE: See the description of *Trichocladium achrasporum* under the perfect state, *Halosphaeria mediosetigera* (p. 297); M. B. Ellis (1976) retains the species in the genus *Culcitalna,* probably because of the occurrence of putative "sporodochia" (Meyers and Moore, 1960); the formation of true sporodochia by *T. achrasporum* is uncertain, although dense accumulations of conidiophores may occur. ETYMOLOGY: From the Greek, *achras* = wild pear tree and *sporos* = seed, spore, in reference to the predominantly pyriform conidial shape.

Zalerion Moore et Meyers

Can. J. Microbiol. **8,** 408, 1962

LITERATURE: Anastasiou (1963a); M. B. Ellis (1976).

CONIDIOPHORES present or obsolete, cylindrical, septate, hyaline or grayish. CONIDIOGENOUS CELLS monoblastic, integrated, terminal, determinate. CONIDIA acrogenous, solitary, helicoid, septate, constricted at the septa, subhyaline to fuscous; diameter and pigmentation of conidial filament more or less equal throughout; conidia coiled regularly or irregularly in three dimensions, simple or rarely branched, often producing a ball or knot of cells. MODE OF LIFE: Saprobic. TYPE SPECIES: *Zalerion nepura* Moore et Meyers [=syn. of *Z. maritimum* (Linder) Anast.]. NOTE: There are two marine and one terrestrial species; it is doubtful if the fungus isolated from stem wounds of trees, *Z. arboricola* Buczacki (1972), is properly placed in *Zalerion; Z. arboricola* has no distinct conidiophores, conidia become verrucose in age, and it forms a second, *Hormiscium*-like conidial state, all characters differing from *Z. maritimum*, type species of the genus. ETYMOLOGY: From the Greek, *zale* = surge of the sea and *erion* = wool, in reference to the habitat of the fungus and the coiled conidia.

Zalerion maritimum (Linder) Anastasiou

Can. J. Bot. **41**, 1136, 1963 (as "*Z. maritima*")

≡*Helicoma maritimum* Linder *in* Barghoorn et Linder, *Farlowia* **1**, 405–406, 1944
=*Helicoma salinum* Linder *in* Barghoorn et Linder, *Farlowia* **1**, 406, 1944
=*Zalerion eistla* Moore et Meyers, *Can. J. Microbiol.* **8**, 413–414, 1962
=*Zalerion nepura* Moore et Meyers, *Can. J. Microbiol.* **8**, 413, 1962
=*Zalerion raptor* Moore et Meyers, *Can. J. Microbiol.* **8**, 415, 1962
=*Zalerion xylestrix* Moore et Meyers, *Can. J. Microbiol.* **8**, 414, 1962

LITERATURE: Anastasiou and Churchland (1969); Barghoorn (1944); Biernacka (1965); Block *et al.* (1973); Brooks *et al.* (1972); Byrne and Eaton (1972); Byrne and Jones (1974, 1975a,b); Catalfomo *et al.* (1972–1973); Churchland and McClaren (1972, 1973, 1976); Cole (1976); Curran (1975); Eaton (1972); Eaton and Dickinson (1976); M. B. Ellis (1976); S. E. J. Furtado *et al.* (1977); Gustafsson and Fries (1956); Henningsson (1974, 1976a,b); Höhnk (1955, 1956, 1967); G. C. Hughes (1968, 1969, 1974, 1975); G. C. Hughes and Chamut (1971); Irvine and Jones (1975); Irvine *et al.* (1972); T. W. Johnson (1956b, 1958b); T. W. Johnson and Sparrow (1961); E. B. G. Jones (1962a,b, 1963c, 1965, 1968b, 1971a, 1972, 1973, 1976); E. B. G. Jones and Irvine (1971); E. B. G. Jones and Jennings (1965); E. B. G. Jones and Le Campion-Alsumard (1970a,b); E. B. G. Jones and Oliver (1964); E. B. G. Jones *et al.* (1972); Kirk and Catalfomo (1970); Kirk *et al.* (1974); Koch (1974); Kohlmeyer (1958a,b, 1959, 1960, 1962a, 1963c,d, 1967, 1971b); Kohlmeyer *et al.* (1959); Kohlmeyer and Kohlmeyer (1964–1969); Meyers (1968a,b, 1971a); Meyers and Reynolds (1963); Meyers *et al.* (1963a,b, 1964a, 1972); Neish (1970); Peters *et al.* (1975); Pisano *et al.* (1964); Poole and Price (1972); Raghukumar (1973); Ritchie (1961); Ritchie and Jacobson (1963); Schaumann (1968, 1969, 1974b,c, 1975a–c); Schmidt (1974); Schneider (1976, 1977); Tubaki (1966, 1969); Tyndall and Kirk (1973); Wilson (1951, 1956a).

Fig. 116a

HYPHAE 2–4 μm in diameter, septate, branched, hyaline to fuscous. CONIDIOPHORES 10–115 μm long, 3–5.5 μm in diameter, micronematous or semimacronematous, slightly increasing in diameter from base to apex,

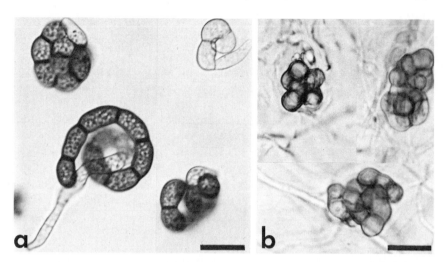

Fig. 116. Coiled conidia of *Zalerion* spp. (a) *Z. maritimum,* immature conidium in the upper right corner (Herb. J.K. 3315). (b) *Z. varium* (type culture). (Both bars = 15 μm; both in bright field.)

simple or rarely ramose, straight or curved, cylindrical, septate, rarely absent, superficial, light to dark fuscous. CONIDIA 19.5–50(–68) μm in diameter, acrogenous, solitary, blastic, coiled in two or three dimensions, forming 1–5(–6) more or less regular coils; conidial filament (3.5–)5.5–12 μm in diameter, 4–27(–35)-septate, slightly or strongly constricted at the septa, fuscous or almost black. MODE OF LIFE: Saprobic. SUBSTRATES: Intertidal and drifting wood and bark, test panels (e.g., *Araucaria* sp., *Betula pubescens, Cryptomeria japonica, Dicorynia paraensis, Fagus sylvatica, Laphira procera, Liriodendron tulipifera, Ocotea rodiaei, Pinus palustris, P. sylvestris, Pinus* sp., *Pseudotsuga douglasii, P. menziesii, Quercus* sp.), submerged leaves, rope, cork, polyurethane foam and wood in cooling towers. RANGE: Atlantic Ocean—Canada [New Brunswick (J. Kohlmeyer, unpublished), Newfoundland, Nova Scotia], Chile, Denmark, France, Germany (Helgoland, Wangerooge, and mainland, F.R.G.), Great Britain (England, Ireland, Scotland, Wales), Norway, Portugal, Puerto Rico, Senegal, Spain (Canary Islands), Sweden, United States [Connecticut, Florida, Maine, Massachusetts, New Jersey (J. Kohlmeyer, unpublished), North Carolina, Rhode Island, Texas]; Baltic Sea—Denmark, Germany (F.R.G. and G.D.R.), Poland, Sweden; Black Sea—Rumania; Indian Ocean—India (Madras); Mediterranean—France, Italy (Linosa and mainland); Pacific Ocean—Australia (New South Wales), Canada (British Columbia), Japan, New Zealand, United States (California, Washington). NOTE: We agree with Anastasiou

(1963a), who considered *Z. eistla, Z. nepura, Z. raptor,* and *Z. xylestrix* to be synonyms of *Z. maritimum;* quantitative physiological characters [as applied by Moore and Meyers (1962)] should not be used to differentiate species; morphological characters are more stable than physiological ones because they are based on complex genetic material, whereas a single physiological feature can change easily by mutation; records of *Z. maritimum* isolated from soil of the Sonoran Desert (Ranzoni, 1968) could not be confirmed because these cultures have not become available for comparison. ETYMOLOGY: From the Latin, *maritimus* = marine, in reference to the habitat of the species.

Zalerion varium Anastasiou

Can. J. Bot. **41**, 1136, 1963 (as "*Z. varia*")

LITERATURE: Churchland and McClaren (1973); Fell *et al.* (1975); Henningsson (1974); G. C. Hughes (1968, 1969); G. C. Hughes and Chamut (1971); Kohlmeyer (1967); Kohlmeyer and Kohlmeyer (1971a); Neish (1970); Newell (1976); Raghukumar (1973); Schaumann (1969, 1975a–c); Schmidt (1974); Shearer (1972b); Tubaki (1968, 1969).

Fig. 116b

HYPHAE septate, branched, immersed, hyaline. CONIDIOPHORES up to 30 μm long, 2–3.5 μm in diameter, micronematous, simple, cylindrical, septate, sometimes absent, superficial, hyaline to light olive-colored. CONIDIA 15–65 × 13.5–56 μm, solitary, irregularly helicoid or coiled in three planes, forming a knot or ball of about 10 to 30 cells; conidial filament lateral, rarely branched or subtending an additional conidium; thick-walled, smooth, brown to dark brown, appearing black in mass; cells 5–13 × 4–10.5 μm; additional complex conidia (or chlamydospores?) composed of up to several hundred cells may be formed in the substrate. MODE OF LIFE: Saprobic. SUBSTRATES: Intertidal and drifting wood and coconut, test panels (e.g., balsa wood), pilings (e.g., *Cryptomeria japonica*), submerged leaves and seedlings of *Rhizophora mangle*. RANGE: Atlantic Ocean—Canada (Newfoundland, Nova Scotia), Gambia, Germany (mainland and Helgoland, F.R.G.), Spain (Canary Islands), Sweden, United States (Florida, Maryland); Baltic Sea—Germany (G.D.R.), Sweden; Indian Ocean—India (Kerala, Madras); Pacific Ocean—Canada (British Columbia), Chile, Guatemala, Japan, Pago Pago. NOTE: Description based on Anastasiou (1963a) and on our examination of the type culture kindly made available to us by Dr. Anastasiou; the species was also collected in inland salt lakes, the Salton Sea (California) on submerged wood of *Tamarix aphylla*, and the Salt Lake (Oahu, Hawaii) on

Prosopis sp. (Anastasiou, 1963a); while *Z. maritimum* is quite distinctive because of its regular coils, *Z. varium* is not so easily identified and may be confused with other known (*Cirrenalia* spp.) or still undescribed species. ETYMOLOGY: From the Latin, *varius* = diversified, various, in reference to the irregularly shaped conidia.

Key to the Marine Species of *Zalerion*
1. Conidial filament turns to produce a terminal, regular spiral; no additional complex conidia formed in the substrate . . . . **Z. maritimum**
1'. Conidial filament turns to a lateral, variable spiral; additional complex conidia composed of up to several hundred cells formed in the substrate . . . . **Z. varium**

3. Tuberculariales
a. Tuberculariaceae

Allescheriella P. Hennings

Hedwigia **36**, 244, 1897

LITERATURE: M. B. Ellis (1971); S. J. Hughes (1951).

HYPHAE interwoven, anastomosing, superficial or immersed, forming cushions. CONIDIOPHORES semimacronematous, mononematous, branched, straight or curved, hyaline to light reddish-brown, smooth. CONIDIOGENOUS CELLS monoblastic, integrated, terminal on stipe and branches or discrete, determinate, cylindrical. CONIDIA acrogenous, solitary, globose, subglobose, obovoid or ellipsoidal, one-celled, thick-walled, smooth, reddish-brown. MODE OF LIFE: Saprobic. SUBSTRATE: Wood. TYPE SPECIES: *Allescheriella crocea* (Mont.) S. J. Hughes. NOTE: Description based on M. B. Ellis (1971); there is one terrestrial and one marine representative. ETYMOLOGY: Named in honor of the German mycologist A. Allescher (1828–1903).

Allescheriella bathygena Kohlm.

Rev. Mycol. **41**, 199–201, 1977

Fig. 117

HYPHAE 2–4 μm in diameter, septate, brown, immersed, forming sporodochia. STROMA 40–50 μm high, 80–130 μm in diameter, pulvinate, partly immersed, pseudoparenchymatous, composed of polygonal, thin-walled cells with large lumina, forming a textura angularis, producing conidia on the surface. CONIDIOGENOUS CELLS about 20 × 4 μm, monoblastic, terminal, determinate, cylindrical, straight, hyaline to light brown,

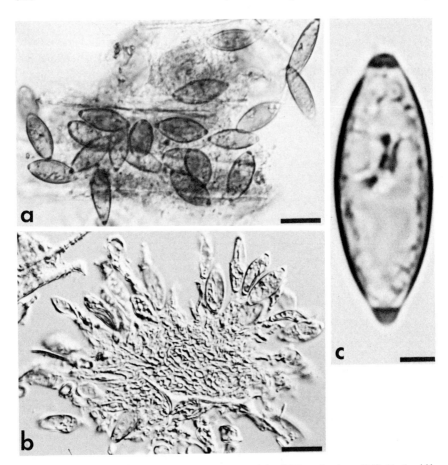

Fig. 117. *Allescheriella bathygena* from wood in 1722 m (isotype IMS, Herb. J.K. 2469). (a) Conidia on the wood surface (bar = 15 μm). (b) 4-μm section through sporodochium bearing conidia (bar = 15 μm). (c) Conidium with dark apices (bar = 3 μm). From Kohlmeyer, *Rev. Mycol.* **41,** 200 (1977), Fig. 25. Reprinted with permission. (a and c in bright field, b in Nomarski interference contrast.)

arising from the upper cells of the stroma. CONIDIA (16.5–) 17–24(–26) × 6.5–9 μm, acrogenous, solitary, ellipsoidal to fusiform with rounded ends, one-celled, thick-walled (especially at the apices), smooth, light brown. MODE OF LIFE: Saprobic. SUBSTRATE: Submerged wood. RANGE: Atlantic Ocean—Deep sea (Tongue of the Ocean near Bahama Islands, 1722 m; known only from the original collection). ETYMOLOGY: From the Greek, *bathys* = deep, deep sea, and the suffix *-genes* = -born, in reference to the origin of the species.

Epicoccum Link ex Schlechtendahl

Link, *Mag. Ges. Naturf. Freunde, Berlin* **7**, 32, 1816; Schlechtendahl,
Synopsis Plantarum Cryptogamarum. . . , Berlin, p. 136, 1824

LITERATURE: M. B. Ellis (1971); Schol-Schwarz (1959); von Arx (1970).

SPORODOCHIA solitary or gregarious, cushionlike, brown to black, superficial. STROMATA globose or hemiglobose. Conidiophores macronematous or semimacronematous, mononematous, forming a dense layer on the surface of the stroma, septate, simple or sometimes branched, short, straight or curved, smooth or verrucose, hyaline to light brown. CONIDIOGENOUS CELLS monoblastic, integrated, terminal, determinate, cylindrical. CONIDIA subglobose to pyriform, muriform, with a basal scar surrounded by a light-colored, protuberant basal cell, verrucose, dark golden brown, solitary, acrogenous. MODE OF LIFE: Saprobic or weakly parasitic. TYPE SPECIES: *Epicoccum purpurascens* Ehrenb. ex Schlecht. (=*E. nigrum* Link). NOTE: Description after M. B. Ellis (1971) and Schol-Schwarz (1959); although species of *Epicoccum* can probably not be considered marine we have included the genus because its representatives are sometimes found on substrates of marine origin, for example, algae washed up on the beach (Sutherland, 1916a) or decaying parts of *Spartina* spp. (Gessner and Kohlmeyer, 1976); incubated substrates from marine habitats often yield *E. purpurascens* probably deriving from terrestrial sources; *E. maritimum* Sutherland is a synonym of *E. purpurascens* (Schol-Schwarz, 1959). ETYMOLOGY: From the Greek prefix, *epi-* = upon, on, and *kokkos* = berry, seed, in reference to the conidia formed on the surface of sporodochia.

Epicoccum spp.

LITERATURE: Gessner and Kohlmeyer (1976); Nicot (1958c); Roth *et al.* (1964); Schol-Schwarz (1959); Siepmann and Johnson (1960); Steele (1967); Sutherland (1916a); Wagner-Merner (1972).

SUBSTRATES: Directly observed on washed-up *Laminaria* sp., also on standing senescent *Spartina alterniflora;* isolated with incubation techniques from seawater, beach sand, and wooden panels; conidia observed in sea foam. RANGE: Atlantic Ocean—Near Bahama Islands, France, Great Britain (England, Orkney Islands), United States (Connecticut, Florida, North Carolina, Virginia); Pacific Ocean—United States (Hawaii).

Tubercularia Tode ex Persoon

Synopsis Methodica Fungorum, Gottingae p. 111, 1801

=*Knyaria* O. Kuntze, *Revisio Generum Plantarum,* Leipzig **2,** pp. 855–856,
   1891
=*Stromateria* Corda, *Icon. Fung.* **1,** p. 5, 1837

LITERATURE: Barron (1968); Kendrick and Carmichael (1973); von Arx
(1970).
SPORODOCHIA cushion-shaped, sessile or short-stalked, erumpent,
smooth or wrinkled, occasionally with marginal hairs, light- or orange-
colored. CONIDIOPHORES forming a compact palisade on the surface of
the sporodochium, elongate, straight, simple or rarely branched, septate,
hyaline. CONIDIOGENOUS CELLS phialidic. CONIDIA ovate, elongate-
cylindrical, rarely globose or naviculate, one-celled, hyaline. MODE OF
LIFE: Saprobic or parasitic. SUBSTRATES: Higher plants, mostly on wood.
TYPE SPECIES: *Tubercularia vulgaris* Tode. NOTE: Description mostly
after Barron (1968); synonyms after Kendrick and Carmichael (1973);
there is 1 marine representative among about 25 terrestrial species.
ETYMOLOGY: From the Latin, *tuberculum* = tubercle, and the suffix
-arius, indicating connection, in reference to the tuberclelike
sporodochia.

Tubercularia pulverulenta Spegazzini

An. Soc. Cient. Argent. **13,** 32–33, 1882

≡*Knyaria pulverulenta* (Spegazzini) O. Kuntze, *Revisio Generum Plantarum,* Leipzig **2,** p. 856. 1891

LITERATURE: Farr (1973).

Fig. 118

SPORODOCHIA 240–400 μm high, 100–400 μm in diameter, cushion-
shaped, sessile or short-stalked, composed of a stromatic tissue that is
restricted to the base of the fructification, forming one to three consecu-
tive layers of conidiophores on the upper surface, superficial or erumpent,
sometimes with the base embedded in decaying fruiting bodies (pycnidia
or ascomata?), fleshy, cream- to orange-colored, solitary, gregarious or
confluent. CONIDIOPHORES 35–40(–80) × 1 μm, filiform, simple, forming
conidia at the apex, hyaline; probably proliferating and forming a second
and third layer on top of the first one. CONIDIA 3–4 × 1.5–2 μm, one-
celled, ellipsoidal, hyaline to light pink in masses, smooth-walled. MODE
OF LIFE: Saprobic (or hyperparasitic?). HOSTS: *Salicornia ambigua, S.
herbacea,* and *S. peruviana.* RANGE: Atlantic Ocean—Argentina (Buenos

Aires), Germany (F.R.G., Island of Sylt). NOTE: Description based on Spegazzini's diagnosis, on our own collection from the Province of Buenos Aires (Herb. J.K. No. 3514, on *S. ambigua*), and on material of Sydow, *Mycotheca germanica* No. 1129 (with *Diplodina salicorniae* Jaap on *S. herbacea;* Herb. B); we were unable to find *T. pulverulenta* in the type material (Herb. LPS, No. 26.804); the capsule contains sketches of this species and inscriptions in Spegazzini's handwriting. ETYMOLOGY: From the Latin, *pulverulentus* = dusty, covered with powder, in reference to the sporodochia covered with conidia.

B. Coelomycetes
1. Sphaeropsidales
a. Sphaerioidaceae

Ascochyta Libert

Plantae cryptogamicae, quas in Arduenna collegit, Leodii, Fasc. **1,**
8, 1830; emend. Saccardo, *Michelia* **1,** 161, 1878

=*Ascoxyta* Lib., *Mém. Soc. Sci. Lille 1829–1830* p. 175, 1831
=*Ascochyta* Lib. sect *Stagonosporoides* Zherbele, *Tr. Vses. Nauchno-Issled. Inst. Zashch. Rast.* **29,** 20, 1971

LITERATURE: Boerema and Bollen (1975); Boerema and Dorenbosch (1973); Brewer and Boerema (1965); Dickinson and Morgan-Jones (1966); Grove (1935); Melnik (1977); Sutton (1973b, 1977); von Arx (1970); Wollenweber and Hochapfel (1936).

PYCNIDIA separate, subglobose to lens-shaped, unilocular, immersed, ostiolate, brown, glabrous or sometimes hairy. CONIDIOGENOUS CELLS holoblastic, annellidic, similar to the inner cells of the peridium, ampulliform or doliiform, seldom short cylindrical. CONIDIA obovoid, becoming two- or more-celled by invaginations of an inner wall layer, smooth-walled, pale brown, without appendages. MODE OF LIFE: Saprobic or parasitic. SUBSTRATES: Higher plants. TYPE SPECIES: *Ascochyta pisi* Libert. NOTE: Description after Boerema and Bollen (1975), Grove (1935), and Sutton (1973b); synonyms after Sutton (1977); among about 30 species (von Arx, 1970) there is *A. salicorniae* described from a salt-marsh plant, *Salicornia;* the identity of others from *Atriplex* and *Halimione* (e.g., *Ascochyta chenopodii* Rostr.) has to be clarified (Grove, 1935). ETYMOLOGY: From the Greek, *askos* = skin (e.g., a bladder) and *chytos* (from *cheo*) = poured out, effused, in reference to the membranaceous pycnidium effusing conidia.

Fig. 118.   Sporodochium (16-μm longitudinal section) of *Tubercularia pulverulenta* from *Salicornia ambigua* (Herb. J.K. 3514; bar = 50 μm; in bright field).

Ascochyta salicorniae P. Magnus *in* Jaap

Schr. Naturwiss. Ver. Schleswig-Holstein **12**, 345, 1902 (non
Ascochyta salicorniae Trotter 1904, later homonym)

≡*Stagonosporopsis salicorniae* (P. Magnus) Diedicke, *Ann. Mycol.* **10**,
141–142, 1912
=*Ascochyta salicorniae* var. *salicorniae-patulae* Trotter, *Ann. Mycol.* **3**,
30, 1905
≡*Ascochyta salicorniae-patulae* (Trotter) Melnik, *Nov. Sist. Niz. Rast.* p.
205, 1975
≡*Ascochyta salicorniae* Trotter, *Ann. Mycol.* **2**, 536, 1904
=*Diplodina salicorniae* Jaap, *Verh. Bot. Ver. Prov. Brandenburg* **49**, 16,
1907

LITERATURE: Grove (1935); Jaap (1907b); Melnik (1977).
Figs. 119a–119d

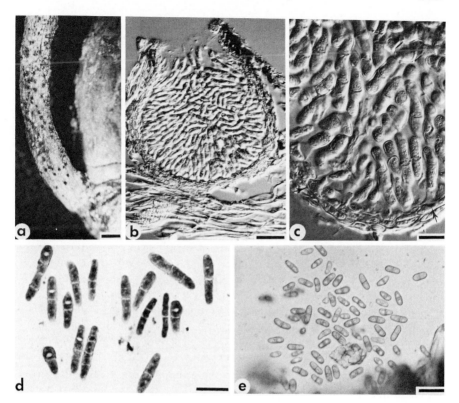

Fig. 119. Marine Coelomycetes. (a) Habit of *Ascochyta salicorniae* on *Salicornia herbacea* (type HBG; bar = 1 mm). (b) Same as a, longitudinal section (8 μm) through pycnidium, mucilage in the ostiole (bar = 25 μm). (c) Same as a, 6-μm section through base of pycnidium with peridium and conidiogenous cells (arrow) (bar = 10 μm). (d) Same as a, one-, two-, and three-septate conidia, stained in violamin (bar = 10 μm). (e) Conidia of *Ascochytula obiones* from *Halimione portulacoides* (type BPI; bar = 15 μm). (a = close-up photograph, b and c in Nomarski interference contrast, d and e in bright field.)

PYCNIDIA 90–180(–230) μm high, 90–210(–275) μm in diameter, sub-globose, ellipsoidal or pyriform, immersed (subepidermal), ostiolate, short papillate or epapillate, coriaceous, thin-walled, brown to black, darkest and thick-walled around the ostiole, solitary and scattered, usu-ally in the lower part of the stem. PERIDIUM 8–12(–34) μm thick, usually composed of two layers of elongate, lenticular, thick-walled cells, dark on the outside, light-colored near the venter, forming a textura angularis. PAPILLAE 25–30 μm high, 45–50 μm in diameter, conical or cylindrical, centric or eccentric, projecting above the surface of the host; ostiole 20–30 μm in diameter; ostiolar canal filled with mucilage and some small hyaline, periphysoid cells broken off the wall. CONIDIOGENOUS CELLS phialidic, flask-shaped, hyaline, originating along the inner wall of the peridium. CONIDIA 10–19(–20) × 4–7 μm, cylindrical with rounded ends to ellipsoidal, one-septate in the middle, rarely with an additional septum in the lower or upper half, very seldom three-septate, not or slightly constricted at the septa, smooth-walled, hyaline to yellowish or light brown, enclosed in an indistinct, irregular, mucilaginous sheath that becomes visible in India ink. MODE OF LIFE: Saprobic (or perthophytic?). HOSTS: Stems or inflorescences of moribund or dead *Salicornia europaea, S. herbacea,* and *S. patula.* RANGE: Atlantic Ocean—Canada [New Brunswick (J. Kohlmeyer, unpublished)], Denmark (Rømø), Germany [Helgoland (J. Kohlmeyer, unpublished) and Sylt, F.R.G.], Great Britain (Wales), United States (Connecticut). NOTE: We examined the holotype material (Herb. HBG), the dimensions of which overlapped those of *A. salicorniae* var. *salicorniae-patulae* and *Diplodina salicorniae;* a mucilaginous sheath around the conidia had not been recorded before for *A. salicorniae;* other specimens examined were the following: Ellis and Everhart, North Amer. Fungi No. 2759 together with *Sphaerella salicorniae* Auersw. (BPI); G. Winter, together with lectotype of *S. salicorniae,* Germany (B); Sydow, Mycotheca Germanica No. 1129, sub *Diplodina salicorniae* (B); collections J. K. 3173 and 3342 (IMS). The fungus on *Salicornia* from inland saline habitats in Germany is identical with marine collections. ETYMOLOGY: Named after the host genus, *Salicornia* (Chenopodiaceae).

Ascochytula (Potebnia) Diedicke

Ann. Mycol. **10,** 141, 1912

≡[*Ascochyta* Lib.] subgen. *Ascochytula* Potebnia, *Ann. Mycol.* **5,** 10, 1907

LITERATURE: Dickinson and Morgan-Jones (1966); Grove (1935); Petrak (1925); Sutton (1977).

Description as in *Ascochyta,* but peridium composed of an outer layer with thick-walled dark cells and an inner layer with small hyaline cells (Dickinson and Morgan-Jones, 1966); conidia light brown (Diedicke, 1912; Grove, 1935). TYPE SPECIES: *Ascochytula obiones* (Jaap) Diedicke. NOTE: Sutton (1977) regards *Ascochytula* as a later synonym of *Pseudodiplodia* (Karst.) Sacc.; we leave the only marine species, *A. obiones,* in *Ascochytula* until its proper generic disposition can be established. ETYMOLOGY: From *Ascochyta* (Sphaeropsidales) and the Latin diminutive ending *-ula,* indicating a relationship between the two genera.

Ascochytula obiones (Jaap) Diedicke

Kryptogamenflora Mark Brandenburg **9**(5), p. 410, 1912 (as *A. obionis*)

≡*Diplodina obionis* Jaap, *Verh. Bot. Ver. Prov. Brandenburg* **47**, 96, 1905

LITERATURE: Dickinson (1965); Dickinson and Morgan-Jones (1966); Dickinson and Pugh (1965b); Diedicke (1912); Grove (1935); Jaap (1907b); Migula (1921).

Fig. 119e

PYCNIDIA 150–200 μm in diameter, ovate, immersed, finally erumpent, ostiolate, papillate or epapillate, thick-walled, olive-brown, on leaves and propagules densely scattered, on stems gregarious or in irregular rows. PERIDIUM composed of many layers of cells; cells of outer stratum pseudoparenchymatous, dark and thick-walled; cells of inner stratum small, hyaline, bearing conidiophores; forming a textura angularis. PAPILLAE short or elongate. CONIDIOGENOUS CELLS 6–7 μm long, 3–4.5 μm in diameter, phialidic, pyriform, short-necked, hyaline, originating along the inner wall of the peridium. CONIDIA 9–11.5 × 3.5–5 μm (in culture 8.5–9 × 3–4 μm), blastic, ellipsoidal to obovate, with an obtuse apex, at first truncate at the base, one-septate, slightly constricted at the septum, upper cell larger than the lower one, smooth-walled, pale yellowish to brownish. CHLAMYDOSPORES thick-walled, one-septate, formed at the tips of vegetative hyphae in culture. MODE OF LIFE: Parasitic and saprobic. HOSTS: Roots, stems, live and moribund leaves of *Halimione portulacoides.* RANGE: Atlantic Ocean—Germany (Amrum, F.R.G.), Great Britain (England); Mediterranean—Yugoslavia [Rab (J. Kohlmeyer, unpublished)]. NOTE: Description based on Dickinson and Morgan-Jones (1966), on examination of type material (BPI), and on our own collection J.K. 2903. ETYMOLOGY: Named after *Obione* (Chenopodiaceae), to which the host, *H. portulacoides,* was formerly assigned.

Camarosporium Schulzer von Müggenburg

Verh. Zool.-Bot. Ges. Wien **20**, 641, 1870

=*Camarosporulum* Tassi, *Bull. Lab. Ort. Bot. Siena* **5**, 63, 1902
= ?*Hyalothyridium* Tassi, *Bull. Lab. Ort. Bot. Siena* **3**, 91, 1900

=?*Hyalothyris* Clements, *Genera of Fungi*, Minneapolis, p. 127, 1909
=*Pseudodichomera* v. Höhn., *Hedwigia* **60**, 186, 1918

LITERATURE: Grove (1937); Sutton (1973b, 1977).

PYCNIDIA solitary or gregarious, subglobose or ellipsoidal, unilocular, immersed or erumpent, ostiolate, papillate or epapillate, thin- or thick-walled, dark brown. CONIDIOGENOUS CELLS holoblastic, annellated, cylindrical. CONIDIA of irregular shapes, muriform, thick-walled, smooth, dark brown. MODE OF LIFE: Saprobic or parasitic. SUBSTRATES: Higher plants. TYPE SPECIES: *Camarosporium quaternatum* (Hazsl.) Schulzer. NOTE: Description after Grove (1937) and Sutton (1973b); synonyms after Sutton (1977); among about 100 terrestrial species there are 3 that have been described from plants growing along the seashore or in coastal salt marshes. Ahmad (1967), Hansford (1954), Koshkelova and Frolov (1973), and Koshkelova *et al.* (1970) described several species of *Camarosporium* from halophytes of inland habitats. ETYMOLOGY: From the Greek, *camara* = vaulted chamber, arch, and *sporos* = seed, spore, in reference to the chambered conidia.

Camarosporium metableticum Trail

Scott. Nat. **8** (New Ser. 2), 267, 1886

≡*Camarographium metableticum* (Trail) Grove, *British Stem- and Leaf-Fungi*, Cambridge Univ. Press, Vol. 2, p. 108, 1937
=*Camarosporium graminicolum* Ellis et Everhart, *Proc. Acad. Nat. Sci. Philadelphia*, p. 161, 1893

LITERATURE: Diedicke (1915); Jaap (1907b); Kohlmeyer and Kohlmeyer (1964–1969); Schmidt (1974).

PYCNIDIA 150–215 µm high, 210–330 µm in diameter, subglobose, immersed, ostiolate, papillate, slightly clypeate, coriaceous, black, solitary. PERIDIUM 12.5–20 µm thick, composed of three or four layers, rarely up to seven layers, of rounded or irregularly ellipsoidal cells with large luminia, forming a textura angularis. PAPILLAE 40–80 µm high, 60–72 µm in diameter, surrounded by a more or less distinct clypeus; ostiolar canal at first filled with a tissue of small cells with thick walls that gelatinize. CONIDIOGENOUS CELLS 5–10 µm long, 2.5 µm in diameter, cylindrical, originating along the inner wall of the pycnidium, almost up to the ostiolar canal, forming conidia singly. CONIDIA (20–)27.5–39 × 11–15.5 µm (excluding apical appendages), ellipsoidal or trapezoidal, muriform, with (3–)5–7(–9) transverse septa, one to four of the central cells with one or rarely two longitudinal septa, without longitudinal septa in the apices, slightly constricted at the septa, fuscous or yellowish-brown, apical cells lighter colored; each apex covered by a caplike, subglobose, gelatinous, indistinctly striate appendage, 8.5–11 μm in diameter, which stains in hematoxylin. MODE OF LIFE: Saprobic. HOSTS: Dead culms and leaves of *Ammophila arenaria, A. baltica,* and *Uniola paniculata* (R. V. Gessner, unpublished; Herb. J. K. 3692); on bark and conifer cones according to Schmidt (1974). RANGE: Atlantic Ocean—Germany (Amrum, Sylt, and mainland, F.R.G.), Great Britain (England, Scotland), United States [New York, North Carolina (R. V. Gessner, unpublished)]; Baltic Sea—Germany (F.R.G. and G.D.R.). NOTE: We examined type material of Trail (one slide from Herb. ABD); the conidia (23–30 × 9–12.5 μm) in this dried-out preparation had shrunk and did not show the apical appendages; but these caps also disappear in slides made from fresh material; other specimens examined were the following (all from Herb. B on *A. arenaria*): three collections of A. Ludwig from northern Germany; Ellis and Everhart, North American Fungi (2nd ser.) No. 2945 from Long Island, New York; C. L. Shear, New York Fungi No. 379. Conidial ontogeny in *Camarosporium* supposedly is holoblastic and annellidic (Sutton, 1973b); we could not detect annellids in conidiogenous cells of *C. metableticum;* also, conidial appendages are not typical for representatives of *Camarosporium;* therefore, *C. metableticum* could probably be assigned to another genus, a step that we prefer not to take because the system of Deuteromycetes remains artificial until the perfect states of the species will be known; *C. metableticum* is not a marine fungus in the strict sense since it occurs on coastal dune plants normally not flooded by seawater; we have included this species because it may be confused with other representatives of the genus that grow on halophytes in salt marshes (e.g., *Atriplex, Halimione, Salicornia,* and *Suaeda*). ETYMOLOGY: From the Greek, *metabletikos* = metabletic, of the nature of barter (Webster's Unabridged Dictionary, 2nd ed.); possibly derived from *metaballo* = to change, vary, in reference to the variable conidia.

Camarosporium palliatum J. Kohlmeyer et E. Kohlmeyer sp. nov.

PYCNIDIA 120–190 μm altis (80–)100–190(–260) μm diametro, subglobosis ad pyriformibus, immersis, ostiolatis, epapillatis, coriaceis, dilute vel atro-brunneis ad nigris, solitariis vel gregariis. OSTIOLIS 10–12 μm diametro, primo cellulis hyalinis, ellipsoideis clausis. PERIDIIS 12–14 μm crassis, ca. 3 stratis cellularum, texturam angularem formatium. PARAPHYSIBUS 30–40 × 3–4 μm, simplicibus, rariter furcatis, filiformibus, hyalinis, nonseptatis. CELLULIS CONIDIOGENIS similis phialidibus, ampullaceis, simplicis, hyalinis, in pariete interno pycnidii undique insertis.

CONIDIIS 20–34 × 9–20 μm, enteroblasticis, monophialidicis, ellipsoideis vel oblongis apicibus obtusis, muriformibus, (3–)5(–6) septis transversis, 3–4 septis longitudinalibus vel obliquis, ad septa leviter constrictis, glabris, primo luteo-brunneis, deinde atro-brunneis, tegumentis mucosis tectis; tegumentis in aqua lente tumescentibus et diffluentibus. SUBSTRATA: plantae mortuae specierum generis *Salicornia*. HOLOTYPUS: On *Salicornia* sp., together with *Mycosphaerella salicorniae* and *Pleospora gaudefroyi*, Villa del Mar, southeast of Bahia Blanca, Prov. Buenos Aires, Argentina, 23 Oct. 1973, leg. J. and E. Kohlmeyer, Herb. J. K. No. 3520a (IMS). ISOTYPUS: Same collecting data, Herb J. K. No. 3520b (NY).

Figs. 120a–120d

PYCNIDIA 120–190 μm high, (80–)100–190(–260) μm in diameter, subglobose to pyriform, immersed, ostiolate, epapillate, coriaceous, light or dark brown to blackish, solitary or gregarious. OSTIOLES 10–12 μm in diameter, at first closed by hyaline, ellipsoidal cells, which are exuded together with the conidia. PERIDIUM 12–14 μm thick, composed of about three layers of thin-walled cells with large lumina, forming a textura angularis. PARAPHYSES 30–40 × 3–4 μm, simple, rarely forked, filiform, hyaline, nonseptate. CONIDIOGENOUS CELLS phialidic, flask-shaped, simple, hyaline, originating along the inner wall of the pycnidium. CONIDIA 20–34 × 9–20 μm (excluding the sheath), enteroblastic, monophialidic, ellipsoidal or oblong with rounded ends, muriform, with (3–)5(–6) transverse and three or four longitudinal or oblique septa, slightly constricted at the septa, smooth, at first yellowish-brown, becoming dark brown, surrounded by a gelatinous sheath, which slowly swells in water and eventually dissolves. MODE OF LIFE: Saprobic (or perthophytic?). HOSTS: Stems and leaves of dead *Salicornia ambigua*, *S. virginica*, *Salicornia* sp. RANGE: Atlantic Ocean—Argentina (Buenos Aires), United States (North Carolina). FURTHER MATERIAL EXAMINED: On *S. ambigua*, saltwater lagoon separated from the ocean, along Highway 229, southeast of Bahia Blanca, Argentina, 23. Oct. 1973, Herb. J. K. No. 3521; on *Salicornia* sp., oil-covered plants, same data as No. 3521, close to Puerto Ing. White, Herb. J.K. No. 3522; on *S. ambigua*, Salinas Las Barrancas, a salt lake west of Bahia Blanca, Argentina, 24. Oct. 1973, Herb. J.K. No. 3523, salinity of water near the plants 32–34.5°/₀₀; on *S. virginica*, Beaufort, North Carolina, 19 May 1966, Herb. J.K. Nos. 1994–1995. NOTE: This species from *Salicornia* spp. is characterized by conidia with usually five transverse septa and a gelatinous sheath; a similar fungus on *Suaeda fruticosa* from England (Herb. Ludwig ex Rhodes *in* B) has smaller conidia; another *Camarosporium* on dry old leaves of *Aster tripolium* (Yugoslavia, Herb. J.K. No. 2906) agrees with *C. palliatum*, except for the coffee-brown

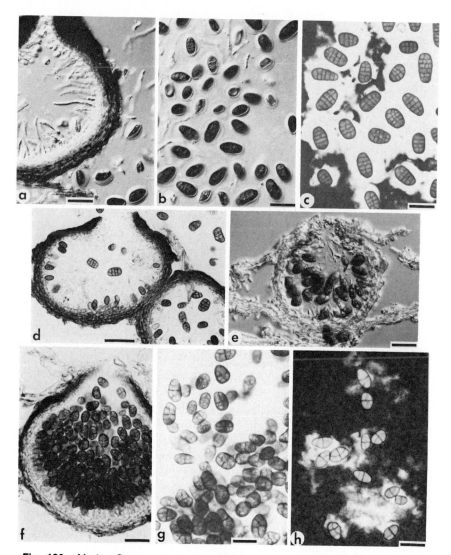

Fig. 120. Marine *Camarosporium* spp. (a) Longitudinal section (16 μm) through pycnidium of *C. palliatum* sp. nov. from *Salicornia* sp., paraphyses along the inner wall (holotype IMS, Herb. J.K. 3520a; bar = 25 μm). (b) Same as a, conidia with gelatinous sheaths (bar = 25 μm). (c) Same as a, conidia in India ink, showing gelatinous sheaths (bar = 25 μm). (d) Same as a, 16-μm section through pycnidia with immature conidia (bar = 50 μm). (e) Longitudinal section (10 μm) through pycnidium of *C. roumeguerii* from *Salicornia herbacea* (type PAD; bar = 20 μm). (f) 12-μm section through pycnidium of *C. roumeguerii* from *Halimione portulacoides* (type of *C. obionis* Jaap in BPI; bar = 25 μm). (g) Same as f, conidia (bar = 15 μm). (h) Sheathless conidia of *C. roumeguerii* in India ink, mucus deriving from pycnidial cavity (type PAD; bar = 20 μm). (a, b, and e in Nomarski interference contrast, the others in bright field.)

color of the conidia in the former. ETYMOLOGY: From the Latin, *palliatus* = clad in a pallium (a coverlet), in reference to the sheath-covered conidia.

Camarosporium roumeguerii Saccardo

Michelia **2**, 112–113, 1880

=*Camarosporium obiones* Jaap, *Verh. Bot. Ver. Prov. Brandenburg* **47**, 97, 1905 (as *C. obionis*)

LITERATURE: Dickinson and Morgan-Jones (1966); Grove (1937); Jaap (1907b); Mameli (1915); Spegazzini (1880); von Höhnel (1911).

Figs. 120e–120h

PYCNIDIA 90–210 μm high, 85–260 μm in diameter, globose to ellipsoidal or lenticular, immersed, ostiolate, short papillate, coriaceous, yellow–brown to olivaceous, darkest around the ostiole, solitary or gregarious; venter filled with a mucilage, especially in the ostiolar canal. PERIDIUM 7–12 μm thick at the base, 12–20 μm at the ostiole, composed of up to four layers of thin-walled cells with large lumina, dark on the outside, almost hyaline near the venter, forming a textura angularis. PARAPHYSES simple, filiform, hyaline, nonseptate. CONIDIOGENOUS CELLS phialidic, flask-shaped, one-celled, hyaline, originating along the inner wall of the pycnidium. CONIDIA 10–20(–22) × 7–13 μm; ontogeny enteroblastic, monophialidic; initially one-celled, hyaline, maturing in the center of the pycnidial venter; mature conidia subglobose, ovoid, ellipsoidal or irregular, muriform, with (one to) three transverse and one or two longitudinal or oblique septa, slightly constricted at the septa, composed of two to eight cells, smooth, gold-, ochraceous-, or olive-brown, without appendages or gelatinous sheaths (irregular floccules of mucus deriving from the pycnidial cavity may adhere to conidia). MODE OF LIFE: Saprobic (or perthophytic?). HOSTS: Stems and leaves of moribund or dead *Atriplex halimus*, *Atriplex* sp., *Halimione portulacoides*, *Salicornia herbacea*, *S. peruviana*, *S. virginica*, and *Suaeda fruticosa*. RANGE: Atlantic Ocean—Argentina (Chubut), Germany (Amrum, F.R.G.), Great Britain (England), United States [Massachusetts (Gessner and Lamore, 1978), North Carolina (J. Kohlmeyer, unpublished)]; Mediterranean—France, Italy (Sardinia); Pacific Ocean—Mexico [Sonora (J. Kohlmeyer, unpublished)]. NOTE: Description based on holotype material of *C. roumeguerii* (PAD), on isotype material of *C. obiones* (BPI), and on our own collections of *Atriplex* and *Suaeda* spp. from Argentina (J.K. No. 3602) and Mexico (J.K. Nos. 3570, 3571); paraphyses have not been reported before for this species, but they are present in all collections; "Sporenträger" (=conidiophores) and "sporophores" mentioned by Jaap

(1905) and Grove (1937), respectively, are in reality paraphyses; there are other *Camarosporium* species besides *C. roumeguerii* on shoreline plants, and the identification is sometimes difficult. ETYMOLOGY: Named in honor of C. Roumeguère (1828–1892), French mycologist.

Key to the Marine Species of *Camarosporium*

1. Conidia predominantly three-septate, usually not longer than 20 μm; without distinctive gelatinous sheaths or appendages . . . . *Camarosporium roumeguerii*
1′. Conidia with five or more transverse septa, 20 μm or longer; with distinctive gelatinous sheaths or appendages (use phase contrast or India ink) 2
    2(1′) Conidia with a gelatinous caplike appendage at each end; mostly on Gramineae in dunes . . . . *Camarosporium metableticum*
    2′(1′) Conidia completely surrounded by a gelatinous sheath; on salt-marsh *Salicornia* spp. *Camarosporium palliatum*

Coniothyrium Corda emend. Saccardo

Corda, *Icon. Fung.* **4**, p. 38, 1840; Saccardo, *Michelia* **2**, 7, 1880

=?*Asteropsis* G. Frag., *Trab. Mus. Cienc. Nat., Madrid, Ser. Bot.* **12**, 50, 1917
=*Clisosporium* Fr., nom. rej., *Syst. Mycol.* **1**, p. XLVII, 1821
=?*Coniothyrinula* Petrak, *Ann. Mycol.* **21**, 2, 1923
=?*Monoplodia* Westd., *Bull. Acad. R. Sci. Belg.,* Ser. 2 **7**, 91, 1859

LITERATURE: Biga *et al.* (1958); Grove (1937); Petrak and Sydow (1927); Sutton (1973b, 1977); von Arx (1970); Wollenweber and Hochapfel (1936).

PYCNIDIA separate, subglobose or depressed, unilocular, immersed, later erumpent or almost superficial, ostiolate, papillate, dark brown, membranaceous. CONIDIOGENOUS CELLS holoblastic, annellidic, doliiform, with one to three annellations, hyaline. CONIDIA cylindrical to oval, one-celled (or also one-septate?), thick-walled, pigmented. MODE OF LIFE: Saprobic or parasitic. SUBSTRATES: Higher plants. TYPE SPECIES: *Coniothyrium palmarum* Corda. NOTE: Description after Grove (1937) and Sutton (1973b); synonyms after Sutton (1977); authors cited above accept between 40 and 125 species in the genus; at least 1 species is known from coastal salt-marsh plants. ETYMOLOGY: From the Greek, *konia* = dust, powder (the conidia), and *thyreos* = large shield (the pycnidium), in reference to the conidia-producing pycnidia.

Coniothyrium obiones Jaap

Schr. Naturwiss. Ver. Schleswig-Holstein **14**, 29, 1907 (as *C. obionis*)

LITERATURE: Dickinson and Morgan-Jones (1966); Dickinson and Pugh (1965b); Grove (1937).

PYCNIDIA 140–190 μm high, 150–230 μm in diameter, subglobose, at first immersed, finally erumpent, ostiolate, short papillate, coriaceous, thin-walled, brown, gregarious. PERIDIUM 10–12 μm thick; on the outside composed of small, brown, thick-walled cells that merge toward the venter with hyaline, thin-walled, flattened cells; cells of inner stratum bearing conidiophores; forming a textura angularis. PAPILLAE about 30 μm in diameter. CONIDIOGENOUS CELLS up to 7 μm long, phialidic, flask-shaped, tapering at the apex, one-celled, hyaline. CONIDIA 4–8 μm long, 3.5–6 μm in diameter, ellipsoidal, ovoid or subglobose, one-celled, smooth- and thick-walled, olivaceous to light brown, originating in basipetal succession on the apex of the conidiogenous cell; hyaline when released, maturing in the pycnidial venter. MODE OF LIFE: Saprobic. HOSTS: Dead stems and moribund propagules of *Halimione por-tulacoides*. RANGE: Atlantic Ocean—Germany (Amrum, F.R.G.), Great Britain (England). NOTE: Description based on Dickinson and Morgan-Jones (1966) and on our examination of the fungus in Jaap "Fungi selecti exsiccati" No. 98; Grove's (1937) suggestion that *C. obiones* was merely an early state of *Ascochytula obiones* was refuted by Dickinson and Morgan-Jones (1966); it is uncertain if *C. atriplicis* Maublanc and *C. halymi* (Cast.) Sacc. described from *Atriplex halimus* are different from *C. obiones*. ETYMOLOGY: Named after *Obione* (Chenopodiaceae), to which the host, *H. portulacoides*, was formerly assigned.

Cytospora Ehrenb. ex Fries

Systema Mycologicum, Gryphiswaldiae **2**, p. 540, 1823

SYNONYMS: See Sutton (1977).
LITERATURE: Defago (1935); Sutton (1973b, 1977); von Arx (1970).
PYCNIDIA eustromatic, complex, immersed, not becoming erumpent; cavity convoluted, divided, with a single ostiole. PARAPHYSES absent. CONIDIOPHORES septate, branched. CONIDIOGENOUS CELLS enteroblas-tic, monophialidic. CONIDIA allantoid, one-celled, hyaline, discharged as cirri or beads. MODE OF LIFE: Parasitic or saprobic. SUBSTRATES: Higher plants. TYPE SPECIES: *Cytospora chrysosperma* (Pers.) Fries. NOTE: De-scription after Sutton (1973b), who lists (1977) 15 synonyms; among about 40 species (von Arx, 1970), mostly conidial states of *Valsa* spp. causing dieback, only *C. rhizophorae* is described from a marine habitat. ETYMOLOGY: From the Greek, *kytos* = cavity, receptacle, and *sporos* = seed, spore, in reference to the conidia-producing pycnidia.

Cytospora rhizophorae J. Kohlmeyer et E. Kohlmeyer

Mycologia **63**, 847–848, 1971

LITERATURE: Kohlmeyer [1969c (sub *Cytospora* sp.)]; Kohlmeyer and Kohlmeyer (1977).

Fig. 121

PYCNIDIA 0.5–0.8 mm high, 1.5–1.7 mm in diameter, of irregular shapes, composed of several locules, immersed under the host cortex, ostiolate, coriaceous, brown, solitary or gregarious, surrounded by thick, white hyphae. LOCULES 370–500 μm high, 100–310 μm in diameter, irregular ampulliform, opening into a common ostiole. PERIDIUM 10–25 μm thick. NECKS cylindrical, central, sometimes clypeate, breaking through the lenticels or the bark of the host; ostiole about 30 μm in

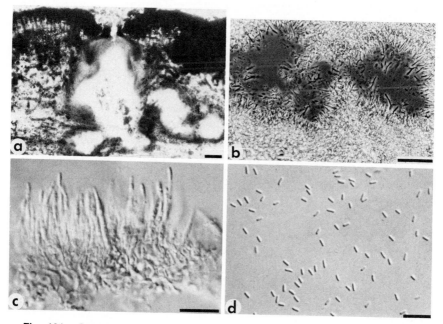

Fig. 121. *Cytospora rhizophorae* from *Rhizophora mangle.* (a) Longitudinal section (48 μm) through pycnidium composed of several locules (bar = 100 μm). (b) Section (4 μm) through locules with conidiophores lining the walls (bar = 50 μm). (c) Conidiophores (bar = 10 μm). (d) Conidia (bar = 15 μm). From Kohlmeyer and Kohlmeyer, *Mycologia* **63**, 849 (1971), Fig. 34. Reprinted with permission. (c is Herb. J.K. 2402, the others are holotype NY, Herb. J.K. 2693; a in bright field, b in phase contrast, c and d in Nomarski interference contrast.)

diameter; canal formed schizogenously. CONIDIOPHORES 13–20 μm long, 1–1.8 μm in diameter, cylindrical, filiform, attenuate, simple or rarely branched, septate, producing conidia at the apex, covering the inner wall of the pycnidia. CONIDIA 3–6 × 1.1–1.5 μm, allantoid or ellipsoid–cylindrical, straight or slightly curved, one-celled, hyaline or light yellowish, discharged as cirri or beads. MODE OF LIFE: Parasitic or perthophytic. HOSTS: Roots and seedlings of *Rhizophora mangle* and *R. racemosa*, causing dieback of young plants. RANGE: Atlantic Ocean—Bahamas (Abaco), Bermuda, Liberia, Mexico (Veracruz), United States (Florida); Pacific Ocean—Guatemala, Papua New Guinea (D. E. Shaw, personal communication), United States [Hawaii (Oahu)]. NOTE: The species is halotolerant and facultative marine. ETYMOLOGY: Named after the host genus, *Rhizophora* (Rhizophoraceae).

Diplodia Fries apud Mont.

Ann. Sci. Nat., Bot. Biol. Vég., Sér. 2 **1**, 302, 1834

= ?*Holcomyces* Lindau, *Verh. Bot. Ver. Prov. Brandenburg* **45**, 155, 1903

LITERATURE: Grove (1937); Sutton (1977); von Arx (1970); R. K. Webster *et al.* (1974); Zambettakis (1954).

PYCNIDIA solitary or gregarious, stromatic or nonstromatic, subglobose, at first immersed, then erumpent, ostiolate, mostly papillate, dark, thick-walled, glabrous or hirsute. PARAPHYSES present or absent. CONIDIOPHORES mostly short cylindrical, straight, obtuse, hyaline, lining the cavity of the pycnidium. CONIDIA mostly over 15 μm long, one-septate, sometimes constricted at the septum, at first hyaline, becoming brown, smooth, striated or granulose; length-to-width ratios 1.5:1 to 6.5:1. MODE OF LIFE: Saprobic and parasitic. SUBSTRATES: Higher plants. TYPE SPECIES: *Diplodia mutila* Fries. NOTE: Description after von Arx (1970), Grove (1937), and R. K. Webster *et al.* (1974); the conidial ontogeny of the type species is not known (Sutton, 1973b); there is 1 marine representative among about 30 species of the genus. ETYMOLOGY: From the Greek, *diploos* = twofold, twin, in reference to the two-celled conidia.

Diplodia oraemaris Linder *in* Barghoorn et Linder

Farlowia **1**, 403, 1944

LITERATURE: Gustafsson and Fries (1956); T. W. Johnson (1956b); T.

W. Johnson and Sparrow (1961); Kohlmeyer (1960); Kohlmeyer and Kohlmeyer (1964–1969).

PYCNIDIA 135–365 μm high, 165–470 μm in diameter, ellipsoidal, immersed or partly immersed, ostiolate, short papillate or epapillate, coriaceous or subcarbonaceous, brownish-black above, subhyaline below, solitary or gregarious. PERIDIUM composed of about six to ten layers of polygonal cells, forming a textura angularis. PAPILLAE conical, short; ostiolar canal at first filled with a hyaline, thin-walled pseudoparenchyma. CONIDIOPHORES 6.5–8.5 × 2.5–3.5 μm, cylindrical or clavate, simple, forming conidia singly at the apex, lining the wall of the pycnidial cavity up to the ostiolar canal. CONIDIA 6–8.5 × 3.5–4.5 μm, ovoid or ellipsoidal, one-septate, not or slightly constricted at the septum, yellowish-brown. MODE OF LIFE: Saprobic. SUBSTRATE: Driftwood. RANGE: Atlantic Ocean—United States (Florida, North Carolina); Pacific Ocean—United States (California, Washington). NOTE: Putative collections of *D. oraemaris* by T. W. Johnson (1956b, 1968) from North Carolina and Iceland are omitted because conidia are clearly larger (8–12 × 5–7 μm) than those of the type; if conidial length would be accepted as a valid character to separate *Diplodia* from *Microdiplodia* Allesch. (=*Diplodia*, fide R. K. Webster *et al.*, 1974), *D. oraemaris* would not belong to *Diplodia;* no taxonomical change is proposed because the position of a Deuteromycete remains doubtful until its sexual state will be known. ETYMOLOGY: From the Latin, *ora* = coast and *mare* = the sea, in reference to the marine habitat of the species.

Macrophoma (Saccardo) Berlese et Vogl.

Atti Soc. Veneto-Trentina Sci. Nat. **10**, 4, 1886

SYNONYMS: See Sutton (1977).

Description as in *Phoma* (p. 530), but conidia are longer; this genus is artificial according to Grove (1935) and marine collections of *Macrophoma* are tentatively assigned to this genus; the only named marine species in the genus, namely, *Macrophoma antarctica* J. Feldmann (1940) from *Curdiea racovitzae*, is a *nomen nudum* based on a *Chadefaudia* (Ascomycetes; J. Kohlmeyer, unpublished). TYPE SPECIES; *Macrophoma sapinea* (Fr.) Petrak. ETYMOLOGY: From the Greek, *makros* = long, large, and *Phoma,* in reference to the large conidia and the resemblance to this genus.

Macrophoma spp.

LITERATURE: Dickinson and Morgan-Jones (1966); Kohlmeyer (1971b); Kohlmeyer and Kohlmeyer (1971a); Schaumann (1969); Steele (1967).

SUBSTRATES: Intertidal wood and mangrove roots (*Avicennia* sp., *Pachira aquatica, Rhizophora mangle*), also isolated from stems and roots of *Halimione portulacoides* and seawater. RANGE: Atlantic Ocean— Germany (Helgoland, F.R.G.), Great Britain (England), United States (Maine); Indian Ocean—South Africa; Pacific Ocean—Guatemala, United States (Hawaii).

Phialophorophoma Linder *in* Barghoorn et Linder

Farlowia **1**, 402, 1944

LITERATURE: Sutton (1973b); Sutton and Pirozynski (1963).

PYCNIDIA separate, subglobose or ellipsoidal, unilocular, immersed, ostiolate, papillate or epapillate, subcarbonaceous, black, glabrous. PERIDIUM pseudoparenchymatous; cells on the outside dark, toward the centrum hyaline and bearing conidiophores. CONIDIOPHORES cylindrical, branched, septate, with terminal and lateral phialides, hyaline, produced on the inner cells of the peridium. CONIDIOGENOUS CELLS monophialidic, enteroblastic. CONIDIA obovoid–ellipsoidal to clavate, one-celled, smooth-walled, hyaline. MODE OF LIFE: Saprobic. TYPE SPECIES: *Phialophorophoma litoralis* Linder. The genus is monotypic. NOTE: Description based mainly on Sutton and Pirozynski (1963). ETYMOLOGY: From the Greek, *phiale* = urne, *-phorus* = -bearing, and *Phoma,* in reference to the phialidlike conidiogenous cells and a similarity to the genus *Phoma.*

Phialophorophoma litoralis Linder *in* Barghoorn et Linder

Farlowia **1,** 403, 1944

LITERATURE: Barghoorn (1944); Cribb and Cribb (1955); Sutton and Pirozynski (1963).

Fig. 122

PYCNIDIA up to 200 μm high (including the neck), 118–200 μm in diameter, subglobose or ellipsoidal, unilocular, immersed, ostiolate, papillate or epapillate, subcarbonaceous, brown to black, glabrous, solitary. PERIDIUM composed of two to five cells, pseudoparenchymatous; cells on the outside dark, toward the centrum hyaline and bearing co-

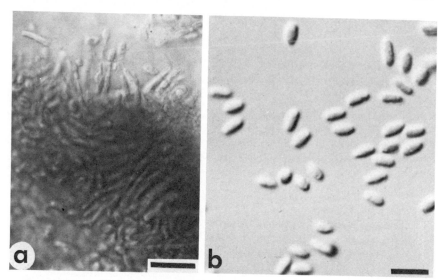

Fig. 122. *Phialophorophoma litoralis* from wood, type slide Barghoorn No. 24, in FH. (a) Conidiophores. (b) Conidia. (Both bars = 10 μm; both in Nomarski interference contrast.)

nidiophores. CONIDIOPHORES 10–25 × 2–2.5 μm, cylindrical, branched, septate, with terminal and lateral phialides, hyaline, produced on the inner cells of the peridium. CONIDIOGENOUS CELLS 7–10.5 × (2.5–)3–3.5 μm, monophialidic, enteroblastic, tapering upward to a cylindrical neck or cylindrical throughout; lateral phialides produced immediately below transverse septa of the conidiophore. CONIDIA 2.5–4.5 × 1.5–2 μm, ellipsoidal, obovoid, or clavate, one-celled, smooth-walled, hyaline. MODE OF LIFE: Saprobic. SUBSTRATES: Intertidal wood and bark of *Quercus* sp., driftwood, dead roots of *Avicennia marina* var. *resinifera*. RANGE: Atlantic Ocean—Great Britain (England), United States (Massachusetts); Pacific Ocean—Australia (Queensland). NOTE: Description after Linder (1944), Sutton and Pirozynski (1963), and our examination of the type slide in Herb. F (Barghoorn No. 24); in contrast to Linder's description, conidiophores in the type material do not show a "flaring cup-shaped collar" (Linder, 1944) but are tapering toward the apex (Fig. 122a); it is possible that this difference is caused by the mounting medium; a report of *P. litoralis* from forest litter in Idaho (Brandsberg, 1969) appears to be doubtful; some marine collections identified as *Phoma* sp. might actually belong to *Phialophorophoma*. ETYMOLOGY: From the Latin, *litoralis* = of the shore, in reference to the occurrence of the species.

Phoma Saccardo

Michelia **2**, 4, 1880

=*Ascochyta* sect. *Phyllostictoides* Zherbele *in Tr. Vses. Nauchno-Issled. Inst. Zashch. Rast.* **29**, 20, 1971
=*Deuterophoma* Petri, *Boll. R. Staz. Patol. Veg., Roma, N.S.* **9**, 396. 1929
=*Leptophoma* von Höhnel, *Sitzungsber. Kais, Akad. Wiss. Wien, Math.-Naturwiss. Kl,* Abt. 1, **124**, 73, 1915
=*Peyronellaea* Goidànich, *Atti Accad. Naz. Lincei, Cl. Sci. Fis., Mat. Nat., Rend.,* Ser. 8 **1**, 451, 1946, ex Togliani, *Ann. Sper. Agrar.,* Ser. 2 **6**, 92, 1952
=*Phoma* Fr. ex Fr., *Syst. Mycol.* **1**, p. LII, 1821; **2**, p. 546, 1823, *nom. rej.*
=*Plenodomus* Preuss, *Linnaea* **24**, 145, 1851
=*Polyopeus* Horne, *J. Bot., London* **58**, 239, 1920
=*Sclerophomella* von Höhnel, *Hedwigia* **59**, 237, 1917
=*Sclerophomina* von Höhnel, *Hedwigia* **59**, 240, 1917
=?*Vialina* Curzi, *Boll. R. Staz. Patol. Veg., Roma, N.S.* **15**, 252, 1935

LITERATURE: Boerema (1969, 1976); Boerema and Bollen (1975); Grove (1935); Sutton (1973b, 1977); von Arx (1970).

PYCNIDIA separate or gregarious, globose, subglobose, ampulliform, obpyriform or irregular, unilocular, immersed or erumpent, ostiolate, papillate or epapillate, thin-walled, dark-colored, glabrous or sometimes hairy or setose; occasionally with more than one ostiole. PERIDIUM pseudoparenchymatous, prosenchymatous, or rarely pseudoscle-renchymatous; cells on the outside dark and thick-walled, toward the centrum hyaline and isodiametric. CONIDIOGENOUS CELLS mono-phialidic, enteroblastic, annellidic, ampulliform, similar to the inner cells of the peridium, hyaline. CONIDIA globose, obovoid, ellipsoidal or clavate, usually once or twice as long as wide, one-celled, but sometimes a secondary septation may result in two- or more-celled conidia, smooth-walled, hyaline, without appendages. MODE OF LIFE: Saprobic or parasitic. SUBSTRATES: Higher and lower plants. TYPE SPECIES: *Phoma herbarum* Westend. NOTE: Description and synonyms after von Arx (1970), Boerema and Bollen (1975), and Sutton (1973b, 1977); there are about 200 species in the genus; we have included 3 species found growing in marine habitats; other, doubtfully marine species of *Phoma* are often isolated by culture methods; Boerema and Bollen (1975) point out that a useful classification of *Phoma* species has to combine the use of morpho-logical characters with growth characteristics *in vitro;* however, most species are insufficiently known and Grove's (1935) definition of the genus still holds true: that *Phoma* "remains very heterogenous and has been used as a kind of waste heap or harbor of refuge for any Coelomycete with colorless and one-celled spores which was not fully explored or understood by its discoverer''; positive identifications of marine *Phoma* species are probably impossible, unless their cultural characteristics are explored or perfect states become known. ETYMOLOGY: From the Greek, *phos* = blister, burn, probably in reference to the pustulelike pycnidia.

Phoma laminariae Cooke et Massee *in* Cooke

Grevillea **18**, 53, 1889

= ?*Diplodina laminariana* Sutherland, *New Phytol.* **15**, 39, 1916 (probably a synonym, but type material absent)

LITERATURE: Grove [1935 (as *Diplodina laminariae,* ex errore)]; T. W. Johnson and Sparrow (1961).

Fig. 123

PYCNIDIA 90–135 μm high, 90–160 μm in diameter, ellipsoidal to sub-globose, immersed or erumpent, ostiolate, short papillate or epapillate, coriaceous, thin-walled, brown to blackish, darkest and thick-walled

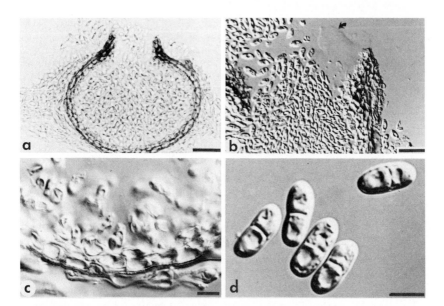

Fig. 123. *Phoma laminariae* from *Laminaria* sp. (a) Longitudinal section (2 μm) through pycnidium (isotype NY; bar = 20 μm). (b) 2-μm section through ostiole filled with mucilage (arrow) (isotype NY; bar = 15 μm). (c) 6-μm section through peridium (holotype K; bar = 5 μm). (d) Septate conidia (holotype K; bar = 5 μm). (a in bright field, the others in Nomarski interference contrast.)

around the ostiole, solitary or gregarious. PERIDIUM 6–8 μm thick, up to 20 μm at the ostiole, usually composed of two layers of elongate, lenticular cells, dark on the outside, light-colored near the venter, forming a textura angularis. PAPILLAE 10–25 μm high, 30–50 μm in diameter, conical; ostiolar canal filled with mucilage. CONIDIOGENOUS CELLS about 6 × 6 μm, conical, hyaline, lining the wall of the pycnidial cavity. CONIDIA 6–10(–12) × 3–4(–4.5) μm, ellipsoidal, mostly one-celled, but also one-septate conidia occurring, not or slightly constricted at the septum, smooth-walled, hyaline. MODE OF LIFE: Saprobic. HOST: Decaying thalli of *Laminaria* sp. RANGE: Atlantic Ocean—Great Britain (England, Scotland, Orkney Islands). NOTE: Description based on our examination of holotype (K) and isotype (NY) material containing besides one-celled conidia also one-septate ones, not mentioned in the original description by Cooke and Massee; conidia of *Phoma* may develop secondary septa (Boerema and Bollen, 1975); therefore, *P. laminariae* can remain in this genus in spite of formation of septate conidia; a collection from wood tentatively identified as *Diplodina laminariana* (Tubaki and Ito, 1973) most probably represents a different species. ETYMOLOGY: Named after the host genus, *Laminaria* (Phaeophyta).

Phoma marina Lind

Danish fungi as represented in the herbarium of E. Rostrup, Nordisk Forlag, Copenhagen, p. 214, 1913 (imperfect or spermogonial state of *Didymosphaeria danica*, see p. 398).

Phoma suaedae Jaap

Schr. Naturwiss. Ver. Schleswig-Holstein **14**, 27, 1907

LITERATURE: Grove (1935); Pirozynski and Morgan-Jones (1968).

PYCNIDIA 150 μm high, 150–230 μm in diameter, subglobose to ellipsoidal, immersed, becoming erumpent, ostiolate, short papillate or epapillate, coriaceous, light to dark brown, darkest around the ostiole, solitary or gregarious. PERIDIUM composed of an outer layer with thick-walled, dark cells and an inner layer with hyaline cells, bearing conidiogenous cells. PAPILLAE about 15 μm in diameter, ostiolar canal about 10 μm in diameter. CONIDIOGENOUS CELLS 5–8 μm long, phialidic, flask-shaped, hyaline, originating along the inner wall of the peridium, similar to the inner cells of the peridium. CONIDIA (5–)6–8(–10) × 3–4(–5) μm, ellipsoidal to subglobose, one-celled, often biguttulate, hyaline (finally somewhat yellowish?), originating in basipetal succession. MODE OF LIFE: Saprobic. HOST: Dead stems and leaves of *Suaeda maritima*. RANGE: Atlantic Ocean—Germany [Amrum and Helgoland (J. Kohlmeyer, unpublished), F.R.G.], Great Britain (England, Wales), United States [Maine (J. Kohlmeyer, unpublished)]. NOTE: Description based on Jaap (1907b), Pirozynski and Morgan-Jones (1968), and our own material (J.K. Nos. 3167, 3168, 3340). ETYMOLOGY: Named after the host genus, *Suaeda* (Chenopodiaceae).

Phoma spp.

LITERATURE: Dickinson and Morgan-Jones (1966); Dickinson and Pugh (1965c); Eaton and Irvine (1972); Eaton and Jones (1971a,b); Gessner and Goos (1973a,b); Gessner and Kohlmeyer (1976); Gold (1959); Goodman (1959); Höhnk (1956); G. C. Hughes (1968, 1969); G. C. Hughes and Chamut (1971); T. W. Johnson (1968); T. W. Johnson *et al.* (1959); E. B. G. Jones (1962a, 1963c, 1968b, 1971a); E. B. G. Jones and Irvine (1971); E. B. G. Jones and Oliver (1964); Kohlmeyer (1960, 1963b, 1966a, 1967, 1968c, 1969c, 1971b, 1976); Kohlmeyer and Kohlmeyer (1971a, 1977); Lee and Baker (1973); Meyers *et al.* (1964a, 1970b); Neish (1970); Nicot (1958a,c); Poole and Price (1972); Pugh (1960); Pugh *et al.* (1963); Quinta (1968); Rai *et al.* (1969); Ritchie (1957); Roth *et al.* (1964); Schaumann (1969); Schmidt (1974); Sivanesan and Manners (1970); Steele (1967); Tubaki and Ito (1973). SUBSTRATES: Isolated from beach and dune sand, salt pans, seawater, sediment, driftwood, intertidal wood, wooden panels (e.g., *Fagus sylvatica, Pinus sylvestris*); isolated from or observed on *Avicennia germinans, Crithmum maritimum* (J. Kohlmeyer, unpublished), *Halimione portulacoides, Hibiscus tiliaceus, Rhizophora mangle, Salicornia herbacea* (J. Kohlmeyer, unpublished), *S. stricta, Spartina alterniflora, S. townsendii, Zostera marina,* and algae [*Fucus vesiculosus* (J. Kohlmeyer, unpublished), *Macrocystis integrifolia, Porolithon onkodes*]. RANGE: Atlantic Ocean—Argentina (Buenos Aires), Bahamas, Bermuda, Canada (New Brunswick, Newfoundland, Nova Scotia), Colombia, France, Germany (Helgoland, F.R.G.), Great Britain (England, Scotland, Wales), Iceland (Surtsey), Liberia, Mexico, Norway (J. Kohlmeyer, unpublished), Panama, Portugal, Spain (Tenerife), Trinidad (J. Kohlmeyer, unpublished), United States [Connecticut, Florida, Georgia (J. Kohlmeyer, unpublished), Maine, Massachusetts, New Jersey, North Carolina, Rhode Island, South Carolina, Virginia]; Baltic Sea—Germany (G.D.R.); Indian Ocean—India (Bengal), South Africa; Pacific Ocean—Canada (British Columbia), Chile, Japan, Peru (Ica), United States [California, Hawaii (Oahu), Washington].

Key to the Marine Species of *Phoma*

1. On algae (*Chondrus* and *Laminaria*) 2
1'. On other substrates 3
    2(1) Saprobic on decaying *Laminaria* spp.; pycnidia separate, not in a stroma
    . . . . *P. laminariae*
    2'(1) Parasitic on *Chondrus crispus;* pycnidia embedded in a stroma (imperfect or
    spermogonial state of *Didymosphaeria danica*) . . . . *P. marina*
3(1') On *Suaeda maritima* . . . . *P. suaedae*
3'(1') On other substrates . . . . *Phoma* sp.

Rhabdospora (Dur. et Mont.) Mont., *nom. cons.*

Sylloge Generum Specierumque Cryptogamarum, Baillière, Paris p. 277, 1856

    =*Septoria* Sacc. sect. *Rhabdospora* Dur. et Mont. apud Mont., *Ann. Sci.
    Nat., Bot. Biol. Vég.*, Sér. 3 **11**, 47, 1849
    =*Filaspora* Preuss, *nom. rej., Linnaea* **26**, 718, 1855

LITERATURE: Grove (1935); Sutton (1977); von Arx (1970).

PYCNIDIA subglobose or depressed, immersed or erumpent, ostiolate, mostly papillate, black or fuscous, thick-walled. CONIDIA bacilliform or filiform, multiguttulate or multiseptate, hyaline. MODE OF LIFE: Saprobic and parasitic. SUBSTRATES: Higher plants. TYPE SPECIES: A lectotype needs to be selected (Sutton, 1977). NOTE: The genus is similar to *Septoria; Rhabdospora* is ill-defined and heterogeneous; more than 100 species have been described, but only 1 marine representative has been named. ETYMOLOGY: From the Greek, *rhabdos* = stick and *sporos* = seed, spore, in reference to the filiform conidia.

Rhabdospora avicenniae J. Kohlmeyer et E. Kohlmeyer

Mycologia **63**, 851, 1971

LITERATURE: Kohlmeyer (1966c, 1976); Kohlmeyer and Kohlmeyer (1965, 1977).

Fig. 124

PYCNIDIA 70–110 μm high, 70–115 μm in diameter, subglobose, immersed or half immersed, ostiolate, papillate or epapillate, coriaceous, dark brown or black, solitary or gregarious, often developing in light-colored spots on the substrate next to ascomata of *Leptosphaeria avicenniae* and *Mycosphaerella pneumatophorae*. PERIDIUM 14–20 μm thick above, 8–12 μm at the base, dark brown, composed of polygonal cells forming a textura angularis. PAPILLAE short or absent; ostioles 6–15 μm in diameter. CONIDIOPHORES 5–12 μm long, 1.2–2.5 μm in diameter, cylindrical or attenuate, simple, forming conidia singly at the apex, lining

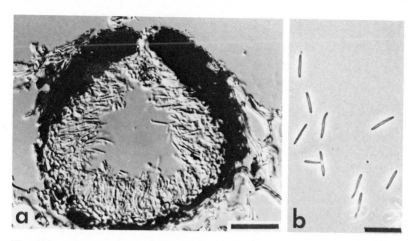

Fig. 124. *Rhabdospora avicenniae* from pneumatophores of *Avicennia germinans* (type IMS, Herb. J.K. 2735a). (a) Longitudinal section (4 μm) through pycnidium. From Kohlmeyer and Kohlmeyer, *Mycologia* **63**, 852 (1971), Fig. 37. Reprinted with permission. (b) Conidia. (Both bars = 25 μm; a in Nomarski interference contrast, b in phase contrast.)

the wall of the pycnidial cavity. CONIDIA 9–12.5 × 1.5–2 μm, botuliform or filiform, one-celled, straight or slightly curved, hyaline. MODE OF LIFE: Saprobic. SUBSTRATES: Bark of pneumatophores and trunks of *Avicennia africana* and *A. germinans*. RANGE: Atlantic Ocean— Bahamas, Bermuda, Colombia, Liberia, United States (Florida). ETYMOLOGY: Named after the host genus, *Avicennia* (Avicenniaceae).

Robillarda Saccardo (*nom. cons. prop.*)

Michelia **2**, 8, 1880 (non *Robillarda* Cast. 1845, *nom. rej. prop.*)

LITERATURE: Cunnell (1958); Morgan-Jones *et al.* (1972); Nag Raj and Morgan-Jones (1972); Nag Raj *et al.* (1972, 1973); Nicot and Rouch (1965); Sutton (1977); von Arx (1970).

PYCNIDIA solitary or gregarious, separate, subglobose, unilocular, immersed, ostiolate, brown, glabrous. CONIDIOGENOUS CELLS holoblastic, ampulliform, hyaline, originating on the inner wall of the peridium. CONIDIA ellipsoidal, one-septate, smooth-walled, hyaline, with a single, branched, apical appendage. MODE OF LIFE: Saprobic and parasitic. SUBSTRATES: Mostly higher plants. TYPE SPECIES: *Robillarda sessilis* (Sacc.) Sacc. NOTE: Description mainly after Morgan-Jones *et al.* (1972); there is one marine representative among about 25 species of the genus.

ETYMOLOGY: Named in honor of the French naturalist Louis Marc Antoine Robillard d'Argentelle (1777–1828).

Robillarda rhizophorae Kohlm.

Can. J. Bot. **47**, 1483, 1969

LITERATURE: Lee and Baker (1972b, 1973); Newell (1976).

PYCNIDIA 89–200 μm high, 248–770 μm long, ellipsoidal, unilocular, at first immersed, later erumpent, epapillate, opening apically with a longitudinal slit, subcoriaceous, black, glabrous, solitary or gregarious, rarely confluent. PERIDIUM 7–24 μm thick, composed of four to six layers of ellipsoidal or irregular, thick-walled cells, dark on the outside, hyaline near the venter, forming a textura angularis. CONIDIOGENOUS CELLS 4.5–8.5 μm high, 2.5–6 μm in diameter, conical or cylindrical, hyaline, originating along the inner wall of the peridium. CONIDIA 9.5–13.5 × 3–4.5 μm, single, ellipsoidal, one-septate in the middle, slightly constricted at the septum, smooth-walled, hyaline, provided with three (or four) apical, radiating appendages; appendages 14–30.5(–38) μm long, 0.8–1.5 μm in diameter at the base, slightly tapering toward the tip, rigid, straight or slightly curved, diverging from a common base, 1.5–2 μm long, 1.5 μm in diameter. MODE OF LIFE: Saprobic. HOST: Seedlings, live and dead prop roots of *Rhizophora mangle*. RANGE: Atlantic Ocean—United States (Florida); Pacific Ocean—United States [Hawaii (Oahu)]. ETYMOLOGY: Named after the host genus, *Rhizophora* (Rhizophoraceae).

Septoria Saccardo (*nom. cons.*)

Syll. Fung. **3**, p. 474, 1884

=*Septaria* Fr., *Systema Mycologicum, Gryphiswaldiae* **1**, p. XL, 1821, *nom. rej.*
=*Septoria* Fr., *Systema Orbis Vegetalis, Lundae* p. 119, 1825, *nom. rej.*
= ?*Spilosphaeria* Rabenh., *Bot. Ztg.* **15**, 173, 1857

LITERATURE: Donk (1968); Grove (1935); Sutton (1973b, 1977).

PYCNIDIA separate, subglobose, ampulliform or ovate, unilocular, immersed, ostiolate, sometimes papillate, glabrous, usually developing in discolored spots of leaves. CONIDIOGENOUS CELLS holoblastic, similar to the inner cells of the peridium, doliiform, ampulliform or elongated, sometimes proliferating. CONIDIA filiform, multiseptate, hyaline, without appendages. MODE OF LIFE: Saprobic or parasitic. SUBSTRATES: Mostly higher plants. TYPE SPECIES: *Septoria cytisi* Desm. NOTE: Description and synonyms mostly after Sutton (1973b). There are 2 representatives from the marine environment among about 1000 terrestrial species; *S. dictyotae* Oudemans from *Dictyota* is a *nomen rejiciendum*. ETYMOLOGY: From the Latin, *septum* = septum and the suffix *-orius*, in reference to the septate conidia.

Septoria ascophylli Melnik et Petrov

Nov. Sist. Niz. Rast. p. 211, 1966

LITERATURE: Kohlmeyer (1968e)

PYCNIDIA 57–80 × 83–115 μm, separate, ampulliform, unilocular, immersed, ostiolate, epapillate, glabrous, becoming pale reddish. PERIDIUM thin. OSTIOLE 20 μm in diameter, round. CONIDIOGENOUS CELLS about 4–8 × 1.5 μm, cylindrical, hyaline. CONIDIA 14.2–25 × 0.6 μm, filiform, nonseptate (?), straight or somewhat curved. MODE OF LIFE: Parasitic (or symbiotic?). HOST: In receptacles of *Ascophyllum nodosum*. RANGE: Arctic Ocean (Barents and White Sea)—U.S.S.R. NOTE: Description based on Melnik and Petrov (1966) and on our examination of type material (holotype LE, isotype IMS); this fungus is probably the spermatial state of *Mycosphaerella ascophylli* Cotton. ETYMOLOGY: Named after the host genus, *Ascophyllum* (Phaeophyta).

Septoria thalassica Spegazzini

An. Mus. Nac. Buenos Aires **20**, 387, 1910

LITERATURE: Farr (1973).

Fig. 125

PYCNIDIA 140–210 μm high, 150–320 μm in diameter, ellipsoidal, unilocular, immersed or erumpent, ostiolate, short papillate, glabrous, black, solitary or gregarious. PERIDIUM 12–16 μm thick, brown. CONIDIA 60–94 × 3–6 μm, filiform, seven (or eight)-septate, straight or curved, hyaline, with a gelatinous, subglobose cap, 2 μm in diameter, at each end. MODE OF LIFE: Saprobic. HOSTS: In wilting leaves and rhizomes of *Distichlis "thalassica"* and *D. spicata*. RANGE: Atlantic Ocean— Argentina (Santa Cruz), United States [Maine (J. Kohlmeyer, unpublished, J.K. 3389)]. NOTE: Description based on our examination of the type material (LPS No. 2944) and on our own collection; the description had to be amended to include septa and gelatinous conidial caps that are present in the type specimen; *S. thalassica* could be included in *Stagonospora*, but a transfer does not appear to be warranted. ETYMOLOGY: From the Latin, *thalassicus* = sea-green (from the Greek, *thalassa* = the sea), in reference to the habitat of the species.

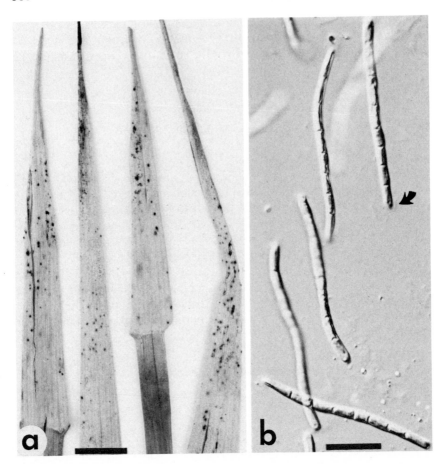

Fig. 125. *Septoria thalassica* from *Distichlis* sp. (type LPS, Herb. Spegazzini 2944). (a) Habit on leaves of the host (bar = 5 mm). (b) Conidia with apical gelatinous caps (arrow) (bar = 20 µm). (a = close-up photograph, b in Nomarski interference contrast.)

Key to the Marine Species of *Septoria*

1. In algae (*Ascophyllum*) . . . . *S. ascophylli*
1'. In Gramineae (*Distichlis*) . . . . *S. thalassica*

Stagonospora (Saccardo) Saccardo (*nom. cons.*)

Syll. Fung. **3**, p. 445, 1884

=?*Diedickella* Petrak, *Ann. Mycol.* **20**, 305, 1922
=*Gymnosphaera* Tassi, *Bull. Lab. Ort. Bot. Siena* **5**, 78, 1902
=*Hendersonia* Berk., *nom. rej.*, *Ann. Mag. Nat. Hist.*, **6**, 430, 1841
=*Hendersonia* Berk. subgen. *Stagonospora* Sacc., *Michelia* **2**, 8, 1880

=*Stagonospora* (Sacc.) Sacc. subgen. *Gymnosphaera* (Tassi) Sacc. et D.
  Sacc., *Syll. Fung.* **18**, p. 361, 1906
=*Stagonosporella* Tassi, *Bull. Lab. Ort. Bot. Siena* **5**, 50, 1902
=?*Stagonosporina* Tassi, *Bull. Lab. Ort. Bot. Siena* **5**, 51, 1902
=?*Tetradia* T. Johnson, *Sci. Proc. R. Dublin Soc.* **10**, 157, 1904

LITERATURE: Cunnell (1956, 1957); Grove (1935); Sutton (1973b, 1977); von Arx (1970).

PYCNIDIA separate, subglobose or ovate, unilocular, immersed, ostiolate, glabrous, thin-walled, often dark around the ostiole. CONIDIOGENOUS CELLS holoblastic, annellidic, originating on the inner wall of the pycnidium. CONIDIA cylindrical, navicular or fusiform, truncate at the base, multiseptate, hyaline, guttulate, without appendages. MODE OF LIFE: Saprobic and parasitic. SUBSTRATES: Mostly higher plants. TYPE SPECIES: *Stagonospora paludosa* (Sacc. et Speg.) Sacc. NOTE: Description and synonyms mostly after Sutton (1973b, 1977); *Stagonospora* is similar to *Septoria;* these genera are separated in von Arx's key (1970, p. 134) as follows: "Spores multiseptate often becoming brownish when ripe, smaller 1-celled spores mostly also present" characterizes *Stagonospora,* whereas the second lead, "Spores hyaline, filiform, fusiform or vermiform" leads to *Septoria;* Sutton (1973b, p. 561) separates the two genera on the basis of the conidiogenous cells, which are annelidic in *Stagonospora,* and "forming solitary conidia" in *Septoria;* more than 100 species of *Stagonospora* have been described, but only 1 marine representative has been named; additional species of *Stagonospora* appear to occur in the marine environment (Gessner and Kohlmeyer, 1976; Gold, 1959; Henningsson, 1974; G. C. Hughes, 1969; E. B. G. Jones, 1963c; E. B. G. Jones and Oliver, 1964). ETYMOLOGY: From the Greek, *stagon* = drop and *sporos* = seed, spore, in reference to the guttulate conidia.

Stagonospora haliclysta Kohlm.

Bot. Mar. **16**, 213, 1973

Figs. 126a–126c

PYCNIDIA 70–160 μm high, 60–150 μm in diameter, subglobose, superficial or half immersed, ostiolate, epapillate, coriaceous, dark brown above, light brown below, solitary or gregarious, rarely confluent. PERIDIUM 12–18 μm thick, composed of five or six layers of ellipsoidal cells, forming a textura angularis. OSTIOLES 25–30 μm in diameter. CONIDIOGENOUS CELLS 2–3 μm high, 4–5 μm in diameter, conoidal, simple, hyaline, forming conidia singly at the apex, lining the wall along the base and side of the pycnidial cavity. CONIDIA 20–27.5 × 3.5–4.5 μm, holoblastic, fusiform with rounded ends, three-septate, not or slightly con-

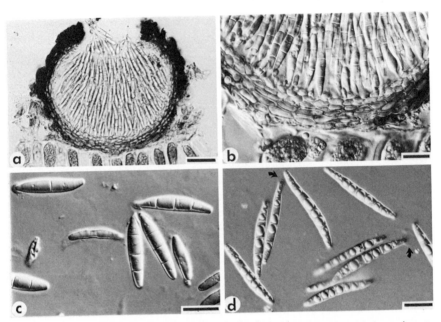

Fig. 126. Marine *Stagonospora* spp. (a) Longitudinal section (8 μm) through pycnidium of *S. haliclysta* on the surface of *Pelvetia canaliculata* (isotype IMS, Herb. J.K. 3151; bar = 25 μm). (b) Same as a, 8-μm section through base of pycnidium with peridium, conidiogenous cells toward the venter (bar = 10 μm). (c) Same as a, mature and immature conidia with a gelatinous cap over one end (bar = 10 μm). (d) Conidia of *Stagonospora* sp. from *Spartina* sp., inconspicuous caps (arrows) over one end (bar = 15 μm). (a in bright field, the others in Nomarski interference contrast.)

stricted at the septa, hyaline, smooth, the upper end covered by a mucilaginous cap; cap 3 μm in diameter, subglobose, dissolving. MODE OF LIFE: Saprobic? HOST: On the surface of attached but damaged *Pelvetia canaliculata*. RANGE: Atlantic Ocean—Norway (known only from the type collection). ETYMOLOGY: From the Greek, *haliklystos* = sea-washed, in reference to the marine intertidal habitat.

Stagonospora sp.

LITERATURE: Gessner (1977); Gessner and Goos [1973b (as *Septoria* sp.)]; Gessner and Kohlmeyer [1976 (as *Stagonospora* sp. II)].

Fig. 126d

PYCNIDIA 140–180 μm high, 110–200 μm in diameter, subglobose or ellipsoidal, immersed or superficial, ostiolate, epapillate or short papillate, coriaceous, light to dark brown, darkest around the ostiole, solitary or gregarious, sometimes confluent. PAPILLA 25 μm high, 35 μm in diameter; ostiole 8–10 μm in diameter. PERIDIUM 12–20 μm thick, composed of four or five layers of ellipsoidal cells with large lumina, forming a textura angularis. CONIDIA 38–61(–76) × (4–)6–7 μm, elongate–cylindrical with rounded apices, (five- or) seven-septate, slightly constricted at the septa, straight or curved, hyaline to light brown, with a hyaline to light yellowish, mucilaginous cap at the upper end. MODE OF LIFE: Saprobic. HOSTS: Old but attached rhizomes, stems and leaves of *Spartina alterniflora, S. patens* and *Spartina* sp. RANGE: Atlantic Ocean—Argentina (Buenos Aires), Canada (New Brunswick, Nova Scotia, Québec), United States (Connecticut, Florida, Maine, New Jersey, North Carolina, Rhode Island, Virginia). NOTE: This species from *Spartina* spp. is similar to *Septoria junci* on *Juncus* spp.; we have examined Desmazière's type material from Herb. PC in which the conidia are 45–75 × 3–4 μm and bear an apical gelatinous cap not mentioned in the original diagnosis; another similar species appears to be *Septoria caricicola,* the seven-septate conidia of which measure 35–55 × 4 μm (Grove, 1935); we have not seen the type specimen of *S. caricicola,* but our collection J.K. 3376 on *Carex paleacea* from Québec (conidia 26.5–62 × 4–5 μm, seven-septate) agrees with this species except for an apical cap, which is not indicated in the type description; at this time, it cannot be decided if the fungi on saltmarsh *Spartina, Juncus,* and *Carex* are identical or not; the matter is even more confused by the fact that imperfect states of *Pleospora vagans* on *Phragmites communis* (Gessner and Kohlmeyer, 1976) and of *Pleospora vagans* var. *spartinae* on *Spartina* (J. Webster and Lucas, 1961) are similar to our *Stagonospora* sp., although conidial appendages appear to be absent; comparisons with isolates from different hosts are required to determine if one or several species are involved.

Key to the Marine Representatives of *Stagonospora*

1. Conidia longer than 30 μm, mostly seven-septate; in *Spartina* spp. . . . . **Stagonospora** sp.
1'. Conidia shorter than 30 μm, mostly three-septate; on *Pelvetia*
· **Stagonospora haliclysta**

b. Excipulaceae

Dinemasporium Léveillé

Ann. Sci. Nat., Bot. Biol. Vég., Sér. 3 **5,** 274, 1846

=*Amphitiarospora* Agnihothrudu, *Sydowia* **16,** 75, 1963
=*Dendrophoma* Saccardo, *Michelia* **2,** 4, 1880
=*Pycnidiochaeta* Camara, *Agron. Lusit.* **12,** 109, 1950

LITERATURE: Grove (1937); Morgan-Jones *et al.* (1972); Sutton (1965, 1973b, 1977); von Arx (1970).

ACERVULI discoid, exposing the conidiogenous layer, superficial, dark brown to black, surrounded by dark setae, solitary or somewhat gregarious. SETAE tapering toward the apex, septate, smooth, brown, divergent. LATERAL EXCIPULUM of acervuli composed of elongated or of pseudoparenchymatous, isodiametric, light brown cells. CONIDIOPHORES septate, branched, hyaline or light brown. CONIDIOGENOUS CELLS phialidic, enteroblastic, cylindrical, tapering toward the apex, smooth, hyaline, producing conidia from an apical meristem. CONIDIA allantoid, botuliform or lentiform, one-celled, smooth, hyaline or light brown, with an unbranched, hairlike setula at each end. MODE OF LIFE: Saprobic. SUBSTRATES: Higher plants. TYPE SPECIES: *Dinemasporium graminum* Léveillé. NOTE: Description after Morgan-Jones *et al.* (1972) and Sutton (1973b); synonyms after Sutton (1977); there is 1 marine representative among about 20 species of the genus. ETYMOLOGY: From the Greek, *di-* = two-, *nema* = filament, thread, and *sporos* = seed, spore, in reference to the conidia bearing two setulae.

Dinemasporium marinum Nilsson

Bot. Not. **110**, 321, 1957

LITERATURE: T. W. Johnson (1968); T. W. Johnson and Sparrow (1961).

ACERVULI 600–1200 μm in diameter, flattened–cupulate, exposing the conidiogenous layer, superficial, grayish-black to black, setose. SETAE up to 800 μm long, 8–10 μm in diameter, needle-shaped, rigid, branched at the base, dark brown, scattered over the surface of the acervuli. LATERAL EXCIPULUM of acervuli composed of loose hyphae. CONIDIOPHORES 40–80 × 2–3.5 μm, septate, simple or slightly branched, elongate–cylindrical, hyaline, standing closely together on the surface of the acervulus. CONIDIOGENOUS CELLS cylindrical, tapering toward the apex, smooth, hyaline, producing conidia at the apex. CONIDIA 11–15 × 2.5–3.5(–6.5) μm, fusoid–allantoid, sometimes straight, slightly attenuate at the apices, one-celled, smooth, hyaline to very slightly greenish, containing two large oil globules, with a simple hairlike setula at each end; setulae 8–10 μm long, straight or obliquely inserted. MODE OF LIFE: Saprobic. SUBSTRATE: Driftwood (*Pinus* sp.). RANGE: Atlantic Ocean—Denmark, Iceland (Surtsey). NOTE: Description based on T. W. Johnson (1968) and S. Nilsson (1957). ETYMOLOGY: From the Latin, *marinus* = marine, in reference to the habitat of the species.

2. Melanconiales
a. Melanconiaceae

Sphaceloma de Bary

Ann. Oenol. **4**, 165, 1874

=*Kurosawaia* Hara, *A List of Japanese Fungi*, p. 172, 1954
=*Manginia* Viala et Pacottet, *Ct. R. Hebd. Seances Acad. Sci.* **139**, 88, 1904
=*Melanobasidium* Maublanc, *Bull. Soc. Mycol. Fr.* **22**, 64, 1906
=*Melanobasis* Clements et Shear, *Genera of Fungi, New York,* p. 370, 1931
=?*Melanodochium* Syd., *Ann. Mycol.* **36**, 310, 1938
=*Melanophora* von Arx, *Verh. K. Ned. Akad. Wet., Afd. Natuurkd., Tweede Sect.* **51**, 71, 1957

LITERATURE: Jenkins and Bitancourt (1941); Sutton (1973b, 1977); von Arx (1970).

HYPHAE in strands or forming a pseudoparenchyma. ACERVULI composed of a stromatic tissue that is restricted to the base of the fructification, forming conidiophores on the upper surface, releasing conidia by regular or irregular splitting of the overlying host tissues, hyaline at the base, usually darker toward the outer surface. CONIDIOPHORES forming a palisade of closely appressed cells that become brown with age, short, cylindrical, pointed at the apex, or broad at the base and tapering toward the tip, often none- or one-septate. CONIDIOGENOUS CELLS polyphialidic, enteroblastic. CONIDIA small, ovoid to oblong–ellipsoidal, one-celled, hyaline; rarely elongate or cylindrical and multiseptate, yellowish or dark; sometimes forming microconidia. MODE OF LIFE: Parasitic. SUBSTRATES: Higher plants and algae. TYPE SPECIES: *Sphaceloma ampelinum* de Bary. NOTE: Description and synonyms after Jenkins and Bitancourt (1941) and Sutton (1973b, 1977); there is 1 marine representative among about 50 terrestrial species. ETYMOLOGY: From the Greek, *sphakelos* = necrosis, and the suffix -*ma,* indicating the result of an action, in reference to the necrotic spots caused by many species of *Sphaceloma.*

Sphaceloma cecidii Kohlm.

J. Elisha Mitchell Sci. Soc. **88**, 255, 1972

LITERATURE: Kohlmeyer (1971a, 1974b).

Fig. 127

ACERVULI 70–140 μm high (excluding covering algal tissue), 200–700

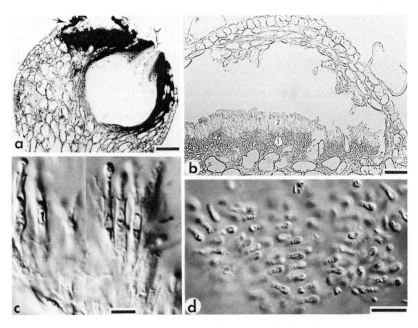

Fig. 127. *Sphaceloma cecidii* in galls caused by *Haloguignardia oceanica* in *Sargassum* sp. (a) Longitudinal section through projection of a gall containing a large ascocarp of *H. oceanica* and an open acervulus of *S. cecidii* (arrow) on the top (bar = 100 μm). (b) 4-μm section through acervulus with pseudoparenchymatous base (1), palisadelike layer of conidiogenous cells (2), and cover formed by algal cells of the gall (bar = 50 μm). (c) Conidiogenous cells with wide mouths forming conidia on "stalks" (bar = 5 μm). (d) Conidia with cylindrical appendages (bar = 10 μm). (a and b from type of *H. oceanica* in C, in bright field; c and d Herb. J.K. 3022, in Nomarski interference contrast.) From Kohlmeyer, *J. Elisha Mitchell Sci. Soc.* **88**, 256 (1972), Figs. 2–5. Reprinted with permission.

μm in diameter, discoid, with circular or elongated outlines, olive-colored, finally black, consisting of a stromatic, pseudoparenchymatous base that is covered by a palisadelike layer of conidiogenous cells; cells of the stroma isodiametric or elongated, at first hyaline, later dark brown; overlying host tissues composed of two to seven layers of cells, rupturing irregularly; the inner part of this dome-shaped algal cover is more or less lined with an inconspicuous, thin crust composed of irregular, thick-walled fungal cells with small lumina; this upper fungal crust was initially connected with the stromatic base of the acervulus and is pushed upward together with the algal cover by the developing conidiogenous cells and conidia. CONIDIOGENOUS CELLS 20–30 × 1.5–2.5 μm, phialidic with funnel-shaped apices, cylindrical or slightly attenuate, simple. CONIDIA 3–4.5(–5) × 1.8–2.5 μm (excluding appendages), enteroblastic, one-celled, ellipsoidal with a truncate base and a basal appendage, hyaline, smooth-walled; appendages 1.8–2.5 × 1–1.5 μm, truncate or short cylindrical, gelatinous. MODE OF LIFE: Hyperparasitic. HOSTS: In galls caused by the Ascomycete, *Haloguignardia irritans* in *Cystoseira osmundacea* and *Halidrys dioica;* by *Haloguignardia oceanica* in *Sargassum fluitans* and *S. natans;* and by *H. tumefaciens* in *Sargassum* sp. RANGE: Atlantic Ocean—Sargasso Sea, United States (North Carolina); Pacific Ocean—Australia (South Australia), United States (California). ETYMOLOGY: From the Greek, *kekidion* = gall, in reference to the substrate, namely, galls produced by other fungi in Phaeophyta.

27. Rejected Names, Doubtful and Excluded Species

Most of the species listed below had previously been included among marine fungi *sensu stricto* (Kohlmeyer and Kohlmeyer, 1971b) that occurred on marine substrates *in situ*. "Molds," however, obtained exclusively by incubation and isolation in the laboratory, are omitted from the following list, as we do not consider them marine until their growth or reproduction, or both, *in situ* have been demonstrated. The majority of the species in the following list have been invalidly published, according to Articles 7–9 and 37 of the International Code of Botanical Nomenclature (Stafleu *et al.*, 1972), since no type material was designated or preserved and new collections of the particular species have not been made.

I. ASCOMYCOTINA

1. *Ceriosporopsis maritima* (Linder) R. Santesson *in* Gustafsson et Fries, *Physiol. Plant.* **9**, 462, 1956
   This combination was not validly published because no basionym was indicated. *Ceriosporopsis maritima* is a *nomen nudum*.

2. *Dermatomeris georgica* Reinsch *in* *Internationale Polarforschung 1882–1883, Die deutschen Expeditionen*, Vol. II, Asher, Berlin, pp. 425–426, 1890
   Inquiries at herbaria B and ER to obtain Reinsch's type material of this and other species have been unsuccessful. *Dermatomeris georgica* collected by H. Will in South Georgia is a lichenized fungus similar to *Turgidosculum complicatulum* except for smaller dimensions of ascocarps,

asci, and ascospores. In the absence of type material, *Dermatomeris* and its incompletely described type species *D. georgica* are considered to be *nomina dubia*.

3. *Didymosamarospora euryhalina* Johnson et Gold, *J. Elisha Mitchell Sci. Soc.* **73**, 104, 1957

This fungus was described from *Juncus roemerianus* from Mullet Pond on Shackleford Banks, North Carolina. Critical features of the fungus, namely, the presence or absence of an apical apparatus in the ascus, uni- or bitunicate condition of the ascus wall, and morphology of the ascocarp venter, are unknown. We were unable to obtain type material, and several searches for the fungus at the type locality and at other sites of *J. roemerianus* in North Carolina have been unsuccessful. The fungus has not been collected by other authors since its original description. *Didymosamarospora* and its type species *D. euryhalina* are considered to be *nomina dubia*.

4. *Entocolax rhodymeniae* Reinsch *in Internationale Polarforschung 1882–1883, Die deutschen Expeditionen,* Vol. II, Asher, Berlin, pp. 399–400, 1890

This fungus from *Rhodymenia* spp. was designated as an Ascomycete by Reinsch; however, the description and illustrations do not permit one to recognize critical characters, and type material is not available. Skottsberg (1923) considers a fungus on *Curdiea reniformis* to be identical with *E. rhodymeniae. Entocolax rhodymeniae* is a *nomen dubium*.

5. *Humarina coccinea* (Crouan) Seaver var. *maritima* (Grélet) Cash *in* Trotter, Saccardo's *Syll. Fung.* **26**, 501, 1972
    ≡*Humaria coccinea* (Crouan) Sacc. var. *maritima* Grélet, *Bull. Soc. Mycol. Fr.* **42**, 204, 1926

This variety was collected on washed-up *Zostera marina* in France. We have not examined this fungus and exclude it from the list of marine species until detailed information becomes available on the occurrence of the fungus.

6. *Lentescospora submarina* Linder *in* Barghoorn et Linder, *Farlowia* **1**, 411–412, 1944
    ≡*Ceriosporopsis submarina* (Linder) Müller *in* Müller et von Arx, *Beitr. Kryptogamenflora Schweiz* **11**(2), 793, 1962

The description of *L. submarina* was apparently based on old material (Kohlmeyer, 1969c), and the type specimen in Herb. F examined by us consists of empty, broken ascocarps. Thus, critical features, like asci and the nature of ascospore appendages, are unknown and we consider *Len-*

tescospora and its type species *L. submarina* to be *nomina dubia*. These names have led to some confusion in the past because fungal collections known now to belong to *Halosphaeria galerita* have been identified often as *L. submarina* (Kohlmeyer, 1969c).

7. *Lophiotrema littorale* Spegazzini, *Michelia* **1**, 466, 1879
   ≡*Lophiostoma appendiculatum* Fuckel var. *littorale* (Speg.)
      Chesters et Bell, *Mycol. Pap.* **120**, 19, 1970

There is no indication that *L. littorale* is a marine fungus and we exclude it from the list of marine species. The type material on *Salix fragilis* derived from an island, the well-known Lido near Venice, and Chesters and Bell reported it from *Salix* sp. from an inland location in New York State. A tentative record by Newell (1976) on seedlings of *Rhizophora mangle* from Florida is possibly a different species because diagnostic features, namely, slitlike ostioles of the ascocarps, are unclear in culture (Newell, 1976).

8. *Maireomyces peyssoneliae* J. Feldmann, *Bull. Soc. Hist. Nat. Afr. Nord* **31**, 165–166, 1940

Maireomyces peyssoneliae from *Peyssonelia squamaria* has been discussed by T. W. Johnson and Sparrow (1961), Meyers (1957), and von Arx and Müller (1954), and the insufficient description of the fungus has been pointed out by these authors. In the absence of type material (Kohlmeyer, 1972b) or other authentic collections, we consider *Maireomyces* and its type species *M. peyssoneliae* to be *nomina dubia*.

9. *Massariella maritima* Johnson, *Mycologia* **48**, 846, 1956

This species was described from wooden panels in North Carolina. The fungus has not been collected again in over 20 years. Since critical features are unknown and type material could not be located, we consider *M. maritima* to be a *nomen dubium*.

10. *Mycaureola dilseae* Maire et Chemin, *C. R. Hebd. Séances Acad. Sci.* **175**, 321, 1922

We have examined the type material from Herb. MPU and found only immature fruiting bodies on the host, *Dilsea edulis*. The description was obviously based on this immature material. Therefore, *Mycaureola* and its type species *M. dilseae* are considered to be *nomina dubia*.

11. *Ophiobolus laminariae* Sutherland, *Trans. Br. Mycol. Soc.* **5**, 147, 1915

This fungus from *Laminaria digitata* was not found again since its original description. Critical morphological features, such as ascocarp and ascus structure, are unknown and type material is absent. Therefore, we consider *O. laminariae* a *nomen dubium*.

12. *Ophiobolus medusa* Ellis et Everhart var. *minor* Ellis et Everhart, *Proc. Acad. Nat. Sci. Philadelphia* **42,** 239, 1891
=*Ophiobolus salina* Meyers, *Mycologia* **49,** 518, 1957

This fungus from *Andropogon muricatus* was reduced to a synonym of *O. herpotrichus* (Fries) Sacc. *in* Roum. et Sacc. by Shoemaker (1976). We exclude the species because it cannot be considered marine.

13. *Pharcidia pelvetiae* Sutherland, *New Phytol.* **14,** 39–41, 1915

This species from *Pelvetia canaliculata* has not been collected again since its description, and type material is absent. The incomplete description of *P. pelvetiae* requires its designation as *nomen dubium*.

14. *Pleospora laminariana* Sutherland, *Trans. Br. Mycol. Soc.* **5,** 260–262, 1916

The species was described from *Laminaria* sp. and is possibly identical with *P. pelvetiae* (Kohlmeyer, 1973b). Critical morphological features are unknown and type material is missing. Therefore, *P. laminariana* must be regarded a *nomen dubium*.

15. *Pleospora purpurascens* R. Santesson *in* Gustafsson et Fries, *Physiol. Plant.* **9,** 462, 1956

This name was published without a description and, thus, *P. purpurascens* is a *nomen nudum*.

16. *Rhaphidophora maritima* Saccardo, *Michelia* **1,** 119, 1878
≡*Halophiobolus maritimus* (Sacc.) Linder *in* Barghoorn et Linder, *Farlowia* **1,** 419, 1944
≡*Linocarpon maritimum* (Sacc.) Petrak, *Sydowia* **6,** 388, 1952
≡*Lulworthia maritima* (Sacc.) Cribb et Cribb, *Univ. Queensl. Pap., Dep. Bot.* **3,** 80, 1955
≡*Ophiobolus maritimus* (Sacc.) Sacc., *Syll. Fung.* **2,** 350, 1883

Walker (1972) examined type material from Herb. PAD and found only empty ascocarps. Isotype specimens from Herb. Magnus in HBG examined by us also bore empty fruiting bodies exclusively. In the absence of asci and ascospores it is not possible to classify *R. maritima* and it has to be considered a *nomen dubium*.

17. *Rosellinia laminariana* Sutherland, *Trans. Br. Mycol. Soc.* **5,** 257–259, 1916

Critical morphological features of this species from *Laminaria* sp. are unknown and type material is absent. Thus, *R. laminariana* is designated as a *nomen dubium*.

18. *Samarosporella pelagica* Linder *in* Barghoorn et Linder, *Farlowia* **1,** 407–408, 1944

The description of this species from submerged wood is based on immature material and critical features cannot be recognized in the type specimen (Kohlmeyer, 1972b). The fungus could belong to any of the marine Ascomycetes with three-septate, appendaged ascospores, but an assignment to a particular species is not possible. *Samarosporella* and its type species *S. pelagica* are *nomina dubia*.

19. *Sphaeria balani* Patouillard *in* Hue, *Bull. Soc. Bot. Fr.* **56,** 322, 1909
Fig. 128

Material of *S. balani* from the Patouillard collection in Herb. FH (Fig. 128) contains an Ascomycete associated with blue–green algae. It derived from *Balanus*-covered rocks in France. Included in the folder is also a letter of E. Bornet dated 24 Jan. 1886, probably addressed to Patouillard, and concerns *S. balani*. The fungus could belong to *Leiophloea,* but a thorough examination is necessary to place and describe it properly. *Sphaeria balani* has never been validly published and is a *nomen nudum*.

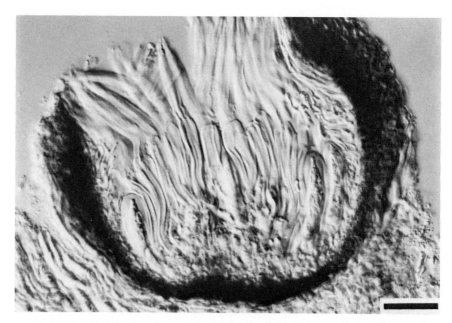

Fig. 128. *"Sphaeria balani,"* a lichenized Ascomycete from Herb. Patouillard (FH); longitudinal section (16 μm) through ascocarp with expanded endoasci of emptied asci (Nomarski interference contrast; bar = 25 μm).

20. *Sphaeria littoralis* Crouan et Crouan, *Florule du Finistère*, p. 29, 1867
    ≡*Ophiobolus littoralis* (Crouan et Crouan) Saccardo, *Syll.*
    Fung. **2,** 349, 1883 (non *O. littoralis* Cribb et Cribb 1956)

We have examined type material (designated as "*Raphidiospora littoralis*" on *Agrostis maritima*) in the Marine Station of the Collège de France at Concarneau. Crouans' original collection does not contain any asci, but some yellow-brownish, filamentous, six- to twelve-septate ascospores, 70–114 × 4–5.5 μm, are present. Bommer and Rousseau (1891) list *O. littoralis* on *Ammophila arenaria* from Belgium; ascospores from their collection are described as being subhyaline, ten- to fourteen-septate, 45–60 × 3 μm. In the absence of asci in the type material it is not possible to apply the name *S. littoralis* to any later collections and, in agreement with Dr. J. Walker (personal communication), we consider it to be a *nomen dubium*.

21. *Sphaeria obionei* Crouan et Crouan, *Florule du Finistère,* p. 25, 1867
    ≡*Physalospora obionis* (Crouan et Crouan) Saccardo, *Syll.*
    Fung. **1,** 448, 1882

The type material at Concarneau examined by us consists of two stems of *Atriplex* sp. (=*Obione*) with small black ascocarps. The ascospores are brown and muriform and resemble those of a *Pleospora*. This fungus is obviously not the species described by the Crouans, and the name *Sphaeria obionei* must be considered a *nomen confusum*.

22. *Sphaeria subsalsa* Crouan et Crouan, *Florule du Finistère,* p. 25, 1867
    ≡*Ceratostomella subsalsa* (Crouan et Crouan) Saccardo,
    Syll. Fung. **1,** 412, 1882

Type material of *S. subsalsa* on *Atriplex* sp. (="*Obione*") in the Crouan herbarium at Concarneau consists of two subglobose, brown ascocarps, 520–550 μm in diameter. The necks are 480–750 μm long and 90–140 μm in diameter. The peridium is 25–30 μm thick. Asci are apparently early deliquescing and absent in the type specimen. A drawing prepared by the Crouans on the paper capsule represents a thin-walled, clavate ascus with eight ascospores. We did not find any paraphyses or pseudoparaphyses. The ascospores are 35–42 × 19–26 μm, broadly ellipsoidal, one-septate, and slightly constricted at the septum and lack appendages. The fungus probably belongs to the Halosphaeriaceae and resembles *Gnomonia salina*. However, we did not find ascospore appendages in the type material of *S. subsalsa,* which could have deteriorated in these over 100-year-old specimens. As the identity of *S. subsalsa* cannot be established with certainty, we consider the name a *nomen dubium*.

23. *Sphaerulina longirostris* R. Santesson *in* Gustafsson et Fries, *Physiol.*
    Plant. **9,** 462, 1956

This name was published without a description and, therefore, *S. lon-girostris* is a *nomen nudum*.

24. *Sphaerulina pedicellata* Johnson, *Mycologia* **48,** 846, 1956

Critical characters of this fungus from submerged wood are unknown, namely, structure of the ascus wall and tip, spore release, and morphology of the ascocarp centrum. The species has been confused frequently with an Ascomycete from *Spartina* spp. (*Buergenerula spartinae*), but is not identical with it (Kohlmeyer and Gessner, 1976). As *S. pedicellata* has become a persistent source of error and type material is not available, the name must be considered to be a *nomen dubium* or *ambiguum*.

25. *Stigmatea pelvetiae* Sutherland, *New Phytol.* **14,** 37–39, 1915

This species, described from *Pelvetia canaliculata* in Scotland, has not been collected again since its description. As critical morphological features are unknown and type material is not available, *S. pelvetiae* is designated as a *nomen dubium*.

26. *Trematosphaeria oraemaris* (Linder) R. Santesson *in* Gustafsson et Fries, *Physiol. Plant.* **9,** 462, 1956

This combination was not validly published because no basionym was indicated. *Trematosphaeria oraemaris* is a *nomen nudum*.

27. *Trematosphaeria purpurascens* R. Santesson *in* Johnson and Sparrow, *Fungi in Oceans and Estuaries,* Cramer, Weinheim, p. 516, 1961

This name was erroneously attributed by Johnson and Sparrow to Santesson (*in* Gustafsson and Fries, 1956). There is no description available, and *T. purpurascens* is considered to be a *nomen nudum*.

28. *Trematosphaeria thalassica* R. Santesson *in* Gustafsson et Fries, *Physiol. Plant.* **9,** 462, 1956

This name was published without a description and, thus, *T. thalassica* is a *nomen nudum*.

II. DEUTEROMYCOTINA

29. *Alternaria maritima* Sutherland, *New Phytol.* **15,** 46–47, 1916

Type material of this fungus from *Laminaria* is not available and the description is not sufficient to classify the species according to modern standards. It appears as if it has become the usage to identify any *Alternaria* isolated from marine habitats as *A. maritima*. In the absence of type material and because *A. maritima* has become a persistent source of error, this name is considered to be a *nomen dubium*.

30. *Aposphaeria boudieri* Rolland, *Bull. Soc. Mycol. Fr.* **12**, 6, 1896
The species is insufficiently known and we have been unable to locate type material. Possibly, this fungus is the spermogonial state of *Halotthia posidoniae* (Ollivier, 1930). We consider *A. boudieri* to be a *nomen dubium*.

31. *Brachysporium helgolandicum* Schaumann, *Helgol. Wiss. Meeresunters.* **25**, 26–27, 1973
We exclude this species from the treatment of marine fungi because its marine origin is uncertain (Schaumann, 1973). The fungus occurred on drift bark collected on the beach of Helgoland. As it was not possible to obtain pure cultures (Schaumann, 1973), information on the variability of *B. helgolandicum* in culture is missing. The fungus is similar to a group of unidentified species in our collections, which occur frequently in marine habitats and which may be related to *Culcitalna* and *Humicola*.

32. *Cephalosporium saulcyanum* Montagne, *Ann. Sci. Nat., Bot. Biol. Vég.,* Sér. 4 **5**, 346–347, 1856
This fungus occurred on *Prasiola orbicularis* (=*Ulva crispa*) in Greenland. The poorly preserved type material in Paris (PC) does not give clues to a marine origin of the collection. It was also examined by W. Gams, who designated it as "*Absidia* sp." We exclude this incompletely known species from the list of marine fungi.

33. *Dictyonema zoophytarum* Reinsch *in Contributiones ad Algologiam et Fungologiam,* Leipzig, p. 95, 1875
This species was described from "zoophytis" (coelenterates?) on marine algae (*Phyllaphora* sp. and *Rhodymenia* sp.). The description and illustrations are insufficient to classify the organism. As type material is unavailable, *Dictyonema* and its type species *D. zoophytarum* are considered to be *nomina dubia*.

34. *Dothichiza litoralis* Petrak, *Ann. Mycol.* **25**, 282–283, 1927
This species on *Polyides* "*lumbricoides*" (=*P. lumbricalis*?, Rhodophyta) was distributed without number as "?Conidien-Pilz" in Rehm's *Ascomycetes exsiccata,* according to Petrak. We were unable to locate the type material and, hence, the species cannot be properly classified. The name *D. litoralis* is a *nomen dubium*.

35. *Fusidium maritimum* Sutherland, *New Phytol.* **15**, 41, 1916
The fungus was described from *Laminaria* sp. and *Pelvetia* sp., collected above the high-tide mark or grown upon incubation of algal thalli in

the laboratory. As type material of *F. maritimum* is absent, the identity cannot be established and the name must be regarded as a *nomen dubium*.

36. *Hypheotrix fucoidea* Piccone et Grunow *in* Piccone, *Nuovo G. Bot. Ital.* **16,** 291, 1884

Professor G. F. Papenfuss kindly drew our attention to this species, which, although described as a blue–green alga, is a filamentous fungus, according to Gomont (1892). As we have not seen any type material, we include *H. fucoidea* in the doubtful species.

37. *Macrophoma antarctica* J. Feldmann, *Bull. Soc. Hist. Nat. Afr. Nord* **31,** 169, 1940

Fruiting bodies of this fungus were first described and illustrated by Gain (1912) as male conceptacles of the host, *Curdiea racovitzae.* Skottsberg (1923) pointed out that these "conceptacles" are perithecia of *Entocolax rhodymeniae.* J. Feldmann (1940) considered the fungus to be sphaeropsidaceous and named it *M. antarctica* without giving a formal description. We have examined Gain's (1912) original material on *C. racovitzae* in Herb. PC and concluded that it is an Ascomycete in the genus *Chadefaudia,* close to *C. gymnogongri* (Kohlmeyer, 1973b). *Macrophoma antarctica* is a *nomen nudum.*

38. *Macrosporium laminarianum* Sutherland, *New Phytol.* **15,** 45–46, 1916

The morphology of this saprobe from *Laminaria* sp. is insufficiently known and type material is absent. Therefore, *M. laminarianum* is considered to be a *nomen dubium*.

39. *Macrosporium pelvetiae* Sutherland, *New Phytol.* **14,** 41–42, 1915

This species, supposedly the imperfect state of *Pleospora pelvetiae,* is incompletely known. In the absence of type material *M. pelvetiae* must be considered to be a *nomen dubium.*

40. *Monodictys austrina* Tubaki *in* Tubaki et Asano, *JARE 1956–1962 Sci. Rep.,* Ser. E **27,** 7, 1965

We exclude this species from the list of marine fungi because its origin is doubtfully marine. It was isolated from soil and unidentified algae in Antarctica. So far, *M. austrina* is known only from the original collection. Conidia appear to be similar to those of *Zalerion varium,* but we were unable to compare the two species because a culture of *M. austrina,* kindly provided by Dr. Tubaki, remained sterile.

41. *Monosporium maritimum* Sutherland, *New Phytol.* **15,** 42, 1916

The species occurs saprobically on decomposing seaweeds. Its identity cannot be established because type material is absent and the description is insufficient. Therefore, *M. maritimum* must be included in the list of *nomina dubia*.

42. *Septoria dictyotae* Oudemans, *Versl. Zittingen Wis- Natuurkd. Afd. K. Akad. Wet. Amsterdam* **3**, 54, 1894
This species from *Dictyota obtusangula* was insufficiently described and has not been collected since its description. We have been unable to locate type material and consider *S. dictyotae* to be a *nomen dubium*.

43. *Sirococcus posidoniae* Rolland, *Bull. Soc. Mycol. Fr.* **12**, 6, 1896
The fungus was described from rhizomes of *Posidonia oceanica*. Ollivier (1928) noted a resemblance of *S. posidoniae* with male conceptacles of *Melobesia membranacea*. Type material could not be located and we consider *S. posidoniae* to be a *nomen dubium*.

44. *Sporotrichum maritimum* Sutherland, *New Phytol.* **15**, 43, 1916
Type material of this insufficiently described species from *Laminaria* sp. is absent. Therefore, *S. maritimum* is considered to be a *nomen dubium*.

45. *Stemphylium codii* Zeller, *Publ. Puget Sound Biol. Stn., Univ. Wash.* **2**, 123, 1918
We have been unable to locate type material of this incompletely known saprobic species from *Codium mucronatum*. *Stemphylium codii* is considered to be a *nomen dubium*.

46. *Stemphylium maritimum* Johnson, *Mycologia* **48**, 844, 1956
It is not possible to classify this species properly because details of conidial ontogeny are unknown. Unfortunately, many authors list *Stemphylium*-like fungi from marine habitats under this name, although the identity cannot be established because type material is not available. We consider *S. maritimum* to be a *nomen dubium*.

47. *Trichocladium constrictum* I. Schmidt, *Biol. Rundsch.* **12**, 102, 1974
There is no description available for this organism. A publication containing descriptions for this and the following two species is in press (I. Schmidt, personal communication). For the time being, *T. constrictum* is a *nomen nudum*.

48. *Trichocladium lignincola* I. Schmidt, *Biol. Rundsch.* **12**, 102, 1974
See the remarks under *T. constrictum* (No. 47). *Trichocladium lignincola* is a *nomen nudum*.

49. *Vesicularia marina* I. Schmidt, *Biol. Rundsch.* **12**, 102, 1974
See the remarks under *T. constrictum* (No. 47). *Vesicularia* and its type species *V. marina* are *nomina nuda*.

28. Yeasts

I. INTRODUCTION

"Yeasts are those fungi which, in a stage of their life cycle, occur as single cells, reproducing by budding or fission." This definition by Kreger-van Rij (1973) embraces a large number of organisms, not belonging to a natural taxonomic group. In this survey of marine-occurring yeasts, we have compiled 177 species, which were isolated from seawater, sediments, plants, animals, and other organic matter in the marine habitat. We have tentatively divided them into an "obligate" and a "facultative" group. "Obligate marine" yeasts are those yeasts that, thus far, have never been collected anywhere *but* in the marine environment, whereas "facultative marine" yeasts are also known from terrestrial habitats. "Obligate marine" species may be confined to marine habitats, especially if they have been collected frequently and exclusively in the sea for several years. Other yeasts listed in this tentative group may be rare and, for that reason, were not found on land, but may be reported from terrestrial areas in the future.

Yeast concentrations have been observed to decrease with increasing remoteness from land (e.g., Ahearn *et al.*, 1968; Hoppe, 1972a; Taysi and van Uden, 1964), and certain yeast species frequently collected in seawater were obtained in highest quantities in the vicinity of heavily polluted areas (Fell and van Uden, 1963). These phenomena could be explained by the lack or the availability of nutrients, respectively. However, such facts could also indicate that the collected organisms were merely contaminants from terrestrial sources, surviving passively in the sea. As in the higher fungi (Chapter 1), the question arises: Are there indigenous marine yeasts? The majority of reports on yeasts from marine environments are based upon indirect collection methods, such as incubation of seawater, sediments, and diverse substrates found in the sea. With such culture techniques, cells may begin to grow *in vitro* that were dormant and would have remained inactive in the marine habitats. Yeast species that have also been found in fruits, soil, domestic animals, and man, to name just a few sources, are most likely not native to estuaries and seas, even if they were isolated from such areas many times. It is more probable that they were washed into marine waters by way of rivers or sewage, or with dust blown seaward by the wind. Observations such as exceptionally high yeast densities following *Noctiluca* blooms in the North Sea (Meyers *et al.*, 1967a) could indicate the presence of indigenous species, but insufficient data did not allow these authors to draw definite conclusions. In addition, the area in question is polluted by sewage disposal and regular passenger-liner traffic (Lüneburg in Gunkel, 1963). Kriss and Rukina (1949) also found plankton blooms in the Black Sea and the Pacific Ocean to be locations of greatest density of yeast populations in the sea. Kriss and Novozhilova (1954) reported that budding yeasts were observed by direct microscopic examination of water samples down to depths of 2000 m. This fact would be evidence for growth of yeasts in seawater; however, the collecting technique with Nansen bottles used by Kriss and co-workers was questioned later when such containers were found to be easily contaminated (Sorokin, 1964).

A truly marine yeast must be able to grow on or in a marine substrate. Incubation of collected substances from the sea cannot prove that this requirement is met; therefore, the organisms must be captured *in vivo* and observed immediately after recovery to ascertain that reproduction has begun *in situ*. Such practices are used with higher fungi from the sea, organisms that are easier to identify because they are more diversified morphologically than yeasts. Investigations of this nature have also been conducted with morphologically indistinctive forms such as bacteria, caught and cultured in the deep sea by Jannasch and Wirsen (1973), and with lower marine fungi by Ulken (1974), who experimented with submerged baits. Direct examination of living marine invertebrates, however, has demonstrated the presence of parasitic and pathogenic yeasts (Fize *et al.*, 1970; Roth *et al.*, 1962; Seki and Fulton, 1969), and, if such species have grown *in situ* in the animal and its native habitat, they could rightly be called indigenous marine.

So far, no physiological clues have been found to explain why a marine-occurring yeast is able to live in this special habitat. Salinity tolerance does not distinguish marine species from terrestrial species, because almost all yeasts can grow in sodium chloride concentrations exceeding those normally present in the sea.

A. Substrates of Marine-Occurring Yeasts

1. Seawater

The majority of yeast collections were obtained by isolation from seawater samples (see Table XXII), a fact that may reflect a preference of investigators for this milieu, rather than the actual distribution of yeasts in the sea. Since the first yeasts were isolated from seawater in 1886 (Fischer and Brebeck, 1894), yeast-containing water samples have been collected in all oceans and seas, from arctic and antarctic waters to the tropics, from littoral habitats to the open ocean, and in the deep sea (e.g., Bhat and Kachwalla, 1955; Fell, 1967, 1976; Kriss, 1963; Kriss *et al.*, 1967; Morris, 1968; van Uden and Fell, 1968). Fell (1976), in his extensive study of yeasts in oceanic regions, noted that the distribution varies from ubiquitous species such as *Debaryomyces hansenii* and its imperfect stage *Torulopsis candida* to those that appear limited by geographical or hydrographical conditions, such as *Leucosporidium antarcticum,* which was only found adjacent to antarctic pack ice.

2. Animals

Animals are a frequent source of marine-occurring yeasts. Half of the species from animals have been recorded from invertebrates and one-third from fish, while birds yielded only 13 of 103 species, and mammals merely 2.

a. Invertebrates. To obtain the isolates, either invertebrates were shaken in sterile seawater and swabs were made from surfaces, sections of the animals were directly placed on culture media, or gut contents, liquid

TABLE XXII
Numbers of Yeast Species Collected in Marine Environments

Seawater		133
Animals		
Invertebrates	55	
Fish	39	
Birds (including bird droppings)	13	
Mammals	2	
	Animals, total	109
Sediments, mud, and littoral sands		47
Algae		45
Plankton		17
Oil slicks		9
Higher plants		5

portions, or internal tissues were taken aseptically. Buck *et al.* (1977), by investigating bivalve shellfish in Long Island Sound, noted that, in general, the liquid portion of the shellfish contained more yeasts than the internal viscera, while Siepmann and Höhnk (1962) reported that about half the number of species found were from the internal parts of the animals and about half from surface swabs.

Yeast populations from conch and spiny lobster on a Bahama Island were studies by Volz *et al.* (1974). They isolated fewer yeasts from the animals than from marine sand and sediment of the same habitat and assumed that the isolates were probably ingested during feeding and did not seem to cause stress to their hosts. Chrzanowski and Cowley (1977) found *Rhodotorula glutinis* and *Torulopsis ernobii* in the guts of the fiddler crab *Uca pugilator* as well as in the soil of its environment. It was speculated that these yeasts might serve as food, but feeding experiments showed that they could not be utilized as a sole food source by the crabs.

In assimilation tests, Siepmann and Höhnk (1962) found strong formation of riboflavin by *Debaryomyces subglobosus* (=*D. hansenii*), a yeast they frequently isolated from internal liquids of invertebrates, and the authors suggested that this yeast may serve as a vitamin source for marine animals.

Pathogens and parasites were discovered by direct examination of invertebrates. The whole body of the amphipod *Podoceras brasiliensis* was invaded by *Rhodotorula minuta* (Roth *et al.*, 1962), and the commercially raised brine shrimp, *Artemia salina,* was parasitized by *Metschnikowia bicuspidata* var. *australis,* a yeast that appears to be equipped with an active predatory mechanism, attacking its host by forcible ascospore discharge (Lachance *et al.*, 1976). Seki and Fulton (1969) showed that the tissues of living marine copepods *(Calanus plumchrus)* were attacked by *Metschnikowia* spp. The authors supposed, however, that the yeasts were of terrestrial origin, because they were found only in winter when the waters of the collecting site, the Strait of Georgia, were strongly influenced by freshwater inflow. In addition, Fize *et al.* (1970) reported a *Metschnikowia* sp. parasitizing living copepods (*Eurytemora velox*) in southern France.

b. Fish. Yeasts associated with fish were isolated from skin, gills, mouths, feces, and gut contents of the animals. Ross and Morris (1965) report the greatest variety and highest number of yeasts from fish skins, while gills produced less and feces much lower yields. In gut contents, van Uden and Castelo-Branco (1963) found high humbers of *Metschnikowia zobellii*. Studies of the intestinal yeast flora of Atlantic fish indicated to the investigators (Fell and van Uden, 1963) that the yeasts were ingested with food and appeared to be restricted both quantitatively and qualitatively to the digestive tract of the fish examined. In the Pacific, van Uden and Castelo-Branco (1963) found certain fish species containing significantly higher numbers of yeast cells than the surrounding seawater, and the authors believe that these yeasts may be able to grow in the intestines of some fish.

c. Birds. Gut and rectal contents of free-living gulls and terns were found to harbor a number of yeasts. Among the most frequently isolated species were *Torulopsis glabrata* and *Candida tropicalis* in birds of the Atlantic Ocean (Portugal: Kawakita and van Uden, 1965) as well as in those of the Pacific Ocean (California: van Uden and Castelo-Branco, 1963). Kawakita and van Uden (1965) suggest that gulls and terns seem to inoculate yeasts with their feces into water bodies all over the world. However, yeasts occurring in gulls were not always found in seawater of the area where the birds were caught (van Uden and Castelo-Branco, 1963), and the authors assumed that low water temperatures can prevent a buildup of detectable yeast populations.

Isolations from shore bird droppings on southern California beaches (Dabrowa *et al.*, 1964) yielded species also occurring in rectal contents of seagulls.

d. Mammals. *Candida tropicalis* was found in the stomachs of marine mammals, namely, a dolphin and a porpoise (Morii, 1973). According to the author, the yeast was probably ingested with indigenous food or seawater. *Rhodosporidium toruloides* was isolated from the intestine of another porpoise that died in captivity (Fell *et al.*, 1970). The above-mentioned three whales are the only marine mammals reported to harbor yeasts. Van Uden and Castelo-Branco (1963), who found no yeasts in intestinal samples from eight California sea lions, say that warm-blooded animals with a high intake of food rich in protein such as fish are, in general, unsuitable hosts for intestinal yeasts.

3. Sediments and Littoral Sands

The investigation of sediments, tideland mud, and littoral sands brought various results. Fell and van Uden (1963) sampled sediments with coring devices and found active yeast populations to be confined to the upper 2 cm of sediment at water depths of 540 m in the Gulf Stream. In shallow Florida waters, however, where strong wave actions and rapid settling of sediments prevail, yeasts were found in coring sample depths to 9 cm. The authors conclude that availability of oxygen is the limiting factor for the growth processes of yeasts within the sediment. Fell *et al.* (1960) reported that, in general, silty muds in their Florida sampling area harbored more yeasts than sandy sediments. They also submerged banana stalks in water and found much greater species diversity in these baits than in sediments;

for instance, in one locality 17 yeast species were isolated from the baits, but only 2 spp. from the sediments beneath them. On the other hand, in the studies of Volz et al. (1974), frequency of isolation and numbers of yeast species were greater in sands and sediments than in two invertebrates of the same area. Similarly, grasses and algae in the investigations of Roth et al. (1962) showed lower cell counts and smaller numbers of species than surrounding waters and sediments of the grass beds.

Meyers et al. (1971a) counted very high concentrations of viable cells in sediments of Spartina alterniflora marshes at the Louisiana coast, much higher than in sediments of adjacent bayous and in water samples. Species of Pichia and Kluyveromyces were predominant and occurred most commonly in the culm-sediment region of the Spartina plants (Ahearn and Meyers, 1972).

4. Algae

Several investigations deal with populations of yeasts on seaweeds. Patel (1975) found that living algae contained lower counts of yeasts compared to counts in the surrounding seawater, but, when decomposition of the plants set in, yeasts in the algal material increased to higher numbers than those found in surrounding seawater. Maximal counts of epiphytic yeasts were reported from rhodophytes and the one chlorophyte investigated by Seshadri and Sieburth (1975), while lower populations on phaeophytes were attributed to the release of inhibitory phenolic substances by these plants. Seasonal variations in numbers of yeasts in surrounding seawater paralleled those on the algae. The great majority of all isolates were strains of Candida, and only a very small fraction consisted of Rhodotorula spp. The authors considered the possibility that the yeasts may utilize exudates of their living hosts.

More numerous studies were made on yeast growth on decomposing algae, which are often found piled in deep masses on rocky shorelines. Bunt (1955) examined the decomposition of giant kelp at Macquarie Island, Antarctica, both in nature and in the laboratory. He observed a wide range of microorganisms, and yeasts were present in large numbers within the kelp tissues. Excrements of seals favored microbial activity, and kelp fly larvae fed on the resulting slime. Kelp fly adults were probably important in spreading inocula of microorganisms. The extensive banks of dead seaweeds were finally converted into a foul-smelling, semiliquid mass that was eventually washed back into the sea. The author assumed that the marine life of Macquarie Island benefits greatly from the kelp food source, and even the sustenance of large populations of seals, penguins, and birds of flight might be assisted by this decomposed plant material.

Van Uden and Castelo-Branco (1963) isolated yeasts from giant kelp in

southern California and found *Metschnikowia zobellii* on all samples
yielding yeasts, except for one.

5. Plankton

Studies on zoo- and phytoplankton revealed more than 20 yeast
species. Fell *et al.* (1973) isolated several *Rhodosporidium* species (Fig.
129) from plankton samples at various water depths in the Pacific Ocean.
Suehiro (1962) detected yeasts on the diatom *Thalassiosira subtilis* in
Japan. He noticed low numbers of yeasts in nature early in the year, but
much higher counts in April when the diatoms began to decay. The same
results were obtained in *Thalassiosira* cultures in flasks. In experiments
with pure cultures of plankton, Suehiro *et al.* (1962) found that apparently
more than 50% of all cell constituents of the cultured phytoplankton were
converted to yeast cells.

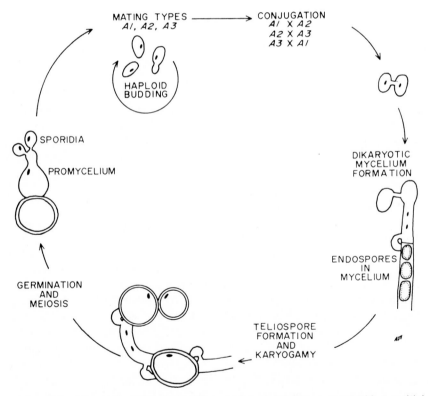

Fig. 129. Life cycle of *Rhodosporidium infirmo-miniatum,* a yeast with a multial-
lelic bipolar incompatibility system. From Fell *et al. Can. J. Microbiol.* **19,** 645 (1973),
Fig. 20. Reproduced by permission of the National Research Council of Canada.

6. Oil Slicks

The responses of yeast populations to oil pollution were investigated by Ahearn and Meyers (1972). Plots of a *Spartina alterniflora* salt marsh in Louisiana were selected as test areas and were saturated with oil. Compared with adjacent control sites, a considerable increase in yeast densities was noticed in the oil-soaked plots, and the predominant yeasts of the marshland were replaced by hydrocarbonoclastic strains, especially *Pichia ohmeri* and *Trichosporon* sp. In the nutrient-rich sediments of the estuary, populations of yeasts continued to increase in the presence of oil. In offshore areas, however, yeast populations declined after an initial increase, perhaps due to a lack of nutrients and vitamins. It is suggested that the tested organisms may have relatively low capacities to decompose crude oil at oil spillage sites.

In the laboratory, Ahearn *et al.* (1971b) tested selected yeasts isolated from oil-polluted habitats for their ability to use hydrocarbons as sole sources of carbon. Most yeasts adhered to the surface of the oil globules, often producing small depressions and, with growing numbers of yeast cells, finally dividing big globules into smaller ones. This process disrupted the surface film and caused emulsification of the oil in the broth. Hyphal cells of a *Trichosporon* sp. were able to penetrate the globule and filled its interior with mycelia, converting it into a small fungal colony with many enclosed small oil droplets. Authors of both aforementioned papers noted that, in general, strains of species isolated from oil-polluted regions exhibited much higher carbonoclastic facilities than the same species from non-polluted areas.

Le Petit *et al.* (1970) studied oil-polluted littoral marine areas in the Mediterranean and found seven species that were able to metabolize hydrocarbon fractions. From non-polluted test sites, only one hydrocarbonoclastic species was isolated. Biodegradation was very slow, and the authors conclude that yeasts probably play only a minor role in the elimination of hydrocarbons from the sea.

7. Higher Plants

Few identified yeast species have been recorded from higher plants of the seashore. In a Louisiana salt marsh, Meyers *et al.* (1975) studied the yeast populations on living *Spartina alterniflora* plants. *Pichia spartinae* and *Kluyveromyces drosophilarum* were found on the outer surfaces of the culm, but the former species is of special interest because it occurred in great concentrations in the plant's intraculm cell liquid and viable tissue. In laboratory experiments, the authors established that *Pichia spartinae* is able to use cellulosic plant breakdown products and they speculated that the yeast could, therefore, play an important part in the cellulosic biodegradation of the marsh grasses.

Yeast populations of *Sporobolomyces roseus* on marsh plants in England were investigated by Pugh and Lindsay (1975). They sampled leaves from drift material and growing plants in the intertidal zone to high points on the beach that were never reached by tides. The tidal washings had an unfavorable effect on the yeasts by significantly reducing cell numbers on the leaves. Leaves of inland plants harbored much higher cell numbers of *Sporobolomyces roseus* than those near the shore. In laboratory experiments, increased salinities retarded growth of the yeast. Therefore, a possible reason for the lower cell counts on shore plants may be the salt depositions on leaves, caused by tides and seawater spray with subsequent drying. Strains from *Halimione portulacoides* were slightly more salt tolerant than those from inland plants. Pugh and Lindsay believe that *Sporobolomyces roseus* probably does not play an important role in the marine environment and that its growth and respiration patterns are more comparable to terrestrial than to marine fungi.

Finally, Newell (1976) mentioned blooms of *Rhodotorula rubra* and *Debaryomyces hansenii* on submerged seedlings of *Rhizophora mangle*.

B. General Comments

In nomenclatural questions, the extensive taxonomic treatise of Lodder (1970) has been the guideline for names and synonyms. The suprageneric classification follows von Arx *et al.* (1977). After the completion of this chapter, it was brought to our attention that a revision of *The Yeasts* (Lodder, 1970) is in preparation, with N. J. W. Kreger-van Rij as editor. Because this volume of *The Yeasts* will correct and update the yeast taxonomy, our text has been limited to citations of literature, sources of marine origin, and range of collections. Descriptions of genera and species were omitted and can be found in the forthcoming detailed account of this group. Names used in the following lists are those applied by collectors of marine-occurring yeasts. It is not within the scope of our treatise to make any revisions or corrections in the yeasts, but we feel that a compilation of all the literature concerning yeasts found in seawater and their substrates and geographical distribution will be useful to researchers not specialized in this field.

We have not listed all synonyms of a species, only those that were used in the literature dealing with marine-occurring yeasts. If, in that literature, a yeast species was mentioned under a synonymous name only (according to Lodder, 1970), it is treated by this name, but a parenthetical note marks it as a synonym [e.g., *Candida silvicola* Shifrine et Phaff 1956 (syn. of

Hansenula holstii Wickerham 1960)]. Yeasts identified to genus only have not been considered.

Keys and technical informations are excluded to avoid duplication with the yeast literature. For keys and general information on yeasts, the books by Barnett and Pankhurst (1974), Lodder (1970), and Phaff *et al.* (1978) and Kreger-van Rij's revision should be consulted. Methods for sampling and isolation of yeasts from marine sources have been presented by Fell (1976), Kriss (1963), and van Uden and Fell (1968), and the reader is referred to these publications.

II. OBLIGATE MARINE YEASTS

A. Ascomycotina
Metschnikowiaceae

Metschnikowia Kamienski

Trav. Soc. Nat. St. Petersbourg (Leningrad) **30**, 363–364, 1899

TYPE SPECIES: *Metschnikowia bicuspidata* (Metschnikoff) Kamienski. ETYMOLOGY: Named in honor of I. I. Metschnikoff (1845–1916), Russian microbiologist.

Metschnikowia bicuspidata (Metschnikoff) Kamienski var. *australis* Fell et Hunter

Antonie van Leeuwenhoek **34**, 369, 1968

LITERATURE: Lachance *et al.* (1976); Miller and van Uden (1970); Pitt and Miller (1970).
ORIGIN: Animals (commercially reared brine shrimp: *Artemia salina*), seawater. RANGE: Atlantic Ocean—Antarctic Sea. ETYMOLOGY: From the Latin, *australis* = southern, in reference to the occurrence of the variety.

Metschnikowia bicuspidata (Metschnikoff) Kamienski var. *californica* Pitt et Miller

Antonie van Leeuwenhoek **36**, 365–366, 1970

ORIGIN: Seawater. RANGE: Pacific Ocean—United States (California). ETYMOLOGY: Named after the origin of the type material.

Metschnikowia krissii (van Uden et Castelo-Branco) van Uden

Rev. Biol. (Lisbon) **3**, 95–96, 1962

≡*Metschnikowiella krissii* van Uden et Castelo-Branco

LITERATURE: Miller and van Uden (1970); Pitt and Miller (1970); van Uden (1967); van Uden and Castelo-Branco (1961).
ORIGIN: Seawater. RANGE: Pacific Ocean—United States (California). ETYMOLOGY: Named in honor of A. E. Kriss, marine microbiologist in the U.S.S.R.

Saccharomycetaceae

Kluyveromyces van der Walt emend. van der Walt

Antonie van Leeuwenhoek **31**, 347, 1965

TYPE SPECIES: *Kluyveromyces polysporus* van der Walt. ETYMOLOGY: Named after A. J. Kluyver (1888–1956), Dutch microbiologist.

Kluyveromyces aestuarii (Fell) van der Walt

Antonie van Leeuwenhoek **31**, 347, 1965

≡*Saccharomyces aestuarii* Fell, *Antonie van Leeuwenhoek* **27**, pp. 29–30, 1961

LITERATURE: Ahearn and Roth (1962); van der Walt (1970); van Uden and ZoBell (1962).
ORIGIN: Marine mud, seawater, varied marine substrates. RANGE: Atlantic Ocean—United States (Florida). ETYMOLOGY: From the Latin, *aestuarium* = estuary, referring to the habitat of the species.

Pichia Hansen

Centralbl. Bakteriol., Parasitenkd., Infektionskr. Hyg., Abt. 2 **12**, 538, 1904

TYPE SPECIES: *Pichia membranaefaciens* Hansen. ETYMOLOGY: Named in honor of Professor Pichi (see Hansen, 1904; no data available).

Pichia spartinae Ahearn, Yarrow et Meyers

Antonie van Leeuwenhoek **36**, 505, 1970

LITERATURE: Ahearn and Meyers (1972); Ahearn *et al.* (1971a,b); Kurtzman and Ahearn (1976); Meyers *et al.* (1971a,b, 1975).

ORIGIN: *Spartina alterniflora*, marsh sediment. RANGE: Atlantic Ocean—United States (Louisiana). NOTE: Spencer *et al.* (1974) reported isolating *P. spartinae* from freshwater; however, Kurtzman and Ahearn (1976), who examined strains of those collections, determined that the strains, in contrast to the type description, failed to grow on osmotic medium or at 37°C; the latter authors, therefore, conclude that the Canadian isolates may represent a different species. ETYMOLOGY: Named after the host genus, *Spartina* (Gramineae).

B. Basidiomycotina

Sporobolomycetaceae

Leucosporidium Fell, Statzell, Hunter et Phaff

Antonie van Leeuwenhoek **35**, 438, 1969

TYPE SPECIES: *Leucosporidium scottii* Fell, Statzell, Hunter et Phaff. ETYMOLOGY: From the Greek, *leuco-* = white, and the Latin *sporidium* = sporidium, referring to the cream color of the cultures of this genus.

Leucosporidium antarcticum Fell, Statzell, Hunter et Phaff

Antonie van Leeuwenhoek **35**, 447, 1969

LITERATURE: Fell (1972, 1976); Fell and Phaff (1970).
ORIGIN: Seawater. RANGE: Atlantic, Indo-Pacific, and Pacific Oceans—Antarctic Sea. ETYMOLOGY: From the Latin *antarcticus* = antarctic, referring to the origin of the type material.

Rhodosporidium Banno

J. Gen. Appl. Microbiol. **13**, 192–193, 1967

TYPE SPECIES: *Rhodosporidium toruloides* Banno. ETYMOLOGY: From the Greek, *rhodo-* = rosy-red, and the Latin, *sporidium* = sporidium, referring to the pink color of the cultures of this genus.

Rhodosporidium bisporidii Fell, Hunter et Tallman

Can. J. Microbiol. **19**, 655, 1973 (as *Rh. bisporidiis*, see *Index of Fungi*, Vol. 4, 225, 1974)

LITERATURE: Fell (1972, 1976).
ORIGIN: Seawater. RANGE: Indian and Pacific Oceans—Antarctic Sea.

ETYMOLOGY: From the Latin, *bi-* = two- and *sporidium* = sporidium, referring to the presence of two distinct sporidial types on the basidium.

Rhodosporidium capitatum Fell, Hunter et Tallman

Can. J. Microbiol. **19**, 655–656, 1973

LITERATURE: Fell (1972, 1974, 1976).
ORIGIN: Seawater. RANGE: Atlantic Ocean—United States (Florida); Indian Ocean—Antarctic Sea; Pacific Ocean—Antarctic Sea, New Zealand. ETYMOLOGY: From the Latin, *capitatus* = with a knoblike head, referring to the enlarged terminal apex of the basidium.

Rhodosporidium dacryoidum Fell, Hunter et Tallman

Can. J. Microbiol. **19**, 656, 1973

LITERATURE: Fell (1972, 1974, 1976).
ORIGIN: Seawater. RANGE: Indian and Pacific Oceans—Antarctic Sea. ETYMOLOGY: From the Greek, *dacryon* = tear, and the diminutive suffix *-idium,* referring to the shape of most of the teliospores.

Rhodosporidium malvinellum Fell et Hunter *in* Fell

In Recent Trends in Yeast Research, D. G. Ahearn, ed. Georgia State University, Atlanta, p. 59, 1970

LITERATURE: Fell (1976); Newell and Hunter (1970).
ORIGIN: Seawater. RANGE: Atlantic, Indian, and Pacific Oceans— Antarctic Sea. ETYMOLOGY: From the Latin, *malvinus* = mauve, and the diminutive suffix *-ellus*, in reference to the mauve color of the culture.

Rhodosporidium sphaerocarpum Newell et Fell

Mycologia **62**, 276, 1970

LITERATURE: Fell (1970); Moore (1972); Newell and Hunter (1970).
ORIGIN: Seawater. RANGE: Atlantic Ocean—Antarctic Sea, Caribbean. ETYMOLOGY: From the Greek, *sphaira* = globe and *carpos* = fruit, in reference to the globose teliospores.

C. Deuteromycotina
Torulopsidaceae

Candida Berkhout

Thesis, Utrecht, pp. 41–42, 1923

TYPE SPECIES: *Candida vulgaris* Berkhout. ETYMOLOGY: From the Latin, *candidus* = shining white, referring to the appearance of the colonies in culture.

Candida glaebosa Komagata et Nakase

J. Gen. Appl. Microbiol. **11**, 262, 1965

LITERATURE: Seshadri and Sieburth (1975); van Uden and Buckley (1970).
ORIGIN: Algae, animals (frozen cuttlefish). RANGE: Pacific Ocean—Japan; Atlantic Ocean—United States (Rhode Island). NOTE: *C. glaebosa* may be a facultative marine species, because it was reported from freshwater in Canada (Saskatchewan) by Spencer *et al.* (1974). ETYMOLOGY: From the Latin, *glaebosus* = lumpy, referring to the tendency of lump formation of cells in this species.

Candida krissii Goto et Iizuka

In Goto, Yamasato, and Iizuka, *J. Gen. Appl. Microbiol.*
20, 313–314, 1974

ORIGIN: Seawater. RANGE: Pacific Ocean. ETYMOLOGY: Named in honor of A. E. Kriss, marine microbiologist in the U.S.S.R.

Candida marina van Uden et ZoBell

Antonie van Leeuwenhoek **28**, 278, 1962

LITERATURE: Seshadri and Sieburth (1975); van Uden (1967); van Uden and Buckley (1970).
ORIGIN: Algae, seawater. RANGE: Indian Ocean—Australia (Queensland); Atlantic Ocean—United States (Rhode Island). ETYMOLOGY: From the Latin, *marinus* = marine, referring to the habitat of the species.

Candida salmonicola Komagata et Nakase

J. Gen. Appl. Microbiol. **11**, 255, 1965

LITERATURE: Seshadri and Sieburth (1975); van Uden and Buckley (1970).
ORIGIN: Algae, animals (frozen salmon). RANGE: Pacific Ocean—Japan; Atlantic Ocean—United States (Rhode Island). ETYMOLOGY: From salmon (pisces) and the Latin, *-cola* = -dweller, in reference to the host of the species.

Candida suecica Rodrigues de Miranda et Norkrans

Antonie van Leeuwenhoek **34**, 115–116, 1968

LITERATURE: Lundström-Eriksson and Norkrans (1968).
ORIGIN: Seawater. RANGE: Atlantic Ocean—Sweden. ETYMOLOGY: From the Latin, *suecicus* = from Sweden, referring to the origin of the type culture.

Cryptococcus Kützing emend. Phaff et Spencer

Proceedings of the Second International Symposium on Yeasts,
Bratislava (Czechoslovakia), pp. 59–65, 1969

TYPE SPECIES: *Cryptococcus neoformans* (Sanfelice) Vuillemin. ETYMOLOGY: From the Greek, *kryptos* = covered, hidden, concealed, and *kokkos* = berry, seed, probably referring to the capsule-covered cells.

Cryptococcus lactativorus Fell et Phaff

Antonie van Leeuwenhoek **33**, 469–470, 1967

LITERATURE: Buck *et al.* (1977); Phaff and Fell (1970).
ORIGIN: Animals (mussel: *Mytilus edulis;* oyster: *Crassostrea virginica;* quahog: *Mercenaria mercenaria*), seawater. RANGE: Atlantic Ocean—Antarctic Sea, United States (Long Island Sound). ETYMOLOGY: From the Latin, *lactosum* = lactose, and the suffix *-vorus* = consuming, referring to the ability of the species to use lactic acid strongly.

Rhodotorula Harrison

Trans. R. Soc. Can., Sect. 5 **22**, 189, 1928

TYPE SPECIES: *Rhodotorula glutinis* (Fres.) Harrison. ETYMOLOGY:

From the Greek, *rhodos* = red, and the genus *Torula* (syn. of *Torulopsis*), in reference to the red pigments of the colonies.

Rhodotorula glutinis var. *salinaria* Hirosawa et Takada 1969

Trans. Mycol. Soc. Jpn. **10**, 38, 1969

LITERATURE: Ito and Takada (1974, 1976); Joho (1970); Matsuda and Takada (1973).
ORIGIN: Seawater. RANGE: Pacific Ocean—Japan. ETYMOLOGY: From the Latin, *salinus* = saline, and the substantival suffix, *-arius,* indicating a place where something is done, in reference to the salt farm where the variety was isolated.

Rhodotorula marina Phaff, Mrak et Williams

Mycologia **64**, 436, 1952

LITERATURE: Fell *et al.* (1960); Goto *et al.* (1974a,b); Volz *et al.* (1974); Yamasato *et al.* (1974).
ORIGIN: Animals (shrimp), seawater, deep-sea and shallow-water sediments. RANGE: Atlantic Ocean—Bahamas, United States (Texas); Pacific Ocean—Japan and open ocean. ETYMOLOGY: From the Latin, *marinus* = marine, referring to the habitat of the species.

Sterigmatomyces Fell

Antonie van Leeuwenhoek **32**, 101, 1966

TYPE SPECIES: *Sterigmatomyces halophilus* Fell. ETYMOLOGY: From the Greek, *sterigmos* = support (Latinized = *sterigma*) and *mykes* = fungus, referring to the development of conidia on sterigmata.

Sterigmatomyces indicus (Fell) Fell

In The Yeasts, J. Lodder, ed., North-Holland Publ., Amsterdam, p. 1232, 1970 (not validly published, see Article 33 of the International Code of Botanical Nomenclature)

≡*Sterigmatomyces halophilus* var. *indicus* Fell, *Antonie van Leeuwenhoek* **32**, 102–103, 1966

ORIGIN: Seawater. RANGE: Atlantic Ocean—Bahamas; Indian Ocean. ETYMOLOGY: From the Latin, *indicus* = Indian, in reference to the collection in the Indian Ocean.

Sympodiomyces Fell et Statzell

Antonie van Leeuwenhoek **37**, 361, 1971

TYPE SPECIES: *Sympodiomyces parvus* Fell et Statzell. The genus is monotypic. ETYMOLOGY: From *sympodium* = sympodium and the Greek, *mykes* = fungus, in reference to the sympodial development of conidia on the conidiophore.

Sympodiomyces parvus Fell et Statzell

Antonie van Leeuwenhoek **37**, 362, 1971

LITERATURE: Fell (1972, 1976).
ORIGIN: Seawater. RANGE: Indian, Indo-Pacific, and Pacific Oceans—Antarctic Sea. ETYMOLOGY: From the Latin, *parvus* = little, small, in reference to the small size of the cells.

Torulopsis Berlese

G. Vitic. Enol. **3**, 54, 1895

TYPE SPECIES: *Torulopsis colliculosa* (Hartmann) Saccardo. ETYMOLOGY: From the Latin, *torulosus* = cylindrical with bulges, and the Greek, *opsis* = aspect, appearance, in reference to the budding cells.

Torulopsis austromarina Fell et Hunter

Antonie van Leeuwenhoek **40**, 309, 1974

LITERATURE: Fell (1972, 1976).
ORIGIN: Seawater. RANGE: Indian and Indo-Pacific Oceans—Antarctic Sea. ETYMOLOGY: From the Latin, *austro-* = south- and *marinus* = marine, in reference to the marine occurrence of the species in the Southern Hemisphere.

Torulopsis haemulonii van Uden et Kolipinski

Antonie van Leeuwenhoek **28**, 78–79, 1962

LITERATURE: Ahearn *et al.* (1968); van Uden (1967); van Uden and Vidal-Leiria (1970).
ORIGIN: Animals (gut contents of fish: *Haemulon sciurus*), seawater. RANGE: Atlantic Ocean—Portugal, United States (Florida). ETYMOLOGY: Named after the host, *Haemulon sciurus* (pisces).

Torulopsis maris van Uden et ZoBell

Antonie van Leeuwenhoek **28**, 281, 1962

LITERATURE: van Uden (1967); van Uden and Vidal-Leiria (1970). ORIGIN: Seawater. RANGE: Pacific Ocean—Australia (near Queensland). ETYMOLOGY: From the Latin, *mare* = the sea, in reference to the habitat of the species.

Torulopsis torresii van Uden et ZoBell

Antonie van Leeuwenhoek **28**, 279–280, 1962

LITERATURE: van Uden (1967); van Uden and Vidal-Leiria (1970). ORIGIN: Seawater. RANGE: Pacific Ocean—Australia (Queensland). ETYMOLOGY: Named after the location of the type collection, the Torres Strait (Australia).

III. FACULTATIVE MARINE YEASTS*

A. Ascomycotina
Metschnikowiaceae

Metschnikowia zobellii (van Uden et Castelo-Branco) van Uden

Rev. Biol. (Lisbon) **3**, 95–96, 1962 (not validly published, see Article 33 of the International Code of Botanical Nomenclature)

≡*Metschnikowiella zobellii* van Uden et Castelo-Branco, *J. Gen. Microbiol.* **26**, 144, 1961

LITERATURE: Fell and Pitt (1969); Fell and van Uden (1963); Pitt and Miller (1970); van Uden (1967); van Uden and Castelo-Branco (1961, 1963).

MARINE ORIGIN: Algae (*Macrocystis pyrifera*), animals (fish: *Atherinopsis affinis littoralis, Trachurus symmetricus*), seawater. RANGE: Pacific Ocean—United States (California).

---

* Terrestrial yeasts isolated from the marine environment. Synonyms according to Lodder (1970), where references can be obtained; full literature citations are also found in the same treatise. When no country is mentioned, collections were made in the open ocean.

Saccharomycetaceae

Debaryomyces globosus Klöcker 1909 [syn. of *Saccharomyces kloeckerianus* van der Walt 1970 (not validly published, see Article 33 of the International Code of Botanical Nomenclature)]

LITERATURE: Kriss *et al.* (1967).
MARINE ORIGIN: Seawater. RANGE: Atlantic Ocean—North Atlantic; Pacific Ocean.

Debaryomyces hansenii (Zopf) Lodder et Kreger-van Rij 1952

> =*Debaryomyces guilliermondii* Dekker (Stelling-Dekker 1931)
> =*Debaryomyces kloeckeri* Guilliermond et Péju 1919
> =*Debaryomyces nicotianae* Giovannozzi 1939
> =*Debaryomyces subglobosus* (Zach) Lodder et Kreger-van Rij 1952

LITERATURE: Ahearn and Roth (1962); Bhat and Kachwalla (1955); Bhat *et al.* (1955); Bruce and Morris (1973); Capriotti (1962a,b); Crow *et al.* (1977); Fell (1967, 1972, 1976); Fell and van Uden (1963); Fell *et al.* (1960); Gezelius and Norkrans (1970); Goto *et al.* (1974a,b); Hoppe (1972a,b); Kriss *et al.* (1967); Lundström-Eriksson and Norkrans (1968); Meyers *et al.* (1967a,b); Newell (1976); Norkrans (1966a,b, 1968, 1969b); Norkrans and Kylin (1969); Picci and Verona (1964); Ross and Morris (1962, 1965); Roth *et al.* (1962); Siepmann and Höhnk (1962); Volz *et al.* (1974); Yamasato *et al.* (1974).
MARINE ORIGIN: Algae (*Ascophyllum nodosum*), animals (octopus: *Octopus vulgaris;* sea cucumber; shrimp eggs; spider crab: *Libinia dubia;* sponge; squid: *Loligo vulgaris;* starfish; fish: numerous species), plankton, *Rhizophora mangle* seedlings, seawater, sediment. RANGE: Atlantic Ocean—Bahamas, Iceland, North Atlantic, Portugal, Scotland, United States (Florida); Baltic Sea—Germany (F.R.G.), Sweden; Black Sea; Indian Ocean—Antarctic Ocean, India, Mauritius, and open ocean; Mediterranean—Italy; Pacific Ocean—Japan and open ocean; southern oceans.

Debaryomyces vanriji (van der Walt et Tscheuschner) Abadie, Pignal et Jacob 1963

LITERATURE: Fell (1972, 1976).
MARINE ORIGIN: Seawater. RANGE: Pacific Ocean—Antarctic Sea.

Debaryomyces vini Zimmermann 1938 [syn. of *Pichia vini* (Zimmermann) Phaff 1956]

LITERATURE: De Queiroz and Macêdo (1972).
MARINE ORIGIN: Seawater. RANGE: Atlantic Ocean—Brazil.

Hanseniaspora uvarum (Niehaus) Shehata, Mrak et Phaff 1955 (see also *Kloeckera apiculata*)

LITERATURE: Fell (1967); Meyers *et al.* (1967a).
MARINE ORIGIN: Seawater. RANGE: Atlantic Ocean—North Atlantic; Indian Ocean.

Hanseniaspora valbyensis Klöcker 1912 (see also *Kloeckera apiculata*)

LITERATURE: Ahearn and Roth (1962); Fell and van Uden (1963); Roth *et al.* (1962).
MARINE ORIGIN: Animals (fish). RANGE: Atlantic Ocean—United States (Florida).

Hansenula anomala (Hansen) H. et P. Sydow 1919

=*Candida pelliculosa* Redaelli 1925

LITERATURE: Ahearn and Roth (1962); Capriotti (1962a); Fell (1967); Fell *et al.* (1960); Fell and van Uden (1963); Kobayashi *et al.* (1953); Roth *et al.* (1962); Suehiro (1963).
MARINE ORIGIN: Animals (clam; fish), seawater, tidal mud. RANGE: Atlantic Ocean—United States (Florida); Indian Ocean—Mauritius; Pacific Ocean—Japan.

Hansenula californica (Lodder) Wickerham 1951

LITERATURE: Phaff *et al.* (1952).
MARINE ORIGIN: Animals (shrimp: *Penaeus setiferus*). RANGE: Atlantic Ocean—Gulf of Mexico.

Hansenula jadinii (A. et R. Sartory, Weill et Meyer) Wickerham 1951

LITERATURE: Meyers *et al.* (1967b).
MARINE ORIGIN: Seawater. RANGE: Black Sea.

Kluyveromyces drosophilarum (Shehata, Mrak et Phaff) van der Walt 1965

LITERATURE: Ahearn and Meyers (1972); Ahearn *et al.* (1971b); Meyers *et al.* (1970a, 1971a,b).
MARINE ORIGIN: Seawater, marsh sediment. RANGE: Atlantic Ocean—United States (Florida, Louisiana).

Pichia bovis van Uden et DoCarmo-Sousa 1957

LITERATURE: Taysi and van Uden (1964).
MARINE ORIGIN: Seawater. RANGE: Atlantic Ocean—Portugal.

Pichia etchellsii Kreger-van Rij 1964

LITERATURE: Lundström-Eriksson and Norkrans (1968); Norkrans (1966a,b).
MARINE ORIGIN: Seawater. RANGE: Atlantic Ocean—Sweden.

Pichia farinosa (Lindner) Hansen 1904

LITERATURE: van Uden and ZoBell (1962).
MARINE ORIGIN: Seawater. RANGE: Pacific Ocean—Torres Strait.

Pichia fermentans Lodder 1932

LITERATURE: Ahearn and Roth (1962); Capriotti (1962a,b); Fell (1967).
MARINE ORIGIN: Algae, seawater. RANGE: Atlantic Ocean—United States (Florida); Indian Ocean.

Pichia fluxuum (Phaff et Knapp) Kreger-van Rij 1964

LITERATURE: Lundström-Eriksson and Norkrans (1968); Norkrans (1966a,b).
MARINE ORIGIN: Seawater. RANGE: Atlantic Ocean—Sweden.

Pichia membranaefaciens (Hansen) Hansen 1904

LITERATURE: Ross and Morris (1962, 1965); Taysi and van Uden (1964).
MARINE ORIGIN: Animals (fish: whiting and unidentified), seawater.
RANGE: Atlantic Ocean—Portugal, Scotland.

Pichia ohmeri (Etchells et Bell) Kreger-van Rij 1964

=*Candida guilliermondii* (Cast.) Langeron et Guerra var.
membranaefaciens Lodder et Kreger-van Rij 1952

LITERATURE: Ahearn and Meyers (1972); Le Petit *et al.* (1970); Picci
and Verona (1964); Seshadri and Sieburth (1975).
MARINE ORIGIN: Algae (Phaeophyta, Rhodophyta), animals (fish:
numerous species), oil slicks, seawater, marsh sediments. RANGE: Atlantic Ocean—United States (Louisiana, Rhode Island); Mediterranean—
Italy, France.

Pichia saitoi Kodama, Kyono et Kodama 1962

LITERATURE: Ahearn and Meyers (1972); Ahearn *et al.* (1971b).
MARINE ORIGIN: Seawater (?), marsh sediments. RANGE: Atlantic
Ocean—United States (Louisiana).

Pichia terricola van der Walt 1957

LITERATURE: Taysi and van Uden (1964).
MARINE ORIGIN: Seawater. RANGE: Atlantic Ocean—Portugal.

Saccharomyces bailii Lindner 1895

LITERATURE: De Queiroz and Macêdo (1972).
MARINE ORIGIN: Seawater. RANGE: Atlantic Ocean—Brazil.

Saccharomyces carlsbergensis Hansen 1908 (syn. of *Saccharomyces uvarum* Beijerinck 1898)

LITERATURE: Capriotti (1962a); Picci and Verona (1964).
MARINE ORIGIN: Animals (fish: *Conger conger, Mugil cephalus,
Sparus aurata*), seawater. RANGE: Atlantic Ocean—United States
(Florida); Mediterranean—Italy.

Saccharomyces cerevisiae Hansen 1883

=*Candida robusta* Diddens et Lodder 1942
=*Saccharomyces ellipsoideus* Hansen 1883

LITERATURE: Artagaveytia-Allende (1960); Capriotti (1962a,b); Fell

(1967); Kawakita and van Uden (1965); Norkrans (1969b); Picci and Verona (1964); Seshadri *et al.* (1966); Suehiro (1963).

MARINE ORIGIN: Animals (clam: *Donax trunculus;* mussel: *Mytilus galloprovincialis;* fish: *Gadus capelanus, Merluccius merluccius, Mullus surmuletus, Scomber scomber, Scomberesox saurus;* gulls: *Larus argentatus, L. fuscus, L. ridibundus;* terns: *Sterna sandvicensis, S. minuta*), seawater, tidal mud. RANGE: Atlantic Ocean—Portugal, Uruguay, United States (Florida); Indian Ocean—India and open ocean; Mediterranean—Italy; Pacific Ocean—Japan.

Saccharomyces delbrueckii Lindner 1901

LITERATURE: Suehiro (1963).
MARINE ORIGIN: Tidal mud. RANGE: Pacific Ocean—Japan.

Saccharomyces exiguus Hansen 1888

LITERATURE: Suehiro (1963).
MARINE ORIGIN: Tidal mud. RANGE: Pacific Ocean—Japan.

Saccharomyces fermentati (Saito) Lodder et Kreger-van Rij 1952

LITERATURE: Volz *et al.* (1974).
MARINE ORIGIN: Sediment. RANGE: Atlantic Ocean—Bahamas.

Saccharomyces florentinus (Castelli) Lodder et Kreger-van Rij 1952

LITERATURE: DeQueiroz (1972).
MARINE ORIGIN: Algae. RANGE: Atlantic Ocean—Brazil.

Saccharomyces fragilis Jörgensen 1909 [syn. of *Kluyveromyces fragilis* (Jörgensen) van der Walt 1965]

LITERATURE: van Uden and Castelo-Branco (1963).
MARINE ORIGIN: Seawater. RANGE: Pacific Ocean—United States (California).

Saccharomyces fructuum Lodder et Kreger-van Rij 1952 (syn. of *Saccharomyces chevalieri* Guilliermond 1914)

LITERATURE: Ahearn and Roth (1962); Bhat and Kachwalla (1955); Fell and van Uden (1963); Fell *et al.* (1960); Shinano (1962).
MARINE ORIGIN: Seawater, sediment. RANGE: Atlantic Ocean—United States (Florida); Indian Ocean—India; Pacific Ocean—North Pacific.

Saccharomyces heterogenicus Osterwalder 1924

LITERATURE: Picci and Verona (1964).
MARINE ORIGIN: Animals (fish: *Saragus saragus*). RANGE: Mediterranean—Italy.

Saccharomyces oleaginosus Santa María 1958

LITERATURE: Kawakita and van Uden (1965).
MARINE ORIGIN: Animals (gulls: *Larus fuscus, L. genei;* terns: *Sterna hirundo, S. minuta*). RANGE: Atlantic Ocean—Portugal.

Saccharomyces rosei (Guilliermond) Lodder et Kreger-van Rij 1952

≡*Debaryomyces rosei* (Guilliermond) Kudriavzev 1954, 1960
≡*Torulaspora rosei* Guilliermond 1913

LITERATURE: Bhat and Kachwalla (1955); Capriotti (1962a); Kriss *et al.* (1967).
MARINE ORIGIN: Seawater. RANGE: Atlantic Ocean—North Atlantic, United States (Florida); Indian Ocean.

Saccharomyces steineri Lodder et Kreger-van Rij 1952 (syn. of *Saccharomyces italicus* Castelli 1938)

LITERATURE: Bhat and Kachwalla (1955).
MARINE ORIGIN: Seawater. RANGE: Indian Ocean—India.

B. Basidiomycotina
Sporobolomycetaceae

Leucosporidium scottii Fell, Statzell, Hunter et Phaff 1969

IMPERFECT STATE: *Candida scottii* Diddens et Lodder 1942.
LITERATURE: Bruce and Morris (1973); Fell (1972, 1974, 1976); Moore (1972).
MARINE ORIGIN: Animals (fish: numerous species), seawater. RANGE: Atlantic Ocean—Scotland; Indo-Pacific and Pacific Oceans—Antarctic Sea. NOTE: Also collected by Goto *et al.* (1969) and J. Sugiyama *et al.* (1967) in an antarctic saline lake.

Rhodosporidium diobovatum Newell et Hunter 1970

LITERATURE: Fell (1970).
MARINE ORIGIN: Seawater. RANGE: Atlantic Ocean—United States (Florida).

Rhodosporidium infirmo-miniatum Fell, Hunter et Tallman 1973

LITERATURE: Fell (1972, 1974, 1976).
MARINE ORIGIN: Seawater, zooplankton. RANGE: Atlantic Ocean—Antarctic Sea; Pacific Ocean—Antarctic Sea, New Zealand.

Rhodosporidium toruloides Banno 1967

LITERATURE: Ahearn and Meyers (1972); Ahearn *et al.* (1971b); Fell (1974); Fell *et al.* (1970); Johnson-Reid and Moore (1972); Newell and Fell (1970); Newell and Hunter (1970).
MARINE ORIGIN: Animals (porpoise), oil slicks, seawater. RANGE: Atlantic Ocean—Bahamas, Gulf of Mexico.

Sporobolomyces albo-rubescens Derx 1930

LITERATURE: Crow *et al.* (1977).
MARINE ORIGIN: Seawater. RANGE: Atlantic Ocean—North Atlantic.

Sporobolomyces hispanicus Palaez et Ramirez 1956

LITERATURE: Fell (1967).
MARINE ORIGIN: Seawater. RANGE: Indian Ocean.

Sporobolomyces holsaticus Windisch 1949

LITERATURE: Kriss and Novozhilova (1954); Novozhilova (1955).
MARINE ORIGIN: Seawater. RANGE: Pacific Ocean.

Sporobolomyces odorus Derx 1930

LITERATURE: Fell (1967).
MARINE ORIGIN: Seawater. RANGE: Indian Ocean.

Sporobolomyces pararoseus Olson et Hammer 1937

LITERATURE: Meyers *et al.* (1967b).
MARINE ORIGIN: Seawater. RANGE: Black Sea.

Sporobolomyces roseus Kluyver et van Niel 1925

LITERATURE: Ahearn *et al.* (1971b); Kriss *et al.* [1967 (as "vars.")];
Meyers *et al.* (1967b); Pugh and Lindsey (1975).
MARINE ORIGIN: *Halimione portulacoides* and terrestrial dune plants,
oil slicks, seawater. RANGE: Atlantic Ocean—England, Gulf of Mexico,
North Atlantic; Black Sea.

Sporobolomyces salmonicolor (Fischer et Brebeck) Kluyver et van Niel 1925

LITERATURE: Fell (1972, 1976); Kriss (1963); Kriss and Novozhilova
(1954); Novozhilova (1955); Novozhilova and Popova (1973); Shinano
(1962).
MARINE ORIGIN: Seawater. RANGE: Atlantic Ocean—Gulf of Guinea,
and open ocean; Black Sea; Indian Ocean—Antarctic Sea; Pacific Ocean.

C. Deuteromycotina
Torulopsidaceae

Candida albicans (Robin) Berkhout 1923

LITERATURE: Ahearn and Roth (1962); Buck *et al.* (1977); Capriotti
(1962b); Combs *et al.* (1971); Crow *et al.* (1977); Dabrowa *et al.* (1964);
Fell (1967); Fell and van Uden (1963); Kawakita and van Uden (1965);
Kishimoto and Baker (1969); Seshadri *et al.* (1966); Suehiro (1960).

MARINE ORIGIN: Algae, animals (crab; mussel: *Mytilus edulis;* oyster: *Crassostrea virginica;* quahog: *Mercenaria mercenaria;* gulls: *Larus fuscus, L. ridibundus),* beach sand, shore bird droppings, seawater, sediment. RANGE: Atlantic Ocean—North Atlantic, Portugal, United States (Florida, Long Island Sound); Indian Ocean—India, and open ocean; Pacific Ocean—Japan, United States (California. Hawaii).

Candida australis Goto, Sugiyama et Iizuka 1969

LITERATURE: Buck *et al.* (1977).
MARINE ORIGIN: Animals (mussel: *Mytilus edulis;* oyster: *Crassostrea virginica;* quahog: *Mercenaria mercenaria).* RANGE: Atlantic Ocean—United States (Long Island Sound).

Candida beechii Buckley et van Uden 1968

LITERATURE: Buck *et al.* (1977); Seshadri and Sieburth (1975).
MARINE ORIGIN: Algae (Phaeophyta, Rhodophyta); animals (mussel: *Mytilus edulis;* oyster: *Crassostrea virginica*; quahog: *Mercenaria mercenaria).* RANGE: Atlantic Ocean—United States (Long Island Sound, Rhode Island).

Candida bogoriensis Deinema 1961

LITERATURE: Seshadri and Sieburth (1975).
MARINE ORIGIN: Algae (Phaeophyta, Rhodophyta), seawater. RANGE: Atlantic Ocean—United States (Rhode Island).

Candida boidinii Ramirez 1953

LITERATURE: Fell *et al.* (1960); Fell and van Uden (1963); Taysi and van Uden (1964).
MARINE ORIGIN: Seawater, sediment. RANGE: Atlantic Ocean—Portugal, United States (Florida).

Candida brumptii Langeron et Guerra 1935

LITERATURE: Seshadri and Sieburth (1975); Suehiro (1963); Suehiro and Tomiyasu (1962); Suehiro *et al.* (1962); Taysi and van Uden (1964).
MARINE ORIGIN: Algae (Chlorophyta, Phaeophyta, Rhodophyta),

plankton, seawater, tidal mud. RANGE: Atlantic Ocean—Portugal, United States (Rhode Island); Pacific Ocean—Japan.

Candida capsuligena (van der Walt) van Uden et Buckley 1970 (not validly published, see Article 33 of the International Code of Botanical Nomenclature)

LITERATURE: Seshadri and Sieburth (1975).
MARINE ORIGIN: Seawater. RANGE: Atlantic Ocean—United States (Rhode Island).

Candida catenulata Diddens et Lodder 1942

LITERATURE: Taysi and van Uden (1964).
MARINE ORIGIN: Seawater. RANGE: Atlantic Ocean—Portugal.

Candida ciferrii Kreger-van Rij 1965

LITERATURE: Seshadri and Sieburth (1975).
MARINE ORIGIN: Algae (Chlorophyta, Phaeophyta, Rhodophyta), seawater. RANGE: Atlantic Ocean—United States (Rhode Island).

Candida claussenii Lodder et Kreger-van Rij 1952

LITERATURE: Seshadri and Sieburth (1975).
MARINE ORIGIN: Algae (Chlorophyta, Phaeophyta, Rhodophyta), seawater. RANGE: Atlantic Ocean—United States (Rhode Island).

Candida conglobata (Redaelli) van Uden et Buckley 1970 (not validly published, see Article 33 of the International Code of Botanical Nomenclature)

LITERATURE: Seshadri and Sieburth (1975).
MARINE ORIGIN: Algae (Chlorophyta). RANGE: Atlantic Ocean—United States (Rhode Island).

Candida curiosa Komagata et Nakase 1965

LITERATURE: Seshadri and Sieburth (1975).
MARINE ORIGIN: Algae (Chlorophyta, Phaeophyta, Rhodophyta), seawater. RANGE: Atlantic Ocean—United States (Rhode Island).

Candida curvata (Diddens et Lodder) Lodder et Kreger-van Rij 1952

LITERATURE: Ahearn and Roth (1962); Fell and van Uden (1963); Fell *et al.* (1960); Volz *et al.* (1974).
MARINE ORIGIN: Animals (conch: *Strombus gigas*), sediment. RANGE: Atlantic Ocean—Bahamas.

Candida diddensii (Phaff, Mrak et Williams) Fell et Meyer 1967

≡*Trichosporon diddensii* Phaff, Mrak et Williams 1952
=*Candida atmosphaerica* Santa María 1959
=*Candida polymorpha* Ohara et Nonomura 1954 (*nomen nudum*)
=*Trichosporon atlanticum* Siepmann et Höhnk 1962

LITERATURE: Buck *et al.* (1977); Fell (1967, 1972, 1976); Fell and Meyer (1967); Goto *et al.* (1974b); Meyers *et al.* (1967a,b); Phaff *et al.* (1952); Siepmann and Höhnk (1962); Volz *et al.* (1974); Yamasato *et al.* (1974).
MARINE ORIGIN: Animals (mussel: *Mytilus edulis;* oyster: *Crassostrea virginica;* quahog: *Mercenaria mercenaria;* shrimp: *Penaeus setiferus* and unidentified shrimp eggs; fish: *Pterolamiops longimanus),* seawater, sediment. RANGE: Atlantic Ocean—Antarctic Sea, Bahamas, Gulf of Mexico, United States (Long Island Sound), North Atlantic; Black Sea; Indian Ocean—Antarctic Sea and elsewhere; Indo-Pacific Ocean—Antarctic Sea; Pacific Ocean—Antarctic Sea and elsewhere.

Candida fabianii Kodama, Kyono, Iida et Onoyama 1964 (syn.
of *Hansenula fabianii* Wickerham 1965)

LITERATURE: Buck *et al.* (1977).
MARINE ORIGIN: Animals (mussel: *Mytilus edulis;* oyster: *Crassostrea virginica;* quahog: *Mercenaria mercenaria*). RANGE: Atlantic Ocean— United States (Long Island Sound).

Candida freyschussii Buckley et van Uden 1968

LITERATURE: Buck *et al.* (1977).
MARINE ORIGIN: Animals (mussel: *Mytilus edulis;* oyster: *Crassostrea virginica;* quahog: *Mercenaria mercenaria*). RANGE: Atlantic Ocean— United States (Long Island Sound).

Candida guilliermondii (Castellani) Langeron et Guerra 1938

=*Candida melibiosi* Lodder et Kreger-van Rij 1952

LITERATURE: Ahearn and Roth (1962); Artagaveytia-Allende (1960); Bhat and Kachwalla (1955); Buck *et al.* (1977); Capriotti (1962a); Combs *et al.* (1971); De Queiroz (1972); De Queiroz and Macêdo (1972); Fell (1967); Fell and van Uden (1963); Fell *et al.* (1960); Meyers *et al.* (1967b); Phaff *et al.* (1952); Picci and Verona (1964); Roth *et al.* (1962); Suehiro (1963); Taysi and van Uden (1964); van Uden (1967); van Uden and Castelo-Branco (1963); Volz *et al.* (1974).

MARINE ORIGIN: Algae, animals (mussel: *Mytilus edulis;* oyster: *Crassostrea virginica;* quahog: *Mercenaria mercenaria;* shrimp: *Penaeus setiferus;* spiny lobster: *Panulirus argus;* squid: *Loligo vulgaris;* fish: numerous species), seawater, sediment, tidal mud. RANGE: Atlantic Ocean—Bahamas, Brazil, Gulf of Mexico, Portugal, Uruguay, United States (Florida, Long Island Sound); Black Sea; Indian Ocean; Mediterranean—Italy; Pacific Ocean—Japan, United States (California).

Candida guilliermondii (Castellani) Langeron et Guerra var. *carpophila* Phaff et Miller 1961

LITERATURE: Seshadri and Sieburth (1975).
MARINE ORIGIN: Algae (Rhodophyta), seawater. RANGE: Atlantic Ocean—United States (Rhode Island).

Candida humicola (Daszewska) Diddens et Lodder 1942

LITERATURE: Artagaveytia-Allende (1960); Hoppe (1972b); Kobayashi *et al.* (1953).
MARINE ORIGIN: Animals (clam), seawater. RANGE: Atlantic Ocean—Uruguay; Baltic Sea—Germany (F.R.G.); Pacific Ocean—Japan. NOTE: Also collected by Goto *et al.* (1969) in an antarctic saline lake.

Candida insectamans Scott, van der Walt et van der Klift apud van der Walt, Scott and van der Klift 1972

LITERATURE: Buck *et al.* (1977).
MARINE ORIGIN: Animals (mussel: *Mytilus edulis;* oyster: *Crassostrea virginica;* quahog: *Mercenaria mercenaria*). RANGE: Atlantic Ocean— United States (Long Island Sound).

Candida intermedia (Ciferri et Ashford) Langeron et Guerra 1938

LITERATURE: Ahearn and Roth (1962); Artagaveytia-Allende (1960);

Fell (1972, 1976); Fell and van Uden (1963); Fell *et al.* (1960); Seshadri and Sieburth (1975); Suehiro (1963); Taysi and van Uden (1964).

MARINE ORIGIN: Algae (Chlorophyta, Phaeophyta, Rhodophyta), seawater, sediment, tidal mud. RANGE: Atlantic Ocean—Portugal, Uruguay, United States (Florida, Rhode Island); Indo-Pacific Ocean—Antarctic Sea; Pacific Ocean—Antarctic Sea, Japan.

Candida krusei (Castellani) Berkhout 1923

LITERATURE: Artagaveytia-Allende (1960); Buck *et al.* (1977); Combs *et al.* (1971); Dabrowa *et al.* (1964); Fell and van Uden (1963); Kawakita and van Uden (1965); Kishimoto and Baker (1969); Meyers *et al.* (1967a); van Uden (1967); van Uden and Castelo-Branco (1963).

MARINE ORIGIN: Animals (crab; mussel: *Mytilus edulis;* oyster: *Crassostrea virginica;* quahog: *Mercenaria mercenaria;* gulls: *Larus fuscus, L. argentatus;* tern: *Sterna hirundo*), beach sand, shore bird droppings, seawater. RANGE: Atlantic Ocean—North Atlantic, Portugal, Uruguay, United States (Florida, Long Island Sound); Pacific Ocean—United States (California, Hawaii).

Candida lambica (Lindner et Genoud) van Uden et Buckley 1970 [not validly published, see Article 33 of the International Code of Botanical Nomenclature; *C. lambica* may be regarded as the imperfect form of *Pichia fermentans* Lodder, according to van Uden and Buckley (1970)]

LITERATURE: Buck *et al.* (1977); Taysi and van Uden (1964).

MARINE ORIGIN: Animals (mussel: *Mytilus edulis;* oyster: *Crassostrea virginica;* quahog: *Mercenaria mercenaria*), seawater. RANGE: Atlantic Ocean—Portugal, United States (Long Island Sound).

Candida lipolytica (Harrison) Diddens et Lodder 1942

LITERATURE: Artagaveytia-Allende (1960); Le Petit *et al.* (1970); Picci and Verona (1964); Ross and Morris [1965 (as "var.")]; Suehiro (1962); Taysi and van Uden (1964).

MARINE ORIGIN: Algae *(Thalassiosira subtilis);* animals (fish: numerous species), oil slicks, seawater. RANGE: Atlantic Ocean—Iceland, North Atlantic, Portugal, Uruguay; Mediterranean—Italy; Pacific Ocean—Japan.

Candida lusitaniae van Uden et Do Carmo-Sousa 1959

LITERATURE: Buck *et al.* (1977); Fell and Meyer (1967); Kawakita and van Uden (1965); Taysi and van Uden (1964).

MARINE ORIGIN: Animals (mussel: *Mytilus edulis;* oyster: *Crassostrea virginica;* quahog: *Mercenaria mercenaria;* gull: *Larus ridibundus*), seawater. RANGE: Atlantic Ocean—Portugal, United States (Long Island Sound).

Candida macedoniensis (Castellani et Chalmers) Berkhout 1923

LITERATURE: Seshadri and Sieburth (1975).
MARINE ORIGIN: Algae (Phaeophyta, Rhodophyta), seawater. RANGE: Atlantic Ocean—United States (Rhode Island).

Candida maritima (Siepmann) van Uden et Buckley 1970 (not validly published, see Article 33 of the International Code of Botanical Nomenclature)

≡*Trichosporon maritimum* Siepmann *in* Siepmann et Höhnk, *Veröff. Inst. Meeresforsch. Bremerhaven* **8**, 89–90, 1962

LITERATURE: Seshadri and Sieburth (1975); Volz *et al.* (1974).
MARINE ORIGIN: Algae (Chlorophyta, Phaeophyta, Rhodophyta), animals (shrimp), seawater, sediment. RANGE: Atlantic Ocean—Bahamas, United States (Rhode Island), open ocean.

Candida melibiosica Buckley et van Uden 1968

LITERATURE: Buck *et al.* (1977); Seshadri and Sieburth (1975).
MARINE ORIGIN: Algae (Chlorophyta, Phaeophyta), animals (mussel: *Mytilus edulis;* oyster: *Crassostrea virginica;* quahog: *Mercenaria mercenaria)*, seawater. RANGE: Atlantic Ocean—United States (Long Island Sound, Rhode Island).

Candida melinii Diddens et Lodder 1942

LITERATURE: Ahearn and Roth (1962); Capriotti (1962a); De Queiroz and Macêdo (1972); Fell and van Uden (1963); Fell *et al.* (1960); Picci and Verona (1964); Volz *et al.* (1974).
MARINE ORIGIN: Animals (fish: *Merluccius merluccius*), seawater, sediment. RANGE: Atlantic Ocean—Bahamas, Brazil, United States (Florida); Mediterranean—Italy.

Candida mesenterica (Geiger) Diddens et Lodder 1942

LITERATURE: Picci and Verona (1964).
MARINE ORIGIN: Animals (octopus: *Octopus vulgaris;* fish: *Gobius cobitis, Mullus surmuletus, Saragus saragus, Trachurus trachurus*). RANGE: Mediterranean—Italy.

Candida mogii Vidal-Leiria 1967

LITERATURE: Buck *et al.* (1977).
MARINE ORIGIN: Animals (mussel: *Mytilus edulis;* oyster: *Crassostrea virginica;* quahog: *Mercenaria mercenaria*). RANGE: Atlantic Ocean—United States (Long Island Sound).

Candida mycoderma (Reess) Lodder et Kreger-van Rij 1952
[syn. of *Candida valida* (Leberle) van Uden et Buckley 1970
or *Candida vini* (Desmazières ex Lodder) van Uden et
Buckley 1970; both species not validly published, see
Article 33 of the International Code of Botanical
Nomenclature]

LITERATURE: Fell (1967); Picci and Verona (1964); Ross and Morris (1962); Suehiro (1963); Taysi and van Uden (1964); van Uden (1967).
MARINE ORIGIN: Animals (fish: gurnard, plaice, *Trachurus trachurus),* seawater, tidal mud. RANGE: Atlantic Ocean—Portugal, Scotland; Indian Ocean; Mediterranean—Italy; Pacific Ocean—Japan.

Candida norvegensis (Dietrichson) van Uden et Farinha ex van Uden et Buckley 1970

LITERATURE: Kawakita and van Uden (1965).
MARINE ORIGIN: Animals (gulls: *Larus fuscus, L. genei;* tern: *Sterna hirundo*). RANGE: Atlantic Ocean—Portugal.

Candida obtusa (Dietrichson) van Uden et Do Carmo-Sousa ex van Uden et Buckley 1970

LITERATURE: Meyers *et al.* (1967a).
MARINE ORIGIN: Seawater. RANGE: Atlantic Ocean—North Atlantic.

Candida parapsilosis (Ashford) Langeron et Talice 1932

=*Candida parapsilosis* var. *intermedia* van Rij et Verona 1949
=*Candida parapsilosis* var. *querci* van Uden et Do Carmo-Sousa 1959

LITERATURE: Ahearn and Meyers (1972); Ahearn and Roth (1962); Ahearn *et al.* (1971b); Buck *et al.* (1977); Capriotti (1962a,b); Combs *et al.* (1971); Dabrowa *et al.* (1964); De Queiroz and Macêdo (1972); Fell (1967, 1972, 1976); Fell and Meyer (1967); Fell and van Uden (1963); Fell *et al.* (1960); Hoppe (1972a,b); Kawakita and van Uden (1965); Kishimoto and Baker (1969); Meyers *et al.* (1967b); Norkrans (1966a,b); Phaff *et al.* (1952); Picci and Verona (1964); Ross and Morris (1962, 1965); Roth *et al.* (1962); Seshadri and Sieburth (1971); Suehiro (1960, 1962); Suehiro and Tomiyasu (1962); Suehiro *et al.* (1962); Taysi and van Uden (1964).

MARINE ORIGIN: Algae *(Chondrus crispus, Fucus vesiculosus, Polysiphonia harveyi, Thalassiosira subtilis, Ulva lactuca,* and others), animals (mussel: *Mytilus edulis;* octopus: *Octopus vulgaris;* oyster: *Crassostrea virginica;* quahog: *Mercenaria mercenaria;* shrimp: *Penaeus setiferus;* squid: *Sepia officinalis;* fish: numerous species; gull: *Larus genei*), beach sand, shore bird droppings, oil slicks, plankton, seawater, sediment. RANGE: Atlantic Ocean—Bahamas, Brazil, Gulf of Mexico, Gulf Stream, Iceland, North Sea, Portugal, Scotland, Sweden, United States (Florida, Long Island Sound, Rhode Island); Baltic Sea—Germany (F.R.G.); Black Sea; Mediterranean—Italy; Indian Ocean—Antarctic Sea; Pacific Ocean—Antarctic Sea, Japan, United States (California, Hawaii).

Candida pelliculosa Redaelli var. *cylindrica* Diddens et Lodder 1942 [syn. of *Hansenula anomala (Hansen)* H. et P. Sydow var. *schneggii* (Weber) Wickerham 1951]

LITERATURE: Buck *et al.* (1977).
MARINE ORIGIN: Animals (mussel: *Mytilus edulis;* oyster: *Crassostrea virginica;* quahog: *Mercenaria mercenaria*). RANGE: Atlantic Ocean—United States (Long Island Sound).

Candida pulcherrima (Lindner) Windisch 1940

=yeast stage of *Metschnikowia pulcherrima* Pitt et Miller 1968 (the ascomycetous stage has not been found in marine habitats)

LITERATURE: Capriotti (1962a); Hoppe (1972a,b); Taysi and van Uden (1964); van Uden (1967).
MARINE ORIGIN: Seawater. RANGE: Atlantic Ocean—Portugal, United States (Florida); Baltic Sea—Germany (F.R.G.).

Candida ravautii Langeron et Guerra 1935

LITERATURE: Buck *et al.* (1977); Seshadri and Sieburth (1975).

MARINE ORIGIN: Algae (Phaeophyta, Rhodophyta); animals (mussel: *Mytilus edulis;* oyster: *Crassostrea virginica;* quahog: *Mercenaria mercenaria*). RANGE: Atlantic Ocean—United States (Long Island Sound, Rhode Island).

Candida reukaufii (Grüss) Diddens et Lodder 1942

=yeast stage of *Metschnikowia reukaufii* Pitt et Miller 1968 (the ascomycetous stage has not been found in marine habitats)

LITERATURE: Artagaveytia-Allende (1960); Shinano (1962).
MARINE ORIGIN: Seawater. RANGE: Atlantic Ocean—Uruguay; Pacific Ocean.

Candida rugosa (Anderson) Diddens et Lodder 1942

=*Candida rugosa* var. *elegans* Dietrichson 1954 (*nomen nudum*)

LITERATURE: Fell (1967); van Uden et Castelo-Branco (1963).
MARINE ORIGIN: Seawater. RANGE: Indian Ocean; Pacific Ocean—United States (California).

Candida sake (Saito et Ota) van Uden et Buckley 1970 (not validly published, see Article 33 of the International Code of Botanical Nomenclature)

≡*Torulopsis sake* (Saito et Ota) Lodder et Kreger-van Rij 1952
=*Candida natalensis* van der Walt et Tscheuschner 1957
=*Candida tropicalis* var. *lambica* (Harrison) Diddens et Lodder 1942
=*Candida vanriji* Capriotti 1958

LITERATURE: Bruce and Morris (1973); Fell (1972, 1976); Fell and Hunter (1968); Fell and Meyer (1967); Seshadri and Sieburth (1975).
MARINE ORIGIN: Algae (Chlorophyta, Rhodophyta), animals (fish: numerous species), seawater. RANGE: Atlantic Ocean—Antarctic Sea, Scotland, United States (Rhode Island); Indo-Pacific Ocean—Antarctic Sea; Pacific Ocean—Antarctic Sea, New Zealand.

Candida salmanticensis (Santa María) van Uden et Buckley 1970 (not validly published, see Article 33 of the International Code of Botanical Nomenclature)

LITERATURE: Seshadri and Sieburth (1975).

Marine Origin: Algae (Chlorophyta, Phaeophyta, Rhodophyta), seawater. Range: Atlantic Ocean—United States (Rhode Island).

Candida scottii Diddens et Lodder 1942

Imperfect State of *Leucosporidium scottii* Fell, Statzell, Hunter et Phaff 1969 (see *L. scottii*).
Note: Also collected by Goto *et al.* (1969) in an antarctic saline lake.

Candida silvicola Shifrine et Phaff 1956 (syn. of *Hansenula holstii* Wickerham 1960)

Literature: Taysi and van Uden (1964).
Marine Origin: Seawater. Range: Atlantic Ocean—Portugal.

Candida sloofii van Uden et Do Carmo-Sousa 1957

Literature: Buck *et al.* (1977).
Marine Origin: Animals (mussel: *Mytilus edulis;* oyster: *Crassostrea virginica;* quahog: *Mercenaria mercenaria*). Range: Atlantic Ocean—United States (Long Island Sound).

Candida solani Lodder et Kreger-van Rij 1952

Literature: Seshadri and Sieburth (1975); Suehiro *et al.* (1962).
Marine Origin: Algae (Chlorophyta, Phaeophyta, Rhodophyta), plankton, seawater. Range: Atlantic Ocean—United States (Rhode Island); Pacific Ocean—Japan.

Candida steatolytica Yarrow 1969

Literature: Buck *et al.* (1977).
Marine Origin: Animals (mussel; *Mytilus edulis;* oyster: *Crassostrea virginica;* quahog: *Mercenaria mercenaria*). Range: Atlantic Ocean—United States (Long Island Sound).

Candida stellatoidea (Jones et Martin) Langeron et Guerra 1939

Literature: Combs *et al.* (1971).
Marine Origin: Seawater. Range: Atlantic Ocean—United States (Long Island Sound).

Candida tenuis Diddens et Lodder 1942

LITERATURE: Ahearn and Roth (1962); Fell (1967); Fell and van Uden (1963); Fell *et al.* (1970); Kawakita and van Uden (1965); Seshadri and Sieburth (1975); Suehiro and Tomiyasu (1962).

MARINE ORIGIN: Algae (Phaeophyta, Rhodophyta), animals (fish: *"Scopelenges alcocki,"* *Coryphaena hippurus;* gull: *Larus ridibundus*), seawater, sediment. RANGE: Atlantic Ocean—Bahamas, Portugal, United States (Rhode Island); Indian Ocean; Pacific Ocean—Japan.

Candida tropicalis (Castellani) Berkhout 1923

=*Trichosporon lodderi* Phaff, Mrak et Williams 1952

LITERATURE: Ahearn and Roth (1962); Bhat and Kachwalla (1955); Bhat *et al.* (1955); Buck *et al.* (1977); Capriotti (1962a,b); Combs *et al.* (1971); De Queiroz (1972); De Queiroz and Macêdo (1972); Fell (1967); Fell-and Meyer (1967); Fell and van Uden (1963); Fell *et al.* (1960); Hoppe (1972a,b); Kawakita and van Uden (1965); Le Petit *et al.* (1970); Meyers *et al.* (1967a,b); Morii (1973); Roth *et al.* (1962); Seshadri *et al.* (1966); Suehiro (1963); Suehiro and Tomiyasu (1962); Suehiro *et al.* (1962); van Uden and Castelo-Branco (1963).

MARINE ORIGIN: Algae *(Sargassum* sp. and others), animals (mussel: *Mytilus edulis;* oyster: *Crassostrea virginica;* quahog: *Mercenaria mercenaria;* shrimp: *Penaeus setiferus;* fish; gulls: *Larus fuscus, L. genei, L. occidentalis;* terns: *Sterna hirundo, S. minuta, S. sandvicensis;* mammals: *Neomeris phocaenoides, Stenella attenuata*), oil slicks, plankton, seawater, sediment, *Thalassia* sp., tidal mud. RANGE: Atlantic Ocean—Bahamas, Brazil, Gulf of Mexico, North Atlantic, Portugal, United States (Long Island Sound, Florida); Baltic Sea—Germany (F.R.G.); Black Sea; Indian Ocean—India, and open ocean; Mediterranean—France; Pacific Ocean—Japan, United States (California).

Candida utilis (Henneberg) Lodder et Kreger-van Rij 1952

LITERATURE: Artagaveytia-Allende (1960); Fell (1972); Kawakita and van Uden (1965).

MARINE ORIGIN: Animals (tern: *Sterna minuta),* seawater. RANGE: Atlantic Ocean—Portugal, Uruguay; southern oceans.

Candida viswanathii Sandhu et Randhawa 1962

LITERATURE: Fell (1967); Fell and Meyer (1967).
MARINE ORIGIN: Animals (fish: *Coryphaena hippurus*). RANGE: Indian Ocean.

Candida zeylanoides (Castellani) Langeron et Guerra 1938

=*Trichosporon piscium* Siepmann *in* Siepmann et Höhnk 1962

LITERATURE: Lundström-Eriksson and Norkrans (1968); Meyers *et al.* (1967a); Norkrans (1966a,b, 1968, 1969a); Ross and Morris (1965); Seshadri and Sieburth (1971, 1975); van Uden (1967).
MARINE ORIGIN: Algae (*Chondrus crispus, Polysiphonia harveyi,* Chlorophyta, Rhodophyta), animals (fish), seawater. RANGE: Atlantic Ocean—North Atlantic, Portugal, Sweden, United States (Rhode Island).

Cryptococcus albidus (Saito) Skinner 1947

≡*Torulopsis albida* (Saito) Lodder 1934
=*Cryptococcus terricolus* Pedersen 1958
=*Torulopsis pseudaeria* Zsolt 1958

LITERATURE: Ahearn and Roth (1962); Buck and Greenfield (1964); Capriotti (1962a,b); Crow *et al.* (1977); Fell (1967, 1972, 1976); Fell and van Uden (1963); Fell *et al.* (1960); Goto *et al.* (1974b); Lundström-Eriksson and Norkrans (1968); Meyers *et al.* (1967b, 1970b); Norkrans (1966a,b, 1968, 1969b); Phaff *et al.* (1952); Ross and Morris (1965); Roth *et al.* (1962); Shinano (1962); Siepmann and Höhnk (1962); Suehiro and Tomiyasu (1962); Suehiro *et al.* (1962); Taysi and van Uden (1964); van Uden and Castelo-Branco (1963); Volz *et al.* (1974); Yamasato *et al.* (1974).
MARINE ORIGIN: Algae, animals (shrimp: *Penaeus setiferus;* fish: *Haemulon aureolineatum* and unidentified), plankton, seawater, sediment. RANGE: Atlantic Ocean—Bahamas, Gulf Stream, North Atlantic, Portugal, Sweden, United States (Florida, Louisiana, Texas); Black Sea; Indian Ocean—Antarctic Sea, India; Indo-Pacific Ocean—Antarctic Sea; Pacific Ocean—Antarctic Sea, Gulf of California, Japan, North Pacific, United States (California), and open ocean. NOTE: Also collected by Goto *et al.* (1969) in an antarctic saline lake.

Cryptococcus albidus (Saito) Skinner var. *diffluens* (Zach *in* Wolfram et Zach) Phaff et Fell 1970 (not validly published, see Article 33 of the International Code of Botanical Nomenclature)

≡*Cryptococcus diffluens* (Zach *in* Wolfram et Zach) Lodder et Kreger-van
Rij 1952

LITERATURE: Ahearn and Roth (1962); Bruce and Morris (1973); Fell
(1972, 1976); Fell and van Uden (1963); Fell *et al.* (1960); Ross and Morris
(1965); Shinano (1962); Suehiro and Tomiyasu (1962); Suehiro *et al.*
(1962).

MARINE ORIGIN: Algae, animals (fish: numerous species), plankton,
seawater, sediment. RANGE: Atlantic Ocean—Bahama Islands, North
Atlantic, Scotland; Pacific Ocean—Antarctic Sea, Japan, and open ocean.
NOTE: Also collected by Sugiyama *et al.* (1967) in an antarctic saline lake.

Cryptococcus hungaricus (Zsolt) Phaff et Fell 1970 (not validly published, see Article 33 of the International Code of Botanical Nomenclature)

LITERATURE: Fell (1976).
MARINE ORIGIN: Seawater. RANGE: Pacific Ocean—Antarctic Sea.

Cryptococcus infirmo-miniatus (Okunuki) Phaff et Fell 1970 (not validly published, see Article 33 of the International Code of Botanical Nomenclature)

≡*Rhodotorula glutinis* (Fres.) Harrison var. *infirmo-miniata* (Okunuki)
Lodder 1934
≡*Rhodotorula infirmo-miniata* (Okunuki) Hasegawa et Banno 1964

LITERATURE: Bruce and Morris (1973); Buck and Greenfield (1964);
Fell (1967, 1972, 1976); Goto *et al.* (1974b); Kriss (1963); Kriss *et al.*
(1952); Lundström-Eriksson and Norkrans (1968); Meyers *et al.* (1967);
Norkrans (1966a,b); Novozhilova (1955); Novozhilova and Popova
(1973); Taysi and van Uden (1964); Volz *et al.* (1974); Yamasato *et al.*
(1974).

MARINE ORIGIN: Animals (spiny lobster: *Panulirus argus;* fish: numerous species), seawater. RANGE: Atlantic Ocean—Bahamas, Gulf of
Guinea, Portugal, Scotland, Sweden; Black Sea; Indian Ocean—
Mauritius; Pacific Ocean—Antarctic Sea, Japan, North Pacific, and open
ocean.

Cryptococcus laurentii (Kufferath) Skinner 1947

≡*Torulopsis laurentii* (Kufferath) Lodder 1934

LITERATURE: Bhat *et al.* (1955); Bruce and Morris (1973); Combs *et al.* (1971); Fell (1967, 1972, 1976); Fell and van Uden (1963); Fell *et al.* (1960); Kriss (1963); Kriss and Novozhilova (1954); Kriss *et al.* (1952); Lundström-Eriksson and Norkrans (1968); Meyers *et al.* (1967b, 1970b); Norkrans (1966a,b); Novozhilova (1955); Picci and Verona (1964); Ranganathan and Bhat (1955); Siepmann and Höhnk (1962); Suehiro (1962, 1963); Suehiro and Tomiyasu (1962); Suehiro *et al.* (1962); Taysi and van Uden (1964); van Uden and Castelo-Branco (1963); Volz *et al.* (1974).

MARINE ORIGIN: Algae (*Thalassiosira subtilis* and others), animals (sponge; fish: *Merluccius merluccius* and others), plankton, submerged banana stalks, seawater, sediment, tidai mud. RANGE: Atlantic Ocean— Bahamas, North Atlantic, Portugal, Scotland, Sweden, United States (Florida, Long Island Sound, Louisiana); Black Sea; Indian Ocean— Antarctic Sea, India, Mauritius, and open ocean; Indo-Pacific Ocean— Antarctic Sea; Mediterranean—Italy; Pacific Ocean—Antarctic Sea, Japan, United States (California).

Cryptococcus laurentii (Kuff.) Skinner var. *flavescens* (Saito) Lodder et Kreger-van Rij 1952

=*Rhodotorula aurea* (Saito) Lodder 1934
=*Rhodotorula penaeus* Phaff, Mrak et Williams 1952 (as *"peneaus"*)

LITERATURE: Ahearn *et al.* (1962); Buck and Greenfield (1964); Fell (1972, 1976); Kriss (1963); Kriss and Novozhilova (1954); Novozhilova (1955).

MARINE ORIGIN: Animals (shrimp: *Penaeus setiferus*), seawater, sediment. RANGE: Atlantic Ocean—Gulf of Mexico; Black Sea; Pacific Ocean—Antarctic Sea, North Pacific.

Cryptococcus laurentii (Kuff.) Skinner var. *magnus* Lodder et Kreger-van Rij 1952

LITERATURE: Fell (1972, 1976).
MARINE ORIGIN: Seawater. RANGE: Indo-Pacific Ocean—Antarctic Sea.

Cryptococcus luteolus (Saito) Skinner 1950

≡*Torulopsis luteola* (Saito) Lodder 1934

LITERATURE: Kriss (1963); Kriss and Novozhilova (1954); Norkrans (1966a,b); Novozhilova (1955); Suehiro *et al.* (1962).

MARINE ORIGIN: Seawater. RANGE: Atlantic Ocean—Sweden; Black Sea; Pacific Ocean—Japan, North Pacific.

Cryptococcus macerans (Frederiksen) Phaff et Fell 1970 (not validly published, see Article 33 of the International Code of Botanical Nomenclature)

LITERATURE: Fell (1972, 1976).
MARINE ORIGIN: Seawater. RANGE: Indian Ocean—Antarctic Sea.

Cryptococcus neoformans (Sanfelice) Vuillemin 1901

≡*Torulopsis neoformans* (Sanfelice) Redaelli 1931

LITERATURE: Kriss (1963); Kriss and Novozhilova (1954); Novozhilova (1955); Suehiro and Tomiyasu (1962); Suehiro *et al.* (1962).
MARINE ORIGIN: Algae, plankton, seawater. RANGE: Black Sea; Pacific Ocean—Japan.

Cryptococcus terreus Di Menna 1954

LITERATURE: Fell and van Uden (1963); van Uden and Castelo-Branco (1963).
MARINE ORIGIN: Seawater. RANGE: Pacific Ocean—United States (California).

Cryptococcus uniguttulatus (Zach *in* Wolfram et Zach) Phaff et Fell 1970 (not validly published, see Article 33 of the International Code of Botanical Nomenclature)

≡*Cryptococcus neoformans* (Sanfelice) Vuillemin var. *uniguttulatus* (Zach *in* Wolfram et Zach) Lodder et Kreger-van Rij 1952

LITERATURE: Ahearn and Roth (1962); Fell *et al.* (1960); Volz *et al.* (1974).
MARINE ORIGIN: Sediment. RANGE: Atlantic Ocean—Bahamas.

Kloeckera apiculata (Reess emend. Klöcker) Janke 1923 (asporogenous form of *Hanseniaspora uvarum* and *H. valbyensis*)

LITERATURE: Capriotti (1962a); Suehiro (1963); Taysi and van Uden (1964); van Uden (1967); van Uden and Castelo-Branco (1963).

MARINE ORIGIN: Animals (fish: *Atherinopsis affinis littoralis*), seawater, tidal mud. RANGE: Atlantic Ocean—Portugal; Pacific Ocean—Japan, United States (California).

Kloeckera lafarii (Klöcker) Janke 1923, 1928 [syn. of *Kloeckera javanica* (Klöcker) Janke var. *lafarii* (Klöcker) Miller et Phaff 1958]

LITERATURE: Taysi and van Uden (1964).
MARINE ORIGIN: Seawater. RANGE: Atlantic Ocean—Portugal.

Rhodotorula aurantiaca (Saito) Lodder 1952

=*Rhodotorula colostri* (Castelli) Lodder 1934
=*Rhodotorula crocea* Shifrine et Phaff 1956

LITERATURE: Ahearn and Roth (1962); Ahearn *et al.* (1962); Fell (1967); Fell and van Uden (1963); Hoppe (1972b); Kriss (1963); Kriss and Novozhilova (1954); Novozhilova (1955); Novozhilova and Popova (1973); Seshadri and Sieburth (1975).
MARINE ORIGIN: Algae (Chlorophyta), seawater. RANGE: Atlantic Ocean—Bahamas, Gulf of Guinea, United States (Rhode Island); Baltic Sea—Germany (G.F.R.); Black Sea; Indian Ocean; Pacific Ocean—North Pacific.

Rhodotorula flava (Saito) Lodder 1934 [syn. of *Cryptococcus flavus* (Saito) Phaff et Fell 1970]

LITERATURE: Artagaveytia-Allende (1960); Suehiro *et al.* (1962).
MARINE ORIGIN: Plankton, seawater. RANGE: Atlantic Ocean—Uruguay; Pacific Ocean—Japan.

Rhodotorula glutinis (Fresenius) Harrison 1928

=*Rhodotorula rufula* (Saito) Harrison 1928
=*Rhodotorula glutinis* (Fresenius) Harrison var. *rubescens* (Saito) Lodder 1934
=*Rhodotorula gracilis* Rennerfelt 1937

LITERATURE: Ahearn and Roth (1962); Ahearn *et al.* (1962); Bruce and Morris (1973); Capriotti (1962a,b); Chrzanowski and Cowley (1977); Combs *et al.* (1971); Fell (1967, 1972, 1976); Fell and van Uden (1963); Fell *et al.* (1960); Goto *et al.* (1974b); Hoppe (1972a,b); Kriss (1963); Kriss and Novozhilova (1954); Kriss *et al.* (1952, 1967); Lundström-Eriksson and Norkrans (1968); Meyers *et al.* (1967b); Newell and Fell (1970); Nicot (1958a,c); Norkrans (1966a,b); Novozhilova (1955); Phaff *et al.* (1952); Ross and Morris (1962, 1965); Roth *et al.* (1962); Seshadri *et al.* (1966); Shinano (1962); Siepmann and Höhnk (1962); Suehiro *et al.* (1962); Taysi and van Uden (1964); van Uden and Castelo-Branco (1963); Volz *et al.* (1974); Yamasato *et al.* (1974).

MARINE ORIGIN: Algae *(Ascophyllum nodosum),* animals (conch: *Strombus gigas;* fiddler crab: *Uca pugilator;* shrimp: *Penaeus setiferus;* spiny lobster: *Panulirus argus;* fish: numerous species), submerged littoral sand, marsh soil, plankton, seawater, sediment. RANGE: Atlantic Ocean—Antarctic Sea, Bahamas, France, Gulf of Mexico, Gulf Stream, Iceland, North Atlantic, Portugal, Scotland, Sweden, United States (Florida, Long Island Sound, South Carolina); Baltic Sea; Black Sea; Indian Ocean—India, Mauritius, and open ocean; Indo-Pacific Ocean— Antarctic Sea; Pacific Ocean—Antarctic Sea, Japan, North Pacific, United States (California), and open ocean.

Rhodotorula glutinis (Fresenius) Harrison var. *dairenensis* Hasegawa et Banno 1958

LITERATURE: Ahearn and Roth (1962); Ahearn *et al.* (1962); Roth *et al.* (1962).
MARINE ORIGIN: Not specified by authors listed. RANGE: Atlantic Ocean.

Rhodotorula graminis Di Menna 1958

LITERATURE: Ahearn and Roth (1962); Ahearn *et al.* (1962); De Queiroz and Macêdo (1972); Fell (1967); Fell and van Uden (1963); Fell *et al.* (1960); Meyers *et al.* (1967b); Roth *et al.* (1962); Seshadri and Sieburth (1975); van Uden and Castelo-Branco (1963).
MARINE ORIGIN: Algae (Chlorophyta, Phaeophyta), seawater, sediment. RANGE: Atlantic Ocean—Brazil, Gulf Stream, United States (Florida, Rhode Island); Black Sea; Indian Ocean—Mauritius, and open ocean; Pacific Ocean—United States (California).

Rhodotorula lactosa Hasegawa 1959

LITERATURE: Seshadri and Sieburth (1971, 1975).
MARINE ORIGIN: Algae *(Ulva lactuca,* Chlorophyta, Phaeophyta), seawater. RANGE: Atlantic Ocean—United States (Rhode Island).

Rhodotorula minuta (Saito) Harrison 1928

LITERATURE: Ahearn and Roth (1962); Ahearn *et al.* (1962); Capriotti (1962b); Combs *et al.* (1971); Fell (1972, 1976); Fell and van Uden (1963); Fell *et al.* (1960); Kobayashi *et al.* (1953); Lundström-Eriksson and Norkrans (1968); Meyers *et al.* (1967b); Norkrans (1966a,b); Ross and Morris (1965); Roth *et al.* (1962); Shinano (1962); Suehiro *et al.* (1962); Volz *et al.* (1974).

MARINE ORIGIN: Animals (clams, fish), plankton, seawater, sediment. RANGE: Atlantic Ocean—Bahamas, Iceland, Sweden, United States (Florida, Long Island Sound); Black Sea; Indo-Pacific Ocean—Antarctic Sea; Pacific Ocean—Antarctic Sea, Japan, and open ocean. NOTE: Also collected by Sugiyama *et al.* (1967) in an antarctic saline lake.

Rhodotorula minuta (Saito) Harrison var. *texensis* (Phaff, Mrak et Williams) Phaff et Ahearn 1970 (not validly published, see Article 33 of the International Code of Botanical Nomenclature)

≡*Rhodotorula texensis* Phaff, Mrak et Williams 1952

LITERATURE: Capriotti (1962b); Fell *et al.* (1960); Siepmann and Höhnk (1962); Volz *et al.* (1974).

MARINE ORIGIN: Animals (shrimp: *Penaeus setiferus*), seawater, sediment. RANGE: Atlantic Ocean—Bahamas, North Atlantic, United States (Florida, Texas). NOTE: Also collected by Goto *et al.* (1969) in an antarctic saline lake.

Rhodotorula pallida Lodder 1934

LITERATURE: Fell (1967); Kriss (1963); Kriss and Novozhilova (1954); Novozhilova (1955); Seshadri *et al.* (1966); Seshadri and Sieburth (1975).

MARINE ORIGIN: Algae (Chlorophyta), seawater. RANGE: Atlantic Ocean—United States (Rhode Island); Black Sea; Indian Ocean—India, and open ocean.

Rhodotorula pilimanae Hedrick et Burke 1951

LITERATURE: Ahearn and Roth (1962); Ahearn *et al.* (1962); Crow *et al.* (1977); Fell (1972, 1976); Meyers *et al.* (1967a,b); Roth *et al.* (1962); Seshadri and Sieburth (1975); Taysi and van Uden (1964).

MARINE ORIGIN: Algae (Phaeophyta), animals (fish), seawater. RANGE: Atlantic Ocean—Bahamas, Helgoland (F.R.G.), North Atlantic, Portugal,

United States (Florida, Rhode Island); Black Sea; Indo-Pacific and Pacific Oceans—Antarctic Sea.

Rhodotorula rubra (Demme) Lodder 1934

=*Rhodotorula mucilaginosa* (Jörgensen) Harrison 1928
=*Rhodotorula mucilaginosa* var. *sanguinea* (Schimon) Lodder 1934
=*Rhodotorula sanniei* (Ciferri et Redaelli) Lodder 1934

LITERATURE: Ahearn and Roth (1962); Artagaveytia-Allende (1960); Bhat *et al.* (1955); Bruce and Morris (1973); Capriotti (1962b); Combs *et al.* (1971); Fell (1967, 1972, 1976); Fell and van Uden (1963); Fell *et al.* (1960); Goto *et al.* (1974a,b); Hoppe (1972a,b); Kobayashi *et al.* (1953); Kriss (1963); Kriss and Novozhilova (1954); Kriss *et al.* (1967); Le Petit *et al.* (1970); Lundström-Eriksson and Norkrans (1968); Meyers *et al.* (1967a,b, 1970b); Newell (1976); Norkrans (1966a,b); Novozhilova (1955); Novozhilova and Popova (1973); Phaff *et al.* (1952); Picci and Verona (1964); Ross and Morris (1965); Roth *et al.* (1962); Seshadri and Sieburth (1971, 1975); Shinano (1962); Siepmann and Höhnk (1962); Suehiro (1962, 1963); Suehiro and Tomiyasu (1962); Suehiro *et al.* (1962); Taysi and van Uden (1964); Volz *et al.* (1974); Yamasato *et al.* (1974).

MARINE ORIGIN: Algae (*Fucus vesiculosus, Thalassiosira subtilis,* and others), animals (clam; conch: *Strombus gigas;* shrimp: *Penaeus setiferus;* fish: *Saragus saragus* and others), oil slicks, plankton, *Rhizophora mangle,* seawater, sediment, tidal mud. RANGE: Atlantic Ocean—Bahamas, Gulf of Guinea, Gulf Stream, Helgoland (F.R.G.), Iceland, North Atlantic, Portugal, Scotland, Sweden, Uruguay, United States (Florida, Long Island Sound, Louisiana, Rhode Island, Texas); Baltic Sea—Germany (F.R.G.); Black Sea; Indian Ocean—India, Mauritius, and open ocean; Indo-Pacific Ocean—Antarctic Sea; Mediterranean—France, Italy; Pacific Ocean—Antarctic Sea, Japan, North Pacific, and open ocean. NOTE: Also collected by Goto *et al.* (1969) and Sugiyama *et al.* (1967) in an antarctic saline lake.

Sterigmatomyces halophilus Fell 1966

LITERATURE: Fell (1967).
MARINE ORIGIN: Seawater. RANGE: Atlantic Ocean—United States (Florida); Indian Ocean.

Torulopsis aeria (Saito) Lodder 1934 [*syn. of Cryptococcus albidus* (Saito) Skinner var. *aerius* Phaff et Fell 1970; not validly published, see Article 33 of the International Code of Botanical Nomenclature)

LITERATURE: Kobayashi *et al.* (1953); Kriss (1963); Kriss *et al.* [1967 (as "vars.")]; Kriss and Novozhilova (1954); Novozhilova (1955); Novozhilova and Popova (1973); Phaff *et al.* (1952).
MARINE ORIGIN: Animals (clam; shrimp: *Penaeus setiferus*), seawater.
RANGE: Arctic Ocean; Atlantic Ocean—Gulf of Guinea, Gulf of Mexico, North Atlantic; Black Sea; Indian Ocean; Pacific Ocean—Japan, and open ocean.

Torulopsis anatomiae Zwillenberg 1966

LITERATURE: Buck *et al.* (1977).
MARINE ORIGIN: Animals (mussel: *Mytilus edulis;* oyster: *Crassostrea virginica;* quahog: *Mercenaria mercenaria*). RANGE: Atlantic Ocean— United States (Long Island Sound).

Torulopsis bovina (van Uden et Do Carmo-Sousa) van Uden et Vidal-Leiria 1970 (not validly published, see Article 33 of the International Code of Botanical Nomenclature)

LITERATURE: Buck *et al.* (1977).
MARINE ORIGIN: Animals (mussel: *Mytilus edulis;* oyster: *Crassostrea virginica;* quahog: *Mercenaria mercenaria*). RANGE: Atlantic Ocean— United States (Long Island).

Torulopsis candida (Saito) Lodder 1934

=*Candida famata* (Harrison) Novák et Zsolt 1961
=*Torulopsis famata* (Harrison) Lodder et Kreger-van Rij 1952
=*Torulopsis minor* (Pollacci et Nannizzi) Lodder 1934

LITERATURE: Ahearn and Roth (1962); Artagaveytia-Allende (1960); Bhat and Kachwalla (1955); Bruce and Morris (1973); Buck *et al.* (1977); Fell and van Uden (1963); Fell *et al.* (1960); Goto *et al.* (1974a); Hoppe (1972b); Kawakita and van Uden (1965); Kriss (1963); Kriss and Novozhilova (1954); Kriss *et al.* (1952, 1967); Le Petit *et al.* (1970); Lundström-Eriksson and Norkrans (1968); Norkrans (1966a,b); Novozhilova (1955); Novozhilova and Popova (1973); Ross and Morris (1962, 1965); Seshadri *et al.* (1966); Shinano (1962); Siepmann and Höhnk (1962); Suehiro (1960); Suehiro and Tomiyasu (1962); Taysi and van Uden (1964); van Uden (1967); Yamasato *et al.* (1974).
MARINE ORIGIN: Algae, animals (mussel: *Mytilus edulis;* oyster: *Crassostrea virginica;* quahog: *Mercenaria mercenaria;* fish: numerous species; tern: *Sterna minuta*), oil slicks, seawater, sediment. RANGE:

Atlantic Ocean—Bahamas, Gulf of Guinea, United States (Long Island Sound), North Atlantic, Portugal, Scotland, Sweden, Uruguay; Baltic Sea; Black Sea; Indian Ocean—India; Mediterranean—France; Pacific Ocean—Japan and open ocean.

Torulopsis cantarelli van der Walt et van Kerken 1961

LITERATURE: Buck *et al.* (1977); Volz *et al.* (1974).
MARINE ORIGIN: Animals (mussel: *Mytilus edulis;* oyster: *Crassostrea virginica;* quahog: *Mercenaria mercenaria*), sediment. RANGE: Atlantic Ocean—Bahamas, United States (Long Island Sound).

Torulopsis dattila (Kluyver) Lodder 1934

LITERATURE: Buck *et al.* (1977); Kriss *et al.* (1967); Le Petit *et al.* (1970); Shinano (1962).
MARINE ORIGIN: Animals (mussel: *Mytilus edulis;* oyster: *Crassostrea virginica;* quahog: *Mercenaria mercenaria*), oil slicks, seawater. RANGE: Atlantic Ocean—North Atlantic, United States (Long Island Sound); Mediterranean—France; Pacific Ocean.

Torulopsis ernobii Lodder et Kreger-van Rij 1952

LITERATURE: Chrzanowski and Cowley (1977).
MARINE ORIGIN: Animals (crab: *Uca pugilator*), marsh soil. RANGE: Atlantic Ocean—United States (South Carolina).

Torulopsis glabrata (Anderson) Lodder et de Vries 1938–1939

LITERATURE: Bhat and Kachwalla (1955); Buck *et al.* (1977); Capriotti (1962a); Kawakita and van Uden (1965); Phaff *et al.* (1952); van Uden and Castelo-Branco (1963).
MARINE ORIGIN: Animals (mussel: *Mytilus edulis;* oyster: *Crassostrea virginica*; quahog: *Mercenaria mercenaria;* shrimp: *Penaeus setiferus;* gulls: *Larus argentatus, L. fuscus, L. genei, L. ridibundus;* tern: *Sterna hirundo*), seawater. RANGE: Atlantic Ocean—Portugal, United States (Florida, Long Island Sound, Texas); Indian Ocean—India; Pacific Ocean—United States (California).

Torulopsis globosa (Olson et Hammer) Lodder et Kreger-van Rij 1952 [syn. of *Citeromyces matritensis* (Santa María) Santa María 1957]

LITERATURE: Buck *et al.* (1977).
MARINE ORIGIN: Animals (mussel: *Mytilus edulis;* oyster: *Crassostrea virginica;* quahog: *Mercenaria mercenaria*), seawater. RANGE: Atlantic Ocean—United States (Long Island Sound).

Torulopsis holmii (Jörgensen) Lodder 1934

LITERATURE: Buck *et al.* (1977); Kriss *et al.* (1967).
MARINE ORIGIN: Animals (mussel: *Mytilus edulis;* oyster: *Crassostrea virginica;* quahog: *Mercenaria mercenaria*), seawater. RANGE: Atlantic Ocean—North Atlantic, United States (Long Island Sound).

Torulopsis inconspicua Lodder et Kreger-van Rij 1952

LITERATURE: Fell (1972); Ross and Morris [1962, 1965 (also as "var.")]; Suehiro (1962); Suehiro *et al.* (1962).
MARINE ORIGIN: Algae *(Thalassiosira subtilis),* animals (fish: numerous species), plankton, seawater. RANGE: Atlantic Ocean—North Atlantic, Scotland; Pacific Ocean—Japan; southern oceans.

Torulopsis ingeniosa Di Menna 1958

LITERATURE: Volz *et al.* (1974).
MARINE ORIGIN: Sediment. RANGE: Atlantic Ocean—Bahamas.

Torulopsis lipofera (den Doren de Jong) Lodder 1934 [syn. of *Lipomyces lipofer* (den Doren de Jong) Lodder et Kreger-van Rij 1952]

LITERATURE: Kriss (1963); Kriss and Novozhilova (1954); Novozhilova (1955); Novozhilova and Popova (1973).
MARINE ORIGIN: Seawater. RANGE: Atlantic Ocean—Gulf of Guinea; Pacific Ocean—North Pacific.

Torulopsis magnoliae Lodder et Kreger-van Rij 1952

LITERATURE: Buck *et al.* (1977).
MARINE ORIGIN: Animals (mussel: *Mytilus edulis;* oyster: *Crassostrea virginica;* quahog: *Mercenaria mercenaria*). RANGE: Atlantic Ocean—United States (Long Island Sound).

Torulopsis norvegica Reiersöl 1958

LITERATURE: Buck *et al.* (1977); Fell (1972, 1976).

MARINE ORIGIN: Animals (mussel: *Mytilus edulis;* oyster: *Crassostrea virginica;* quahog: *Mercenaria mercenaria*), seawater. RANGE: Atlantic Ocean—United States (Long Island Sound); Indian Ocean; Indo-Pacific Ocean; southern oceans.

Torulopsis pintolopesii van Uden 1952

LITERATURE: Kawakita and van Uden (1965).
MARINE ORIGIN: Animals (gulls: *Larus argentatus, L. fuscus, L. genei, L. ridibundus*). RANGE: Atlantic Ocean—Portugal.

Torulopsis pinus Lodder et Kreger-van Rij 1952

LITERATURE: Buck *et al.* (1977).
MARINE ORIGIN: Animals (mussel: *Mytilus edulis;* oyster: *Crassostrea virginica;* quahog: *Mercenaria mercenaria*). RANGE: Atlantic Ocean—United States (Long Island Sound).

Torulopsis pulcherrima (Lindner) Sacc. 1906 (syn. of *Metschnikowia pulcherrima* Pitt et Miller 1968)

LITERATURE: Kriss (1963); Kriss and Novozhilova (1954); Kriss *et al.* (1952); Novozhilova (1955).
MARINE ORIGIN: Seawater. RANGE: Black Sea; Pacific Ocean—North Pacific.

Torulopsis sphaerica (Hammer et Cordes) Lodder 1934 [imperfect form of *Kluyveromyces lactis* (Dombrowski) van der Walt 1965]

LITERATURE: Shinano (1962).
MARINE ORIGIN: Seawater. RANGE: Pacific Ocean.

Torulopsis stellata (Kroemer et Krumbholz) Lodder 1932

LITERATURE: Buck *et al.* (1977).
MARINE ORIGIN: Animals (mussel: *Mytilus edulis;* oyster: *Crassostrea virginica;* quahog: *Mercenaria mercenaria*). RANGE: Atlantic Ocean—United States (Long Island Sound).

Torulopsis versatilis (Etchells et Bell) Lodder et Kreger-van Rij 1952

LITERATURE: De Queiroz (1972).
MARINE ORIGIN: Algae. RANGE: Atlantic Ocean—Brazil.

Trichosporon arenicola Lima et Queiroz 1972

MARINE ORIGIN: Beach sand. RANGE: Atlantic Ocean—Brazil.

Trichosporon behrendii Lodder et Kreger-van Rij 1952 [syn. of
Endomycopsis burtonii (Boidin, Pignal, Lehodey, Vey et Abadie)
Kreger-van Rij 1970; not validly published, see Article 33 of the
International Code of Botanical Nomenclature)

LITERATURE: Suehiro (1960).
MARINE ORIGIN: Algae. RANGE: Pacific Ocean—Japan.

Trichosporon capitatum Diddens et Lodder 1942

LITERATURE: Capriotti (1962a); Taysi and van Uden (1964).
MARINE ORIGIN: Seawater. RANGE: Atlantic Ocean—Portugal, United
States (Florida).

Trichosporon cutaneum (De Beurmann, Gougerot et Vaucher) Ota 1926

> =*Trichosporon infestans* (Moses et Vianna) Ciferri et Redaelli 1935
> =*Trichosporon cutaneum* var. *penaeus* Phaff, Mrak et Williams 1952 (as
> *"peneaus"*)

LITERATURE: Ahearn and Roth (1962); Buck *et al.* (1977); Capriotti
(1962b); Fell (1972, 1976); Fell and van Uden (1963); Fell *et al.* (1960);
Hoppe (1972b); Kishimoto and Baker (1969); Phaff *et al.* (1952); Roth *et
al.* (1962); Siepmann and Höhnk (1962); Suehiro (1960, 1963); Suehiro and
Tomiyasu (1962); Suehiro *et al.* (1962); Taysi and van Uden (1964); Volz
et al. (1974).
MARINE ORIGIN: Algae (*Ascophyllum nodosum* and unidentified), ani-
mals (mussel: *Mytilus edulis;* oyster: *Crassostrea virginica;* quahog: *Mer-
cenaria mercenaria;* shrimp: *Penaeus setiferus;* fish), beach sand,
plankton, seawater, sediment, tidal mud. RANGE: Atlantic Ocean—
Bahamas, Gulf of Mexico, North Atlantic, Portugal, United States
(Florida, Long Island Sound); Baltic Sea—Germany (F.R.G.); Indian and
Indo-Pacific Oceans—Antarctic Sea; Pacific Ocean—Japan, United States
(Hawaii); southern oceans.

Trichosporon pullulans (Lindner) Diddens et Lodder 1942

LITERATURE: Bruce and Morris (1973); Suehiro (1963).
MARINE ORIGIN: Animals (fish: numerous species), tidal mud. RANGE: Atlantic Ocean—Scotland; Pacific Ocean—Japan.

IV. APPENDIX*

Aureobasidium mansonii (Cast.) Cooke 1962

LITERATURE: Roth *et al.* (1964); Steele (1967).
MARINE ORIGIN: Coastal sands, seawater. RANGE: Atlantic Ocean— United States (Florida); Pacific Ocean—Phoenix Islands.

Aureobasidium pullulans (DeBary) Arnaud 1918

≡*Pullularia pullulans* (DeBary) Berkhout 1923

LITERATURE: Capriotti (1962a,b); Crow *et al.* (1977); Fell *et al.* (1960); Kobayashi *et al.* (1953); Meyers *et al.* (1967a,b); Phaff *et al.* (1952); Ross and Morris (1962); Roth *et al.* (1964); Siepmann and Höhnk (1962); Steele (1967).
MARINE ORIGIN: Animals (clam; shrimp: *Penaeus setiferus;* starfish; fish: *Haemulon aureolineatum, Lactophrys quadricornuta,* salmon, and unidentified), coastal sands, seawater, sediment. RANGE: Atlantic Ocean—Gulf of Mexico, Gulf Stream, North Atlantic, Scotland, United States (Florida); Black Sea; Pacific Ocean—Japan, Phoenix Islands, United States (Hawaii).

Pullularia fermentans Wynne et Gott var. *saccharofermentans*
Gladoch 1969

MARINE ORIGIN: Animals (fish: *Trachurus* sp.). RANGE: Atlantic Ocean—Africa.

---

* The genus *Aureobasidium* (syn. *Pullularia*) belongs to the Hyphomycetes, but its species are often confused with yeasts because of yeastlike budding of primary conidia. Because they are frequently isolated from the marine environment, we include *A. mansonii,* *A. pullulans,* and *Pullularia fermentans* var. *saccharofermentans* with the asporogenous yeasts. *Aureobasidium* was included in a recent monograph by Hermanides-Nijhof (1977).

Glossary*

acervulus A pseudoparenchyma or aggregation of hyphae without distinct wall or ostiole; producing conidia on the upper surface.

acrogenous Borne at apices.

acropetal Developing from below and toward the apex (e.g., in a chain of conidia where the youngest conidium is at the tip).

acropleurogenous At the end and on the sides.

agar A dry, gelatinlike phycocolloid obtained from red algae, used to prepare culture media into gels.

aleuriospore A conidium developed from the blown-out tip of a conidiogenous cell or hyphal branch from which it secedes with difficulty [term rejected by Kendrick (1971b)].

algicolous Living on algae.

alginate The sodium salt of alginic acid, produced in the cell walls of certain brown algae (e.g., *Ascophyllum, Fucus, Laminaria,* and *Macrocystis*).

allantoid Sausage-shaped.

ampulliform Flask-shaped; basal part swollen.

amyloid Stained blue in iodine (Melzer's reagent).

anastomosing Fusing of hyphae, producing an irregular network.

annellated Conidiogenous cell provided with an annellation or ringlike structure, remaining attached after secession of holoblastically produced conidia.

annulus A ring-shaped apical apparatus in certain asci.

antheridium The male gametangium.

aphysoclastic An ascus type that releases the spores without rupture of an outer, less extensible portion of the wall and stretching of an inner zone of the wall (J. S. Furtado and Olive, 1970). *See also* physoclastic.

apical apparatus A refractive ring-, plate-, or fish-trap-like structure in the apex of asci of certain Ascomycetes, sometimes staining blue in IKI (Iodine–Potassium–Iodide).

apiosporous Septate in the lower part (e.g., certain ascospores).

apothecium Cup- or disklike ascocarp with exposed asci (e.g., in the Discomycetes).

appressorium A swelling, attaching a hypha or germ tube to the substrate.

aquatic fungi Marine and freshwater fungi.

arenicolous Sand-inhabiting (living among or on grains of sand).

* Definitions follow mainly Ainsworth *et al.* (1971), M. B. Ellis (1971), Kendrick (1971b), Snell and Dick (1971), and Stearn (1966).

ascocarp Fruiting body of an Ascomycete, producing asci with ascospores. *See also* apothecium; cleistothecium; perithecium.

ascogenous Producing asci.

ascogonium A cell or group of cells in Ascomycetes fertilized by a sexual act.

ascospore A spore formed in an ascus.

ascosporogenesis The development of an ascospore from an ascospore rudiment to the mature structure.

ascus The saclike reproductive cell of Ascomycetes; the formation of usually eight ascospores is preceded in the young ascus by karyogamy and meiosis.

bacilliform Rodlike.

barophilic Able to grow at high pressures.

basidiocarp Fruiting body of a Basidiomycete, producing basidia with basidiospores.

basidiospore A spore formed on a basidium.

basidium The reproductive cell of Basidiomycetes; the formation of basidiospores is preceded by karyogamy and meiosis.

basipetal Developing toward the base (e.g., in a chain of conidia where the youngest conidium is at the base).

benthic Living on the bottom of the sea.

bitunicate An ascus with two walls; we use the term in a functional sense and define a bitunicate ascus as one that ejects the spores forcibly by rupturing of an inelastic outer wall layer and stretching of an inner layer ("jack-in-a-box" mechanism). *See also* physoclastic; unitunicate.

blastic A type of conidial development in which the conidial initial is markedly enlarged before it is delimited by a septum.

blastoconidium Holoblastic conidium.

botuliform Sausagelike; cylindrical with rounded ends.

bulbil A small sclerotium consisting of a small number of cells.

caespitose In tufts or groups, not grown together.

carbonaceous Charcoal-like, easily broken, dark-colored.

catenate, catenulate Forming chains.

catenophysis A persistent chain of utricular, thin-walled cells formed by the vertical separation of the pseudoparenchyma in the centrum of certain Ascomycetes [e.g., some Halosphaeriaceae (Kohlmeyer and Kohlmeyer, 1971a)].

cellulolytic Cellulose decomposing.

cephalothecoid An arrangement of cells of the peridium in ascocarps of Cephalothecaceae and members of other unrelated groups; the peridium is composed of plates of radiating cells that originate from a number of meristematic regions (Hawksworth and Booth, 1974).

cheiroid Shaped like a hand.

chitinoclastic Decomposing chitin.

chlamydospore A thick-walled, intercalary or terminal asexual, nondeciduous spore.

cicatrized Bearing scars (e.g., conidiogenous cells).

cirrus Spores forced out of a fruiting body in a tendril-like mass.

clamp A small semicircular hyphal protuberance, making a connection between the two cells that result from cell division, and arching over the septum between these adjoining cells.

cleistothecial Having an enclosed ascocarp (cleistothecium).

cleistothecium A closed ascocarp without ostiole, rupturing irregularly at maturity to release the ascospores.

clypeate   Having a clypeus.

clypeoid   Similar to a clypeus.

clypeus   A shieldlike stromatic growth around the ostiole of an ascocarp, sometimes covering several fruiting bodies.

conceptacle   A cavity containing reproductive organs (e.g., in algae, or used for pycnidia).

conidiogenesis   The development of conidia.

conidiogenous cell   A cell from which, or within which, conidia are produced.

conidiophore   A hypha bearing conidiogenous cells, which produce conidia.

conidium   An asexual reproductive unit of the Deuteromycotina (imperfect fungi).

convoluted   Brainlike.

coprophilous   Dung-inhabiting.

coriaceous   Leatherlike.

cortex   In algae: the outer cell layer or tissue of an algal thallus.

corticolous   Bark-inhabiting.

crozier   The hook of an ascogenous hypha, prior to the formation of asci.

cupulate   Cuplike.

cyphelloid   Similar to *Cyphella* (Aphyllophorales), having disk- or cup-shaped basidiocarps.

cystidium   A sterile end of a hypha in the hymenium of a Basidiomycete, usually projecting beyond the basidia.

cystocarp   The female reproductive structure in red algae.

deciduous   Easily falling off at maturity, not persistent (e.g., conidia).

determinate   Having a definite limit (e.g., growth ceasing in conidiophores or conidiogenous cells at or before the beginning of conidiogenesis).

detritivore   An animal feeding chiefly on detritus (organic debris).

discrete   Having a distinctive shape, different from the supporting vegetative cells (e.g., conidiogenous cells). *See also* integrated *(ant.)*.

doliiform   Barrel-like.

dothideaceous   Having the asci in locules of a stroma (e.g., in *Dothidea*).

driftwood   Floating or loose wood on the shore (G. C. Hughes, 1968). *See also* intertidal wood.

echinulate   Having spines or small pointed processes (e.g., spores).

ectoascus   Outer wall of the ascus. *See also* bitunicate.

elater   Coiled appendage of spores of horsetails (*Equisetum* spp.), assisting in spore dispersal.

endoascus   Inner wall of the ascus. *See also* bitunicate.

endopsammon   Microenvironment between grains of sand in marine beaches.

enteroblastic   A type of conidial development in which the outer wall layer of the conidiogenous cell does not contribute to the formation of the conidial wall.

epapillate   Without a papilla (in ascocarps, pycnidia, and spermogonia).

epiphyte   A plant attached to another plant, using it for support without parasitizing it.

epispore   The first wall layer of a developing ascospore.

epithecium   A layer above the asci (e.g., formed by the tips of paraphyses in disks of certain Discomycetes).

erumpent   Breaking through the surface of the substrate.

étang   Salt pond connected to the ocean (in France).

excipulum   The marginal tissues of an apothecium adjacent to the hypothecium.

exospore   The final outer wall layer of an ascospore.

foliicolous   Leaf-inhabiting.

geniculate   Bent like a knee.
germ pore   A thin area in the wall of a spore through which a germ tube may come out.
gleba   The spore-producing tissue of closed fruiting bodies (e.g., in Gasteromycetes).

halophyte   A plant capable of tolerating high amounts of salt, usually 0.5% or more NaCl.
hapteron   A basal, usually cylindrical, outgrowth of the holdfast of certain algae.
haustorium   A side branch of a hypha serving as an organ of attachment and absorption of nutrients, especially inside a living cell of the host.
helicoid   Spiral-like, coiled, or curved; especially conidia of some Deuteromycetes.
hirsute   Provided with long hairs.
holoblastic   A type of conidial development in which all layers of the conidiogenous cell participate in the formation of the conidial wall.
hydrocarbonoclastic   Decomposing hydrocarbons.
hydrocaulus   The erect stemlike structures of hydrozoan colonies.
hydrorhiza   The rootlike network of hydrozoa covering the substrate.
hymenium   The spore-producing layer of a fruiting body (e.g., formed by basidia in Basidiomycetes or by asci in Ascomycetes, often with interspersed sterile filaments between them).
hyperparasite   A parasite growing on another parasite.
hypersaprobe   A saprobic fungus growing on another, dead saprobic fungus.
hypha   A filament of fungi, forming the mycelium.
hyphopodium   A hyphal modification, consisting of a short lateral, often lobed outgrowth, adhering firmly to the substrate and producing narrow filaments that penetrate the cell of the host; a specialized form of an appressorium (Goos and Gessner, 1975).
hypostroma   The footlike base of a stroma under an ascocarp.
hypothecium   The hyphal layer under the hymenium of an apothecium.
hysterothecium   An elongate fruiting body that opens apically by a long slit, exposing the hymenium.

integrated   Being morphologically indistinguishable from vegetative cells (e.g., conidiogenous cells). *See also* discrete (*ant.*).
interascicular tissue   Sterile pseudotissue, often filamentous, between asci in ascocarps of certain Ascomycetes.
intercalary   Between two cells.
interstitium   The space between grains of sand on the shore.
intertidal wood   Permanently fixed wood in the intertidal zone (G. C. Hughes, 1968). *See also* driftwood.
involucrellum   Tissue of the upper part of ascocarps of some lichenized Ascomycetes (e.g., in certain species of *Pharcidia*).
isocryme   A line connecting points having the same mean temperature for a specified coldest time of the year.
isothere   A line connecting points having the same mean summer temperature.
isotherm   A line connecting points having the same temperature at a given time.

lageniform   With a swollen base and tapering toward the top; flask-shaped.
laminarin   A reserve polysaccharide in many Phaeophyta.
lenticel   A raised, usually elliptical, pore in the rind of woody plants, filled with loosely arranged cells and allowing exchange of gases (e.g., in prop roots of mangroves).

lentiform   Lens-shaped.

lichen   A symbiosis between algae and fungi, usually resulting in an association that is morphologically different from both partners, usually producing specific lichen substances.

lignicolous   Wood-inhabiting.

loculus, locule   A cavity (e.g., in a stroma).

lysigenous   Formed by the breakdown of cells (e.g., in the formation of an ostiolar canal). *See also* schizogenous *(ant.).*

macronematous   Being morphologically different from vegetative hyphae (e.g., conidiophores).

mangal   The mangrove vegetation.

manglicolous   Living on mangrove trees.

mangrove   Intertidal trees of the tropics and subtropics, belonging to the genera *Avicennia, Rhizophora,* and others.

medulla   The central tissue under the cortex of algae.

meristem   A type of conidial development in which hyphae or conidiogenous cells become converted into conidia.

mesospore   Wall layer between epispore and plasmalemma of a developing ascospore.

microconidium   The smaller conidium of a fungus, which also produces macroconidia; sometimes functioning as a spermatium.

micronematous   Being morphologically similar to vegetative hyphae (e.g., conidiophores).

monoblastic   A type of conidial development in which a holoblastic conidiogenous cell blows out at only one point.

mononematous   Being solitary, not fused (e.g., conidiophores).

monophialidic   Phialidic conidiogenous cells having one opening.

muriform   Spores having transverse and longitudinal septa.

mycelium   A network of fungal filaments (hyphae).

mycobiont   The fungal partner in a lichen.

mycology   The science dealing with fungi.

mycophycobiosis   An obligate symbiotic association between a systemic marine fungus and a macroalga in which the habit of the alga dominates.

mycorrhiza   A symbiotic association between fungi and roots of higher plants.

mycostasis   The inhibition of fungal growth by some chemical or physical agent.

mycota   The fungal population of a particular area.

nomen confusum   A name based on a nomenclatural type consisting of discordant elements.

nomen dubium   A name of uncertain application.

nomen nudum   A name published without a description; therefore, being not validly published.

obpyriform   The reverse of pear-shaped.

obturbinate   The reverse of top-shaped.

ontogeny   The whole course of development during the life history of an individual.

operculus   A little lid at the ascus apex (e.g., in *Orcadia ascophylli*).

ostiolar canal   Cavity in the papilla or neck of a fruiting body, opening to the outside with a pore, the ostiole.

ostiolate   Provided with an ostiole.

ostiole, ostiolum   A pore in the peridium of a fruiting body (ascocarp or pycnidium) through which spores are released.

paraphysis   A sterile filament between or around the asci in ascocarps of Discomycetes and Pyrenomycetes, being attached at the base, or growing downward as an apical paraphysis.

paraphysoid   Similar to paraphyses.

parasite   An organism growing on a living host and obtaining its food from it.

parietal   Formed along the wall.

pathogen   A parasite causing disease in its host.

pedunculate   Having a little stalk.

percurrent   A type of vegetative proliferation of conidiophores or conidiogenous cells in which each successive tip grows through the previous apex.

peridium   The wall of fruiting bodies (ascocarps, basidiocarps, pycnidia, spermogonia).

periphysis   A short filament in the ostiolar canal of ascocarps, pycnidia, and spermogonia; periphyses often occluding the ostiole.

perithecium   The bottle-shaped ascocarp of Pyrenomycetes.

perthophyte   An organism living on dead tissues of living hosts.

perthophytic   Having the mode of life of a perthophyte.

phialide   A conidiogenous cell that produces conidia in basipetal succession through one opening or several openings; the wall does not participate in the formation of conidia.

phragmosporous   A spore with two or more transverse septa.

phycobiont   The algal partner in a lichen.

physoclastic   An ascus that releases the spores by rupture of an outer, less extensible portion of the wall and stretching of an inner zone of the wall (J. S. Furtado and Olive, 1970). *See also* aphysoclastic.

pit connection   A cytoplasmic strand connecting two adjoining cells through a pit in the walls.

plectenchyma   A thick tissue in which the hyphae have grown together, intertwining and adhering to each other.

pleomorphic   Having more than one independent spore stage in the life cycle.

pleurogenous   Formed on the side.

pneumatophore   Negatively geotropic aerating roots of mangrove trees (e.g., in the genus *Avicennia*).

polyblastic   A type of conidial development in which a holoblastic conidiogenous cell blows out at more than one point.

polytretic   The formation of enteroblastic conidia by protrusion of the inner wall of the conidiogenous cell through several channels in the outer wall.

primary marine fungi   Fungal species derived from marine ancestors and remaining in the marine environment (e.g., members of Spathulosporales).

proliferate   The successive development of new structures within the old wall.

prop root   An adventitious aerial root that serves as an anchoring and aerating organ of mangrove trees (e.g., in *Rhizophora* spp.); prop roots are usually arched and up to 2 m long.

prosenchyma   A plectenchyma in which the hyphal elements are recognizable as hyphae.

protocysts   The inflated bulbous tips of ramified sterile hyphae in the venter of immature basidiocarps of *Nia vibrissa* (Doguet, 1969).

protoperithecial initial   The first recognizable stage of an ascocarp, developing into a protoperithecium.

protoperithecium   The immature stage of an ascocarp.

pseudoclypeus   A clypeus in which parts of the host's tissue are enclosed in the fungal stroma.

pseudoparaphysis   A sterile filament between the asci in ascocarps of Loculoascomycetes, usually growing downward in the locule before the formation of asci.

pseudoparenchyma A plectenchyma in which the hyphal elements are not recognizable as hyphae.

pseudoseptum A membrane formed by the protoplasm or a vacuole and appearing like a cell wall.

pseudostroma A stroma in which fungal cells and host tissue are mixed.

psychrophilic Fungi able to grow at temperatures below 10°C, with an optimum below 20°C.

pulvillus Turgescent swelling cushion in the ostiole of *Turgidosculum ulvae,* closing the opening and preventing penetration of water into the ascocarp cavity; also, part of an apical apparatus in certain asci.

pulvinate Cushionlike.

pulvinus A cushion.

pycnidium A fruiting body of Sphaeropsidales (Deuteromycotina), producing conidia, often globose or bottle-shaped.

ramo-conidium The apical part of a conidiophore, which becomes detached and functions as a conidium.

ramose Branched.

receptacle In fungi (e.g., Laboulbeniales), an organ that bears reproductive structures (e.g., a stalk); in algae (Fucales), the specialized fertile area of branches containing conceptacles.

reniform Kidney-shaped.

repand Having a wavy edge and turned back.

resupinate Covering the substrate (without stalk) with the hymenium facing outward.

reticulate Netlike.

rhizome An underground stem that serves as a means of perennation and vegetative propagation [e.g., in sea grasses (*Posidonia, Zostera,* etc.)].

rhizoplane The root surface of higher plants.

rhizosphere The zone of the soil immediately surrounding the roots.

rostrate Having a beak.

saprobe An organism utilizing dead organic substrates as food.

schizogenous Formed by splitting or cracking (e.g., in the formation of an ostiolum in ascocarps and pycnidia). *See also* lysigenous *(ant.).*

scleroplectenchymatous Plectenchymatous, with thickened cell walls to produce a hard tissue.

sclerotium A firm resting body consisting of a hardened mass of hyphae, often with a dark rind.

scolecosporous Having long, worm- or threadlike spores.

scutellate Shaped like a small plate or shield.

sea grass Marine phanerogams of the genera *Posidonia, Thalassia, Zostera,* etc.

secondary marine fungi Fungal species derived from terrestrial ancestors that have migrated secondarily into the marine environment (e.g., members of Loculoascomycetes).

senescent Growing old.

setose Bearing bristles or hairs.

setula A fine hairlike appendage at the apex of a conidium.

sigmoid S-shaped.

smut A disease caused by members of Ustilaginales; or the fungus itself.

soft rot Deterioration of wood by hyphae of higher fungi (Ascomycetes and Deuteromycetes), which attack predominantly the unlignified layers of the cell walls, causing characteristic tunnels inside the walls.

sorus   The erumpent spore mass of certain fungi (e.g., smuts).

spermatiophore   A structure producing spermatia.

spermatium   A nonmotile male gamete, uniting with the trichogyne (e.g., in Laboulbeniales and Spathulosporales).

spermatophyte   A seed plant.

spermogonium   A fruiting body producing spermatia, similar to the conidia-producing pycnidia.

sporodochium   An erumpent asexual fructification in which conidia originate on a cushion-like cluster of short conidiophores; the latter are usually seated on a stroma (e.g., Tuberculariaceae).

sporogenous   Producing spores.

staurosporous   Having star-shaped spores (with three or more arms).

sterigma   A cell process (e.g., a pedicel on basidia supporting a spore).

stroma   A mass of vegetative hyphae in, on, or under which fructifications are produced.

subiculate   Provided with a subiculum.

subiculum   A crust- or wool-like growth of hyphae under fruiting bodies.

symbiont   An organism that lives in a state of symbiosis.

symbiosis   A living together of dissimilar organisms (e.g., in lichens).

sympodial   A type of vegetative proliferation of conidiophores or conidiogenous cells in which, after production of a terminal spore on the main axis, a succession of apices is produced, each of which originates below and to one side of the previous apex.

systemic   Distributed throughout an organism.

tetraradiate   Having four branches (e.g., certain conidia).

textura angularis   A hyphal tissue composed of short polyhedral cells without intercellular spaces; the separate hyphae are not distinguishable.

textura epidermoidea   A hyphal tissue composed of long cells without interhyphal spaces; hyphae with their walls united, forming a membranous tissue; the separate hyphae are distinguishable.

textura intricata   A hyphal tissue composed of long cells with interhyphal spaces; hyphae with their walls not united; the separate hyphae are distinguishable.

thalassopsammon   The microfauna between grains of sand in marine habitats; includes representatives of most invertebrate classes.

thalloconidium   A conidium produced by thallic conidiogenous cells; the conidium initial enlarges after delimitation by a septum.

thallus   The vegetative body of fungi and lower plants.

tomentose   Having a cover of densely matted, woolly hairs.

tretic   The formation of enteroblastic conidia by protrusion of the inner wall of the conidiogenous cell through one or more channels in the outer wall.

trichogyne   The receptive hypha of the female organ (e.g., in Ascomycetes and Rhodophyta).

trigonous   Three-angled.

turbinate   Top-shaped.

umbonate   Having an umbo, that is, a central raised knob.

uncinate   Hooked.

unitunicate   An ascus with one wall; we use the term in a functional sense (as opposed to the bitunicate ascus) and define a unitunicate ascus as one that releases the spores without the separation of an outer and inner wall layer; electron microscopy has shown that walls of so-called unitunicate asci are composed of several layers. *See also* aphysoclastic.

valsoid Having groups of ascocarps with their necks pointing inward, as in the genus *Valsa*.
verrucose Having small warts or rounded processes.
verruculose Delicately warty.

yeast A unicellular, budding fungus; yeasts are phylogenetically heterogenous, belonging to the Ascomycetes, Basidiomycetes, or Deuteromycetes.

Bibliography

Abdel-Fattah, H. M., Moubasher, A. H., and Abdel-Hafez, S. I. (1977). Studies on mycoflora of salt marshes in Egypt. I. Sugar fungi. *Mycopathology* **61,** 19–26.

Ahearn, D. G., and Meyers, S. P. (1972). The role of fungi in the decomposition of hydrocarbons in the marine environment. *In* "Biodeterioration of Materials" (A. H. Walters and E. H. Hueck-van der Plas, eds.), pp. 12–18. Applied Science, London.

Ahearn, D. G., and Roth, F. J. (1962). Vitamin requirements of marine-occurring yeasts. *Dev. Ind. Microbiol.* **3,** 163–173.

Ahearn, D. G., Roth, F. J., Jr., and Meyers, S. P. (1962). A comparative study of marine and terrestrial strains of *Rhodotorula. Can. J. Microbiol.* **8,** 121–132.

Ahearn, D. G., Roth, F. J., Jr., and Meyers, S. P. (1968). Ecology and characterization of yeasts from aquatic regions of South Florida. *Mar. Biol.* **1,** 291–308.

Ahearn, D. G., Meyers, S. P., Crow, S., and Berner, N. H. (1971a). Effect of oil in Louisiana marshland yeast populations. *Bacteriol. Proc.* **71,** 35.

Ahearn, D. G., Meyers, S. P., and Standard, P. G. (1971b). The role of yeasts in the decomposition of oils in marine environments. *Dev. Ind. Microbiol.* **12,** 126–134.

Ahmad, S. (1967). Contributions to the fungi of West Pakistan—VI. *Biologia (Lahore)* **13,** 15–42.

Ahmadjian, V. (1967). A guide to the algae occurring as lichen symbionts: isolation, culture, cultural physiology and identification. *Phycologia* **6,** 127–160.

Ainsworth, G. C. (1968). The number of fungi. *In* "The Fungi" (G. C. Ainsworth and A. S. Sussman, eds.), Vol. 3, pp. 505–514. Academic Press, New York.

Ainsworth, G. C., James, P. W., and Hawksworth, D. L. (1971). "Ainsworth & Bisby's Dictionary of the Fungi, Including the Lichens," 6th ed. Commonw. Mycol. Inst., Kew, Surrey, England.

Ainsworth, G. C., Sparrow, F. K., and Sussman, A. S., eds. (1973). "The Fungi," Vol. 4A. Academic Press, New York.

Alderman, D. J. (1973). Fungal infection of crawfish (*Palinurus elephas*) exoskeleton. *Trans. Br. Mycol. Soc.* **61,** 595–597.

Alderman, D. J. (1976). Fungal diseases of marine animals. *In* "Recent Advances in Aquatic Mycology" (E. B. G. Jones, ed.), pp. 223–260. Wiley, New York.

Aleem, A. A. (1952). Sur la présence de *Melanopsamma tregoubovii* Ollivier (Pyrénomycète) dans la manche occidentale (Parasite de *Dilophus fasciola* (Roth) Howe). *Bull. Lab. Marit. Dinard* **36,** 21–24.

Allaway, A. E., and Jennings, D. H. (1970a). The influence of cations on glucose uptake by the fungus *Dendryphiella salina. New Phytol.* **69,** 567–579.

Allaway, A. E., and Jennings, D. H. (1970b). The influence of cations on glucose transport and metabolism by, and the loss of sugar alcohols from, the fungus *Dendryphiella salina*. *New Phytol.* **69**, 581–593.

Allaway, A. E., and Jennings, D. H. (1971). The effect of cations on glucose utilization by, and on the growth of, the fungus *Dendryphiella salina*. *New Phytol.* **70**, 511–518.

Anastasiou, C. J. (1961). Fungi from salt lakes. I. A new species of *Clavariopsis*. *Mycologia* **53**, 11–16.

Anastasiou, C. J. (1963a). The genus *Zalerion* Moore et Meyers. *Can J. Bot.* **41**, 1135–1139.

Anastasiou, C. J. (1963b). Fungi from salt lakes. II. Ascomycetes and Fungi Imperfecti from the Salton Sea. *Nova Hedwigia* **6**, 243–276.

Anastasiou, C. J., and Churchland, L. M. (1968). An *Olpidiopsis* parasitic on a marine fungus. *Syesis* **1**, 81–85.

Anastasiou, C. J., and Churchland, L. M. (1969). Fungi on decaying leaves in marine habitats. *Can. J. Bot.* **47**, 251–257.

Andrews, J. H. (1976). The pathology of marine algae. *Biol. Rev. Cambridge Philos. Soc.* **51**, 211–253.

Apinis, A. E. (1964). On fungi isolated from soils and *Ammophila* debris. *Kew Bull.* **19**, 127–131.

Apinis, A. E., and Chesters, C. G. C. (1964). Ascomycetes of some salt marshes and sand dunes. *Trans. Br. Mycol. Soc.* **47**, 419–435.

Arnold, R. H. (1967). The *Nectriella* element of *"Hyalodothis"*. *Mycologia* **59**, 246–254.

Artagaveytia-Allende, R. C. (1960). Levaduras aisladas frente a las costas del Uruguay. *Atti Ist. Bot. Univ. Lab. Crittogam., Pavia* [5] **18**, 1–4.

Bandoni, R. J. (1975). Significance of the tetraradiate form in dispersal of terrestrial fungi. *Rep. Tottori Mycol. Inst. (Jpn.)* **12**, 105–113.

Barghoorn, E. S. (1944). II. Biological Aspects. *Farlowia* **1**, 434–467.

Barghoorn, E. S., and Linder, D. H. (1944). Marine fungi: their taxonomy and biology. *Farlowia* **1**, 395–467.

Bärlocher, F., and Kendrick, B. (1973). Fungi in the diet of *Gammarus pseudolimnaeus* (Amphipoda). *Oikos* **24**, 295–300.

Bärlocher, F., and Kendrick, B. (1975a). Leaf-conditioning by microorganisms. *Oecologia* **20**, 359–362.

Bärlocher, F., and Kendrick, B. (1975b). Assimilation efficiency of *Gammarus pseudolimnaeus* (Amphipoda) feeding on fungal mycelium or autumn-shed leaves. *Oikos* **26**, 55–59.

Barnett, J. A., and Pankhurst, R. J. (1974). "A New Key to the Yeasts." Am. Elsevier, New York.

Barr, M. E. (1972). Preliminary studies on the Dothideales in temperate North America. *Contrib. Univ. Mich. Herb.* **9**, 523–638.

Barr, M. E. (1976a). Perspectives in the Ascomycotina. *Mem. N.Y. Bot. Gard.* **28**, 1–8.

Barr, M. E. (1976b). *Buergenerula* and the Physosporellaceae. *Mycologia* **68**, 611–621.

Barr, M. E. (1977). *Magnaporthe, Telimenella,* and *Hyponectria* (Physosporellaceae). *Mycologia* **69**, 952–966.

Barron, G. L. (1968). "The genera of Hyphomycetes from Soil." Williams & Wilkins, Baltimore, Maryland.

Bauch, R. (1936). *Ophiobolus kniepii,* ein neuer parasitischer Pyrenomycet auf Kalkalgen. *Pubbl. Staz. Zool. Napoli* **15**, 377–391.

Becker, G. (1959). Biological investigations on marine borers in Berlin-Dahlem. *In* "Marine Boring and Fouling Organisms" (D. L. Ray, ed.), pp. 62–83. Univ. of Washington Press, Seattle.

Becker, G. (1961). Holzbeschädigung durch *Sphaeroma hookeri* Leach (Isopoda) an der französischen Mittelmeerküste. *Z. Angew. Zool.* **48,** 333–339.

Becker, G., and Kohlmeyer, J. (1958a). Holzzerstörung durch Meerespilze in Indien und ihre besondere Bedeutung für Fischereifahrzeuge. *Arch. Fischereiwiss.* **9,** 29–40.

Becker, G., and Kohlmeyer, J. (1958b). Deterioration of wood by marine fungi in India and its special significance for fishing-crafts. *J. Timber Dryers' Preserv. Assoc. India* **4,** 1–10.

Becker, G., Kampf, W.-D., and Kohlmeyer, J. (1957). Zur Ernährung der Holzbohrasseln der Gattung *Limnoria*. *Naturwissenschaften* **44,** 473–474.

Beer, R. (1920). On a new species of *Melanotaenium* with a general account of the genus. *Trans. Br. Mycol. Soc.* **6,** 331–343.

Benjamin, R. K. (1971). "Introduction and Supplement to Roland Thaxter's Contribution Towards a Monograph of the Laboulbeniaceae." Cramer, Lehre.

Benjamin, R. K. (1973). Laboulbeniomycetes. *In* "The Fungi" (G. C. Ainsworth, F. K. Sparrow, and A. S. Sussman, eds.), Vol. 4A, pp. 223–246. Academic Press, New York.

Bergen, L., and Wagner-Merner, D. T. (1977). Comparative survey of fungi and potential pathogenic fungi from selected beaches in the Tampa Bay area. *Mycologia* **69,** 299–308.

Berlese, A. N. (1888). Monografia dei generi *Pleospora. Clathrospora* e *Pyrenophora*. *Nuovo G. Bot. Ital.* **20,** 5–176 and 193–260.

Berlese, A. N. (1894). "Icones Fungorum," Vol. 1. Abellini.

Bhat, J. V., and Kachwalla, N. (1955). Marine yeasts off the Indian Coast. *Proc. Indian Acad. Sci., Sect. B* **41,** 9–15.

Bhat, J. V., Kachwalla, N., and Mody, B. N. (1955). Some aspects of the nutrition of marine yeasts and their growth. *J. Sci. Ind. Res.* **14,** 24–27.

Biernacka, I. (1965). Badania pali wyciągniętych na ląd z trzech przybrzeżnych akwenów Bałtyku. *Sylwan* **109,** 7–17.

Biga, M. L. B., Ciferri, R., and Bestagno, G. (1958). Ordinamento artificiale delle specie del genere *Coniothyrium* Corda. *Sydowia* **12,** 258–320.

Blanchard, D. C., and Syzdek, L. D. (1970). Mechanism for the water-to-air transfer and concentration of bacteria. *Science* **170,** 626–628.

Blanchard, D. C., and Syzdek, L. D. (1974). Bubble tube: apparatus for determining rate of collection of bacteria by an air bubble rising in water. *Limnol. Oceanogr.* **19,** 133–138.

Block, J. H., Catalfomo, P., Constantine, G. H., Jr., and Kirk, P. W., Jr. (1973). Triglyceride fatty acids of selected higher marine fungi. *Mycologia* **65,** 488–491.

Boerema, G. H. (1969). The use of the term *forma specialis* for *Phoma*-like fungi. *Trans. Br. Mycol. Soc.* **52,** 509–513.

Boerema, G. H. (1976). The *Phoma* species studied in culture by Dr. R. W. G. Dennis. *Trans. Br. Mycol. Soc.* **67,** 289–319.

Boerema, G. H., and Bollen, G. J. (1975). Conidiogenesis and conidial septation as differentiating criteria between *Phoma* and *Ascochyta*. *Persoonia* **8,** 111–144.

Boerema, G. H., and Dorenbosch, M. M. J. (1973). The *Phoma* and *Ascochyta* species described by Wollenweber and Hochapfel in their study on fruit-rotting. *Stud. Mycol.* **3,** 1–50.

Bommer, E., and Rousseau, M. (1891). Contributions à la flore mycologique de Belgique. *Bull. Soc. R. Bot. Belg.* **29,** 205–302.

Booth, C. (1957). Studies of Pyrenomycetes: I. Four species of *Chaetosphaeria,* two with *Catenularia* conidia. II. *Melanopsamma pomiformis* and its *Stachybotrys* conidia. *Mycol. Pap.* **68,** 1–27.

Booth, C. (1958). The genera *Chaetosphaeria* and *Thaxteria* in Britain. *The Naturalist*, pp. 83–90.

Børgesen, F. (1913). The marine algae of the Danish West Indies. Part 1. Chlorophyceae. *Dan. Bot. Ark.* **1**, 1–158.

Borut, S. Y., and Johnson, T. W., Jr. (1962). Some biological observations on fungi in estuarine sediments. *Mycologia* **54**, 181–193.

Bose, S. K. (1961). Studies on *Massarina* Sacc. and related genera. *Phytopathol. Z.* **41**, 151–213.

Böttger, M. (1967). Die Flechten der Insel Helgoland. *Veroeff. Inst. Meeresforsch. Bremerhaven* **10**, 247–259.

Böttger, M. (1969). Die Flechten der Inseln Neuwerk und Scharhörn (Elbmündung). *Veroeff. Inst. Meeresforsch. Bremerhaven* **11**, 293–302.

Boyd, D. A. (1901). Fungi (Microscopic). *In* "Fauna, Flora, and Geology of the Clyde Area," pp. 61–77. Br. Assoc. Adv. Sci., Glasgow.

Boyd, D. A. (1909). Some recent additions to the fungus-flora of the Clyde Area. *Glasgow Nat.* **1**, 110–115.

Boyd, D. A. (1911). Mycological Notes. *Glasgow Nat.* **4**, 14–18.

Boyd, D. A. (1916). Notes on the microfungi of the Kyles of Bute District. *Glasgow Nat.* **8**, 1–8.

Brandsberg, J. W. (1969). Fungi isolated from decomposing conifer litter. *Mycologia* **61**, 373–381.

Brewer, J. G., and Boerema, G. H. (1965). Electron microscope observations on the development of pycnidiospores in *Phoma* and *Ascochyta*. *Proc. K. Ned. Akad. Wet., Ser. C* **68**, 86–97.

Brodo, I. M. (1977). Lichenes Canadenses exsiccati: fascicle II. *Bryologist* **79**, 385–405.

Brooks, R. D. (1975). The presence of dolipore septa in *Nia vibrissa* and *Digitatispora marina*. *Mycologia* **67**, 172–174.

Brooks, R. D., Goos, R. D., and Sieburth, J. M. (1972). Fungal infestation of the surface and interior vessels of freshly collected driftwood. *Mar. Biol.* **16**, 274–278.

Brown, J. C. (1958a). British records, 17. *Asteromyces cruciatus* Moreau. *Trans. Br. Mycol. Soc.* **41**, 64.

Brown, J. C. (1958b). Soil fungi of some British sand dunes in relation to soil type and succession. *J. Ecol.* **46**, 641–664.

Bruce, J., and Morris, E. O. (1973). Psychrophilic yeasts isolated from marine fish. *Antonie van Leeuwenhoek* **39**, 331–339.

Bruun, A. F. (1957). Deep sea and abyssal depths. *In* "Treatise on Marine Ecology and Paleoecology." I. Ecology. *Mem., Geol. Soc. Am.* **67**, 641–672.

Buck, J. D., and Greenfield, L. J. (1964). Calcification in marine-occurring yeasts. *Bull. Mar. Sci. Gulf Caribb.* **14**, 239–245.

Buck, J. D., Bubucis, P. M., and Combs, T. J. (1977). Occurrence of human-associated yeasts in bivalve shellfish from Long Island Sound. *Appl. Environ. Microbiol.* **33**, 370–378.

Buczacki, S. T. (1972). *Zalerion arboricola*, a new helicosporous hyphomycete from conifer stems. *Trans. Br. Mycol. Soc.* **59**, 159–161.

Bultman, J. D., and Ritchie, D. D. (1976). Inhibition of fungal growth and reproduction by obtusaquinone and some cinnamylphenols. *In* "Proceedings of a Workshop on the Biodeterioration of Tropical Woods: Chemical Basis for Natural Resistance" (J. D. Bultman, ed.), pp. 57–65. Nav. Res. Lab., Department of the Navy, Washington, D.C.

Bultman, J. D., and Southwell, C. R. (1972). A preliminary investigation of the marine borer resistance of the tropical wood *Dalbergia retusa*. *Nav. Res. Lab. Rep.* **7416**, 1–14.

Bultman, J. D., Felsenstein, D., and Ritchie, D. D. (1973). The inhibitory effects of obtusaquinone on the growth and reproduction of two marine fungi. *Nav. Res. Lab. Rep.* **7653**, 1–10.

Bunt, J. S. (1955). The importance of bacteria and other microorganisms in the sea-water at Macquarie Island. *Aust. J. Mar. Freshwater Res.* **6**, 60–65.

Byrne, P. J., and Eaton, R. A. (1972). Fungal attack of wood submerged in waters of different salinity. *Int. Biodeterior. Bull.* **8**, 127–134.

Byrne, P. J., and Jones, E. B. G. (1974). Lignicolous marine fungi. *Veroeff. Inst. Meeresforsch. Bremerhaven, Suppl.* **5**, 301–320.

Byrne, P., and Jones, E. B. G. (1975a). Effect of salinity on spore germination of terrestrial and marine fungi. *Trans. Br. Mycol. Soc.* **64**, 497–503.

Byrne, P. J., and Jones, E. B. G. (1975b). Effect of salinity on the reproduction of terrestrial and marine fungi. *Trans. Br. Mycol. Soc.* **65**, 185–200.

Capriotti, A. (1962a). Yeasts of the Miami, Florida, area. II. From the Miami River. *Arch. Mikrobiol.* **41**, 147–153.

Capriotti, A. (1962b). Yeasts of the Miami, Florida, area. III. From sea water, marine animals and decaying materials. *Arch. Mikrobiol.* **42**, 407–414.

Carlucci, A. F., and Williams, P. M. (1965). Concentration of bacteria from sea water by bubble scavenging. *J. Cons., Cons. Int. Explor. Mer* **30**, 28–33.

Carriker, M. R., Smith, E. H., and Wilce, R. T. (1969). Penetration of calcium carbonate substrates by lower plants and invertebrates. An international multidisciplinary symposium. *Am. Zool.* **9**, 629–1020.

Carroll, F. E., and Carroll, G. C. (1973). Senescence and death of the conidiogenous cell in *Stemphylium botryosum* Wallroth. *Arch. Mikrobiol.* **94**, 109–124.

Catalfomo, P., Block, J. H., Constantine, G. H., and Kirk, P. W., Jr. (1972–1973). Choline sulfate (ester) in marine higher fungi. *Mar. Chem.* **1**, 157–162.

Cavaliere, A. R. (1966a). Marine Ascomycetes: Ascocarp morphology and its application to taxonomy. I. *Amylocarpus* Currey, *Ceriosporella* gen. nov., *Lindra* Wilson. *Nova Hedwigia* **10**, 387–398.

Cavaliere, A. R. (1966b). Marine Ascomycetes: Ascocarp morphology and its application to taxonomy. II. Didymosporae. *Nova Hedwigia* **10**, 399–424.

Cavaliere, A. R. (1966c). Marine Ascomycetes: Ascocarp morphology and its application to taxonomy. IV. Stromatic species. *Nova Hedwigia* **10**, 438–452.

Cavaliere, A. R. (1968). Marine fungi of Iceland: A preliminary account of Ascomycetes. *Mycologia* **60**, 475–479.

Cavaliere, A. R. and Alberte, R. S. (1970). Fungi in animal shell fragments. *J. Elisha Mitchell Sci. Soc.* **86**, 203–206.

Cavaliere, A. R., and Johnson, T. W., Jr. (1966a). Marine Ascomycetes: Ascocarp morphology and its application to taxonomy. III. A revision of the genus *Lulworthia* Sutherland. *Nova Hedwigia* **10**, 425–437.

Cavaliere, A. R., and Johnson, T. W., Jr. (1966b). Marine Ascomycetes: Ascocarp morphology and its application to taxonomy. V. Evaluation. *Nova Hedwigia* **10**, 453–461.

Cavaliere, A. R., and Markhart, A. H., III (1972). Marine fungi of Iceland: Calcareophilous forms. *Surtsey Prog. Rep.* **6**, 20–22.

Chadefaud, M. (1969). Une interprétation de la paroi des ascospores septées, notamment celles des *Aglaospora* et des *Pleospora*. *Bull. Soc. Mycol. Fr.* **85**, 145–157.

Chadefaud, M. (1975). L'origine "para-floridéenne" des Eumycètes et l'archétype ancestral de ces champignons. *Ann. Sci. Nat., Bot. Biol. Vég.* [12] **16**, 217–247.

Chapman, V. J. (1960). "Salt Marshes and Salt Deserts of the World." Wiley (Interscience), New York.

Chapman, V. J. (1974). Salt marshes and salt deserts of the world. *In* "Ecology of Halophytes" (R. J. Reimold and W. H. Queen, eds.), pp. 3–19. Academic Press, New York.

Chapman, V. J. (1976). "Mangrove Vegetation." Cramer, Vaduz, Liechtenstein.

Cheng, L., ed. (1976). "Marine Insects." North-Holland Publ., Amsterdam.

Chesters, C. G. C., and Bull, A. T. (1963a). The enzymic degradation of laminarin. 1. The distribution of laminarinase among microorganisms. *Biochem. J.* **86,** 28–31.

Chesters, C. G. C., and Bull, A. T. (1963b). The enzymic degradation of laminarin. 2. The multicomponent nature of fungal laminarinases. *Biochem. J.* **86,** 31–38.

Chesters, C. G. C., and Bull, A. T. (1963c). The enzymic degradation of laminarin. 3. Some effects of temperature, pH and various chemical reagents of fungal laminarinases. *Biochem. J.* **86,** 38–46.

Chidambaram, P., Mathur, S. B., and Neergaard, P. (1973). Identification of seed-borne *Drechslera* species. *Friesia* **10,** 165–207.

Chrzanowski, T. H., and Cowley, G. T. (1977). Response of *Uca pugilator* to diets of two selected yeasts. *Mycologia* **69,** 1062–1068.

Chupp, C. (1953). "A Monograph of the Fungus Genus *Cercospora*." Chupp, Ithaca, New York.

Church, A. H. (1919). Thalassiophyta and the subaerial transmigration. *Bot. Mem.* **3,** 1–95.

Churchland, L. M., and McClaren, M. (1972). The effect of Kraft pulp mill effluents on the growth of *Zalerion maritimum*. *Can. J. Bot.* **50,** 1269–1273.

Churchland, L. M., and McClaren, M. (1973). Marine fungi isolated from a Kraft pulp mill outfall area. *Can. J. Bot.* **51,** 1703–1710.

Churchland, L. M., and McClaren, M. (1976). Growth of filamentous marine fungi in a continuous culture system. *Can. J. Bot.* **54,** 893–899.

Clapp, W. F., and Kenk, R. (1963). "Marine borers. An Annotated Bibliography." Off. Nav. Res., Department of the Navy, Washington, D.C.

Clements, F. E., and Shear, C. L. (1931). "The Genera of Fungi," Wilson Co., New York.

Cole, G. T. (1976). Conidium ontogeny in marine hyphomycetous fungi: *Asteromyces cruciatus* and *Zalerion maritimum*. *Mar. Biol.* **38,** 147–158.

Colon, M. (1940). "Morphologie, Physiologie et Écologie de Deux Champignons des Dunes, *Asteromyces cruciatus* Moreau et *Monopodium uredopsis* Delacroix." Mém. Dipl. Et. sup., Fac. Sc. Caen, Reims, Imp. du Nord-est.

Combs, T. J., Murchelano, R. A., and Jurgen, F. (1971). Yeasts isolated from Long Island Sound. *Mycologia* **63,** 178–181.

Conners, I. L. (1967). An annotated index of plant diseases in Canada. *Can., Dep. Agric., Res. Br., Publ.* **1251.**

Conover, J. T., and Sieburth, J. M. (1964). Effect of *Sargassum* distribution on its epibiota and antibacterial activity. *Bot. Mar.* **6,** 147–157.

Cooke, M. C. (1890). New British fungi. *Grevillea* **18,** 73–74.

Corbaz, R. (1957). Recherches sur le genre *Didymella* Sacc. *Phytopathol. Z.* **28,** 375–414.

Corbett, N. H. (1965). Micro-morphological studies on the degradation of lignified cell walls by Ascomycetes and Fungi Imperfecti. *J. Inst. Wood Sci.* **14,** 18–29.

Corlett, M. (1973). Surface structure of the conidium and conidiophore of *Stemphylium botryosum*. *Can. J. Microbiol.* **19,** 392–393.

Corlett, M. (1975). Observations and comments on the *Pleospora* centrum type. *Nova Hedwigia* **24,** 347–366.

Corte, A. M. (1975). Osservazioni sul genere *Lulworthia* Suth. e sui suoi rapporti con *Limnoria* Menzies e segnalazioni di altre specie. *G. Bot. Ital.* **109,** 227–237.

Cotton, A. D. (1909). Notes on marine Pyrenomycetes. *Trans. Br. Mycol. Soc.* **3,** 92–99.

Courtois, H. (1963). Mikromorphologische Befallssymptome beim Holzabbau durch Moder-fäulepilze. *Holzforsch. Holzverwert.* **15,** 88–101.

Cowley, G. T. (1973). Variations in soil fungus populations in a South Carolina salt marsh. *In* "Estuarine Microbial Ecology" (L. H. Stevenson and R. R. Colwell, eds.), pp. 441–454. Univ. of South Carolina Press, Columbia.

Cribb, A. B., and Cribb, J. W. (1955). Marine fungi from Queensland-I. *Univ. Queensl. Pap., Dep. Bot.* **3,** 77–81.

Cribb, A. B., and Cribb, J. W. (1956). Marine fungi from Queensland-II. *Univ. Queensl. Pap., Dep. Bot.* **3,** 97–105.

Cribb, A. B., and Cribb, J. W. (1960a). Marine fungi from Queensland - III. *Univ. Queensl. Pap., Dep. Bot.* **4,** 39–44.

Cribb, A. B., and Cribb, J. W. (1960b). Some marine fungi on algae in European herbaria. *Univ. Queensl. Pap., Dep. Bot.* **4,** 45–48.

Cribb, A. B., and Cribb, J. W. (1969). Some marine fungi from the Great Barrier Reef area. *Queensl. Nat.* **19,** 118–120.

Cribb, A. B., and Herbert, J. W. (1954). Three species of fungi parasitic on marine algae in Tasmania. *Univ. Queensl. Pap., Dep. Bot.* **3,** 9–13.

Crouan, P. L., and Crouan, H. M. (1867). "Florule du Finistère." Klincksieck, Paris and Brest.

Crow, S. A., Bowman, P. I., and Ahearn, D. G. (1977). Isolation of atypical *Candida albicans* from the North Sea. *Appl. Environ. Microbiol.* **33,** 738–739.

Cundell, A. M., and Mitchell, R. (1977). Microbial succession on a wooden surface exposed to the sea. *Int. Biodeterior. Bull.* **13,** 67–73.

Cunnell, G. J. (1956). Some pycnidial fungi on *Carex. Trans. Br. Mycol. Soc.* **39,** 21–47.

Cunnell, G. J. (1957). *Stagonospora* spp. on *Phragmites communis* Trin. *Trans. Br. Mycol. Soc.* **40,** 443–455.

Cunnell, G. J. (1958). On *Robillarda phragmitis* sp. nov. *Trans. Br. Mycol. Soc.* **41,** 405–412.

Curran, P. M. T. (1975). Lignicolous marine fungi from Ireland. *Nova Hedwigia* **26,** 591–596.

Cutter, J. M., and Rosenberg, F. A. (1972). The role of cellulolytic bacteria in the digestive processes of the shipworm. II. Requirement of bacterial cellulase in the digestive system of Teredine borers. *In* "Biodeterioration of Materials" (A. H. Walters and E. H. Hueck-van der Plas, eds.), pp. 42–51. Applied Science, London.

Dabrowa, N., Landau, J. W., Newcomer, V. D., and Plunkett, O. A. (1964). A survey of tide-washed coastal areas of Southern California for fungi potentially pathogenic to man. *Mycopathol. Mycol. Appl.* **24,** 137–150.

Damon, S. C. (1952). Type studies in *Dictyosporium, Speira,* and *Cattanea. Lloydia* **15,** 110–124.

Dapper, H. (1967). Über den Abbau von Kork unter verschiedenen Pflanzengesellschaften. *Oecol. Plant.* **2,** 125–138.

Dapper, H. (1969). Der Abbau von Kork in Erde unter Gewächshausbedingungen. *Ber. Dtsch. Bot. Ges.* **82,** 573–576.

David, H. M. (1943). Studies in the autecology of *Ascophyllum nodosum* Le Jol. *J. Ecol.* **31,** 178–198.

Davidson, D. E. (1973). Mucoid sheath of *Lulworthia medusa. Trans. Br. Mycol. Soc.* **60,** 577–579.

Davidson, D. E. (1974a). The effect of salinity on a marine and a freshwater Ascomycete. *Can. J. Bot.* **52,** 553–563.

Davidson, D. E. (1974b). Wood-inhabiting and marine fungi from a saline lake in Wyoming. *Trans. Br. Mycol. Soc.* **63**, 143–149.

Dawson, E. Y. (1949). Contributions toward a marine flora of the southern California channel islands, I-III. *Occ. Pap. Allan Hancock Found.* **8**, 1–57.

Dawson, E. Y. (1956). "How to Know the Seaweeds." W. C. Brown, Dubuque, Iowa.

Dean, R. C. (1976). Cellulose and wood digestion in the marine mollusk *Bankia gouldi* Bartsch. *In* "Proceedings of the Third International Biodegradation Symposium" (J. M. Sharpley and A. M. Kaplan, eds.), pp. 955–965. Applied Science, London.

DeBaun, R. M., and Nord, F. F. (1951). The resistance of cork to decay by wood-destroying molds. *Arch. Biochem. Biophys.* **33**, 314–319.

De Bertoldi, M., Lepidi, A. A., and Nuti, M. P. (1972). Classification of the genus *Humicola* Traaen. I. Preliminary reports and investigations. *Mycopathol. Mycol. Appl.* **46**, 289–304.

De Bertoldi, M., Lepidi, A. A., and Nuti, M. P. (1973). The significance of DNA base composition in classification of *Humicola* and related genera. *Trans. Br. Mycol. Soc.* **60**, 77–85.

Défago, G. (1935). De quelques Valsées von Höhnel parasites des arbres à noyau dépérissants. *Beitr. Kryptogamenflora Schweiz* **8**, 1–109.

Deighton, F. C. (1965). Proposal for the conservation of the generic name *Pleospora* Rabenhorst ex Cesati & De Notaris. *Regnum Veg.* **40**, 13–14.

Demoulin, V. (1974). The origin of Ascomycetes and Basidiomycetes. The case for a red algal ancestry. *Bot. Rev.* **40**, 315–345.

Den Hartog, C. (1970). "The Sea-Grasses of the World." North-Holland Publ., Amsterdam.

Denison, W. C., and Carroll, G. C. (1966). The primitive Ascomycete: A new look at an old problem. *Mycologia* **58**, 249–269.

Dennis, R. W. G. (1964). The fungi of the isle of Rhum. *Kew Bull.* **19**, 77–131.

Dennis, R. W. G. (1968). "British Ascomycetes." Cramer, Lehre.

De Queiroz, L. A. (1972). Análise quanti-qualitativa de leveduras isoladas de algas marinhas. I. *Publ. Inst. Micol., Recife* **677**, 1–13.

De Queiroz, L. A., and Macêdo, S. J. (1972). Análise quanti-qualitativa de leveduras isoladas de aguas marinhas. I. *Publ. Inst. Micol., Recife* **676**, 1–17.

Desmazières, J. B. H. J. (1849). "Plantes Cryptogames de France," 2nd ed., No. 1778. Lille.

Desmazières, J. B. H. J. (1850). "Plantes Cryptogames de France," 1st ed., Fasc. 42. Lille.

de Vries, G. A. (1952). "Contribution to the knowledge of the genus *Cladosporium* Link ex Fr." Hollandia Press, Baarn.

Dickinson, C. H. (1965). The mycoflora associated with *Halimione portulacoides*. III. Fungi on green and moribund leaves. *Trans. Br. Mycol. Soc.* **48**, 603–610.

Dickinson, C. H., and Morgan-Jones, G. (1966). The mycoflora associated with *Halimione portulacoides*. IV. Observations on some species of Sphaeropsidales. *Trans. Br. Mycol. Soc.* **49**, 43–55.

Dickinson, C. H., and Pugh, G. J. F. (1965a). Use of a selective cellulose agar for isolation of soil fungi. *Nature (London)* **207**, 440–441.

Dickinson, C. H., and Pugh, G. J. F. (1965b). The mycoflora associated with *Halimione portulacoides*. I. The establishment of the root surface flora of mature plants. *Trans. Br. Mycol. Soc.* **48**, 381–390.

Dickinson, C. H., and Pugh, G. J. F. (1965c). The mycoflora associated with *Halimione portulacoides*. II. Root surface fungi of mature and excised plants. *Trans. Br. Mycol. Soc.* **48**, 595–602.

Diedicke, H. (1912). Die Abteilung Hyalodidymae der Sphaeropsideen. *Ann. Mycol.* **10**, 135–152.

Diedicke, H. (1915). Pilze. *In Kryptogamenflora Mark Brandenburg* **9**, 1–962.

Dietrich, R., and Höhnk, W. (1958). Über das Öl des submers lebenden Pilzes *Cerato-stomella* spec. (Sphaeriales, Ascomycetes). *Veroeff. Inst. Meeresforsch. Bremerhaven* **5**, 135–142.

Dixon, M. (1968). *Trichocladium pyriformis* sp. nov. *Trans. Br. Mycol. Soc.* **51**, 160–164.

Dodge, C. W. (1973). "Lichen Flora of the Antarctic Continent and Adjacent Islands." Phoenix Publ., Canaan, New Hampshire.

Doguet, G. (1962a). *Digitatispora marina*, n.g., n.sp., Basidiomycète marin. *C. R. Hebd. Séances Acad. Sci.* **254**, 4336–4338.

Doguet, G. (1962b). Recherches sur les noyaux des basides du *Digitatispora marina*. *Bull. Soc. Mycol. Fr.* **78**, 283–290.

Doguet, G. (1963). Basidiospores anormales chez le *Digitatispora marina*. *Bull. Soc. Mycol. Fr.* **79**, 249–252.

Doguet, G. (1964). Influence de la température et de la salinité sur la croissance et la fertilité du *Digitatispora marina* Doguet. *Bull. Soc. Fr. Physiol. Vég.* **10**, 285–292.

Doguet, G. (1967). *Nia vibrissa* Moore et Meyers, remarquable Basidiomycète marin. *C. R. Hebd. Séances Acad. Sci., Sér. D* **265**, 1780–1783.

Doguet, G. (1968). *Nia vibrissa* Moore et Meyers, Gastéromycète marin. I. Conditions générales de formation des carpophores en culture. *Bull. Soc. Mycol. Fr.* **84**, 343–351.

Doguet, G. (1969). *Nia vibrissa* Moore et Meyers, Gastéromycète marin. II. Développement des carpophores et des basides. *Bull. Soc. Mycol. Fr.* **85**, 93–104.

Donk, M. A. (1962). Confusion. *Taxon* **11**, 120–122.

Donk, M. A. (1964). A conspectus of the families of Aphyllophorales. *Persoonia* **3**, 199–324.

Donk, M. A. (1968). Report of the committee for fungi and lichens 1964–1969. *Taxon* **17**, 578–581.

Doty, M. S. (1947). The marine algae of Oregon. Part I. Chlorophyta and Phaeophyta. *Farlowia* **3**, 1–65.

Drechsler, C. (1923). Some graminicolous species of *Helminthosporium*. I. *J. Agric. Res.* **24**, 641–740.

Dring, D. M. (1973). Gasteromycetes. *In* "The Fungi" (G. C. Ainsworth, F. K. Sparrow, and A. S. Sussman, eds.), Vol. 4B, pp. 451–478. Academic Press, New York.

Durán, R. (1973). Ustilaginales. *In* "The Fungi" (G. C. Ainsworth, F. K. Sparrow, and A. S. Sussman, eds.), Vol. 4B, pp. 281–300. Academic Press, New York.

Durieu de Maisonneuve, C., and Montagne, J. F. C. (1869). Pyrenomycetes Fr. *In* "Exploration Scientifique de l'Algérie, Botanique" (J. Bory de Saint-Vincent and C. Durieu de Maisonneuve, eds.), pp. 443–608, Paris.

Eaton, R. A. (1972). Fungi growing on wood in water cooling towers. *Int. Biodeterior. Bull.* **8**, 39–48.

Eaton, R. A., and Dickinson, D. J. (1976). The performance of copper chrome arsenic treated wood in the marine environment. *Mater. Org., Beih.* **3**, 521–529.

Eaton, R. A., and Irvine, J. (1972). Decay of untreated wood by cooling tower fungi. *In* "Biodeterioration of Materials" (A. H. Walters and E. H. Hueck-van der Plas, eds.), pp. 192–200. Applied Science, London.

Eaton, R. A., and Jones, E. B. G. (1971a). The biodeterioration of timber in water cooling towers. I. Fungal ecology and the decay of wood at Connah's Quay and Ince. *Mater. Org.* **6**, 51–80.

Eaton, R. A., and Jones, E. B. G. (1971b). The biodeterioration of timber in water cooling towers. II. Fungi growing on wood in different positions in a water cooling system. *Mater. Org.* **6**, 81–92.

Elliott, J. S. B. (1930). The soil fungi of the Dovey salt marshes. *Ann. Appl. Biol.* **17**, 284–305.

Ellis, J. B., and Everhart, B. M. (1885). New fungi. *J. Mycol.* **1**, 42–44.

Ellis, J. B., and Everhart, B. M. (1892). "North American Pyrenomycetes." Ellis & Everhart, Newfield, New Jersey.

Ellis, J. P. (1976). British *Microthyrium* species and similar fungi. *Trans. Br. Mycol. Soc.* **67**, 381–394.

Ellis, M. B. (1971). "Dematiaceous Hyphomycetes." Commonw. Mycol. Inst., Kew, Surrey, England.

Ellis, M. B. (1976). "More Dematiaceous Hyphomycetes." Commonw. Mycol. Inst., Kew, Surrey, England.

Eltringham, S. K. (1971). Marine borers and fungi. *In* "Marine Borers, Fungi and Fouling Organisms of Wood" (E. B. G. Jones and S. K. Eltringham, eds.), pp. 327–337. Org. Econ. Coop. Dev., Paris.

Eriksson, O. (1964). *Nectriella laminariae* n.sp. in stipes of a *Laminaria. Sven. Bot. Tidskr.* **58**, 233–236.

Eriksson, O. (1967a). On graminicolous Pyrenomycetes from Fennoscandia. 2. Phragmosporous and scolecosporous species. *Ark. Bot.* [2] **6**, 381–440.

Eriksson, O. (1967b). Studies on graminicolous Pyrenomycetes from Fennoscandia. *Acta Univ. Ups.* **88**, 1–16.

Eriksson, O. (1973). *Orbilia marina,* an over-looked Discomycete on members of Fucales. *Sven. Bot. Tidskr.* **67**, 208–210.

Escobar, G. A., McCabe, D. E., and Harpel, C. W. (1976). *Limnoperdon,* a floating Gasteromycete isolated from marshes. *Mycologia* **68**, 874–880.

Estee, L. M. (1913). Fungus galls on *Cystoseira* and *Halidrys. Univ. Calif., Berkeley, Publ. Bot.* **4**, 305–316.

Farr, M. L. (1973). "An Annotated List of Spegazzini's Fungus Taxa," Vol. 2, Cramer, Lehre.

Fassatiová, O. (1967). Notes on the genus *Humicola* Traaen. II. *Česká Mykol.* **21**, 78–89.

Fazzani, K., and Jones, E. B. G. (1977). Spore release and dispersal in marine and brackish water fungi. *Material Org.* **12**, 235–248.

Feldmann, G. (1957). Un nouvel Ascomycète parasite d'une algue marine: *Chadefaudia marina. Rev. Gén. Bot.* **64**, 140–152.

Feldmann, G. (1959). Une ustilaginale marine, parasite du *Ruppia maritima* L. *Rev. Gén. Bot.* **66**, 35–39.

Feldmann, J. (1931). Contribution à la flore algologique marine de l'Algérie. Les algues de Cherchell. *Bull. Soc. Hist. Nat. Afr. Nord* **22**, 179–254.

Feldmann, J. (1932). Sur la répartition dans la Méditerranée occidentale, du *Melanopsamma tregoubovii* Ollivier var. *Cystoseirae* Oll. Pyrénomycète parasite du *Cystoseira abrotanifolia* C. Ag. *Rev. Algol.* **6**, 225–226.

Feldmann, J. (1937). Sur les gonidies de quelques *Arthopyrenia* marins. *Rev. Bryol. Lichénol.* **10**, 64–73.

Feldmann, J. (1938). Le *Blodgettia confervoides* Harv. est-il un lichen? *Rev. Bryol. Lichénol.* **11**, 155–163.

Feldmann, J. (1940). Une nouvelle espèce de Sphéropsidée parasite d'une algue marine. *Bull. Soc. Hist. Nat. Afr. Nord* **31**, 167–170.

Feldmann, J. (1954). Inventaire de la flore marine de Roscoff. Algues, Champignons, Lichens et Spermatophytes. *Trav. Stn. Biol. Roscoff, Suppl.* **6**, 1–152.

Fell, J. W. (1967). Distribution of yeasts in the Indian Ocean. *Bull. Mar. Sci.* **17**, 454–470.

Fell, J. W. (1970). Yeasts with heterobasidiomycetous life cycles. *In* "Recent Trends in Yeast Research" (D. G. Ahearn, ed.), pp. 49–66. Georgia State Univ., Atlanta.

Fell, J. W. (1974a). Distributions of yeasts in the water masses of the southern oceans. *In* "Effect of the Ocean Environment of Microbial Activities" (R. R. Colwell and R. Y. Morita, eds.), pp. 510–523. Univ. Park Press, Baltimore, Maryland.

Fell, J. W. (1974b). Heterobasidiomycetous yeasts *Leucosporidium* and *Rhodosporidium*. Their systematics and sexual incompatibility systems. *Trans. Mycol. Soc. Jpn.* **15**, 316–323.

Fell, J. W. (1976). Yeasts in oceanic regions. *In* "Recent Advances in Aquatic Mycology" (E. B. G. Jones, ed.), pp. 93–124. Wiley, New York.

Fell, J. W., and Hunter, I. L. (1968). Isolation of heterothallic yeast strains of *Metschnikowia* Kamienski and their mating reactions with *Chlamydozyma* Wickerham spp. *Antonie van Leeuwenhoek* **34**, 365–376.

Fell, J. W., and Master, I. M. (1973). Fungi associated with the degradation of mangrove (*Rhizophora mangle* L.) leaves in south Florida. *In* "Estuarine Microbial Ecology" (L. H. Stevenson and R. R. Colwell, eds.), pp. 455–465. Univ. of South Carolina Press, Columbia.

Fell, J. W., and Master, I. M. (1975). Phycomycetes (*Phytophthora* spp. nov. and *Pythium* sp. nov.) associated with degrading mangrove (*Rhizophora mangle*) leaves. *Can. J. Bot.* **53**, 2908–2922.

Fell, J. W., and Meyer, S. A. (1967). Systematics of yeast species in the *Candida parapsilosis* group. *Mycopathol. Mycol. Appl.* **32**, 177–193.

Fell, J. W., and Phaff, H. J. (1970). *Leucosporidium* Fell, Statzell, Hunter et Phaff. *In* "The Yeasts" (J. Lodder, ed.), pp. 776–802. North-Holland Publ., Amsterdam.

Fell, J. W., and Pitt, J. I. (1969). Taxonomy of the yeast genus *Metschnikowia:* a correction and a new variety. *J. Bacteriol.* **98**, 853–854.

Fell, J. W., and van Uden, N. (1963). Yeasts in marine environments. *In* "Symposium on Marine Microbiology" (C. H. Oppenheimer, ed.), pp. 329–340. Thomas, Springfield, Illinois.

Fell, J. W., Ahearn, D. G., Meyers, S. P., and Roth, F. J., Jr. (1960). Isolation of yeasts from Biscayne Bay, Florida and adjacent benthic areas. *Limnol. Oceanogr.* **5**, 366–371.

Fell, J. W., Phaff, H. J., and Newell, S. Y. (1970). *Rhodosporidium* Banno. *In* "The Yeasts" (J. Lodder, ed.), pp. 803–814. North-Holland Publ., Amsterdam.

Fell, J. W., Hunter, I. L., and Tallman, A. S. (1973). Marine basidiomycetous yeasts (*Rhodosporidium* spp. n.) with tetrapolar and multiple allelic bipolar mating systems. *Can. J. Microbiol.* **19**, 643–657.

Fell, J. W., Cefalu, R. C., Master, I. M., and Tallman, A. S. (1975). Microbial activities in the mangrove (*Rhizophora mangle*) leaf detrital system. *In* "The Biology and Management of Mangroves" (G. Walsh, S. Snedaker, and H. Teas, eds.), Vol. 2, pp. 661–679. Univ. of Florida, Gainesville.

Fenchel, T. M., and Riedl, R. J. (1970). The sulfide system: a new biotic community underneath the oxidized layer of marine sand bottoms. *Mar. Biol.* **7**, 255–268.

Ferdinandsen, C., and Winge, O. (1920). A *Phyllachorella* parasitic on *Sargassum*. *Mycologia* **12**, 102–103.

Fischer, B., and Brebeck, C. (1894). "Zur Morphologie, Biologie and Systematik der Kahmpilze, der *Monilia candida* Hansen und des Soorerregers." Fischer, Jena.

Fize, A. (1960). Présence de deux espèces de champignons pyrénomycètes dans les sables littoraux du Golfe d'Aigues-Mortes. *Vie Milieu* **11**, 675–677.

Fize, A., Manier, J. F., and Maurand, J. (1970). Sur un cas d'infestation du Copépode *Eurytemora velox* (Lillj) par une levure du genre *Metschnikowia* (Kamienski). *Ann. Parasitol. Hum. Comp.* **45**, 357–363.

Fletcher, A. (1975a). Key for the identification of British marine and maritime lichens. I. Siliceous rocky shore species. *Lichenologist* **7**, 1–52.

Fletcher, A. (1975b). Key for the identification of British marine and maritime lichens. II. Calcareous and terricolous species. *Lichenologist* **7**, 73–115.

Fletcher, A. (1976). Nutritional aspects of marine and maritime lichen ecology. *In* "Lichenology: Progress and Problems" (D. H. Brown, D. L. Hawksworth, and R. H. Bailey, eds.), pp. 359–384. Academic Press, New York.

Furtado, J. S., and Olive, L. S. (1970). Ascospore discharge and ultrastructure of the ascus in *Leptosphaerulina australis*. *Nova Hedwigia* **19**, 799–823.

Furtado, S. E. J., and Jones, E. B. G. (1976). The performance of *Dalbergia* wood and *Dalbergia* extractives impregnated into pine and exposed in a water cooling tower. *In* "Proceedings of a Workshop on the Biodeterioration of Tropical Woods: Chemical Basis for Natural Resistance" (J. D. Bultman, ed.), pp. 41–56. Nav. Res. Lab., Department of the Navy, Washington, D.C.

Furtado, S. E. J., Jones, E. B. G., and Bultman, J. D. (1977). The effect of certain wood extractives on the growth of marine micro-organisms. *C. R. Congr. Int. Corros. Mar. Salissures, 4th, 1976,* pp. 195–201.

Furuya, K., and Udagawa, S. (1973). Coprophilous Pyrenomycetes from Japan III. *Trans. Mycol. Soc. Jpn.* **14**, 7–30.

Furuya, K., and Udagawa, S. (1975). Two new species of cleistothecial Ascomycetes. *J. Jpn. Bot.* **50**, 249–254.

Gaertner, A. (1968). Niedere, mit Pollen köderbare marine Pilze diesseits und jenseits des Island-Färöer-Rückens im Oberflächenwasser und im Sediment. *Veroeff. Inst. Meeresforsch. Bremerhaven* **11**, 65–82.

Gain, L. (1912). La flore algologique des régions antarctiques et subantarctiques. *In* "Deuxième Expédition Antarctique Française (1908–1910), Commendée par le Dr. Jean Charcot". Masson, Paris.

Gambogi, P. (1969). Micromiceti nuovi o rari per l'Italia, isolati da terreno di bosco del litorale tirrenico. *Humicola parvispora* n. sp. e *Gonytrichum macrocladum* (Sacc.) Hughes var. *terricola* n. var. *G. Bot. Ital.* **103**, 33–46.

Gäumann, E. (1964). "Die Pilze," 2nd ed. Birkhaeuser, Basel.

Gessner, R. V. (1976). *In vitro* growth and nutrition of *Buergenerula spartinae,* a fungus associated with *Spartina alterniflora*. *Mycologia* **68**, 583–599.

Gessner, R. V. (1977). Seasonal occurrence and distribution of fungi associated with *Spartina alterniflora* from a Rhode Island estuary. *Mycologia* **69**, 477–491.

Gessner, R. V., and Goos, R. D. (1973a). Fungi from decomposing *Spartina alterniflora*. *Can. J. Bot.* **51**, 51–55.

Gessner, R. V., and Goos, R. D. (1973b). Fungi from *Spartina alterniflora* in Rhode Island. *Mycologia* **65**, 1296–1301.

Gessner, R. V., and Kohlmeyer, J. (1976). Geographical distribution and taxonomy of fungi from salt marsh *Spartina*. *Can. J. Bot.* **54**, 2023–2037.

Gessner, R. V., and Lamore, B. J. (1978). Fungi from Nantucket salt marshes and beaches. *Rhodora* **80**, 581–586.

Gessner, R. V., Goos, R. D., and Sieburth, J. M. (1972). The fungal microcosm of the internodes of *Spartina alterniflora*. *Mar. Biol.* **16**, 269–273.

Gezelius, K., and Norkrans, B. (1970). Ultrastructure of *Debaryomyces hansenii*. *Arch. Mikrobiol.* **70**, 14–25.

Ginns, J., and Malloch, D. (1977). *Halocyphina,* a marine Basidiomycete (Aphyllophorales). *Mycologia* **69**, 53–58.

Gladoch, M. (1969). Nowa odmiana *Pullularia fermentans* Wynne et Gott. *Zesz. Nauk. Wyzsz. Szk. Roln. Szczecinie* **30**, 99–104.

Gold, H. S. (1959). Distribution of some lignicolous Ascomycetes and Fungi Imperfecti in an estuary. *J. Elisha Mitchell Sci. Soc.* **75**, 25–28.

Golubic, S., Brent, G., and Le Campion, T. (1970). Scanning electron microscopy of endolithic algae and fungi using a multipurpose casting-embedding technique. *Lethaia* **3**, 203–209.

Golubic, S., Perkins, R. D., and Lukas, K. J. (1975). Boring microorganisms and microborings in carbonate substrates. *In* "The Study of Trace Fossils" (R. W. Frey, ed.), pp. 229–259. Springer-Verlag, Berlin and New York.

Gomont, M. (1892). Monographie des Oscillariées (Nostocacées homocystées). *Ann. Sci. Nat., Bot. Biol. Vég.* [7] **15**, 263–368.

Goodman, P. J. (1959). The possible role of pathogenic fungi in die-back of *Spartina townsendii* Agg. *Trans. Br. Mycol. Soc.* **42**, 409–415.

Goos, R. D., and Gessner, R. V. (1975). Hyphal modifications of *Sphaerulina pedicellata:* Appressoria or hyphopodia? *Mycologia* **67**, 1035–1038.

Goto, S., Sugiyama, J., and Iizuka, H. (1969). A taxonomic study of Antarctic yeasts. *Mycologia* **61**, 748–774.

Goto, S., Yamasato, K., and Iizuka, H. (1974a). Identification of yeasts isolated from the Pacific Ocean. *J. Gen. Appl. Microbiol.* **20**, 309–316.

Goto, S., Ohwada, K., and Yamasato, K. (1974b). Identification of yeasts isolated from sea water and sediment in Aburatsubo Inlet. *J. Gen. Appl. Microbiol.* **20**, 317–322.

Grove, W. B. (1935). "British Stem and Leaf Fungi (Coelomycetes)," Vol. 1. Cambridge Univ. Press, London and New York.

Grove, W. B. (1937). "British Stem and Leaf Fungi (Coelomycetes)," Vol. 2, Cambridge Univ. Press, London and New York.

Gunkel, W. (1963). Daten zur Bakterienverteilung in der Nordsee. *Veroeff. Inst. Meeresforsch. Bremerhaven, Suppl.* pp. 80–89.

Gunn, C. R., and Dennis, J. V. (1971). Ocean journeys by mangrove seedlings. *Shore & Beach* **39**, 19–22.

Gunn, C. R., and Dennis, J. V. (1976). "World Guide to Tropical Drift Seeds and Fruits." A Demeter Press Book. Quadrangle/The New York Times Book Co., New York.

Gustafsson, U., and Fries, N. (1956). Nutritional requirements of some marine fungi. *Physiol. Plant.* **9**, 462–465.

Hafellner, J., and Poelt, J. (1976). Die Gattung *Karschia*—Bindeglied zwischen bitunicaten Ascomyceten und lecanoralen Flechtenpilzen? *Plant Syst. Evol.* **126**, 243–254.

Hansen, E. C. (1904). Grundlinien zur Systematik der Saccharomyceten. *Centralbl. Bakteriol., Parasitenkd., Infektionskr. Hyg., Abt. 2* **12**, 529–538.

Hansford, C. G. (1954). Australian fungi. II. New records and revisions. *Proc. Linn. Soc. N.S.W.* **79**, 97–141.

Harding, G. C. H. (1973). Decomposition of marine copepods. *Limnol. Oceanogr.* **18**, 670–673.

Hariot, P. (1889). Algues, Champignons. *In* "Mission Scientifique du Cap Horn, 1882–1883," Vol. 5, pp. 3–109 and pp. 173–200.

Harvey, W. H. (1858). Nereis Boreali-Americana: or contributions to the history of the marine algae of North America. Part III-Chlorospermeae. *Smithson. Contrib. Knowl.* **10**, 1–119.

Hawksworth, D. L. (1974). "Mycologist's Handbook." Commonw. Mycol. Inst., Kew, Surrey, England.

Hawksworth, D. L. (1976). Looking at lichens. *Nat. Hist. Book Rev.* **1**, 8–15.

Hawksworth, D. L., and Booth, C. (1974). A revision of the genus *Zopfia* Rabenh. *Mycol. Pap.* **135**, 1–38.

Heald, E. J., and Odum, W. E. (1969). The contribution of mangrove swamps to Florida fisheries. *Proc. Gulf Caribb. Fish. Inst., 22nd Annu. Sess.* pp. 130–135.

Hedjaroude, G. A. (1968). Etudes taxonomiques sur les *Phaeosphaeria* Miyake et leurs formes voisines (Ascomycètes). *Sydowia* **22**, 57–107.

Hennebert, G. L. (1962). *Wardomyces* and *Asteromyces*. *Can. J. Bot.* **40**, 1203–1216.

Hennebert, G. L. (1971). Pleomorphism in Fungi Imperfecti. *In* "Taxonomy of Fungi Imperfecti" (B. Kendrick, ed.), pp. 202–223. Univ. of Toronto Press, Toronto.

Henningsson, M. (1974). Aquatic lignicolous fungi in the Baltic and along the west coast of Sweden. *Sven. Bot. Tidskr.* **68**, 401–425.

Henningsson, M. (1976a). Degradation of wood by some fungi from the Baltic and the west coast of Sweden. *Mater. Org., Beih.* **3**, 509–519.

Henningsson, M. (1976b). Studies on aquatic lignicolous fungi from Swedish coastal waters. *Acta Univ. Ups.* **386**, 1–15.

Henssen, A., and Jahns, H. M. (1974). "Lichenes. Eine Einführung in die Flechtenkunde." Thieme, Stuttgart.

Hermanides-Nijhof, E. J. (1977). *Aureobasidium* and allied genera. *Stud. Mycol., Baarn* **15**, 141–177.

Höhnk, W. (1952). Studien zur Brack-und Seewassermykologie. I. *Veroeff. Inst. Meeresforsch. Bremerhaven* **1**, 115–125.

Höhnk, W. (1954a). Studien zur Brack-und Seewassermykologie. IV. Ascomyceten des Küstensandes. *Veroeff. Inst. Meeresforsch. Bremerhaven* **3**, 27–33.

Höhnk, W. (1954b). Von den Mikropilzen in Watt und Meer. *Abh. Naturwiss. Ver. Bremen* **33**, 407–429.

Höhnk, W. (1955). Studien zur Brack-und Seewassermykologie. V. Höhere Pilze des submersen Holzes. *Veroeff. Inst. Meeresforsch. Bremerhaven* **3**, 199–227.

Höhnk, W. (1956). Studien zur Brack-und Seewassermykologie. VI. Über die pilzliche Besiedlung verschieden salziger submerser Standorte. *Veroeff. Inst. Meeresforsch. Bremerhaven* **4**, 195–213.

Höhnk, W. (1958). Mykologische Notizen: I. Mikropilze im Eis. *Veroeff. Inst. Meeresforsch. Bremerhaven* **5**, 193–194.

Höhnk, W. (1959). Ein Beitrag zur ozeanischen Mykologie. *Dtsch. Hydrogr. Z., Reihe B* No. 3, pp. 81–87.

Höhnk, W. (1961). A further contribution to the oceanic mycology. *Rapp. P.-V. Réun., Cons. Int. Explor. Mer* **149**, 202–208.

Höhnk, W. (1962a). Hinweise auf die Rolle der marinen Pilze im Stoffhaushalt des Meeres. *Kiel. Meeresforsch.* **18**, 145–150.

Höhnk, W. (1962b). Daten zur Verbreitung und Ökologie mariner Pilze. *Zentralbl. Bakteriol., Parasitenkd., Infektionskr. Hyg.* **184**, 278–287.

Höhnk, W. (1967). Über die submersen Pilze an der rumänischen Schwarzmeerküste nahe Constanza. *Veroeff. Inst. Meeresforsch. Bremerhaven* **10**, 149–158.

Höhnk, W. (1968). Zur Entfaltung der marinen Mykologie. *Ber. Dtsch. Bot. Ges.* **81**, 380–390.

Höhnk, W. (1969). Über den pilzlichen Befall kalkiger Hartteile von Meerestieren. *Ber. Dtsch. Wiss. Komm. Meeresforsch.* **20**, 129–140.

Höhnk, W. (1972). Fungi. *In* "Research Methods in Marine Biology" (C. Schlieper, ed.), pp. 142–155. Univ. of Washington Press, Seattle.

Holligan, P. M., and Jennings, D. H. (1972a). Carbohydrate metabolism in the fungus *Dendryphiella salina*. I. Changes in the levels of soluble carbohydrates during growth. *New Phytol.* **71**, 569–582.

Holligan, P. M., and Jennings, D. H. (1972b). Carbohydrate metabolism in the fungus *Dendryphiella salina*. II. The influence of different carbon and nitrogen sources on the accumulation of mannitol and arabitol. *New Phytol.* **71**, 583–594.

Holligan, P. M., and Jennings, D. H. (1972c). Carbohydrate metabolism in the fungus *Dendryphiella salina*. III. The effect of the nitrogen source on the metabolism of [1-^{14}C]- and [6-^{14}C]-Glucose. *New Phytol.* **71**, 1119–1133.

Holm, L. (1952). Taxonomical notes on Ascomycetes. II. The herbicolous Swedish species of the genus *Leptosphaeria* Ces. et De Not. *Sven. Bot. Tidskr.* **46**, 18–46.

Holm, L. (1953). Taxonomical notes on Ascomycetes. III. The herbicolous Swedish species of the genus *Didymella* Sacc. *Sven. Bot. Tidskr.* **47**, 520–525.

Holm, L. (1957). Etudes taxonomiques sur les Pléosporacées. *Symb. Bot. Ups.* **14**, 1–188.

Holm, L. (1975). Nomenclatural notes on Pyrenomycetes. *Taxon* **24**, 475–488.

Holmgren, P. K., and Keuken, W. (1974). "Index Herbariorum," 6th ed., Part I, No. 92. Int. Bur. Plant Taxon. Nomenclature, Utrecht.

Hooker, J. D. (1847). The Botany. The Antarctic Voyage of H. M. Discovery Ships *Erebus* and *Terror* in the years 1839–1843. I. Flora Antarctica. Part I. Botany of Lord Auckland's Group and Campbell's Island. Part II. Botany of Fuegia, The Falklands-Kerguelen's Land. London.

Hooker, J. D., and Harvey, W. H. (1845). Algae Antarcticae, being characters and descriptions of the hitherto unpublished species of Algae, discovered in Lord Auckland's Group, Campbell's Islands, Kerguelen's Land, Falkland Islands, Cape Horn and other southern circumpolar regions, during the voyage of H. M. discovery ships *Erebus* and *Terror*. *J. Bot., London* **4**, 249–276 and 293–298.

Hoppe, H. G. (1972a). Untersuchungen zur Ökologie der Hefen im Bereich der westlichen Ostsee. *Kiel. Meeresforsch.* **28**, 54–77.

Hoppe, H. G. (1972b). Taxonomische Untersuchungen an Hefen aus der westlichen Ostsee. *Kiel. Meeresforsch.* **28**, 219–226.

Hopper, B. E., and Meyers, S. P. (1966). Aspects of the life cycle of marine nematodes. *Helgol. Wiss. Meeresunters.* **13**, 444–449.

Howe, M. A. (1905). Phycological studies—I. New Chlorophyceae from Florida and the Bahamas. *Bull. Torrey Bot. Club* **32**, 241–252.

Huang, L. H. (1973). *Zopfiella flammifera*, a new species from Nigerian soil. *Mycologia* **65**, 690–694.

Hue, Abbé (1909). Le *Mastoidea tessellata* Hook. fil. et Harv. *Bull. Soc. Bot. Fr.* **56**, 315–322.

Hughes, G. C. (1968). Intertidal lignicolous fungi from Newfoundland. *Can. J. Bot.* **46**, 1409–1417.

Hughes, G. C. (1969). Marine fungi from British Columbia: occurrence and distribution of lignicolous species. *Syesis* **2**, 121–140.

Hughes, G. C. (1974). Geographical distribution of the higher marine fungi. *Veroeff. Inst. Meeresforsch. Bremerh., Suppl.* **5**, 419–441.

Hughes, G. C. (1975). Studies of fungi in oceans and estuaries since 1961. I. Lignicolous, caulicolous and foliicolous species. *Oceanogr. Mar. Biol. Annu. Rev.* **13**, 69–180.

Hughes, G. C., and Chamut, P. S. (1971). Lignicolous marine fungi from southern Chile, including a review of distributions in the southern hemisphere. *Can. J. Bot.* **49**, 1–11.

Hughes, S. J. (1951). Studies on micro-fungi. VII. *Allescheriella crocea, Oidium simile*, and *Pellicularia pruinata. Mycol. Pap.* **41**, 1–17.

Hughes, S. J. (1952). *Trichocladium* Harz. *Trans. Br. Mycol. Soc.* **35**, 152–157.

Hughes, S. J. (1958). Revisiones Hyphomycetum aliquot cum appendice de nominibus rejiciendis. *Can. J. Bot.* **36**, 727–836.

Hughes, S. J. (1969). New Zealand Fungi. 13. *Trichocladium* Harz. *N.Z. J. Bot.* **7**, 153–157.

Ingold, C. T. (1942). Aquatic Hyphomycetes of decaying alder leaves. *Trans. Br. Mycol. Soc.* **25**, 339–417.

Ingold, C. T. (1971). "Fungal Spores. Their Liberation and Dispersal." Oxford Univ. Press (Clarendon), London and New York.

Ingold, C. T. (1972). *Sphaerobolus:* The story of a fungus. *Trans. Br. Mycol. Soc.* **58,** 179–195.

Ingold, C. T. (1975a). Convergent evolution in aquatic fungi: the tetraradiate spore. *Biol. J. Linn. Soc.* **7,** 1–25.

Ingold, C. T. (1975b). An illustrated guide to aquatic and water-borne Hyphomycetes (Fungi Imperfecti) with notes on their biology. *Freshwater Biol. Assoc., Sci. Publ.* **30,** 1–96.

Ingold, C. T. (1976). The morphology and biology of freshwater fungi excluding Phycomycetes. *In* "Recent Advances in Aquatic Mycology" (E. B. G. Jones, ed.), pp. 335–357. Wiley, New York.

Iqbal, S. H., and Webster, J. (1973). The trapping of aquatic Hyphomycete spores by air bubbles. *Trans. Br. Mycol. Soc.* **60,** 37–48.

Irvine, J., and Jones, E. B. G. (1975). The effect of a copper-chrome-arsenate preservative and its constituents on the growth of aquatic micro-organisms. *J. Inst. Wood Sci.* **7,** 1–5.

Irvine, J., Eaton, R. A., and Jones, E. B. G. (1972). The effect of water of different ionic composition on the leaching of a water borne preservative from timber placed in cooling towers and in the sea. *Mater. Org.* **7,** 45–71.

Ito, N., and Takada, H. (1974). Adenine-dependent growth of marine yeast, *Rhodotorula glutinis* var. *salinaria* under sodium chloride: hypertonic condition. *Bot. Mag.* **87,** 293–300.

Ito, N., and Takada, H. (1976). Latent period for obligate halophilic growth of a marine yeast, *Rhodotorula. Trans. Mycol. Soc. Jpn.* **17,** 144–148.

Jaap, O. (1905). Verzeichnis zu meinem Exsiccatenwerk "Fungi selecti exsiccati," Serien I-IV (Nummern 1-100), nebst Bemerkungen. *Verh. Bot. Ver. Prov. Brandenburg* **47,** 77–99.

Jaap, O. (1907a). Zweites Verzeichnis zu meinem Exsiccatenwerk "Fungi selecti exsiccati," Serien V - VIII (Nummern 101-200) nebst Beschreibungen neuer Arten und Bemerkungen. *Verh. Bot. Ver. Prov. Brandenburg* **49,** 7–29.

Jaap, O. (1907b). Weitere Beiträge zur Pilzflora der nordfriesischen Inseln. *Schr. Naturwiss. Ver. Schleswig-Holstein* **14,** 15–33.

Janex-Favre, M.-C. (1965). Sur le Pyrénomycète lichénicole *Pharcidia gyrophorarum* Zopf et la position systématique du genre *Pharcidia* Körber. *C. R. Hebd. Séances Acad. Sci.* **261,** 4803–4806.

Jannasch, H. W., and Wirsen, C. O. (1973). Deep-sea microorganisms: in situ response to nutrient enrichment. *Science* **180,** 641–643.

Jannasch, H. W., and Wirsen, C. O. (1977). Microbial life in the deep sea. *Sci. Am.* **236,** 42–52.

Jannasch, H. J., Wirsen, C. O., and Taylor, C. D. (1976). Undecompressed microbial populations from the deep sea. *Appl. Environ. Microbiol.* **32,** 360–367.

Jansson, B.-O. (1968). Studies on the ecology of the interstitial fauna of marine sandy beaches. Ph.D. Thesis, University of Stockholm, Stockholm, Sweden.

Jenkins, A. E., and Bitancourt, A. A. (1941). Revised descriptions of the genera *Elsinoe* and *Sphaceloma. Mycologia* **33,** 338–340.

Jennings, D. H., and Aynsley, J. S. (1971). Compartmentation and low temperature fluxes of potassium in mycelium of *Dendryphiella salina. New Phytol.* **70,** 713–723.

Jensen, J. R., and Sguros, P. L. (1971a). Cellulose degradation by filamentous marine fungi. *Bacteriol. Proc.* **71,** 140.

Jensen, J. R., and Sguros, P. L. (1971b). Characteristics of the cellulolytic enzyme complex in certain filamentous marine fungi. *Q. J. Fla. Acad. Sci.* **34,** 11.

Jewell, T. R. (1974). A qualitative study of cellulose distribution in *Ceratocystis* and *Europhium*. *Mycologia* **66**, 139–146.

Johnson, R. G. (1974). Particulate matter at the sediment-water interface in coastal environments. *J. Mar. Res.* **32**, 313–330.

Johnson, T. W., Jr. (1956a). Marine fungi. I. *Leptosphaeria* and *Pleospora*. *Mycologia* **48**, 495–505.

Johnson, T. W., Jr. (1956b). Marine fungi. II. Ascomycetes and Deuteromycetes from submerged wood. *Mycologia* **48**, 841–851.

Johnson, T. W., Jr. (1956c). Ascus development and spore discharge in *Leptosphaeria discors*, a marine and brackish-water fungus. *Bull. Mar. Sci. Gulf Caribb.* **6**, 349–358.

Johnson, T. W., Jr. (1958a). Marine fungi. IV. *Lulworthia* and *Ceriosporopsis*. *Mycologia* **50**, 151–163.

Johnson, T. W., Jr. (1958b). Some lignicolous marine fungi from the North Carolina coast. *J. Elisha Mitchell Sci. Soc.* **74**, 42–48.

Johnson, T. W., Jr. (1963a). Some aspects of morphology in marine Ascomycetes: *Halosphaeria* Linder. *Nova Hedwigia* **6**, 67–81.

Johnson, T. W., Jr. (1963b). Some aspects of morphology in marine Ascomycetes: *Corollospora* Werdermann. *Nova Hedwigia* **6**, 83–93.

Johnson, T. W., Jr. (1963c). Some aspects of morphology in marine Ascomycetes: *Amylocarpus* Currey, *Herpotrichiella* Petrak and *Torpedospora* Meyers. *Nova Hedwigia* **6**, 157–168.

Johnson, T. W., Jr. (1963d). Some aspects of morphology in marine Ascomycetes: *Ceriosporopsis* Linder. *Nova Hedwigia* **6**, 169–178.

Johnson, T. W., Jr. (1967). The estuarine mycoflora. *In* "Estuaries" (G. H. Lauff, ed.); Publ. No. 83, pp. 303–305. Am. Assoc. Adv. Sci., Washington, D.C.

Johnson, T. W., Jr. (1968). Aquatic fungi of Iceland: Introduction and preliminary account. *J. Elisha Mitchell Sci. Soc.* **84**, 179–183.

Johnson, T. W., Jr., and Anderson, W. R. (1962). A fungus in *Anomia simplex* shell. *J. Elisha Mitchell Sci. Soc.* **78**, 43–47.

Johnson, T. W., Jr., and Cavaliere, A. R. (1963). Some aspects of morphology in marine Ascomycetes: *Remispora* Linder. *Nova Hedwigia* **6**, 179–198.

Johnson, T. W., Jr., and Gold, H. S. (1959). A system for continual-flow sea-water cultures. *Mycologia* **51**, 89–94.

Johnson, T. W., Jr., and Howard, K. L. (1968). Aquatic fungi of Iceland: Species associated with algae. *J. Elisha Mitchell Sci. Soc.* **84**, 305–311.

Johnson, T. W., Jr., and Sparrow, F. K., Jr. (1961). "Fungi in Oceans and Estuaries." Cramer, Weinheim.

Johnson, T. W., Jr., Ferchau, H. A., and Gold, H. S. (1959). Isolation, culture, growth and nutrition of some lignicolous marine fungi. *Phyton* **12**, 65–80.

Johnson-Reid, J. A., and Moore, R. T. (1972). Some ultrastructural features of *Rhodosporidium toruloides* Banno. *Antonie van Leeuwenhoek* **38**, 417–435.

Joho, M. (1970). Effect of alkaline earth metals on heat treated marine yeast *Rhodotorula glutinis* var. *salinaria*. *Mem. Ehime Univ., Nat. Sci., Ser. B* **6**, 173–179.

Joly, P. (1964). Le genre *Alternaria*. *Encycl. Mycol.* **33**, 1–250.

Jones, E. B. G. (1962a). Marine fungi. *Trans. Br. Mycol. Soc.* **45**, 93–114.

Jones, E. B. G. (1962b). *Haligena spartinae* sp. nov., a Pyrenomycete on *Spartina townsendii*. *Trans. Br. Mycol. Soc.* **45**, 245–248.

Jones, E. B. G. (1962c). Some wood-inhabiting marine Pyrenomycetes. *Trans. Br. Mycol. Soc.* **45**, 587–588.

Jones, E. B. G. (1963a). Marine fungi II. Ascomycetes and Deuteromycetes from submerged wood and drift *Spartina*. *Trans. Br. Mycol. Soc.* **46**, 135–144.

Jones, E. B. G. (1963b). British Records: *Halosphaeria mediosetigera* Cribb et Cribb, *Ceriosporopsis calyptrata* Kohlm. *Trans. Br. Mycol. Soc.* **46**, 460–462.

Jones, E. B. G. (1963c). Observations on the fungal succession on wood test blocks submerged in the sea. *J. Inst. Wood Sci.* **11**, 14–23.

Jones, E. B. G. (1964). *Nautosphaeria cristaminuta* gen. et sp. nov., a marine Pyrenomycete on submerged wood. *Trans. Br. Mycol. Soc.* **47**, 97–101.

Jones, E. B. G. (1965). *Halonectria milfordensis* gen. et sp. nov., a marine Pyrenomycete on submerged wood. *Trans. Br. Mycol. Soc.* **48**, 287–290.

Jones, E. B. G. (1968a). Marine fungi. *Curr. Sci.* **37**, 378–379.

Jones, E. B. G. (1968b). The distribution of marine fungi on wood submerged in the sea. *In* "Biodeterioration of Materials" (A. H. Walters and J. J. Elphick, eds.), pp. 460–485. Elsevier, Amsterdam.

Jones, E. B. G. (1971a). The ecology and rotting ability of marine fungi. *In* "Marine Borers, Fungi and Fouling Organisms of Wood" (E. B. G. Jones and S. K. Eltringham, eds.), pp. 237–258. Organ. Econ. Coop. Dev., Paris.

Jones, E. B. G. (1971b). Aquatic fungi. *In* "Methods in Microbiology" (J. R. Norris, D. W. Ribbons, and C. Booth, eds.), Vol. 4, pp. 335–365. Academic Press, New York.

Jones, E. B. G. (1972). The decay of timber in aquatic environments. *Br. Wood Pres. Assoc. Annu. Conv.* pp. 1–18.

Jones, E. B. G. (1973). Marine fungi: spore dispersal, settlement and colonization of timber. *Proc. Int. Congr. Mar. Corros. Fouling, 3rd, 1972* pp. 640–647.

Jones, E. B. G. (1976). Lignicolous and algicolous fungi. *In* "Recent Advances in Aquatic Mycology" (E. B. G. Jones, ed.), pp. 1–51. Wiley, New York.

Jones, E. B. G., and Eaton, R. A. (1969). *Savoryella lignicola* gen. et. sp. nov. from water-cooling towers. *Trans. Br. Mycol. Soc.* **52**, 161–165.

Jones, E. B. G., and Irvine, J. (1971). The role of fungi in the deterioration of wood in the sea. *J. Inst. Wood Sci.* **29**, 31–40.

Jones, E. B. G., and Irvine, J. (1972). The role of marine fungi in the biodeterioration of materials. *In* "Biodeterioration of Materials" (A. H. Walters and E. H. Hueck-van der Plas, eds.), pp. 422–431. Applied Science, London.

Jones, E. B. G., and Jennings, D. H. (1964). The effect of salinity on the growth of marine fungi in comparison with non-marine species. *Trans. Br. Mycol. Soc.* **47**, 619–625.

Jones, E. B. G., and Jennings, D. H. (1965). The effect of cations on the growth of fungi. *New Phytol.* **64**, 86–100.

Jones, E. B. G., and Le Campion-Alsumard, T. (1970a). The biodeterioration of polyurethane by marine fungi. *Int. Biodeterior. Bull.* **6**, 119–124.

Jones, E. B. G., and Le Campion-Alsumard, T. (1970b). Marine fungi on polyurethane covered plates submerged in the sea. *Nova Hedwigia* **19**, 567–582.

Jones, E. B. G., and Oliver, A. C. (1964). Occurrence of aquatic Hyphomycetes on wood submerged in fresh and brackish water. *Trans. Br. Mycol. Soc.* **47**, 45–48.

Jones, E. B. G., and Stewart, R. J. (1972). *Tricladium varium,* an aquatic Hyphomycete on wood in water cooling towers. *Trans. Br. Mycol. Soc.* **59**, 163–167.

Jones, E. B. G., and Ward, A. W. (1973). Septate conidia in *Asteromyces cruciatus. Trans. Br. Mycol. Soc.* **61**, 181–186.

Jones, E. B. G., Byrne, P., and Alderman, D. J. (1971). The response of fungi to salinity. *Vie Milieu, Suppl.* **22**, 265–280.

Jones, E. B. G., Kühne, H., Trussell, P. C., and Turner, R. D. (1972). Results of an international cooperative research programme on the biodeterioration of timber submerged in the sea. *Mater. Org.* **7**, 93–118.

Kampf, W.-D., Becker, G., and Kohlmeyer, J. (1959). Versuche über das Auffinden und den

Befall von Holz durch Larven der Bohrmuschel *Teredo pedicellata* Qutrf. *Z. Angew. Zool.* **46**, 257–283.

Kappen, L., and Lange, O. L. (1972). Die Kälteresistenz einiger Makrolichenen. *Flora (Jena)* **161**, 1–29.

Karsten, P. (1873). Mycologia Fennica 2. *Bidr. Finl. Nat. Folk* **23**, 1–252.

Kawakita, S., and van Uden, N. (1965). Occurrence and population densities of yeast species in the digestive tracts of gulls and terns. *J. Gen. Microbiol.* **39**, 125–129.

Kendrick, B., ed. (1971a). "Taxonomy of Fungi Imperfecti." Univ. of Toronto Press, Toronto.

Kendrick, B. (1971b). Conclusions and recommendations. *In* "Taxonomy of Fungi Imperfecti" (B. Kendrick, ed.), pp. 253–262. Univ. of Toronto Press, Toronto.

Kendrick, W. B., and Bhatt, G. C. (1966). *Trichocladium opacum. Can. J. Bot.* **44**, 1728–1730.

Kendrick, W. B., and Carmichael, J. W. (1973). Hyphomycetes. *In* "The Fungi" (G. C. Ainsworth, F. K. Sparrow, and A. S. Sussman, eds.), Vol. 4A, pp. 323–509. Academic Press, New York.

Kingham, D. L., and Evans, L. V. (1977). The *Pelvetia/Ascophyllum—Mycosphaerella* inter-relationship. *Br. Phycol. J.* **12**, 120.

Kirby, C. J., and Gosselink, J. G. (1976). Primary production in a Louisiana Gulf coast *Spartina alterniflora* marsh. *Ecology* **57**, 1052–1059.

Kirk, P. W., Jr. (1966). Morphogenesis and microscopic cytochemistry of marine pyrenomycete ascospores. *Nova Hedwigia Beih.* **22**, 1–128.

Kirk, P. W., Jr. (1969). Isolation and culture of lignicolous marine fungi. *Mycologia* **61**, 174–177.

Kirk, P. W., Jr. (1970). Neutral red as a lipid fluorochrome. *Stain Technol.* **45**, 1–4.

Kirk, P. W., Jr. (1976). Cytochemistry of marine fungal spores. *In* "Recent Advances in Aquatic Mycology" (E. B. G. Jones, ed.), pp. 177–192. Wiley, New York.

Kirk, P. W., Jr., and Catalfomo, P. (1970). Marine fungi: The occurrence of ergosterol and choline. *Phytochemistry* **9**, 595–597.

Kirk, P. W., Jr., Catalfomo, P., Block, J. H., and Constantine, G. H., Jr. (1974). Metabolites of higher marine fungi and their possible ecological significance. *Veroeff. Inst. Meeresforsch. Bremerhaven, Suppl.* **5**, 509–518.

Kishimoto, R. A., and Baker, G. E. (1969). Pathogenic and potentially pathogenic fungi isolated from beach sands and selected soils of Oahu, Hawaii. *Mycologia* **61**, 537–548.

Klement, O., and Doppelbaur, H. (1952). Über die Artberechtigung einiger mariner Arthopyrenien. *Ber. Dtsch. Bot. Ges.* **65**, 166–174.

Knebel, G. (1963). Monographie der Algenreihe der Prasiolales, insbesondere von *Prasiola crispa. Hedwigia* **75**, 1–120.

Knudsen, J. (1961). The bathyal and abyssal *Xylophaga* (Pholadidae, Bivalvia). *Galathea Rep.* **5**, 163–209.

Kobayashi, Y., Tsubaki, K., and Soneda, M. (1953). Marine yeasts isolated from Little-neck Clam. *Bull. Natl. Sci. Mus., Tokyo* **33**, 47–52.

Koch, J. (1974). Marine fungi on driftwood from the west coast of Jutland, Denmark. *Friesia* **10**, 209–250.

Kohlmeyer, J. (1956). Über den Cellulose-Abbau durch einige phytopathogene Pilze. *Phytopathol. Z.* **27**, 147–182.

Kohlmeyer, J. (1958a). Beobachtungen über mediterrane Meerespilze sowie das Vorkommen von marinen Moderfäule-Erregern in Aquariumszuchten holzzerstörender Meerestiere. *Ber. Dtsch. Bot. Ges.* **71**, 98–116.

Kohlmeyer, J. (1958b). Holzzerstörende Pilze im Meerwasser. *Holz Roh- Werkst.* **16,** 215–220.

Kohlmeyer, J. (1959). Neufunde holzbesiedelnder Meerespilze. *Nova Hedwigia* **1,** 77–99.

Kohlmeyer, J. (1960). Wood-inhabiting marine fungi from the Pacific Northwest and California. *Nova Hedwigia* **2,** 293–343.

Kohlmeyer, J. (1961a). Pilze von der nördlichen Pazifik-Küste der U.S.A. *Nova Hedwigia* **3,** 80–86.

Kohlmeyer, J. (1961b). Hypersaprophytismus eines Salzwasser-Ascomyceten. *Ber. Dtsch. Bot. Ges.* **74,** 305–310.

Kohlmeyer, J. (1962a). Halophile Pilze von den Ufern Frankreichs. *Nova Hedwigia* **4,** 389–420.

Kohlmeyer, J. (1962b). *Corollospora maritima* Werderm.: Ein Ascomycet. *Ber. Dtsch. Bot. Ges.* **75,** 125–127.

Kohlmeyer, J. (1962c). Über *Pleospora gaudefroyi* Patouillard. *Willdenowia* **3,** 315–324.

Kohlmeyer, J. (1963a). Zwei neue Ascomyceten-Gattungen auf *Posidonia*-Rhizomen. *Nova Hedwigia* **6,** 5–13.

Kohlmeyer, J. (1963b). Parasitische und epiphytische Pilze auf Meeresalgen. *Nova Hedwigia* **6,** 127–146.

Kohlmeyer, J. (1963c). Répartition de champignons marins (Ascomycetes et Fungi Imperfecti) dans la Méditerranée. *Rapp. Comm. Int. Explor. Sci. Mer Méditerr.* **17,** 723–730.

Kohlmeyer, J. (1963d). Fungi marini novi vel critici. *Nova Hedwigia* **6,** 297–329.

Kohlmeyer, J. (1963e). The importance of fungi in the sea. *In* "Symposium on Marine Microbiology" (C. H. Oppenheimer, ed.), pp. 300–314. Thomas, Springfield, Illinois.

Kohlmeyer, J. (1964). Pilzfunde am Meer. *Z. Pilzkd.* **30,** 43–51.

Kohlmeyer, J. (1966a). Ecological observations on arenicolous marine fungi. *Z. Allg. Mikrobiol.* **6,** 94–105.

Kohlmeyer, J. (1966b). Ascospore morphology in *Corollospora*. *Mycologia* **58,** 281–288.

Kohlmeyer, J. (1966c). Neue Meerespilze an Mangroven. *Ber. Dtsch. Bot. Ges.* **79,** 27–37.

Kohlmeyer, J. (1967). Intertidal and phycophilous fungi from Tenerife (Canary Islands). *Trans. Br. Mycol. Soc.* **50,** 137–147.

Kohlmeyer, J. (1968a). A new *Trematosphaeria* from roots of *Rhizophora racemosa*. *Mycopathol. Mycol. Appl.* **34,** 1–5.

Kohlmeyer, J. (1968b). The first Ascomycete from the deep sea. *J. Elisha Mitchell Sci. Soc.* **84,** 239–241.

Kohlmeyer, J. (1968c). Marine fungi from the tropics. *Mycologia* **60,** 252–270.

Kohlmeyer, J. (1968d). Dänische Meerespilze (Ascomycetes). *Ber. Dtsch. Bot. Ges.* **81,** 53–61.

Kohlmeyer, J. (1968e). Revisions and descriptions of algicolous marine fungi. *Phytopathol. Z.* **63,** 341–363.

Kohlmeyer, J. (1969a). Deterioration of wood by marine fungi in the deep sea. *In* "Materials Performance and the Deep Sea", *Am. Soc. Test. Mater., Spec. Tech. Publ.* **445,** 20–29.

Kohlmeyer, J. (1969b). The role of marine fungi in the penetration of calcareous substances. *Am. Zool.* **9,** 741–746.

Kohlmeyer, J. (1969c). Marine fungi of Hawaii including the new genus *Helicascus*. *Can. J. Bot.* **47,** 1469–1487.

Kohlmeyer, J. (1969d). Ecological notes on fungi in mangrove forests. *Trans. Br. Mycol. Soc.* **53,** 237–250.

Kohlmeyer, J. (1969e). Perithecial hairs with phialides in *Spathulospora phycophila*. *Mycologia* **61,** 1012–1015.

Kohlmeyer, J. (1970). Ein neuer Ascomycet auf Hydrozoen im Südatlantik. *Ber. Dtsch. Bot. Ges.* **83**, 505–509.

Kohlmeyer, J. (1971a). Fungi from the Sargasso Sea. *Mar. Biol.* **8**, 344–350.

Kohlmeyer, J. (1971b). Annotated check-list of New England marine fungi. *Trans. Br. Mycol. Soc.* **57**, 473–492.

Kohlmeyer, J. (1972a). Marine fungi deteriorating chitin of hydrozoa and keratin-like annelid tubes. *Mar. Biol.* **12**, 277–284.

Kohlmeyer, J. (1972b). A revision of Halosphaeriaceae. *Can. J. Bot.* **50**, 1951–1963.

Kohlmeyer, J. (1972c). Parasitic *Haloguignardia oceanica* (Ascomycetes) and hyperparasitic *Sphaceloma cecidii* sp. nov. (Deuteromycetes) in drift *Sargassum* in North Carolina. *J. Elisha Mitchell Sci. Soc.* **88**, 255–259.

Kohlmeyer, J. (1973a). Spathulosporales, a new order and possible missing link between Laboulbeniales and Pyrenomycetes. *Mycologia* **65**, 614–647.

Kohlmeyer, J. (1973b). Fungi from marine algae. *Bot. Mar.* **16**, 201–215.

Kohlmeyer, J. (1974a). On the definition and taxonomy of higher marine fungi. *Veroeff. Inst. Meeresforsch. Bremerhaven, Suppl.* **5**, 263–286.

Kohlmeyer, J. (1974b). Higher fungi as parasites and symbionts of algae. *Veroeff. Inst. Meeresforsch. Bremerhaven, Suppl.* **5**, 339–356.

Kohlmeyer, J. (1975a). New clues to the possible origin of Ascomycetes. *BioScience* **25**, 86–93.

Kohlmeyer, J. (1975b). Revision of algicolous *Zignoella* spp. and description of *Pontogeneia* gen. nov. (Ascomycetes). *Bot. Jahrb.* **96**, 200–211.

Kohlmeyer, J. (1976). Marine fungi from South America. *Mitt. Inst. Colombo-Alemán Invest. Cient. "Punta de Betin"* **8**, 33–39.

Kohlmeyer, J. (1977). New genera and species of higher fungi from the deep sea (1615–5315 m). *Rev. Mycol.* **41**, 189–206.

Kohlmeyer, J., and Gessner, R. V. (1976). *Buergenerula spartinae* sp. nov., an Ascomycete from salt marsh cordgrass, *Spartina alterniflora. Can. J. Bot.* **54**, 1759–1766.

Kohlmeyer, J., and Kohlmeyer, E. (1964). "Synoptic Plates of Higher Marine Fungi," 2nd ed. Cramer, Weinheim.

Kohlmeyer, J., and Kohlmeyer, E. (1964–1969). "Icones Fungorum Maris." Cramer, Weinheim and Lehre.

Kohlmeyer, J., and Kohlmeyer, E. (1965). New marine fungi from mangroves and trees along eroding shorelines. *Nova Hedwigia* **9**, 89–104.

Kohlmeyer, J., and Kohlmeyer, E. (1966). On the life history of marine Ascomycetes: *Halosphaeria mediosetigera* and *H. circumvestita. Nova Hedwigia* **12**, 189–202.

Kohlmeyer, J., and Kohlmeyer, E. (1971a). Marine fungi from tropical America and Africa. *Mycologia* **63**, 831–861.

Kohlmeyer, J., and Kohlmeyer, E. (1971b). "Synoptic Plates of Higher Marine Fungi. An Identification Guide for the Marine Environment," 3rd ed. Cramer, Lehre.

Kohlmeyer, J., and Kohlmeyer, E. (1972a). Is *Ascophyllum nodosum* lichenized? *Bot. Mar.* **15**, 109–112.

Kohlmeyer, J., and Kohlmeyer, E. (1972b). A new genus of marine Ascomycetes on *Ulva vexata* Setch. et Gard. *Bot. Jahrb.* **92**, 429–432.

Kohlmeyer, J., and Kohlmeyer, E. (1972c). Permanent microscopic mounts. *Mycologia* **64**, 666–669.

Kohlmeyer, J., and Kohlmeyer, E. (1975). Biology and geographical distribution of *Spathulospora* species. *Mycologia* **67**, 629–637.

Kohlmeyer, J., and Kohlmeyer, E. (1977). Bermuda marine fungi. *Trans. Br. Mycol. Soc.* **68**, 207–219.

Kohlmeyer, J., Becker, G., and Kampf, W.-D. (1959). Versuche zur Kenntnis der Ernährung der Holzbohrassel *Limnoria tripunctata* und ihre Beziehung zu holzzerstörenden Pilzen. *Z. Angew. Zool.* **46**, 457–489.

Kohlmeyer, J., Schmidt, I., and Nair, N. B. (1967). Eine neue *Corollospora* (Ascomycetes) aus dem Indischen Ozean und der Ostsee. *Ber. Dtsch. Bot. Ges.* **80**, 98–102.

Kölliker, A. (1860a). Über das ausgebreitete Vorkommen von pflanzlichen Parasiten in den Hartgebilden niederer Thiere. *Z. Wiss. Zool.* **10**, 215–232.

Kölliker, A. (1860b). On the frequent occurrence of vegetable parasites in the hard tissues of the lower animals. *Q. J. Microsc. Sci.* **8**, 171–187.

Koponen, H., and Mäkelä, K. (1975). *Leptosphaeria* s. lat. (*Keissleriella, Paraphaeosphaeria, Phaeosphaeria*) on Gramineae in Finland. *Ann. Bot. Fenn.* **12**, 141–160.

Korf, R. P. (1973). Discomycetes and Tuberales. *In* "The Fungi" (G. C. Ainsworth, F. K. Sparrow, and A. S. Sussman, eds.), Vol. 4A, pp. 249–319. Academic Press, New York.

Koshkelova, E. N., and Frolov, I. P. (1973). "Mikoflora podgornoi raviniy kopetdaga i tsentral'nykh Karakumov (Mikromitsetiy)" (Mycoflora of the mountainous plains of Kopet Dagh and Central Kara Kum). Ashkhabat.

Koshkelova, E. N., Frolov, I. P., and Dzhuraeva, S. (1970). "Mikoflora Badkhiza, Karabilya i yuzhnoi Chasti Murgabskogo Oazisa" (Mycoflora of Badkhyz, Karabil and the southern part of the Murgab Oasis). Ashkhabad.

Kreger-van Rij, N. J. W. (1973). Endomycetales, basidiomycetous yeasts, and related fungi. *In* "The Fungi" (G. C. Ainsworth, F. K. Sparrow, and A. S. Sussman, eds.), Vol. IV A, pp. 11–32. Academic Press, New York.

Kremer, B. P. (1973). Untersuchungen zur Physiologie von Volemit in der marinen Braunalge *Pelvetia canaliculata*. *Mar. Biol.* **22**, 31–35.

Kriss, A. E. (1963). "Marine Microbiology. (Deep Sea)." Oliver & Boyd, Edinburgh and London.

Kriss, A. E., and Novozhilova, M. I. (1954). Yavlyayutsya li drosheviye organismi obitatelyami morei i okeanov? (Are yeast organisms inhabitants of seas and oceans? Translation by Jean S. ZoBell). *Mikrobiologia* **23**, 669–683.

Kriss, A., and Rukina, E. A. (1949). Microbiology of the Black Sea. *Mikrobiologia* **18**, 141.

Kriss, A. E., Rukina, E. A., and Tikhonenko, A. S. (1952). Rasprostraneniye droshevikh organizmov v morye. *Akad. Nauk Soyuza S.S.R. Zh. Obshch. Biol.* **13**, 233–242.

Kriss, A. E., Mishustina, I. E., Mitskevich, N., and Zemtsova, E. V. (1967). "Microbial Population of Oceans and Seas." Arnold, London.

Kühne, H. (1971). The identification of wood-boring crustaceans. *In* "Marine Borers, Fungi and Fouling Organisms of Wood" (E. B. G. Jones and S. K. Eltringham, eds.), pp. 65–88. Organ. Econ. Coop. Dev., Paris.

Kurtzman, C. P., and Ahearn, D. G. (1976). Sporulation in *Pichia spartinae*. *Mycologia* **68**, 682–685.

Lachance, M. A., Miranda, M., Miller, M. W., and Phaff, H. J. (1976). Dehiscence and active spore release in pathogenic strains of the yeast *Metschnikowia bicuspidata* var. *australis:* possible predatory implication. *Can. J. Microbiol.* **22**, 1756–1761.

Lamb, I. M. (1948). Antarctic pyrenocarp lichens. *'Discovery' Rep.* **25**, 1–30.

Lamb, I. M. (1963). "Index Nominum Lichenum." Ronald, New York.

Lane, C. E. (1961). The teredo. *Sci. Am.* **204**, 132–142.

Larsen, P. (1952). Studies in Danish Pyrenomycetes. *Dan. Bot. Ark.* **14**, 1–61.

Le Campion-Alsumard, T. (1970). Etude qualitative et quantitative des salissures biologiques de plaques expérimentales immergées en pleine eau. 2. Etude préliminaire de quelques Pyrénomycètes marins récoltés sur des plaques de polyuréthane. *Tethys* **1**, 715–718.

Lee, B. K. H., and Baker, G. E. (1972a). An ecological study of the soil microfungi in a Hawaiian mangrove swamp. *Pac. Sci.* **26,** 1–10.

Lee, B. K. H., and Baker, G. E. (1972b). Environment and the distribution of microfungi in a Hawaiian mangrove swamp. *Pac. Sci.* **26,** 11–19.

Lee, B. K. H., and Baker, G. E. (1973). Fungi associated with the roots of red mangrove, *Rhizophora mangle. Mycologia* **65,** 894–906.

Leightley, L. E., and Eaton, R. A. (1977). Mechanisms of decay of timber by aquatic micro-organisms. *Br. Wood Pres. Assoc. Annu. Conv.* pp. 1–26.

Leonard, K. J. (1976). Synonymy of *Exserohilum halodes* with *E. rostratum,* and induction of the ascigerous state, *Setosphaeria rostrata. Mycologia* **68,** 402–411.

Le Petit, J., N'Guyen, M. H., and Devèze, L. (1970). Etude de l'intervention des levures dans la bio-dégradation en mer des hydrocarbures. *Ann. Inst. Pasteur, Paris* **118,** 709–720.

Lepidi, A. A., Nuti, M. P., de Bertoldi, M., and Santulli, M. (1972). Classification of the genus *Humicola* Traaen: II. The DNA base composition of some strains within the genus. *Mycopathol. Mycol. Appl.* **47,** 153–159.

Letrouit-Galinou, M. A. (1969). Les algues des lichens. *Mém. Soc. Bot. Fr.* pp. 35–77.

Levy, J. (1965). The soft rot fungi: Their mode of action and significance in the degradation of wood. *Adv. Bot. Res.* **2,** 323–357.

Lewis, D. H. (1973). Concepts in fungal nutrition and the origin of biotrophy. *Biol. Rev. Cambridge Philos. Soc.* **48,** 261–278.

Lewis, J. R. (1964). "The Ecology of Rocky Shores." English Univ. Press, London.

Liese, W. (1970). Ultrastructural aspects of woody tissue disintegration. *Annu. Rev. Phytopathol.* **8,** 231–258.

Liese, W., and Cote, W. A., Jr. (1960). Electron microscopy of wood: Results of the first ten years of research. *Proc. World For. Congr., 5th, 1960* Vol 2, pp. 1288–1298.

Lind, J. (1913). "Danish Fungi as Represented in the Herbarium of E. Rostrup." Nordisk Forlag, Copenhagen.

Lindau, G. (1898). Bau und Entwicklungsgeschichte von *Amylocarpus encephaloides* Curr. *Verh. Bot. Ver. Prov. Brandenburg* **40,** XXIV–XXV.

Lindau, G. (1899). Über Entwicklung und Ernährung von *Amylocarpus encephaloides* Curr. *Hedwigia* **38,** 1–19.

Linder, D. H. (1944). I. Classification of the marine fungi. *In* E. S. Barghoorn and D. H. Linder, Marine fungi: their taxonomy and biology. *Farlowia* **1,** 401–433.

Lloyd, L. S., and Wilson, I. M. (1962). Development of the perithecium in *Lulworthia medusa* (Ell. et Ev.) Cribb et Cribb, a saprophyte on *Spartina townsendii. Trans. Br. Mycol. Soc.* **45,** 359–372.

Lockwood, J. L. (1977). Fungistasis in soils. *Biol. Rev. Cambridge Philos. Soc.* **52,** 1–43.

Lodder, J. (1970). "The Yeasts. A Taxonomic Study," 2nd ed. North-Holland Publ., Amsterdam.

Lucas, M. T. (1963). Culture studies on Portuguese species of *Leptosphaeria* I. *Trans. Br. Mycol. Soc.* **46,** 361–367.

Lucas, M. T., and Webster, J. (1964). Conidia of *Pleospora scirpicola* and *P. valesiaca. Trans. Br. Mycol. Soc.* **47,** 247–256.

Lucas, M. T., and Webster, J. (1967). Conidial states of British species of *Leptosphaeria. Trans. Br. Mycol. Soc.* **50,** 85–121.

Lindqvist, N. (1969). *Tripterospora* (Sordariaceae s. lat., Pyrenomycetes). *Bot. Not.* **122,** 589–603.

Lundqvist, N. (1972). Nordic Sordariaceae s. lat. *Symb. Bot. Ups.* **20,** 1–374.

Lundström-Eriksson, A., and Norkrans, B. (1968). Studies on marine-occurring yeasts:

relations to inorganic nitrogen compounds, especially hydroxylamine. *Arch. Mikrobiol.* **62**, 373–383.

Lutley, M., and Wilson, I. M. (1972a). Development and fine structure of ascospores in the marine fungus *Ceriosporopsis halima. Trans. Br. Mycol. Soc.* **58**, 393–402.

Lutley, M., and Wilson, I. M. (1972b). Observations on the fine structure of ascospores of marine fungi: *Halosphaeria appendiculata, Torpedospora radiata* and *Corollospora maritima. Trans. Br. Mycol. Soc.* **59**, 219–227.

Luttrell, E. S. (1951). Taxonomy of the Pyrenomycetes. *Univ. Mo. Stud.* **24**, 1–120.

Luttrell, E. S. (1973). Loculoascomycetes. *In* "The Fungi" (G. C. Ainsworth, F. K. Sparrow, and A. S. Sussman, eds.), Vol. 4A, pp. 135–219. Academic Press, New York.

Malacalza, L., and Martínez, A. (1971). Ascomycetes marinos de Argentina. *Bol. Soc. Argent. Bot.* **14**, 57–72.

Malloch, D., and Cain, R. F. (1971). New cleistothecial Sordariaceae and a new family, Coniochaetaceae. *Can. J. Bot.* **49**, 869–880.

Malloch, D., and Cain, R. F. (1972). New species and combinations of cleistothecial Ascomycetes. *Can. J. Bot.* **50**, 61–72.

Mameli, E. (1915). Sulla flora micologica della Sardegna. Seconda contribuzione. *Atti Ist. Bot. Univ. Pavia* [2] **14**, 1–13.

Mason, E. (1928). Note on the presence of mycorrhiza in the roots of salt marsh plants. *New Phytol.* **27**, 193–195.

Matsuda, H., and Takada, H. (1973). Salt-tolerance of mutants of marine yeast, *Rhodotorula glutinis* var. *salinaria* induced by ultraviolet light. *Trans. Mycol. Soc. Jpn.* **14**, 137–143.

Mattick, F. (1953). Lichenologische Notizen. 1–5. *Ber. Dtsch. Bot. Ges.* **66**, 263–276.

Maxwell, G. S. (1968). Pathogenicity and salinity tolerance of *Phytophthora* sp. isolated from *Avicennia resinifera* (Forst F.)—some initial investigations. *Tane* **14**, 13–23.

Meadows, P. S., and Anderson, J. G. (1966). Micro-organisms attached to marine and freshwater sand grains. *Nature (London)* **212**, 1059–1060.

Meadows, P. S., and Anderson, J. G. (1968). Micro-organisms attached to marine sand grains. *J. Mar. Biol. Assoc. U.K.* **48**, 161–175.

Melnik, V. A. (1977). "Opredelitel gribov roda *Ascochyta* Lib." (Monograph of Fungi of the Genus *Ascochyta* Lib.). Acad. Sci. SSSR, Leningrad.

Melnik, V. A., and Petrov, Yu. E. (1966). Novyi vid griba s morskoi buroi Vodorosli *Ascophyllum nodosum* (L.) Le Jolis. (De specie nova fungi in alga marina *Ascoph. nodos.* (L.) Le J. (Phaeoph.) inventa notula). *Nov. Sist. Niz. Rast.* pp. 211–212.

Menzies, R. J., Zaneveld, J. S., and Pratt, R. M. (1967). Transported turtle grass as a source of organic enrichment of abyssal sediments off North Carolina. *Deep-Sea Res.* **14**, 111–112.

Meyers, S. P. (1953). Marine fungi in Biscayne Bay, Florida. *Bull. Mar. Sci. Gulf Caribb.* **2**, 590–601.

Meyers, S. P. (1954). Marine fungi in Biscayne Bay, Florida. II. Further studies of occurrence and distribution. *Bull. Mar. Sci. Gulf Caribb.* **3**, 307–327.

Meyers, S. P. (1957). Taxonomy of marine Pyrenomycetes. *Mycologia* **49**, 475–528.

Meyers, S. P. (1966). Variability in growth and reproduction of the marine fungus, *Lulworthia floridana. Helgol. Wiss. Meeresunters.* **13**, 436–443.

Meyers, S. P. (1968a). Degradative activities of filamentous marine fungi. *In* "Biodeterioration of Materials" (A. H. Walters and J. J. Elphick, eds.), pp. 594–609. Elsevier, Amsterdam.

Meyers, S. P. (1968b). Observations on the physiological ecology of marine fungi. *Bull. Misaki Mar. Biol. Inst., Kyoto Univ.* **12**, 207–225.

Meyers, S. P. (1969). Thalassiomycetes XI. Further studies of the genus *Lindra* with a description of *L. marinera,* a new species. *Mycologia* **61**, 486–495.

Meyers, S. P. (1971a). Isolation and identification of filamentous marine fungi. *In* "Marine Borers, Fungi and Fouling Organisms of Wood" (E. B. G. Jones and S. K. Eltringham, eds.), pp. 89–113. Org. Econ. Coop. Dev., Paris.

Meyers, S. P. (1971b). Development in the biology of filamentous marine fungi. *In* "Marine Borers, Fungi and Fouling Organisms of Wood" (E. B. G. Jones and S. K. Eltringham, eds.), pp. 217–235. Org. Econ. Coop. Dev., Paris.

Meyers, S. P., and Hopper, B. E. (1966). Attraction of the marine nematode, *Metoncholaimus* sp., to fungal substrates. *Bull. Mar. Sci.* **16**, 142–150.

Meyers, S. P., and Hopper, B. E. (1967). Studies on marine fungal-nematode associations and plant degradation. *Helgol. Wiss. Meeresunters.* **15**, 270–281.

Meyers, S. P., and Hoyo, L. (1966). Observations on the growth of the marine Hyphomycete *Varicosporina ramulosa. Can. J. Bot.* **44**, 1133–1140.

Meyers, S. P., and Kohlmeyer, J. J. (1965). *Varicosporina ramulosa* gen. nov. sp. nov., an aquatic Hyphomycete from marine areas. *Can. J. Bot.* **43**, 915–921.

Meyers, S. P., and Moore, R. T. (1960). Thalassiomycetes II. New genera and species of Deuteromycetes. *Am. J. Bot.* **47**, 345–349.

Meyers, S. P., and Reynolds, E. S. (1957). Incidence of marine fungi in relation to woodborer attack. *Science* **126**, 969.

Meyers, S. P., and Reynolds, E. S. (1958). A wood incubation method for the study of lignicolous marine fungi. *Bull. Mar. Sci. Gulf Caribb.* **8**, 342–347.

Meyers, S. P., and Reynolds, E. S. (1959a). Effects of wood and wood products on perithecial development by lignicolous marine Ascomycetes. *Mycologia* **51**, 138–145.

Meyers, S. P., and Reynolds, E. S. (1959b). Growth and cellulolytic activity of lignicolous Deuteromycetes from marine localities. *Can. J. Microbiol.* **5**, 493–503.

Meyers, S. P., and Reynolds, E. S. (1959c). Cellulolytic activity in lignicolous marine Ascomycetes. *Bull. Mar. Sci. Gulf Caribb.* **9**, 441–455.

Meyers, S. P., and Reynolds, E. S. (1960a). Occurrence of lignicolous fungi in northern Atlantic and Pacific marine localities. *Can. J. Bot.* **38**, 217–226.

Meyers, S. P., and Reynolds, E. S. (1960b). Cellulolytic activity of lignicolous marine Ascomycetes and Deuteromycetes. *Dev. Ind. Microbiol.* **1**, 157–168.

Meyers, S. P., and Reynolds, E. S. (1963). Degradation of lignocellulose material by marine fungi. *In* "Symposium on Marine Microbiology" (C. H. Oppenheimer, ed.), pp. 315–328. Thomas, Springfield, Illinois.

Meyers, S. P., and Scott, E. (1967). Thalassiomycetes X. Variation in growth and reproduction of two isolates of *Corollospora maritima. Mycologia* **59**, 446–455.

Meyers, S. P., and Scott, E. (1968). Cellulose degradation by *Lulworthia floridana* and other lignicolous marine fungi. *Mar. Biol.* **2**, 41–46.

Meyers, S. P., and Simms, J. (1965). Thalassiomycetes VI. Comparative growth studies of *Lindra thalassiae* and lignicolous Ascomycete species. *Can. J. Bot.* **43**, 379–392.

Meyers, S. P., and Simms, J. (1967). Thalassiomycetes IX. Comparative studies of reproduction in marine Ascomycetes. *Bull. Mar. Sci.* **17**, 133–148.

Meyers, S. P., Prindle, B., and Reynolds, E. S. (1960). Cellulolytic activity of marine fungi. Degradation of ligno-cellulose material. *Tappi* **43**, 534–538.

Meyers, S. P., Feder, W. A., and Tsue, K. M. (1963a). Nutritional relationships among certain filamentous fungi and a marine nematode. *Science* **141**, 520–522.

Meyers, S. P., Prindle, B., Kamp, K., and Levi, J. U. (1963b). Degradation of manila cordage by marine fungi: Analysis of breaking strength tests. *Tappi* **46**, 164A–167A.

Meyers, S. P., Feder, W. A., and Tsue, K. M. (1964a). Studies of relationships among nematodes and filamentous fungi in the marine environment. *Dev. Ind. Microbiol.* **5**, 354–364.

Meyers, S. P., Kamp, K. M., Johnson, R. F., and Shaffer, D. L. (1964b). Thalassiomycetes IV. Analysis of variance of ascospores of the genus *Lulworthia*. *Can. J. Bot.* **42**, 519–526.

Meyers, S. P., Orpurt, P. A., Simms, J., and Boral, L. L. (1965). Thalassiomycetes VII. Observations on fungal infestation of turtle grass, *Thalassia testudinum* König. *Bull. Mar. Sci.* **15**, 548–564.

Meyers, S. P., Ahearn, D. G., Gunkel, W., and Roth, F. J., Jr. (1967a). Yeasts from the North Sea. *Mar. Biol.* **1**, 118–123.

Meyers, S. P., Ahearn, D. G., and Roth, F. J., Jr. (1967b). Mycological investigations of the Black Sea. *Bull. Mar. Sci.* **17**, 576–596.

Meyers, S. P., Nicholson, M. E., and Ahearn, D. G. (1970a). Amine compounds as nitrogen sources for yeasts. *Bacteriol. Proc.* **70**, 32.

Meyers, S. P., Nicholson, M. L., Rhee, J., Miles, P., and Ahearn, D. G. (1970b). Mycological studies in Barataria Bay, Louisiana, and biodegradation of oyster grass, *Spartina alterniflora*. *Coastal Stud. Bull., Spec. Sea Grant Issue* No. 5, pp. 111–179.

Meyers, S. P., Ahearn, D. G., and Miles, P. C. (1971a). Characterization of yeasts in Barataria Bay. *Coastal Stud. Bull.* **6**, 7–15.

Meyers, S. P., Miles, P., and Ahearn, D. G. (1971b). Occurrence of pulcherrimin producing yeasts in Louisiana marshland sediments. *Bacteriol. Proc.* **71**, 36.

Meyers, S. P., Chung, S. L., and Ahearn, D. G. (1972). Biodegradation of cellulosic substrates by marine fungi. *In* "Biodeterioration of Materials" (A. H. Walters and E. H. Hueck-van der Plas, eds.), pp. 121–128. Applied Science, London.

Meyers, S. P., Ahearn, D. G., Alexander, S. K., and Cook, W. L. (1975). *Pichia spartinae,* a dominant yeast of the *Spartina* salt marsh. *Dev. Ind. Microbiol.* **16**, 262–267.

Migula, W. (1921). "Kryptogamen-Flora von Deutschland, Deutsch-Österreich und der Schweiz," Vol. III, Part 4, No. 1, Bermühler, Berlin.

Miller, M. W., and van Uden, N. (1970). *Metschnikowia* Kamienski. *In* "The Yeasts" (J. Lodder, ed.), pp. 408–429. North-Holland Publ., Amsterdam.

Mix, A. J. (1949). A monograph of the genus *Taphrina*. *Univ. Kans. Sci. Bull.* **33**, 3–167.

Miyabe, K., and Tokida, J. (1948). Black-dots disease of *Gloiopeltis furcata* Post. et Rupr. caused by a new ascomycetous fungus. *Bot. Mag.* **61**, 116–118.

Montagne, J. F. C. (1856). "Sylloge Generum Specierumque Cryptogamarum." Baillière et Fils, Paris.

Moore, R. T. (1958). Deuteromycetes I: The *Sporidesmium* complex. *Mycologia* **50**, 681–692.

Moore, R. T. (1972). Ustomycota, a new division of higher fungi. *Antonie van Leeuwenhoek* **38**, 567–584.

Moore, R. T., and Meyers, S. P. (1962). Thalassiomycetes III. The genus *Zalerion*. *Can. J. Microbiol.* **8**, 407–416.

Moorhouse, J., and de Bertoldi, M. (1975). Electrophoretic characteristics of enzymes as a taxonomic criterion in the genus *Humicola*. *Mycotaxon* **3**, 109–118.

Moreau, F., and Moreau, Mme. F. (1941). Première contribution à l'étude de la microflore des dunes. *Rev. Mycol.* **6**, 49–94.

Morgan-Jones, G., Nag Raj, T. R., and Kendrick, B. (1972). Icones generum coelomycetum IV. *Univ. Waterloo Biol. Ser.* **6**, 1–42.

Morgan-Jones, J. F. (1953). Morpho-cytological studies of the genus *Gnomonia*. I. *Gnomonia intermedia* Rehm: its development in pure culture. *Sven. Bot. Tidskr.* **47,** 284–308.

Morgan-Jones, J. F. (1958). Morpho-cytological studies of the genus *Gnomonia*. II. The asexual stage of *Gnomonia ulmea* (Schw.) Thüm. *Sven. Bot. Tidskr.* **52,** 363–372.

Morgan-Jones, J. F. (1959). Morpho-cytological studies of the genus *Gnomonia*. III. Early stages of perithecial development. *Sven. Bot. Tidskr.* **53,** 81–101.

Morii, H. (1973). Yeasts predominating in the stomach of marine little toothed whales. *Bull. Jpn. Soc. Sci. Fish.* **39,** 333.

Morris, E. O. (1968). Yeasts of marine origin. *Oceanogr. Mar. Biol.* **6,** 201–230.

Moss, S. T., and Jones, E. B. G. (1977). Ascospore appendages of marine Ascomycetes: *Halosphaeria mediosetigera*. *Trans. Br. Mycol. Soc.* **69,** 313–315.

Mounce, I., and Diehl, W. W. (1934). A new *Ophiobolus* on eelgrass. *Can. J. Res.* **11,** 242–246.

Moustafa, A. F. (1975). Osmophilous fungi in the salt marshes of Kuwait. *Can. J. Microbiol.* **21,** 1573–1580.

Moustafa, A. F., and Al-Musallam, A. A. (1975). Contribution to the fungal flora of Kuwait. *Trans. Br. Mycol. Soc.* **65,** 547–553.

Moustafa, A. F., Sharkas, M. S., and Kamel, S. M. (1976). Thermophilic and thermotolerant fungi in the desert and salt-marsh soils of Kuwait. *Norw. J. Bot.* **23,** 213–220.

Müller, E. (1950). Die schweizerischen Arten der Gattung *Leptosphaeria* und ihrer Verwandten. *Sydowia* **4,** 185–319.

Müller, E. (1951). Die schweizerischen Arten der Gattungen *Clathrospora, Pleospora, Pseudoplea* und *Pyrenophora*. *Sydowia* **5,** 248–310.

Müller, E., and von Arx, J. A. (1962). Die Gattungen der didymosporen Pyrenomyceten. *Beitr. Kryptogamenflora Schweiz* **11,** 1–922.

Müller, E., and von Arx, J. A. (1973). Pyrenomycetes: Meliolales, Coronophorales, Sphaeriales, *In* "The Fungi" (G. C. Ainsworth, F. K. Sparrow, and A. S. Sussman, eds.). Vol. 4A, pp. 87–132. Academic Press, New York.

Munk, A. (1956). On *Metasphaeria coccodes* (Karst.) Sacc., and other fungi probably related to *Massarina* Sacc. (Massarinaceae nov. fam.). *Friesia* **5,** 303–308.

Munk, A. (1957). Danish Pyrenomycetes. A preliminary flora. *Dan. Bot. Ark.* **17,** 1–491.

Muraoka, J. S. (1966a). "Deep-ocean Biodeterioration of Materials," Part III, Tech. Rep. R 428, pp. I–IV and 1–47. U.S. Nav. Civ. Eng. Lab., Port Hueneme, Calif.

Muraoka, J. S. (1966b). "Deep-ocean Biodeterioration of Materials," Part V, Tech. Rep. R 495, pp. I–IV and 1–46. U.S. Nav. Civ. Eng. Lab., Port Hueneme, California.

Nagai, M. (1940). Marine algae of the Kurile Islands. I. *J. Fac. Agric., Hokkaido Imp. Univ.* **46,** 1–138.

Nag Raj, T. R., and Morgan-Jones, G. (1972). Nomina conservanda proposita. Proposal for the conservation of the generic name *Robillarda* Saccardo. *Taxon* **21,** 535–536.

Nag Raj, T. R., Morgan-Jones, G., and Kendrick, B. (1972). Genera coelomycetarum. IV. *Pseudorobillarda* gen. nov., a generic segregate of *Robillarda* Sacc. *Can. J. Bot.* **50,** 861–867.

Nag Raj, T. R, Morgan-Jones, G., and Kendrick, B. (1973). *Pseudorobillarda* Nag Raj et al., a later homonym of *Pseudorobillarda* Morelet. *Can. J. Bot.* **51,** 688–689.

Nair, N. B. (1970). The problem of timber destroying organisms along the Indian coasts. *Proc. Int. Congr. Mar. Corros. Fouling, 2nd, 1968* pp. 1–7.

Neergaard, P. (1945). "Danish Species of *Alternaria* and *Stemphylium*. Taxonomy, Parasitism and Economical Significance," Munksgaard, Copenhagen.

Neish, G. (1970). Lignicolous marine fungi from Nova Scotia. *Can. J. Bot.* **48,** 2319–2322.

Newell, S. Y. (1973). Succession and role of fungi in the degradation of red mangrove seedlings. *In* "Estuarine Microbial Ecology" (L. H. Stevenson and R. R. Colwell, eds.), pp. 467–480. Univ. of South Carolina Press, Columbia.

Newell, S. Y. (1976). Mangrove fungi: The succession in the mycoflora of red mangrove (*Rhizophora mangle* L.) seedlings. *In* "Recent Advances in Aquatic Mycology" (E. B. G. Jones, ed.), pp. 51–91. Wiley, New York.

Newell, S. Y., and Fell, J. W. (1970). The perfect form of a marine-occurring yeast of the genus *Rhodotorula. Mycologia* **62**, 272–281.

Newell, S. Y., and Fell, J. W. (1975). Preliminary experimentation in the development of natural food analogues for culture of detritivorous shrimp. *Univ. Miami Sea Grant, Tech. Bull.* **30**, I–X, 1–115.

Newell, S. Y., and Hunter, I. L. (1970). *Rhodosporidium diobovatum* sp. n., the perfect form of an asporogenous yeast (*Rhodotorula* sp.). *J. Bacteriol.* **104**, 503–508.

Nicoli, R. M., and Russo, A. (1974). Le genre *Humicola* Traaen et les genres voisins (Hyphomycetes). *Nova Hedwigia* **25**, 737–798.

Nicot, J. (1958a). Remarques sur la mycoflore des sols sableux immergés à marée haute. *C.R. Hebd. Séances Acad. Sci.* **246**, 451–454.

Nicot, J. (1958b). Une moisissure arénicole du littoral atlantique: *Dendryphiella arenaria* sp. nov. *Rev. Mycol.* **23**, 87–99.

Nicot, J. (1958c). Quelques micromycètes des sables littoraux. *Bull. Soc. Mycol. Fr.* **74**, 221–235.

Nicot, J., and Rouch, J. (1965). Développement et sporogénèse d'une sphaeropsidale isolée d'un sol de vignoble toulousain. *Ann. Sci. Nat., Bot. Biol. Vég.* [12] **6**, 769–780.

Nilsson, S. (1957). A new Danish fungus, *Dinemasporium marinum. Bot. Not.* **110**, 321–324.

Nilsson, S. (1964). Freshwater Hyphomycetes. Taxonomy, morphology and ecology. *Symb. Bot. Ups.* **18**, 1–130.

Nilsson, T. (1973). Studies on wood degradation and cellulolytic activity of microfungi. *Stud. For. Suec.* **104**, 1–40.

Nilsson, T. (1974a). Formation of soft rot cavities in various cellulose fibres by *Humicola alopallonella* Meyers and Moore. *Stud. For. Suec.* **112**, 1–30.

Nilsson, T. (1974b). The degradation of cellulose and the production of cellulase, xylanase, mannanase and amylase by wood-attacking microfungi. *Stud. For. Suec.* **114**, 1–61.

Nilsson, T. (1974c). Microscopic studies on the degradation of cellophane and various cellulosic fibres by wood-attacking microfungi. *Stud. For. Suec.* **117**, 1–32.

Nolan, R. A. (1972). *Asteromyces cruciatus* from North America. *Mycologia* **64**, 430–433.

Norkrans, B. (1966a). On the occurrence of yeasts in an estuary off the Swedish westcoast. *Sven. Bot. Tidskr.* **60**, 463–482.

Norkrans, B. (1966b). Studies on marine occurring yeasts: growth related to pH, NaCl concentration and temperature. *Arch. Mikrobiol.* **54**, 374–392.

Norkrans, B. (1968). Studies on marine occurring yeasts: respiration, fermentation, and salt tolerance. *Arch. Mikrobiol.* **62**, 358–372.

Norkrans, B. (1969a). Hydroxylamine as the sole nitrogen source for growth of some *Candida* sp. *Acta Chem. Scand.* **23**, 1457–1459.

Norkrans, B. (1969b). The sodium and potassium contents of yeasts differing in halotolerance, at various NaCl concentrations in the media. *Antonie van Leeuwenhoek* **35**, Suppl. Yeast Symp., G31-G32.

Norkrans, B., and Kylin, A. (1969). Regulations of the potassium to sodium ratio and of the osmotic potential in relation to salt tolerance in yeasts. *J. Bacteriol.* **100**, 836–845.

Norris, J. N. (1971). Observations on the genus *Blidingia* (Chlorophyta) in California. *J. Phycol.* **7**, 145–149.

Norris, J. N., and Abbott, I. A. (1972). Some new records of marine algae from the R/V Proteus cruise to British Columbia. *Syesis* **5**, 87–94.

Norris, J. R., Ribbons, D. W., and Booth, C., eds. (1971). "Methods in Microbiology," Vol. 4. Academic Press, New York.

Novozhilova, M. I. (1955). The quantitative characteristics, species composition and distribution of yeast-like organisms in the Black Sea, the Sea of Okhotsk and in the Pacific Ocean. *Tr. Inst. Mikrobiol., Akad. Nauk SSSR* **4**, 155–195.

Novozhilova, M. I., and Popova, L. E. (1973). The yeast of the Gulf of Guinea. *Tr. Inst. Okeanol., Akad. Nauk SSSR* **95**, 168–179.

Nylander, W. (1884). Lichenes novi e Freto Behringii. *Flora (Jena)* **67**, 211–223.

Odum, E. P., and De La Cruz, A. A. (1967). Particulate organic detritus in a Georgia salt marsh-estuarine ecosystem. *In* "Estuaries" (G. H. Lauff, ed.), Publ. No. 83, pp. 383–388. Am. Assoc. Adv. Sci., Washington, D.C.

Odum, W. E., and Heald, E. J. (1972). Trophic analyses of an estuarine mangrove community. *Bull. Mar. Sci.* **22**, 671–738.

Ollivier, G. (1926). *Thalassoascus tregoubovi* (nov. gen., nov. sp.) pyrénomycète marin, parasite des Cutlériacées. *C. R. Hebd. Séances Acad. Sci.* **182**, 1348–1349.

Ollivier, G. (1928). Contribution à la connaissance de la flore marine des Alpes-Maritimes. *Bull. Inst. Océanogr.* **522**, 1–8.

Ollivier, G. (1930). Etude de la flore marine de la Côte d'Azur. *Ann. Inst. Océanogr. (Paris)* [N.S.] **7**, 53–173.

Orpurt, P. A., Meyers, S. P., Boral, L. L., and Simms, J. (1964). Thalassiomycetes V. A new species of *Lindra* from turtle grass, *Thalassia testudinum* König. *Bull. Mar. Sci. Gulf Caribb.* **14**, 405–417.

Ostenfeld-Hansen, C. (1897). Contribution à la flore de l'île Jan-Mayen. *Bot. Tidsskr.* **21**, 18–32.

Padgett, D. E., and Lundeen, C. V. (1977). A variable-medium culture chamber for fungi. *Mycologia* **69**, 835–837.

Padhye, A. A., Pawar, V. H., Sukapure, R. S., and Thirumalachar, M. J. (1967). Keratinophilic fungi from marine soils of Bombay, India: I. *Hind. Antibiot. Bull.* **10**, 138–141.

Parguey-Leduc, A. (1966). Recherches sur l'ontogénie et l'anatomie comparée des ascocarpes des Pyrénomycètes ascoloculaires. (Première partie: Les ascocarpes des Pyrénomycètes ascoloculaires bituniqués). *Ann. Sci. Nat., Bot. Biol. Vég.* [12] **7**, 505–690.

Parguey-Leduc, A. (1967). Recherches sur l'ontogénie et l'anatomie comparée des ascocarpes des Pyrénomycètes ascoloculaires. (Seconde partie: Les ascocarpes des Pyrénomycètes ascoloculaires unituniqués). *Ann. Sci. Nat., Bot. Biol. Vég.* [12] **8**, 1–110.

Park, D. (1972a). Methods of detecting fungi in organic detritus in water. *Trans. Br. Mycol. Soc.* **58**, 281–290.

Park, D. (1972b). On the ecology of heterotrophic micro-organisms in fresh-water. *Trans. Br. Mycol. Soc.* **58**, 291–299.

Patel, K. S. (1975). The relationship between yeasts and marine algae. *Proc. Indian Acad. Sci., Sect. B* **82**, 25–28.

Patouillard, N. (1886). "Tabulae Analyticae Fungorum," 2nd Sér. Klincksieck, Paris.

Pawar, V. H., and Thirumalachar, M. J. (1966). Studies on halophilic soil fungi from Bombay. *Nova Hedwigia* **12**, 497–508.

Pawar, V. H., Padhye, A. A., and Thirumalachar, M. J. (1963). Isolation of *Monosporium apiospermum* from marine soils in Bombay. *Hind. Antibiot. Bull.* **6**, 50–53.

Pawar, V. H., Rahalkar, P. W., and Thirumalachar, M. J. (1965). *Cladosporium wernecki* Horta isolated from marine habitat. *Hind. Antibiot. Bull.* **8**, 19–20.

Pawar, V. H., Mathur, P. N., and Thirumalachar, M. J. (1967). Species of *Phoma* isolated from marine soils in India. *Trans. Br. Mycol. Soc.* **50**, 259–265.

Penot, M., and Penot, M. (1977). Quelques aspects originaux des transports à longue distance dans le thalle de *Ascophyllum nodosum* (L.) Le Jolis (Phaeophyceae, Fucales). *Phycologia* **16**, 339–347.

Perkins, R. D., and Halsey, S. D. (1971). Geologic significance of microboring fungi and algae in Carolina shelf sediments. *J. Sediment. Petrol.* **41**, 843–853.

Petch, T. (1938). British Hypocreales. *Trans. Br. Mycol. Soc.* **21**, 243–305.

Peters, J. E., Catalfomo, P., Constantine, G. H., Jr., and Kirk, P. W., Jr. (1975). Free amino acids in higher marine fungi. *J. Pharm. Sci.* **64**, 176–177.

Petersen, H. E. (1935). Preliminary report on the disease of the eelgrass (*Zostera marina* L.). *Rep. Dan. Biol. Stn. Min. Shipping Fisheries* **40**, 3–8.

Petersen, R. H. (1962). Aquatic Hyphomycetes from North America. I. Aleuriosporae (Part 1), and key to the genera. *Mycologia* **54**, 117–151.

Petersen, R. H. (1963). Aquatic Hyphomycetes from North America. II. Aleuriosporae (Part 2) and Blastosporae. *Mycologia* **55**, 18–29.

Petrak, F. (1925). Mykologische Notizen. *Ann. Mycol.* **23**, 1–143.

Petrak, F. (1952). Über die Gattungen Gaeumannomyces v. Arx et Olivier, *Halophiobolus* Linder und *Linocarpon* Syd. *Sydowia* **6**, 383–388.

Petrak, F., and Sydow, H. (1927). Die Gattungen der Pyrenomyzeten, Sphaeropsideen und Melanconieen. I. Teil. Die phaeosporen Sphaeropsideen und die Gattung *Macrophoma. Beih. Rep. Spec. Nov Veg.* **42**, 1–551.

Phaff, H. J., and Fell, J. W. (1970). *Cryptococcus* Kützing emend. Phaff et Spencer. *In* "The Yeasts" (J. Lodder, ed.), pp. 1088–1145. North-Holland Publ., Amsterdam.

Phaff, H. J., Mrak, E. M., and Williams, O. B. (1952). Yeasts isolated from shrimp. *Mycologia* **44**, 431–451.

Phaff, H. J., Miller, M. W., and Mrak, E. M. (1978). "The Life of Yeasts. Their Nature, Activity, Ecology, and Relation to Mankind." 2nd ed. Harvard Univ. Press, Cambridge, Massachusetts.

Picard, F. (1908). Sur une Laboulbéniacée marine (*Laboulbenia marina* n. sp.) parasite d'*Aepus robini* Laboulbène. *C. R. Séances Soc. Biol.* **65**, 484–486.

Picci, G. (1966). Sulla micoflora presente nelle strutture in legno soggette all'azione dell'acqua di mare. *Ric. Sci.* **36**, 153–157.

Picci, G., and Verona, O. (1964). Sulla flora blastomicetica di alcuni animali marini. *Ric. Sci.* **33**, 85–93.

Pirozynski, K. A., and Morgan-Jones, G. (1968). Notes on microfungi. III. *Trans. Br. Mycol. Soc.* **51**, 185–206.

Pirozynski, K. A., and Shoemaker, R. A. (1971). Some Coelomycetes with appendaged conidia. *Can. J. Bot.* **49**, 529–541.

Pisano, M. A., Mihalik, J. A., and Catalano, G. R. (1964). Gelatinase activity by marine fungi. *Appl. Microbiol.* **12**, 470–474.

Pitt, J. I., and Miller, M. W. (1970). The parasexual cycle in yeasts of the genus *Metschnikowia. Mycologia* **62**, 462–473.

Pitts, G. Y., and Cowley, G. T. (1974). Mycoflora of the habitat and midgut of the fiddler crab, *Uca pugilator. Mycologia* **66**, 669–675.

Pollack, F. G. (1967). A simple method for preparing dried reference cultures. *Mycologia* **59**, 541–544.

Poole, N. J., and Price, P. C. (1972). Fungi colonizing wood submerged in the Medway estuary. *Trans. Br. Mycol. Soc.* **59**, 333–335.

Porter, C. L., and Zebrowski, G. (1937). Lime-loving molds from Australian sands. *Mycologia* **29**, 252–257.

Prince, J. S., and Kingsbury, J. M. (1973). The ecology of *Chondrus crispus* at Plymouth, Massachusetts. I. Ontogeny, vegetative anatomy, reproduction, and life cycle. *Am. J. Bot.* **60**, 956–963.

Printz, H. (1964). Die Chaetophoralen der Binnengewässer. *Hydrobiologia* **24**, 1–376.

Pugh, G. J. F. (1960). The fungal flora of tidal mud-flats. *In* "The Ecology of Soil Fungi" (D. Parkinson and J. S. Waid, eds.), pp. 202–208. Liverpool Univ. Press, Liverpool.

Pugh, G. J. F. (1962a). Studies on fungi in coastal soils. I. *Cercospora salina* Sutherland. *Trans. Br. Mycol. Soc.* **45**, 255–260.

Pugh, G. J. F. (1962b). Studies on fungi in coastal soils. II. Fungal ecology in a developing salt marsh. *Trans. Br. Mycol. Soc.* **45**, 560–566.

Pugh, G. J. F. (1966). Cellulose-decomposing fungi isolated from soils near Madras. *J. Indian Bot. Soc.* **45**, 232–241.

Pugh, G. J. F. (1968). A study of fungi in the rhizosphere and on the root surfaces of plants growing in primitive soils. *In* "Methods of Study in Soil Ecology" (J. Phillipson, ed.), pp. 159–164. UNESCO, Paris.

Pugh, G. J. F. (1974). Fungi in intertidal regions. *Veroeff. Inst. Meeresforsch. Bremerhaven, Suppl.* **5**, 403–418.

Pugh, G. J. F., and Hughes, G. C. (1975). Epistolae mycologicae V. Keratinophilic fungi from British Columbia coastal habitats. *Syesis* **8**, 297–300.

Pugh, G. J. F., and Lindsey, B. I. (1975). Studies of *Sporobolomyces* in a maritime habitat. *Trans. Br. Mycol. Soc.* **65**, 201–209.

Pugh, G. J. F., and Mathison, G. E. (1962). Studies on fungi in coastal soils. III. An ecological survey of keratinophilic fungi. *Trans. Br. Mycol. Soc.* **45**, 567–572.

Pugh, G. J. F., and Mulder, J. L. (1971). Mycoflora associated with *Typha latifolia*. *Trans. Br. Mycol. Soc.* **57**, 273–282.

Pugh, G. J. F., and Nicot, J. (1964). Studies on fungi in coastal soils. V. *Dendryphiella salina* (Sutherland) comb. nov. *Trans. Br. Mycol. Soc.* **47**, 263–267.

Pugh, G. J. F., and Williams, G. M. (1968). Fungi associated with *Salsola kali*. *Trans. Br. Mycol. Soc.* **51**, 389–396.

Pugh, G. J. F., Blakeman, J. P., Morgan-Jones, G., and Eggins, H. O. W. (1963). Studies on fungi in coastal soils. IV. Cellulose decomposing species in sand dunes. *Trans. Br. Mycol. Soc.* **46**, 565–571.

Quinta, M. L. (1968). Natural destruction of the mycological flora of sea salt. *Food Technol.* **22**, 1599–1601.

Radford, A. E., Ahles, H. E., and Bell, C. R. (1968). "Manual of the Vascular Flora of the Carolinas." Univ. of North Carolina Press, Chapel Hill.

Raghukumar, S. (1973). Marine lignicolous fungi from India. *Kavaka* **1**, 73–85.

Rai, J. N., and Chowdhery, H. J. (1975). *Hemisartorya*, a new genus of cleistothecial Ascomycetes with *Aspergillus* state. *Kavaka* **3**, 73–76.

Rai, J. N., and Chowdhery, H. J. (1976). Cellulolytic activity and salinity relationship of some mangrove swamp fungi. *Nova Hedwigia* **27**, 609–617.

Rai, J. N., and Tewari, J. P. (1963). On some isolates of the genus *Preussia* Fuckel from Indian soils. *Proc. Indian Acad. Sci., Sect. B* **57**, 45–55.

Rai, J. N., Tewari, J. P., and Mukerji, K. G. (1969). Mycoflora of mangrove mud. *Mycopathol. Mycol. Appl.* **38**, 17–31.

Ranganathan, B., and Bhat, J. V. (1955). Growth response of a marine yeast, *Cryptococcus laurentii,* to thiamine. *J. Indian Inst. Sci.* **37,** 154–159.

Ranzoni, F. V. (1968). Fungi isolated in culture from soils of the Sonoran Desert. *Mycologia* **60,** 356–371.

Raper, J. R. (1968). On the evolution of fungi. *In* "The Fungi" (G. C. Ainsworth and A. S. Sussman, eds.), Vol. 3, pp. 677–693. Academic Press, New York.

Ray, D. L. (1959a). Nutritional physiology of *Limnoria. In* "Marine Boring and Fouling Organisms" (D. L. Ray, ed.), pp. 46–61. Univ. of Washington Press, Seattle.

Ray, D. L. (1959b). Some properties of cellulase from *Limnoria. In* "Marine Boring and Fouling Organisms" (D. L. Ray, ed.), pp. 372–396. Univ. of Washington Press, Seattle.

Ray, D. L. (1959c). Marine fungi and wood borer attack. *Proc. Am. Wood-Preserv. Assoc.* **55,** 147–153.

Ray, D. L., and Julian, J. R. (1952). Occurrence of cellulase in *Limnoria. Nature (London)* **169,** 32–33.

Ray, D. L., and Stuntz, D. E. (1959). Possible relation between marine fungi and *Limnoria* attack on submerged wood. *Science* **129,** 93–94.

Reed, M. (1902). Two new ascomycetous fungi parasitic on marine algae. *Univ. Calif., Berkeley, Publ. Bot.* **1,** 141–164.

Rehm, A., and Humm, H. J. (1973). *Sphaeroma terebrans:* A threat to the mangroves of southwestern Florida. *Science* **182,** 173–174.

Reichenbach-Klinke, H. H. (1956). Über einige bisher unbekannte Hyphomyceten bei verschiedenen Süsswasser-und Meeresfischen. *Mycopathol. Mycol. Appl.* **7,** 333–347.

Reimold, R. J., and Queen, W. H., eds. (1974). "Ecology of Halophytes." Academic Press, New York.

Reinsch, P. F. (1890). Zur Meeresalgenflora von Süd-Georgien. *In* "Internationale Polarforschung 1882–1883: Die deutschen Expeditionen und ihre Ergebnisse," Vol. 2, pp. 366–449. Asher, Berlin.

Reisinger, O. (1968). Remarques sur les genres *Dendryphiella* et *Dendryphion. Bull. Soc. Mycol. Fr.* **84,** 27–51.

Reisinger, O., and Guedenet, J. C. (1968). Morphologie ultrastructurale et critères taxinomiques chez les Deuteromycètes. I. Les parois sporales chez *Dendryphiella oinosa* (Berk. et Curt.) Reisinger. *Bull. Soc. Mycol. Fr.* **84,** 19–26.

Renn, C. E. (1936). The wasting disease of *Zostera marina.* I. A phytological investigation of the diseased plant. *Biol. Bull. (Woods Hole, Mass.)* **70,** 148–158.

Rheinheimer, G. (1971). "Mikrobiologie der Gewässer." Fischer, Jena.

Rheinheimer, G. (1977). Bakteriologisch-ökologische Untersuchungen in Sandstränden an Nord- und Ostsee. *Bot. Mar.* **20,** 385–400.

Riedl, H. (1961). Die Arten der Gattung *Mycoporellum* Müll. Arg. *sensu* A. Zahlbruckner. Catal., nebst Bemerkungen zum System dothidealer Flechten. *Sydowia* **15,** 257–287.

Riedl, H. (1962). Vorstudien zu einer Revision der Gattung *Arthopyrenia* Mass. *sensu amplo.* I. *Sydowia* **16,** 263–274.

Riedl, H. (1968). Beobachtungen an *Pleospora hookeri* (Borr.) Keissler und einigen weiteren *Pleospora*-Arten. *Sydowia* **22,** 395–402.

Ristanović, B., and Miller, C. E. (1969). Salinity tolerances and ecological aspects of some fungi collected from fresh-water, estuarine and marine habitats. *Mycopathol. Mycol. Appl.* **37,** 273–280.

Ritchie, D. (1954). A fungus flora of the sea. *Science* **120,** 578–579.

Ritchie, D. (1957). Salinity optima for marine fungi affected by temperature. *Am. J. Bot.* **44,** 870–874.

Ritchie, D. (1961). Interrelation of temperature, nutrition, and osmotic pressure as factors affecting growth of *Zalerion eistla*. *Bacteriol. Proc.* **61**, 39.

Ritchie, D. (1968). Invasion of some tropical timbers by fungi in brackish waters. *J. Elisha Mitchell Sci. Soc.* **84**, 221–226.

Ritchie, D., and Jacobsohn, M. K. (1963). The effects of osmotic and nutritional variation on growth of a salt-tolerant fungus, *Zalerion eistla*. *In* "Symposium on Marine Microbiology" (C. H. Oppenheimer, ed.), pp. 286–299. Thomas, Springfield, Illinois.

Rodrigues, J., Sguros, P. L., and White, J. L. (1970). Dehydrogenase patterns in filamentous marine fungi. *Proc. Fla. Acad. Sci.* **34**, 1.

Rogerson, C. T. (1970). The hypocrealean fungi (Ascomycetes, Hypocreales). *Mycologia* **62**, 865–910.

Rosenberg, F. A., and Breiter, H. (1969). The role of cellulolytic bacteria in the digestive processes of the shipworm. 1. Isolation of some cellulolytic micro-organisms from the digestive system of teredine borers and associated waters. *Mater. Org.* **4**, 147–159.

Rosenberg, F. A., and Cutter, J. (1973). The role of cellulolytic bacteria in the digestive processes of the shipworm. *Proc. Int. Congr. Mar. Corros. Fouling, 3rd, 1972* pp. 778–796.

Rosinski, M. A., and Campana, R. J. (1964). Chemical analysis of the cell wall of *Ceratocystis ulmi. Mycologia* **56**, 738–744.

Ross, S. S., and Morris, E. O. (1962). Effect of sodium chloride on the growth of certain yeasts of marine origin. *J. Sci. Food Agric.* **13**, 467–475.

Ross, S. S., and Morris, E. O. (1965). An investigation of the yeast flora of marine fish from Scottish coastal waters and a fishing ground off Iceland. *J. Appl. Bacteriol.* **28**, 224–234.

Rostrup, E. (1884). Underjordiske Svampe i Danmark. *Medd. Bot. Foren. Kjoebenhavn* **1**, 102–106.

Rostrup, E. (1888). Mykologiske Meddelelser. *Medd. Bot. Foren. Kjoebenhavn* **2**, 84–93.

Roth, F. J., Jr., Ahearn, D. G., Fell, J. W., Meyers, S. P., and Meyer, S. A. (1962). Ecology and taxonomy of yeasts isolated from various marine substrates. *Limnol. Oceanogr.* **7**, 178–185.

Roth, F. J., Jr., Orpurt, P. A., and Ahearn, D. G. (1964). Occurrence and distribution of fungi in a subtropical marine environment. *Can. J. Bot.* **42**, 375–383.

Sachs, J. (1874). "Lehrbuch der Botanik," 4th ed. Engelmann, Leipzig.

Saitô, T. (1952). The soil fungi of a salt marsh and its neighbourhood. *Ecol. Rev. (Sendai)* **13**, 111–119.

Santesson, R. (1939). Amphibious pyrenolichens I. *Ark. Bot.* **29A**, 1–68.

Savile, D. B. O. (1968). Possible interrelationships between fungal groups. *In* "The Fungi" (G. C. Ainsworth and A. S. Sussman, eds.), Vol 3, pp. 649–675. Academic Press, New York.

Savile, D. B. O. (1973). Review of: C. T. Ingold, "Fungal spores: Their liberation and dispersal," *Mycologia* **75**, 259–263.

Savory, J. G. (1954a). Breakdown of timber by Ascomycetes and Fungi Imperfecti. *Ann. Appl. Biol.* **41**, 336–347.

Savory, J. G. (1954b). Damage to wood caused by micro-organisms. *J. Appl. Bacteriol.* **17**, 213–218.

Săvulescu, T. (1947). "Herbarium Mycologicum Romanicum." Fasc. 27, No. 1318. Bucharest.

Schafer, R. D. (1966). Survival ability of *Limnoria* on a protein-deficient diet. *Wasmann J. Biol.* **24**, 109–115.

Schafer, R. D., and Lane, C. E. (1957). Some preliminary observations bearing on the nutrition of *Limnoria. Bull. Mar. Sci. Gulf Caribb.* **7,** 289–296.

Schaumann, K. (1968). Marine höhere Pilze (Ascomycetes und Fungi imperfecti) aus dem Weser-Ästuar. *Veroeff. Inst. Meeresforsch. Bremerhaven* **11,** 93–117.

Schaumann, K. (1969). Über marine höhere Pilze von Holzsubstraten der Nordsee-Insel Helgoland. *Ber. Dtsch. Bot. Ges.* **82,** 307–327.

Schaumann, K. (1972). *Corollospora intermedia* (Ascomycetes, Halosphaeriaceae) vom Sandstrand der Insel Helgoland (Deutsche Bucht). *Veroeff. Inst. Meeresforsch. Bremerhaven* **14,** 13–22.

Schaumann, K. (1973). *Brachysporium helgolandicum* nov. sp., ein neuer Deuteromycet auf Treibborke im Meer. *Helgol. Wiss. Meeresunters.* **25,** 26–34.

Schaumann, K. (1974a). Zur Verbreitung saprophytischer höherer Pilzkeime in der Hochsee. Erste quantitative Ergebnisse aus der Nordsee und dem NO-Atlantik. *Veroeff. Inst. Meeresforsch. Bremerhaven, Suppl.* **5,** 287–300.

Schaumann, K. (1974b). Experimentelle Untersuchungen zum Einfluss des Salzgehaltes und der Temperatur auf das Mycelwachstum höherer Pilze aus dem Meer- und Brackwasser. *Veroeff. Inst. Meeresforsch. Bremerhaven, Suppl.* **5,** 443–474.

Schaumann, K. (1974c). Experimentelle Untersuchungen zur Produktion und Aktivität cellulolytischer Enzyme bei höheren Pilzen aus dem Meer- und Brackwasser. *Mar. Biol.* **28,** 221–235.

Schaumann, K. (1975a). Vergleich der Pilzbesiedlung treibender und ortsfester Holzsubstrate im Meer bei Helgoland (Deutsche Bucht). *Veroeff. Inst. Meeresforsch. Bremerhaven* **15,** 13–26.

Schaumann, K. (1975b). Ökologische Untersuchungen über höhere Pilze im Meer- und Brackwasser der Deutschen Bucht unter besonderer Berücksichtigung der holzbesiedelnden Arten. *Veroeff. Inst. Meeresforsch. Bremerhaven* **15,** 79–182.

Schaumann, K. (1975c). Marine Pilzfunde von der Norwegischen Rinne, der Barents-See und von den Küsten Westafrikas und der Kanarischen Inseln. *Veroeff. Inst. Meeresforsch. Bremerhaven* **15,** 183–194.

Scheinpflug, H. (1958). Untersuchungen über die Gattung *Didymosphaeria* Fuckel und einige verwandte Gattungen. *Ber. Schweiz. Bot. Ges.* **68,** 325–386.

Schlichting, H. E. (1971). A preliminary study of the algae and protozoa in seafoam. *Bot. Mar.* **14,** 24–28.

Schmidt, I. (1967). Über das Vorkommen mariner Ascomyceten und Fungi imperfecti in der Ostsee und einigen angrenzenden Boddengewässern. *Nat. Naturschutz Mecklenburg* **5,** 115–126.

Schmidt, I. (1969a). *Carbosphaerella pleosporoides* gen. nov. et spec. nov. und *Cirrenalia fusca* spec. nov., zwei neue marine Pilzarten von der Ostseeküste. *Feddes Repert.* **80,** 107–112.

Schmidt, I. (1969b). *Corollospora intermedia,* nov. spec., *Carbosphaerella leptosphaeriodes,* nov. spec. und *Crinigera maritima,* nov. gen., nov. spec., 3 neue marine Pilzarten von der Ostseeküste. *Nat. Naturschutz Mecklenburg* **7,** 5–14.

Schmidt, I. (1974). Höhere Meerespilze der Ostsee. *Biol. Rundsch.* **12,** 96–112.

Schneider, J. (1976). Lignicole marine Pilze (Ascomyceten und Deuteromyceten) aus zwei Ostseeförden. *Bot. Mar.* **19,** 295–307.

Schneider, J. (1977). Fungi. *Ecol. Stud.* **25,** 90–102.

Schol-Schwarz, M. B. (1959). The genus *Epicoccum* Link. *Trans. Br. Mycol. Soc.* **42,** 149–173.

Schultz, D. M., and Quinn, J. G. (1973). Fatty acid composition of organic detritus from *Spartina alterniflora. Estuarine Coastal Mar. Sci.* **1,** 177–190.

Schuster, R. (1966). Hornmilben(Oribatei) als Bewohner des marinen Litorals. *Veroeff. Inst. Meeresforsch. Bremerhaven, Suppl.* **2**, 319–327.

Seifert, K. (1964). Über den Abbau der Holzsubstanz durch die Bohrassel *Limnoria. Holz Roh- Werkst.* **22**, 209–215.

Seki, H., and Fulton, J. (1969). Infection of marine copepods by *Metschnikowia* sp. *Mycopathol. Mycol. Appl.* **38**, 61–70.

Seshadri, R., and Sieburth, J. M. (1971). Cultural estimation of yeasts on seaweeds. *Appl. Microbiol.* **22**, 507–512.

Seshadri, R., and Sieburth, J. M. (1975). Seaweeds as a reservoir of *Candida* yeasts in inshore waters. *Mar. Biol.* **30**, 105–117.

Seshadri, R., Krishnamurthy, K., and Ramamurthy, V. D. (1966). Bacteria and yeasts in marine and estuarine waters of Portonovo (S. India). *Bull. Dep. Mar. Biol. Oceanogr., Univ. Kerala* **2**, 5–11.

Setchell, W. A. (1926). Phytogeographical notes on Tahiti. II. Marine vegetation. *Univ. Calif., Berkeley, Publ. Bot.* **12**, 291–324.

Sguros, P. L., and Simms, J. (1963a). Role of marine fungi in the biochemistry of the oceans. II. Effect of glucose, inorganic nitrogen, and tris (hydroxymethyl)aminomethane on growth and pH changes in synthetic media. *Mycologia* **55**, 728–741.

Sguros, P. L., and Simms, J. (1963b). Role of marine fungi in the biochemistry of the oceans. III. Growth factor requirements of the Ascomycete *Halosphaeria mediosetigera. Can. J. Microbiol.* **9**, 585–591.

Sguros, P. L., and Simms, J. (1964). Role of marine fungi in the biochemistry of the oceans. IV. Growth responses to seawater inorganic macroconstituents. *J. Bacteriol.* **88**, 346–355.

Sguros, P. L., Meyers, S. P., and Simms, J. (1962). Role of marine fungi in the biochemistry of the oceans. I. Establishment of quantitative technique for cultivation, growth measurement and production of inocula. *Mycologia* **54**, 521–535.

Sguros, P. L., Rodrigues, J., and White, J. L. (1970). Enzymatic indicators of carbohydrate metabolism in filamentous marine fungi. *Bacteriol. Proc.* **70**, 147.

Sguros, P. L., Simms, J., and Rodrigues, J. (1971). Utilization of sole carbon and nitrogen sources by certain marine fungi. *Bacteriol. Proc.* **71**, 36.

Sguros, P. L., Rodrigues, J., and Simms, J. (1973). Role of marine fungi in the biochemistry of the oceans. V. Patterns of constitutive nutritional growth responses. *Mycologia* **65**, 161–174.

Shearer, C. A. (1972a). Tentative outline for inventory of division mycota: *Corollospora pulchella* (fungus). *Chesapeake Sci.* **13**, Suppl., S171–S172.

Shearer, C. A. (1972b). Fungi of the Chesapeake Bay and its tributaries. III. The distribution of wood-inhabiting Ascomycetes and Fungi Imperfecti of the Patuxent River. *Am. J. Bot.* **59**, 961–969.

Shearer, C. A., and Crane, J. L. (1971). Fungi of the Chesapeake Bay and its tributaries. I. Patuxent River. *Mycologia* **63**, 237–260.

Shearer, C. A., and Crane, J. L. (1977). Fungi of the Chesapeake Bay and its tributaries. VI. *Trichocladium achrasporum,* the imperfect state of *Halosphaeria mediosetigera,* a marine Ascomycete. *Mycologia* **69**, 1218–1223.

Shearer, C. A., and Miller, M. (1977). Fungi of the Chesapeake Bay and its tributaries. V. *Aniptodera chesapeakensis* gen. et. sp. nov. *Mycologia* **69**, 887–898.

Shinano, H. (1962). Studies on yeasts isolated from various areas of the North Pacific. *Bull. Jpn. Soc. Sci. Fish.* **28**, 1113–1122.

Shoemaker, R. A. (1962). *Drechslera* Ito. *Can. J. Bot.* **40**, 809–836.

Shoemaker, R. A. (1976). Canadian and some extralimital *Ophiobolus* species. *Can. J. Bot.* **54**, 2365–2404.

Sieburth, J. M. (1968). The influence of algal antibiosis on the ecology of marine microorganisms. *Adv. Microbiol. Sea* **1**, 63–94.

Sieburth, J. M., and Conover, J. T. (1965). *Sargassum* tannin, an antibiotic which retards fouling. *Nature (London)* **208**, 52–53.

Sieburth, J. M., Brooks, R. D., Gessner, R. V., Thomas, C. D., and Tootle, J. L. (1974). Microbial colonization of marine plant surfaces as observed by scanning electron microscopy. *In* "Effect of the Ocean Environment on Microbial Activies" (R. R. Colwell and R. Y. Morita, eds.), pp. 418–432. Univ. Park Press, Baltimore, Maryland.

Siepmann, R. (1959a). Ein Beitrag zur saprophytischen Pilzflora des Wattes der Wesermündung. I. Systematischer Teil. *Veroeff. Inst. Meeresforsch. Bremerhaven* **6**, 213–282.

Siepmann, R. (1959b). Ein Beitrag zur saprophytischen Pilzflora des Wattes der Wesermündung. Zweiter Teil. *Veroeff. Inst. Meeresforsch. Bremerhaven* **6**, 283–301.

Siepmann, R., and Höhnk, W. (1962). Über Hefen und einige Pilze (Fungi imp., Hyphales) aus dem Nordatlantik. *Veroeff. Inst. Meeresforsch. Bremerhaven* **8**, 79–97.

Siepmann, R., and Johnson, T. W., Jr. (1960). Isolation and culture of fungi from wood submerged in saline and fresh waters. *J. Elisha Mitchell Sci. Soc.* **76**, 150–154.

Simmons, E. G. (1967). Typification of *Alternaria, Stemphylium,* and *Ulocladium*. *Mycologia* **59**, 67–92.

Simmons, E. G. (1969). Perfect states of *Stemphylium*. *Mycologia* **61**, 1–26.

Sivanesan, A., and Manners, J. G. (1970). Fungi associated with *Spartina townsendii* in healthy and "die-back" sites. *Trans. Br. Mycol. Soc.* **55**, 191–204.

Skottsberg, C. (1923). Botanische Ergebnisse der schwedischen Expedition nach Patagonien und dem Feuerlande 1907–1909. - IX. Marine algae 2. Rhodophyceae. *K. Sven. Vetenskapsakad. Handl.* **63**, 1–70.

Sleeter, T. D., Boyle, P. J., Cundell, A. M., and Mitchell, R. (1978). Relationships between marine microorganisms and the wood-boring isopod, *Limnoria tripunctata*. *Mar. Biol.* **45**, 329–336.

Smith, A. L. (1908). New or rare microfungi. *Trans. Br. Mycol. Soc.* **3**, 111–123.

Smith, A. L., and Ramsbottom, J. (1915). Is *Pelvetia canaliculata* a lichen? *New Phytol.* **14**, 295–298.

Snell, W. H., and Dick, E. A. (1971). "A Glossary of Mycology." Harvard Univ. Press, Cambridge, Massachusetts.

Sorokin, J. I. (1964). A quantitative study of the microflora in the central Pacific Ocean. *J. Cons., Cons. Int. Explor. Mer* **29**, 25–40.

Southwell, C. R., Bultman, J. D., Forgeson, B. W., and Hummer, C. W. (1970). Biological deterioration of wood in tropical environments. Part 2—Marine borer resistance of natural woods over long periods of immersion. *Nav. Res. Lab. Rep.* **7123**, 1–44.

Sparrow, F. K. (1937). The occurrence of saprophytic fungi in marine muds. *Biol. Bull. (Woods Hole, Mass.)* **73**, 242–248.

Sparrow, F. K. (1973). The peculiar marine Phycomycete *Atkinsiella dubia* from crab eggs. *Arch. Mikrobiol.* **93**, 137–144.

Spegazzini, C. (1880). Fungi Argentini. *An. Soc. Cient. Argent.* **10**, 145–168.

Spencer, J. F. T., Gorin, P. A. J., and Gardner, N. R. (1974). Yeasts isolated from some lakes and rivers of Saskatchewan. *Can. J. Microbiol.* **20**, 949–954.

Stafleu, F. A., Bonner, C. E. B., McVaugh, R., Meikle, R. D., Rollins, R. C., Ross, R., Schopf, J. M., Schulze, G. M., de Vilmorin, R., and Voss, E. G., eds. (1972). International Code of Botanical Nomenclature. *Regnum Veg.* **82**, 1–426.

Stearn, W. T. (1966). "Botanical Latin." Hafner, New York.

Steele, C. W. (1967). Fungus populations in marine waters and coastal sands of the Hawaiian, Line, and Phoenix islands. *Pac. Sci.* **21**, 317–331.

Stevens, R. B., ed. (1974). "Mycology Guidebook." Univ. of Washington Press, Seattle.

Stevenson, J. R. (1961). Absence of chitinase in *Limnoria*. *Nature (London)* **190**, 463.

Stolk, A. C. (1955). *Emericellopsis minima* sp. nov. and *Westerdykella ornata* gen. nov., sp. nov. *Trans. Br. Mycol. Soc.* **38**, 419–424.

Størmer, L. (1977). Arthropod invasion of land during late Silurian and Devonian times. *Science* **197**, 1362–1364.

Subramanian, C. V. (1971). "Hyphomycetes. An Account of Indian Species, Except Cercosporae." Indian Counc. Agric. Res., New Delhi.

Subramanian, C. V., and Jain, B. L. (1966). A revision of some graminicolous helminthosporia. *Curr. Sci.* **35**, 352–355.

Subramanian, C. V., and Raghukumar, S. (1974). Ecology of higher fungi in soils of marine and brackish environments in and around Madras. *Veroeff. Inst. Meeresforsch. Bremerhaven, Suppl.* **5**, 377–402.

Suckow, R., and Schwartz, W. (1960). Die Mikrobenassoziationen der Seegrasbänke. *Z. Allg. Mikrobiol.* **1**, 71–77.

Suehiro, S. (1960). Studies on the yeasts developing in putrefied marine algae. *Sci. Bull. Fac. Agric., Kyushu Univ.* **17**, 443–449.

Suehiro, S. (1962). Studies on the marine yeasts II. Yeasts isolated from *Thalassiosira subtilis* (marine diatom) decayed in flasks. *Sci. Bull. Fac. Agric., Kyushu Univ.* **20**, 101–105.

Suehiro, S. (1963). Studies on the marine yeasts III. Yeasts isolated from the mud of tideland. *Sci. Bull. Fac. Agric., Kyushu Univ.* **20**, 223–227.

Suehiro, S., and Tomiyasu, Y. (1962). Studies on the marine yeasts V. Yeasts isolated from seaweeds. *J. Fac. Agric., Kyushu Univ.* **12**, 163–169.

Suehiro, S., Tomiyasu, Y., and Tanaka, O. (1962). Studies on the marine yeasts IV. Yeasts isolated from marine plankton. *J. Fac. Agric., Kyushu Univ.* **12**, 155–161.

Sugiyama, J., Sugiyama, Y., Iizuka, H., and Torii, T. (1967). "Report of the Japanese Summer Parties in Dry Valleys, Victoria Land, 1963–1965," Vol. IV, Part 2, Antarct. Rec. No. 28, pp. 2247–2256. Natl. Sci. Mus., Tokyo.

Sutherland, G. K. (1915a). New marine fungi on *Pelvetia*. *New. Phytol.* **14**, 33–42.

Sutherland, G. K. (1915b). Additional notes on marine Pyrenomycetes. *New Phytol.* **14**, 183–193.

Sutherland, G. K. (1915c). New marine Pyrenomycetes. *Trans. Br. Mycol. Soc.* **5**, 147–155.

Sutherland, G. K. (1916a). Marine Fungi Imperfecti. *New Phytol.* **15**, 35–48.

Sutherland, G. K. (1916b). Additional notes on marine Pyrenomycetes. *Trans. Br. Mycol. Soc.* **5**, 257–263.

Sutton, B. C. (1965). Typification of *Dendrophoma* and a reassessment of *D. obscurans*. *Trans. Br. Mycol. Soc.* **48**, 611–616.

Sutton, B. C. (1973a). Hyphomycetes from Manitoba and Saskatchewan, Canada. *Mycol. Pap.* **132**, 1–143.

Sutton, B. C. (1973b). Coelomycetes. *In* "The Fungi" (G. C. Ainsworth, F. K. Sparrow, and A. S. Sussman, eds.), Vol. 4A, pp. 513–582. Academic Press, New York.

Sutton, B. C. (1976). *Angiopoma* Lév., 1841, an earlier name for *Drechslera* Ito, 1930. *Mycotaxon* **3**, 337–380.

Sutton, B. C. (1977). Coelomycetes VI. Nomenclature of generic names proposed for Coelomycetes. *Mycol. Pap.* **141**, 1–253.

Sutton, B. C., and Pirozynski, K. A. (1963). Notes on British Microfungi. I. *Trans. Br. Mycol. Soc.* **46**, 505–522.

Sutton, B. C., and Pirozynski, K. A. (1965). Notes on Microfungi. II. *Trans. Br. Mycol. Soc.* **48,** 349–366.

Swart, H. J. (1958). An investigation of the mycoflora in the soil of some mangrove swamps. *Acta Bot. Neerl.* **7,** 741–768.

Swart, H. J. (1963). Further investigation of the mycoflora in the soil of some mangrove swamps. *Acta Bot. Neerl.* **12,** 98–111.

Swart, H. J. (1970). *Penicillium dimorphosporum* sp. nov. *Trans. Br. Mycol. Soc.* **55,** 310–313.

Szaniszlo, P. J., and Mitchell, R. (1971). Hyphal wall compositions of marine and terrestrial fungi of the genus *Leptosphaeria*. *J. Bacteriol.* **106,** 640–645.

Szaniszlo, P. J., Wirsen, C., Jr., and Mitchell, R. (1968). Production of a capsular polysaccharide by a marine filamentous fungus. *J. Bacteriol.* **96,** 1474–1483.

Talbot, P. H. B. (1973). Aphyllophorales I: General characteristics; telephoroid and cupuloid families. *In* "The Fungi" (G. C. Ainsworth, F. K. Sparrow, and A. S. Sussman, eds.), Vol. 4B, pp. 327–349. Academic Press, New York.

Taysi, I., and van Uden, N. (1964). Occurrence and population densities of yeast species in an estuarine-marine area. *Limnol. Oceanogr.* **9,** 42–45.

Thaxter, R. (1896). Contribution towards a monograph of the Laboulbeniaceae. Part I. *Mem. Am. Acad. Arts Sci.* **12,** 187–429.

Tresner, H. D., and Hayes, J. A. (1971). Sodium chloride tolerance of terrestrial fungi. *Appl. Microbiol.* **22,** 210–213.

Trussel, P. C., and Jones, E. B. G. (1970). Protection of wood in the marine environment. *Int. Biodeterior. Bull.* **6,** 3–7.

Tubaki, K. (1961). On some fungi isolated from the Antarctic materials. *Publ. Seto Mar. Biol. Lab.* **14,** 3–9.

Tubaki, K. (1966). Marine fungi from Japan. Lignicolous I. *Trans. Mycol. Soc. Jpn.* **7,** 73–87.

Tubaki, K. (1967). An undescribed species of *Heleococcum* from Japan. *Trans. Mycol. Soc. Jpn.* **8,** 5–10.

Tubaki, K. (1968). Studies on the Japanese marine fungi. Lignicolous group II. *Publ. Seto Mar. Biol. Lab.* **15,** 357–372.

Tubaki, K. (1969). Studies on the Japanese marine fungi, lignicolous group (III), algicolous group and a general consideration. *Annu. Rep. Inst. Ferment., Osaka* **4,** 12–41.

Tubaki, K., and Asano, I. (1965). Additional species of fungi isolated from the Antarctic materials. *JARE 1956–1962 Sci. Rep., Ser. E.* **27,** 1–12.

Tubaki, K., and Ito, T. (1973). Fungi inhabiting in brackish water. *Rep. Tottori Mycol. Inst. (Jpn.)* **10,** 523–539.

Turner, R. D. (1971). Identification of marine wood-boring mollusks. *In* "Marine Borers, Fungi and Fouling Organisms of Wood" (E. B. G. Jones and S. K. Eltringham, eds.), pp. 17–64. Org. Econ. Coop. Dev., Paris.

Turner, R. D. (1973). Wood-boring bivalves, opportunistic species in the deep sea. *Science* **180,** 1377–1379.

Tutin, T. G. (1934). The fungus on *Zostera marina. Nature (London)* **134,** 573.

Tyndall, R. W., and Kirk, P. W., Jr. (1973). Factors in seawater affecting spore germination in marine lignicolous fungi. *Va. J. Sci.* **24,** 136.

Udagawa, S., and Furuya, K. (1972). *Zopfiella pilifera,* a new cleistoascomycete from Japanese soil. *Trans. Mycol. Soc. Jpn.* **13,** 255–259.

Udagawa, S., and Furuya, K. (1974). Notes on some Japanese Ascomycetes. XIII. *Trans. Mycol. Soc. Jpn.* **15,** 206–214.

Udagawa, S., and Horie, Y. (1974). Notes on some Japanese Ascomycetes. XII. *Trans. Mycol. Soc. Jpn.* **15,** 105–112.

Udagawa, S., and Takada, M. (1974). Notes on some Japanese Ascomycetes. XI. *Trans. Mycol. Soc. Jpn.* **15**, 23–29.

Ulken, A. (1970). Phycomyceten aus der Mangrove bei Cananéia (São Paulo, Brasilien). *Veroeff. Inst. Meeresforsch. Bremerhaven* **12**, 313–319.

Ulken, A. (1972). Physiological studies on a Phycomycete from a mangrove swamp at Cananéia, São Paulo, Brazil. *Veroeff. Inst. Meeresforsch. Bremerhaven* **13**, 217–230.

Ulken, A. (1974). Chytridineen im Küstenbereich. *Veroeff. Inst. Meeresforsch. Bremerhaven, Suppl.* **5**, 83–104.

Ulken, A. (1975). Phycomyceten und *Actinoplanes* sp. aus Mangrovesümpfen von Hawaii. *Veroeff. Inst. Meeresforsch. Bremerhaven* **15**, 27–36.

van den Hoek, C. (1975). Phytogeographic provinces along the coasts of the northern Atlantic Ocean. *Phycologia* **14**, 317–330.

van der Walt, J. P. (1970). *Kluyveromyces* van der Walt emend. van der Walt. In "The Yeasts" (J. Lodder, ed.), pp. 316–378. North-Holland Publ., Amsterdam.

van Uden, N. (1967). Occurrence and origin of yeasts in estuaries. In "Estuaries" (G. H. Lauff, ed.), Publ. No. 83, pp. 306–310. Am. Assoc. Adv. Sci., Washington, D.C.

van Uden, N., and Buckley, H. (1970). *Candida* Berkhout. In "The Yeasts" (J. Lodder, ed.), pp. 893–1087. North-Holland Publ., Amsterdam.

van Uden, N., and Castelo-Branco, R. (1961). *Metschnikowiella zobellii* sp. nov. and *M. krissii* sp. nov., two yeasts from the Pacific Ocean pathogenic for *Daphnia magna*. *J. Gen. Microbiol.* **26**, 141–148.

van Uden, N., and Castelo-Branco, R. (1963). Distribution and population densities of yeast species in Pacific water, air, animals, and kelp off southern California. *Limnol. Oceanogr.* **8**, 323–329.

van Uden, N., and Fell, J. W. (1968). Marine yeasts. *Adv. Microbiol. Sea* **1**, 167–201.

van Uden, N., and Vidal-Leiria, M. (1970). *Torulopsis* Berlese. In "The Yeasts" (J. Lodder, ed.), pp. 1235–1308. North-Holland Publ., Amsterdam.

van Uden, N., and ZoBell, C. E. (1962). *Candida marina* nov. spec., *Torulopsis torresii* nov. spec. and *T. maris* nov. spec., three yeasts from the Torres Strait. *Antonie van Leeuwenhoek* **28**, 275–283.

Vargo, G. A., Hargraves, P. E., and Johnson, P. (1975). Scanning electron microscopy of dialysis tubes incubated in flowing seawater. *Mar. Biol.* **31**, 113–120.

Vembu, D., and Sguros, P. L. (1972). Citric acid cycle and glyoxylate bypass in glucose-grown, filamentous marine fungi. *Bacteriol. Proc.* **72**, 166.

Verona, O., and Benedek, T. (1965). Iconographia Mycologia. *Mycopathol. Mycol. Appl., Suppl.* **12**, A276.

Vestergren, T. (1900). Verzeichnis nebst Diagnosen und kritischen Bemerkungen zu meinem Exsiccatenwerke "Micromycetes rarioies selecti," Fasc. VII–X. *Bot. Not.*, pp. 27–44.

Vitellaro-Zuccarello, L. (1973). Ultrastructure of the byssal apparatus of *Mytilus galloprovincialis*. I. Associated fungal hyphae. *Mar. Biol.* **22**, 225–230.

Volz, P. A., Jerger, D. E., Wurzburger, A. J., and Hiser, J. L. (1974). A preliminary survey of yeasts isolated from marine habitats at Abaco Island, The Bahamas. *Mycopathol. Mycol. Appl.* **54**, 313–316.

von Arx, J. A. (1949). Beiträge zur Kenntnis der Gattung *Mycosphaerella*. *Sydowia* **3**, 28–100.

von Arx, J. A. (1970). "The Genera of Fungi Sporulating in Pure Culture." Cramer, Lehre.

von Arx, J. A. (1973). Ostiolate and nonostiolate Pyrenomycetes. *Proc. K. Ned. Acad. Wet., Ser. C* **76**, 289–296.

von Arx, J. A., and Müller, E. (1954). Die Gattungen der amerosporen Pyrenomyceten. *Beitr. Kryptogamenflora Schweiz* **11**, 1–434.

von Arx, J. A., and Müller, E. (1975). A re-evaluation of the bitunicate Ascomycetes with keys to families and genera. *Stud. Mycol.* **9**, 1–159.

von Arx, J. A., Rodrigues de Miranda, L., Smith, M. T., and Yarrow, D. (1977). The genera of yeasts and the yeast-like fungi. *Stud. Mycol.* **14**, 1–42.

von Höhnel, F. (1911). Fragmente zur Mykologie. XIII. *Sitzungsber. Kais. Akad. Wiss. Wien, Math.-Naturwiss. Kl.* **70**, 1–106.

von Keissler, K. (1937). Pyrenulaceae. *In* "L. Rabenhorst's Kryptogamenflora" (A. Zahlbruckner, ed.), Vol. 9, Sect. I, Part 2, pp. 1–421.

Vuillemin, P. (1912). "Les Champignons. Essay de Classification." Doin, Paris.

Wagner, D. T. (1965). Developmental morphology of *Leptosphaeria discors* (Saccardo and Ellis) Saccardo and Ellis. *Nova Hedwigia* **9**, 45–61.

Wagner, D. T. (1969). Ecological studies on *Leptosphaeria discors,* a graminicolous fungus of salt marshes. *Nova Hedwigia* **18**, 383–396.

Wagner-Merner, D. T. (1972). Arenicolous fungi from the south and central Gulf Coast of Florida. *Nova Hedwigia* **23**, 915–922.

Wainio, E. A. (1903). Lichens. *In* "Résultats du Voyage de S.Y. Belgica en 1897-1898-1899 sous le Commandement de A. de Gerlache de Gomery. Expédition Antarctique Belge," pp. 1–46. Rapp. Sci. Bot. Buschmann, Anvers.

Waisel, Y. (1972). "Biology of Halophytes." Academic Press, New York.

Walker, J. (1972). Type studies on *Gaeumannomyces graminis* and related fungi. *Trans. Br. Mycol. Soc.* **58**, 427–457.

Walker, J. (1975). Mutual responsibilities of taxonomic mycology and plant pathology. *Annu. Rev. Phytopathol.* **13**, 335–355.

Walsh, G. E. (1967). An ecological study of a Hawaiian mangrove swamp. *In* "Estuaries" (G. H. Lauff, ed.), Publ. No. 83, pp. 420–431. Am. Assoc. Adv. Sci., Washington, D.C.

Walsh, G. E. (1974). Mangroves: A review. *In* "Ecology of Halophytes" (R. J. Reimold and W. H. Queen, eds.), pp. 51–174. Academic Press, New York.

Webber, E. E. (1966). Fungi from a Massachusetts salt marsh. *Trans. Am. Microsc. Soc.* **85**, 556–558.

Webber, E. E. (1970). Marine Ascomycetes from New England. *Bull. Torrey Bot. Club* **97**, 119–120.

Webber, F. C. (1967). Observations on the structure, life history and biology of *Mycosphaerella ascophylli*. *Trans. Br. Mycol. Soc.* **50**, 583–601.

Weber, W. A., and Shushan, S. (1959). Lichens of the Queen Charlotte Islands, Canada, collected in 1957 by Dr. Herman Persson. *Sven. Bot. Tidskr.* **53**, 299–306.

Webster, J. (1955). *Hendersonia typhae,* the conidial state of *Leptosphaeria typharum*. *Trans. Br. Mycol. Soc.* **38**, 405–408.

Webster, J. (1959). Experiments with spores of aquatic Hyphomycetes. I. Sedimentation and impaction on smooth surfaces. *Ann. Bot. (London), New Ser.* **23**, 595–611.

Webster, J. (1965). The perfect state of *Pyricularia aquatica*. *Trans. Br. Mycol. Soc.* **48**, 449–452.

Webster, J. (1969). The *Pleospora* state of *Stemphylium triglochinicola*. *Trans. Br. Mycol. Soc.* **53**, 478–482.

Webster, J., and Lucas, M. T. (1961). Observations on British species of *Pleospora*. II. *Trans. Br. Mycol. Soc.* **44**, 417–436.

Webster, R. K., Hewitt, W. B., and Bolstad, J. (1974). Studies on *Diplodia* and *Diplodia*-like fungi. VII. Criteria for classification. *Hilgardia* **42**, 451–464.

Wehmeyer, L. E. (1948). The developmental pattern within the genus *Pleospora* Rab. *Mycologia* **40**, 269–294.

Wehmeyer, L. E. (1952). The genera *Leptosphaeria, Pleospora,* and *Clathrospora* in Mt. Rainier National Park. *Mycologia* **44,** 621–655.

Wehmeyer, L. E. (1961). "A World Monograph of the Genus *Pleospora* and its Segregates." Univ. of Michigan Press, Ann Arbor.

Wehmeyer, L. E. (1963). Some Himalayan Ascomycetes of the Punjab and Kashmir. *Mycologia* **55,** 309–336.

Weise, W. (1976). Aufwuchs von Bakterien auf Quarzkörnern. *In* "Jahresbericht 1975," (G. Hempel, ed.), pp. 62–63. Inst. Meeresk., Univ. Kiel.

Weresub, L. K. (1974). *Papulaspora sepedonioides. In* "Fungi Canadenses," No. 27. Natl. Mycol. Herb., Biosyst. Res. Inst., Res. Branch, Agric. Can., Ottawa, Ontario.

Weresub, L. K., and LeClaire, P. M. (1971). On *Papulaspora* and bulbilliferous Basidiomycetes *Burgoa* and *Minimedusa. Can. J. Bot.* **49,** 2203–2213.

White, W. L., and Downing, M. H. (1953). *Humicola grisea,* a soil-inhabiting cellulolytic Hyphomycete. *Mycologia* **45,** 951–963.

Wicks, S. R. (1974). Presence of *Azotobacter* in marine sand beaches. *Fla. Sci.* **37,** 167–169.

Wilcox, W. W. (1970). Anatomical changes in wood cell walls attacked by fungi and bacteria. *Bot. Rev.* **36,** 1–28.

Wilson, I. M. (1951). Notes on some marine fungi. *Trans. Br. Mycol. Soc.* **34,** 540–543.

Wilson, I. M. (1954). *Ceriosporopsis halima* Linder and *Ceriosporopsis cambrensis* sp. nov.: two Pyrenomycetes on wood. *Trans. Br. Mycol. Soc.* **37,** 272–285.

Wilson, I. M. (1956a). Some marine fungi on wood. *Trans. Br. Mycol. Soc.* **39,** 386.

Wilson, I. M. (1956b). Some new marine Pyrenomycetes on wood or rope: *Halophiobolus* and *Lindra. Trans. Br. Mycol. Soc.* **39,** 401–415.

Wilson, I. M. (1960). Marine fungi: A review of the present position. *Proc. Linn. Soc. London* **171,** 53–70.

Wilson, I. M. (1963). Development of the ascospore and its appendages in the marine fungus *Ceriosporopsis. Nature (London)* **200,** 601.

Wilson, I. M. (1965). Development of the perithecium and ascospores of *Ceriosporopsis halima. Trans. Br. Mycol. Soc.* **48,** 19–33.

Wilson, I. M., and Knoyle, J. M. (1961). Three species of *Didymosphaeria* on marine algae: *D. danica* (Berlese) comb. nov., *D. pelvetiana* Suth. and *D. fucicola* Suth. *Trans. Br. Mycol. Soc.* **44,** 55–71.

Wiltshire, S. P. (1938). The original and modern concepts of *Stemphylium. Trans. Br. Mycol. Soc.* **21,** 211–239.

Winge, Ö. (1923). The Sargasso Sea, its boundaries and vegetation. *In Rep. Dan. Oceanogr. Exped. Medit.* **3,** 1–34.

Winter, G. (1887). Exotische Pilze IV. *Hedwigia* **26,** 6–18.

Wirsen, C. O., and Jannasch, H. W. (1975). Activity of marine psychrophilic bacteria at elevated hydrostatic pressures and low temperatures. *Mar. Biol.* **31,** 201–208.

Withers, R. G., Farnham, W. F., Lewey, S., Jephson, N. A., Haythorn, J. M., and Gray, P. W. G. (1975). The epibionts of *Sargassum muticum* in British waters. *Mar. Biol.* **31,** 79–86.

Wolff, T. (1970). The concept of the hadal or ultra-abyssal fauna. *Deep-Sea Res.* **17,** 983–1003.

Wollenweber, H. W., and Hochapfel, H. (1936). Beiträge zur Kenntnis parasitärer und saprophytischer Pilze. I. *Phomopsis, Dendrophoma, Phoma,* und *Ascochyta,* und ihre Beziehung zur Fruchtfäule. *Z. Parasitenkd.* **8,** 561–605.

Wright, E. P. (1876). *Blodgettia confervoides* Harvey, from a new station (Bermuda). *Q. J. Microsc. Sci.* [N.S.] **16,** 342.

Wright, E. P. (1880). *Blodgettia. Q. J. Microsc. Sci.* [N.S.] **20,** 111.

Wright, E. P. (1881). On *Blodgettia confervoides* of Harvey, forming a new genus and species of fungi. *Trans. R. Ir. Acad.* **28,** 21–26.

Yamada, Y. (1934). The marine Chlorophyceae from Rynkyn, especially from the vicinity of Nawa. *J. Fac. Sci., Hokkaido Imp. Univ., Ser. 5* **3,** 33–88.

Yamasoto, K., Goto, S., Ohwada, K., Okuno, D., Araki, H., and Iizuka, H. (1974). Yeasts from the Pacific Ocean. *J. Gen. Appl. Microbiol.* **20,** 289–307.

Yonge, C. M. (1927). The absence of a cellulase in *Limnoria. Nature (London)* **119,** 855.

Young, D. N. (1977). A note on the absence of flagellar structures in spermatia of *Bonnemaisonia. Phycologia* **16,** 219–222.

Zambettakis, C. (1954). Recherches sur la systématique des Sphaeropsidales-Phaeodidymae. *Bull. Soc. Mycol. Fr.* **70,** 219–350.

Zanefeld, J. S. (1969). "Iconography of Antarctic and sub-Antarctic Benthic Marine Algae," Part I. Cramer, Lehre.

Zebrowski, G. (1936). New genera of Cladochytriaceae. *Ann. Mo. Bot. Gard.* **23,** 553–564.

Zinn, D. J. (1967). Between grains of sand. *Underwater Nat.* **4,** 4–8.

Zinn, D. J. (1968). A brief consideration of the current terminology and sampling procedures used by investigators of marine interstitial fauna. *Trans. Am. Microsc. Soc.* **87,** 219–225.

ZoBell, C. E. (1968). Bacterial life in the deep sea. *Bull. Misaki Mar. Biol. Inst., Kyoto Univ.* **12,** 77–96.

Zschacke, H. (1925). Die mitteleuropäischen Verrucariaceen. IV. *Hedwigia* **65,** 46–64.

Organism Index

Names of hosts are those used by the various authors in connection with a particular fungus. Hosts for which no authorities could be found are followed by a question mark (?). Invalid names, e.g., unpublished herbarium names, are listed in quotation marks.

A

Abaphospora Kirschstein, 425

Abies Mill., 113, 251, 255, 257, 268, 273, 295, 299, 304, 326

A. alba Mill., 113, 465

A. firma Sieb. et Zucc., 113, 257, 275, 327, 465, 479

Absidia van Tiegh., 553

Abyssomyces Kohlm., 214, 345–347

A. hydrozoicus Kohlm., 43, 44, 138, 165, 202, 346–347

Acer L., 113, 225, 314

Acmaea Rathke *in* Eschscholtz, 358

Actinospora megalospora Ingold, 32

Aepus robini Laboulbène, 139, 227, 228

Aesculus hippocastanum L., 113, 225

Aglaozonia Zanard., 11, 57, 65, 448, 449

A. chilosa Falkenb., 449

A. parvula (Grev.) Zanard., 449

Agonomycetaceae, 215, 217, 464–466

Agonomycetales, 215, 217, 464–466

Agropyron, see Agropyrum

Agropyrum Gaertn., 85, 88, 476

A. junceiforme (Löwe) Löwe, 85, 283, 419, 422

A. pungens (Pers.) Roem. et Schult., 85, 283, 293, 309, 317, 318, 399, 422, 435

Agrostis L., 90

A. maritima?, 551

Alaria Grev., 57, 195

A. esculenta (L.) Grev., 64, 340

Allescheriella P. Hennings, 216, 509–510

A. bathygena Kohlm., 43, 120, 205, 509–510

A. crocea (Montagne) S. J. Hughes, 509

Alnus Ehrh., 113, 225, 257, 304

A. glutinosa (L.) Gaertn., 113, 255

A. rubra Bong., 106

Alternaria Nees von Esenbeck ex Fries, 49, 81, 83, 84, 99, 106, 114, 115, 117, 120, 210, 215, 437, 473–474, 503, 552

A. alternata (Fries) Keissler, 88, 89, 473

A. fasciculata Cooke et Ellis, 473

A. geophila Daszewska, 473

A. humicola Oudemans, 473

A. maritima Sutherland, 132, 473, 552–553

A. tenuis Nees ex Pers., 106, 473

A. tenuissima (Kunze ex Pers.) Wiltshire, 473

Ammophila Host., 201, 435

A. arenaria (L.) Link, 85, 434, 455, 476, 479, 488, 519, 551

A. arenaria × *Calamagrostis epigejos*, 435

A. baltica Lind, 85, 519

Amphididymella Petrak, 425

Amphisphaeria biturbinata (Durieu et Montagne) Sacc., 446

A. maritima Linder, 430

A. posidoniae (Durieu et Montagne *in* Montagne) Cesati et DeNotaris, 404, 405

Subject Index